Surgery and Arthroscopy of the Knee

Second European Congress
of Knee Surgery and Arthroscopy
Basel, Switzerland, 29 Sept.–4 Oct. 1986

Edited by We. Müller and W. Hackenbruch

With 432 Figures and 144 Tables

Springer-Verlag Berlin Heidelberg New York
London Paris Tokyo

Dr. med. Werner Müller

Privatdozent der Universität Basel, Chefarzt Orthopädische Klinik, Kantonsspital Bruderholz, 4101 Bruderholz BL, Switzerland

Dr. med. Werner Hackenbruch

Chefarzt Orthopädische Abteilung, Regionalspital, 4900 Langenthal, Switzerland

ISBN-13: 978-3-642-72784-9 e-ISBN-13: 978-3-642-72782-5

DOI: 10.1007/978-3-642-72782-5

Library of Congress Cataloging-in-Publication Data.
European Congress of Knee Surgery and Arthroscopy (2nd : 1986 : Basel, Switzerland) Surgery and arthroscopy of the knee.
Includes index.
1. Knee – Surgery – Congresses. 2. Arthroscopy – Congresses. I. Müller, Werner, 1933– . II. Hackenbruch, W. (Werner), 1939– . III. Title. [DNLM: 1. Arthroscopy – congresses. 2. Knee – surgery – congresses. W3 EU844 2nd 1986s/WE 870 E89 1986s] RD561.E95 1986 617.582 88-4357

Softcover reprint of the hardcover 1st edition 1988

Typesetting: Brühlsche Universitätsdruckerei, Giessen
Offsetprinting: Saladruck, Berlin. Bookbinding: Lüderitz & Bauer, Berlin
2124/3020-543210

Preface

The need for greater international collaboration is becoming increasingly clear. Practically every nation or people has, in addition to its own characteristics and history, its own medical tradition. Methods and ways of thinking that are successfully established in one place may not be accepted in another for some time.

The medical tradition and training in different countries can take quite different forms. What would our present knowledge be

- Without the creative, spontaneous ideas and contributions from the romance countries?
- Without the opportunity of research, as it is practised on a large scale in the USA, where so many young Europeans took their first steps in experimental science?
- Without the conscientious clinical work done in Central Europe and the valuable experience gained there?
- Without, finally, the sober and critical scientific approach, as fostered in the cool heads of Northern Europe for generations?

None of these qualities is sufficient without the others, but together they are a source of strength. Ultimately, we hope, through sensible concentration of efforts (while maintaining variety), to overcome the lack of an integrated approach in experimental science and to prevent repetition of painful mistakes. In this regard, the first two ESKA congresses have brought us great advances. To meet, to learn, and to make constructive progress are our goals, and it is with these in view that selected contributions from the remarkable second ESKA congress (29 September to 4 October 1986, Basel) have been collected in this stimulating and forward-looking book.

Bruderholz and Langenthal,
January 1988

Werner Müller
Werner Hackenbruch

Table of Contents

Surgical Anatomy and Biomechanics of the Knee

The Theory of Movement and the Modern Mode of Thought in Biology
A. Menschik 3

The Mechanics of the Knee in the Sagittal Plane
J. O'Connor, T. Shercliff and J. Goodfellow 12

The Evaluation of Ligamentous Knee Joint Instability in the Gait Laboratory
J. U. Baumann 31

Two-Bundle Ligamentoplasty in Anterior Laxities of the Knee
J. L. Rhenter 37

Technique and Model of Rehabilitation in Anterior Cruciate Ligament Repair of the Knee: A Review of 175 Cases
R. Feinstein, B. Segesser, P. Jenoure and P. Michel 43

Functional Rehabilitation in Anterior Cruciate Ligament Reconstruction (4–7-Year Results)
U. Munzinger, T. Drobny and A. Arnold 44

Anatomy and Biomechanics of the Anteroposterolateral Femorotibial Structures in the Iliotibial Band System
P. Lobenhoffer, C. Krettek, P. Posel and S. Witt 45

Mechanoreceptors of Some Anatomical Structures of the Human Knee
G. Cerulli, A. Ceccarini, P. F. Alberti, A. Caraffa and G. Caraffa . 50

The Function of Hamstrings: A Pathogenic Hypothesis of Femoropatellar Osteoarthritis
P. Blaimont, P. Klein, M. Alameh and P. van Elegem 55

Anatomy of the Popliteus Muscle and Posterolateral Structures
C. Fabbriciani and M. Oransky 58

Femoropatellar Pathology and Rotational and Torsional Abnormalities of the Inferior Limbs: The Use of CT Scan
J. L. Lerat, B. Moyen, M. Bochu and O. Galland 61

Anterior Cruciate Ligament and Anterior Instability

Grading the Pivot Shift: An Objective System with Treatment Implications
R. P. Jakob, H. U. Stäubli and J. Deland 69

The Jerk Test in External Rotation in Anterior Cruciate Ligament-Deficient Knees
J. Y. Dupont 71

Translation and Rotation in Knee Instability: A Prospective Stress Radiographic Analysis with the Knee in Extension
H. U. Stäubli, R. P. Jakob and B. Noesberger 82

The Diagnostic and Prognostic Value of the "Active Radiologic Lachman"
H. Dejour, P. Chambat, G. Walch and R. Ranger 84

Measure of Laxities by Stress Radiography and by KT 1000 Arthrometer
J. L. Lerat, B. Moyen, L. Dupré la Tour, E. Mainetti, J. J. Lalain and E. Brunet-Guedj 85

Stress Radiographic Comparison of Drawer Symptoms Measured in 90° Versus 15° of Knee Flexion (the Lachman Position)
B. F. Iversen, J. Stürup and K. Jacobsen 91

Unrecognized Abnormal Osseous Metabolic Activity about the Knee of Patients with Symptomatic Anterior Cruciate Ligament Deficiency
S. F. Dye 93

Stresses Applied to the Knee in Meniscal and Ligamentous Disorders – A Study of the Medial Tibiofemoral Joint
D. D'Hondt, P. Frain, C. Fontaine and C. Argenson 95

Instability of the Anterior Cruciate Deficient Knee
J. Kärrholm, Göran Selvik, L.-G. Elmqvist and L. I. Hansson . . . 104

The Reliability of the Genucom Knee Analysis System
C. L. Highgenboten and A. Jackson 107

Isolated Tears of the Medial Collateral Ligament of the Knee – Operative or Conservative Treatment?
H. Seiler, M. Kayser, H. Niemeyer and V. Bühren 111

Early Mobilization of Isolated Partial Ruptures of the Medial Collateral Ligament: A Randomized Comparison of Bracing and Elastic Wrapping
M. Lundberg and P. Hamberg 112

Congenital Absence of Cruciate Ligaments: Clinical, Radiological and Arthroscopic Aspects
P. Schlepckow and H.-U. Ernst 116

Surgical Versus Non-surgical Treatment of Knee Ligament Injuries
M. Bauer and R. Sandberg 121

Evaluation of Knee Ligament Injuries
Y. Tegner, J. Lysholm and J. Gillquist 123

Comparative Results of 163 Anterior Cruciate Ligament Injuries Managed by Repair/Reinsertion, Primary Augmentation, or Reconstruction
R. Biedert, We. Müller, W. Hackenbruch and R. Baumgartner . . 130

The Anterior Cruciate Deficient Knee Treated by a Modified Marshall-Macintosh Repair
G. N. Fernandez and J. A. Robertson 142

Substitution of the Anterior Cruciate Ligament Using the Torn Meniscus
E. Lais, P. Hertel, J. Wortmann, R. Schrempf and H. J. Barrach . 145

Recent Rupture of the Anterior Cruciate Ligament: Clinical Results of Treatment by Reinsertion and Allogenic Augmentation
O. Kwasny, R. Schabus, M. Wagner and H. Plenk 155

Autogenic Versus Polylactate-Coated Carbon-Fiber Augmentation in Anterior Cruciate Reconstruction
J. Witvoet, P. Christel and F. Gaucher 160

ACL Reconstruction Procedure with Autogenous Graft under Arthroscopic Control
J.-C. Imbert 168

Treatment for Anterior Cruciate Ligaments Insufficiency and Combined Meniscus Injuries in Athletes
N. Kanazawa, A. Saito, T. Itagaki, Y. Tomoyasu and K. Sato . . . 170

Late Results after Primary Repair of Anterior Cruciate Ligament Tears
W. Kipfer, P. Ballmer, B. Grünig, H. U. Stäubli, R. Zehnder and R. P. Jakob 173

Five to Ten-Year Follow-up after Primary Suture of Anterior Cruciate Ligament Tears
T. Strand and A. Mølster 181

Anterior Cruciate Ligament Repair Combined with Tibial Valgisation Osteotomy in Medial Osteoarthritis with Insufficient Anterior Cruciate Ligament
H. Bereiter, P. Giger and A. Gächter 182

The Natural History of ACL Ruptures
J. Y. Dupont, C. Scellier and D. Chaudières 185

A Comparative Analysis of Four Prosthetic Anterior Cruciate Ligaments in the Goat
N. P. Thomas and I. G. Turner 187

A Comparative Analysis of the Mechanical Properties of Four Prosthetic Anterior Cruciate Ligaments
N. P. Thomas, K. Wright, I. G. Turner and I. W. Nelson 191

Mechanical and Biological Considerations in Prosthetic Cruciate Replacement
B. B. Seedhom, K. Fujikawa and P. J. Atkinson 197

Anterior Cruciate Ligament Reconstruction Using the Leeds Keio Terylene Ligament
J. B. Hamilton, G. K. Sefton, R. B. Smith and T. J. Cain 199

Reconstruction of the Anterior Cruciate Ligament Using a Ligament Prosthesis
J. Gillquist and M. Odensten 201

Reconstruction of the Anterior and Posterior Cruciate Ligaments of the Knee under Arthroscopy. First Results
J. Bahuaud, J. Rebour, J. Mayer, D. Besse, F. Mautalen and H. Baspeyre 204

Short-Term Results in Anterior Cruciate Ligament Reconstruction with a Dacron Prosthesis
T. Wredmark 213

Anterior Cruciate Ligament Reconstruction in the Chronically Unstable Knee Using an Expanded Polytetrafluoroethylene (PTFE) Prosthetic Ligament
S. K. Ahlfeld, R. L. Larson and H. R. Collins 216

The Use of Proflex in the Repair of Anterior Cruciate Ligament Lesions
M. Allard, J. Bahuaud. L. Barbat, C. Chaix, G. Gacon, F. Gougeon, R. Ferro and Ch. Mansat 225

The Inherent Mechanical Properties of Allograft Fascia Lata
M. J. Gibbons, D. L. Butler, F. R. Noyes and T. Schlegal 235

Alternative Operative Procedure for Anterior Cruciate Ligament Replacement with a Composite Graft for Early Functional Treatment
R. Schabus, O. Kwasny, M. Wagner and H. Plenk Jr. 237

Technique, Complications, and Results of Arthroscopic Anterior Cruciate Ligament Reconstructions with Bioprosthetic Ligaments
S. I. Springer 243

Synovitis after Anterior Cruciate Ligament Reconstruction Effusion: Confusion
W. L. Lanzer, A. Gown and C. Piening 246

Posterior Cruciate Ligament and Posterior Instability

Diagnosis of Posterior Cruciate Ligament Insufficiency. A Prospective Clinical and Stress Radiographical Analysis with the Knee in Extension
H. U. Stäubli and R. P. Jakob 251

Isometric Points of the Posterior Cruciate Ligament
E. S. Grood, M. S. Hefzy, T. L. Lindenfeld and F. R. Noyes 252

Posterior Instability of the Knee. Surgical Treatment, Rehabilitation and Results
We. Müller 254

Treatment of Acute and Chronic Injuries of the Posterior Cruciate Ligament
H. Oshiro 269

Experimental Investigations of Strees Production on the Cruciate Ligaments after Transpatellar Steinmann's Nail Fixation
R. Weinstabl, H. Kern and M. Wagner 271

The Transpatellar Steinmann Pin as a Postoperative Safeguard after Reconstruction of the Posterior Cruciate Ligament or after Trillat's Operation
E. Orthner, O. Kwasny, R. Schabus and M. Wagner 275

Experiences with Reconstruction of the Posterior Cruciate Ligament Using a Free Graft of the Rectus Muscle Tendon Combined with a Patella Bone Graft
Z. G. Harnach and V. Pokorný 279

Treatment of Posterior Cruciate Ligament Injuries: Long-Term Effects
J. Lysholm, M. Lindh and J. Gillquist 283

Surgical Stabilisation of Serious Posterior Instability of the Knee: A Preliminary Report on Personal Techniques
J.-L. Meystre 287

Surgical Treatment of Chronic Posterolateral Laxity of the Knee: Procedure, Indications and Early Results
J. H. Jaeger and J. R. Meyer 291

Treatment of a Delayed, Nonreducible, Chronic Posterior Dislocation of the Tibia with Posterior Instability after Elsewhere Internal Fixation of a Femoral Shaft Fracture
P. Lauber, A. Kentsch and We. Müller 296

Arthroscopic Operations on the Meniscus

Follow-up of More than 150 Meniscal Repairs up to 13 Years Postoperatively
A. Schreiber and M. Rodriquez 305

Meniscus Repair by Arthroscopy
J. Defrere, A. Franckart and R. Bassleer 306

Combined Arthroscopic-Open Repair of Meniscal Injuries
S. C. Marshall 317

Electron Microscopical Observations on Sutured Meniscus Tears
A. Schmid, F. Schmid and T. Tiling 328

Four Years' Experience with Arthroscopic Meniscal Suture – Where Are We?
R. P. Jakob, H. U. Stäubli and K. Zuber 335

Surgical Repair of Chronic Meniscal Injury: Plastic Meniscus Surgery
P. L. Melanotte and M. Baldovin 336

Arthroscopic Management of Degenerative Meniscal Tears in Patients Over 60
G. Stroz 341

Results of Posterior Horn Resection of the Medial Meniscus
P. Hertel 345

CO_2 Laser Meniscectomy under Arthroscopy
K. Inoue, K. Mikanagi, H. Morita and T. Namiki 349

Results of Arthroscopic Knee Operations in Out-Patients Under Local Anaesthesia
H. Sieczewicz and H. Henkemeyer 350

Pathological Examination of Meniscus Fragments: The Effect of Arthroscopic Excision
A. Fisseler, J. Witt, J. Krämer and K.-M. Müller 352

Results of Partial Arthroscopic Meniscectomies in Anterior Cruciate Ligament Deficient Knees
P. Aglietti, R. Buzzi and A. Pisaneschi 358

Arthroscopic Meniscectomie on Patients with Untreated Rupture of the Anterior Crucuiate Ligament
A. Georgoulis, E. Lais and P. Hertel 360

Total Meniscectomy in Children: A Long-Term Follow-up Study
P. Abdon, A. J. G. Swanson and M. S. Turner 366

Meniscectomy in Senior Citizens
T. Hoogland 369

The Consequence of Meniscectomy in 147 Athletes: A 14.5-Year (Median) Prospective Longitudinal Study
U. Jørgensen, S. Sonne-Holm, F. Lauridsen and A. Rosenklint . . 371

Meniscus Reattachment or Allotransplantation? An Experimental Comparison on Sheep
K. Wulff, B. Gunterberg, G. Moberger and P. Phillips 378

Meniscus Transplantation – Experimental Study, Clinical Report, Arthroscopic Findings
K. A. Milachowski, K. Weismeier, C. J. Wirth and D. Kohn 380

Antiphlogistics Enhance Recovery after Arthroscopic Meniscectomy
M. Bauer, D. Ogilvie-Harris and P. Corey 389

Problems on the Patella and Extensor Apparatus

Classification of Cartilage Damage
E. W. Morscher 397

Quantitative Structural Analysis of Articular Cartilage: A Comparison of Weight-Bearing and Less-Weight-Bearing Areas in the Femoral Condyle of Young Adult Rabbits
P. S. Eggli, E. B. Hunziker, T. Naef and R. K. Schenk 407

Comparative Morphological Examination of Experimental Patellar Dislocation
I. Földes, J. Gyarmati Jr., I. Kiss, J. Csongor, M. Kern and G. Deseö 415

Experimental Investigations into the Reaction of Articular Cartilage and Synovial Membrane to Microtraumatical Impulse Loading
H. Fengler, H. Schmidt and K.-J. Schulze 422

Electron Microscope Findings After Cartilage Shaving
A. Schmid, F. Schmid and T. Tiling 426

A New Method of Depiciting the Relationship between Patella and the Femur
W. Sloan 433

The Three-Dimensional Tracking Pattern of the Patella in the Human Knee Joint and the Effects of Surgical Interventions
A. van Kampen, R. Huiskes, L. Blankevoort and T. J. G. Van Rens . 434

Epidemiology of Chondromalacia (CHM)
K. Franke 446

The Natural History of Anterior Knee Pain in Japanese Adolescents
A. Fujimoto, Y. Mori, Y. Kuroki, R. Yamamoto, H. Hino and H. Okumo 447

Patellofemoral Dysplasia
H. A. Tomairek, N. A. Mahmoud, A. M. El-Eleimi, N. Galal and B. El-Serwy 452

Chondropathia Patellae – A 5-Year Follow-Up Study after Surgical Treatment
H. Kiefer, W. Mutschler and J. M. Friedrich 462

Patella Realignment via Lateral Release: Results of Surgery and Subsequent Aetiological Investigation
W. D. Stanish, P. K. Smith, L. F. Pow and J. Kozey 470

Longitudinal Osteotomy of the Patella after Morscher
M. Pećina 471

Patellectomy as a Salvage Procedure
G. Blatter, R. W. Jackson and O. Bayne 476

Patellectomy: Long-Term Results
G. Brunelli, G. Stefani, F. Brunelli and C. Milini 486

The Biomechanical Analysis of Chronic Patellar Tendinitis and Treatment with Eccentric Loading
W. D. Stanish, H. Lamb and S. Curwin 493

Jumper's Knee Ultrasonography
D. Fritschy 495

Is Patellar Tendinitis a Surgical Affection?
G. Saillant, J. P. Benazet, G. Gagna, F. Combelles and R. Roy-Camille 498

The Effect of Five Different Patella Braces on Performance and Pain in an Isokinetic Quadriceps Torque Test
U. Lindberg, J. Gillquist and J. Lysholm 499

Degenerative Arthritis and Osteotomy

Zed Osteotomy with Trapezoidal Resection in Internal Rotations of the Tibia and Secondary Static Arthrosis of the Knee
G. Brunelli, F. Brunelli, B. Battiston and P. A. Guizzi 503

Unicompartmental Replacement of the Knee: A Long-Term Study Comparing the Results with Tibial Osteotomy
N. S. Broughton, J. H. Newman and R. A. J. Baily 510

Knee Prosthesis

Total Knee Replacement: The Gallannaugh Semi-Constrained Linked Prosthesis
S. C. Gallannaugh 513

Ten Year's Experience with the Total Condylar Knee Prosthesis
M. Soudry M. Iusim, A. Menachem and D. G. Mendes 516

PCA Unicompartmental Knee Endoprosthesis
A. Lindstrand and A. Stenström 519

Assessment of Results Following Total Knee Replacement: A Comparison of Systems
R. Binazzi, M. Soudry, L. Mestriner and J. N. Insall 521

Total Knee Replacement with Customized Prosthesis
R. Binazzi, M. Soudry, L. Mestriner and J. N. Insall 525

Knee Stability after Total Knee Replacement with the Biomet Prosthesis as Evaluated by Stress Radiography
B. F. Iversen, K. Jacobsen, N. Laursen and P. Lyndrup 529

Unicompartmental Arthroplasty with the Oxford Meniscal Knee Prosthesis
J. Goodfellow and J. O'Connor 533

Micromotion of Uncemented Freeman-Samuelson Knee Prosthesis: A Roentgen Stereophotogrammetric Analysis
B. E. J. Albrektsson, L. Ryd, P. Herberts, A. Lindstrand and G. Selvik 536

Application of Bench-Mounted Saws for Precision Replacement Arthroplasty of the Arthritic Knee – The Questor Systems
T. D. V. Cooke, Y. Harada, G. Saunders, D. Siu, H. Wevers and Y. Yoshioka 542

Incidence and Form of Knee Deformity Associated Osteoarthritis
T. D. V. Cooke, D. Siu, B. Fisher and J. Rudan 556

Arthroplasty of the Knee Joint with the PCA (Porous Coated Anatomic) Endoprosthesis
H. Jenny 570

The Anatomical Endoprosthesis System for the Knee Joint with Metal Spongiosa Surface Structures Allowing Biological Fixations
W. Thomas 579

The Alloarthroplasty of the Knee Combined with Bone Reconstruction
R. Hoos, H. J. Refior and H. Stürz 585

Patellar Complications with the Total Condylar Posteriorly Stabilized Knee Prosthesis
P. Aglietti, R. Buzzi and A. Pisaneschi 589

The Failure of a Semi-Constrained Total Knee Prosthesis
M. I. B. Besser 591

Revision Arthroplasty Total-Knee Implants in Infected PCA
M. Rodriguez and A. Papandreou 595

Arthrodesis of the Knee after Infected Total Knee Prosthesis
K. H. Müller 596

The Value of Cruciate Ligaments in an Unconstrained Knee Prosthesis
J. L. Lerat, B. Moyen, L. P. Jacomy, G. Py and J. J. Lalain . . . 597

Endoprosthetic Replacement of the Knee in Rheumatic Diseases
K. Tillmann 601

Prosthetic Replacements of Joint Surfaces and the Influence on Activity and Evolution of Rheumatoid Disease
R. Neumann 602

Total Knee Arthroplasty in the Treatment of Rheumatoid Arthritis and Other Forms of Osteoarthritis
F.-W. Hagena and G. O. Hofmann 603

Unconstrained Knee Replacement in Patients with Rheumatoid Arthritis
C. T. Trepte, W. Noack and B. Gondolph-Zink 612

Our Experiences with Kinematic Total Knee Replacements
F. Kerschbaumer, R. Siorpaes, R. Hingshammer, M. Eichenauer and M. Kofler 615

Autologous and Homologous Bone Grafts in Fixation of Knee-Joint Prostheses
G. Waertel, D. Wessinghage and E. Kißlinger 618

Knee Arthroplasty with GSB and Engelbrecht's and Tönnis' Sledge Prosthesis in Rheumatoid Arthritis and Osteoarthritis
E. Kißlinger, D. Wessinghage and J. Zacher 626

Cementless Knee Arthroplasty in Rheumatoid Arthritis
G. Bontemps 640

Condylar Anatomical Knee Joint Implants
R. K. Miehlke 644

Innovations and Free Communications

Magnetic Resonance Imaging of the Knee: Normal and Pathological Imaging and Correlations with Arthroscopy
M. Denti, C. Bonizzoni, V. Ramondetta and G. Peretti 649

Diagnosis of Ligamentous and Meniscal Lesions of the Knee Joint with Magnetic Resonance Imaging (MRI)
M. Rodriguez, G. v. Schulthess and N. Friederich 655

The Treatment of Hemophilic Knee Arthropathy and Inferior Limb Discrepancies by the "Ilizarov" Method
G. Torri, G. Bagnoli, E. Lozej, P. Cerea, G. Mistò, R. Ventura and V. Pietrogrande . . . 656

Antibiotic Absorption under Tourniquet
D. P. Johnson, G. C. Bannister, J. M. Auchincloss and J. H. Newman . . . 657

Pigmented Villonodular Synovitis of the Knee: Advantages of Arthroscopy
B. Locker, J. Beguin, C. Vielpeau and G. Loyau . . . 661

Villous Lipomatous Proliferation of the Synovia (Lipoma Arborescens)
T. Hallel and S. Lew . . . 666

Rheumatoid Arthritis: A Result of Different Pathomechanisms
H. G. Fassbender . . . 670

A Scientific Basis for Surgery in Rheumatoid Arthritis
T. D. V. Cooke . . . 682

Arthroscopic Knee Synovectomy
E. S. Bittar . . . 688

Synovectomy Applied to Inflammatory Rheumatic Diseases of the Knee under Arthroscopic Control
M. Moraldo, R. Schleberger and A. Hedtmann . . . 690

Radical Versus Partial Synovectomy: Pathophysiological Considerations
K. Tillmann and U. Dümmer . . . 695

Rehabilitation after Knee Injuries – Importance of Pain Relief
E. Eriksson and I. Arvidsson . . . 696

Recent Results on Post-traumatic Osteoarthrosis after Internal Fixation of Fractures of the Tibial Plateau
H.-M. Märklin and U. Holz . . . 697

Traumatic Loss of a Condyle of the Femur: Replacement by an Ipsilateral Patellar Graft
V. Bühren, H. Seiler and H. Niemeyer . . . 701

Electron Microscopical Studies on Human Meniscal Tissue Preserved for Transplantation
A. Schmid, F. Schmid and T. Tiling . . . 703

Ultrastructural Analysis of the Cruciate Ligaments and of Tissue Used for their Plastic Repair
H. Arnold-Schmiebusch, W. Hackenbruch and J. Staubesand . . . 708

What is a Meniscal Insufficiency?
M. F. Fischmeister 713

Reconstruction of the Lateral Femoral Condyle with Homologous Bone Graft after a Third-Degree Compound Comminuted Fracture in Polytrauma after Motorcycle Accident
P. Lauber, A. Kentsch and We. Müller 717

Subject Index 729

Surgical Anatomy and Biomechanics of the Knee

The Theory of Movement and the Modern Mode of Thought in Biology

A. Menschik

Lorenz Böhler Unfallkrankenhaus, Vienna, Austria

When one searches for the most comprehensive criterion specific to fauna, particularly in reference to vertebrates, the phenomenon of movement appears to be the most characteristic feature. The vital functions of acquiring and consuming food, of flight and of aggression are closely related to this phenomenon. Every organic function – the central nervous system, the neural system, respiration, blood circulation, muscular function, metabolism, the optical organs and the ears – is constructed for movement.

From the earliest days, walking and running fascinated the human mind. In medicine and biology today a separate branch of science has established itself, biomechanics, the main area of research for which is the phenomenon of motion. As Amtmann of Würzburg has observed, biomechanics sees its main task in registering the movement of single segments or of the entire body, in comprehending the "how" kinematographically and in mathematically describing the orbits of moving curves, or marking points, without primarily asking for the cause, the "what" of the motion. This dualistic mode of thinking – to deal with the "how" mathematically and to delegate the "what" to natural sciences – is a typical medieval way of thinking, which has long since been overcome by Newtonian and post-Newtonian physics.

This dualistic mode of thought goes back to the Arab mathematician and philosopher Ibnu'l Haitam Alhazan (965–1038), who, in his work *On Light*, divided the mathematical sciences from the natural sciences in the following way: the treatment of the "how" of the radiation of light requires the mathematical sciences, and the "what" of light belongs to the natural sciences. Seen from a modern standpoint, for example, Johannes Kepler (1571–1630) failed in his mathematical description of the plantary movements, insofar as he limited his observations to the "how". Only Isaac Newton (1643–1727) was able to explain the cause, the "what", of the planets' movements, by applying the law of gravitation in a mathematical-physical way.

That the dualistic mode of thought characteristic of today's biomechanics has not led to substantial results is demonstrated by the great number of highly equipped biomechanical laboratories throughout the world which have not been able to take a single step towards clarifying, for example, the knee joint – despite some 50 years of intensive activity by numerous technicians, computer experts, mathematicians, biologists and medical personnel. The simplest functions of the knee joint – the rolling and gliding movement and the orthogonal transmission of forces – remain today inadequately explained phenomena of "nature" for biomechanics. Biomechnics itself is unable to define the elementary features of knee

Surgery and Arthroscopy of the Knee
2nd Congress of the European Society
Editors: Müller/Hackenbruch

movements, the joint faces. A discussion of the myriad inadequacies and contradictions of modern medicine and biology in relation to motion and the causes of motion would lead, however, far beyond the bounds of the present paper.

Let us return, rather, to our main topic, that of the theory of movement and a modern mode of thought in biology. Any consistently reproducible movement must be encompassed by certain laws, regardless of whether it is a technical or a biological dynamic system. Consequently, the substantive structural principle, the "what", must be at the centre of attention, for it is the cause of the motion, the "how", and not vice versa.

In considering the knee joint, for example, we must at the outset confront the still undefined central term of the dynamic system, the joint face. Because of the different appearances of the tibial and femoral joint faces an incongruence seems to arise. However, when the knee joint is examined in flexion and extension, it appears that the joint faces have in every position a constant point of contact or "identity point". Flexion and extension therefore involve a sequence of identity points. When two joint faces consist of reproducible identity point sequences, they correlate very highly and cannot therefore be called incongruent, as is nevertheless still done in contemporary literature. The joint faces of the tibial and femoral condyles have an "identity" in the sense of an identical sequence of points, which appears in the two different phenomena, while the different appearances correspond to each other to the highest degree – since they consist of identical sequences of points.

If one stabilizes the thigh and moves the lower leg, an identity point sequence appears, but one also discovers that the tibial joint face envelops that of the thigh. When the motion is reversed – lower leg stabilized and thigh moved – the joint face of the thigh envelops that of the lower leg. Two joint faces that envelop each other while in a reproducible movement consisting of identity point sequences constitute "envelopes". Thus we can define the joint faces of the knee joint as envelopes. Envelopes, however, are purely geometric formations. Surprisingly, we discover that the biological substance, the cells of the system, embody or "realize" a purely geometric formation.

The discovery that the joint faces of the knee joint are envelopes clarifies two trivial functions of the knee joint which had previously been inexplicable. The first of these is the rolling and gliding movement of the knee joint. This was described by the Weber brothers 150 years ago as a consequence of the envelopes' movement. Envelopes in motion execute a rolling and gliding movement that can be described mathematically. Therefore, the rolling and gliding motion appears to be a geometric-kinematic result, rather than a biological one.

The second function is that of the orthogonal transmission of forces at the points of exertion of the joint faces, which was postulated by Benninghoff early this century and on which he based his theory of cartilaginous function. He reasoned that "almost frictionless joint faces can receive only orthogonal pressure (or force)". This has, in fact, been proved now by envelope kinematics. Envelopes have a common face tangent to their exertion points. The normal (line) to the generating (line) at the exertion point and the enveloped face fit into each other and cross the momentary angle point of the system. When forces are applied to this system, these forces are transposed orthogonally via normal curves at the exertion point,

from the enveloped to the generating face or vice versa. Also here, the orthogonal transmission of forces at the exertion points of the joint faces is not a biological but a geometric-kinematic consequence.
Understanding the concept of "envelope" also helps to explain the function of the cruciate ligaments, for they are the control system of the envelopes. Envelopes need a permanent control system for their consistently reproducible sequence of movements. The envelopes have only an indirect relation to the control system, which is the reason why there are four differently formed joint faces at the knee joint that nevertheless display a uniform dynamic system. Much more importantly, the discovery that the joint faces are envelopes makes unmistakably clear that the biological dynamic systems – independently of their content – embody geometric-kinematic relations which must be discovered little by little.
Still our modern medical-biological view regards geometry as something that has been introduced from "outside", projected by the human mind upon the biological dynamic systems for the sake of explanatory utility, but having scarcely anything to do with the biological reality itself.
Gradually the relativistic mode of thought in contemporary physics is infiltrating the medical and biological sciences. By applying geometric concepts to physical occurrences in a completely new way we become ever more aware that we ourselves – and every living being – have emerged out of this physical space-time world. Biological dynamic systems can thereby come to be seen as embodying force-free relations which we analyse in terms of dynamic geometry or kinematics. Biological dynamic systems are therefore self-realizing geometric-kinematic laws that must be discovered little by little. Only after analytical clarification of the structural principle of the dynamic systems can questions as to improvement (or optimization) be put.
We lack the space in this paper to discuss the interaction of cruciate and collateral ligaments. In this matter I refer to Müller (1981), who treats the subject in a masterful and lucid fashion.
The validity of observations on kinematic relations in the resting system to those in the active system applies to all joint rectangles, regardless of their dimensions and material. This fact was of special difficulty when structurally evaluating the control system, i.e. the twisted (oblique) joint rectangle, consisting of the cruciate ligaments and their points of attachment.
If the length of the cruciate ligaments is given in a joint rectangle, every dimension of the joint rectangle would be possible, as far as allowed by the length of the contributing ligaments. However, within a biological dynamic system, e.g. the knee joint, the whole system is clearly determined when the lengths of the cruciate ligaments are given. The length of the cruciate ligaments corresponds to certain other dimensions, such as the distance to their insertion points at the lower leg, the three-dimensional transposed distance to the insertion points at the thigh, and the height and width of the fossa intercondylica. If the length of the cruciate ligaments –the steering arms of the joint rectangle – determine the dimensions of the twisted joint rectangle, there must exist a geometric transformation system which, at the base of the cruciate ligaments, determines inter alia the height of the fossa intercondylica.

The question remains as to which transformation system induces this geometrical interdependence among the various parameters. While it is clear that genetic material is responsible for form and function, we still must ask how the genetic information is transposed and according to what laws, especially since the joint faces, as envelopes, are purely geometrical phenomena.
When in motion, the ends of all osseous fixated filaments of ligamentary systems that span the joint gap (collateral and cruciate ligaments) describe cones with spherical bases while the filaments themselves describe spherical envelopes, that is, each filament of a ligamentary system describes an arc.
We can therefore conclude that the arc is the only force-less structural element for self-realization of biological dynamic systems. Rulers, protractors and other such tools for comprehending technical dynamic systems are not available here, since biological dynamic systems realized themselves without any help from "outside".
This leads to the trivial but substantial question as to whether nature can divide the distance between two points in three-dimensional space symmetrically only by arcs; that is, is it possible to fix two points on paper and to bisect the fictional line between them exactly without using a ruler and pair of compasses. This question was the subject of a book by the Italian mathematician Lorenzo Mascheroni (1750–1800) entitled *Geometria del compasso* (1797). Here he developed a method of solving geometric problems with only a pair of compasses and without a ruler or protractor. The analytic basis of arc geometry is the inverse transformation system with the elementary equation: $r{:}\bar{r} = \pm c^2$ (Fig. 1).
If on a straight line lies, besides a fixed point O,a movable pair of points P and $\bar{P}$, the product of $OP \cdot \bar{O}\bar{P}$ is of a constant value c^2; one says that P and $\bar{P}$ are of inverse likeness to one another with reference to the fixed point O as a pole or a centre and c^2 as the potential of that inversion (hyperbolic inversion). The inverse transformation system is valid for a plane as well as for three-dimensional space, for circle and angle and also for length. Plane or spatial formations of one kind are transformed into formations of another kind according to the relevant laws.

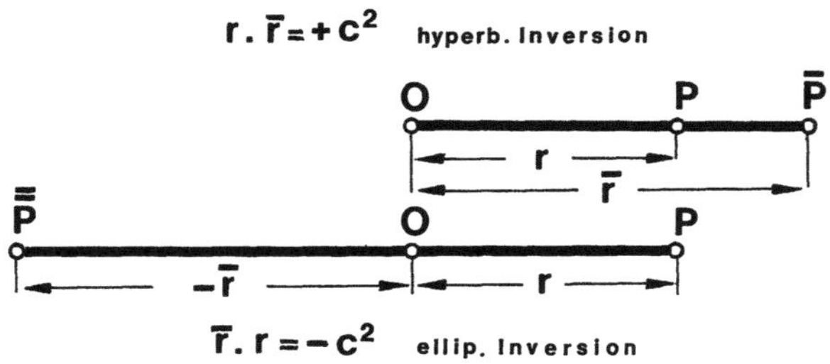

Fig. 1. The point *O* is the center of inversion or the pole with the potency of c^2. If the points *P* and $\bar{P}$ are situated on the same side of the polebeam from the centre one talks of hyperbolic inversion. The potency then is positive ($r \cdot \bar{r} = +c^2$. If the inverse points *P* and $\bar{P}$ lie on different sides of the center *O* one talks of elliptic inversion. Then the potency is negative ($r \cdot \bar{r} = -c^2$). Inversion is true to circle and angle (congruity) and is also valid for the three dimensional space (Plücker 1834)

The inverse transformation system does not need any system of coordinates – a fact of great importance for biological dynamic systems. The relation $r \cdot \bar{r} = -c^2$ (elleptic inversion) shows that complex numbers of the type $z = x + iy$ are also involved in the inverse transformation system. If it is actually the case that the only structural element free of force is the arc or the inverse transformation system for self-realization of biological dynamic systems, this must be expressed by the kinematic relation between the resting and active systems and in the relation between the lower leg and thigh, because the dynamic system realizes itself in the algorithm, the inverse transformation. This means that the kinetostatic relationships between resting and moving systems, between lower leg and thigh, must be realized by the relation $r \cdot \bar{r} = c^2$.
If the cruciate ligaments, together with their insertion points (the control system, the joint rectangle) is drawn on a plane showing the lateral view in such a way that the insertion points of the cruciate ligaments describe circles when moving (Fig. 2), then the anterior cruciate ligament shows that the point X* is the insertion point at the thigh, while the point X is the insertion point at the lower leg.

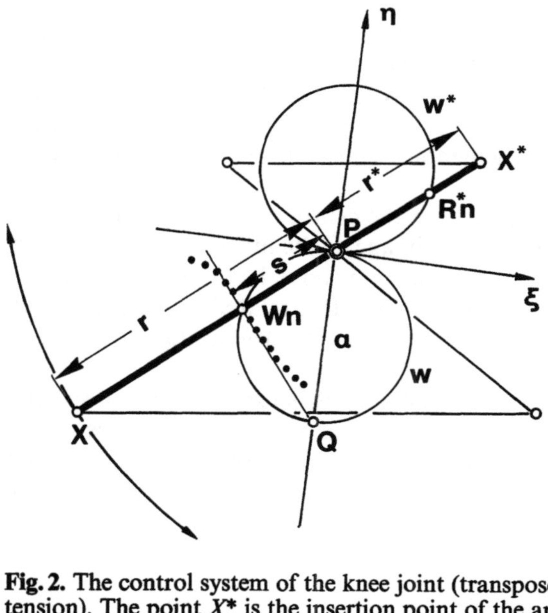

Fig. 2. The control system of the knee joint (transposed joint rectangle for flexion and extension). The point *X** is the insertion point of the anterior cruciate ligament at the thigh while the point *X* is the insertion point at the lower leg. The point *P* is the momentary center, the pivot of the knee joint, at the given moment of movement. On the entire pole beam that carries the anterior cruciate ligament a single point *Wn* exists that commands the following properties: When the path point *X* moves to the left on its circular path, then the point *Wn* describes a curve with the aperture of the curvature pointing downwards. If the point moves to the right then the point *Wn* describes a curve with its aperture pointing upwards. The direction of the curvature changes at the point *Wn*. The centre point of flexion therefore lies in infinity. The return points of all path points at this very moment of motion are situated on a circle-line, the so-called return circle, that touches the rolling (or trundle) tangent *ζ* at the momentary center *P*

The point P (crossing point of the anterior and posterior cruciate ligaments) is the momentary centre of rotation, the momentary axis P of the knee joint.
The insertion point X describes while in motion an arc at the lower leg, whose centre of curvature X* is the insertion point at the thigh. On the line which bears the anterior cruciate ligament (v) only one single point (Wn) exists, which possesses most interesting qualities. When the path point (X) moves on its path to the left (Fig. 2), the point Wn describes an orbit with the curve opening downwards. When the point X moves to the right, the point Wn describes an orbit with the opening upwards, that is, at the point Wn it changes the direction of its curvature. The consequence is that the orbit has no curvature at the point Wn. The centre of flexion, or curvature, of the point Wn therefore lies at infinity. The point Wn is described as a point of inflection, or turning point, of its path.
On every pole beam that crosses the momentary centre P such a turning point Wn exists. The entirety of all such points lies on a circle that passes through the momentary centre P and is centred on the trundle normal. This circle is called the inflexion, or turning circle w. If one examines at a given position in the system the relation of these four points of the anterior cruciate ligament – the origin X* at the thigh and at the insertion point X at the lower leg, the momentary centre P and the inflection point Wn – the following results are observed. Multiplying the length of the distance XWn (the length of distance between insertion point X at the lower leg and the inflection point Wn) by the total length of the anterior cruciate ligament XX* shows that the product $XWn \cdot XX^*$ is equal to the square $(XP)^2$ of the distance between the path-point X and the momentary centre P. The relation $XWn \cdot XX^* = (XP)^2$ is equal to the elementary equation $r \cdot \bar{r} = c^2$ of the inverse transformation system.
The equation $XWn \cdot XX^* = (XP)^2$ states that the centre of flexion X* (origin of the anterior cruciate ligament) is the inverse image of the inflection point Wn, or the inflection point Wn is the mirror image, the inverse likeness, of the centre point of path curvature (X*) under the potential $(XP)^2$, related to the path point X (insertion of the anterior cruciate ligament at the lower leg) as the centre of the inversion.
This proves that the geometry of the arc – the only structural element of self-realization in biological dynamic systems which is free of force, the inverse transformation system, the algorithm with the elementary equation $r \cdot \bar{r} = \pm c^2$ – forms the geometric base of *all* dynamic systems that command a momentary centre and that cause consistently reproducible movements.

Summary

What is different in the view of biological dynamic systems in reference to movement and the cause of movement?

1. Every consistently reproducible movement is determined by kinematic-geometric laws.
2. The question is not over the "how" of motion but the "what", that is, over the cause, the structural principle of the dynamic system.
3. Therefore the dynamic system must be examined kinetostatically while in resting mode and the control system must be developed a priori; that is, the prin-

ciple of construction cannot be found by experiment nor can it be discovered or invented. It must be developed, discovered a priori. The dynamic system should be examined in a position which requires a minimum of geometric-mathematical effort. In the case of the knee joint this would mean that position at which the tibial plane is parallel to the fossa intercondylica.

4. The discovery that the joint faces of the knee are envelopes and thus a purely geometric formation shows us that what has been regarded as impossible until now – the biological substance itself, the total amount of cell formation – realizes a purely geometric formation in its construction. The rolling and gliding motion and the orthogonal transmission of forces at the exertion points of the knee joint, e.g. the retropositio femoris and tibiae, the retroversio of the tibial plane, and the valgus of the knee, are geometric consequences and by no means biological ones.
5. Because of the self-realization of the biological dynamic systems without outside influence there must exist a transformation system that relates all parameters to one another. This transformation system – hitherto unknown in biology – is the inversion with the basic equation $r \cdot \bar{r} = \pm c^2$, which itself is given in the relationship between resting and moving systems. This means that without the inversion laws there would not be a dynamic system with a momentary centre that causes consistently reproducible movements, either in physics nor in biology.
6. Life means movement; especially for vertebrates, life means consistently reproducible movements and self-realization of dynamic systems; without movement the functions of the central nervous system, the neural system, and other organic systems would be meaningless. The reproducible movement itself with a rotation axis as momentary centre is bound to geometric transformation systems, in this case to that of inversion.

   The biological dynamic systems realize themselves in developing within themselves the laws of motion; therefore, all parameters that comprise these laws must themselves be subject directly or indirectly to the principle of order, to the inversion.

Therefore, inversion can be seen as a law of nature, as the principle of order of living beings (particularly of vertebrates), with the basic relationship $r . \bar{r} = \pm c^2$. The integration of the laws of inversion with kinematic-kinetostatic laws offers a new method of operation which can, to use Buchner's term, be referred to as "biometry". By using this method a new way is offered to search deeper into the sphere of yet unknown biological dynamic systems.

The new working mode of biometry opens many new questions, which up to now could not be asked. This is of particular importance because preceding every scientific perception is a question.

Appendix

On every pole beam that passes through the momentary centre P., the actual pair of points $X \rightarrow X^*$ form two projective lines of points. Their flight points are those intersection points, different to the momentary centre P of the pole beam with the turning circle. The double points (fixed points) are united in the momentary centre P. Therefore the projectivity is parabolic.

In geometry projective point lines are distinguished by double-relation similarity. Four points of one line and the corresponding points of the other lines have the same double relation.
The double relation (A B C D) of four points A, B, C, D of a straight line produces the following relation:

$$(ABCD)=\frac{AC}{BC}:\frac{AD}{BD}$$

When applying the laws of projective geometry to the respective points $X \rightarrow X^*$ of the flexion relation (as an example the anterior cruciate ligament is taken (Fig. 2), with its origin X^* at the thigh and its insertion point X at the lower leg, the point P is the momentary axle of the knee joint: the points Wn and Rn^* are the turning points, $Wn^{*\infty}$ and Rn^{∞} are the centre points of flexion of the turning points, at infinity. The following relations result:

$$P \leftrightarrow P,\ Wn \leftrightarrow Wn^{*\infty},\ Rn^{\infty} \leftrightarrow Rn^*,\ X \leftrightarrow X^*$$

forming thereof the double relation:

$$(PWnR_n^{\infty}X)-(PW_n^{*\infty}R_n^*X^*) \Rightarrow \overline{WnX}:\overline{PX}=\overline{PR_n^*}:\overline{PX^*}\ (\overline{PR_n^*}=\overline{WnP})$$
$$\Rightarrow \overline{WnX}:\overline{WnP}=\overline{PX}:\overline{PX^*} \Rightarrow \overline{WnX}:\overline{PX}=\overline{PX}:\overline{X^*X}$$

$$\boxed{\overline{XWn}\cdot\overline{XX^*}=(\overline{PX})^2}$$

The path point X is the centre and $(XP)^2$ the potential of the inversion. The turning point Wn and the centre point of path flexion X^* are then the inverse images of each other: they are also called conjugate points.
From the inversion equation $WnX \cdot XX^*=(PX)^2$ the Euler-Savary equation $\frac{1}{r}+\frac{1}{r^*}=\frac{1}{s}$ constant (lens equation) on a line can easily be deduced as follows:
The lens equation states that the total of the reciprocal values of the distances between the path point X and its centre point of flexion X^* from the momentary centre P has a constant value s^{-1}; "s" means the distance of the turning point Wn from the momentary centre P.
If the distance XP is denoted r and $\overline{X^*P}=r^*$ and $\overline{WnP}=s$, then we have (Fig. 3): $\overline{XWn}=(r-s)$ and $\overline{XX^*}=(r+r^*)$. Substituting these expressions into

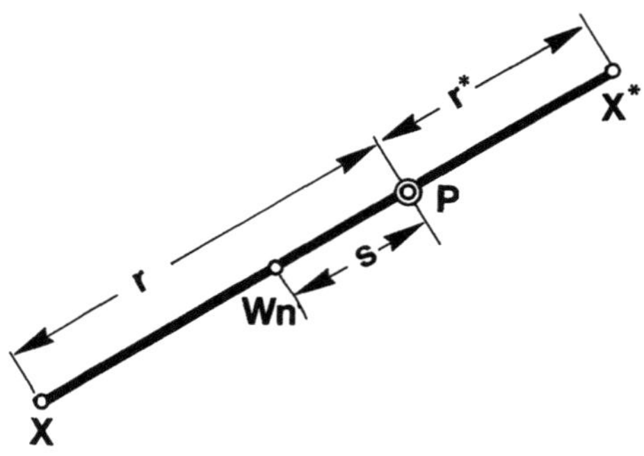

Fig. 3. Considering only the anterior cruciate ligament, the four points *X, Wn, P, X** are related to each other: By multiplying the distance *X Wn* by the total length of the anterior cruciate ligament *X X** the resulting product is equal to the square of the distance *X P* $[XWn \cdot XX^*=(XP)^2]$. That means, the insertion point X^* of the anterior cruciate ligament at the thigh is the inverse mirror image of the return point *Wn* – or vice versa – considering the path point *X* as the centre of inversion under the potency $(XP)^2$

$\overline{XWn} \cdot \overline{XX^*} = (\overline{XP})^2$ leads to:

$$(r-s)(r+r^*)=r^2$$

$$r^2-sr+rr^*-sr=r^2$$

$$rr^*=sr+sr^*$$

Dividing throughout by srr* gives

$$\frac{1}{s}=\frac{1}{r^*}+\frac{1}{r}$$

The following equation is valid for all points X→X* satisfying the flexion relation at a given moment of motion:

$$\frac{1}{r}+\frac{1}{r^*}=\frac{1}{\alpha \sin p} \text{ because } s=\alpha \sin p \text{ where}$$

α = diameter of the turning circle
p = angle between one line and the trundle tangent

References

Adler A (1890) Theorie der Mascheroni'schen Konstruktionen. Wiener Akademie, vol 99 IIa

Gruson JP (1825) L. Mascheroni's Gebrauch des Zirkels. Schlesinger'sche Buch- und Musikhandlung, Berlin

Jank W (1974) Das menschliche Knie als Gelenkviereck. Anzeiger d. math.-naturwiss. Klasse der Öst. Akademie d. Wissenschaft 10:157–162

Menschik A (1974) Mechanik des Kniegelenkes: Part 1. Z Orthop 112:481–495

Menschik A (1975) Mechanik des Kniegelenkes: Part 2. Z Orthop 113:388–400

Menschik A (1974) Mechanik des Kniegelenkes: Part 3. F. Sailer, Wien

Menschik A (1976) Die Synoviapumpe des Kniegelenkes. Z Orthop 114:89–94

Menschik A (1976) Injuries of the ligaments and their repair. In: Chapchal G (ed) Seventh international symposium on topical problems in orthopaedic surgery, Lucerne. Thieme, Stuttgart

Menschik A (1979) Kinematik und Endoprothetik des Kniegelenkes. Orthopädie-Technik 5:61–64

Menschik A (1981) Einführung in die Kinematik des Kniegelenkes unter Berücksichtigung allgemeiner Gesichtspunkte. Allg Probl Chirurg Orthop 15:9–18

Menschik A (1984) Grundsätzliches zur Kinematik und Selbstverwirklichung der unbekannten biologischen Bewegungssysteme unter besonderer Berücksichtigung des Kniegelenkes. H Unfallheilk 167:23–48

Menschik A (1987) Biometrie, das Konstruktionsprinzip des Kniegelenkes, des Hüftgelenkes, der Beinlänge und der Körpergröße. Springer, Berlin Heidelberg New York (to be published)

Müller W (1982) Kniegelenk. Springer, Berlin Heidelberg New York

Schmidt H (1950) Die Inversion und ihre Anwendung. Oldenbourg, München

Wunderlich W (1970) Ebene Kinematik, vol 1. Hochschul-Taschenbücher 447–447/a, Mannheim

The Mechanics of the Knee in the Sagittal Plane.

Mechanical Interactions Between Muscles, Ligaments and Articular Surfaces

J. O'Connor[1], T. Shercliff[1], and J. Goodfellow[2]

[1] Department of Engineering Science, University of Oxford, UK
[2] Nuffield Orthopaedic Centre, Oxford, UK

Introduction

Dr. Mueller's book (1983) cites 404 references to clinical and experimental investigations of the human knee. This is only a portion of the growing volume of published experimental work that demands, for its understanding, the development of theoretical models of the joint. These models could provide a basis for the design of experiments and the interpretation of their results. The work reported in this paper is one such development. It represents a series of hypotheses about the functional anatomy of the joint. Our object is to calculate the magnitudes of the forces transmitted by the articular surfaces, ligaments and muscle tendons at the human knee in the presence of external load and as a result of antagonistic muscle action. The results of such calculations can subsequently be compared to actual measurements in order to test the validity of the hypotheses. In this paper, we describe only the response in the sagittal plane to loads applied to the lower limb. Before we can proceed with the study of the mechanics, it is first necessary to determine the position and direction of the lines of action of the muscle, ligament and contact forces. We will start with a study of the geometry of the joint.

Geometry of the Knee

Both the geometric and the mechanical analyses are centred on the behaviour of the four-bar linkage comprising the femur, the tibia and the two cruciate liga-

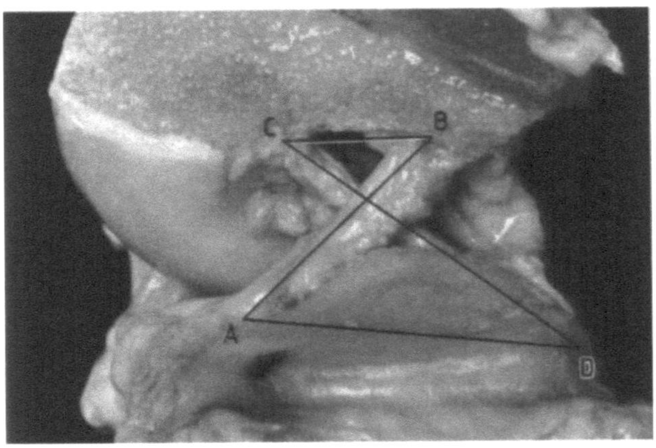

Fig. 1. A human knee with the lateral femoral condyle removed exposing the cruciate ligaments. A diagram of the cruciate linkage *ABCD* is superimposed

Surgery and Arthroscopy of the Knee
2nd Congress of the European Society
Editors: Müller/Hackenbruch

ments. A description of the knee joint in terms of the behaviour of this linkage was first attempted by Strasser (1917) and was further developed by Kapandji (1970), Menschik (1974), Huson (1974), Goodfellow and O'Connor (1978) and O'Connor et al. (1981). Figure 1 shows a human knee with the lateral femoral condyle removed, exposing the cruciate ligaments. Four lines, representing the four elements of the cruciate linkage, are shown. The line AB represents the neutral fibre of the anterior cruciate ligament, the line CD is the neutral fibre of the posterior cruciate ligament. BC is the femoral link joining the points of attachment of the cruciate ligaments to the femur. The tibial link AD joins the attachment points of the ligaments on the tibia.

Angles of the Ligaments

During flexion or extension of the joint, the angle between the tibial and femoral links changes by an amount equal to the change in the angle of flexion. The angles between each of the ligaments and each of the bone links can be calculated at each position of flexion. The algebraic formulae describing the ligament angles formed the basis of a computer program that was used to draw the linkage in various positions of flexion. Figure 2a shows the linkage in 15 such positions, with the tibial link fixed. Figure 2b similarly shows the linkage in the same 15 positions of flexion but with the femoral link fixed. With increasing flexion, the anterior cruciate ligament swings about 40° clockwise about its tibial attachment point and approaches the tibial plateau. The posterior cruciate ligament swings about the same distance, clockwise about its attachment point, departing from the tibial plateau (see Fig. 2). Both ligaments rotate about 100° counter clockwise about their attachments to the femur.

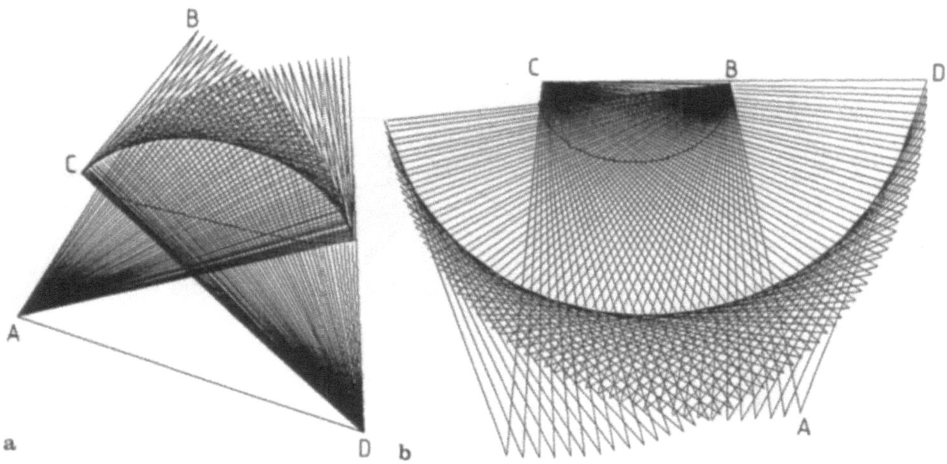

Fig. 2a, b. The four-bar linkage *ABCD* in 15 superimposed positions of flexion. **a** Fixed-tibial link *AB*. *BC* marks the position of the femoral link in full extension. The tibial centrode, the track of the flexion axis relative to the tibia, is included. **b** Fixed-femoral link *BC*. *AD* marks the position of the tibial link at 140° of flexion. The femoral centrode, the track of the flexion axis relative to the femur, in included

The Instant Centre of the Linkage, the Flexion Axis

An important feature of the linkage, on which most of the subsequent geometric and mechanical theory is based, is that the point of intersection of the neutral fibres of the two ligaments is the *instant centre* of the joint. The bones flex and extend relative to each other about an axis that passes through the instant centre. We shall refer to the instant centre as the flexion axis.

Because the geometry of the linkage changes with position, the flexion axis of the joint moves relative to each of the bones. Curves marking the track of the flexion axis relative to each of the bones are included in Fig. 2. These curves are called the *centrodes* of the linkage. The tibial centrode is approximately a straight line. The femoral centrode is an elliptical curve. Figure 3 shows the linkage in four positions with the tibial link fixed. The tibial centrode is attached to the fixed tibia. The femoral centrode is attached to and moves with the flexing femur. The two curves are seen to roll upon each other without slip. The distances between the corresponding pairs of contact points on the two centrodes are precisely equal. If the articular surfaces of the bones coincided exactly with the centrodes, they would roll without slip upon each other during flexion and extension while allowing the neutral fibres of the two ligaments to remain isometric over the flexion range. This result demonstrates the intimate relationship between the geometry of the ligaments and the geometry of the articular surfaces of a joint. Strasser (1917), Menschik (1974) and Hudson (1974) used geometric constructions, drawing the linkage in various positions, to obtain the shapes of the centrodes. In the present work, the position of the intersection of the neutral fibres of the ligaments was calculated relative to both the tibial and the femoral links, the formulae then being used in the computer program to calculate the shapes of the centrodes. The results of Fig. 3 demonstrate Kapandji's (1970) statement that: "The shape of the femoral condyles is geometrically determined by the lengths of the cruciates,

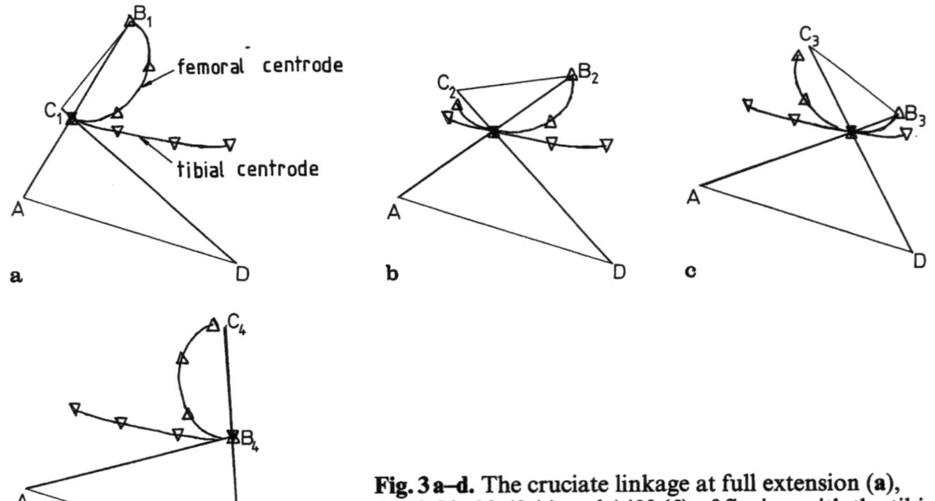

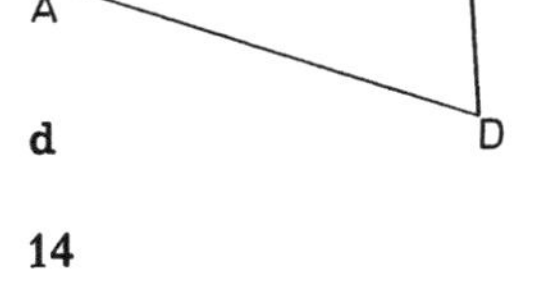

Fig. 3 a–d. The cruciate linkage at full extension (**a**), 46.7° (**b**), 93.4° (**c**) and 140° (**d**) of flexion with the tibial link *AD* and the tibial centrode fixed. The femoral centrode rolls without slip on the tibial centrode

their length ratio, and the arrangement of their insertions." In Figs. 2 and 3 as well as the the subsequent figures, the lengths of the four links, the coordinates of points of insertion of ligaments and tendons on the two bones and other parameters of the model were estimated from studies of cadaver specimens and therefore represent an approximation of the geometry of the human knee in the sagittal plane.

Geometry of the Articular Surfaces

The articular surfaces of the human knee do not coincide with its centrodes but lie distal to the flexion axis. We can choose the shape and location of the articular surface on one of the bones and calculate the compatible shape and location of the articular surface on the other. The two surfaces together then allow the bones to flex and extend upon each other while keeping the neutral fibres of the cruciate ligaments at constant length.

Menschik (1974) and Huson (1974) have shown that the femoral surface compatible with a flat tibial plateau co-incident with the tibial link is the envelope of that link when the linkage is drawn in many positions with the femur fixed, as in Fig. 2b. Equally, the tibial surface compatible with a flat femoral surface co-incident with the femoral link is the envelope shown in Fig. 2a. These examples demonstrate the truth of the assertion of Goodfellow and O'Connor (1978) that there is an infinite number of pairs of possible articular surfaces compatible with a single cruciate linkage. We use a somewhat more general technique to construct compatible articular surfaces. Figure 4a shows a concave tibial surface, which represents the medial tibial plateau. Assuming the tibial surface to be fixed and the femur moving, where would the surface of the femur make contact with the surface of the tibia? Because the flexion axis of the joint passes through the intersection point 1, all points on the femur must move on circular tracks centred at 1. If contact between the femur and the tibia were to occur at point F1, then that point would have to move tangential to its circular track and would tend to interpenetrate the tibia. Similarly, if F2 were the contact point, moving on its circular track, it would separate from the femur. If interpenetration or separation of the bones is to be avoided, contact must occur at point F, where the common normal to the articular surfaces passes through the flexion axis. With contact at F, the surfaces of the bones merely slide relative to each other.

The position of point F (Fig. 4) relative to the femoral link can be calculated, thus giving the shape of the femoral condyle that is compatible with the chosen shape of the tibial plateau. The curves representing the femoral condyles, as calculated using these methods, which are appropriate for convex and concave tibial plateaux are shown in Fig. 5. It can be seen that the distances between successive contact points on the femur are larger than those on the tibia. Therefore, during flexion, the femur slides forward while rolling backward on the tibia. During extension, the femur slides backwards while rolling forwards on the tibia. Such a pattern of rolling and sliding movements in the human knee was described by W. Weber and E. Weber (1836). It occurs because the articular surfaces of the bones lie distal to the flexion axis. Were they to lie proximal to the flexion axis, the articular surface of the femur would then be shorter than that of the tibia and the directions of rolling and sliding of the femur on the tibia would be the same.

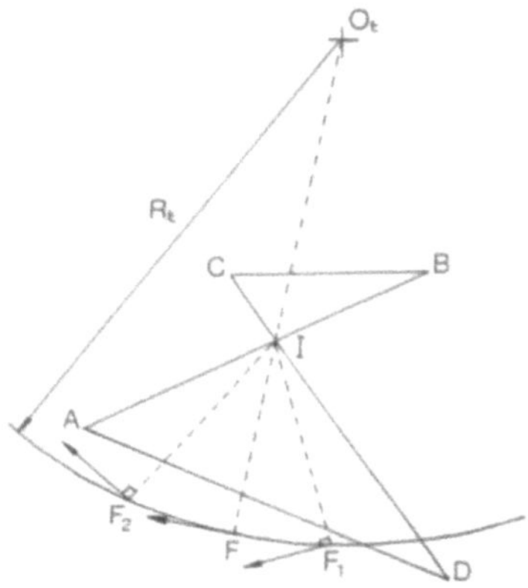

Fig. 4. Diagram showing that the point of contact *F* between the articular surfaces lies on the common normal through the flexion axis

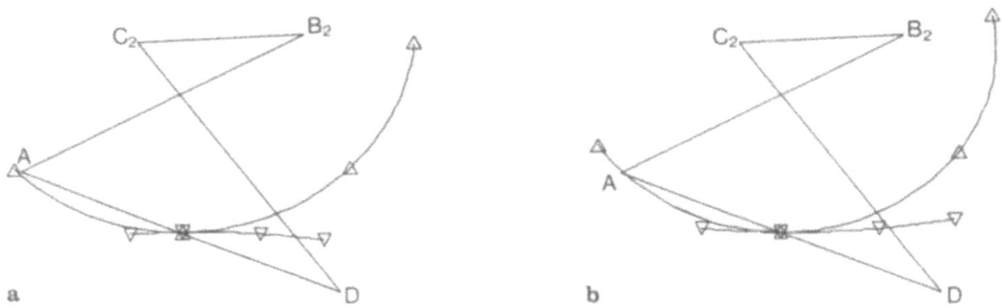

Fig. 5 a, b. Femoral articular surfaces calculated for convex (**a**) and concave (**b**) tibial surfaces

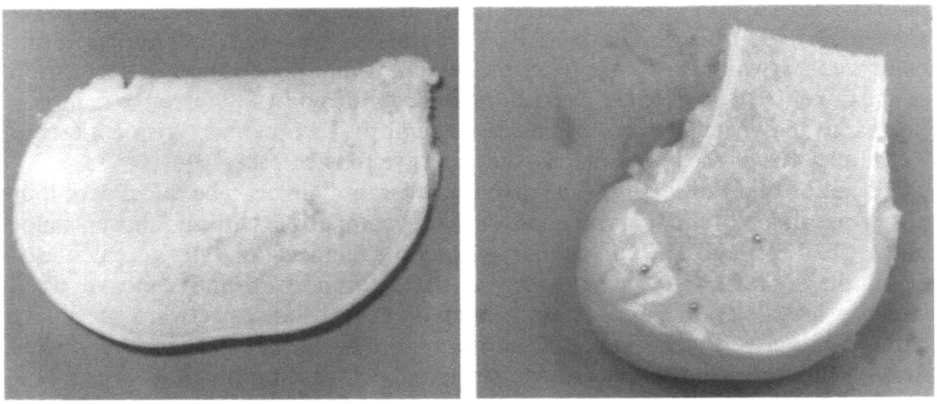

Fig. 6 a, b. Parasagittal sections of a human femur: **a** through the lateral condyle and **b** through the sulcus of the trochlear groove

In further development of the model, the tibial plateau is taken to be flat, as a compromise between the convex lateral and the concave medial plateaux of the human knee. The perpendicular to the flat tibial plateau, at its point of contact with the femur, passes through the flexion axis. We must now add the two long bones to their articular surfaces.
Dealing first with the femur, Fig. 6 shows a section through the sulcus of the trochlear groove of a human femur. The curve defining the sulcus, which guides the patella over the femur near extension, is not continuous with the curve defining the tibial facets of the femoral condyles, in contrast to the section through the condyle (Fig. 6a). In Fig. 7a, the anterior aspect of the femoral condyle is taken to be a circle, centred at the point 0, representing the shape of the sulcus. The long axes of the bones are assumed to coincide at full extension. The flat tibial plateau slopes downwards and backwards from the axis of the tibia at an angle of 13.5°. In each of the following diagrams, the model is drawn with the tibial plateau in a horizontal position. Figure 7b shows the femur drawn in a number of positions of flexion on a fixed tibia, demonstrating the phenomenon of roll back very clearly.

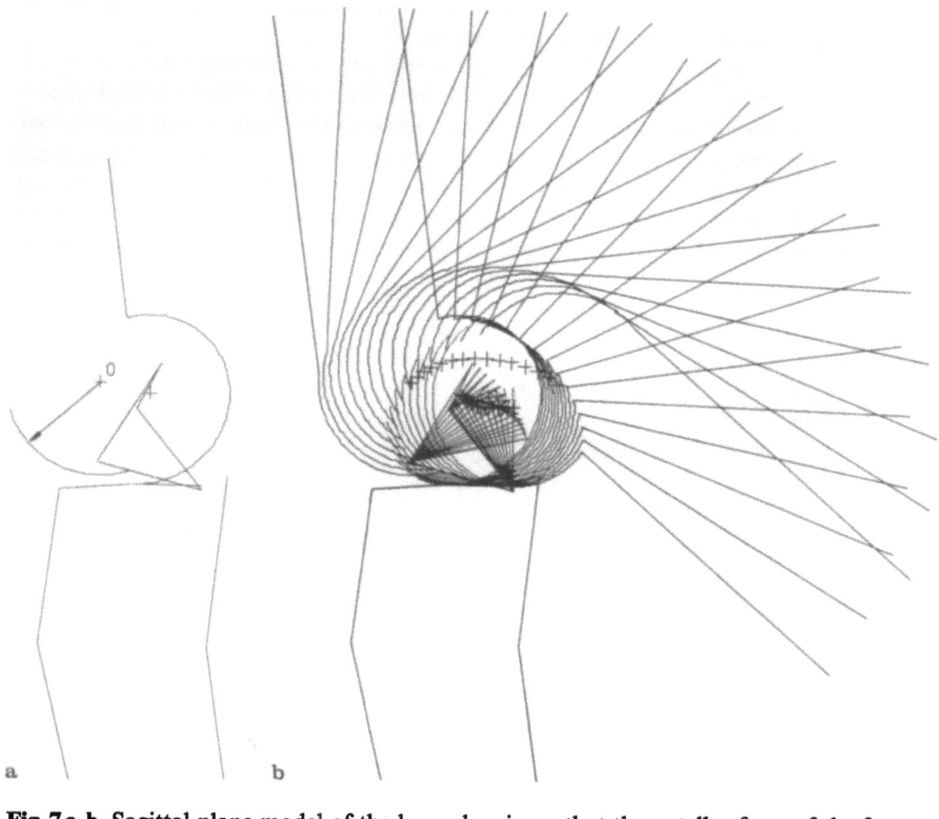

Fig. 7a, b. Sagittal plane model of the knee showing **a** that the patellar facet of the femur is taken to be a circle centered at 0 and **b** the femur drawn in a number of superimposed flexion positions to demonstrate rollback

Non-turning Loads

If the neutral fibres of the cruciate ligaments represent the lines of action of the resultant tensile forces transmitted by those ligaments, then the directions of those lines of action relative to both bones vary continuously over the flexion range (see Figs. 2, 7 b). We have argued that the common normal to the articular surfaces at their point of contact passes through the flexion axis. The articular surfaces of synovial joints exhibit very low friction. They can transmit only compressive stress from one to the other. The resultant force transmitted between them must therefore lie along the common normal at their point of contact and must pass through the flexion axis. The line of action of the tibio-femoral contact force is therefore perpendicular to the tibial plateau and passes through the point of contact. It moves backwards on the tibia during flexion and forwards again during extension.

Since the lines of action of the two ligament forces and the contact force all pass through the flexion axis, those forces have zero moment, that is, zero turning effect about that axis. Some combination of these three forces is then sufficient to balance any external load applied to the tibia with a line of action also passing through the flexion axis. Such loads are called *non-turning* loads. Muscle action is not required to balance non-turning loads.

We give some simple examples of the response to non-turning loads in Fig. 8. Figure 8 a shows a load applied through the foot with a line of action passing perpendicular to the tibial plateau through the flexion axis. Such a load can be balanced at the knee joint by the compressive contact force C alone, without ligament or muscle action. In Fig. 8 b, the load W is applied to the tibia precisely along the line of the anterior cruciate ligament. It is balanced at the knee joint by a tension force A in that ligament alone, without muscle or contact forces.

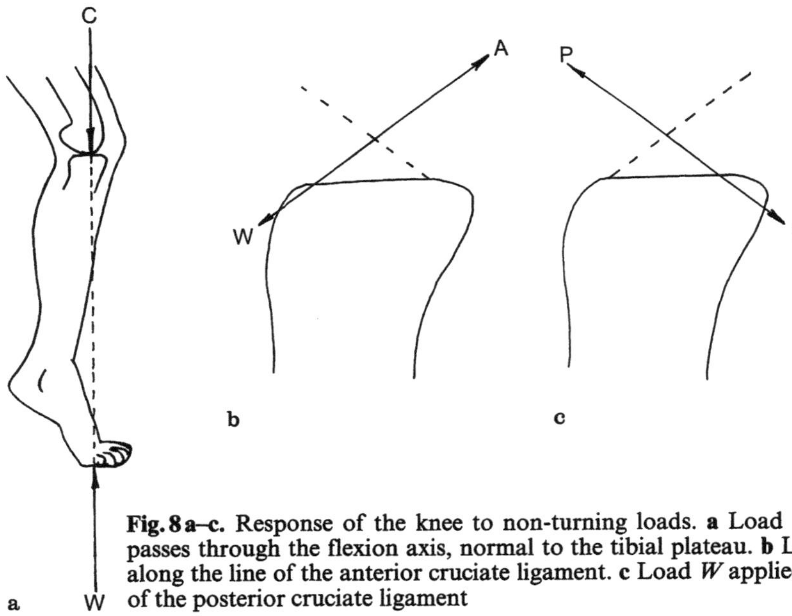

Fig. 8 a–c. Response of the knee to non-turning loads. **a** Load W, whose line of act passes through the flexion axis, normal to the tibial plateau. **b** Load W applied precis along the line of the anterior cruciate ligament. **c** Load W applied precisely along the l of the posterior cruciate ligament

Figure 8c shows, similarly, a load applied to the tibia along the line of the posterior cruciate ligament, balanced by a tension force in that ligament alone. The forces *C*, *A* and *P* are each equal to the load *W*.

The examples in Fig. 9 are simulations of the drawer test. In this test, the examiner slides the tibia either forwards or backwards on the femur, keeping the flexion angle fixed. Since flexion or extension are specifically inhibited, the effective line of action of the force applied by the examiner to the tibia must pass through the flexion axis.

In Fig. 9a, an anteriorly directed load *W* applied to the tibia is balanced by tension force *A* in the anterior cruciate ligament and to tibio-femoral contact force *N*. In Fig. 9b, the posteriorly directed load *W* applied to the tibia is balanced by a tension force *P* in the posterior cruciate ligament and the tibio-femoral contact force. In each case, the two forces are sufficient to balance the external load: The component of the ligament force parallel to the plateau is equal to the load; the component of the ligament force normal to the plateau is equal to the contact force. The magnitudes of the ligament and contact forces are proportional to the applied loads and can be calculated by a simple parallelogram of forces, as indicated in the diagrams.

In Figs. 9a and b, values of the ligament and contact forces, expressed as proportions of the applied load, are plotted against the angle of flexion. The ligament and contact forces induced by the drawer test vary with position because the directions of the ligaments vary. In the anterior drawer test, Fig. 9a, the anterior

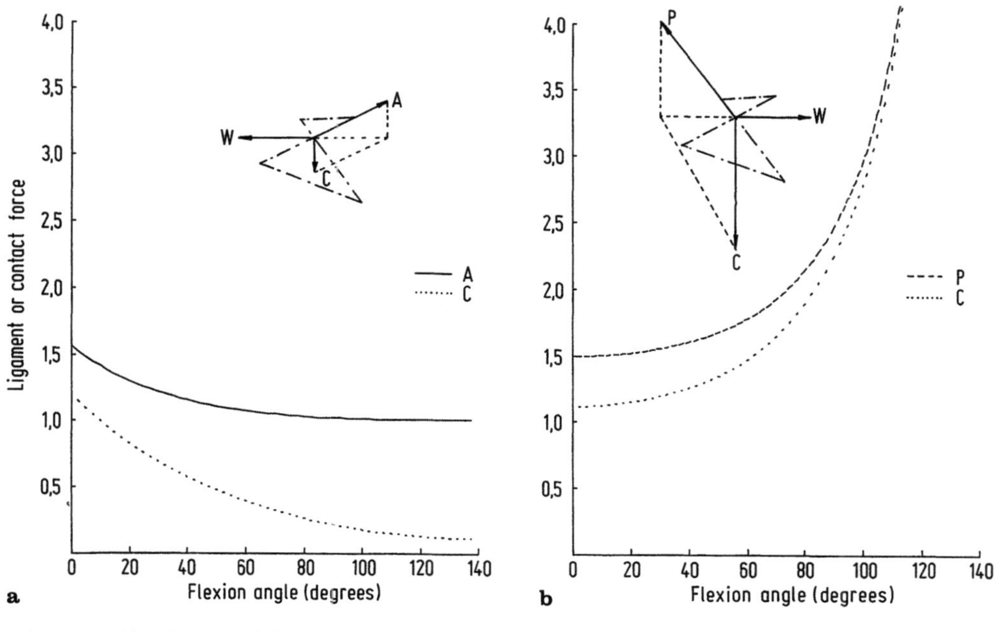

Fig. 9 a, b. Simulation of the drawer test. Calculated values of ligament and contact forces plotted against the angle of flexion. **a** Response to anterior load. **b** Response to posterior load. The curves give the values of the forces divided by the applied load

cruciate and contact forces both diminish with increasing flexion as the anterior cruciate ligament moves towards the direction of the applied load. In the posterior drawer test, Fig. 9b, the posterior cruciate and contact forces both increase with increasing flexion as the posterior cruciate moves away from the direction of the applied load.
In the examples shown in Figs. 8 and 9, it was possible to balance loads applied to the tibia through the flexion axis by some combination of ligament and contact forces without muscle action. When the line of action of the load does not pass through the flexion axis, it tends either to flex or extend the joint. Muscle action is then essential for equilibrium.

Geometry of the Muscle Tendons

In order to be able to balance the flexing or extending effect of a load that does not pass through the flexion axis, it is essential that the lines of action of the tendons of the flexor and extensor muscles should similarly not pass through the flexion axis. A tension force in a tendon lying anterior to the flexion axis will induce extension or resist flexion. Such a force in a tendon lying posterior to the flexion axis will induce flexion or resist extension. In the sagittal plane, we model the tendons of the extensors (quadriceps) and the flexors (hamstrings or gastrocnemius) as single straight-line elements. These lines may be taken to represent the lines of action of the resultant tension forces transmitted by the tendons to the bones.

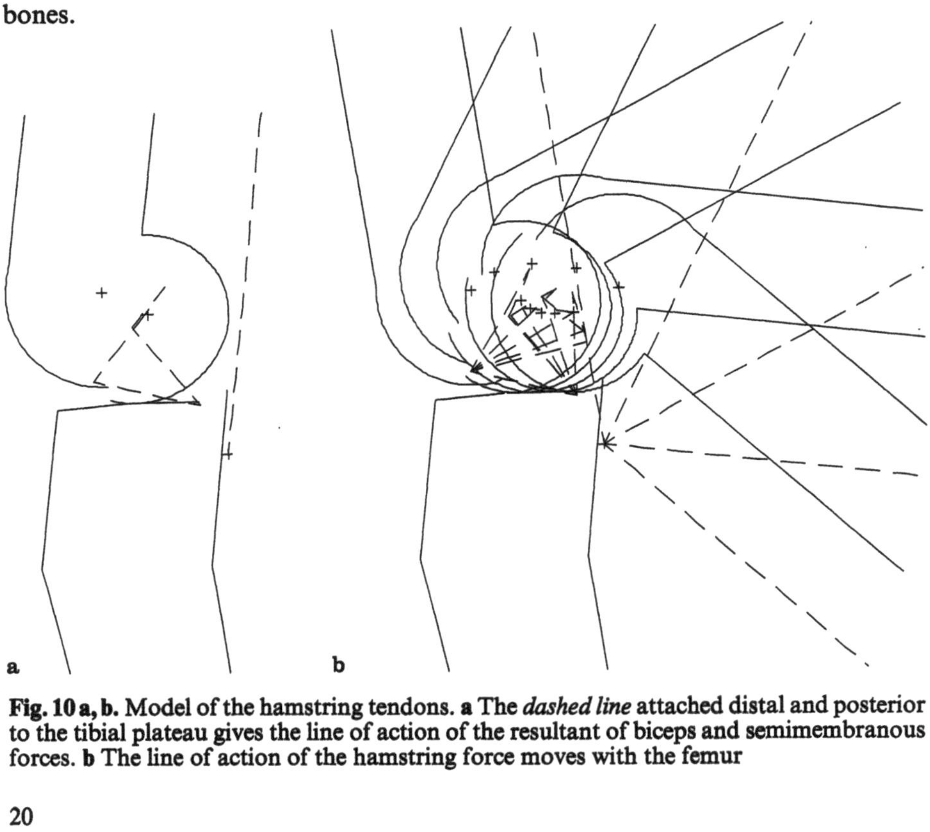

Fig. 10 a, b. Model of the hamstring tendons. **a** The *dashed line* attached distal and posterior to the tibial plateau gives the line of action of the resultant of biceps and semimembranous forces. **b** The line of action of the hamstring force moves with the femur

Hamstrings

The hamstring tendons, mainly those of the biceps and semimembranosus, are modelled as single-line elements attached to the posterior tibia somewhat distal to the joint cleft (Fig. 10a). The tendon is assumed to lie parallel to the axis of the femur so that its direction, relative to the tibia, changes continuously during flexion (Fig. 10b).

Gastrocnemius

Distally, gastrocnemius inserts into the Achilles tendon; proximally, the tendon arises from the back of the femur. When near full extension, the tendon wraps

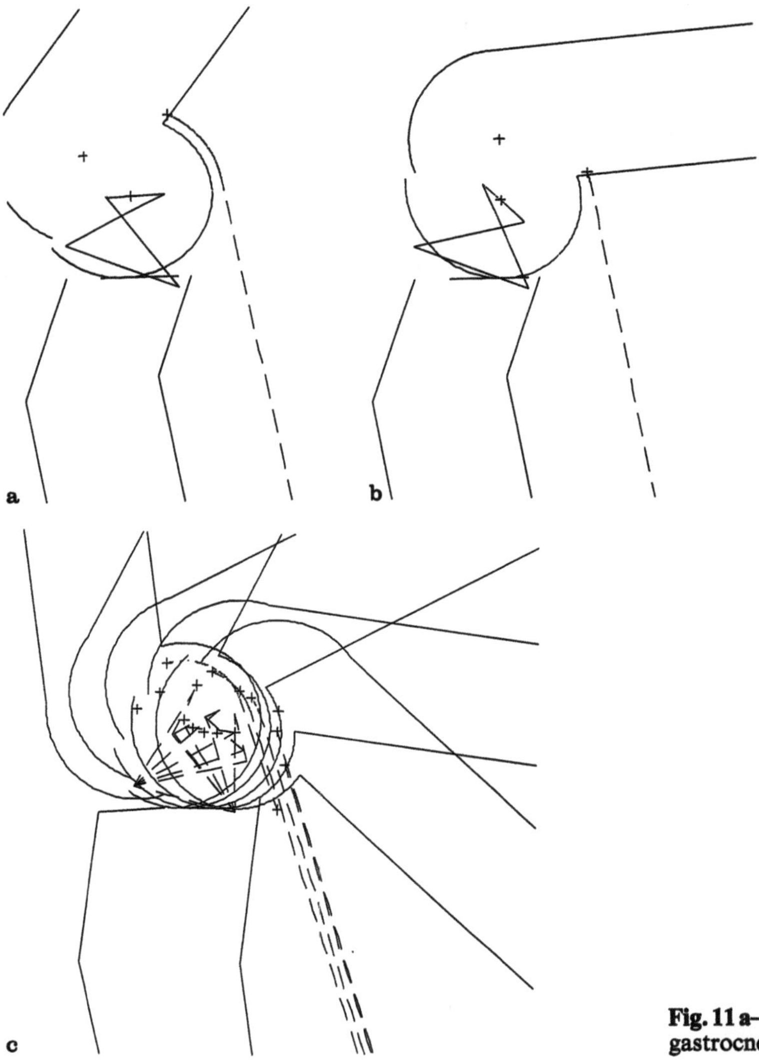

Fig. 11 a–c. Model of the gastrocnemius tendons

around the posterior aspect of the femur (Fig. 11 a). The tendon has finite thickness and is shown lying somewhat posterior to the surface of the femoral condyle. At higher flexion angles, the tendon loses contact with the condyle and hangs from the back of the femur (Fig. 11 b). Although the gastrocnemius muscles lie generally parallel to the tibia, the line of action of the tendon as it traverses the knee joint changes somewhat with increasing flexion due to the roll back of the femur on the tibia (Fig. 11 c).

The Patella-Femoral Joint

Figure 12 a shows a free-body diagram of the patella. In the sagittal plane, the patella is loaded by three forces; tension in the quadriceps and patellar tendons as well as the compressive patella-femoral contact force. For equilibrium, the lines of action of these three forces must be co-planar and must meet at a single point. We call that point the patellar force centre. In the simplest model of the patella-femoral joint, we assume that the length of the patellar tendon from the tibial tubercle to the patellar force centre is constant and we assume that the patellar force centre moves on a circle whose centre coincides with that of the patellar facet of the femur (the point 0 in Fig. 12 b). The quadriceps tendon is taken to lie parallel to the femur. At high flexion angles, the quadriceps tendon wraps around the front of the femur, forming the tendo-femoral joint (Fig. 12 c). At high angles of flexion, therefore, compressive forces are applied to the femur at the knee joint at three distinct places: at the tibio-femoral joint, at the patella-femoral joint and at the tendo-femoral joint. The patella-femoral and, more particularly, the tendo-femoral forces help to eliminate the bending moments induced in the femur by the tibio-femoral force.

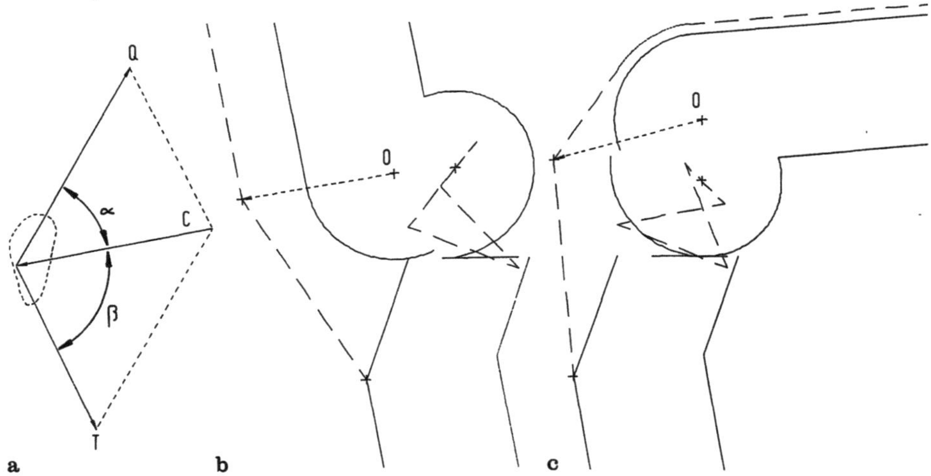

Fig. 12. a Free-body diagram of the patella. The line of action of the quadriceps force Q and the patellar tendon force T coincide with that of the patella-femoral force. C The angles α and β are not equal. **b, c** Model of the quadriceps mechanism

The forces in the quadriceps and the patellar tendons are not equal because the angles between each of the tendons and the line of action of the patella-femoral contact force are not equal. We can calculate the ratio of the forces in the two tendons by considering equilibrium of the patella. The force ratio is plotted against flexion angle in Fig. 13 and is compared there with measurements made by Bishop and Denham (1977) and by Ellis et al. (1981). The calculation describes the measurement quite well. This comparison is a test not only of a model of the patella-femoral joint, but also of the modelling for the cruciate linkage because the cruciate linkage controls the roll back of the femur on the tibia and therefore the inclination of the patella tendon. In Fig. 14a, the femur is shown flexing on a fixed tibia with the quadriceps mechanism in place. Due to roll back, the inclination of the patella tendon relative to the tibia changes by about 30°. The patella tendon pulls the tibia forwards near extension; it pulls the tibia backwards when the knee is highly flexed. Figure 14b shows X-rays of a human knee at full extension and 90° of flexion. A wire threaded along the patellar tendon points anteriorly in extension and posteriorly at 90°. The computer model of Fig. 14a reproduces this effect quite well.

In summary, the lines of action of the three principle muscle tendons spanning the joint cleft of the knee change their inclinations relative to the tibia in a systematic and predictable fashion during flexion and extension.

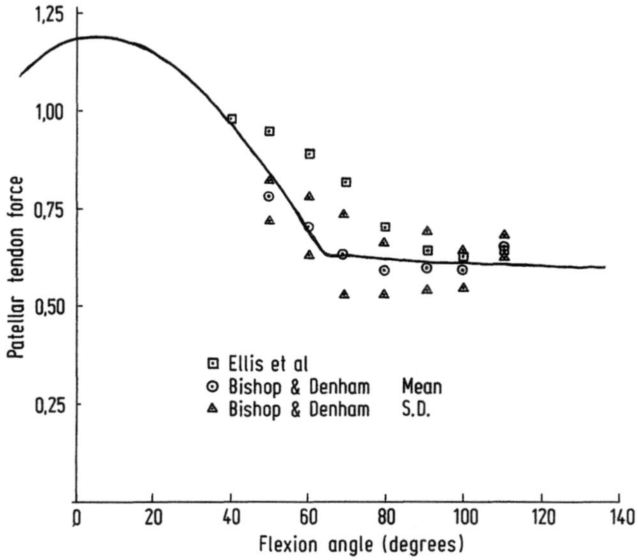

Fig. 13. The *solid curve* shows the variation with flexion angle of the ratio of patellar tendon to quadriceps tendon force. The *points* represent measurements by Bishop and Denham (1977) and Ellis et al. (1981)

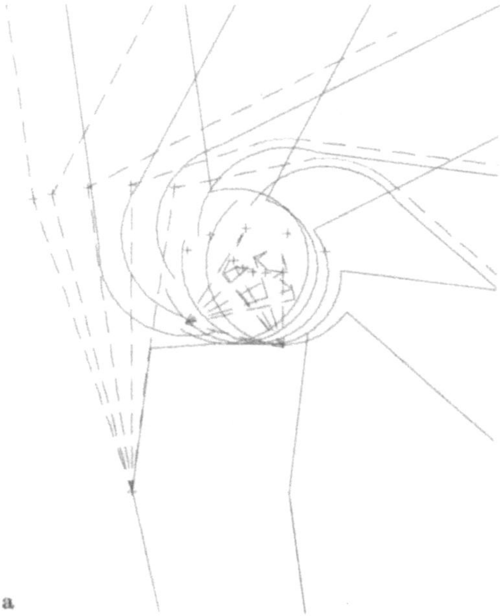

a

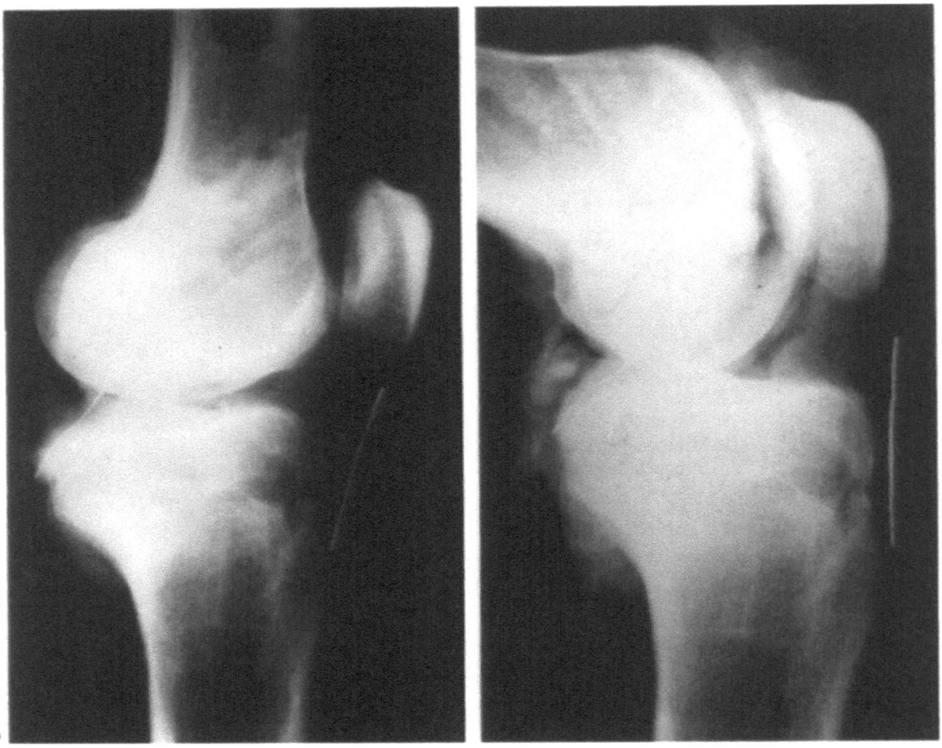

b c

Fig. 14. a Demonstration of changing direction of the patellar tendon. **b, c** Human knee with wire placed along the patellar tendon

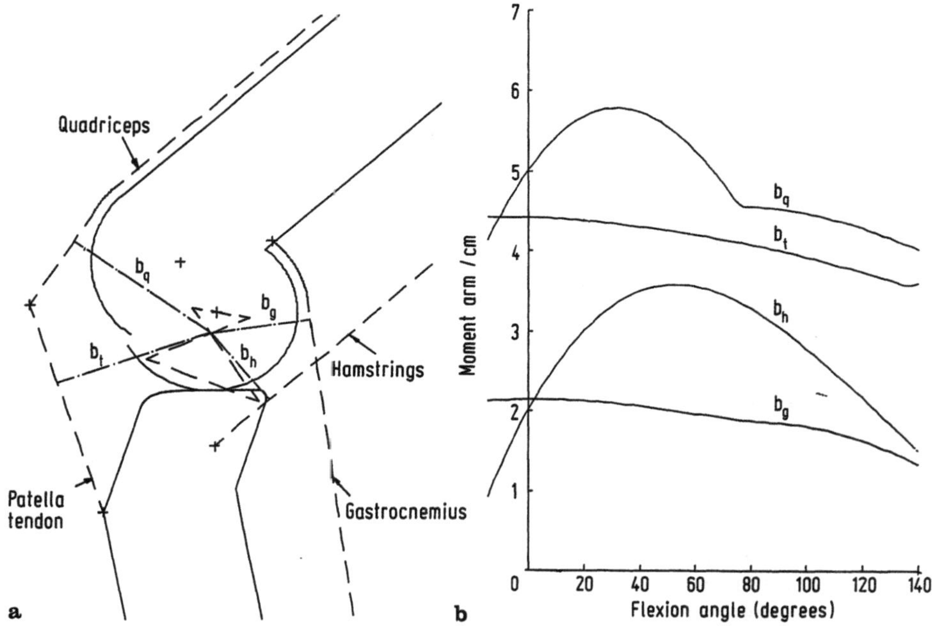

Fig. 15 a, b. Muscle lever arms. **a** The lever arms of the three principal muscles about the flexion axis. **b** The calculated values of the lever arms plotted against flexion angle

Muscle Lever Arms

The power of a muscle to induce or resist flexion or extension of the bones about the flexion axis depends on the length of its intrinsic lever arm. Figure 15a show diagrams of the lever arms of the three principle muscles tendons and the perpendicular distances from the flexion axis to the patellar, hamstring and gastrocnemius tendons, respectively. The lengths of the lever arms are plotted against the flexion angle in Fig. 15b. They vary with flexion angle for two reasons: (a) because the directions of the tendons change and (b) because the position of the flexion axis changes.

Flexing and Extending Loads

The modelling of the knee joint in the sagittal plane is now complete. It enables us to calculate the position and direction of the lines of action of the tibio-femoral contact force and of the tension forces in each of the muscle tendons and cruciate ligaments. With this information, it is then possible to calculate the magnitudes of the forces transmitted by these structures when external loads, tending to flex or extend the knee joint, are applied to the tibia. Even when the line of action of the load relative to the tibia is fixed, the relationships between the load and the muscle, ligament and contact forces change with position because of the changing lines of action of those forces and the changing lever arms of the muscle tendons.

Single Muscle Action

We shall consider first situations in which the turning effect of a flexing load is precisely balanced by a tension force in either the patellar tendon, the hamstrings or the gastrocnemius tendons. The tibia can then be held in equilibrium in the presence of external load by a single muscle force plus a single ligament force plus the tibio-femoral contact force. We need to define the circumstances in which a particular ligament acts with a particular muscle to balance a particular external load.

As seen from Fig. 15b, the lever arms available to the muscle tendons vary in magnitude between about 2 and 4 cm. When external load is applied through the sole of the foot, the external lever arm available to that load may be as much as the length of the shank, about 40 cm. Since the turning effect (distance × force) of the muscle force is equal to that of the load, the muscle forces may be 10–20 times larger than the applied load. Since the inclination of each muscle tendon relative to the tibia varies with position, it will rarely be true that the component of the muscle force parallel to the plateau and tending to slide the tibia in one direction is precisely equal to the component of the load in the opposite direction. In the absence of antagonistic muscle action, equilibrium can be achieved only by means of a tension force in one or the other of the cruciate ligaments. The anterior cruciate ligament pulls the tibia backwards and upwards, the posterior cruciate ligament pulls the tibia forwards and upwards. In any one situation, the difference between the components of the muscle force and the load parallel to the plateau will require either a backwards pull or a forwards pull from the anterior or posterior cruciate ligament, respectively, for equilibrium.

A specific example is shown in Fig. 16. A horizontal load W tending to flex the knee is balanced by tension T in the patellar tendon. In Fig. 16a, the inclination

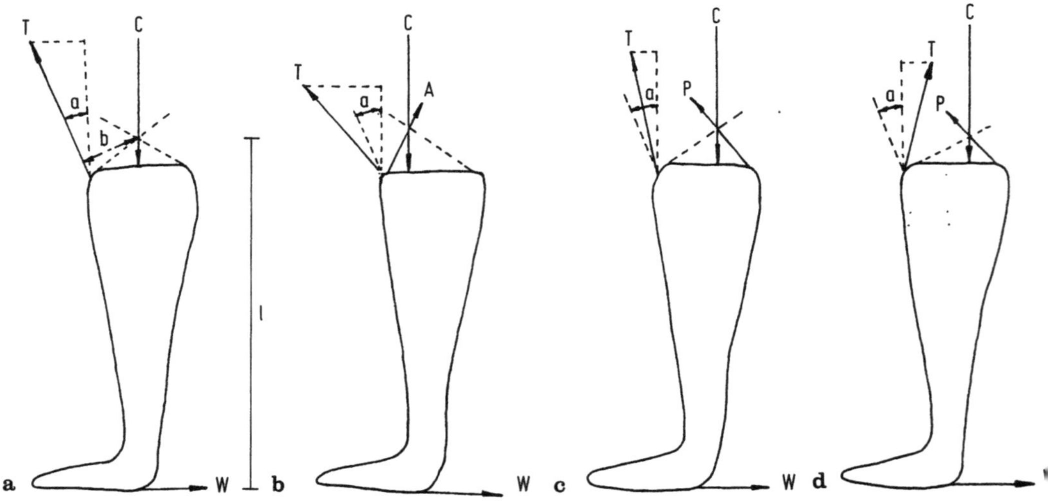

Fig. 16a–d. Critical tendon angle α. **a** Turning load W balanced by a contact force and tendon force when the tendon angle $= \alpha$. **b** Tendon angle $> \alpha$, a force is also required in the anterior cruciate ligament. **c** Tendon angle $< \alpha$ and **d** in the opposite direction to α, a force is required in the posterior cruciate ligament

of the tendon to the tibial plateau is such that the component of the muscle force parallel to the plateau is precisely equal to the load. No ligamentous action is then needed. We call the angle between the tendon and the normal to the tibial plateau in these circumstances the critical tendon angle α. In Fig. 16 b, the angle of the tendon exceeds its critical value; the component of the tendon force parallel to the plateau is larger than the load. Equilibrium is then achieved by means of the backward pull of the anterior cruciate ligament. In Fig. 16c, with higher angles of flexion, the tendon angle is now less than its critical value. The component of the muscle force parallel to the plateau is now less than that of the load and is augmented by the forwards pull of the posterior cruciate ligament. The backwards push of the load is balanced by the combined forward pulls of the muscle and the posterior cruciate ligament. Relatively large forces are needed in the posterior cruciate in Fig.16d where the tendon force has a backward-directed component in the same direction as the load.

The previous paragraph suggests that when the flexing or extending effect of an external load is balanced by a single muscle group, a tension in the anterior cruciate ligament is required for equilibrium near full extension and a tension in the posterior cruciate ligament is required in the flexed joint. The critical angle of flexion at which the cruciate ligaments interchange their roles is that at which the muscle tendon is inclined at its critical value to the tibial plateau.

As examples of these muscle and ligament interactions, we present the results of the calculations of muscle, ligament and contact forces for a simulation of quadriceps strengthening exercises (Fig. 17) where weight suspended at the end of the tibia is balanced by tension in the quadriceps mechanism. Since the muscle, ligament and contact forces are all proportional to load, we present results in non-dimensional form, where the ratio of each of the forces divided by the load is plotted against flexion angle (Fig. 18). Because of the mechanical advantage enjoyed by the load (Fig. 17), the muscle and contact forces are very large, in places reaching 18 times the value of the applied load. It can be seen that the anterior cruciate ligament is under tension up to about 95° of flexion, the posterior cruciate liga-

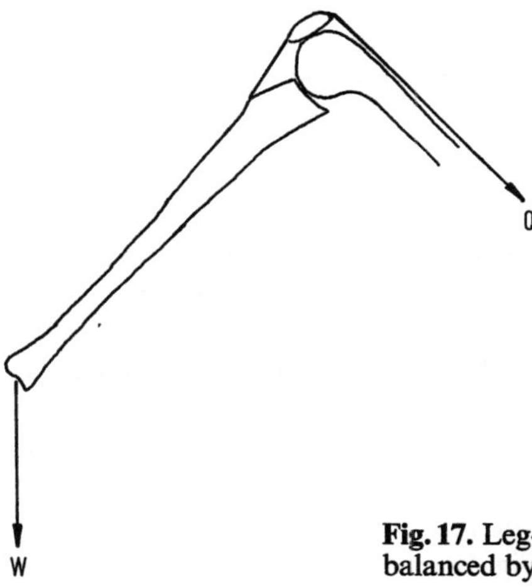

Fig. 17. Leg-lift experiment. Weight suspended from the tibia balanced by tension in the quadriceps muscles

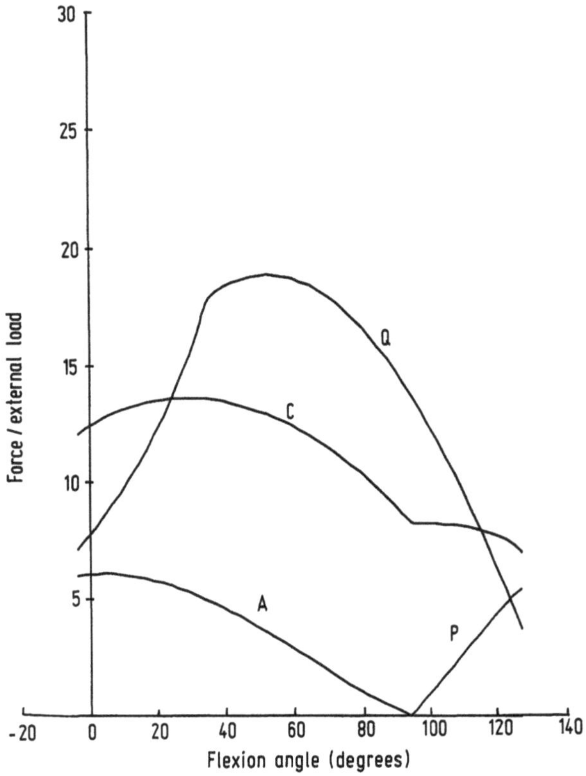

Fig. 18. Quadriceps and cruciate ligament action in the leg-lift experiments. Forces in the quadriceps tendon *Q*, anterior cruciate ligament *A*, posterior cruciate *P* and at the tibio-femoral contact *C* are plotted against flexion angle. Each curve gives the value of the force divided by the load

ment thereafter. When the hamstrings are balancing extending loads, we find that the ligaments interchange their roles at about 20° of flexion. When the gastrocnemius muscles are balancing an extending load, the point of interchange is at 100° of flexion.

Evidence will be presented elsewhere that demonstrates that the values of the muscle and the ligament forces calculated in Fig. 18 are in reasonable agreement with measurements made on cadaveric specimens. We have not included in this figure the effects of the posterior capsule in limiting extension.

Antagonistic Muscle Action

In the previous example, we calculated the values of three forces at the knee required for equilibrium of the tibia in the presence of flexing or extending loads: the tibio-femoral contact force, one cruciate ligament force and one muscle force, with the remaining muscles and ligaments unloaded. In this section, we discuss equilibrium of the tibia under the action of the tibio-femoral force and *two* muscle forces, with the remaining muscle and all ligaments unloaded. This example defines the circumstances in which antagonistic (or agonistic) muscle action can protect the ligaments.

The dashed lines in Fig. 19 are the antagonistic muscle solution for the configuration of Fig. 17. The solid lines are those of Fig. 18 and are included for compari-

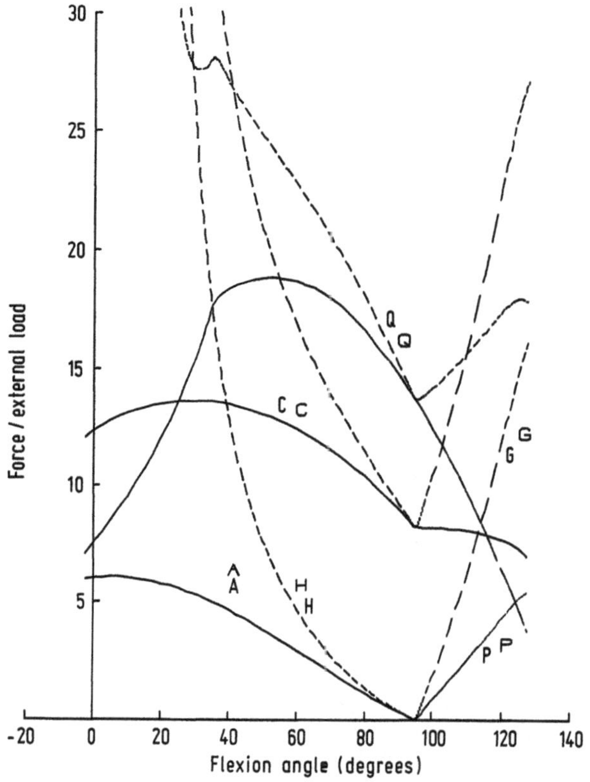

Fig. 19. Antagonistic muscle action. The *dashed lines H* and *G* show the values of the forces in the hamstrings and gastrocnemius tendons. The *solid lines* are those of Fig. 18

son. Near full extension and at high angles of flexion, the required muscle and contact forces become very large compared with the single muscle solution. Within these limits, it will be seen that the muscle and contact forces in the antagonistic solution are always larger than those of the single muscle solution, except at the critical flexion angle of 95° where quadriceps and contact forces alone are sufficient for equilibrium and where the antagonistic and single muscle solutions are identical. It will be seen that the price paid for using antagonistic action to protect the ligaments is an increase in the muscle and contact forces. Conversely, ligament action serves to minimise the muscle and contact forces necessary to balance any external load.

The reason that antagonistic action without ligament forces is not possible near extension or in full flexion is that, in these extremes, all three muscles pull the tibia in one direction only; near extension, they pull the tibia forwards parallel to its plateau, in full flexion they pull it backwards. This is in contrast to the cruciate ligaments; the anterior cruciate pulls the tibia backwards and the posterior cruciate pulls it forwards so that one of them is always available to provide a component of force parallel to the plateau.

Clinical Applications

This paper has described the basic elements of a computerised model of the knee joint in the sagittal plane. Development of the model continues. In particular, final values of the parameters of the model are still being considered in order to bring the results of the calculations into the best possible agreement with measurements.

We are using results calculated from the model to guide clinical practice in three separate ways. In operations intended to repair or replace one of the ligaments of the knee, the calculations can define the position on each bone of ligamentous fibres that would remain isometric during flexion and extension. It is then possible to calculate the length change patterns in non-isometric fibres. Examples such as Fig. 18 can then be used to help define regimes of activity in which the repaired or replaced ligament is placed under minimal tension. Calculations leading to Fig. 6 can help to define the shapes of articular surfaces that are compatible with the ligaments of the knee joint, with application to the design of total knee prostheses.

Acknowledgements. The work reported here was supported by the Arthritis and Rheumatism Council, London. Mr. David Fitzpatrick and Mr. John Mooney helped with the production of the diagrams.

References

Bishop RED, Denham RA (1977) A note on the ratio between tensions in the quadriceps tendon and the infra-patellar ligament. Eng Med 6:53

Goodfellow J, O'Connor J (1978) The mechanics of the knee and prosthesis design. J Bone Joint Surg [Br] 60:358–369

Huson A (1974) Biomechanische Probleme des Kniegelenks. Orthopade 3:119–126

Kapandji IA (1970) The physiology of the joints, vol II. Churchill Livingstone, Edinburgh

Menschik A (1974) Mechanik des Kniegelenkes, Teil I. Z Orthop 112:481–495

Müller W (1983) The knee. Springer, Berlin Heidelberg New York

O'Connor J, Goodfellow J, Biden E (1981) Designing the human knee. In: Stokes IAF (ed) Mechanical factors and the skeleton. Libbey, London

Weber W, Weber E (1836) Mechanik der menschlichen Gehwerkzeuge. Göttingen

The Evaluation of Ligamentous Knee Joint Instability in the Gait Laboratory

J. U. Baumann

Neuro-Orthopaedic Department, Children's Hospital, 4005 Basel, Switzerland

Introduction

In normal daily life, with the exception of sport activities, ligamentous knee instability may be sufficiently compensated for by muscles and osseous structures. Where residual or recurrent instability following one or several operative interventions persists, the indication for a further procedure is difficult to determine. The examination of the patient's gait can help the physician to judge the functional importance of the impairment and is a valuable reference for further planning and assessment of treatment.

During a period of 3 years, 16 patients with injured knee ligaments were evaluated systematically in order to detect even minor gait disturbances during walking at normal speeds on even ground. In all patients, the initial injury had occurred years before, and all had undergone one or several operations prior to examination. The tests were in addition to thorough clinical examination.

Methods

Gait movements were registered by cinephotography on 16-mm film with an exposure time of 1/500 s or less, and camera speeds of either 50 or 100 images/s. Straight views from front and back as well as lateral views from the left and right were taken while the patients walked over a pair of transparent force plates (Kistler Z4305). In addition, a series of steps was recorded by a pair of high speed cameras moving with the patients on rails while they walked over a distance of 8–10 m. This allowed the evaluation of step-by-step variability at a walking speed that was normal for each person. The position of markers indicating joint axes of hip, knee, and talocrural joints as well as the corresponding axes of the body segments were digitized and are represented graphically. Angular values for flexion and extension of hip, knee, and ankle were measured from the lateral film images and were averaged over three steps. Varus-valgus angles and transverse rotation at the knee joint were evaluated from the frontal views while transverse rotation of the feet could be measured from pictures through the transparent force plates from below. Two typical examples are presented.

Results

Case No. 1 (FB 848)

Injury. This 21-year-old female patient suffered an injury to the right knee while skiing at the age of 16 years. As a result, she underwent a series of operations includ-

Surgery and Arthroscopy of the Knee
2nd Congress of the European Society
Editors: Müller/Hackenbruch

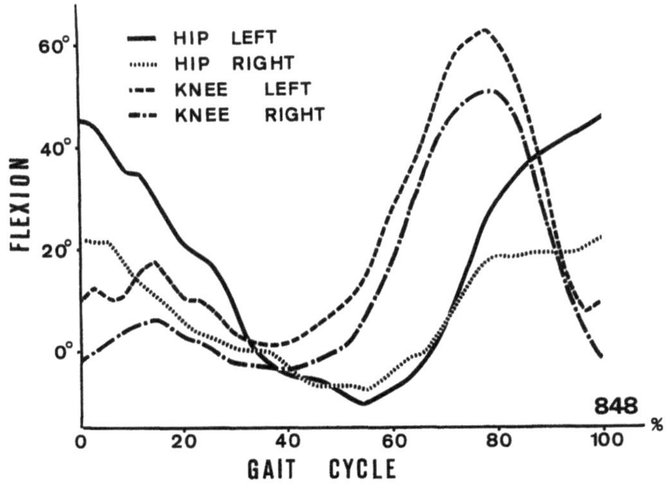

Fig. 1. Average angular values of three steps for flexion extension of the right and left hip and knee joints. Marked reduction of right hip and knee flexion during stance and swing phases. Extended knee gait on the right in presence of anterior cruciate ligament insufficiency, combined with almost normal floor reaction forces

ing medial meniscectomy, replacement of the anterior curciate ligament using the McIntosh procedure, and a semimembranosus transfer. The patient complained about instability of the right knee even when walking on flat ground. Pain occurred regularly in the evening while she was standing, sitting, or lying, but rarely when walking. There was a moderate degree of general ligamentous laxity. The passive range of motion of the knee joints with the hips extended permitted 140° of flexion bilaterally, while extension reached 15° beyond neutral on the right and 10° on the left side. Examination revealed posterolateral and posteromedial instability as well as anterolateral rotatory instability of the right, but normal stability of the left knee.

Gait Evaluation. Gait evaluation from the marker positions on the *lateral* film frames showed pathological overextension of the right knee on initial contact of the heel (Fig. 1). Knee flexion during contact response in an early stance phase was missing on the right but was normal on the left side. During propulsion in the late stance phase, the right knee was regularly extended to neutral while the left one maintained a normal flexion of 3°–4°. Knee flexion in the swing phase reached 57° on the right and a normal value of 62° on the left. *Frontal* views showed a shift of the right knee into 3°–4° of varus during early stance. On the left side, a normal valgus angulation was maintained throughout the gait cycle. The pictures taken *vertically* from below, through the glass top force plates, showed external rotation of the longitudinal axis of the foot by an average of 5° on the right and 6° on the left side.

Force Plate Measurements. The *vertical force-time record* shows a normal heel-strike response bilaterally, as well as normal and equal vertical force action during the phases of loading and breaking of the steps. In the second half of stance, during propulsion, floor pressure by the impaired right leg is 20% less than by the left one. The short initial period of propulsion upon loading of the heel is followed by a bilateral breaking shear force of equal intensity. Propulsion by the impaired right leg is reduced by 10% when comparing average values of five steps. The frontal shear forces initially show a normal and equal amount of medial shear, which stabilizes the foot and prepares the leg for the exertion of lateral shear throughout most of the stance phase. On the right side, this medial shear force is maintained for an abnormally long time, 0.07 s on the average. The temporal development and amplitude of the lateral shear force connected with foot loading and breaking is almost normal bilaterally. The second maximum of lateral shear, which is connected with propulsion, is slightly reduced in amplitude on the right.

Summary. For this patient with global instability of the right knee and generalized joint laxity, the comparison of leg movements and foot-to-ground force transmission provides different but complementary information. The movements of the knee and hip joint of the right leg are considerably disturbed. The right leg is unusually extended in early stance, i.e., during initial loading and breaking. There is also exaggerated knee extension during propulsion in late stance and diminished flexion during the swing phase. The foot-to-ground force records, in contrast, show almost normal force transmission during the first half of the stance phase. Only the prolonged period of inital medial shear indicates a reaction to anterolateral knee instability. During the second half of the stance phase, which corresponds to propulsion, force transfer by the right foot is reduced by 10%–20% in comparison to the left foot, which reduces the moment leading to the loading of the posterolateral structures of the knee.
In this patient, gait function, as evaluated from floor reaction forces, is mainly affected by posterolateral instability, but is only mildly affected by the lack of anterolateral stabilization. The patient is capable of partially compensating for the global ligamentous knee instability on the right by muscle activity and corresponding knee movements. The main insufficiency remaining concerns the posterolateral knee structures and the propulsive action of the right leg. The anterolateral and anteromedial instability can be compensated when walking on even ground and in a straight direction.

Case 2 (BA 795)

Injury. This 27-year-old male patient suffered an injury to the right knee at the age of 25 years. A posterior cruciate ligament tear was diagnosed and operated on. He was examined for posterior instability of the right knee following a new injury. On clinical examination, he measured 182 cm in height and weighed 80 kg. His musculature was well developed. A mild varus deformity of both knees was more pronounced on the healthy left side.

Gait Evaluation. The patient's showed only minor deviations from normal. In the *lateral* view, the amplitude and temporal development of hip flexion/extension was normal bilaterally. The knee angles at initial contact of the foot measured

16°–18° bilaterally and were followed by further normal flexion of 12°–14° during weight acceptance and by breaking in the early stance phase bilaterally. During propulsion in the second half of the stance phase, both knees were extended to neutral. In the swing phase, maximum knee flexion reached 72°–74° on both sides. Flexion/extension movements at the ankle joints were normal.
The *frontal* images showed neutral rotation of both thighs and knees throughout the stance phase. During the swing phase, there was a normal smooth movement of the right leg. During forward swing of the left leg, an increase in initial external and subsequent internal transverse rotation of the femur was observed. The *vertical* view through the force plates showed external rotation of the longitudinal axis on the right of 0°–5° and internal rotation of the left foot axis by 0°–5°.

Force Plate Measurements. The speed of walking varied between 1.43 and 1.73 m/s.

The duration of the stance phase on the right was 5%–10% longer than on the left. The *vertical* force record (see Fig. 2) showed a normal symmetrical heel spike. Subsequently however, floor pressure by the right foot during the period of weight acceptance and breaking was diminished by 20% in comparison to the left foot. In the second half of stance, during propulsion, the floor pressure by the impaired right leg was even mildly stronger than the normal left one. The *sagittal* force component showed a diminution of the breaking shear force by 30% on the right in comparison to the left side. In contrast, propulsion in the second half of stance was 30% stronger on the right than on the left. The *frontal* shear forces, which are a measure for mediolateral stabilization, showed normal initial medial shear bilaterally. Its duration was slightly longer on the right than on the left. At an average of four steps, lateral shear during breaking on the right was reduced, while the lateral component of propulsive force transmission remained normal bilaterally throughout the second half of the stance phase.

Summary. This patient with impaired function of the right posterior cruciate ligament and a mild posterior drawer sign demonstrated only minimal alterations of knee movements during walking on level ground. Force plate measurements, however, demonstrated serious impairment in force transmission during the first half of the stance phase, reducing the breaking forward shear force by 30%, vertical loading by 20%, and the lateral shear by 70%. Force transmission in the second half of stance, during propulsion by the right leg, exceeds that of the healthy left side. This shifts the task of breaking towards the leg with the intact posterior cruciate ligament.

Fig. 2 a–c. Pathological floor reaction forces following a posterior cruciate ligament injury of the right knee, combined with normal walking movements. **a** Vertical floor reaction for left and right leg. Increased irregularity of vertical floor reaction and reduction of floor pressure during early stance phase on the right. **b** Sagittal shear forces for left and right leg. Reduction of breaking shear force, increased propulsion, positive values, for the right leg. **c** Frontal shear forces between foot and floor. Impairment of laterally directed shear force during entire stance phase on the right

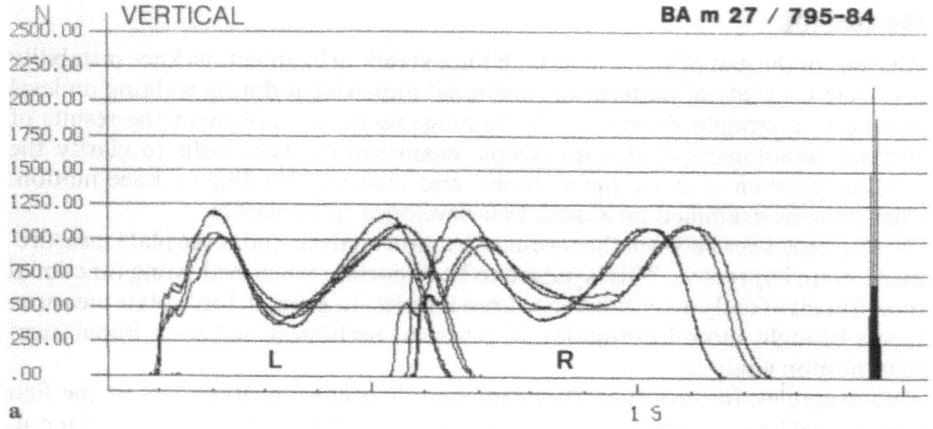
N
VERTICAL
BA m 27 / 795-84
2500.00
2250.00
2000.00
1750.00
1500.00
1250.00
1000.00
750.00
500.00
250.00
.00
L
R
1 S

a

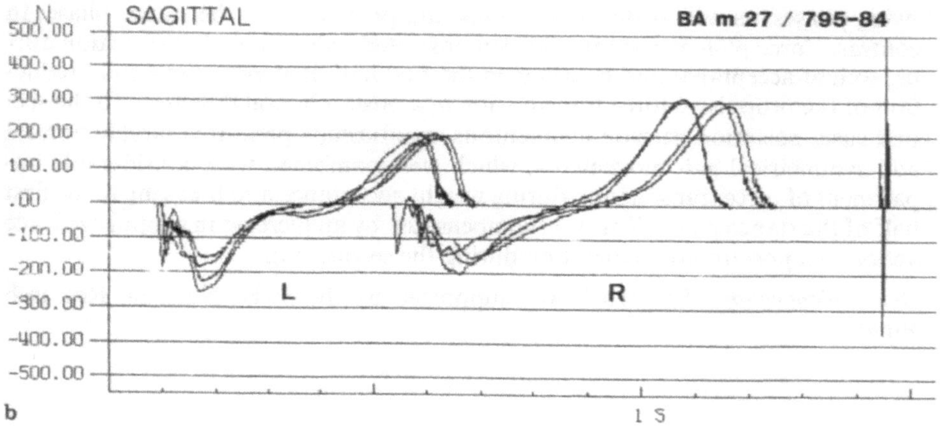
N
SAGITTAL
BA m 27 / 795-84
500.00
400.00
300.00
200.00
100.00
.00
-100.00
-200.00
-300.00
-400.00
-500.00
L
R
1 S

b

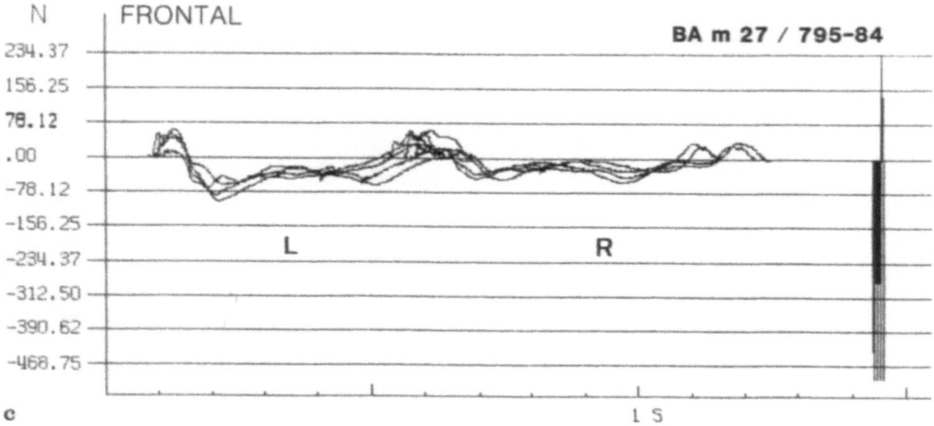
N
FRONTAL
BA m 27 / 795-84
234.37
156.25
78.12
.00
-78.12
-156.25
-234.37
-312.50
-390.62
-468.75
L
R
1 S

c

Discussion

Analysis of the gait of 16 patients with long-standing ligamentous knee instability disclosed a variety of patterns of functional impairment during walking on level ground in a straight direction. The findings usefully supplement the results of clinical, radiological, and arthroscopic examination. They help to clarify the relation between osseous, ligamentous, and muscular guiding of knee motion. Ther persons examined possessed well-developed musculature.

The discrepancies between the results of motion analysis and force plate measurements were impressive. This is judged to be important when evaluating functional impairment of daily activities due to knee injuries. In general, force plate measurements brought more information of practical usefulness in testing impairment than motion analysis.

Two examples are presented that demonstrate opposing situations. In the first case, global knee instability is primarily due to an insufficiency of the anterior cruciate ligament. The motion of knee and hip on the injured side was considerably altered, favoring a more extended position of the knee during both stance and swing phases, as well as during breaking and propulsion in the stance phase. In contrast, force plate measurements demonstrated only minimal dysfunction during weight acceptance and breaking in the first half of stance, and a mild reduction of the propulsive force transmission in a posterolateral direction. In the second case, posterior cruciate ligamentous insufficiency permitted largely normal and symmetrical gait movements, which were combined with considerable impairment of force transmission during weight acceptance and breaking in the first half of the stance phase. This was compensated by an increase in propulsive force action in a posterolateral direction during the second half.

Acknowledgement. This work was supported by the Swiss National Research Fund.

Two-Bundle Ligamentoplasty in Anterior Laxities of the Knee

J. L. Rhenter

Centre Hospitalier Régional and Universitaire de Sainte-Etienne,
Hôpital de Bellevue, Boulevard Pasteur, 42023 St Etienne Cedex, France

In the surgical treatment of anterior laxities of the knee, single-bundle ligamentoplasties afford a functional outcome that generally allows recuperation of physical abilities not unlike those possessed before injury and surgical intervention. However, in other cases, the anatomical outcome is not very satisfactory; persisting anterior loosening, outcome of a jerk test and so forth can be observed.

In addition to problems like poor healing of the transplant or a possible faulty positioning, another reason for residual laxity is no doubt played by the anatomical insufficiency of the single-bundle plasty, which reproduces only a part of the anterior cruciate ligament of the knee. The realizatioin of a more anatomical plasty of the anterior cruciate ligament would no doubt enable this problem to be reduced by producing a better quality of function in the middle and long term phases of recovery. This is what we attempted to do by performing a *two-bundle plasty,* which, to our knowledge, has not been mentioned in any publication so far.

Summary of Knee Kinematics and Physiology

The anterior cruciate ligament (ACL) plays an important role in knee kinematics and physiology. Flexion and rotation are two fundamental and combined knee movements that provide perfect stability in spite of bone parts that are hardly what one would call congruous.

As regards kinematics, we should bear in mind the following points:

1. *The Shape of the Bone Parts.* The lower extremity of the femur displays two femur condyles that are asymmetrical with regard to the three planes of space. In the frontal plane the external condyle is behind the internal condyle during extension. In flexion, both condyles gradually adopt a horizontal position; this is the correction of the physiological valgus. Their long axis diverges from front to back.

   In the horizontal plane the external condyle has the shape of a bean with three segments and the internal condyle has only one long axis. In the sagittal plane the radius of curvature decreases towards the back and is longer for the external condyle. Thus, the greatest surface produced by the external condyle is involved in the automatic rotation of the knee. At 90° flexion, the femur and tibia lie in the same sagittal plane; their projection in the frontal plane is equal.

Surgery and Arthroscopy of the Knee
2nd Congress of the European Society
Editors: Müller/Hackenbruch

In the upper extremity of the tibia, in the frontal plane, the sockets are symmetrical and concave. In the sagittal plane, the external socket is convex and the internal socket is concave. In the tibiofemoral joint the clump of tibial spines and the intercondyloid opening from a modified simple joint and create two compartments, an internal compartment, whose shape assures stability, and an external one, whose function, because of its shape, is to assure mobility.

2. *The Cruciate Ligaments and the Flexion-Rotation Axes.* The crossing-over of the instant axes of flexion and rotation produces a kind of spiral, a geometrical point at the instant centers of flexion-rotation. This central knee joint is located in the joint opening and is on an internal projection of the cruciate ligament crossover.
   The shape of the bone parts and the cruciate ligaments allow a harmonious kinematics of flexion-rotation, thereby guaranteeing the distribution between slipping and rolling. Without this balance, increases in localized constraints and anterior or posterior incomplete dislocation of the femur caused by a camshaft effect could result. (The asynchronism between slipping and rolling when a rupture of the ACL occurs is responsible for jerk when doing the jerk test.)
   With cruciate ligaments the external rotation uncrosses cruciate ligaments and loosens joint surfaces. The neutral rotation position is an intermediate position. Internal rotation crosses the cruciates and approximates joint surfaces. In external flexion-rotation, there is an anteroposterior and frontal play produced by loosening of the cruciates. In extension there is minimal bilateral play.
3. *The Meniscus.* The menisci are true articular fibrocartilaginous joints and largely make up for residual play. However, they are not numerous enough to assure stability in external rotation, a position where joint decoaptation is important.
4. *The Popliteal Muscle.* Functioning as a posteroexternal *strap*, this muscle increases the external joint coaptation and limits external rotation.
5. *Lateral Ligaments.* As a lateral strap, the lateral ligaments play a role in joint coaptation. When they are stretched in extension, coaptation is ideal; when not stretched and in 0° to 90° flexion, coaptation is inadequate.
6. *Posterior Supports: Internal and External.* The two posterior supports are relatively symmetrical and in close connection with the corresponding meniscus, and each is under the control of a particular muscle. For the internal posterior support (IPS) the semimembranosus and its trifurcation exercise control, and for the external posterior support (EPS) the popliteal and its trifurcation. Because of this relative symmetry, the action of the posterior supports will, on the one hand, be completed, and on the other hand, improve the action of the cruciate ligaments. Thus, in internal rotation the action of the IPS predominates over that of the EPS, by taking part in the tightening of the cruciates and by strengthening coaptation of the internal compartment. The EPS, through contraction of the popliteal muscle, maintains the coaptation of the external

compartment and thereby protects the ligaments, whose action it completes (especially in the case of associated constraint).
In external rotation, the EPS, stretched by the popliteal muscle, assists the cruciate ligaments, ensuring stability of the articular center by coaptation of the external compartment and limitation of the external rotation. The IPS, stretched by the semimembranosus, plays a role in the breaking of the external rotation and is in opposition to the valgus.

7. *Meniscal Kinematics*. The attachments of the external meniscus are loose, and those of the internal meniscus are firm. During flexion, the stretching of the external meniscus is greater (1–15 cm) than that of the internal meniscus (0.5 cm). During rotation, their movements are inversed. Thus, the external meniscus ensures mobility and the internal meniscus ensures stability.
The anatomical connections between each meniscus and each posterior support ensures a dynamic tibiofemoral coaptation by means of the menisci under the control of each internal and external posterior support.
Thus, the internal meniscus, in flexion-extension, plays a part in the centering of the knee and distributes the constraints over the cartilage when slipping and rolling, directed by the cruciate ligaments, occurs. In internal rotation, it plays a part in coaptation transmission through tightening of the cruciates. In external rotation, it relieves the EPS by limiting, through its posterior horn, the projection of the internal tibial shelf.
The external meniscus, in flexion-extension, plays a part in the centering of the knee. In internal rotation, it limits, by means of its posterior horn, the projection of the external tibial shelf. In external rotation, it limits the backward movement of the external tibial shelf.

Rupture of the Anterior Cruciate Ligament and Its Consequences

Rupture of the ACL is frequently encountered, especially in sports traumatology; an ACL lesion seriously impairs knee kinematics. The mechanisms of an ACL rupture are now well known. In true varus, the ACL breaks at the end of the movement. This is not true in true valgus, valgus in external rotation, valgus in internal rotation, posteroanterior sagittal shocks, and hyperextension. Associated lesions determine the immediate complimentary gravity that is added to the ACL rupture.
Normally, the articular center of the knee allows for the smooth accomplishment of flexion and extension. Because of this "guiding" action by the capsular ligament system, the mapping of the speed vectors of the tibiofemoral point of tangency exactly reproduces the casing of the condylar contour. Thus, there is no cam effect and no constraint of the tibiofemoral cartilage due to a movement of flexion-extension. In the normal knee, we should take into account only those constraints that are due to the need to fight against gravity.
ACL rupture *disrupts* this balance, rendering the center of articulation unstable during flexion-extension-rotation, thereby creating abnormal cartilaginous constraints. As a result, the instant flexion-rotation centers tend to be displaced to

the rear for the flexion centers, or to the rear and to the inside for the rotation centers. This gives rise to disorders of two kinds:

1. *Asynchronism* between slipping and rolling during flexion: the speed vectors of the tibiofemoral point of tangency no longer reproduce the casing of the condylar contour. It follows, therefore, that abnormal constraints of the cartilage occur through shearing, a prelude to arthrosis, as well as meniscal wear and tear (lesions).
2. *Overload* of the peripheral capsular ligament elements occurs because of these abnormal constraints. This is the beginning of the distension of the posterolateral, internal, and external capsular ligament components. Because of their greater mobility and the constraint in physiological varus during monopodal support in walking and running, the components controlling the external compartments distend more rapidly, especially on a suitable morphotype, a genu varum for example.

Thus, by means of this process of *auto-aggravation* an initially isolated rupture of the ACL becomes generalized and results in a general anterior laxity, most often interspersed with recurrent meniscal accidents.

Explanation for the Good Functional But Poor Anatomical Results After Single-Bundle Plasties

Almost all the plasties used in single bundle are associated with coaptative plasties, which are either internal, external, or both. Thus, in the active-passive plasties of Bousquet, the semitendinous actively fights against anterior instability when the subject is walking downwards, even in the case of distension of the passive transplantation.

Certain external movements also increase frontal and horizontal external coaptation. The free-transplantation-type plasties, the Müller pedicled type, or the MacIntoch type, are almost always associated with an external plasty. Finally, even when distended, we can imagine that certain passive transplantations have a protective role, especially in extreme movements. However, in such cases the vicious circle of auto-aggravation by distension of the ACL is not broken, and secondary degradation can be observed: arthrosis, meniscal lesion, and the reapparance of instability.

ACL Two-Bundle Plasties

The ACL is made up of three bundles: the anterointernal, the posteroexternal (both well characterized), and a small intermediate bundle between these two. Not every bundle is stretched in the same way at each degree of flexion, but there is always one bundle, or at least a part of its fibers that is stretched at each degree of flexion. Thus, from 60° to 20°, the posterior fibers of each bundle have a tendency to distend and then to become tense again in extension. From 20° to 60°, the anterior fibers of each bundle have a tendency to distend and then to become tense again in flexion. This may be explained by the area of ACL condylar insertion in "boomerang" form that is displaced during flexion.

This, in our opinion, seems to justify a two-bundle plasty, because it takes into account the anatomical boomerang insertion of the ACL on the external condyle.

This plasty uses, for the anterointernal bundle, a portion of the extensor system and for the posteroexternal bundle, the semitendinous tendon. The two condylar holes are clearly separated. In the beginning of our research it was not easy to identify two separate tibial holes (we were successful in only 3 cases out of 40). Now, we have achieved a 100% success rate.
There are some associated procedures that give balance to the external compartment by coapting it in the horizontal and frontal planes. This can be accomplished with a retention plasty of the popliteal by means of a fascia lata band or by an anteroexternal plasty by the terminal of the anterointernal bundle. Other associated procedures, such as meniscectomy, meniscal reinsertion, or antivarus osteotomy, depend on the nature of the lesion.
This intervention may be performed using two classical, widely spaced, lateropatellar approaches. We prefer, in order to reduce separation, an approach with a stippled cutaneous incision. Thus, a complete surgical operation on anterior laxities, even complex surgery, may be realized with hardly any traumatism. Moreover, it is quite possible to realize transplantations under arthroscopy with video (8 cases out of 40 were performed using this stipple technique).

Treatment of the First 40 Cases

Each case was examined according to a standard record card. Examination of laxities was comparative whenever possible. Dynamic photos were used in two-thirds of the preoperative cases and in postoperative cases at the time of check-up; they are now being used for postoperative comparative clinical examination.
The average follow-up examination occurred 1 year and 6 months after the operation (ranged from 8 months to 5 years 9 months). The average age was 24 years (ranged from 16 to 41 years). There were 7 women and 33 men (3 bilateral cases counted twice).
In all cases except four, the subjects were atheletic: 22 entered competitions and 14 practiced sport in their free time. Football and skiing were the most common sports practiced.
The operative indication was functional impairment except in four cases of recent ligament rupture (intervention circa 20 days after injury). In the majority of the cases the accident occurred during sports (36 out of 40). In five cases the injury leads one to suppose that there is a two-stroke rupture of the ACL.
The preoperative functional impairment was dominated by instability in sports practice and sometimes in daily life. Pain was rare and was related to a lesion of the meniscus or to the onset of arthrosis after meniscectomy.
The procedures carried out included internal meniscectomy (five total, five partial), external meniscetomy (three total, five partial), reinsertion and meniscal suture (15 internal, 12 external, 20 of 40 very good – "normal" knees), and antivarus – tibial osteotomy (eight cases). Criteria were wear of the internal condyle and replacement of the so-called "mechanical" axis at 0. No osteotomy is conducted if the cartilage is healthy.

Results

There was a follow-up examination after an average of 18 months. The functional results were good in 39 out of 40 cases: stable, painless, dry knees, and the resumption of sports activities at the same level as before the injury.

The anatomical results are the ones of greatest interest. Very good was defined as normal active and passive mobility, no laxity, no hydrarthrosis, no amyotrophy; good as mobility between 135 and 140, no laxity, no hydrarthrosis, amyotrophy <1.5 cm; and poor as all others, in particular laxity (whatever its form) or positivity on one or several dynamics tests. Very good and good results were obtained for 28 of the 40 patients; the results were poor in the remaining 12 cases, i.e., 70% very good and good results and 30% poor anatomical results.

Among the complications which may appear, we noted one cutaneous necrosis (2 cm^2) on an external scar, which healed over in 2 months, but no phlebitis, mobilization under general anesthesia, or algodystrophy.

Conclusion. This two-bundle plasty of the ACL seems to be interesting, particularly from the anatomical point of view.

Technique and Model of Rehabilitation in Anterior Cruciate Ligament Repair of the Knee: A Review of 175 Cases

R. Feinstein, B. Segesser, P. Jenoure, and P. Michel

Praxisklinik Rennbahn und physiotherapeutisches Institut Rennbahn für Orthopädie und Sportmedizin, CH-4132 Muttenz

We have followed up 175 cases of anterior cruciate ligament repair of the knee done in the past 5 years using an identical surgical technique. The best results in a matched group of sports patients arise from an active rehabilitation and functional treatment based on a removable Neofract-Cast system, electromyostimulation, subsequent treatment on a continuous passive knee motion device, training in endurance and mobility and an intensive week of physiotherapy starting 6 weeks after the operation.
In the second phase of the rehabilitation program, the main priority is given to isokinetic training, cyclic, and progressive, sport-specific training until full recovery of the capacity to participate in sports is achieved.

Surgery and Arthroscopy of the Knee
2nd Congress of the European Society
Editors: Müller/Hackenbruch

Functional Rehabilitation in Anterior Cruciate Ligament Reconstruction (4–7-Year Results)

U. Munzinger, T. Drobny, and A. Arnold

Klinik Wilhelm Schulthess, Neumünsterallee 3, 8008 Zürich, Switzerland

The use of rehabilitation therapy following reconstruction of the anterior cruciate ligament is controversial. The purpose of this clinical retrospective study is to determine the effect of using different femoral pathways for intra-articular autografts (bony femoral channel versus "through" the top) and to determine whether functional rehabilitation improves the efficacy of the procedure.

Material and Methods

In 100 patients with chronic failure of the anterior cruciate ligament, a reconstructive procedure utilizing a free bone-ligamentum patellae-bone graft was performed. A bony tibial tunnel was drilled medially from the tuberositas in the direction of the intercondylar eminentia in all patients. During femur surgery two groups of patients were formed. In the first group we used a transcondylar tunnel; in the second a chiseling of the over the top region was carried out. An additional, lateral, We-Müller extra-articular procedure was done in all patients; 30% of patients underwent medial side reconstruction. Limited active exercises were begun the first postoperative day, locomotion with full load was allowed after 50 days, and a brace was used for a minimum of 6 months.

Follow-up studies were carried out according to the very detailed R.P. Jakob scheme. Apart from the usual criteria special attention was paid to axial and rotatory circumstances. Pre- and postoperative X-ray films were used to differentiate between translatory movements affecting either the medial or lateral compartment.

Results

Patellofemoral pain was dramatically reduced by functional treatment. An average mobility (130/0/2°) was achieved when employing active measures in a much shorter time. Of considerable importance is neuromuscular coordination and continuous physical exercise. No disadvantages for stability due to active extension in the early phase (at 20° for the first 50 days, subsequently to 0°) were noted, if special consideration was paid to PNF criteria.

Measurements of translation affecting the medial and lateral compartments in pre- and postoperative dynamical X-ray films (additionally compared to the non-affected knee) showed a slight increase in translation in the latter.

In approximately 50%, increased posterolateral translation was observed. Stability results were more convincing in patients with a graft passed "through" the top than in patients who received a femoral channel.

Surgery and Arthroscopy of the Knee
2nd Congress of the European Society
Editors: Müller/Hackenbruch

Anatomy and Biomechanics of the Anteroposterolateral Femorotibial Structures in the Iliotibial Band System

P. Lobenhoffer [1], C. Krettek [1], P. Posel [2], and S. Witt [2]

[1] Unfallchirurgische Klinik, Medizinische Hochschule, Hannover, FRG
[2] Anatomische Anstalt, Ludwig-Maximilians-Universität, Munich, FRG

The distal iliotibial tract is a principal structure of the lateral side of the knee. Whereas extensive procedures involving this structure have been proposed to increase knee stability and to augment an intraarticular reconstruction [1, 3], the descriptions of the functional anatomy of the distal tract remain inconsistent [4, 5, 10]. This study deals with the detailed anatomy and function of the distal-femoral fixation of the iliotibial tract.

Methods

We analyzed 20 freshly removed and 80 formaldehyde-fixed knee joints. Preoperated joints and those with obvious injuries were excluded from this study. In all joints, the femoral insertions of the iliotibial tract were dissected and documented. Their extension and the distance from the insertions to anatomically defined points of the lateral side of the knee was measured. The 20 freshly removed joints were then fixed in an apparatus. The quadriceps mechanism was held under a constant tension of 10 kp. The joints were then gradually flexed from 0° to 90° of flexion. Every 30° the distances from each of the femoral tract insertions to two points marking the delimitations of Gerdy's tubercle were measured. In 10 joints, the tract insertions were mobilized with a little bone block and were connected to strain transducers (Kistler Co.) via a bone tunnel. Tension measurements were performed during extension/flexion of the knee as well as during manually induced varus-valgus force. Drawer tests in 20° and 90° of flexion were also performed.

Results

Anatomy

Supracondylar Tract Insertion. A pronounced fiber bundle running from proximal-lateral to distal-medial was found in 93% of all dissections. It was localized at the transition of the lateral condyle/femoral shaft and was formed by fibers converging in the form of an arc from the tract and septum. A constant feature of this structure was its relationship to the lateral superior genicular artery. The branch ran proximal to the insertion from dorsal to ventral (Fig. 1). The ligament formed by the fibers described above had a maximal transverse diameter of 13 ± 5 mm, the bone insertion itself had a length of 12 ± 6 mm on average in lateral projection.

Proximal Insertion near the Septum. In 73% of the dissections, an additional tract insertion point was found in the dorsolateral part of the distal femoral shaft and

Surgery and Arthroscopy of the Knee
2nd Congress of the European Society
Editors: Müller/Hackenbruch

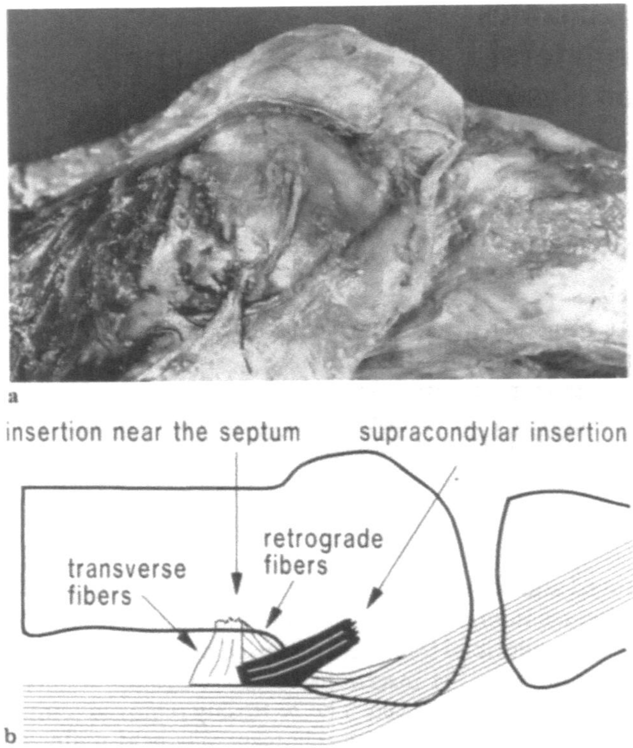

Fig. 1 a, b. Dissection and scheme drawing of the distal femoral fixation of the iliotibial tract

was a prolongation of the intermuscular septum, but was clearly delimited by this septum. The fibers radiating from the distal and proximal tract (see below) displayed a parallel, almost transverse course between the superficial tract and the femur. The maximum breadth of this two-dimensionally arranged structure was on average 10 ± 3 mm, and the bone insertion itself had a size of 9 ± 4 mm. All measurements in the sagittal plane were highly correlated with each other ($r = 0.7$ to 0.8).

Retrograde Fiber Tracts. A closer view of the insertion point just described revealed another fiber tract inserting here. In all of the 73 knees, there was an arcuate fiber arrangement spanning the articular cleft with distal insertions to the Gerdy's tubercle and proximal fixation to the point just discussed. As these fibers have a reversed course in relation to the other insertions of the iliotibial tract to the femur, we have named them the *retrograde fiber tracts* (Fig. 1).

Isometrics Study

Supracondylar Insertion. When measuring the distance between the supracondylar insertion and Gerdy's tubercle, a marked distance increase was found in knee flexion amounting to a maximum of 18% (Fig. 2).

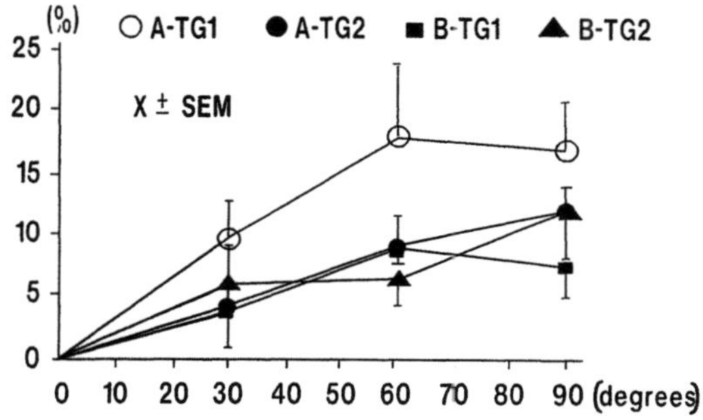

Fig. 2. Distance measurements between the delimitations of the supracondylar insertion of the tract (A, B) and the ventral and dorsal border of Gerdy's tubercle (TG1, TG2)

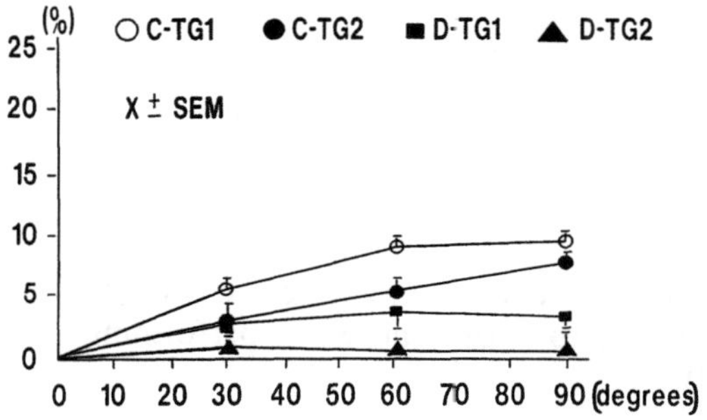

Fig. 3. Distance measurements between the delimitations of the insertion near the septum of the tract (C, D) and the ventral and dorsal border of Gerdy's tubercle (TG1, TG2)

Insertion near the Septum. The distance measurements showed only small variations with a mean variation of 6% (Fig. 3).

Tension Measurements

Supracondylar Insertion. The supracondylar insertion showed a mild tension increase at 30° of flexion of the knee. There was a clear tension increase in internal and external rotation as well as valgus stress of the joint. Varus stress, anterior and posterior drawer, as well as the Lachman test did not increase the tension in the supracondylar insertion (Table 1).

Table 1. Relative tension measurements in the distal femoral fixations of the iliotibial tract

	Supracondylar insertion	Insertion near septum
Extension	−	+
Flexion	(+)	−
Varus	−	+
Valgus	+	−
Internal rotation	+	+
External rotation	+	+
Anterior drawer	−	+
Posterior drawer	−	+
Lachman	−	+

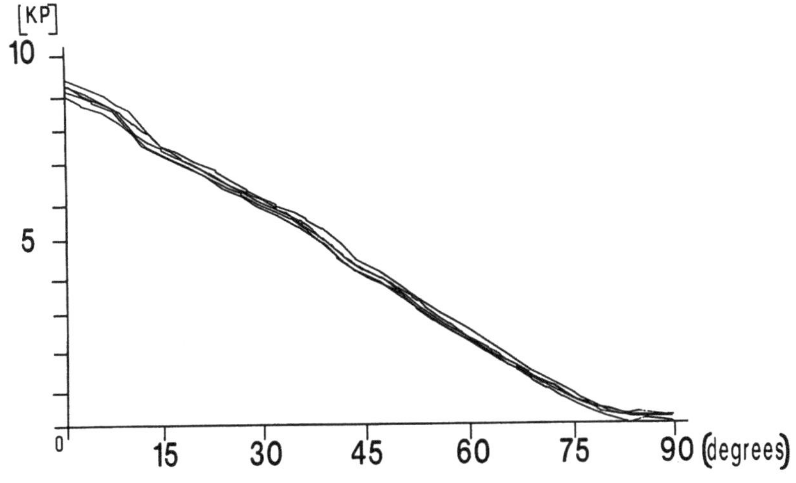

Fig. 4. Tension registration in the insertion near the septum in knee motion

Insertion near the Septum. The insertion near the septum was stressed in extension of the knee, in internal as well as external rotation, and when performing the drawer and the Lachman tests (Fig. 4).

Discussion

The supracondylar insertion was first described and named by Kaplan [6]. This structure did not seem to contribute to lateral knee stability. First, it was formed exclusively by fibers with a course from proximal to distal, which does not seem suitable for function at the knee joint. Second, the distance between supracondylar insertion and Gerdy's tubercle varied significantly during knee motion. Therefore, a static fixation of fibers between these two points seems impossible. Third, our tension measurements showed no tension increase in this structure in varus, drawer, or Lachman stress. In contrast, these fibers are suitable to support the tension-band function of the tract on the thigh [6] and to enable force transmission to the distal femur.

Fibers with a different orientation are inserted near the septum. First, bundles connect the superficial tract layer to the femur. Like the intermuscular septum, they fix the tract to the femur at a defined distance and prevent its ventral dislocation. On the other hand, the retrograde fibers suspended between this insertion point and the Gerdy's tubercle seem appropriate to stabilize the lateral side of the knee. The kinematics of knee joint function requires fibers from dorsal-femoral to ventral-tibial for this purpose [8]. Our own measurements showed only minimal distance change between these two points, which lie in the range defined for ligaments with a static fixation [2]. Lesions and hematomas of this region have been described in operations on extensive capsule ligament injuries of the knee [9]. We conclude that the arcuate fiber tracts in the dorsal tract, which we have made visual, correspond to the "lig. femoro-tibiale laterale anterius" described first by Müller [9] and the "capsulo-osseous layers" of Terry [10]. As these fibers can be separated from the dorso-lateral structures only by sharp dissection, they also are involved in the posterolateral stabilization of the knee joint. Our tension measurements have demonstrated the function of the tract fixations near the septum in knee stabilization. There is a tension increase in ventral dislocation of the tibia, in varus stress, as well as in posterior and posterolateral stress. The cranial insertion point of these fibers has a constant anatomical localization. This spectific localization, which is far proximal and dorsal, should be considered in knee surgery.

References

1. Andrews JR, Sanders RA, Mori B (1985) Surgical treatment of anterolateral rotatory instability. Am J Sports Med 13(2):112–119
2. Bartel DL, Marshall DWM, Schieck RA, Wang JB (1977) Surgical repositioning of the medial collateral ligament. J Bone Joint Surg [Am] 59:107–116
3. Ellison AE (1979) Distal iliotibial-band transfer for anterolateral rotatory instability of the knee. J Bone Joint Surg [Am] 61:330–337
4. Hassler H, Jacob RP (1981) Ein Beitrag zur Ursache der anterolateralen Instabilitaet des Kniegelenks. Arch Orthop Trauma Surg 98:45–50
5. Jacob RP, Hassler H, Staeubli HU (1981) Observations of rotatory instability of the lateral compartment of the knee. Acta Orthop Scand 191(Suppl):52
6. Kaplan EB (1958) The iliotibial tract. Clinical and morphological significance. J Bone Joint Surg [Am] 40:817–832
7. Krackow KA, Brooks RL (1983) Optimization of knee ligament position for lateral extraarticular reconstruction. Am J Sports Med 11(5):293–302
8. Menschik A (1974) Mechanik des Kniegelenks, Teil 1. Z Orthop 112:481–495
9. Müller W (1982) Das Knie. Form, Funktion und ligamentaere Wiederherstellungschirurgie. Springer, Berlin Heidelberg New York
10. Terry GC, Hughston JC, Norwood LA (1986) The anatomy of the iliopatellar band and iliotibial tract. Am J Sports Med 14(1):39–45

Mechanoreceptors of Some Anatomical Structures of the Human Knee

G. Cerulli [1], A. Ceccarini [1], P. F. Alberti [2], A. Caraffa [1], and G. Caraffa [1]

[1] Department of Orthopaedic Surgery, and
[2] Department of Pathologic Anatomy, University of Perugia, Italy

Although recent studies have given us a broad basis for understanding the biomechanisms of the meniscus and anterior cruciate ligament, the high incidence of lesions of these structures has promulgated many investigations concerning the microanatomy and, in particular, the innervation. The studies conducted on the meniscal innervation are those from Laudicella (1947), Loffredo (1948) and Gardner (1948) in human fetal menisci, and from Wilson (1969) and Freeman and Wyke (1967) on animals.

It was not until 1976 that O'Connor found free nerve fibres in the anterior and posterior horn of dog menisci; in 1978 O'Connor and McConnoughey discovered that two corpuscular endings were located in the posterior horn of cat menisci. None was found in the meniscal body (O'Connor 1976; O'Connor and McConnoughey 1978). The importance of these studies and the necessity to identify clearly the complex mechanisms underlying meniscal function are now beyond any doubt.

Concerning the anterior cruciate ligament, there have been no studies since Schultz's (1984) and our investigations. In fact, the current understanding of the physiology of the anterior cruciate ligament suggests that it is solely a passive stabiliser of the knee joint.

Materials and Methods

We conducted the research on cadavers and subdivided the investigation into three sections: the first for the meniscus, the second for the anterior cruciate ligament (ACL) and the third for the patellar and semitendinous tendons.

Slides were prepared in horizontal serial sections from the numerous samples extracted from each anatomical preparation. Staining in block was performed according to Ruffini's gold chloride method as modified by O'Connor and Gonzales (1979). Each of these was immersed in a solution containing three parts lemon juice and one part 88% formic acid conserved in a light-protected container and maintained at room temperature.

After 10 min this solution was substituted by another of 1% gold chloride and incubated in the dark for 10 min. The sample was then immersed in 25% formic acid for 30 min, and after rinsing three times in ethanol at 70% for a period of 10 min each time, the sample was transferred to a glycerol solution for 24 h. After having been consolidated into a paraffin mould, sections of 8–10 µm were prepared and observed under a light microscope.

Surgery and Arthroscopy of the Knee
2nd Congress of the European Society
Editors: Müller/Hackenbruch

Meniscus

The study was conducted on eight cadavers, seven male and one female, with ages ranging from 35 to 70 years. The total number of menisci involved was 16 medial menisci and 16 lateral menisci.

Results. Free nerve fibres were located in all of the menisci examined. These findings were made primarily in the meniscal body, although corpuscular endings were also seen in the anterior and posterior horns.

We have classified the articular nerve endings in accordance with Freeman and Wyke into four categories: Type I or Ruffini's receptors (Fig. 1), type II or Pacini's receptors (Fig. 2), type III receptors (Fig. 3) and type IV receptors or free endings (Fig. 4). Nevertheless, we have found it useful to subdivide the anatomical

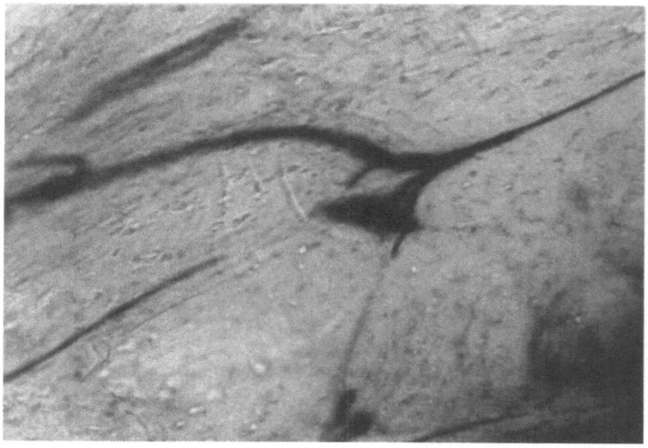

Fig. 1. Type I receptor or Ruffini's corpuscle. Gold chloride, ×200

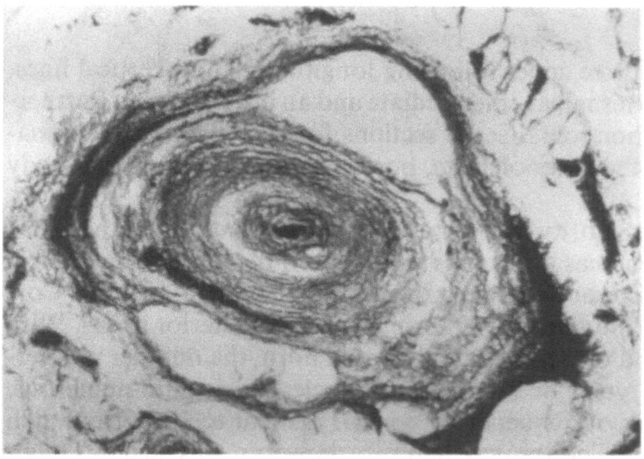

Fig. 2. Type II receptor or Pacini's corpuscle in transverse section, in typical onion bulb shape. Gold chloride, ×200

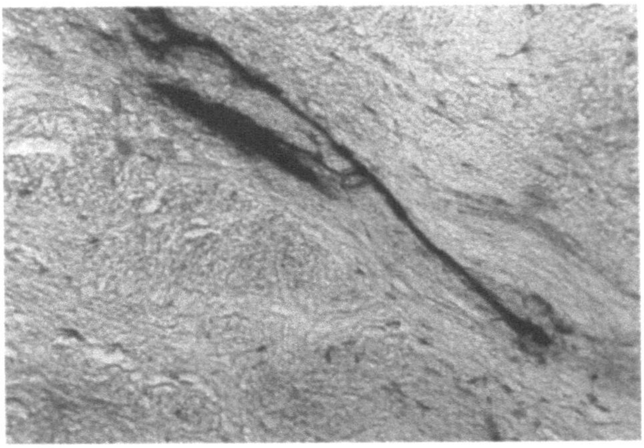

Fig. 3. Type III receptor (Golgi-like). Gold chloride, ×200

Fig. 4. Type IV receptor or free endings. Gold chloride, ×200

areas of the meniscus into three zones following longitudinal and vertical lines. This division results in an external, an intermediate and an internal zone. Furthermore, while preparing the horizontal serial sections for each meniscus, an anatomical distinction among the anterior horn, posterior horn and meniscal body was maintained.

Analysis disclosed that the majority of free nerve fibres was located on the external third of the meniscus whereas a minority was seen on the intermediate third. There was no substantial difference between the medial and lateral menisci. Corpuscular endings were more numerous on the anterior and posterior horns. With regard to the quantity of end organs observed in each horn, the number was certainly not over, while the type I or Ruffini's corpuscles were quite numerous. These were followed in order of frequency by type III (similar to Golgi) and then type II or Pacini's corpuscles. The absolute majority, however, was held by Golgi-Mazzoni ends, which are a Pacinian type of corpuscle, the difference being that they are smaller in size and arranged in clusters. These corpuscles are classified

according to the morphology of the fibre and the sheath and can be included in types I, II or III.
With the onset of age, variations in the quantity and quality of articular innervation may be seen. These age-related changes include a reduction in volume of fibrocartilaginous tissue and a numerical reduction of nerve endings of both the corpuscle and free-ending types.
Unfortunately, we are not able to provide information concerning sex-related distributions since our study involved predominantly male subjects.

Conclusions. The results of our research have provided evidence of both free and corpuscular nerve structure located in the horns as well as the meniscal body. The majority of free-ending fibres was seen on the external third and intermediate area of the body. This area corresponds to the so-called central fibrocartilaginous portion nominated by Freeman and Wyke. The majority of end organs at the level of the anterior and posterior horns consisted of Golgi-Mazzoni and type I, II, and III corpuscles.
In view of this actual data, the role of the meniscus as a passive fibrocartilaginous structure has now become obsolete. It has induced us to reconsider the meniscus as a storehouse of proprioceptive information for coordinating movements and regulating muscle tone.

Anterior Cruciate Ligament

Our investigation was conducted on ten fresh cadavers (eight males, two females) of people age 30–70 years (mean 48). Total number of ACLs was 20.

Results. Free termination and some corpusculated elements were noted in each of the cruciate ligaments examined. They were morphologically divided into groups according to Freeman and Wyke (1967).
Most of these formations were located in the middle third of the ligament at its femoral insertion (Fig. 5). Free terminations constituted the highest percentage, followed by type II, type I and type III corpuscles at much the same values.

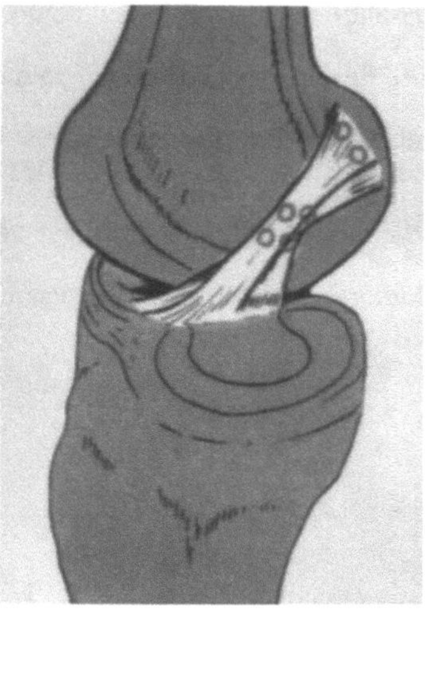

Fig. 5. Map of distribution of free and corpuscular endings in the anterior cruciate ligament of the human knee

Conclusions. The strong presence of free and corpusculated endings on the ACL support the view that it, along with other capsuloligamentous knee structures, is not just a passive stabiliser of the joint but is also the site of proprioceptive information. In light of these data, it is an open question whether instability of the knee after injury to the ACL is solely due to loss of the mechanical function provided by the ligament, or whether interruption of the afferent impulses derived from its nerve structures may also be in some way involved.
The structure most frequently used in the biological intra-articular reconstruction of the ACL (patellar and semitendinous tendons) do not have the same anatomy and proprioceptive innervation.
Perhaps this type of procedure can be used to reconstruct the mechanical propriety but not proprioceptivity of the ACL.

Patellar and Semitendinous Tendons

The investigation was conducted on ten cadavers, five males and five females, aged from 35 to 68 years (mean 45) for a total 20 patella (central third) and 20 semitendinous tendons.

Results. In all patellar and semitendinous tendons we found a very few free terminations, localised at the bone insertion, and no single corpuscular endings.

References

Cerulli G, Ceccarini A, Alberti PF, Caraffa A (1982) Studio sulle terminazioni nervose del menisco nell'uomo. Il Ginocchio 5:133–141

Cerulli G, Ceccarini A, Alberti PF, Caraffa G (1986) Studi sulla propriocettività del legamento crociato anteriore del ginocchio umano. It J Sports Traumatol 8:49–52

Freeman MAR, Wyke BD (1964) Articular contributions to limb muscle reflexes. I. An electromyographic study of the ankle joint mechanoreceptors upon reflex activity in the gastrocnemius muscle of the cat. J Physiol (Lond) 171:20–21

Freeman MAR, Wyke BD (1965) The innervation of the ankle joint. Nature 207:196

Laudicella V (1947) Sulla innervazione dei menischi del ginocchio. Rass Biol Umana 26:195

Loffredo C (1948) Le espansioni nervose del menisco articolare del ginocchio dell'uomo. Quad Anat Prat 3:39

O'Connor BL (1976) The histological structure of dog knee menisci with comments on its possible significance. Am J Anat 147:407–418

O'Connor BL, McConnaughey JS (1978) The structure and innervation of cat knee menisci and their relation to a "sensory hypothesis" of meniscal function. Am J Anat 153:431–442

O'Connor BL, Gonzales J (1979) Mechanoreceptors of the medial collateral ligament of the cat knee joint. J Anat 129:719

Schultz RA, Miller DC, Kerr CS, Micheli L (1984) Mechanoreceptors in human cruciate ligaments. J Bone Joint Surg [Am] 66:1072

The Function of Hamstrings: A Pathogenic Hypothesis of Femoropatellar Osteoarthritis

P. Blaimont, P. Klein, M. Alameh, and P. van Elegem

Institut Médico-Chirurgical d'Ixelles,
Département d'Orthopédie-Traumatologie,
Rue Jean Paquot, 63, 1050 Bruxelles, Belgium

Since the end of the last century there has been controversy concerning the function of the hamstrings. By 1867 Duchenne de Boulogne spoke of a paradoxical action of extension which Lombard (1907) later described in the following manner: "The anterior vector stretches the knee, but flexes the hip; the hamstrings bend the knee and stretch the hip. Their actions should cancel one another; however, they complete one another." In opposition to this Storck (1932) and more recently Maquet (1955) categorically deny the knee extension function of the hamstrings. This is because, says Maquet (1984), "the muscle being located posteriorly to the centre of knee flexion can only be a knee-bender." This discussion continues today: are the hamstrings agonists or antagonists to the quadriceps?
If one defers to the vectorialists, the patella would appear to be a quite surprising bone. Various authors, including Maquet, presume that at 45° flexion the patella presses against the femoral condyles with a pressure of 400 kg. Knowing the contact area of the patella at this angle, the constraint would be about 60 kg/cm^2. Disbelieving such a prodigy, we have conducted a controlled study by pressing with an experimental apparatus a series of patellae against their condyles. Fracture occurred at pressures of 250–280 kg. Consequently, the "mathematical" patella would tolerate constraints passing beyond the resistance of the "biological" patella. Our patellae should fracture, if one believes the vectorialists, every time we bend our knee at 45°. A further paradox lies in the fact that when one goes up stairs, the hamstrings contract at the same time as the quadriceps. Because the hamstrings are posterior to the centre of knee flexion, they should normally be in opposition to the action of the quadriceps and even increase the veneering of the patella.
We shall seek to demonstrate that the hamstrings are truly agonists to the quadriceps during the movement of picking something up, with feet on the ground.

Electromyographic Arguments. When climbing steps or pushing on bicycle pedals, we observe synchronous anterior and posterior muscle contraction. Also in the case of a weight lifter raising a barbell, there is synchronous contracture anteriorly and posteriorly. The realisation of extension under difficult conditions for the knee requires synchronous anterior and posterior muscle contraction. How does one realise extension under these conditions if the quadriceps oppose, on one hand, gravity and, on the other, the posterior muscles?

Anatomical Proof. Using drawings made from large radiographs of the lower extremities we observe that during the passage from semi-bent position to upright position the points of muscular insertion are brought closer together, a fact which

Surgery and Arthroscopy of the Knee
2nd Congress of the European Society
Editors: Müller/Hackenbruch

proves that the electromyographic activity corresponds to a contraction with shortening of the muscle.

Faradic Stimulation. When a faradic discharge is applied to a subject in half-sitting position, the subject receives an impulse toward the upright position. Above 60° flexion, however, the same stimulation produces an impulse toward flexion.

Analogical Model. Comprising the analogical model (Fig. 1) are three articulated parts which simulate pelvis, femur and tibia, and hamstrings and quadriceps (represented by ropes in Fig. 1). The distance between the ropes and the centres of the joints represent reality: thus, the lever arm of the hamstrings is more important at the hip than at the knee below 60° knee flexion. By stretching only the hamstrings, we obtain an extension of the pelvis. If we stabilise the pelvis by applying

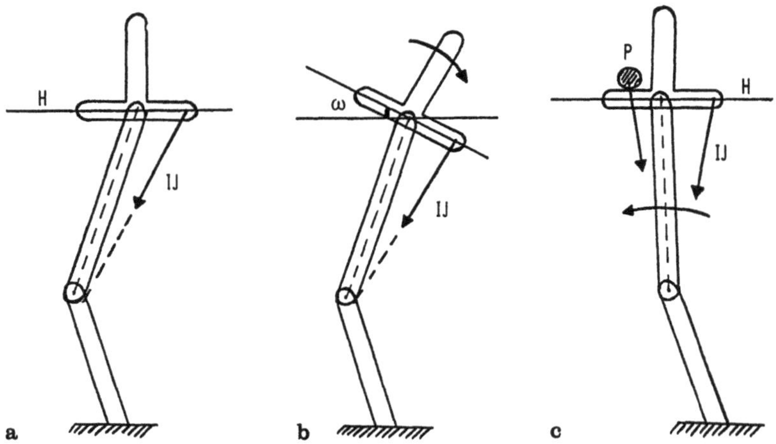

Fig. 1 a–c. Analogical model

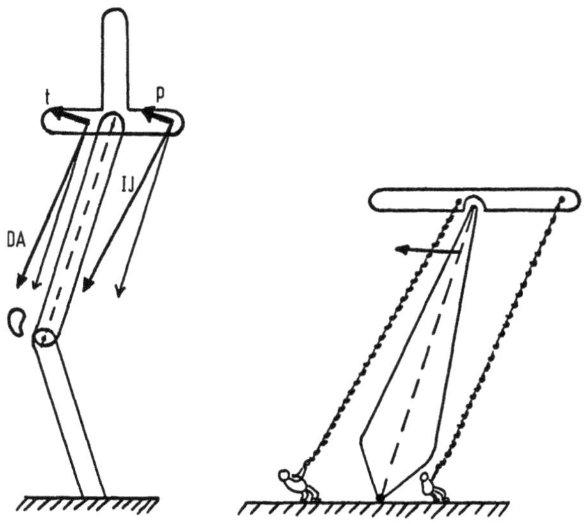

Fig. 2. Vectorial interpretation of the phenomenon

pressure from above to below, anteriorly to the femoral head, we obtain an extension of the knee.

Vectorial Interpretation of the Phenomenon. We accept the vectorial analysis as far as it illustrates phenomena which are demonstrated by physical reality, and this seems to us true in this case (Fig. 2). The anterior vector and the hamstrings are oblique to the mechanical axis of the femur. Their pulling force is composed of one moment which is parallel to this axis (this force equilibrates the pelvis), and one directed forwards from behind. The anterior and posterior muscles thus constitute a synergistic system of "push-pull". During the passage from crouching to standing positions, the patella remains less loaded by the action of the gastrocnemius, which pulls the knee back.

Discussion

The weakness in – or the contracture of – the hamstrings lead to an overloading of the anterior system, and thereby of the patella. This is demonstrated by several conditions of patellar disease. In the case, for example, of chondropathies of cyclists or skiers, in which the patella is well aligned, the patellar disease does not respond to surgical operation. Physical examination shows a contracture of the posterior muscles; several sessions of stretching these muscles is sufficient to see this pain disappear. Also, in the case of incipient femoropatellar osteoarthritis the posterior muscles are too weak.
During the last 6 years we have treated with success the early forms of this disease by developing the tonicity of the gastrocnemius and the hamstrings, as well as by mental training for descending stairs in a "dynamic" rather than in a "static" manner.

Conclusion

We may conclude from these observations the following points:

- The hamstrings are invaluable partners to the quadriceps.
- When the foot is fixed on the ground, the hamstrings must be considered posterior extensors (0°–60° knee flexion).
- The weakness or the contracture of the hamstrings leads to an overloading of the patella.
- Postoperative physical therapy after surgery on the patella must therefore aim at harmonising the anterior and posterior muscles rather than merely at improving the tonicity of the quadriceps.

Anatomy of the Popliteus Muscle and Posterolateral Structures

C. Fabbriciani and M. Oransky

Clinica Ortopedica dell'Università Cattolica S. Cuore, Roma, Italy

The popliteus muscle is the most important structure on the lateral side of the knee. In the early years of this century this was studied by many authors [for instance, in the fundamental monography of Fürst (1903)]. Thereafter, however, it was neglected for a long time. At present its importance is well-recognised.

Its muscular fibres originate from the medial side just behind the medial lateral ligament and the semimembranous tendon to which the fibres are strictly connected. These fibres end in a complex aponeurotic structure that we divide here into three sections (capsular, meniscal, fibular) and in a strong tendon. The aponeurotic structure is very important, and knowledge about it allows us to understand the original function of the popliteus muscle.

The popliteal capsular fibres form the most superficial layer of aponeurotic expansion of the popliteus muscle. They end on the posterior lateral capsule; the contraction of the popliteus muscle contributes to determining tension on the posterior lateral capsule.

The expansion of popliteus aponeurosis on the lateral meniscus forms the popliteal meniscal fibres which have been described by Last (1950). These fibres insert into the inferior edge of the meniscus. However, if the popliteus tendon is detached from the femoral condyle, two different bundles of fibres appear which extend from the tendon to the meniscus. The composition of popliteal meniscal fibres is very complex, although they are generally described as a single bundle. At the beginning of our anatomical investigation we described two terminal expansions of the popliteal meniscal fibres; at present we no longer consider this description correct. While they can be divided into two bundles – superior and inferior – there is also another bundle that we call inferolateral. This originates at the popliteus tendon and reaches the edge of the lateral meniscus at the level of the anterior edge of the hiatus. The connections between the popliteus muscle and the meniscus can also be observed in the embryo, as our histological studies of 12-week-old fetuses have shown.

In the posterior horn of the lateral meniscus it is possible to observe the distal insertion of the meniscal femoral ligaments of Humphry and Wrisberg. These ligaments perform a special function: they help the lateral meniscus to hold a correct position and strengthen the posterior cruciate ligament (PCL). We must stress the anatomical continuity of the popliteal meniscal fibres, lateral meniscus and the PCL. Through the popliteal meniscal fibres the popliteus muscle plays a role in the active control of the lateral meniscus as it is drawn back: in this way a forward shift of the meniscus is prevented, and squeezing of the meniscus between femur

Surgery and Arthroscopy of the Knee
2nd Congress of the European Society
Editors: Müller/Hackenbruch

and tibia is avoided. The displacement causes a traction of Wrisberg's and Humphry's ligaments and consequently that of the PCL.

The popliteal fibular fibres form the inferior portion of popliteus expansions. Although these fibres are very important, they are usually incorrectly described or even ignored, not only in anatomy textbooks but also in specialist papers. These fibres are usually very strong and are attached to the fibular head behind the external lateral ligament. Popliteal fibular fibres are the only fibres that have a bony insertion. The functional role of popliteal fibular fibres is very important, for they contribute to the control of tibial external rotation and recurvatum and also give a good holding position to the popliteus tendon. We have observed that an experimental cut in the popliteal fibular fibres alone is enough to produce posterolateral instability.

The functional role of the popliteus tendon is also of great importance; its best known function is the internal rotation of the femur on the tibia. Also important, however, is the restraining of external rotation. The action of flexing the muscle is not very important – not only as shown by electromyographic studies, but also in consideration that in the flexion the rotation axis of the lateral femoral condyle passes near the femoral tendon insertion.

The external lateral ligament crosses the popliteus tendon at the level of its middle third, resembling a rounded cord, and is thus unlike the medial lateral ligament, which is flat. The fibres twist each other in forming it, and the posterior fibres at the level of femoral condyle become anterior at the fibular head. The ligament, which is independent from the capsule throughout its length, ends on the fibula in the biceps expansions with which it is wrapped. It may be reinforced by a small additional bundle that must not be confused with the short lateral ligament, a particular kind of arcuate ligament.

In the popliteus muscle it is possible to observe a variation which consists of an additional part of the muscle. This was first described by Gruber in 1875 who discovered this variation in eight cases of the 250 dissections he performed (Gruber 1875). This type of muscle was named "musculus popliteus biceps". In our anatomical investigation we found one case of popliteus biceps: the additional part of the muscle was in the superior position as related to the proximal insertion of the tendon of the lateral gastrocnemius.

The other component of the posterolateral corner of the knee is the arcuate complex. Anatomical descriptions of this as found in classical textbooks of anatomy are incomplete, inaccurate, often contradictory and cause genuine confusion. Although the Nomina Anatomica specifies the term "ligamentum popliteum arcuatum", the terms actually adopted are quite numerous. The discrepancies in description and appellation are justified by the extreme anatomical variability of the area. In order to understand the anatomy of the region correctly it is necessary to differentiate those findings based on the presence or absence of fabella in the knee.

In the first (the one most frequently observed) group the arcuate ligement shows total of 40 fresh corpses).

In the first the one most frequently observed group the arcuate ligament shows a thickened external edge along its entire length. The lateral inferior geniculate vessels always pass under the arcuate ligament. This finding contrasts with the description by Kaplan (1961) that these vessels normally run above it. In some cases the arcuate ligament consists of a very thin band of fibrous tissue without the structural and morphological features of a true ligament. In a second group the most external fibres or arcuate ligament differentiate in a distinct ligamentous structure that is parallel to the arcuate ligament: this is the "short external lateral ligament". The lateral inferior geniculate vessels pass over the arcuate ligament and under the short external lateral ligament. We should point out that in earlier nomenclature the short external lateral ligament was confused with the external arch of the arcuate ligament.

In order to understand the functional role of the arcuate complex we conducted an experimental study on cadavers. Both before and after sectioning the arcuate complex, we carried out the more important clinical tests to evaluate posterolateral instabilities, and only then did we section the other structures of the lateral compartment. Only sectioning of the arcuate complex produces no signs of laxity. A slight increase on clinical tests is obtained by the additional cutting of the external lateral ligament. Laxity increases after cutting the popliteus tendon; we want to emphasise the definite increase of the positivity on clinical tests after cutting the popliteal fibular fibres.

When the fabella is present (in 8% of our observations) the arcuate ligament usually consists of a strong ligament with features resembling those of the external lateral ligament. These structures should be called a fabellofibular ligament and extend from the fabella to the head of the fibula, passing, as does the arcuate ligament, above the tendon of the popliteus muscle. This structure is stronger and more solid than that of the arcuate ligament. The fabellofibular ligament is a special type of arcuate ligament but is not a structure that is present at the same time as the arcuate ligament (as has been asserted by various authors). The incidence of the fabellofibular ligament is limited to cases of osseous fabella: only in this case is it actually a ligament. In fibrocartilaginous fabella this ligament is thin or absent. The functional role of the fabellofibular ligament, even if it is difficult to define, is certainly more important than the arcuate ligament. In fact, the fabellofibular ligament is a strong ligamentous structure and it can plausibly be suggested to play a role in limiting hyperextension and external rotation.

References

Basmajian JV, Lovejoy JF (1961) Functions of the popliteus muscle in man. J Bone Joint Surg [Am] 53:557–562

Fabbriciani C, Oransky M, Zoppo U (1982a) Il muscolo popliteo: studio anatomico. Arch Ital Anat Embriol 87(3):203–217

Fabbriciani C, Oransky M, Zoppi U (1982b) Il legamento popliteo arcuato e lesue varianti. Ital J Sports Traumatol 4(3):171–178

Fürst CM (1903) Der Musculus Popliteus und seine Sehne. Acta Regiae Societatis Physiographicae Lundensis

Gruber W (1875) Über den Musculus Popliteus Biceps. Arch Anat Physiol Wiss Med 599–605

Higgins H (1894) The popliteus muscle. J Anat 569

Kaplan EB (1961) The fabellofibular and short lateral ligaments of the knee joint. Surg Gynecol Obstet 104:346

Last RJ (1950) The popliteus muscle and the lateral meniscus. J Bone Joint Surg [Br] 32:93–99

Lovejoy JF, Harden TP (1971) Popliteus muscle in man. Anat Rec 169:727–730

Femoropatellar Pathology and Rotational and Torsional Abnormalities of the Inferior Limbs: The Use of CT Scan

J. L. Lerat [1], B. Moyen [1], M. Bochu [2], and O. Galland [1]

[1] Service de chirurgie orthopédique et traumatologique et de médicine du sport, Hôpital Edouard Herriot, F-69374 Lyon, France
[2] Service de Radiologie, Pavillon B, Hôpital Edouard Herriot, 69374 Lyon, France

The CT Scan in Knee Joint Examination

Since 1979 the CT scan has been in common use for femoropatellar pathology. So far it represents the only way to study the femoropatellar joint in the range of 0°–20° flexion, in which traditional tangential X-rays are insufficient. The CT scan can also be used to explore the knee in hyperextension.

Materials and Methods

We studied 60 normal adult knees (i.e. without any knee pathology) and 400 pathologic knees. Among the pathologic knees two groups were examined, those with patellar chondropathy (defined by pain on the anterior aspect of the knee with or without effusion and without any symptom of patellar instability) and those with patellar instability (with either one or several forms of patellar instability or with true recurrent patellar luxation).
Patients were asked to lie supine on the X-ray table, with feet positioned against a vertical support and well stabilised. Sometimes the feet were placed according to the previously determined step angle. The knees were in extension. For the standard examination nine X-ray views were taken: two each at the hip level (one in centre of the head of the femur, one at the base of the neck), at the level of the femoral condyles (one 2 cm from the joint line, one through the middle of the patella) and through the upper extremity of the tibial epiphysis (one under the level of the plateau, one at the level of the tibial tubercle), one at the level of the ankle malleoli and one through the foot including the metatarsal heads and calcaneum. In addition to this, a dynamic CT scan was made with the knee in flexion (10°–20°), the tibia in external rotation and quadriceps in contraction.

Results

Inferior Femoral Epiphysis Anatomy. Only the CT scan permits one simultaneously to view the femoral trochlea and the posterior aspect of the femoral condyles. With it we can therefore measure the lateral trochlear slope in relation to the posterior femoral condylar plane. In so doing we have observed that the lateral trochlear slope increases from the proximal to the distal. We have averaged two measurements of this slope, one at 2 cm above the knee joint and the other in the middle of the patella. Table 1 presents the means of these measurements

Surgery and Arthroscopy of the Knee
2nd Congress of the European Society
Editors: Müller/Hackenbruch

Table 1. Lateral trochlear slope by knee type

	Normal	Patellar chondropathy	Patellar instability	Permanent patellar luxation
Lateral trochlear slope	22.5° ± 5.5° (*n*=51)	21.3° ± 8.4° (*n*=82)	19.4° ± 7.6° (*n*=54)	12.1° ± 6.4° (*n*=12)
Distance TA–TG (mm)	12.1° ± 4.2° (*n*=63)	11.6° ± 7° (*n*=278)	17° ± 9.6° (*n*=117)	20.7° (*n*=12)
Femorotibial rotation in extension	3.2° ± 3.7° (*n*=60)	5.48° ± 5.9° (*n*=190)	7.2° ± 5.4° (*n*=96)	13.6° ± 8.5° (*n*=18)
Femoral torsion	10.9° ± 7.1° (*n*=69)	17° ± 11.7° [a] (*n*=231)	16.1° ± 9.4° [b] (*n*=122)	22.3° ± 15.8° (*n*=12)
External tibial torsion	32.8° ± 7.8° (*n*=60)	34.1° ± 11° (*n*=193)	36.1° ± 9.8° (*n*=94)	28.1° ± 14.4° (*n*=12)
Sum of femoral and tibial torsions	44.5° ± 13° (*n*=62)	56.5° ± 18.2° (*n*=216)	59.6° ± 16.1° (*n*=120)	50.4° (*n*=12)
Sub malleolar torsion	9.4° ± 5.5° (*n*=60)	11.6° ± 7° (*n*=174)	12.1° ± 7.4° (*n*=72)	–

[a] $P<0.001$.
[b] $P<0.005$.

in each of four types of knees. The only statistically significant difference is between normal knees and those with permanent patellar luxation ($P<0.001$). These measurements were taken in relation to the posterior femoral condylar plane. However, during walking it is necessary to take into account the femoral antetorsion, the external tibial torsion and the step angle; here the knee looks inward and the lateral trochlear plane lies at an average of 35.5° to the frontal plane.

Patellar Shape. CT scans give more precise information than do conventional X-rays concerning patellar shape and thickness of the cartilage and bony trabeculation.

Femoropatellar Congruence. The CT scan provides important information on the static and dynamic aspects of the femoropatellar joint and of the knee in extension and flexion from 0° to 20°. It is of great importance here to control the exact position of the knee in the sagittal plane and in particular to avoid having the knee in hyperextension. In extension, patellar subluxation is common, particularly when the quadriceps is contracted. In slight flexion any patella is normally located in the trochlear groove. If this is not the case, or if the lateral subluxation is increased in tibial external rotation, the patella is considered as abnormal and subluxed. Arthrography-CT scans using iodine dye or air can furnish valuable information on the thickness of the cartilage and on the presence of fissurations and ulcerations. One can also detect the presence of a medial patellar plica. The measurement of the bone mineral content is also important in assessing overload and in providing objective base lines before surgery.

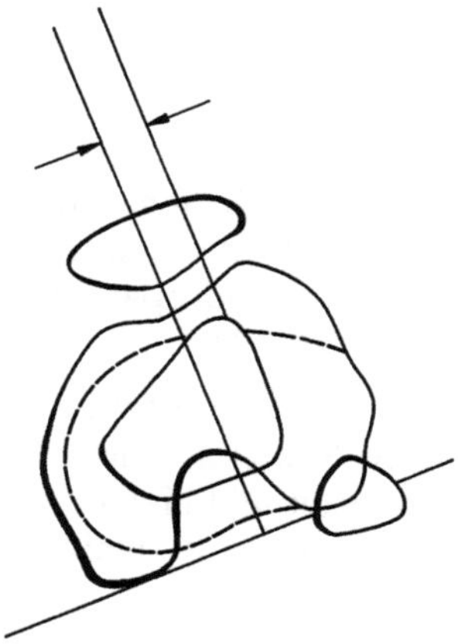

Fig. 1. The measurement of the distance TA–TG by superposition of two views passing through the femoral condyles and the tibial anterior tuberosity

Upper Tibial Extremity. Two views are necessary in examining the upper tibial extremity. The first is through the tibial epiphysis just beneath the articular surface; this permits drawing the posterior tibial plateau landmarks. By convention we consider this line as the transversal axis of the tibial plateau. The second CT scan is at the level of the upper part of the anterior tibial tuberosity. This view passes just above the peroneotibial joint. By superposition it is possible to calculate the distance between the deepest part of the trochlear groove (TG) and the midpart of the tibial anterior tuberosity (TA). This distance represents the potential tracking of the patella, but it is not similar to the Q angle. For these measurements lines are drawn from the two landmarks perpendicular to the posterior condylar plane (Fig. 1). The respective measurements are indicated in Table 1 for three types of knees. Statistically significant differences ($P<0.005$) were found between unstable knees, on the one hand, and both normal knees and chondropathic knees, on the other.
It is essential to take into account the TA–TG distance in the pre-operative examination when a medial tibial tuberosity transposition is planned. In practice we always take care to have a minimal post-operative distance of 6 mm. Depending on the importance of the medial transposition, this distance may also be important in verifying the secondary effect of medial tibial tuberosity transposition on the medial femorotibial joint.

Femorotibial Rotation in Extension. None of the conventional X-ray techniques is able to calculate the femorotibial rotation in extension. Using the CT scan we choose as landmarks the posterior tibial plateau line and the posterior femoral condylar line. The tibia is in external rotation in relation to the femur when the knee is maintained in full extension. Table 1 describes the angles of the various types of knees. There is a statistically significant difference in normal knees ($P<0.001$), on the one hand, and those, on the other, with either patellar instability or permanent patellar luxation. When the extension increases, as is possible in lax knees, the femorotibial rotation and the TA–TG distance increase. In some cases we have observed that a hyperextension of 15° is associated with an increase of 13° in the femorotibial rotation and an increase of 7 mm of the TA-TG distance. When extension increases, patellar position also changes in height. In this position all these actions help to explain the patellar instability.

The CT Scan in Measuring Torsional Abnormalities of the Inferior Limbs

Femoral Torsion. The angle of femoral torsion is measured between the femoral neck axis and the posterior femoral condylar line. There is a statistically significant difference here between normal knees, on the one hand, and chondropathic knees ($P<0.001$) and unstable knees ($P<0.005$), on the other (Table 1).

External Tibial Torsion. The angle of external tibial torsion is measured between the posterior tibial plateau line and the axis of the ankle malleoli. Successive views from the proximal to the distal show that the major part of the external tibial torsion comes from the proximal part of the tibia. There are no statistically significant differences among the various types of knees (Table 1).

Femoral Torsion and External Tibial Torsion: The Concept of Horizontal Morphotype. The external tibial torsion is always greater than the femoral torsion; the difference between the two figures is referred to as the tibiofemoral index (TFI). A high TFI indicates that the external tibial torsion is important, while a low TFI indicates that the two torsions (femoral and tibial) are either small or important. Table 2 shows the relationship of various TFI levels to the three types of knees analysed here, presenting the frequency of observation for each. A high TFI appears to be more frequent in patellar pathology and a low TFI more frequent in normal knees. A low TFI associated with patellar pathology is related to large femoral and tibial torsions. In this morphotype the knee is constrained between two large and inverse torsions.
The sum of femoral and tibial torsions (SFTT) for the various knee types is indicated in Table 1. A statistically significant difference ($P<0.005$) was found between the normal knees and those with patellar chondropathy. TFI and SFTT are important factors in assessing torsional osteotomies.

Submalleolar Torsion. The angle of submalleolar torsion is measured between the malleolar axis and the foot axis. The foot axis is the line extending from the head of the third metatarsal bone and the posterior midpoint of the calcaneum. There are no statistically significant differences among the three groups (Table 1). One can observe that the submalleolar torsion always extends inwardly. In considering torsional osteotomies TFI and SFTT must be taken into account, as are the submalleolar torsion and the step angle. A torsional osteotomy at one level (femoral or tibial) can, for example, influence the position of the foot during walking and also the position of the femoral neck axis in the horizontal plane. The hori-

Table 2. Tibiofemoral index by knee type

	Normal (%)	Patellar chondropathy (%)	Patellar instability (%)
Small (<24°)	45	26	18
Intermediate (24°–32°)	45	54	46
Large (32°–42°)	10	16	20
Major (>42°)	0	4	15

zontal plane is not the only one which must be analysed, however, in order to explain patellar pathology. The frontal plane, the best known and best analysed, and the sagittal plane must be carefully examined by clinical and radiological means. A three-dimensional morphotype must therefore be conceptualised here. Various groups can be distinguished, each of which can represent a potential factor of patellar pathology. Additional factors are also important, such as ligamentous weight, muscular weight, height of the subject and position of the patella (either alta or infera).

Conclusion

The horizontal plane is essential to patellar pathology. Certain of its abnormalities are surgically correctable. Three interdependent planes must always be considered; for example, when the knee progresses from extension to recurvatum, the femorotibial rotation increases, and this subsequently increases the varus. This study of the horizontal plane has demonstrated that the factors involved in surgical assessment include the following: femoral antetorsion, tibial torsion, TA–TG distance, lateral trochlear slope, ligamentous laxity or contracture and muscular unbalance. We must not, of course, forget the two other planes (frontal and sagittal), which may be influenced by any modification in the horizontal plane and which may also need to be corrected. Any decision as to surgery must therefore be made with extreme prudence.

Anterior Cruciate Ligament and Anterior Instability

Grading the Pivot Shift: An Objective System with Treatment Implications

R. P. Jakob, H. U. Stäubli, and J. Deland

Orthopedic Department, University of Berne, Inselspital, Berne, Switzerland

Accurately grading the pivot shift and other laxity tests with a rotatory component is problematic in the clinical setting. Old systems have used the clinical impression of the severity of the shift. By identifying the presence of the pivot shift in specific tibial rotations, a more objective grading system is possible. Such a system, along with measuring the displacement of the medial versus lateral tibial plateaus in a nonblocked drawer test, helps quantify laxities which differ between the medial and lateral sides. In particular, clinically significant levels of laxities that have been described as anterolateral, anteromedial and posterolateral are identified. The anterior laxity of each individual tibial plateau was measured roentgenographically under anesthesia in 21 patients who had anterior cruciate ligament injuries. These measurements could be statistically correlated with the type and degree of laxity defined by the grading system. These tests have two ad-

Table 1. The grading system

Grade	criteria for grading	Usual correlations		
		Lachman	Average values for non-blocked anterior drawer	
			Medial	Lateral
I	Small sliding shift with tibia in internal rotation	6–9 mm	5 mm	12 mm
II	a) Shift present with tibia in internal and neutral rotation b) Shift with a definitive clunk with tibia in internal rotation Usually, both a + b are present, but either defines a grade II	10–15 mm	10 mm	18 mm
III	Shift with tibia in internal, neutral and external rotation a) Shift most severe in internal rotation b) Shift progressively more severe in neutral and external rotation and appearance of reversed pivot shift.	>15 mm	15 mm	22 mm

Surgery and Arthroscopy of the Knee
2nd Congress of the European Society
Editors: Müller/Hackenbruch

vantages. First, the clinician has a more objective measurement that can be used as a standard by different knee examiners. Second, the examiner gains an evaluation of the different laxities which occur in the absence of an anterior cruciate ligament. This evaluation can help make for more accurate treatment.

The Jerk Test in External Rotation in Anterior Cruciate Ligament-Deficient Knees

J. Y. Dupont

Clinique du Marais, 11bis rue Barbette, 75003, Paris, France

Introduction

The findings of Noyes [2] and Markolf [21, 22] have shown the anterior cruciate ligament (ACL) to be the primary stabilizer against anterior subluxation of the tibia, and to be effective regardless of the degree of flexion. It is also a stabilizer in limiting internal as well as external rotation (IR, ER). The two cruciates are intimately linked with the shape of the femoral condyles. This can be represented by the model of the "four bars crossed linkage" by Menschik and Muller [23]. The degree of harmonious gliding and rolling varies according to the flexion angle throughout the entire range of motion [9, 16, 23]. The lateral tibial plateau is convex (Figs. 1, 2). Its summit is posterior to the summit of convexity of the lateral condyle [16].

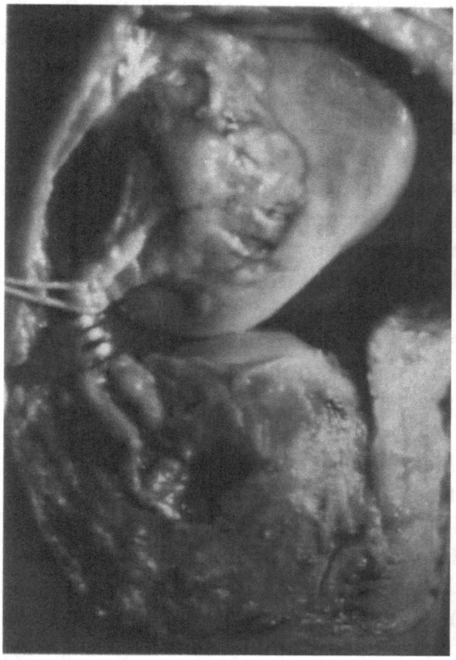

Fig. 1. Lateral compartment dissection on a cadaveric knee showing the convexity of the tibial plateau

Fig. 2. Sagittal section of the lateral compartment

Surgery and Arthroscopy of the Knee
2nd Congress of the European Society
Editors: Müller/Hackenbruch

Damage to the ACL results in abnormal movements in the knee, not only in one axis or in one plane, but invariably in all three dimensions [22]. The femur must be considered as the fixed anatomical element around which tibial movements occur. The most obvious pathological displacement of the tibia is a forward gliding. This can be expressed by the classical anterior drawer test at 90° flexion [12] or by the Lachman test at 20° flexion. Markolf has shown in vivo that the maximum degree of laxity occurs between 30° and 60° flexion, and between 15° and 30° external rotation [21, 22].
The ACL-deficient knee loses its harmonious gliding-rolling mechanics [23]. Allum [1] and Frain [9] have demonstrated that gliding occurs in the first 30° of flexion, followed by a sudden rolling. At the same time, the centre of rotation displaces medially and slightly posteriorly [23]. The lever of arm between this centre and the lateral peripheral structures is augmented, and a slight lateral hypermobility appears.
Abnormal movements of the knee around a longitudinal axis also occur, for the ACL is a restraint to both IR and ER. The anterior drawer test in ER has been popularized by Slocum [26]. It is due to the addition of posteromedial damage to the ACL deficiency. A relation cannot be hypothesized between this test and the functional instability which patients experience in twisting conditions during sports.
A major advance in the understanding of the mechanics of the knee came with the discovery of the sudden subluxation of the lateral tibial plateau under the femoral condyle at 30° flexion. Evoked by Palmer [24], it was first described by Lemaire [17], then popularized by Galway and MacIntosh [11]. All these authors related that test to the incompetence of the ACL. Further descriptions appeared with modifications by Hughston [13], Losee [20] and Slocum [27]. It is now commonly accepted [3, 6–8, 10, 18–20, 28] that the jerk test or pivot-shift test consists of a sudden displacement at about 30° flexion of the lateral tibial plateau under the femoral condyle (both convex), and is related to a deficient ACL and to damage of posterolateral structures. No such phenomenon exists in the medial compartment because the medial tibial plateau is concave.
Galway [11] does not emphasize the need for IR of the tibia to provoke the pivot shift, but for Lemaire and others this is essential, and the instability in internal rotation summarizes the pathophysiology of ACL-deficient knees. The displacement of the tibia can either be described from a position of reduction in flexion to a position of subluxation in extension (pivot-shift, Lemaire test), or from a subluxed to a reduced position (jerk test).
There is a need for a reference plane, which is the bicondylar plane. In normal knees, this is tangent to the plane of the tibial plateaux. In ACL-deficient knees, during the jerk test phenomenon, lateral X-ray views in the subluxated position show the anterior displacement of the lateral tibial plateau. The medial tibial plateau keeps its normal relationship with its femoral condyle (Fig. 3).
The value of this test, as compared with the anterior drawer and the Lachman tests, is in the close relationship between functional instability, such as giving way, and the examination. The patient often recognizes the feeling at the time when the examiner suddenly dislocates the knee [11]. This relationship has led surgeons to invent surrogate lateral extra-articular procedures, trying functionally to sta-

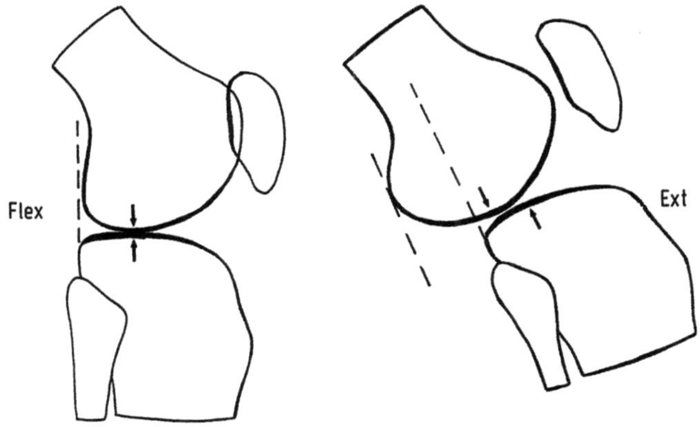

Fig. 3. Positions of the lateral condyle and tibial plateau in the pivot-shift test in internal rotation

bilize the knee simply by preventing it from being lax in IR. Other laxity tests were therefore neglected, and the anterior drawer remained positive. These procedures (Lemaire [17], MacIntosh (unpublished data), Ellison [6]) are not truly reconstructions, but substitutions. They produce a tenodesis effect, stabilizing the knee in ER, and preventing IR. The question remains, however, as to what happens to the pivot-shift phenomenon when the tibia is externally rotated. This led me to create a new laxity test, in 1979, and to utilize it for patients with ligamentous disorders, both acute and chronic. It shall be referred to as the jerk test in ER.

The Jerk Test in External Rotation

As in the case of the classical pivot-shift test in IR, the test in ER can be regarded from a position of reduction in flexion to one of subluxation in extension. I prefer, however, to test it from extension to flexion. The patient is more relaxed, and the test therefore easier to apply; indeed, it is most preferable to perform the test under general anaesthesia.

The position of the examiner's hands is important (Fig. 4). One is placed posterolaterally against the fibular head and applies a valgus stress, as well as forward displacement, of the tibia under the femur. The limb is lifted up from the examination table so that the weight of the thigh helps to keep the tibia anteriorly subluxated. The other hand holds the patient's ankle and provides ER to the leg. At the same time, it determines the flexion-extension angle of the knee, beginning from complete extension angle of the knee, beginning from complete extension and progressing towards flexion. This latter hand must not be placed over the lateral condyle, for this can produce falsely negative results. Also, an unrelaxed patient, a poor examination technique, or an inexperienced examiner can lead to unsatisfactory results. This test is somewhat difficult to perform until the examiner thoroughly understands the manoeuvre. This was also stressed regarding the original pivot-shift test when it was first described.

I have not attempted to grade the test according to the sensation of jerk that it induces; this sensation can be varied simply by applying more or less valgus stress.

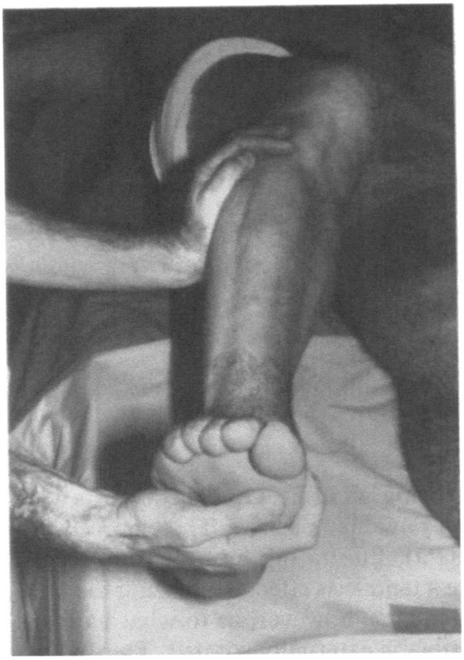

Fig. 4. Position of the examiner's hands in the technique of the pivot-shift test in external rotation

Rather, I have considered only whether it was positive or negative. In some cases the test was only slightly positive; in other cases, however, the knee could almost be dislocated in ER, due to a cam effect of the lateral tibial plateau. This was unexpected because all other laxity tests were only "normally" positive, such as a + + Lachman test, and a positive jerk test in IR.

A close inspection of the knee when performing the test clearly demonstrates that several abnormal movements of the knee occur simultaneously and before the sudden reduction at 30° flexion. The tibia displaces medially as well as anteriorly with an attempt to correct excessive ER by spontaneous internal rotation. In this forward global subluxation, both anterior margins of the tibial plateaux move anteriorly.

X-ray exposures and examination under image intensifier allow us to understand the phenomenon (Fig. 5). In the initial degrees of flexion (the starting position), the tibia grossly displaces anteriorly in a gliding movement of both compartments. The distance between the femoral condyles (the reference plane) and the posterior margin of both plateaux is increased (Fig. 6). At 30° flexion or so, the entire tibia suddenly reduces under the femur. The difference to classical pivot-shift or jerk in IR lies here in the anterior displacement of the medial compartment. The necessary and sufficient lesions to warrant this test are a deficient ACL and damage (rupture or stretching) to both posteromedial and posterolateral corners.

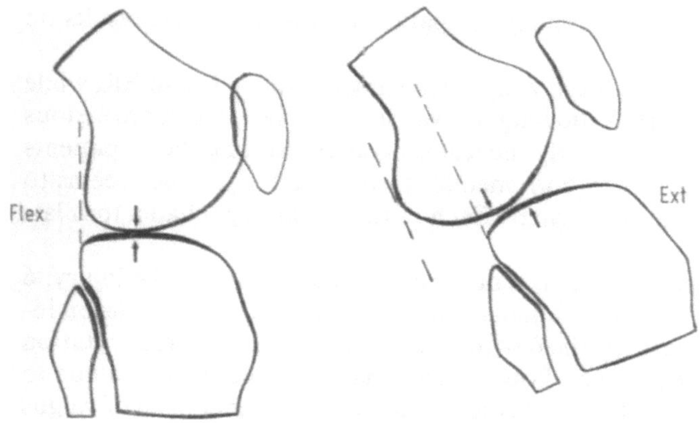

Fig. 5. Positions of the lateral condyle and tibial plateau in the pivot-shift test in external rotation

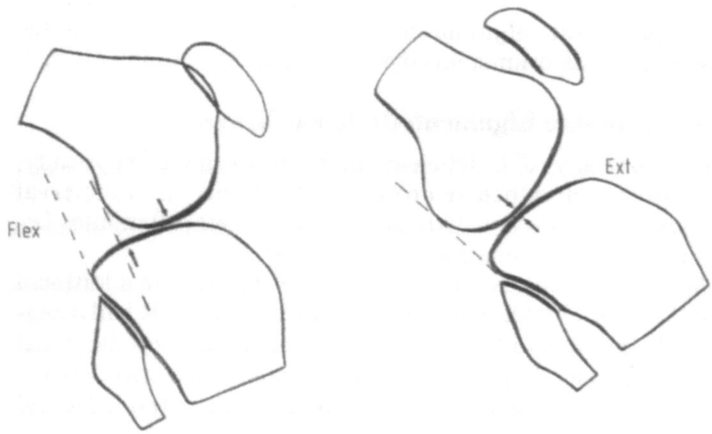

Fig. 6. Positions of the lateral condyle and tibial plateau in the reverse pivot-shift test

Clinical Value of the Jerk Test in External Rotation

I have used this test since 1979 in all types of ligamentous injuries. It has always proven negative in posterior cruciate ligament as well as medial collateral ligament (MCL) and lateral collateral ligament (LCL) injuries, acute and chronic, whether isolated or associated. It was positive only in cases of ACL deficiency.

Acute Anterior Cruciate Ligament Injuries

Arthroscopic examinations included 54 patients with acute haemarthrosis, showing recent damage to the ACL, either isolated or in addition to meniscal and/or peripheral lesions, including 23 MCL [4, 5]. Their laxity was analysed under anaesthesia. Of these patients, 12 showed no gross jerk either in IR or ER, but only some degree of minor gliding of the tibia. Twenty-five had a positive pivot-shift in IR with a negative test in ER. Finally, 17 patients showed positive results on both tests.

In 34 cases, ACL ruptures were total, with 14 showing a positive test in ER, while 20 ruptures were partial, with 3 showing a positive test. Thus, the more serious the ACL rupture, the more frequently the test is positive. Among the 17 patients with a positive test in ER, 14 had both menisci torn posteriorly, which seems to indicate damage to the posterior corners. Each of the remaining 3 had a torn lateral meniscus.

In 24 cases, the patient was unable to remember the mechanism of the injury: 6 had a positive test in ER. Of the 6 patients who experienced a hyperextension injury, none had a positive test. Of the 6 with a varus extension – internal rotation injury, only one had a positive test. Four patients had incurred direct trauma to the knee. None had a positive test. A further 14 described a mechanism of valgus flexion and external rotation. Of these, 10 had a positive test in ER.

The test in ER apparently reproduces the mode of injury which is the most frequent, although it has been denied as a cause of ACL rupture without concomitant MCL rupture. In spite of the valgus mechanism, I did not find any relationship between this test, the valgus trauma mechanism and damage to the MCL.

Chronically Anterior Cruciate Ligament-Deficient Knees

A total of 375 cases of chronically ACL-deficient knees were examined physically, with and without anaesthesia, and then arthroscopically. There was an interval of from 1 month to more than 20 years, with an average of 3 years 9 months between the initial sprain and the examination (Fig. 7) [5, 24].

Of these cases, 352 (94%) had a positive test in IR: 222 (59%) with an additional test in ER and 130 in IR alone (37%). None of those with a test in ER had a negative test in IR. I correlated the two tests and the time span between the initial injury and the examination. The frequency of positive tests varied with this period. In the period between 1 and 3 months, 86% showed a positive test in IR and

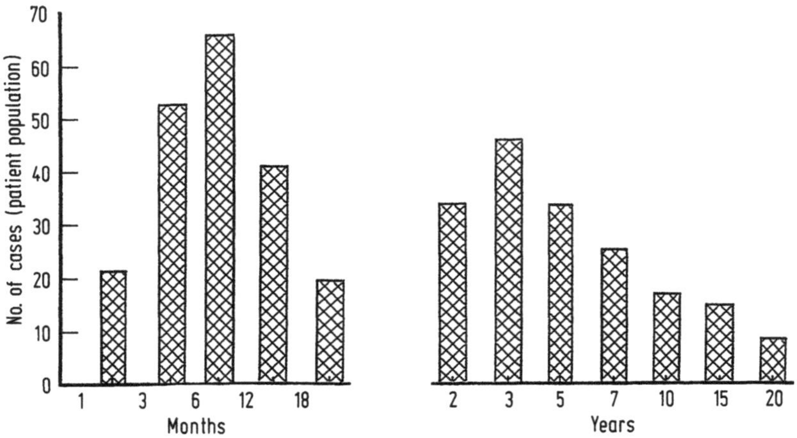

Fig. 7. Delay between initial injury and arthroscopy in 375 ACL-deficient knees (1 month to 40 years; mean 3.9 years)

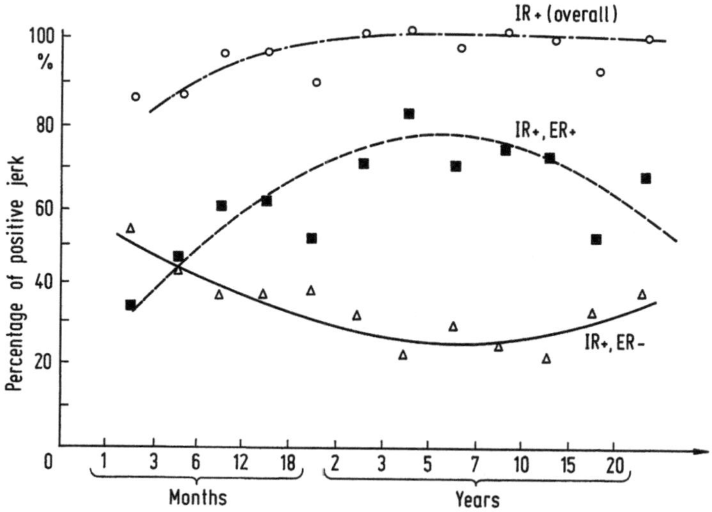

Fig. 8. Percentage of positivity of jerk tests in IR and ER in 375 ACL-deficient knees

35% in ER. In the period after 2 years, the corresponding figures are 97% and 65% (Fig. 8). I conclude from this that laxity increases after the initial ACL injury, due to secondary stretching of the peripheral restraints. A similar relationship was discovered regarding the anterior drawer and the Lachman test, but no correlation was found between time and medial laxity in valgus.

A high correlation was found between the jerk test in ER and the Lachman test: 88% of the patients with a Lachman test of 10 mm or more had a positive test in ER. Each indicates lesions to the ACL and both posteromedial and posterolateral corners.

The specificity of the jerk test in ER is perfect, as it means in all cases a torn or deficient ACL. However, its sensitivity is poor, for 41% of the patients with chronic ACL insufficiency and 69% of acute ACL injuries had a negative test. A positive test indicates a gross global anterior laxity, which does not occur in all the ACL-deficient knees. The specificity of the test in ER is better than that in IR: we encountered cases of falsely positive results in IR, in hyperlax teenagers or some rare lateral meniscal tears with normal ACL. The test in IR is more sensitive: in my experience, this was positive in 78% of acute and 94% of chronic ACL injuries. This is not as high as other reported findings; the difference in our findings may be due to our inclusion of even the mildest ACL partial ruptures or distensions.

The relationship in the case of chronic ACL insufficiancy between laxity tests and such functional disability as giving way was ascertained according to a special instability grading [25]. The strongest relationship was found with the test in ER, in which 97% of cases with such a test were functionally unstable, compared to 86% of cases with positive test in IR alone. The difference between instability scores in both groups was highly significant ($p=0.001$).

Discussion

This test differs from the Losee test [19], which describes the reduction of the pivot shift when the tibia is displaced from a position of IR to a position of ER at less than 30° flexion. It differs also from the reverse pivot shift described by Jacob [15], in which the tibia is posteriorly subluxated in flexion and ER, and reduced in extension. This tests posterolateral laxity, whereas our test examines global anterior laxity. In case of doubt, X-ray examination in a displaced and reduced position explains the type of laxity (Fig. 9). Radiological and clinical examination can therefore be used to distinguish true posterolateral laxities and global anterior laxities. In the former, the reverse pivot shift, when present, is associated with a Hughston recurvatum test [14], a posterior drawer in ER, a varus laxity, and an external hyper-rotation of the tibia. This is a rare case, and may be either isolated or associated with a PCL-deficient knee. We have observed on non-operated knees no such test associated with a torn ACL and an intact PCL. It is our feeling that posterolateral laxities are not associated with ACL insufficiency. In global anterior laxities, on the other hand, the jerk test in ER is associated with a Lachman test, an anterior drawer and a jerk test in IR. This is a very frequent condition. Damaged posterolateral structures allow an abnormal anterior excursion of the lateral tibial plateau which does not correlate with posterolateral excursion of the tibia.

The only, very rare cases in which positive posterolateral laxity tests could be observed on ACL-deficient knees were among four patients who had undergone an extra-articular procedure. Posterolateral laxity was iatrogenic and may have been caused by the forced position of external rotation of the knee at the time of surgery.

The jerk test in ER does not belong to any current knee laxity classification. I propose modifying these classifications and using the term "global anterior laxity" for those cases in which tests both in IR and ER are positive. This can be graded according to the magnitude of the anterior displacement at the Lachman test; for

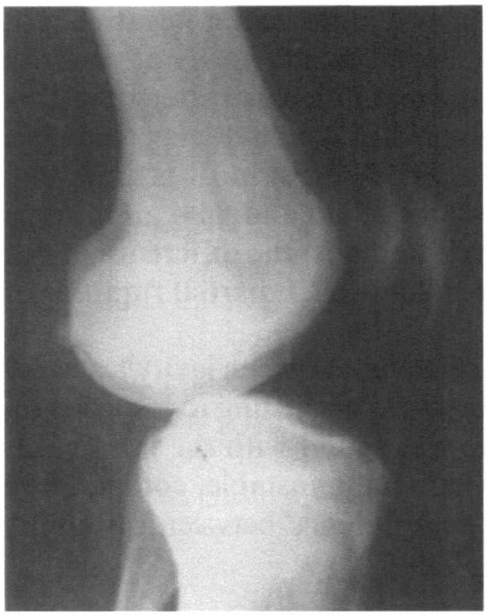

Fig. 9. Radiological examination of the knee position in the pivot-shift test in external rotation and in subluxated position. Note that both tibial plateaus are anteriorly displaced

example, "+" would signify 5 mm, "++", 10 mm, "+++", 15 mm. Cases in which the test in IR is positive and the test in ER is negative could be termed "internal rotation anterior laxity". One must bear in mind that some patients are likely to change their classification over the years following their initial ACL injury. The terms "laxity" and "instability" should definitely not be used interchangeably.

Therapeutic Considerations

The goal of treatment lies in stabilizing the patient functionally by preventing the knee from pursuing abnormal movements. If one analyses only the possibilities of jerk tests, the three following types of case can be distinguished: neither test positive (around 5% of patients); positive test in IR, negative test in ER (35%, defining internal rotational anterior laxity); both tests positive (60% defining global anterior laxity).

In the first of these cases, I found no patient whose knee was unstable. There was no surgical indication, and the patients were treated conservatively.

In the case of internal rotatory laxity, functional instability may lead to surgical treatment. Then, a lateral extra-articular palliative tenodesis can prevent the knee from subluxating, but it must be solid enough to resist progressive stretching due to the absence of primary restraint.

In the case of global anterior laxity, functional instability appears very frequently and is related to the level of patients' activity. If surgical treatment is indicated, it should aim to stabilize the knee in some position of rotation. The treatment chosen would be an isometric intra-articular reconstruction, such that the graft would mimic the ACL in its physiological role.

A lateral extra-articular tenodesis is contraindicated in such a case for two reasons. First, it would not prevent the knee from subluxating in ER and a functional failure could occur; and, secondly, it would not prevent the other abnormal movements reappearing by secondary stretching of the tenodesis in the months thereafter.

Extra-articular procedures are often unsuccessful in middle and long term follow-ups, even though precise data on this is lacking in the literature. In my experience, such patients have a negative jerk test in IR, a positive jerk test in ER, and a positive Lachman test. The conjunction of a peripheral posteromedial tenodesis has also been advocated. It is likely that, due to the lever arm phenomenon, these procedures are unable to stabilize the knee perfectly. Most surgeons who have used these procedures have abandoned them without having published their results.

The presence of a positive result to the jerk test in ER is a clear contraindication to any extra-articular procedure and, if surgery is indicated, an intra-articular reconstruction should be performed. The criteria for an excellent anatomical result after reconstruction is the disappearance of all response to laxity tests. When this goal is attained, the functional result is always excellent.

Conclusions

The jerk test in ER needs to be considered as an important addition to the set of laxity tests for ACL-deficient knees for the following reasons.

1. It reproduces the most frequent mechanism of ACL ruptures.
2. It is pathognomonic of ACL deficiency; though not sensitive, it is a test for global laxity and is a sign of severity.
3. There is a stronger relationship between its presence and functional instability than in the case of the jerk test or pivot-shift test in IR.
4. A positive result is a contraindication for lateral extra-articular procedures.
5. It must be differentiated from the "reverse pivot shift", which is a test for posterolateral laxity.

References

1. Allum R, Jones D, Mowbray MAS, Galway HR (1984) Triaxial electrogoniometric examination of the pivot-shift sign for rotatory instability of the knee. Clin Orthop 183:144–146
2. Butler DL, Noyes FR, Grood ES (1980) Ligamentous restraints to anterior-posterior drawer in the human knee. J Bone Joint Surg [Am] 624:259–270
3. Chatard JC (1979) Fréquence et valeur séméiologique du ressaut rotatoire du compartiment externe du genou dans le cadre du syndrôme d'insuffisance du ligament croisé antérieur. Thèse, Université Saint-Etienne
4. Chaudieres D (1986) Les lésions dans les ruptures récentes die ligament croisé antérieur. Thèse, Université Paris-Ouest
5. Dupont JY, Scellier C, Chaudieres D (1986) Les lésions intra-articulaires et leur évolutivité au cours des ruptures récentes et anciennes du ligament croisé antérieur. Rapport préliminaire. Acta Orthop Belg 52:488–515
6. Ellison AE (1980) The pathogenesis and treatment of anterolateral rotatory instability. Clin Orthop 147:51–55
7. Fetto JF, Marshall JL (1979) Injury to the anterior cruciate ligament producing the pivot-shift sign. J Bone Joint Surg [Am] 61:710–713
8. Fowler PJ (1980) The classification and early diagnosis of knee joint instability. Clin Orthop 147:15–21
9. Frain P, Fontaine C, D'Hondt D Constraintes du genou par dérangement ménisco-ligamentaire. Etude de l'articulation condylo-tibiale interne. Méthode cinématique expérimentale. Rev Chir Orthop 70(5):361–370
10. Fukubayashi T, Torzilli PA, Sherman MF, Warren RF (1982) An in vitro biomechanical evaluation of anterior-posterior motion of the knee. J Bone Joint Surg [Am] 64:258–264
11. Galway HR, MacIntosh DL (1980) The lateral pivot-shift: a symptom and sign of anterior cruciate insufficiency. Clin Orthop 147:45–50
12. Hughston JC, Andrews JR, Cross MJ, Moschi A (1976) Classification of knee ligament instabilities. Part 1: the medial compartment and cruciate ligaments. J Bone Joint Surg [Am] 58:159–172
13. Hughston JC, Andrews JR, Cross MJ, Moschi A (1976) Classification of knee ligament instabilities. Part 2: the lateral compartment. J Bone Joint Surg [Am] 58:173–179
14. Hughston JC, Norwood LA (1980) The posterolateral drawer test and external rotational recurvatum test for posterolateral rotatory instability of the knee. Clin Orthop 147:82–87
15. Jakob RP, Hassler H, Staeubli HU (1981) Observations on rotatory instability of the lateral compartment of the knee. Acta Orthop Scand [Suppl] 191(52):1–31
16. Kapandji IA (1974) Physiologie articulaire. Fascicule II. Maloine, Paris, pp 72–135
17. Lemaire M (1967) Ruptures anciennes du ligament croisé antérieur du genou. Fréquence, clinique, traitement. J Chir (Paris) 93:311–320

18. Losee RE (1983) Concepts of the pivot-shift. Clin Orthop 172:45–51
19. Losee RE, Johnson TR, Southwick WO (1978) Anterior subluxation of the lateral tibial plateau. A diagnostic test and operative repair. J Bone Joint Surg [Am] 60:1015–1030
20. Lucie RS, Wiedel JD, Messner DG (1984) The acute pivot-shift: clinical correlation. Am J Sports Med 12:189–191
21. Markolf KL (1985) A summary of knee stability measurements. AAOS Symposium on sports medicine. The knee. Mosby, St Louis, pp 87–93
22. Markolf KL, Kochan A, Amstutz HC (1984) Measurement of knee stiffness and laxity in patients with documented absence of the anterior cruciate ligament. J Bone Joint Surg [Am] 66:242–252
23. Muller W (1984) The knee. Springer, Berlin Heidelberg New York Tokyo, p 314
24. Palmer I (1938) On the injury to the ligaments of the knee joint. Acta Orthop Scand [Suppl 53]:282
25. Scellier C (1986) Evolutivité des lésions après rupture du ligament croisé antérieur. Thèse, Université Amiens
26. Slocum DB, Larson RL (1968) Rotatory instability of the knee. J Bone Joint Surg [Am] 50:211–225
27. Slocum DB, James SL, Larson RL (1976) Clinical test for anterolateral rotatory instability of the knee. Clin Orthop 118:63–69
28. Wang CJ, Walker PS (1974) Rotatory laxity of the human knee joint. J Bone Joint Surg [Am] 56:161–169

Translation and Rotation in Knee Instability: A Prospective Stress Radiographic Analysis with the Knee in Extension

H. U. Stäubli, R. P. Jakob, and B. Noesberger

Orthopaedic Department, University of Berne, Inselspital, 3010 Berne, Switzerland

Several studies have shown that the Lachman test and the pivot shift test have a high diagnostic accuracy in the clinical evaluation of acute anterior cruciate ligament (ACL) insufficiencies. Controversy still exists however, whether an acute ACL rupture leads to an abnormal translational or rotational laxity of the knee.

Material and Methods

In a prospective stress radiographic analysis, medial and lateral compartment femorotibial displacements were measured under anaesthesia at high force levels (200 N). The previously described stress radiographic measurements technique was used to measure individually the laxity of the medial and lateral compartment of the knee at 10° of flexion.
The stress radiographic measurements were compared with the integrity of the cruciate ligaments at surgery, using a probe from 138 patients above the age of 16 years, 138 knees were entered into the final analysis. All had well-documented clinical records, including anterior stress radiograms and known integrity of the cruciate ligaments. Thus 138 anterior and 138 posterior stress radiograms were analyzed.

Results

Anterior stress radiographic measurements in the ACL-intact and in the ACL-deficient groups yielded the following results:

1. In the ACL-intact population ($n=53$), the anterolateral displacement measured 4.0 ± 3.2 mm, and the anteromedial displacement measured 2.8 ± 2.3 mm, with an arithmetic mean of 3.4 ± 2.0 mm.
2. In the ACL-deficient population ($n=85$), the anterolateral displacement measured $15.8 + 4.6$ mm, and the anteromedial displacement measured 9.9 ± 5.3 mm, with an arithmetic mean of 12.8 ± 4.1 mm.

Discussion

The radiographic measurement technique of Jacobsen and Torzilli with the knee at 90° of flexion was modified to assess compartmental translational and rotational femorotibial motion. Anterior stress radiograms of 85 ACL-deficient knees were compared with a group of 53 knees with an intact ACL.

Surgery and Arthroscopy of the Knee
2nd Congress of the European Society
Editors: Müller/Hackenbruch

With an intact ACL it is possible to demonstrate stress radiographically as a physiologically combined translational and rotational anterior knee motion at large force levels.

In an unstable knee with an insufficiency of the ACL, it is possible to demonstrate stress radiographically at 10° of flexion as a significant increase of a combined translational and rotational anterior knee laxity.

A substantial insufficiency of the ACL led to a significant combined anterolateral and anteromedial femoral-tibial subluxation ($P<0.001$) in comparison with the ACL-intact control group.

The Diagnostic and Prognostic Value of the "Active Radiologic Lachman"

H. Dejour, P. Chambat, G. Walch, and P. Ranger

Clinique de Chirurgie Orthopédique et Traumatologie,
Centre Hospitalier Lyon Sud – pavillon 3A, 69310 Pierre-Benite, France

We feel that it is very important now to quantify anterior laxity in order to be able to evaluate our results in an objective way. We suggest a very simple way to do this without resort to any sophisticated instrumentation.

An isolated complete rupture of the anterior cruciate ligament will give an anterior drawer near the complete extension; that is what we call the Lachman test. This drawer means a laxity that can be reproduced by a single contraction of the quadriceps which we have tried to quantify by the "active radiologic Lachman". A lateral X-ray of the knee in extension with a contraction of the quadriceps is taken. The anterior subluxation of the tibial plateau in relation to the condyles is found in millimeters. In a normal knee, the mean value in 3.2 mm (range from 0 to 6 mm). There are some rules that we have to follow very closely to give a precise repetition of the results. The aim is twofold:

First it helps clarify the diagnosis; indeed the mean value of the active radiologic Lachman for a rupture of the anterior cruciate ligament goes up to 11 mm, and there is always a difference of 5 mm as compared with the opposite side. We studied 150 chronic and 30 acute anterior laxities seen and examined prior to operation (since March 1984). This showed an excellent correlation with both the rupture of the anterior cruciate ligament and the associated injuries: previous meniscectomy, peripheral tear at the posterior horn of the medial meniscus, or a strain of the posteromedial corner which increases the anterior laxity.

Second, it gives an evaluation for the prognosis, by an objective estimate of the laxity after the surgery. So we concluded that even if it was easy to obtain a good result for stability by an intra- and extra-articular plasty, on the other hand it was much more complicated to eliminate the laxity in extension. We then reviewed 150 operations on the anterior cruciate ligament with a follow-up of 3 years. The remaining laxity was measured by the "active radiologic Lachman". The values that we found correlated closely with both the clinical examination (pivot shift, jerk test, etc.) and the degeneration of the femorotibial articulation. From this last correlation it appears that if the dynamic tests reflect the instability, the importance of the anterior laxity is the most faithful indication of the extent of arthrosis evolution.

Surgery and Arthroscopy of the Knee
2nd Congress of the European Society
Editors: Müller/Hackenbruch

Measure of Laxities by Stress Radiography and by KT 1000 Arthrometer

J. L. Lerat, B. Moyen, L. Dupre La Tour, E. Mainetti, J. J. Lalain, and E. Brunet-Guedj

Service de Chirurgie Orthopédique et de Médicine du Sport,
Hôpital Edouard Herriot, 69374 Lyon Cedex 08, France

Numerous studies have shown the Lachman test at 20° flexion to have a higher diagnostic accuracy in the clinical evaluation of anterior cruciate ligament (ACL) insufficiencies than does the anterior drawer test of the knee at 90° flexion [1, 7, 12, 13, 15, 16]. Several new clinical tests improve our diagnosis of sagittal instability, but these clinical methods are unsuccessful in determining accurately the degree of laxity [2, 4, 14]. The most reliable and reproducible assessments of sagittal laxity to date have been by radiographic methods [3, 5, 8–11]. The objectives of this study are to document the effectiveness of several stress-radiographic techniques in assessing sagittal laxity of the knee and to compare this with the results obtained with the KT 1000 arthrometer.

Material and Methods

A total of 371 knees were tested among patients ranging in age from 17 to 35 years. The knees were categorised as follows: 180 normal knees, unharmed by traumatism and without symptoms (134 males, 46 females); 125 knees with ruptured ACL as verified by arthrotomy or by arthroscopy (102 males, 23 females); 66 knees operated on by several procedures of intra-articular reconstruction of the ACL (47 males, 19 females). Three radiographic tests were carried out on these knees. These were passive anterior stress X-ray, active anterior stress X-ray and the stress-radiographic measurement technique. In addition the knees were examined with the KT 1000 arthrometer.

Passive Anterior Stress X-Rays. The patient is placed on the X-ray table in the decubitus dorsal position, with hips on the edge of the table and lower limbs off the table. The tested knee is placed on a special support; the thigh is free, the knee flexed at 20° and the leg placed on a rigid plastic gutter supported from the floor (Fig. 1). The flexing angle of 20° is determined by the fixed gradient of the splint. The leg is attached to the device by straps. A load of 9 kg is applied to the thigh by a strap just above the patella, weighting down the femoral condyles while the upper extremity of the tibia is fixed. This manoeuvre reproduces the drawer of the Trillat-Lachmann test.

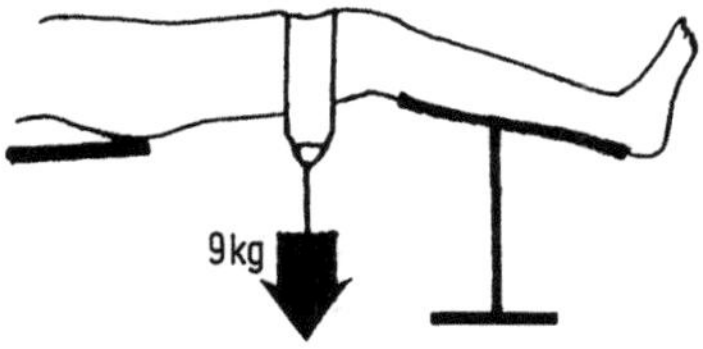

Fig. 1. Passive anterior stress X-rays. Notice the position of the rigid plastic gutter supporting the leg, and the application of the 9-kg load above the patella

Surgery and Arthroscopy of the Knee
2nd Congress of the European Society
Editors: Müller/Hackenbruch

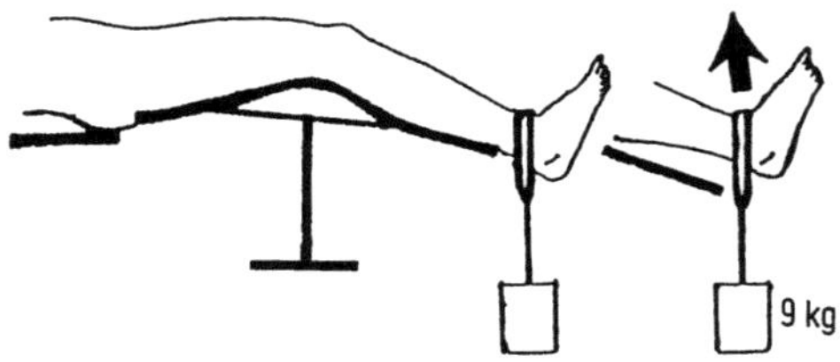

Fig. 2. Active anterior stress X-rays. Notice that the knee is supported by a gutter at 30° of flexion. The 9-kg load is applied to the ankle. The patient is asked to lift the load

Active Anterior Stress X-Rays. The support is placed differently than in the above case. The thigh and the leg rest on a gutter supported from the floor (Fig. 2). The knee is flexed at 30°. A load of 9 kg is applied to the ankle. The subject lifts the load so as just to detach the leg from its support. The thigh rests in contact with the apparatus. The flexion of the knee thus passes from 30° to 20°, and the X-ray is taken in this position. The opposite limb rests on a stool at a lower level in order to prevent superposition on the X-ray. After 5 s of preloading a lateral radiographic stress view is obtained. The X-ray film is placed near the femoral condyle, and the distance between the X-ray source and the film is held constant (1 m). In positioning the patient it is important to avoid excessive rotation.

Stress-Radiographic Measurement Technique. The measuring of the drawer is achieved by taking the posterior femoral medial condyle and the posterior medial tibial plateau as landmarks. Jacobsen [5, 6] identified the medial femoral condyle as smaller and rounder, with the condylopatellar sulcus (limiting groove) slightly anterior to the middle of the curve of the condyle, while the limiting groove of the lateral femoral condyle is situated in the middle. The tibial plateaus can be distinguished easily. The medial tibial plateau is larger and more squared, while the lateral tibial plateau is smaller and rounder. Parallel to the sagittal posterior part of the tibial diaphysis two base lines are drawn tangential to the most posterior aspect of the medial femoral condyle and medial tibial plateau. The distance between the tangential lines of the medial compartment of the knee is measured using a graduated transparent template. All measurements were recorded and rounded to the nearest 1.0 mm. All knees were examined by two examiners in a standard manner. The values were not corrected for magnification.

Measurement of the Anterior Drawer with KT 1000 Arthrometer. The two inferior limbs are placed on the arthrometer support system. This consists of a thigh support platform 11 cm in height which places the knees symmetrically in 25° ± 5° flexion. A footrest positions the feet in 15°–25° of symmetrical external rotation. The patient is asked to relax. The arthrometer is applied to the leg so that the patellar sensor pad stabilizes the patella within the femoral trochlea. A firm, constant pressure is applied to the patellar sensor pad and maintained throughout the test. The measurement reference position is obtained by applying a 89 N posterior load and then releasing it several times. The instrument dial is then set to zero. An anterior pull is applied to the handle and the displacement is read from the dial at both the 67 N (first "beep") and 89 N (second "beep") levels. In this study only the reading at the 89 N level was taken into account. After each pull force cycle a posterior push force is applied, and the force-sensing handle is released to confirm this; the needle of the dial is returned to zero within a margin of 0.5 mm. The average of three reproducible measurements to within 0.5 mm was recorded.

Results

The three radiographic methods described above were applied to measure the importance of the anterior laxity of the knees (Table 1). Measurement error was assessed both in terms of inter- and intra-observer differences. As regards inter-observer differences, 47 randomly selected knees were remeasured by a second examiner; the mean error was 1.5 ± 1.6 mm. To assess intra-observer differences, each observer remeasured 58 randomly selected X-rays; the mean error was 0.7 ± 0.9 mm. In 49 cases here the difference between the figures of the two observers was less than 1 mm, and in nine cases the difference was greater than 1 mm.

Normal Knees. No significant difference was found between right and left sides, between sexes nor between the passive and active radiological anterior drawer. However, a significant difference was found between the values of the anterior drawer given by the KT 1000 arthrometer and those given by radiological methods ($P < 0.0001$). Analysis of the two radiological methods showed a significant correlation between the active and the passive method ($r = 0.62$, $P < 0.0001$), but no significant correlation was found between the radiological methods and the KT 1000 arthrometer (with the active drawers, $r = 0.10$; with the passive drawer, $r = 0.04$).

Anterior Cruciate Deficient Knees. No significant difference was found between the two sexes, between the average values of the passive and active radiological anterior drawer nor between the radiological methods and the KT 1000 arthrometer ($P < 0.0001$). Analysis of the two radiological methods showed a significant correlation between the active and passive method ($r = 0.38$, $P < 0.0001$) and between the passive drawer and KT 1000 ($r = 0.24$, $P < 0.05$). No significant correlation, however, was found between the active radiological method and the KT 1000 arthrometer ($r = 0.15$).

Surgically Treated. No significant difference was found between the two sexes. The average values of the passive drawer and the active drawer were statistically comparable. A significant difference was found between the values of the drawer

Table 1. Comparison of the three methods of measuring anterior laxity in Shace groups of knees

	Normal knees ($n = 180$)	ACL-ruptured knees ($n = 125$)	Operated knees ($n = 66$)
Passive dynamic radiography (mm)	3.1 ± 1.9	10.1 ± 3.1	5.8 ± 3.2
Active dynamic radiography (mm)	3.3 ± 2.0	10.8 ± 2.7	6.4 ± 3.0
KT 1000 arthrometer (mm)	5.7 ± 4.8	10.6 ± 3.1	7.7 ± 2.7

given by the KT 1000 arthrometer and those given by the two radiological methods ($P<0.005$). Analysis of the two radiological methods showed a significant correlation between the passive and the active radiological anterior drawer ($r=0.84$, $P<0.0001$). No significant correlation was found between the radiological methods and the KT 1000 arthrometer (with the active drawer, $r=0.18$; with the passive drawer, $r=0.02$).

Discussion

Anterior physiological laxity may vary from one case to another but not systematically with the sex of the patient. Measurement of the pathological laxity of one knee can be regarded as correct only when compared with the other. Active and passive radiological measurements of the drawer are comparable in the groups of knees. The two radiological methods provide the same information with the same criteria of measurement and produced a permanent radiographical document.

The passive anterior stress X-ray method presents the following advantages: a fixed angle of flexion; a fixed load having always the same application point and the same moment of action; an immobilised knee at rest without muscular contraction which could jeopardise the test; the possibility of studying the knees of cadavers with the same criteria.

The active radiological method, conversely, shows the following disadvantages: the angle of flexion is more difficult to control; the rotation is more difficult to control, as it can vary during active extension; it may be painful in acute cases and shortly after operation, thus reducing its usefulness; the moment arm and the quadriceps acting force may vary.

We measured the anterior laxity on the medial knee compartment, which is the more stable. As Hooper and Staeubli have shown [3, 15], it is possible by stress X-rays to measure the anterior drawer of the lateral compartment. Here the figures are much higher. The passive stress X-ray drawer is 9.8 ± 3.9 mm for normal knees, 17.5 ± 7.2 mm for anterior cruciate ligament deficient knees and 11.7 ± 4.7 mm for operated on knees. The active stress X-ray drawer is 8.6 ± 4.2 mm for normal knees, 13.3 ± 5.3 mm for anterior cruciate ligament-deficient knees and 10.5 ± 6.4 mm for operated-on knees. The lateral compartment is more subject to rotation, and that which is measured on lateral stress X-rays represents merely translation. Nevertheless, the rotational abnormalities of the lateral compartment are important to appreciate, while its objective radiological measurements seem, in our view, less valid.

The absolute values of the drawer, measured radiologically and by arthrometer, do not differ greatly in specifically abnormal knees. In fact, we measured the same phenomenon but under different conditions. With the arthrometer global laxity is measured from two landmarks: the anterior part of the patella and the anterior part of the tibial tuberosity, both anatomically localised in the mid part of the knee. The arthrometer measures anterior laxity from a position of rest, which does not always correspond with the radiological "zero" position but which could be a position of posterior drawer. The radiological "zero" position is the position at which the two base lines tangential to the most posterior aspect of the medial femoral condyle and medial tibial plateau are superimposed. From this "zero"

Table 2. Posterior drawer in three groups of knees, measured by posterior passive stress X-rays studies

	Normal knees ($n=39$)	ACL-ruptured knees ($n=20$)	Operated knees ($n=22$)
Posterior drawer (mm)	2.3 ± 3.7	3.2 ± 3.2	3.2 ± 2.9

position we tested the posterior passive laxity in 39 normal knees and in 42 anterior cruciate insufficient knees. The knee is flexed at 20°, and the thigh is placed in a horizontal gutter supported from the floor, while the foot is fixed. A 9-kg load is applied at the level of the anterior tibial tuberosity. The manoeuvre reproduces the posterior drawer at 20° knee flexion.

Table 2 presents the results, showing that some knees had a significant physiological posterior drawer. Anterior cruciate insufficient knees had a higher but not significant posterior drawer. These observations may explain why the KT 1000 arthrometer tends to give a higher reading than do stress-radiographic techniques.

Conclusion

Our tests on normal knees and on anterior cruciate ligament-insufficient knees, operated and non-operated, have proven dynamic radiography to be a precise evaluation method, both in its active and its passive forms. The passive method, in contrast to the active method, has the advantage of being able to be reproduced in a comparative manner during the evolution of knee laxity. It is usable for acute knees and also in the postopertive period without any interference by pain. The same method can therefore be used to analyse a freshly operated knee and a chronic knee with comparative documents which may be kept and readily verified.

The KT 1000 arthrometer has the advantage of being easy to use in daily practice without exposure to X-rays. Its inconvenience lies in its not leaving any documents. These tests permit additional diagnosis which complete the clinical tests; they are primarily post-operative control tests. We must conclude, therefore that, given the present state of our knowledge, dynamic radiography and the arthrometer appear to complement one another.

References

1. Butler DL, Noyes FR, Grood ES (1980) Ligamentous restraints to anterior-posterior drawer in the human knee: a biomechanical study. J Bone Joint Surg [Am] 62:259
2. Galway HR, MacIntosh DL (1980) The lateral pivot shift: a symptom and sign of anterior cruciate ligament insufficiency. Clin Orthop 147:45–50
3. Hooper GJ (1986) Radiological assessment of anterior cruciate ligament deficiency. J Bone Joint Surg [Br] 68:292–296
4. Hughston JC, Andrews FR, Cross MJ, Moschi A (1976) A classification of knee ligament instabilities: parts I and II. J Bone Joint Surg [Am] 58:159–179

5. Jacobsen K (1981) Gonylazometry. Acta Orthop Scand [Suppl] 52:194
6. Jacobsen K (1976) Stress radiographical measurement of the antero-posterior, medial and lateral stability of the knee joint. Acta Orthop Scand 47:335–344
7. Kennedy JC, Fowler PJ (1971) Medial and anterior instability of the knee: an anatomical and clinical study using stress machines. J Bone Joint Surg [Am] 53:1257–1270
8. Leven H (1977) Determination of sagittal instability of the knee joint. Acta Radiol [Diagn] (Stockh) 18:689–697
9. Leven H (1978) Radiologic determination of rotational instability of the knee joint. Acta Radiol [Diagn] (Stockh) 19:599–608
10. McPhee IB, Fraser JG (1980) Stress radiography in acute ligamentous injuries of the knee. Injury 81(12):383–388
11. Malcom LL, Daniel DM, Stone ML, Sachs R (1985) The measurement of anterior knee laxity after ACL reconstructive surgery. Clin Orthop 196:35–41
12. Markolf KL, Mensch JS, Amstutz HC (1976) Stiffness and laxity of the knee – the contributions of the supporting structures: a quantitative in vitro study. J Bone Joint Surg [Am] 58:583–594
13. Markolf KL, Kochan A, Amstutz HC (1984) Measurement of knee stiffness and laxity in patients with documented absence of the anterior cruciate ligament. J Bone Joint Surg [Am] 66:242–253
14. Slocum DB, James SL, Larson RL, Singer KM (1976) Clinical test for antero-lateral rotatory instability of the knee. Clin Orthop 118:63–69
15. Staeubli HU, Jakob RP, Noesberger R (1984) Anterio-posterior knee instability and stress radiography: a prospective biomechanical analysis with the knee in extension. Biomechanics: current interdisciplinary research. Nijhoff, The Hague
16. Torzilli PA, Greenberg RL, Insall J (1981) An in vivo biomechanical evaluation of antero-posterior motion of the knee: roentgenographic measurement technique, stress machine, and stable population. J Bone Joint Surg [Am] 63:960–968

Stress Radiographic Comparison of Drawer Symptoms Measured in 90° Versus 15° of Knee Flexion (the Lachman Position)

B. F. Iversen, J. Stürup, and K. Jacobsen

Gammel Strandvej 94, 3050 Humlebaek, Denmark

In recent years tests for assessing anterior laxity in knees, executed with the knee flexed to 15°, have emerged. In cases of rupture of a cruciate ligament, the blocking effect of the posterior horn of the meniscus as well as the effect of spasms in the hamstring muscles are minimized in this position. The aim of this study was to quantitate the differences between drawer symptoms measured in 15° and 90° of flexion using stress radiography and to assess the effect of muscular defence on drawer symptoms.

Material and Methods

The patients investigated comprised two groups. Group 1 (11 patients, 22 knees) had no history of knee injuries. Group 2 (16 patients, 16 knees) had arthroscopically verified ruptures of the anterior cruciate ligament (ACL). X-rays were obtained with the knees flexed to 15° as well as 90°, unloaded and with forward or backward traction of 220 N. The patients in group 2 also had radiograms taken under simultaneous forward traction and contraction of the musculature of the thigh, thus simulating muscular defence. The laxity was calculated according to Jacobsen. For statistical analysis, the Wilcoxon test for paired data was used.

Results

Decreasing the knee flexion from 90° to 15° significantly increased the anterior drawer symptoms. This applied both to injured ($P<0.01$) and uninjured ($P<0.05$) knees. In 90° of flexion, contracting the muscles of the thigh (thus simulating muscular defence) had the potential of totally eliminating the forward drawer symptom and even displacing the tibial condyles beyond the (unloaded) resting position. In 15° of flexion, the effect of muscular defence was significantly less ($P<0.002$) than in 90° of flexion. The within-patient variation between the knees in the uninjured group was $\leqq 3$ mm of 90°, and $\leqq 4.5$ mm at 15° of flexion.

Discussion

The quantitative data obtained in this study support the clinical studies in favour of the Lachman position (knee flexed to 15°) for testing for the anterior drawer symptom. In this position, the drawer symptoms measured are greater than at 90°, and the effect of muscular defence is considerably lesser. However, it should be

Surgery and Arthroscopy of the Knee
2nd Congress of the European Society
Editors: Müller/Hackenbruch

emphasized that testing at 15° often requires an assistant to secure the patient's thigh.

Summary

In recent years the Lachman test for detecting the anterior drawer symptom has been introduced. This test is executed with the knee flexed to 15°. The anterior drawer symptom is found to be significantly greater, and the effect of muscular defence is significantly smaller at 15° of flexion than at 90°.

Unrecognized Abnormal Osseous Metabolic Activity About the Knee of Patients with Symptomatic Anterior Cruciate Ligament Deficiency

S. F. Dye

Orthopaedic Surgery Service, Letterman Army Medical Center,
San Francisco, California, USA

Introduction

A primary treatment goal of young adults with symptomatic anterior cruciate ligament deficient knees is the prevention or amelioration of early degenerative arthritis. Current methods of documenting the development of degenerative changes about the knee include roentgenographic evaluation and direct intra-articular examination. At our institution over the past 5 years we have gained experience in utilizing technetium 99m bone scans in the evaluation of patients with patellofemoral symptoms and found this technique to indicate often an objective osseous abnormality in the presence of normal roentgenographic and arthroscopic findings. The purpose of this study was to evaluate a group of patients with anterior cruciate ligament deficient knees with technetium 99m bone scans in order to delineate possible sites of abnormal osseous metabolic activity about the knee, as compared with arthroscopic and roentgenographic findings.

Methods

Twenty-seven individuals (19 males and 8 females) between the ages of 18 and 39 years with symptomatic anterior cruciate ligament insufficiency and positive Lachman and pivot shift tests were included in this study. Besides a standard history and physical examination, evaluation of these patients included AP and lateral roentgenographs and scintigraphic evaluation with technetium 99m MDP generating AP and lateral static projections 3 hours after injection. In addition 18 of the 27 patients had documentation of intra-articular pathology by direct examination.

Results

Of these patients 74% (20/27) revealed scintigraphic evidence of abnormally increased osseous metabolic activity of the medial compartment, 52% (14/27) of the lateral compartment, and 48% (13/27) of the patellofemoral compartment. This compares with arthroscopic evidence of pathology (either torn meniscus or chondromalacia) in 50% (9/18) in the medial compartments, 28% (5/18) in the lateral compartments, and 33% (6/18) in the patellofemoral compartments of those patients who had arthroscopy. Plain roentgenographs revealed abnormality in only 26% (7/27) of the medial compartments, 19% (5/27) of the lateral compartments, and 7% (2/27) of the patellofemoral compartments.

Surgery and Arthroscopy of the Knee
2nd Congress of the European Society
Editors: Müller/Hackenbruch

Conclusions

(a) A heretofore unreported and unappreciated osseus abnormality (increased osseous metabolic activity) was documented in the medial, lateral, and patellofemoral compartments of young adults with symptomatic anterior cruciate deficient knees. (b) This study indicates that most patients with symptomatic anterior cruciate ligament insufficiency have established widespread abnormal osseous metabolic activity, often in the presence of normal roentgenographs. We postulate that this process may represent the initial phase in the development of early degenerative arthritis in many patients. (c) The data presented have implications for interpretation of the effect on the knee of the altered mechanics associated with anterior cruciate ligament insufficiency. For example, the medial compartment appears to be less able to withstand the pathological increase in mobility resulting from anterior cruciate ligament deficiency than the lateral compartment, which seems to be structurally designed for greater mobility and, therefore, is more tolerant of the increases in motion.

Stresses Applied to the Knee in Meniscal and Ligamentous Disorders – A Study of the Medial Tibiofemoral Joint

D. D'Hondt [1], P. Frain [2], C. Fontaine [3], and C. Argenson [1]

[1] Department of Orthopaedics-Traumatology, CHU Nice, Hôpital Saint Roch, 5 rue Pierre Dévoluy, 06000 Nice, France
[2] 49 avenue Gustave Dron, 59200 Tourcoing, France
[3] Department of Orthopaedics-Traumatology B, CHU Lille, 59000 Lille, France

Using a geometric and kinematic approach, we would like to consider the rolling-gliding behaviour of the medial condyle and variations which occur after section of different ligaments and after meniscectomy, demonstrating the presence of stresses which do not appear in normal knee joints. Previous studies of this rolling-gliding behaviour have often been contradictory because only by determining a geometric centre and an instantaneous centre of rotation for every point on the curve of the medial condyle surface is it possible to define these concepts clearly.

Theoretical Geometric and Kinematic Bases

To determine the geometric centre of the medial condyle, the curve of the surface is compared to: (a) a logarithmic spiral arch whose centre is the point of attachment of the medial ligament: I_f (Fig. 1); (b) a series of circular arcs. These circles are known as osculating circles. Their centre is the centre of the curve at a given point. To locate the curve centre for a point T on the surface of the medial condyle, the tangent at T and then the perpendicular at T to the latter are drawn. First $I_f T$ and then the perpendicular at I_f to this segment are drawn. The intersection of these lines locates the geometric centre 0 (Fig. 2).

Thus, for each point on the curve of the medial condyle surface, there is a corresponding curve centre, and the series of curve centres forms Fick's evolute (Fig. 1). Generally, the geometric centre is different from the centre of rotation.

To determine the instantaneous centre of rotation C for a movement of flexion-extension with a point on the curve of the medial condyle surface, it is necessary to locate two (or three) fixed points A and B on the lower end of the femur. In our study, these points were located in five different positions: in extension and flexion at 30°, 60°, 90°, 120° and 150° respectively.

The centre $C01$ is the point of intersection of the perpendiculars brought to the middle of the displacing vectors $A01$ and $B01$ between positions 0 and 1 (Fig. 3). By repeating this procedure, the centres $C12$, $C23$, $C34$, and $C45$ are located. It is known that the centre $C1$ at the midpoint of $C01$–$C12$ is the instantaneous centre of rotation for the position of flexion at 30°, $C2$ for this position at 60°, $C3$ for this position at 90° and $C4$ for this position at 120°. From each point C,

Surgery and Arthroscopy of the Knee
2nd Congress of the European Society
Editors: Müller/Hackenbruch

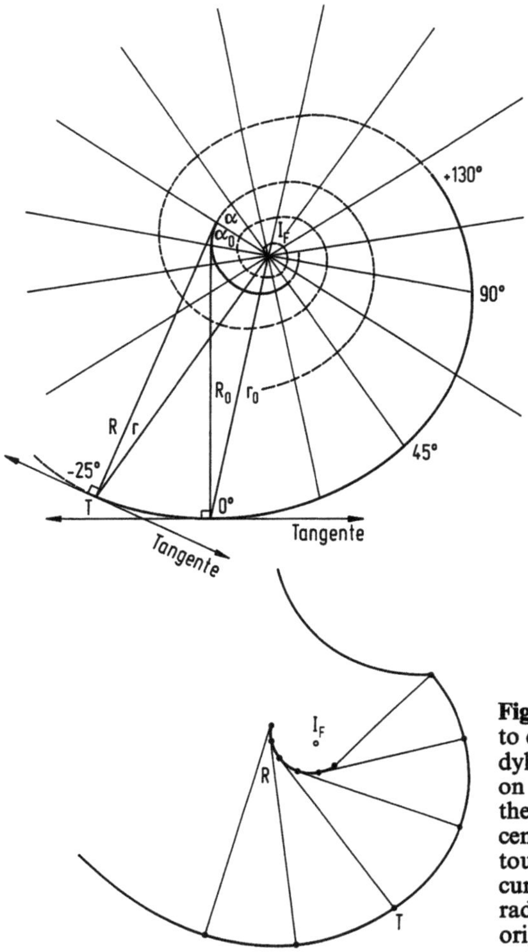

Fig. 1. *Top*, a mathematical model adapted to describe the contour of the medial condyle: it is a logarithmic spiral arch centred on I_F, the point of insertion on the femur for the medial ligament. *Bottom*, locus of curve centres of the surface of the condylar contour. Each point of the condyle has its own curve centre which is unique and specific. *R*, radius of curve at *T*; *r*, vector radius at *T*; *O*, origin

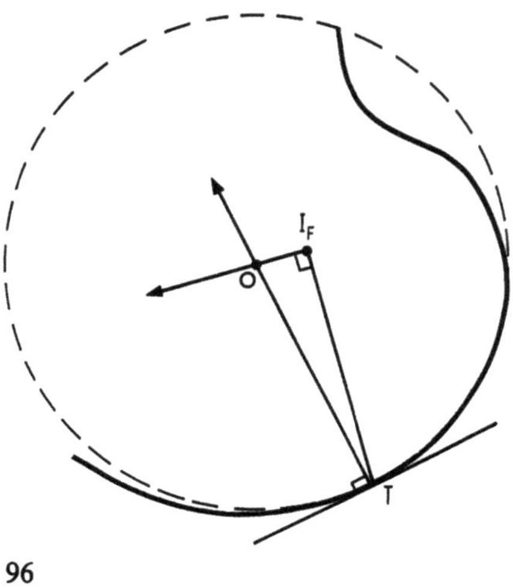

Fig. 2. Construction of curve centre *O* from a point *T* on the medial condyle. I_F, femoral insertion for the medial ligament

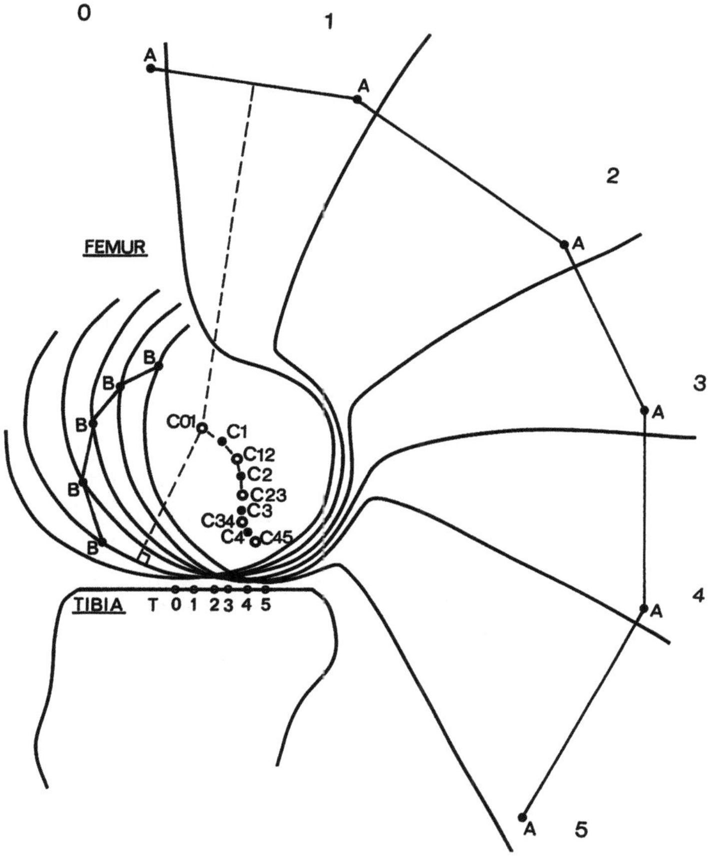

Fig. 3. A tracing of the instantaneous centres of rotation of the knee joint on the outlines of X-rays taken in the profile position at 0°, 30°, 60° and 120°. Concomitant positioning of the reciprocal tangential points on the condyle and the tibial plateau is also shown

one may trace the velocity vector to the tangential point *T* by tracing the perpendicular to the radius, *CT* (our diagrams are intended for an angular velocity of 30°/s).

Before considering the behaviour of the medial condyle, let us begin by examining the behaviour of a wheel whose geometric centre is 0 which is tangent to the ground at *T*, and which turns around an instantaneous centre of rotation *C* under conditions of weightlessness (Fig. 4).

Four situations are thus possible, depending on the respective positions of 0 and *C*:

1. The geometric centre and kinetic centre *C* are the same: the velocity vector is tangent to the circle at *T*, and there is pure gliding motion.
2. The kinetic centre *C* is at point *T*: the velocity vector at *T* is zero, the vector at 0 is parallel to the ground, and there is pure rolling motion.

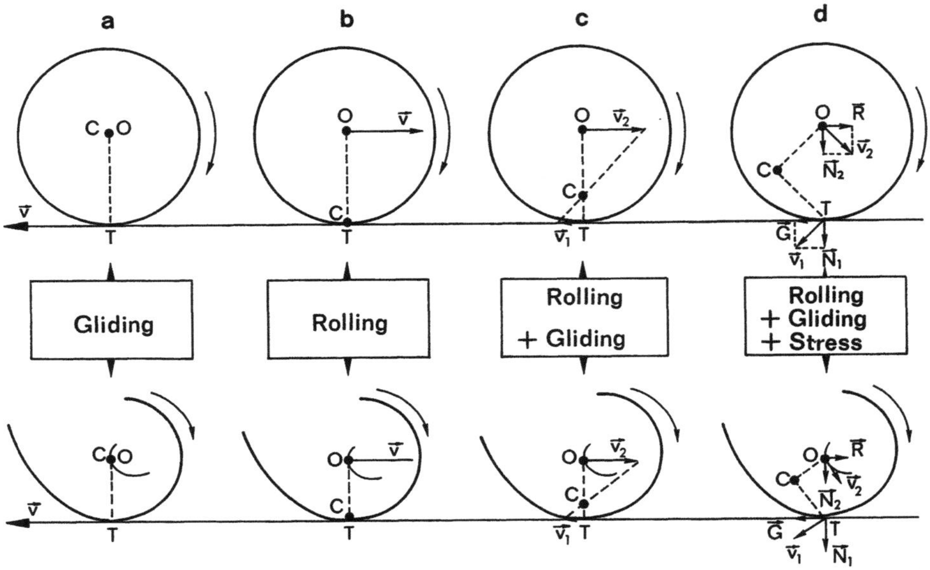

Fig 4 a–d. Instantaneous speeds at *O* and at *T*, according to centre of rotation of a circle (*top*) and of the condyle (*below*) during the motion of flexion. **a** If *O* (centre of circle) and *C* (instantaneous centre) are the same, gliding occurs. **b** If *C* and *T* (point of contact of circle with ground) are the same, rolling occurs. **c** If *C* is located on *OT*, gliding and rolling occur. **d** If *C* is located outside of *OT*, gliding, rolling and stress occur

3. The kinetic centre *C* is located on 0*T*: a vector *V*1 is drawn at *T*, producing a gliding motion, and *V*2 at 0 causes rolling. The ratio *C*0:*CT* provides the proportions for rolling and gliding.
4. *C* is located outside of 0*T*: vectors *V*1 and *V*2 are no longer parallel to the ground. Their separation determines a vector *R* at 0, and *G* at *T*. The normal vertical components *N*1 and *N*2 to the ground represent the stress which originates at the tangential point.

The same reasoning applies for the motion of the spiral condyle, considering the osculating circle as the curve of the surface of the medial condyle at the tangential point. The geometric centre 0 here is the curve centre at *T* (Fig. 4).

Material and Methods

In this study, 20 knee joints from fresh cadavers were examined. X-ray studies in the strict profile position of the medial condyle-tibial joint were made in extension and in flexion at 30°, 60°, 90°, 120° and 150°. Three metallic markers inserted into the femur enabled us to trace the instantaneous centre of rotation *C*. The femoral and tibial insertions of the cruciate ligaments were thus located. Flexion of the knee joint was accomplished passively and then against a transpatellar resistance of 6 kg, to simulate the force of the quadriceps muscle. On the outlines of the X-rays, the centres of rotation *C*, the geometric centres 0 and the tangential femorotibial points were drawn.

Results

The Normal Knee

Behaviour of the Cruciate Ligaments
Radiographic determination of the sagittal component of the anterior cruciate ligament shows that it decreases in length during flexion, while the length of the posterior cruciate ligament increases. This confirms the notion that the anterior cruciate ligament is held in extension and that the posterior cruciate ligament is held in flexion.

Behaviour of the Condylotibial Interface at the Tangential Point
A simple approach to the rolling-gliding behaviour may be obtained by comparing the tangential points *T0–T5* on the medial condyle and on the tibial plateau, showing that there is a large proportion of sliding at the beginning and at the end of flexion. Figure 5 summarizes the results. Kinematic study of the normal knee, which is much more accurate, shows that the centres of rotation *C*1, *C*2, *C*3 and *C*4 are located on the radius of the curve at the tangential points called *T*1, *T*2, *T*3 and *T*4. There is thus rolling and gliding motion without stress, and velocity vectors are tangent to the curve at 01 and *T*1, for example.
Figure 6 summarizes the proportions of rolling and gliding motion: flexion begins with gliding, then rolling progressively develops, to attain 50% between 60° and 90°. At the end of flexion, rolling decreases, giving way to a gliding motion.

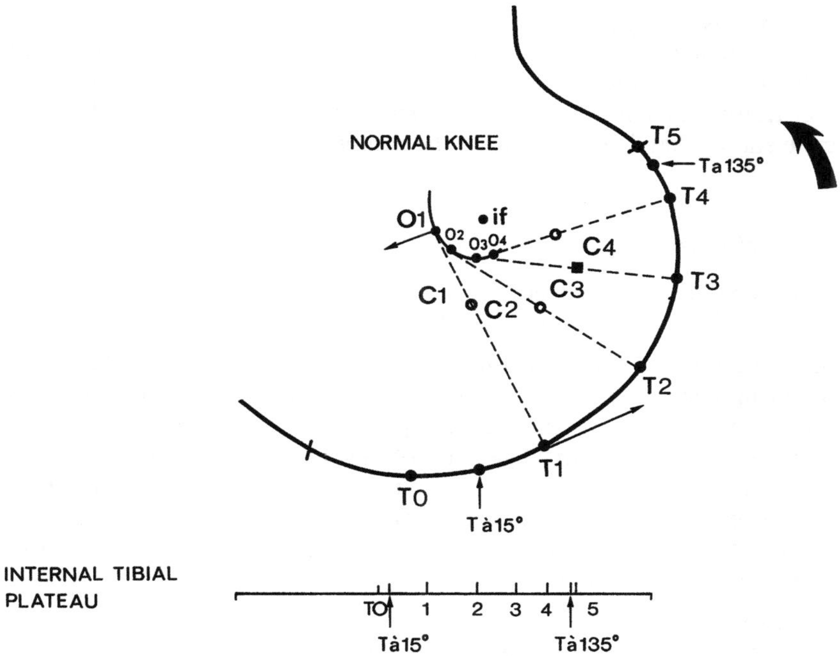

Fig. 5. Kinetic centres of rotation of the normal knee. *if*, femoral insertion of medial ligament; *O*, curve centre; *C*, centre of rotation; *T*, femorotibial tangential point

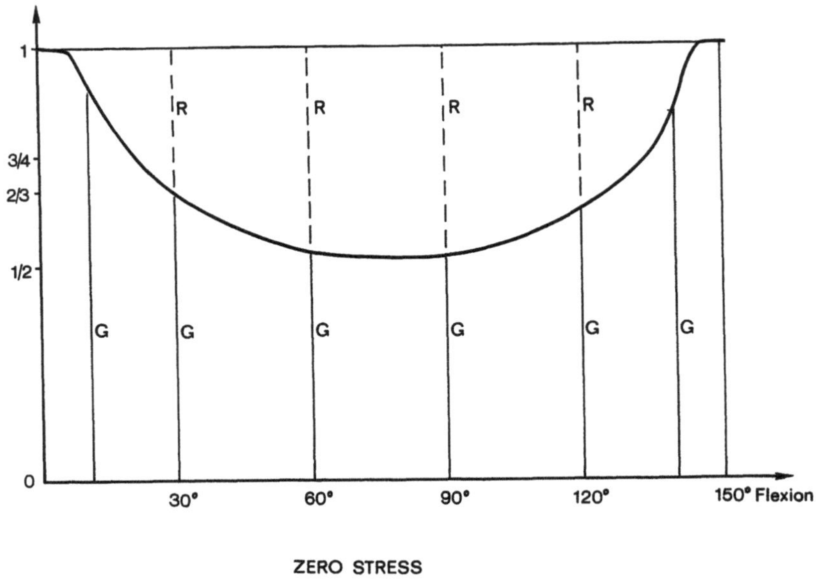

Fig. 6. Proportions of rolling (*R*) and gliding (*G*) during flexion of the normal knee joint. For example, at 30°, it may be seen that 2/3 of the motion is gliding and 1/3 is rolling. Stress is zero

That flexion begins with gliding is explained by the anterior cruciate and medial ligaments being under maximum tension during extension, preventing rolling of the condyle during the first degrees of flexion. Then, the gradual tension of the posterior cruciate ligament during flexion combined with relaxation of the anterior cruciate and medial ligaments draws the condyle back and causes increasing rolling.

The Abnormal Knee

Behaviour of the Cruciate Ligaments

Measurement of the sagittal component of the cruciate ligaments shows that:

1. Section of the anterior cruciate ligaments produces only very minimal forward sliding during flexion against resistance. To obtain a clear sliding effect, it is necessary to associate either meniscectomy or section of the patellar flaps or, especially, section of the medial ligaments of the knee. This sliding effect is corrected at about 60° by the tension of the posterior cruciate ligament, which abruptly brings back the tibial plateau under the condyle.
2. Section of the posterior cruciate ligament causes a net posterior sliding effect from the first degrees of flexion.

Behaviour of the Joint Interface

Kinematic study of the joint is of greater interest:

After meniscectomy (Fig. 7), the centres of rotation are no longer on the radii of the curve, and there is a resultant stress $N1 + N2$. The velocity vector is no longer

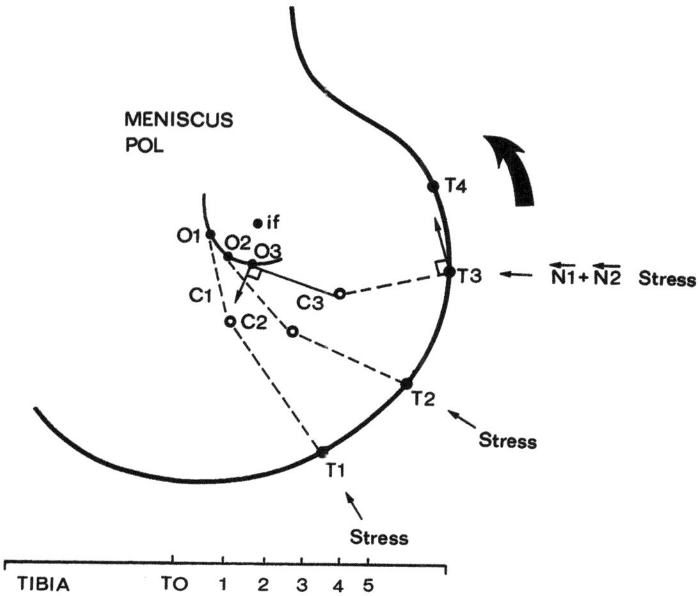

Fig. 7. Centres of rotation after internal meniscectomy and section of the posterior oblique ligament (POL). $N_1 + N_2$, stress vector (at an angular speed of 30°/s). For definitions of other symbols, see legend of Fig. 5

tangent to the curve of the medial condyle surface and penetrates the latter. A study of the knee undergoing resistance shows that these disturbances increase. First, an increase in rolling motion is observed at the beginning of flexion, then an increase in gliding after 60°, as is shown in Fig. 8, at the bottom of which the effect of stresses at 30°, 60° and 90° is represented.

Section of Hughston's posterior oblique ligament (POL) does not change the sagittal centres of rotation. On the other hand, section of the medial ligaments displaces the centres of rotation and thus causes stresses to appear throughout movement, thus disturbing the rolling-gliding motion. Flexion begins with a large amount of rolling. Associated meniscectomy aggravates these disturbances even more.

Section of the anterior cruciate ligament does not produce any change. Only a dynamic study appears to show a very slight change in the sagittal centres of rotation. We know that disturbances in rotation should occur, but they cannot be detected by the techniques employed here.

Section of the posterior cruciate ligament produces stresses which remain weak and accentuate gliding, as compared with the normal knee. Probably, the contraction of the quadriceps muscle, which normally causes a forward-sliding movement of the tibia under the femur, opposes the posterior-sliding movement produced by section of the posterior cruciate ligament that partially protects the tibia from too great kinetic disturbances.

Section of the posterior capsule does not change kinetics at the joint interface.

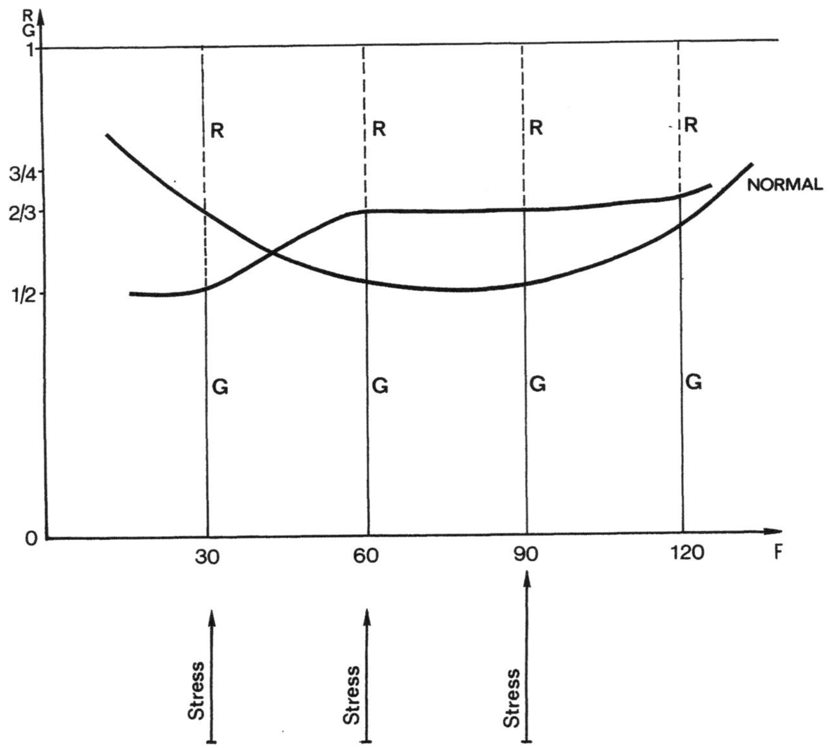

Fig. 8. Proportions of rolling and gliding motion after internal meniscectomy and section of the POL

A Study of the Abnormal Knee

Examination of the knee with anterointernal slack movement, combined with prior section of the anterior cruciate ligament, distension of the medial ligaments and posterior oblique ligament and severe osteoarthritis of the internal joint compartment with maintenance of a laminated meniscus, shows severe disturbances in the centres of rotation.

The rolling-gliding curve becomes inverted, as compared with the normal one, and in all positions we have demonstrated major stress.

Conclusion

Although the shape of the medial condyle is that of a spiral arch, we have not observed any lifting effects. Thus, during flexion-extension, if one draws a series of velocity vectors, the contour for the medial condyle is obtained (Fig. 9). The ligamentous meniscocapsule orients movement by placing the centre of rotation at each moment on the curve radius. We have located these centres exactly.

Section of ligaments and examination of the abnormal knee show that the series of velocity vectors at the tangential point forms an envelope which is different from that of the condyle (Fig. 10). Movement is poorly directed, and stresses develop which cause the onset of osteoarthritis.

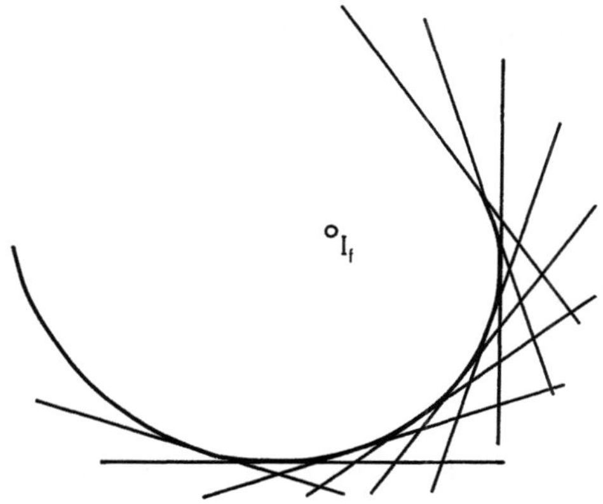

Fig. 9. Normal knee. A tracing of the vectors for velocity, which indicates the direction of the displacement from the femorotibial tangential point, reproduces the envelope of the condylar contour exactly. This means that the tibial tangential point at each moment follows a direction which is tangent to the medial condyle. There cannot be any lifting or stress effects

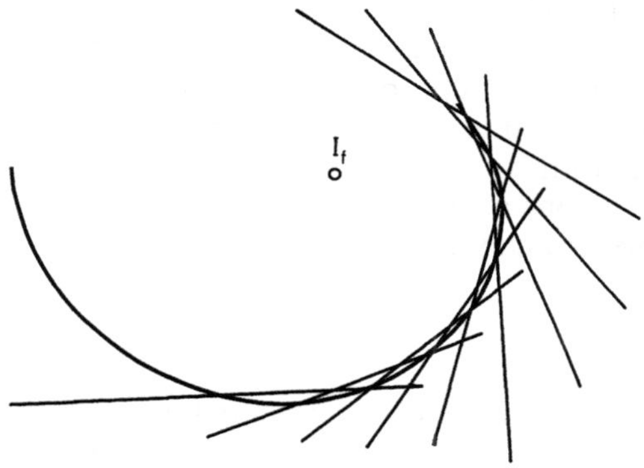

Fig. 10. Abnormal knee. A tracing of the vectors for velocity at the femorotibial tangential point shows that their direction is no longer tangential, as in Fig. 9, but intersects the condylar curve. The envelope thus drawn does not reproduce the condylar contour and is located inside it. It is thus the condyle itself which guides movement, and the tibia follows, while pressing on it. There is a lifting effect, and stress arises here

Instability of the Anterior Cruciate Deficient Knee

J. Kärrholm[1], G. Selvik[2], L.-G. Elmqvist[1], and L. I. Hansson[1]

[1] Department of Orthopaedic Surgery, University Hospital in Umeå, 901 85 Umeå, Sweden
[2] Department of Anatomy, University of Lund, 223 65 Lund, Sweden

Introduction

The effect of ligamentous injuries on the stability of the knee joint has been measured or estimated in clinical and anatomical investigations. In order to assess knee instability under reproducible conditions, external transducers (Markolf et al. 1978) or radiographic techniques (Jacobsen 1981; Torzilli et al. 1981) have been introduced to register the effect of standardized forces. An increased mean sagittal laxity in a group of patients with a previous rupture of the anterior cruciate ligament (ACL) has been established in some investigations, but the diagnostic accuracy has been variable. The purpose of this study was to measure the three-dimensional in vivo translations and rotations of the tibia in the ACL deficient knee using roentgen stereophotogrammetric analysis.

Materials and Methods

In a consecutive series of 25 men the sagittal stability was measured on average 3 years and 6 months after injury. A tear of the ACL had been confirmed at injury or later on by arthroscopy. Three patients had, in addition, had an injury of the medial collateral ligament, seven of the medial meniscus and seven of the lateral meniscus. Total or partial meniscectomy had been performed in nine patients, suture of the meniscus in one, and suture of the medial collateral ligament also in one patient.

During preoperative arthroscopy or under local anaesthesia three to six tantalum balls were implanted in the distal femur and proximal tibia bilaterally. Three to six weeks later the roentgen stereophotogrammetric examination was performed with the knee in a Plexiglas calibration cage. The femur was stabilized in an adjustable metal frame. All patients were examined in the following positions: (1) extended knee (2) 30° flexion and (3) anterior traction of the tibia (150 N), and 18 with (4) posterior traction of the tibia (80 N).

In the testing procedure a sling was placed 6–7 cm distal to the knee joint and anterior traction with 150 N was applied. The foot was held in the neutral position. Posterior stability was investigated applying loads of 80 N both medially and laterally.

Absolute motions of the two bone segments (distal femur, proximal tibia) and relative tibial motions using the distal femur as a fixed bone segment were calculated. Tibial translations were measured as the mean translations of the two tips of the tibial intercondylar eminence using the computer program "point transfer"

Surgery and Arthroscopy of the Knee
2nd Congress of the European Society
Editors: Müller/Hackenbruch

in KINEMA. To maintain a constant position of the tibia in relation to the cardinal axes between the positions of extended and flexed unloaded knee, the tibia and femur were rotated using the computer program rotate.

Results

Reproducibility of the Tests

Repeated examinations were performed in 13 patients 1–12 months after the initial examination. The standard deviations of sagittal translations between the flexed unloaded position and the loaded positions varied between 0.8 and 1.2 mm. The reproducibility of rotations varied between 1.6° and 5.0° depending on the type of rotation measured.

Translation of the Tibial Intercondylar Eminence

We investigated the translations at anterior traction along the cardinal axes and the vectorial sum of these translations. At anterior traction all patients showed an increased anterior displacement of the injured knee (mean 11.2 ± 2.6 mm) in relation to the intact one (mean 3.6 ± 1.9 mm). Both the intact and injured knees showed minimal average distal and lateral displacement, and the difference was insignificant. On the intact side the anterior translations amounted to 6.8 mm. On the injured side the smallest value was 6.8 mm and the highest 16.9 mm. The difference between the injured and the intact side was on average 7.6 mm.
The anterior-posterior (A-P) laxity manifested a similar pattern but the separation between the intact and injured knees was more distinct, with a mean difference of 8 mm (intact side 2.5–7.2 mm, injured side 8.6–17.9 mm).

Tibial Rotations at Anterior and Posterior Load

We also looked at the rotations occurring between extended knee and flexed loaded knee with anterior traction. The average knee flexion was about 30° on the intact and injured sides, and increased slightly when anterior traction was applied. All but one case showed an internal rotation on the intact side and an internal or external rotation on the injured one. The internal rotation averaged 4.3° less on the injured side ($p < 0.01$). Reduced adduction was also recorded on the injured side, with an average difference of 3.4° ($p < 0.001$).
Posterior traction in relation to the extended position resulted more often in an external rotation on the injured side than on the intact one. The average difference was about 4° ($p < 0.01$).

Summary and Conclusions

Roentgen stereophotogrammetric analysis provides a complete kinematic analysis of movements occurring during a test situation. An individual reference position can be determined in each knee and can be used at later examination.

Absence of the anterior cruciate ligament implied an increase in anterior laxity of on average 7.6 mm (range 3.3–13.8 mm). The A-P laxity manifested a still better separation of the injured knees from the intact ones.
The rotatory asymmetry of the ACL deficient knees compared with the intact ones indicates that the ACL not only counteracts anterior translation but also stabilizes the knee in internal rotation and adduction when a sagittal force is applied to the tibia.

References

Jacobsen K (1981) Goylaxometry. Stress radiographic measurement of passive stability in the knee joint of normal subjects and patients with ligament injuries. Acta Orthop Scand [Suppl] 194:1

Markolf KL, Graff-Radford A, Amstutz HC (1978) In vivo knee stability. A quantitative assessment using an instrumented clinical testing apparatus. J Bone Joint Surg [Am] 60:664–674

Torzilli PA, Greenberg RL, Insall J (1981) An in vivo biomechanical evaluation of anterior – posterior motion of the knee. J Bone Joint Surg [Am] 63:960–968

The Reliability of the Genucom Knee Analysis System

C. L. Highgenboten [1] and A. Jackson [2]

[1] 7777 Forest Lane, Bldg. 4, S 106, Dallas Texas 75230, USA
[2] North Texas State University Denton, Texas, USA

The Genucom knee analysis system is a relatively new computerized tool for the evaluation of the knee (Faro Inc 1984). This system has been developed to aid the orthopedic community in the diagnosis and treatment of knee ligament injuries. It combines a biomechanical digitizer, an electrogoniometer, a microcomputer, and a variety of common and unique knee laxity test protocols to provide a clinical description of total static knee ligament behavior. The system generates a detailed test report which includes both graphical and numerical values.
If the Genucom knee analysis system is to become a universally accepted component of the evaluation of the knee and subsequent diagnosis of injury or abnormality, then the reliability and validity of the system's results must be estimated through appropriate scientific methods. The purpose of the present investigation was to determine the reliability of the results that would be generated from the Genucom system for a battery of knee laxity tests that normally could be performed by an orthopedic practitioner or his assistants. In a normal motor performance test situation, the sources of measurement error are: the subject, the test equipment and protocol, and the test administrator (Baumgartner and Jackson 1982). These same sources of measurement error are present in knee laxity tests conducted using the Genucom. The knee analysis system may have electronic or computer inaccuracies, the test administrator may be inconsistent in performing the laxity tests or in running the system, the subject may consciously or unconsciously resist the test manipulations. Any reliability determinations should cover all three primary sources of error.

Methods

Subjects

The sample of the study was 20 subjects, 11 males and 9 females, with a mean age of 31.17 ± 2.48 years, a mean weight of 68.94 ± 3.42 kg, and a mean height of 171.02 ± 1.91 cm. All subjects were volunteers who had signed informed consent. Both legs were tested on each subject ($n = 40$ for subsequent data analysis).

Procedures

The procedures established for the Genucom system in conducting a knee evaluation (Faro Inc 1984) were followed as closely as possible. A summary is given below:

Surgery and Arthroscopy of the Knee
2nd Congress of the European Society
Editors: Müller/Hackenbruch

1. The subject's permanent data file was established using the system's computer and stored on magnetic media.
2. The subject's test leg was secured with the system's restraints and an electrogoniometer was attached to the lower leg.
3. Seven anatomical landmarks were digitized and recorded.
4. A soft tissue compensation was completed.
5. The system's speed test protocol was utilized to perform two independent trials of nine knee laxity tests which yielded 23 separate values for the reliability estimation (Table 1).
6. The subject's test results were permanently stored on magnetic media.
7. A printed copy of both trials of each test was obtained and the data for statistical analysis were gathered from this printed output.

Table 1. Knee laxity tests and derived test values

Test description	Variable label
Anterior/posterior drawer test – leg angle 90°	
Anterior displacement	AA1
Posterior displacement	AP1
Anterior/posterior drawer test – leg angle 90° with 10° external rotation	
Anterior displacement	AA2
Posterior displacement	AP2
Anterior/posterior drawer test – leg angle 90° with 10° internal rotation	
Anterior displacement	AA3
Posterior displacement	AP3
Anterior/posterior drawer test – leg angle 30°	
Anterior displacement	AA4
Posterior displacement	AP4
Anterior/posterior drawer test – leg angle 30°	
Medial compartment	
Anterior displacement	AA5
Posterior displacement	AP5
Lateral compartment	
Anterior displacement	AA6
Posterior displacement	AP6
Anterior/posterior drawer test – leg angle 30°	
Medial compartment	
Anterior displacement	AA7
Posterior displacement	AP7
Lateral compartment	
Anterior displacement	AA8
Posterior displacement	AP8
Varus/valgus stress test – leg angle 20°	
Valgus	VL1
Varus	VR1
Varus/valgus stress test – leg angle 0°	
Valgus	VL2
Varus	VR2
Recurvatum/screw home	
Point of hyperextension	RH1
Point of external rotation	RE1
Amount of external rotation	RA1

Data Analysis

Descriptive statistics, including the mean and standard error of the mean for both trials of each test, were calculated. The zero order correlations were calculated between the results of the two trials for all tests to provide reliability estimates. A t-test for correlated means was also performed to determine if any significant differences existed between the trials.

Results

The means, standard error of the means, and the test-retest correlations for the 23 test values are presented in Table 2. An examination of the correlation values reveals that 10 of the coefficients exceed 0.90, nine were between 0.80 and 0.89, and only four were between 0.70 and 0.79. The t-tests for correlated means indicated no significant differences ($P > 0.05$) between any of the pairs of the independent trials. The similarity of the means and the standard errors, and the finding that no trial differences exceeded 1 mm of displacement, indicated that a high degree of consistency can be achieved between trials in terms of average test results and variability.

Table 2. Means, standard errors of means, and test-retest correlations

Variable label	Trial 1		Trial 2		
	Mean	S. E.	Mean	S. E.	*r*
AA1	7.92	0.51	8.01	0.47	0.91
AP1	2.89	0.26	2.95	0.26	0.79
AA2	7.05	0.44	6.89	0.43	0.90
AP2	2.81	0.28	2.81	0.27	0.80
AA3	6.11	0.54	6.19	0.48	0.89
AP3	2.35	0.24	2.51	0.29	0.70
AA4	5.59	0.39	5.76	0.41	0.82
AP4	2.97	0.28	3.19	0.31	0.76
AA5	7.35	0.43	7.05	0.46	0.91
AP5	2.49	0.31	2.46	0.34	0.85
AA6	8.78	0.46	8.51	0.47	0.95
AP6	4.02	0.36	3.81	0.38	0.92
AA7	4.81	0.38	4.86	0.45	0.78
AP7	1.97	0.32	2.00	0.28	0.85
AA8	6.41	0.54	6.64	0.53	0.83
AP8	4.18	0.44	4.18	0.42	0.90
VL1	5.21	0.34	5.54	0.33	0.88
VR1	7.02	0.52	7.08	0.53	0.93
VL2	3.22	0.37	3.16	0.29	0.87
VR2	3.86	0.28	4.06	0.33	0.87
RH1	6.94	0.62	7.06	0.63	0.98
RE1	24.78	1.63	24.78	1.65	0.99
RA1	12.38	1.52	12.03	1.38	0.90

Discussion

The results of this study indicate that generally reliable data for the knee laxity tests administered in this study can be achieved using the Genucom knee analysis system. This contention is supported by the observation that 19 of the possible 23 reliability estimates exceeded 0.80 and the other four exceeded 0.70. Reliability in this study is defined as the consistency between trials. The numerical similarity and lack of significant differences between the trial means and standard error of the means is further evidence of between trial consistency. The system's rapid test protocol was used in this study which does not allow immediate inspection of test results. In the clinical situation the practitioner will be able to view the test results on the computer monitor and redo an obviously poor test before saving the results for subsequent analysis, which should improve reliability. It is also important to note that the Genucom, the test administrator, and the subject all contributed to the measurement error and true score variance that was observed. The laboratory technician who administered the tests in this study spent approximately 20 h practice time learning to run the Genucom test protocol. Any potential user should be prepared to spend the preparation time necessary for reliable testing in the clinical situation. It is also important to have the subject relax and not resist the test manipulations.

Further research with the Genucom could examine the validity of the system for use as a diagnostic tool for knee injuries. This would involve comparing presurgery diagnoses made using the Genucom with the exact surgical findings of injury. Another potentially valuable effort would be the testing of large numbers of non-symptomatic individuals and the establishment of normative values for a variety of knee laxity tests. This could aid in the recognition of excessive knee laxity from normal levels. A third area of study would be to examine the Genucom as a tool for the evaluation of the results of ligament repairs and reconstruction.

References

Baumgartner T, Jackson A (1982) Measurement for evaluation in physical education. Brown, Dubuque, pp 91–108

Faro Inc (1984) The Genucom knee analysis system. Faro Inc, Montreal

Glass G, Hopkins K (1984) Statistical methods in education and psychology. Prentice Hall, Englewood Cliffs, pp 240–242

Isolated Tears of the Medial Collateral Ligament of the Knee – Operative or Conservative Treatment?

H. Seiler, M. Kayser, H. Niemeyer, and V. Bühren

Chirurgische Universitätsklinik – Abteilung Unfallchirurgie,
6650 Homburg/Saar, FRG

Isolated tears of the medial collateral ligament (MCL) are a common injury. Recently, nonoperative, functional treatment has been advocated when lesions of the menisci and cruciates can be excluded. It is the purpose of this paper to compare the results obtained in three groups of patients with a new grade 2 or 3 valgus instability in flexion of the knee after operation (A), cast immobilisation (B) and purely functional treatment (C).

Material and Methods

Between 1974 and 1980, 53 knees were operated upon (A); between 1980 and 1983, 43 knees were treated either by cast immobilisation (B) (20) or by functional means (C) (23) for isolated tears of the MCL. Included are only cases where no bony avulsions of the ligaments were visible. Arthroscopy had been done in most cases of B and C, and arthrotomy in A, to exclude concomitant lesions of the cruciate ligaments and menisci. Up to now 31 knees of A, 15 of B and 17 of C have been followed up. Mean postoperative interval was 4.2 years in A, and 2.1 years in B and C.

Results

Average time of treatment was 9 (A), 7½ (B) and 6½ (C) weeks. Three infections and one deep thrombosis were the complications of the operative procedures. Results were good in 74%, satisfactory in 24% and poor in 2%. Without operative treatment 84% were good, the rest, satisfactory (Marshall 1977). The difference between cast and functional treatment was not striking, but advantages for the latter are evident.

Summary and Conclusion

Of 96 patients with 96 knees suffering from an isolated tear of the MCL, 63 knees were followed up after a mean interval of 2.1 or 4.2 years. Treatment had been operative in 31, cast immobilisation in 15 and functional in 17 cases. By arthroscopy concomitant lesions were excluded. Because of lack of complications, shorter time of treatment and better results, conservative procedures are recommended, particularly functional ones.

Surgery and Arthroscopy of the Knee
2nd Congress of the European Society
Editors: Müller/Hackenbruch

Early Mobilization of Isolated Partial Ruptures of the Medial Collateral Ligament: A Randomized Comparison of Bracing and Elastic Wrapping

M. Lundberg and P. Hamberg

Department of Orthopaedic Surgery, University Hospital, 581 85 Linköping, Sweden

Rupture of the medial collateral ligament (MCL) is common after knee trauma. The treatment of these ruptures has long been discussed. Partial ruptures have been treated surgically or in a cast, followed by long rehabilitation.

During the last decade there has been a change of attitudes in favour of nonoperative treatment. Ellsasser et al. (1974) had better results from conservative treatment than from surgery. Studies by Fetto and Marshall (1978), Hastings (1980), Derscheid et al. (1981) and Holden et al. (1983) all show good results from conservative treatment. However, these studies differ concerning the selection of patients, diagnostic criteria or treatment, and therefore general conclusions are difficult to draw.

The need for extensive rehabilitation after partial MCL rupture seems limited. Lysholm et al. (1982) showed that medial instability in the knee joint has little importance for its function.

This prospective investigation concerns itself with the early mobilization of isolated partial MCL ruptures, using elastic wrapping or a premade derotation-type brace.

Methods

In this study all patients with a knee trauma in which a meniscus or ligamentous injury was suspected were taken in for examination under anaesthesia and diagnostic arthroscopy. The ruptures that were disclosed were graded according to Marshall (1978). In grade I, there is pain over the MCL but no significant instability at 30° of flexion. In grade II, medial instability is moderate and the end point is firm. All patients had X-rays taken to exclude bony fragments.

When injury other than MCL rupture had been excluded by arthroscopy, the patients were randomized into two groups. In the first group the patients had early mobilization with elastic wrapping. The wrapping was used until postoperative swelling had gone down and MCL pain had faded. The patients of the other group had early mobilization in a premade derotation type brace. The brace was worn continuously for 6 weeks and thereafter on strenuous activity for another 6 weeks.

This ECKO brace limits abduction and rotation but allows a full range of motion (Tegner, personal communication). All patients had the same rehabilitation pro-

Surgery and Arthroscopy of the Knee
2nd Congress of the European Society
Editors: Müller/Hackenbruch

gramme. Full weight-bearing was allowed as soon as the initial pain had disappeared.

The patients of both groups were followed up after 3 and 18 months. Clinical stability, isokinetic quadriceps and hamstring strength, and knee function according to the Lysholm score were evaluated. The time until return to work and return to the previous level of activity was recorded.

An additional objective measurement of medial stability was done at 18 months. Using a sliding shoe, the patient's knees were given a 75-N valgus stress at 30° of flexion. The femoral condyles were fixed in a rigid lock and the patients were positioned for blocking of femoral rotation. The movements at the level of the joint line were measured for both knees. Standard error of the method was less than 1 mm at joint line level. Instability was considered present when the difference in laxity between the knees exceeded two standard deviations compared with a reference group, i.e. more than 3 mm.

Results

During the study, 53 patients, 39 men and 14 women, mean age 25 years (range 13–43 years) were found to have an isolated partial tear of the MCL. Forty-two had sports-related injuries, most of which resulted from soccer or alpine skiing. Twenty-seven patients were randomized to early mobilization with elastic wrapping and 26 were randomized to wearing a brace.

At follow-up, after 3 and 18 months, no statistically significant differences could be discovered between the two groups of patients.

At 3 months, 43 of 45 patients had normal or almost normal knee function. Eight patients were not able to participate in the evaluation. The mean Lysholm score was 97 points (range 80–100). Mean quadriceps strength was 98% (±10%). Thirty-seven patients were stable on clinical examination and eight had a slight instability compared with the unaffected knee.

Eighteen months of follow-up have been completed in 25 patients, of whom 12 had used a brace. Twenty-four of the 25 patients had normal or almost normal knee function. The total mean Lysholm score was 95 points (range 83–100). One patient had a score of only 46 points because of severe patellofemoral osteoarthritis. Quadriceps and hamstring strength were unchanged compared with the 3-month follow-up. As mentioned above, there were no statistical differences between the patients wearing a brace and those with elastic wrapping.

On machine testing for medial laxity, the injured knee in six patients showed significant instability, i.e. the increase in medial laxity was two standard deviations or more compared with the normal material. However, no increase in laxity exceeded 5 mm. Of the six patients with moderate instabilities, five had used a brace during rehabilitation.

The average sick leave was 2–3 weeks (range 1–6 weeks). Many of the patients were students who could return to studies shortly after the trauma. Return to the previous level of activity occurred after an average of 11 weeks (range 1–18 weeks).

Discussion

In this study, most patients with an isolated partial MCL rupture soon regained almost normal knee function. The good results from the 3-month follow-up were repeated at 18 months.

Earlier studies have dealt mostly with professional sportsmen and their capacity to achieve an early return to competition. Our patients are mainly nonprofessionals on a physical fitness programme, with less experience of efficient training programmes. In this study, arthroscopy gives us a precise description of the injury and excludes cruciate or meniscus lesions in all patients.

The brace that was used did not shorten the rehabilitation or reduce the resulting degree of instability.

Most patients in this study have normal knee function and a stable knee after early mobilization. This indicates that early mobilization with an unlimited range of motion provides adequate conditions for the healing of isolated partial tears of the MCL and that there is no need for a brace in the rehabilitation of these injuries.

References

American Medical Association Committee on the medical aspects of sports (1968) Standard nomenclature of athletic injuries, pp 99, 101

Balkfors B (1982) The course of knee-ligament injuries. Acta Orthop Scand [Suppl] 198

Derscheid GL, Garrick JG (1981) Medial collateral ligament injuries in football. Nonoperative treatment management of grade I and grade II sprains. Am S Sports Med 9:365–368

Ellsasser JC, Reynolds FC, Omohundro JR (1974) The non-operative treatment of collateral ligament injuries of the knee in professional football players. J Bone Joint Surg [Am] 56:1185–1190

Fetto JF, Marshall JL (1978) Medial collateral ligament injuries of the knee: a rationale for treatment. Clin Orthop 132:206–218

Frank C, Woo SLY, Amiel D, Harwood F, Gomez M, Akeson W (1983) Medial collateral ligament healing. A multidisciplinary assessment in rabbits. Am J Sports Med 11:379–389

Hastings DE (1980) The non-operative management of collateral ligament injuries of the knee joint. Clin Orthop 147:22–28

Hofmann AA, Wyatt RWB, Bourne MH, Daniels AU (1984) Knee stability in orthotic knee braces. Am J Sports Med 12:371–374

Holden DL, Eggert AW, Butler JE (1983) The nonoperative treatment of grades I and II medial collateral ligament injuries to the knee. Am J Sports Med 11:340–343

Hughston JC, Eilers AF (1973) The role of the posterior oblique ligament in repairs of acute medial (collateral) ligament tears of the knee. J Bone Joint Surg [Am] 55:923–940

Hunter SC, Marascalco R, Hughston JC (1983) Disruption of the vastus medialis obliquus with medial knee ligament injuries. Am J Sport Med 11:427–431

Indelicato PA (1983) Non-operative treatment of complete tears of the medial collateral ligament of the knee. J Bone Joint Surg [Am] 65:323–329

Inoue M, McGurk-Burleson E, Hollis M, Woo SLY (1987) Treatment of the medial collateral ligament injury: I. The importance of anterior cruciate ligament on the varus-valgus knee laxity. Am J Sports Med 15:15–21

Kennedy JC, Fowler PJ (1971) Medial and anterior instability of the knee. J Bone Joint Surg [Am] 53:1257–1270

Lysholm J, Gillquist J, Liljedahl S-O (1982) Long term results after treatment of acute knee injuries. Acta Orthop Scand 53:109–118

Marshall JL, Rubin RM (1980) Knee ligament injuries – a diagnostic and therapeutic approach. Clin Orthop 146:78–83
O'Connor GA (1979) Collateral ligament injuries of the joint. Am J Sports Med 7:209–210
O'Donoghue DH (1950) Surgical treatment of fresh injuries to the major ligaments of the knee. J Bone Joint Surg [Am] 32:721–738
O'Donoghue DH (1955) An analysis of end results of surgical treatment of major injuries to the ligaments of the knee. J Bone Joint Surg [Am] 37:1–12
Sandberg R, Balkfors B, Nilsson B, Westlin N (1987) Operative versus non-operative treatment of recent injuries to the ligaments of the knee. J Bone Joint Surg [Am] (to be published)
Shoemaker SC, Markolf KL (1982) In vivo rotatory knee stability. J Bone Joint Surg [Am] 64:208–216
Slocum DB, Larson RL (1968) Rotatory instability of the knee. J Bone Joint Surg [Am] 50:211–225
Tegner Y (1985) Cruciate ligament injuries in the knee. Evaluation and rehabilitation. Linköping University, Medical Dissertations No. 203:65–73
Woo SLY, Inoue M, McGurk-Burleson E, Gomez MA (1987) Treatment of the medial collateral ligament injury: II. Structure and function of canine knees in response to differing treatment regimens. Am J Sports Med 15:22–29

Congenital Absence of Cruciate Ligaments: Clinical, Radiological and Arthroscopic Aspects

P. Schlepckow and H.-U. Ernst

Orthopaedic Department, University of Freiburg,
Hugstetter Str. 55, 7800 Freiburg im Breisgau, FRG

Pronounced congenital limb deformities such as femoral defects of the third degree (Aitken 1969; Amstutz 1969; Blauth 1967; Henkel et al. 1978) and tibial dysplasia (Jones et al. 1978), require, because of extreme shortening of the leg, operative or orthotic treatment.

Congenital absence of cruciate ligaments is in this respect of minor importance and was therefore not mentioned in earlier literature (Panting and Williams 1978; Hamanishi 1980; Gillespie and Torodo 1983). However, in the last 5 years under the influence of arthroscopy, cases of congenital absence of cruciate ligaments have been described which were only associated with insignificant syndromes of deformity (Johansson and Aparisi 1982, 1983, Thomas et al. 1985). Mainly patients with congenital leg shortening who had no complaints of instability and were active in sports were involved. The list of deformity syndromes where congenital absence of cruciate ligaments has been described includes congenital femoral defects (Gillespie and Torodo 1983; Johansson and Aparisi 1983; Thomas et al. 1985), congenital dislocation of the knee as a single phenomenon or in connection with Larsen's syndrome (Larsen et al. 1950; Niebauer and King 1960; Katz et al. 1967) and also the thrombocytopenia-absent-radius (TAR) syndrome (Schoenecker et al. 1984) and highly pronounced forms of tibial dysplasia (Thomas et al. 1985). In the Anglo-American terminology the following terms are used to describe femoral defects: "proximal femoral focal deficiency", "congenital short femur" and "distal femoral focal deficiency". Arthroscopic findings have been published in the last few years by Johansson and Aparisi (1983) and Thomas, Jackson and Aichroth (1985).

We report here on two young patients with symptoms of instability. The first, a 27-year-old man, was, until a soccer injury, asymptomatic and active in sports. A shortening of the leg of 5 cm was tolerated without compensation. Besides the nonexistence of the fifth toe there were no further deformities. Arthroscopically both cruciate ligaments were missing.

The second, a deaf-mute 12-year-old boy, was instable on exercise. The left knee joint exhibited genu varum and recurvatum, but there was no leg shortening or other skeletal deformity. A subtotal congenital absence of cruciate ligaments was diagnosed.

In the case of the total aplasia, above all posterior and lateral instability was obvious. In the case of the subtotal anterior cruciate ligament (ACL) aplasia, mainly medial and lateral instability existed.

Congenital absence of cruciate ligaments coincides with typical radiological secondary changes of the bones visible in survey exposures (Figs. 1 and 2). We found

Surgery and Arthroscopy of the Knee
2nd Congress of the European Society
Editors: Müller/Hackenbruch

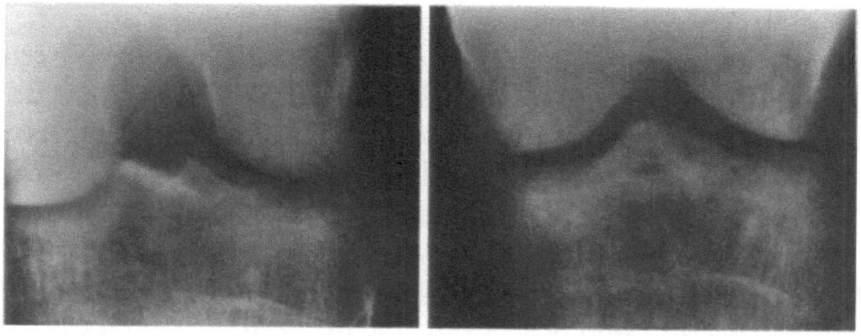

Fig. 1. Total cruciate ligament aplasia. Intercondylar notch is only rudimentally expressed

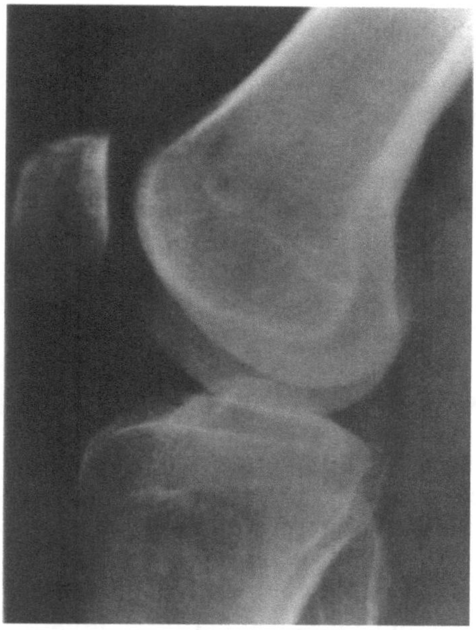

Fig. 2. Blumensaat's line is missing in case of total cruciate aplasia

a cone-shaded enlargement of eminentia intercondylaris with flattening of the tubercle, hypoplasia of the lateral femoral condyle, dysplasia of the proximal fibula, and, in the sagittal plane, absence of Blumensaat's line. The intercondylar view showed the most obvious changes: the intercondylar notch, the space where the cruciate ligaments move, was only rudimentally expressed. The disturbances of the fossa are so typical that they allow, as Johansson and Aparisi (1983) have shown, differentiation into anterior and posterior cruciate aplasia.
In the second case, with subtotal ACL aplasia, radiologically we diagnosed flattened eminentia humps. The fossa was fully shaped, but somewhat plump. The lateral femoral condyle was hypoplastic, compared with that on the opposite side. Obvious was the width of the lateral joint line, accompanied by the existence of a voluminous discoid meniscus.

X-rays in a standing position showed a varus and recurvatum curvature. Studying the embryonic development can give clues to the connection between form and function. The intercondylar structures and joint surfaces, which develop from the 9th week of pregnancy, are classified into two phases by most authors: The first stage of formation, which depends on the potential of self-differentiation of the joints, is followed by further forming by the mechanical influence of muscular activity and joint movement (Tillmann 1974). At this early stage, the cruciate ligaments and menisci already appear as pronounced structures, whereas the characteristic delineation of the femoral condyles is not yet noticeable (McDermott 1943; Gray and Gardner 1953; Kaplan 1955). Thus the damage in regard to congenital absence of cruciate ligaments must have occurred in the 9th or 10th week of pregnancy. The ligaments are primarily affected, but with malfunction of the joint, secondary osseous changes were inevitable, and were noticed in the X-rays.

Congenital absence of cruciate ligaments in connection with pronounced femoral and tibial defects must, however, originate in some perturbation of the formation of the extremities during the 4th and 5th weeks of embryogenesis, much earlier. Exogenous noxious factors during embryogenesis were not evident from the case histories of our two patients. Similar changes, mainly of the intercondylar notch, are also known to us in cases of posttraumatic conditions. The arthroscopic pictures (Figs. 3 and 4) demonstrate, in case of complete cruciate ligament aplasia, the empty, hypoplastic fossa, enabling the view into the hollow of the knee. Rudimentary ligaments are seen on the roof of the fossa as well as in the region of the cone-shaped eminentia. The view from intercondylary shows that the posterior horn of medial meniscus turns far towards the front, leading to a king of ring, as Noble (1975) noticed in one case. The lateral meniscus is shaped regularly and the whole joint shows no signs of degenerative changes. Upon rotation and translation movements, the condyles are arrested by the enlarged eminentia, suggesting the appearance of bony compensation of the instability.

With the second patient, arthrotomy was performed. A ruptured, voluminous, lateral discoid meniscus was present, the ACL exhibited macroscopically more the structure of fibrocartilage, took an atypical course, and narrowed progressively toward dorsal. Here, too, no cartilage damage was evident.

Decisions on therapeutic action must take into account the changed osseous conditions. Therefore we do not consider intraarticular reconstruction, such as ligament replacement, to be of value. Synergistically helpful extraarticular structures, such as the posterolateral area, could be strengthened. We did not operate in the case of the total aplasia. In the second case the discoid meniscus was reduced in size and ACL replacement was carried out according to the technique of Werner Müller (1982).

Congenital absence of cruciate ligaments is, in connection with limb deformities, frequently the source of knee instabilities, but has so far rarely been recognized in cases of minor lesions. When leg length discrepancies are clinically observed, roentgenograms of the knee should be made, as because of typical changes the diagnosis can already be established. Arthroscopy should be performed mainly in order to recognize and treat secondary changes and damage, e.g. a meniscus tear.

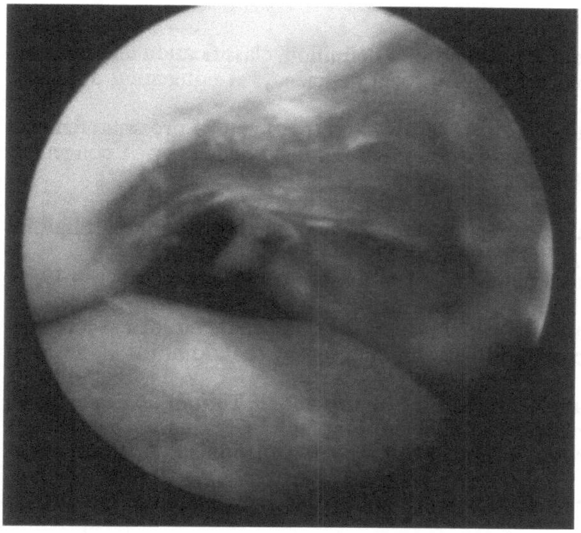

Fig. 3. Missing cruciate ligaments in a hypoplastic notch; arthroscopic view from anterolateral, medial femoral condyle on the right

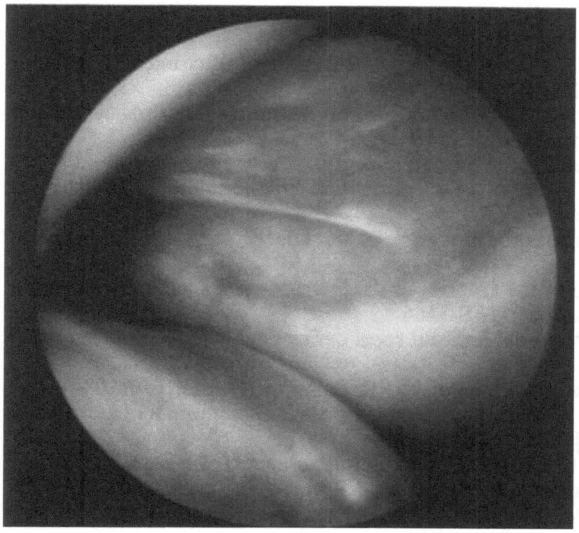

Fig. 4. Arthroscopic view from anterolateral to the medial femoral condyle with rudimentary posterior cruciate ligament on the roof of the fossa

References

Aitken GT (1969) Proximal femoral focal deficiency – definition, classification and management. In: Aitken GT (ed) Proximal femoral focal deficiency. A congenital anomaly. Natl Acad Sci, Washington, pp 1–22

Amstutz HC (1969) The morphology, natural history and treatment of proximal femoral focal deficiencies. In: Aitken GT (ed) Proximal femoral focal deficiency. A congenital anomaly. Natl Acad Sci, Washington, pp 50–76

Blauth W (1967) Der kongenitale Femurdefekt. Z Orthop 103 [Suppl]

Gillespie R, Torodo IP (1983) Classification and management of congenital abnormalities of the femur. J Bone Joint Surg [Br] 65(5):557–568

Gray DJ, Gardner E (1953) Prenatal development of the human knee and superior tibiofibular joints. Am J Anat 92:219–276

Hamanishi C (1980) Congenital short femur, clinical, genetic and epidemiological comparison of the natural occurring condition with that caused by thalidomid. J Bone Joint Surg [Br] 62:307–320

Henkel HL, Willert HG, Gressmann C (1978) Eine internationale Terminologie zur Klassifikation angeborener Gliedmaßenfehlbildungen. Empfehlungen einer Arbeitsgruppe der International Society of Prothetics and Orthotics. Arch Orthop Trauma Surg 93:1–19

Johansson E, Aparisi T (1982) Congenital absence of the cruciate ligaments. Clin Orthop 162:108–111

Johansson E, Aparisi T (1983) Missing cruciate ligament in congenital short femur. J Bone Joint Surg [Am] 65:1109–1115

Jones D, Barnes J, Lloyd-Roberts GC (1978) Congenital aplasia and dysplasia of the tibia with intact fibula. J Bone Joint Surg [Br] 60:31–39

Kaplan EB (1955) The embryology of the menisci of the knee joint. Bull Hosp Jt Dis Orthop Inst 16:111–124

Katz MP, Grogomo BJS, Sepes KC (1967) Etiology and treatment of congenital dislocation of the knee. J Bone Joint Surg [Br] 49:112–120

Larsen LJ, Schottstaedt ER, Bost FC (1950) Multiple congenital dislocations associated with characteristic facial abnormality. J Pediatr 37:574

Mc Dermott LJ (1943) Development of the human knee joint. Arch Surg 46:705–719

Müller W (1982) Das Knie. Form, Funktion und ligamentäre Wiederherstellungschirurgie. Springer, Berlin Heidelberg New York

Niebauer JJ, King DE (1960) Congenital dislocation of the knee. J Bone Joint Surg [Am] 42:207

Noble J (1975) Congenital absence of the anterior cruciate ligament associated with a ring meniscus. J Bone Joint Surg [Am] 57:1165–1166

Panting AL, Williams PF (1978) Proximal femoral focal deficiency. J Bone Joint Surg [Br] 60:46–52

Schoenecker PL, Cohn AK, Sedgwick WG, Manske PR, Salafsky I, Millar EA (1984) Dysplasia of the knee associated with the syndrome of thrombocytopenia and absent radius. J Bone Joint Surg [Am] 66:421–427

Thomas NP, Jackson AM, Aichroth PM (1985) Congenital absence of the anterior cruciate ligament: a common component of knee dysplasia. J Bone Joint Surg [Br] 67:572–575

Tillmann B (1974) Zur funktionellen Morphologie der Gelenkentwicklung. Orthop Praxis 12(10):691–697

Surgical Versus Non-surgical Treatment of Knee Ligament Injuries

M. Bauer and R. Sandberg

Department of Orthopaedic Surgery, Malmö General Hospital, Malmö, Sweden

Methods

In a prospective series 200 unselected cases of fresh knee ligament injuries were randomly allocated either to surgical treatment or to treatment with a cylinder cast only. Included were patients over age 15, with no previous knee injury and a medial and/or an anterior sagittal instability.
Surgical treatment consisted of primary ligamentous suture followed by cylinder cast fixation for 6 weeks. The *non-surgical group* received a cylinder cast after arthroscopy and stability tests had been performed under anaesthesia. In both groups immediate full weight-bearing was allowed. The rigid cast was worn for 6 weeks. The same post-operative regimen was applied in both groups under the supervision of a qualified physiotherapist. Isometric muscle exercises were instituted while in the cast and after removal as long as an effusion persisted, followed by active dynamic exercises. Ranges of motion, muscular strength and endurance (Cybex testing) and stability (Sylvin caliper) were measured at regular intervals. Subjective discomfort was estimated with the help of the Lysholm knee score. *Follow-up* was performed in all cases after at least 1 year and in half the group after 30 months.

Material

In all, 500 patients (133 male), were included. Mean age was 29 (range 15–61 years). Some 75% were athletic injuries, 57% were soccer injuries. The following instabilities were encountered:

	Medial (MCL)	Sagittal (MCL)	Combination (ACL/MCL)
Surgical group	20	54	26
Non-surgical group	24	47	29

Results

The return to physical activity and to work was significantly faster for both medial collateral ligament injuries and anterior cruciate ligament injuries which were treated non-surgically. For the combination injuries no difference could be demonstrated. The functional result at 1 year as expressed by the Lysholm score ($M \pm SD$):

Surgery and Arthroscopy of the Knee
2nd Congress of the European Society
Editors: Müller/Hackenbruch

	MCL		ACL		ACL/MCL	
	n		*n*		*n*	
Surgical group	20	98±5	54	90±13	26	90± 9
Non-surgical group	24	87±5	47	89±13	29	89±11

The sagittal stability at 1 year expressed as the anterior drawer (difference injured minus control im mm):

ACL	*n*	Internal rotation	Neutral	External rotation
Surgical group	54	0.8±1.6	1.7±1.7	0.9±1.9
Non-surgical group	47	1.6±1.5	2.7±2.1	2.0±2.0

After on average 30 months five patients in the non-surgical and three patients in the surgical group had a late reconstruction for instability. In the group of isolated MCL injuries no patient developed instability. In the first 100 patients who were seen at 3 years there was no deterioration when compared with the primary results as regards to Lysholm score and stability.

Conclusions

Non-surgical therapy leads to a faster recovery and quicker return to work and to sports activities. Surgical treatment leads to better stability. Both non-surgical and surgical treatment give the same good functional result.

Evaluation of Knee Ligament Injuries

Y. Tegner [1], J. Lysholm [2], and J. Gillquist [2]

[1] Department of Orthopaedic Surgery, Central Hospital, 961 85 Boden, Sweden
[2] Department of Orthopaedic Surgery, University Hospital, 581 85 Linköping, Sweden

Our target when treating patients with knee ligament injuries should be to restore knee function. This could mean operative or conservative therapy. To measure whether we have succeeded in our aims or not, we need an evaluation system that measures knee function. The system needs to be reproducible and objective, so that comparisons between different treatments are possible.
Several attempts have been made to create good evaluation systems for patients with knee injuries. None of these methods has been able to include all the aspects of the problems (Kettelkamp and Thompson 1975; Larson 1972; Lysholm and Gillquist 1982; Marshall et al. 1977).
In 1980 the World Health Organization published a manual of classification relating to the consequences of disease (WHO 1980). It is stated therein that the effects of illness are: (1) Loss of an anatomical structure or function – *impairment;* (2) departure from the norm in terms of performance of the individual – *disability;* (3) limitations in activity for the individual – *handicap*. In patients with cruciate ligament injury there is a loss of an anatomical structure resulting in impairment of knee function. They are disabled because they experience symptoms of loss of function. This results in a handicap, because they can no longer be as active as they wish (Fig. 1).

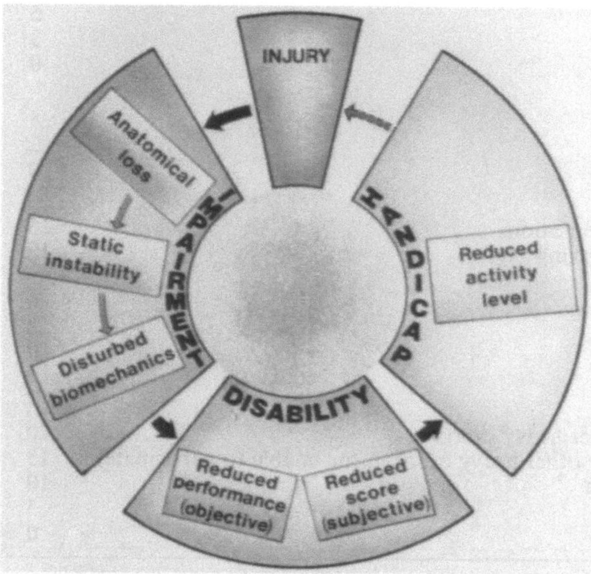

Fig. 1. A schematic drawing of the WHO classification

Surgery and Arthroscopy of the Knee
2nd Congress of the European Society
Editors: Müller/Hackenbruch

We have developed an evaluation system based on the WHO classification. This system consists of several different items in order to evaluate all the aspects of the knee injury.

Methods of Evaluation

Impairment

To evaluate impairment, all patients whom we suspect of having a cruciate injury undergo arthroscopy and stability testing under anaesthesia. To measure strength we use a Cybex isokinetic dynamometer, and we calculate the isokinetic strength at 30°/s and 180°/s and the isometric strength at 60° of knee flexion.

Disability

To measure disability we use a knee score scale and a performance test. The Lysholm knee score consists of eight different items, all relevant to knee function. It has earlier been shown that the score gives a good view of the patient's disability (Lysholm and Gillquist 1982). The items are limp, support, locking, instability, pain, swelling, stair-climbing, and squatting (Table 1; Tegner and Lysholm 1985a). The maximum score is 100 points; above 84 is considered good and below 65 poor.

The performance test consist of four different activities that all put stress on the knee in sport-like situations (Tegner et al. 1986a): (1) Running in a figure-of-eight for two laps, a total of 40 m. Timing is done electrically with an ECG writer and photoelectric cells (Fig. 2a). The photoelectric cells are so situated that the run-

Table 1. Lysholm knee score scale

Limp	
None	5
Slight and/or periodical	3
Severe and/or constant	0
Support	
None	5
Stick or crutch	2
Weight-bearing impossible	0
Locking	
No locking and no catching sensations	15
Catching sensations but no locking	10
Locking	
Occasionally	6
Frequently	2
Locked joint on examination	0
Instability	
No giving way	25
Rarely during athletics or other heavy exertion	20
Frequently during athletics or other heavy exertion (or unable to participate)	15
Occasionally in daily activities	10
Often in daily activities	5
At every step	0

Table 1 (continued)

Pain	
None	25
Inconstant and slight during heavy exertion	20
Marked during heavy exertion	15
Marked on or after walking more than 2 km	10
Marked on or after walking less than 2 km	5
Constant	0
Swelling	
None	10
On heavy exertion	6
On normal exertion	2
Constant	0
Stair-climbing	
No problems	10
Slightly impaired	6
One step at a time	2
Impossible	0
Squatting	
No problems	5
Slightly impaired	4
Not beyond 90°	2
Impossible	0

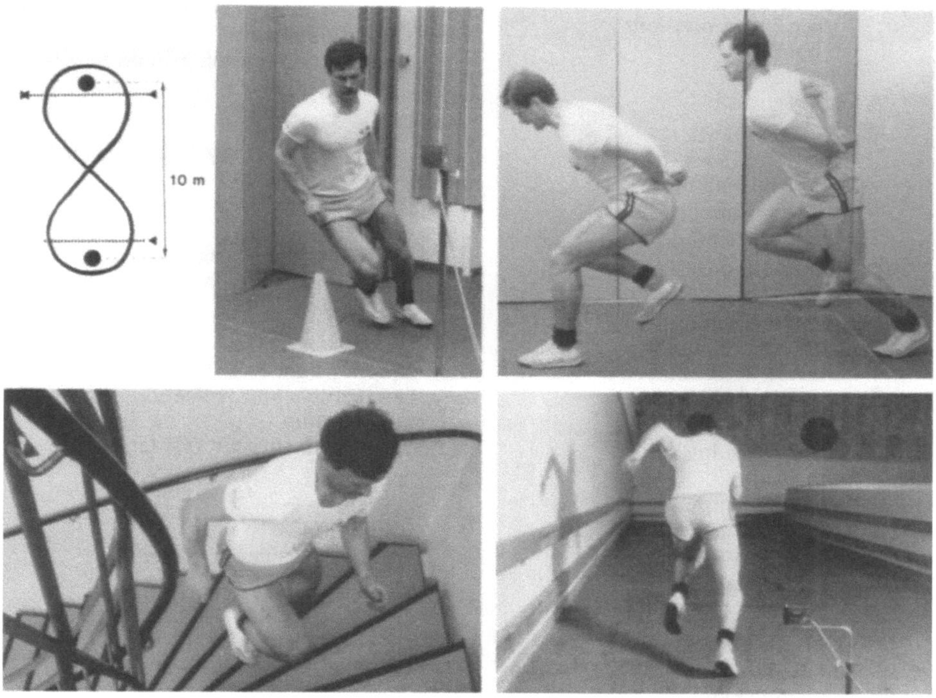

Fig. 2. a Figure-of-eight with photoelectric cells to separate running straight forward and in the turn. **b** The hop. **c** Running up stairs. **d** Running up the indoor slope

ning time straight forward and in the turn can be measured separately. (2) A one-leg hop performed by the patients with their hands on the back (Fig. 2b). We measure both hop-length and the quotient between the best results for the injured and the noninjured leg. (3) Running up and down a spiral staircase of 25 steps on time (Fig. 2c). (4) Running up and down a 55-m indoor slope on time (Fig. 2d).

Handicap

The patient's handicap is graded using an activity level scale (Table 2; Tegner and Lysholm 1985a). In this scale, activities in sports and daily life are graded numerically between 0 and 10. An elite soccer player will be graded 10, a good jogger level 6 and the lazy overfed individual 1. Level 0 is when patients are on sick leave because of knee problems.

Table 2. Activity scale

10. Competitive sports
    - Soccer, national or international level

9. Competitive sports
    - Soccer, lower divisions
    - Ice hockey
    - Wrestling
    - Gymnastics

8. Competitive sports
    - Bandy
    - Squash or badminton
    - Athletics (jumping etc.)
    - Downhill skiing

7. Competitive sports
    - Tennis
    - Athletics (running)
    - Motocross or speedway
    - Handball or basketball

   Recreational sports
    - Soccer
    - Bandy or ice hockey
    - Squash
    - Athletics (jumping)
    - Cross-country

   Recreational and competitive orienteering

6. Recreational sports
    - Tennis or badminton
    - Handball or basketball
    - Downhill skiing
    - Jogging at least 5 times weekly

5. Work
    - Heavy (e.g. building, forestry)

   Competitive sports
    - Cycling
    - Cross-country skiing

   Recreational sports
    - Jogging on uneven ground at least twice weekly

4. Work
    - Moderately heavy (e.g. lorry driving, office cleaning)

   Recreational sports
    - Cycling
    - Cross-country skiing
    - Jogging on even ground at least twice weekly

3. Work
    - Fairly light (e.g. nursing)

   Competitive and recreational sports
    - Swimming

   Walking in rough forest terrain

2. Work
    - Light

   Recreation

   Walking on uneven ground

1. Work
    - Sedentary

   Recreation

   Walking on even ground

0. Sick leave or disability pension
    - Because of knee problems

Results

The different parts of our system were first evaluated separately in 145 patients and 66 healthy soccer players.

The functional score was compared with the Marshall scoring scale. The Marshall system includes items measuring impairment, disability and handicap on the same scale. We were able to show that the Lysholm score gave a more differentiated picture of the patient's disability and, further, that the Marshall score may jeopardize the functional evaluation of the patient's problems (Tegner and Lysholm 1985 a). The performance test was evaluated by comparing healthy soccer players and athletes with injuries of the anterior ligament (ACL). The uninjured soccer players showed significantly better results in all parts of the test. Even with normal thigh muscle strength, most of the patients could not perform normally in the test (Tegner et al. 1986 a).

The activity grading scale was constructed by letting 43 patients with ACL injury grade a number of activities according to how troublesome they were to perform.

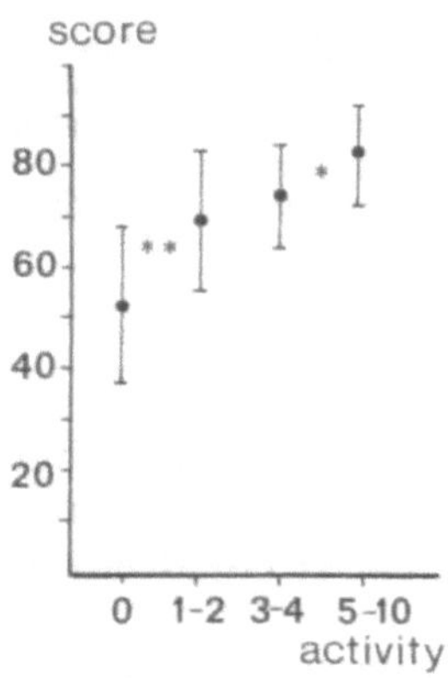

Fig. 3. The relation between the functional score and the activity level. Significant differences in functional score existed between patients at different activity levels

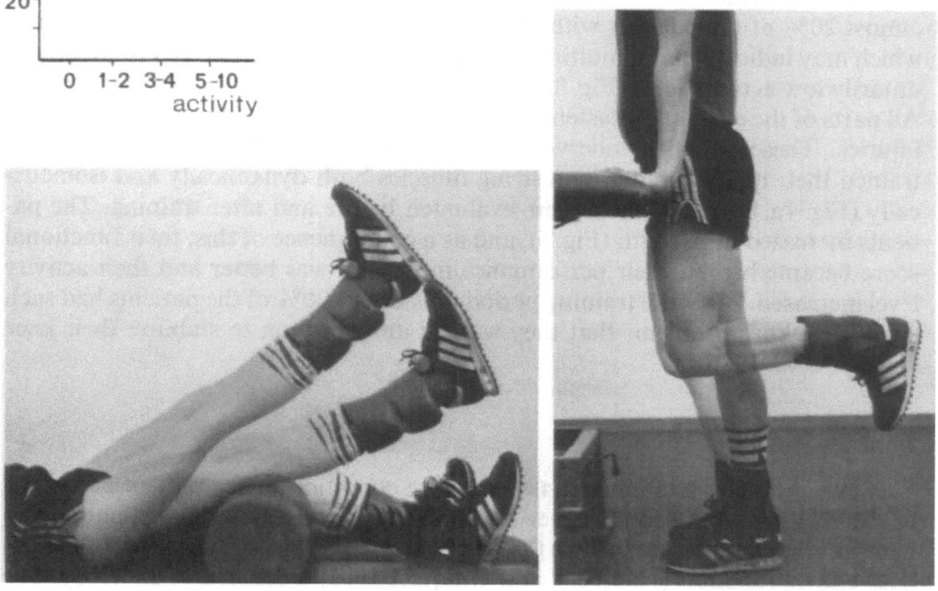

Fig. 4. a Strengthening exercise for the quadricep muscle. b Strengthening exercise for the hamstring muscles

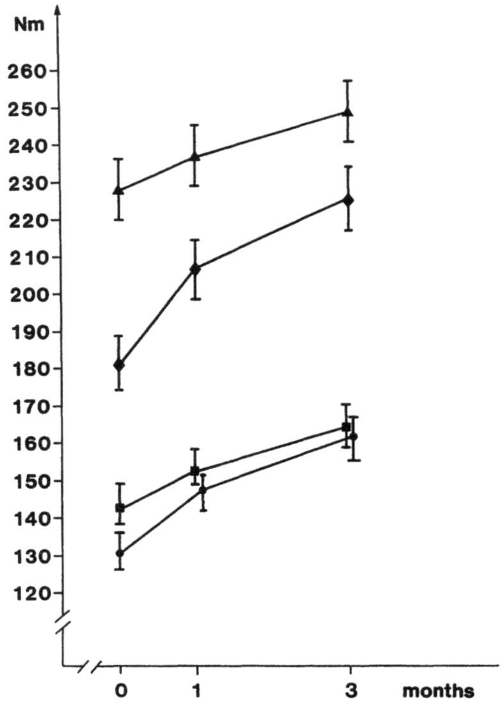

Fig. 5. The increase in strength for hamstring and quadricep muscles as measured with the Cybex isokinetic dynamometer. ▲ quadriceps strength, noninjured leg; ◆ quadriceps strength, injured leg; ■ hamstring strength, noninjured leg; ● hamstring strength, injured leg

A good correlation existed between the Lysholm score and the activity score, but almost 20% of the patients with an activity level below 3 had high knee scores, which may indicate that limitations in knee function may be masked by an unvoluntarily low activity level (Fig. 3; Tegner and Lysholm 1985 a).

All parts of the evaluation system were used for 52 patients with cruciate ligament injuries. These patients underwent a 3-month strength training period. They trained their quadricep and hamstring muscles both dynamically and isometrically (Fig. 4 a, b). All of them were evaluated before and after training. The patients increased in strength (Fig. 5), and as a consequence of this, their functional score became higher, their performance in the test was better and their activity level increased. After the training period only about 20% of the patients had such significant knee problems that they wanted an operation to stabilize their knee (Tegner et al. 1986 b).

Discussion

This evaluation system aims at assessing the condition of the patient according to the WHO classification (Tegner 1985 b). After having used this evaluation systems in other studies, we believe that it gives us a very good indication of the patient's knee function (Hamberg and Gillquist 1983; Odensten et al. 1983). The knee score is useful for evaluation of the disability before and after treatment in the same group of patients or for comparison of treatments in randomized studies. The functional performance test is a good complement to the scoring scale for grading disability. It can also be used for guiding rehabilitation after treatment, as normalization of the performance in the test should be achieved before the patient returns to full activity. The activity scale was constructed because different activities make distinctively different demands on the knee. The main advantage of the activity scale is not to compare different patients but to note changes in activity level in the same patient at different times. With our scale, the preinjury, present and described activity levels can easily be defined.

The disability and handicap after a cruciate ligament injury can be significantly reduced with a strength training period. The results obtained by training as sole therapy can be used as a baseline result when operative therapy is evaluated, as any operation should surpass this result.

References

Hamberg P, Gillquist J (1984) Knee function after arthroscopic meniscectomy. Acta Orthop Scand 55:172

Kettelkamp DB, Thompson C (1975) Development of a knee scoring scale. Clin Orthop 107:93

Larson RL (1972) Rating sheet for knee function. In: Smillie IS (1974) Disease of the knee joint. Churchill-Livingston, Edinburgh, p 29

Lysholm J, Gillquist J (1982) Evaluation of knee ligament surgery results with special emphasis on use of a scoring scale. Am J Sports Med 10:150

Marshall JL, Fetto JF, Bottero PM (1977) Knee ligament injuries. Clin Orthop 123:115

Odensten M, Tegner Y, Lysholm J, Gillquist J (1983) Knee function and muscle strength following distal ileotibial band transfer for antero-lateral rotatory instability. Acta Orthop Scand 54:924

Tegner Y, Lysholm J (1985 a) Rating systems in the evaluation of knee ligament injuries. Clin Orthop 198:43

Tegner Y (1985 b) Cruciate ligament injuries. Evaluation and rehabilitation. Med dissertation no 203, Linköping University

Tegner Y, Lysholm J, Lysholm M, Gillquist J (1986 a) A performance test to monitor rehabilitation and evaluate anterior cruciate ligament injuries. Am J Sports Med 14:156

Tegner Y, Lusholm J, Lysholm M, Gillquist J (1986 b) Strengthening exercises for old cruciate ligament tears. Acta Orthop Scand 57:130

WHO (1980) International classification of impairments, disability and handicaps. World Health Organization, Geneva

Comparative Results of 163 Anterior Cruciate Ligament Injuries Managed by Repair/Reinsertion, Primary Augmentation, or Reconstruction

R. Biedert, We. Müller, W. Hackenbruch, and R. Baumgartner

Department of Orthopedic Surgery, Kantonsspital, 4101 Bruderholz, Switzerland

In recent years we have seen a marked improvement in the outcomes of anterior cruciate ligament (ACL) operations performed at our center. To learn the reasons for this, we conducted a retrospective study of patients operated upon at our facility between 1981 and 1984. It was hoped that the findings of this review would lead to discoveries that would advance future operative techniques.

Materials and Methods

Between 1981 and 1984, 600 operations were performed on the ACL at our district hospital in Bruderholz. Every third case was randomly selected for review. Thirty-seven of these 200 cases were excluded for various reasons: concomitant injury of the posterior cruciate ligament, severe posterolateral instability necessitating popliteal bypass, or unusual combined lesions.
The 163 cases selected for evaluation had been operated on by 10 different surgeons in our department. Follow-ups were performed over a three-month period by the same examiner. Results were recorded in a questionnaire, which included data on the nature and mechanism of the injury, limitation of daily living activities, limitation of sporting activities, subjective complaints, the patient's general condition, etc. Also documented were the comparative stabilities of the operated and nonoperated knees and the surgical outcomes.
The data were evaluated by a statistician working with a computer.

Timing of Follow-Up Examination. The patients had undergone surgery from 12 to 48 months before returning for follow-up. The average follow-up period was 29.4 months.

Age of Patients. The patients ranged in age from 15 to 64 years, with an average age of 27.2 years.

Evaluation

The data filled more than 5000 computer pages and basically covered preoperative documentation, the operation report, and observations derived from the patient's history and our special follow-up. A detailed analysis of this data will be presented elsewhere [3]. Here we shall highlight some of the major findings, considering them in terms of three main criteria: 1. Type of injury; 2. surgical procedure; 3. general results.

Surgery and Arthroscopy of the Knee
2nd Congress of the European Society
Editors: Müller/Hackenbruch

Table 1. Criteria for evaluating outcomes

Results

1. Type of injury, instability

2. Surgical procedure

Objective findings
- Lachman test 0°–30°
- Anterior drawer 60°–90°
- Pivot shift

Subjective findings
- Giving way
- Effusion
- Pain on effort
- Reduction in performance

Functional test
- Figure 8 run

3. General results

Special emphasis was placed on the results of eight objective and subjective tests when evaluating the outcomes (Table 1).

Types of Injury

The great majority of ACL ruptures were sustained during sporting activities. Comparatively small percentages were due to vehicular accidents (mostly involving motorcyclists), job-related injuries, and miscellaneous causes (Table 2).

Table 2. Sources of injury in the follow-up population (n = 163)

Sports	139 (=85.3%)
Vehicular	12 (= 7.3%)
Job-related	6 (= 3.7%)
Other	6 (= 3.7%)

Results by Type of Injury and Associated Instability

The 163 patients were divided into 7 groups based on the patterns of injury found at operation (Table 3). The corresponding instabilities were described in terms of the four-quadrant scheme of W. Mueller (Table 4).
Groups B and C (+ D) had combined injuries involving the "semimembranosus corner" of the knee.
The results by type of injury are summarized in Table 5 for the eight findings that we consider most pertinent.

Importance of the Type of Injury and Instability. It is apparent that the more structures involved in the injury, the more difficult it is to achieve an optimum end result. The prognosis is most favorable for the "isolated" ACL injury (Group A). But in that group only 32% were found to have full stability between 0° and 30°, and 6% between 60° and 90°. This suggests that the diagnosis in these patients was incomplete or was made grossly, causing coexisting lesions of the agonists (especially the posterior oblique ligament) to be missed.
In Group C with peripheral lesions of the medial meniscus and posterior oblique ligament, only 21% had a negative Lachman test. The findings in the range from 60° to 90° were comparable to those in the other groups. This confirms that the semimembranosus corner acts synergistically with the posterior horn of the meniscus as the knee moves into extension.
The complex instabilities of Groups F and G, which include injuries to lateral structures, have the least favorable prognosis and show significantly poorer results for most of the criteria listed.

Results by Surgical Procedure

To compare and correctly analyze the results of the operations, it was necessary to divide the patients into three groups:
Fresh injuries: Patients in this group had no prior history of knee problems and underwent surgery within three weeks after sustaining their injury.
Chronic instabilities: These patients sustained their injuries four weeks or more before undergoing surgery. They had not had a previous operation for ACL instability.
Previously operated instabilities: Patients in this group had a residual instability following a minimum of one or maximum of three previous operations on the ACL.

Table 3. Seven groups of injury combinations, with numbers of patients affected (n)

A:	n = 32	ACL "isolated"
B:	n = 22	ACL + PMCL
C:	n = 19	ACL + PMCL + peripheral lesion of medial meniscus
D:	n = 36	ACL + MCL + PMCL
E:	n = 23	ACL + MCL
F:	n = 20	ACL + ALFTL + popliteus corner/arcucate complex
G:	n = 11	ACL + ALFTL + popliteus corner/arcuate complex + PMCL

Table 4. Four-quadrant scheme for documenting instability

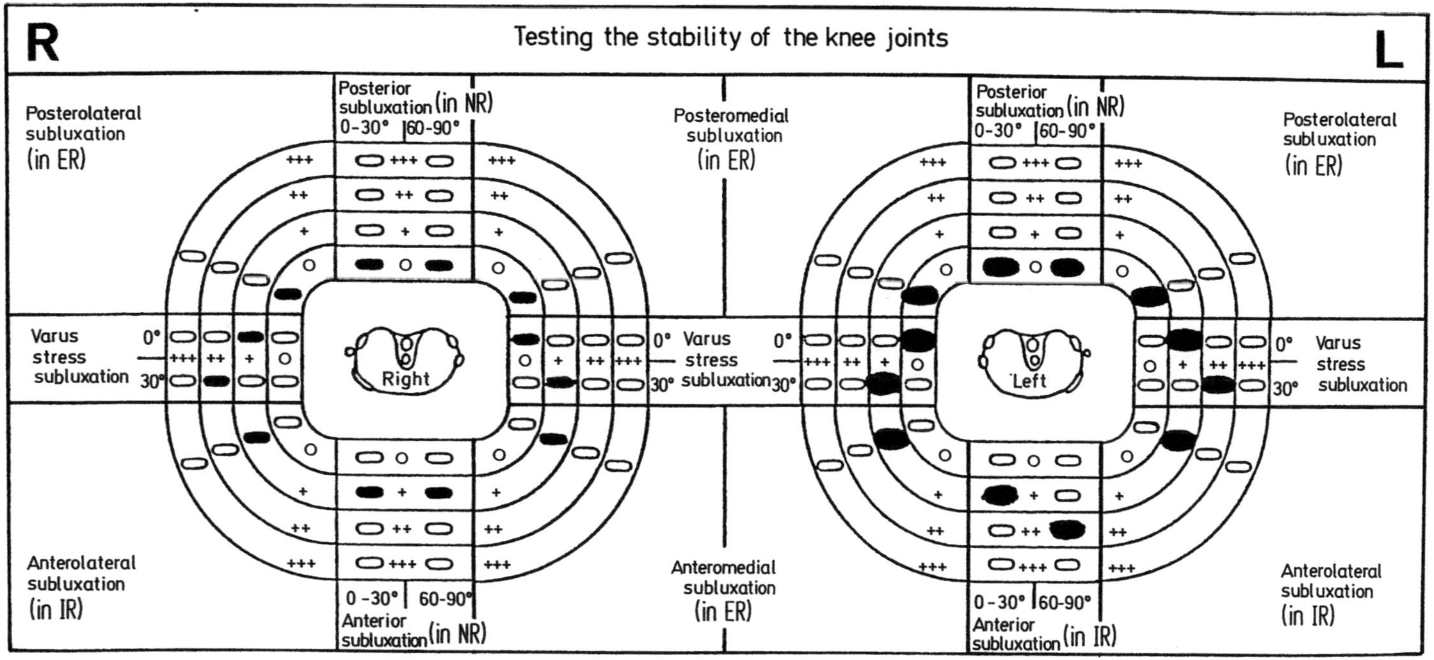

Pivot shift:	Negative / + / + + / + + + / not tested / not testable / other pivot shifts
Reverse pivot shift:	Negative / + / + + / + + + / not tested / not testable /
Other snapping or clicking signs:	Yes / no
"Impingement" anteriorly:	Hard / soft
"Impingement" posteriorly:	Hard / soft
Effusion:	Yes / no
Thickening of capsule:	None / mild / severe

Table 5. Results by type of injury, instability

Group		A	B	C	D	E	F	G

	N=	32	22	19	36	23	20	11
		Data in %						
Lachman 0–30°	0	32	32	21	36	33	17	9
	+	56	57	79	50	54	72	82
	++	9	11	0	14	13	11	9
	+++	3	0	0	0	0	0	0
Drawer 60–90°	0	6	3	6	3	9	0	0
	+	75	82	84	83	76	78	91
	++	19	15	10	14	15	22	9
	+++	0	0	0	0	0	0	0
Pivot shift	Neg	60	57	58	64	59	39	45
	+	33	30	42	17	26	44	36
	++	7	13	0	19	15	11	19
Giving way	Pre	85	80	79	73	67	90	91
	Post	9	13	11	14	11	22	9
Effusion	None	91	89	95	86	89	79	91
	Occ	3	0	0	0	0	5	0
	Reg	0	0	0	0	0	5	0
	Exer	6	11	5	14	11	11	9
Pain on effort	None	67	59	53	58	59	78	91
	Occ	33	38	47	40	39	17	9
	Pers	0	3	0	2	2	5	0
Red. perform.	None	87	80	89	74	78	82	91
	Occ	13	19	11	26	22	13	9
	Sev	0	1	0	0	0	5	0
Figure 8 run	r+1	94	95	90	94	92	78	91
	1 side	6	4	5	3	4	0	0
	Imp	0	1	5	3	4	22	9

Table 6. Breakdown of operations

Fresh: (*n*=91; 56%)	(H, I):	ACL isolated: 12 R=5, A=7
	(K, L):	ACL+ALFTL: 19 R=8, A=11
	(M, N):	ACL+semi: 19 R=6, A=13
	(O, P):	ACL+ALFTL+semi: 28 R=14, A=14
	(Q, R):	ACL+arc/pop: 13 R=11, A=2
Chronic: (*n*=49; 30%)	(S):	ACL isolated: 2
	(T):	ACL+ALFTL: 22
	(U):	ACL+semi: 12
	(V):	ACL+ALFTL+semi: 5
	(W):	ACL+arc/pop: 8
Previously operated: (*n*=23; 14%)	(X):	ACL isolated: 1
	(Y):	ACL+ALFTL: 8
	(Z):	ACL+semi: 7
	(AA):	ACL+ALFTL+semi: 5
	(BB):	ACL+arc/pop: 2

The percentage distribution of these groups is shown in Table 6.

Comments on Group 1 Patients with Fresh Injuries

Two subgroups of these patients were identified:
Patients undergoing a simple *repair of reinsertion* of the ACL (R = repair, Groups H, K, M, O, and Q).
Patients undergoing *primary augmentation* of the ACL with a patellar tendon graft (central one-third, with proximal and distal bone blocks, free or with fat pad pedicle) (A = augmentation, Groups I, L, N, P, and R).
The following additional procedures were used to protect the ACL during healing:

a) Repair of the ALFTL (anterolateral femorotibial ligament, Kaplan system), Groups K and L.
b) Repair of the semimembranosus corner (PMCL, posteromedial capsule, posterior horn of medial meniscus, capsular expansion of semimembranosus muscle), Groups M and N.
c) Combined repair of the ALFTL and semimembranosus corner, Groups O and P.
d) Repair of the popliteus corner (without popliteal bypass), Groups Q and R.

The outcomes for the 91 patients with fresh ligamentous injuries are summarized in Table 7.

Discussion and Evaluation of the Results

Lachman Test. Clearly, the best stability in the range from 0° to 30° was achieved by primary augmentation of the ACL with a patellar tendon graft combined with repair of the ALFTL and semimembranosus corner (64% absolutely stable). The results were least favorable in the groups undergoing repair of the popliteus corner and arcuate complex.

Table 7. Results by surgical procedure: fresh injuries

Group		H	I	K	L	M	N	O	P	Q	R
		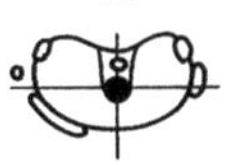	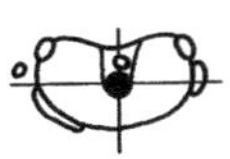		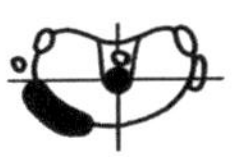		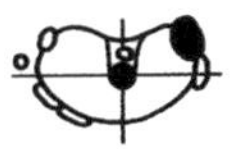	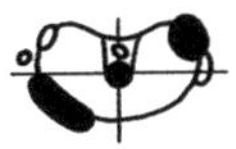	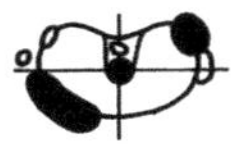		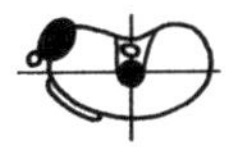
	$N=$	5	7	14	14	11	2	6	13	8	11
		Data in %									
Lachman 0–30°	0	20	43	26	27	32	38	36	64	11	0
	+	20	57	62	54	60	54	50	36	72	50
	++	60	0	12	19	8	8	14	0	17	50
	+++	0	0	0	0	0	0	0	0	0	0
Drawer 60–90°	0	20	28	12	9	16	8	14	22	0	0
	+	20	72	76	73	68	84	86	78	78	50
	++	60	0	12	18	16	8	0	0	22	50
	+++	0	0	0	0	0	0	0	0	0	0
Pivot shift	Neg	40	57	26	82	32	69	36	78	17	50
	+	60	43	62	9	56	8	50	22	77	50
	++	0	0	12	9	12	23	14	0	6	0
Giving way	Pre	80	71	72	81	76	78	71	86	72	100
	Post	20	0	12	9	28	8	14	0	22	0
Effusion	None	100	72	100	100	100	92	100	71	88	100
	Occ	0	14	0	0	0	0	0	0	0	0
	Reg	0	0	0	0	0	0	0	0	6	0
	Exer	0	14	0	0	0	8	0	29	6	0
Pain on effort	None	40	57	62	72	68	54	36	57	67	100
	Occ	60	43	38	28	32	38	64	43	33	0
	Pers	0	0	0	0	0	8	0	0	0	0
Red. perform.	None	100	86	88	82	84	69	79	86	72	50
	Occ	0	14	12	18	16	31	21	14	22	50
	Sev	0	0	0	0	0	0	0	0	0	0
Figure 8 run	r+1	100	100	88	91	84	92	93	86	78	50
	1 side	0	0	12	9	16	8	0	7	0	0
	Imp	0	0	0	0	0	0	7	7	22	50

Table 8. Results by surgical procedure: chronic instabilities

Group		S	T	U	V	W
			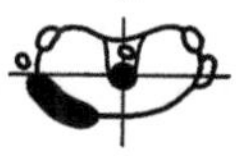		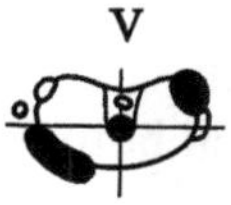	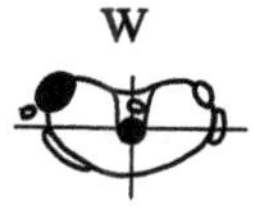
	N=	2	22	12	5	8
		Data in %				
Lachman 0–30°	0	0	27	25	40	0
	+	100	69	67	60	100
	++	0	4	8	0	0
	+++	0	0	0	0	0
Drawer 60–90°	0	0	4	0	0	0
	+	100	87	83	100	88
	++	0	9	17	0	12
	+++	0	0	0	0	0
Pivot shift	Neg	50	72	83	80	76
	+	50	24	17	20	24
	++	0	4	0	0	0
Giving way	Pre	0	96	92	100	100
	Post	0	4	8	0	12
Effusion	None	100	92	84	100	64
	Occ	0	4	8	0	12
	Reg	0	0	0	0	0
	Exer	0	4	8	0	24
Pain on effort	None	100	60	50	80	64
	Occ	0	36	42	20	24
	Pers	0	4	8	0	12
Red. perform.	None	100	72	83	80	76
	Occ	0	28	17	20	24
	Sev	0	0	0	0	0
Figure 8 run	r+1	100	86	83	80	76
	1 side	0	14	17	0	24
	Imp	0	0	0	20	0

Pivot Shift. Primary augmentation (maximum of 82% negative pivot shift) showed significantly better values compared with repair or reinsertion of the ACL alone (maximum 40% negative). Repair of the ALFTL emerges as an important factor. But repair of the ALFTL combined with ACL repair is not sufficient to produce a good result. A significant reduction in pivot shift can be achieved only by concurrent augmentation of the ACL. A similar percentage reduction in pivot shift can be obtained by repair of the semimembranosus corner (compared with an isolated ACL repair).

"Giving Way." Patients undergoing primary augmentation of the ACL reported significantly less giving way than patients undergoing a direct repair (0%–9% versus 12%–28%). The most striking improvement, from 86% before surgery to

0% afterward, was achieved by augmentation combined with repair of the ALFTL and semimembranosus corner.

Figure 8 Run. All patients with an isolated ACL rupture were able to run a figure-of-eight course in both directions without disability. Patients having the greatest difficulty with the course were those who required repair of the popliteus corner and arcuate complex.

Pain on Effort. On average, patients undergoing a primary augmentation had less exertional pain than the group undergoing direct repair. This result was further improved by concurrent repair of the ALFTL.

Reduction in Performance. There were no instances of severe performance reduction in any of the groups evaluated. An average of 20% of patients reported occasional disability.

Effusion. Only 1 of the 91 patients had chronic effusion. No significant differences were noted among the different surgical procedures in this regard.

Chronic Instabilities

All the patients with chronic instabilities underwent ACL reconstruction using a free or vascularized patellar tendon graft with proximal and distal bone blocks. The individual results are shown in Table 8.

Discussion and Evaluation of the Results

Stability. Anterior stability in the 0°–30° and 60°–90° ranges was generally poorer in this group than in the patients with more recent injuries. This is attributable to the progression of anterior instability over time and the associated loosening of secondary restraints. The outcome could be markedly improved, however, by the selective repair of these structures.

Pivot Shift. No significant differences in pivot shift are apparent between the groups with fresh and chronic instabilities.

"Giving Way." No patient who underwent a patellar tendon graft combined with repair of the ALFTL and semimembranosus corner complained of a giving-way sensation postoperatively. All the patients expressed that complaint before surgery.

Pain on Effort. Pain of effort was a much more constant finding in these patients than in the freshly operated group. This is attributed to secondary chondropathies and degenerative changes in the cartilage and menisci, which were more prevalent in patients with longer-standing injuries.

Figure 8 Run. Far fewer patients were able to run the complete figure-of-eight course than in the freshly operated group.

Patients with Previous Operations

These patients comprised the smallest group ($n = 23$), so the results should be interpreted simply as trends (Table 9).

Table 9. Results by surgical procedure: previously operated instabilities

Group		X	Y	Z	AA	AB
					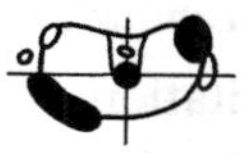	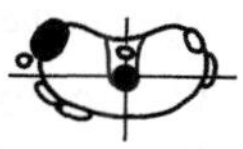
	N=	1	8	7	5	2
		Data in %				
Lachman 0–30°	0	0	12	14	20	0
	+	100	64	72	60	100
	++	0	24	14	20	0
	+++	0	0	0	0	0
Drawer 60–90°	0	0	0	0	0	0
	+	100	76	72	80	100
	++	0	24	28	20	0
	+++	0	0	0	0	0
Pivot shift	Neg	100	76	72	60	100
	+	0	24	14	20	0
	++	0	0	14	20	0
Giving way	Pre	100	88	86	80	100
	Post	0	12	0	0	0
Effusion	None	100	88	72	80	50
	Occ	0	0	14	0	50
	Reg	0	0	0	0	0
	Exer	0	12	14	20	0
Pain on effort	None	100	44	58	40	50
	Occ	0	44	42	60	50
	Pers	0	12	0	0	0
Red. perform.	None	100	50	86	80	100
	Occ	0	50	14	20	0
	Sev	0	0	0	0	0
Figure 8 run	r+1	100	76	86	80	50
	1 side	0	24	14	20	50
	Imp	0	0	0	0	0

General Results

Attitude toward Operation. The vast majority of the patients, 95.8%, would undergo surgery again; 2.4% would not, and 1.8% would consent to the surgery under different conditions.

Use of the Knee Joint. In 98% of cases, use of the knee joint was markedly improved for the activities of daily living. Improvement in more strenuous activities was present in 97%.

Condition of the Knee Joint. The patients rated the condition of their knee as follows:
- 75% greatly improved
- 13% equivalent to unoperated side

- 11% better
- 1% worse (1 patient)

Postoperative Limitation in Sporting Activities.

- 53% no limitation
- 39% mild limitation
- 8% severe limitation (all patients with previous operations)

Fat Pad Fibrosis. In the cases where a patellar tendon graft with a fat pad pedicle was used to reconstruct the ligament, 9% of patients complained of pain referrable to fibrosis of the fat pad. Four percent of these patients required reoperation (partial arthroscopic resection of the fat pad).

Complications

The following postoperative complications were encountered:

Hematoma formation	5%	(2 of these 8 patients required operative revision)
Infrapatellar nerve entrapment	3%	(neuroma excision was performed later in 2 of these 5 patients)
Infection	1.2%	(resolved completely with antibiotic treatment)

Social Status. No patient required disability compensation, and 98% were able to return to their former work.

Range of Motion. The following ranges of motion were determined:

- Flexion to 10° or less = 14 patients (9%)
- Flexion past 140° = 135 patients (83%)
- Extension loss of 5° or less = 10 patients (6%)
- Recurvatum of 5° = 27 patients (16%)
- Recurvatum of 10° = 1 patients (0.6%)
- Recurvatum of 20° = 1 patient (0.6%)

At the time of follow-up, no patient was found to have more than a 10° loss of flexion or extension in the operated knee.

Conclusions

1. The more extensive the instability-producing injury, the more difficult it is to obtain a good late result. However, the definitive outcome depends less on the combination of injuries than on a complete diagnostic workup and adequate operation.
2. Clearly the best results in fresh ACL injuries are achieved by primary augmentation or reconstruction with a patellar tendon graft combined with the repair of all injured structures (ALFTL and semimembranosus corner).
3. Primary augmentation is of significantly greater value in improving pivot shift than is repair or reinsertion of the ACL. The type of ligamentous tear (midsubstance, proximal, distal) is not a significant factor.

4. Repair of the ALFTL and, to a lesser extent, the semimembranosus corner significantly decreases the incidence of pivot shift, but only when combined with augmentation.
5. The results in all groups (fresh injury, chronic instability, previous surgery) were poorest when there was a coexisting lesion of the popliteus corner and arcuate complex.
6. Knee stability in freshly operated patients is better than in patients with chronic instabilities, whose stability, in turn, is better than in patients who have had previous ACL surgery.

The Anterior Cruciate Deficient Knee Treated by a Modified Marshall-Macintosh Repair

G. N. Fernandez and J. A. Robertson

Lord Major Treloar Hospital, Alton, Hampshire GU 34 1RJ, UK

The University Hospital of Southampton provides an accident service for a population of 500 000. All patients in this series underwent surgery at the Lord Mayor Treloar Orthopaedic Hospital which acts as a regional referral centre for a population of 1.5 million. The use of fascia lata for reconstruction of the anterior cruciate ligament (ACL) was first described in 1917 by Hey Groves – a former surgeon at this hospital.

Our indications for this procedure include a history of instability during simple daily activity together with the clinical findings of a positive pivot shift, jerk test and Lachmann's sign. We have found the precise mechanical definition of the instability unhelpful.

The operation is a mixture of Marshall's and Macintosh's ideas with some modifications: A distally based strip of fascia lata is prepared and passed under the lateral collateral ligament as described by Macintosh. We have found it unnecessary to pass the graft under a subperiosteal tunnel; instead, we suture the graft to the collateral ligament with a 1 vicryl suture whilst maintaining firm external rotation of the tibia. The graft is then passed anterior to the lateral intermuscular septum, "over the top" of the femoral condyle, and through the posterior capsule into the intercondylar notch, where it is retrieved through a medial arthrotomy and drawn through a tibial drill hole which runs from the site of the ACL remnant to exit just medial to the tibial tubercle. The graft is then drawn tight and secured with a Landanger staple and the tensor fascia lata closed as described by Marshall.

Our series of 25 patients comprised 23 men and two women with an average age of 27. The left knee was affected in 18 cases and the right knee in seven. Soccer accounted for 18 of the injuries, motorcycle accidents for two, physical assault for two, skiing, dancing and squash for the remaining three. The mechanism of injury was reported as a pure twisting force in 11 cases, whilst six reported twisting with an added valgus component and one a varus component. Pure valgus and varus forces accounted for four cases. The mechanism of injury was unknown in the remaining three. The time from injury to surgery ranged from 10 months to 15 years with an average of 5 years. Fourteen patients had previously undergone medial meniscectomy and three a lateral meniscectomy. Pes anserinus transfer had been performed on two of these. Under anaesthesia all demonstrated a markedly positive Lachmann's test together with a positive pivot shift and jerk test. All patients underwent arthroscopy prior to surgery, when degenerative changes were noted in 14 (56%). Four of these patients had changes affecting all three compartments of the knee, six had changes in two compartments and four had changes in one compartment.

Surgery and Arthroscopy of the Knee
2nd Congress of the European Society
Editors: Müller/Hackenbruch

The average length of follow-up was 23 months with a range of 9 months to 4 ½ years. All the surgery was performed by the senior author (J. A. R.), and the patients were evaluated by the junior author, who questioned and examined 21 patients. The remaining four patients were abroad but were able to submit the relevant information by post. Subjective assessment was obtained using the Cincinnati Sports Clinic questionnaire, which gave a pre- and postoperative knee score based on the symptoms of pain, swelling, giving way and functional activity level, with a maximum score of 100. The most significant objective factor was the presence or absence of the pivot shift sign, although residual instability, restricted range of movement, quadriceps wasting and effusion were also looked for.

Immediate postoperative pain was reported as severe by four patients, and one stated he would not go through surgery again despite an excellent functional and pain-free result. Only one patient developed long-term discomfort and stiffness, and this was attributed to early degenerative changes in his patellofemoral joint which were seen at arthroscopy 3 years after surgery. His arthroscopy demonstrated the intact revascularised intra-articular portion of his graft.

Prospectively, all 25 patients reported giving way as a "severe problem" to the extent that they were unable to turn or twist while walking without the knee giving way. Postoperatively, 19 reported no further episodes of giving way, whilst the remaining six reported "occasional giving way during strenuous sports" such as squash. One patient in this group had a stable knee for 18 months until he tripped on uneven ground and sustained a partial tear of his graft. His knee has only given way twice since and his pivot shift has remained negative even under anaesthetic.

With regard to functional recovery, preoperatively 21 patients were unable to partake in any sport and the remaining four were severely restricted to activities such as cycling and swimming. Following surgery all were able to resume sporting activities to a variable extent. Three were able to return to the same contact sports at their preinjury performance level: TH returned to rugby and hockey, RA to playing soccer twice at weekends (having been reliant on a stick for walking preoperatively) and SS returned to club hockey despite weighing 95 kg; 11 returned to the same sports, such as squash tennis, riding and ice skating, but at a lower performance level. Eight took up different and less strenuous sports such as cycling, running and table tennis. The remaining three were able to partake in more restricted activities such as swimming.

Increase in the knee scores following surgery could be seen. The average score preoperatively was 30 (out of 100) with a range from 18 to 55. This rose after surgery to an average of 83 with a range of 50–100. High scores are still obtained 4 ½ years after surgery.

Examination of the knees produced the following findings: Lachmann's test remained positive in all patients. The pivot shift test remained negative in all, but four gave the impression that pivot shift would occur with muscular relaxation. Anterior draw was abolished in internal rotation in all, but was present in neutral and external rotation in 12. No patient had joint line opening on valgus or varus stressing in full extension, but at 30° flexion eight demonstrated both medial and lateral joint line opening, two medial joint line opening only and five lateral joint line opening. No attempt was made to quantify the degree of opening. All patients had a full range of movement and none had an effusion. Measurable wasting of the quadriceps was present in three patients.

In conclusion, the high incidence of meniscus injuries (68%) and degenerative changes (56%) in these patients is similar to that found in other series. These are thought to be a consequence of the recurrent instability rather than of the original injury.

This operation has been shown to improve function in the anterior curciate deficient knee with chronic rotary instability. The improvement has been maintained for periods up to 4½ years with only one patient showing any deterioration, and this occurred after significant trauma to the knee. All patients were able to return to sports and all but one would undergo the operation again.

We found the Cincinnati knees score to correlate well with the overall clinical picture: those with degenerative changes obtained an average score of 80 compared to a score of 87 in the nondegenerate group, although it should be noted that the three patients with changes in all three compartments had individual scores of 70, 87 and 90. The four patients with suspected pivot shift after surgery had an average score of 67 compared to that of 80 in the remainder. Other forms of residual instability did not preclude a high score, although the lower scores tended to be associated with other residual instabilities. Factors of no prognostic significance when related to the final knee score included the age and sex of the patient, the time lapse between injury and surgery, the side and mode of injury, and previous meniscectomy or ligament reconstruction.

Although these are still early results, we feel that they represent a yardstick against which any artificial ligament replacement should be judged.

Substitution of the Anterior Cruciate Ligament Using the Torn Meniscus

E. Lais [1], P. Hertel [1], J. Wortmann [1], R. Schrempf [2], and H.J. Barrach [2]

[1] Section of Trauma, Department of Surgery, Rudolf-Virchow-Krankenhaus, 1000 Berlin 65, FRG
[2] Institut für Toxikologie und Embryonalpharmakologie der FU Berlin, 1000 1000 Berlin 33, Garystraße 5, FRG

Introduction

Currently, the indication for a surgical correction of chronic anterior instabilities of the knee joint is a large subject area because of the extensive number of associated secondary lesions [1, 2, 6, 8]. The unsatisfactory results of extra-articular plasty and the different resiliences of various inserted transplanted tissues have turned attention once again to the ruptured meniscus, which is not suitable for refixation, as a substitute for the anterior cruciate ligament (ACL) [3, 10, 12, 13, 15].

Material and Methods

Between 1982 and 1986 in 61 instances of a ventral pedunculated bucket handle rupture or lobe rupture of a meniscus, the ACL was reinforced or replaced. This method was exclusively used where a refixation of the meniscus did not seem useful any longer. Except in some cases, an arthroscopy was performed under anaesthesia and the bucket handle was frequently cut off dorsally by employing arthroscopy (Fig. 1). One either uses the bucket handle or, in cases of shorter

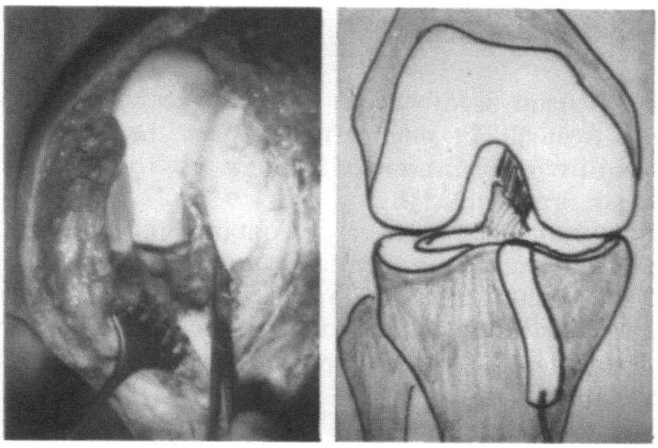

a b

Fig. 1 a, b. Typical anatomy in a case of anterior instability: **a** the proximal insertion of the anterior cruciate ligament is empty and there is only a short distal stump left; **b** the bucket handle is dorsally dissected

Surgery and Arthroscopy of the Knee
2nd Congress of the European Society
Editors: Müller/Hackenbruch

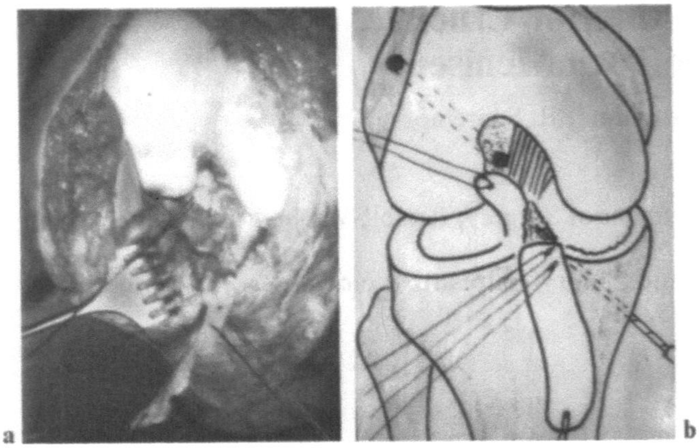

Fig. 2 a, b. Fixation of the ventral pedunculated part of the meniscus by strong transosseous Vicryl sutures in a deep bone groove in the ventral part of the eminentia ant.; the free proximal part was guided over the top and transosseously through the lateral condyle (together with the short distal stump of the anterior cruciate ligament)

vertical ruptures or lobe ruptures, a resection which was carried out in a way that preserved a sufficient solid margin. With the joint open, the ventral pedunculated part of the meniscus is fixed by strong Vicryl (polyglactin) transosseous sutures in a deep bone groove located in the area of the anterior eminence.
The free proximal part of the graft is divided into two portions: one is guided transosseously through the lateral condyle and the other over the top of the lateral condyle (Fig. 2). The fixation sutures for both portions are tied over a bone bridge on the lateral femur. The insertion in the inner side of the lateral condyle is cleaned and, if necessary, a notch plasty is carried out. In 22 cases a stump of the ACL, grown on "en nourisson" at the posterior cruciate ligament (PCL), were prepared and used as a posterolateral bundle.

Results

Forty-two male and 19 female patients aged between 16 and 57 years, with an average age of 30 years, were treated. The time period between the accident causing the instability and operative treatment ranged from 3 weeks to 15 years. The main accidental cause was sports (Table 1).
In six of the patients arthroscopy had already been carried out but only five were diagnosed as suffering a loss of the cruciate ligament. One test arthrotomy was carried out without a result, and in another a lateral meniscus was refixed by a medial arthrotomy after another accident. At that time the damage to the cruciate ligament was apparently underestimated. Two of our patients were already conventionally meniscectomised (Table 2).
It is interesting that in 46 patients signs of instability appeared shortly after the accident. On examination under anaesthesia intra-operatively, a positive Lachmann test and a positive jerk test were always found, whereas the anterior drawer and the varus:valgus instability were expressed at varying strengths. Concerning

Table 1. Accidents resulting in anterior cruciate ligament reconstruction

Accidents	Number of cases
Sport	39
Soccer	17
Skiing	9
School	2
Handball	2
Other	9
Daily events	14
Traffic accident	6
Motorcycle	4
Industrial accident	2

Table 2. Number and type of earlier operations in these patients ($n=61$)

Arthroscopy	6	
Arthrotomy	2	(1 refixation of the meniscus)
Meniscectomy (conventional)	2	
	10	

Table 3. Effects of instability ($n=61$)

	Number of cases
Synovitis	61
Tear of the contralateral meniscus	12
Refixation	4
Resection	8
Damage to the cartilage	
Femoral condyle	49
Patella	36

side-effects, synovitis always occurred. In 12 cases the contralateral meniscus was also damaged, in 49 cartilage damage was indicated at the femur condyles, and damage of the patellofemoral joint was recorded in 36 cases (Table 3). Postoperative complications were not seen in seven patients, for whom the time between the accident and the operation was only between 3 weeks and 4 months. A posteromedial capsule doubling after Hughston [7] was carried out 21 times because of a clearly increased medial opening capacity. In one case a popliteus bypass was carried out, in a second a lateral tract plasty and in a third a bony rupture of the lateral collateral ligament, and the popliteals tendon was refixed by osteosynthe-

Table 4. Complications

	Number of cases
Infection	2 (1 *Staphylococcus aureus*)
Loosening of the graft	1
Rupture of the graft (sudden fall in the phase of mobilisation)	1

Table 5. Results scored according to Müller [15] ($n=61$)

	Number of cases
Very good	17
Good	26
Moderate	10
Poor	2

Table 6. Results scored according to Lysholm and Gillquist [13] ($n=61$)

	Number of cases
Very good	18
Good	20
Moderate	15
Poor	2

sis. Two cases were complicated by precocious infection; arthrotomy was carried out again and drainage was laid for lavage and suction which allowed healing to begin. Both patients were followed up. One of them already plays football again without any trouble. The results in the other patient remain moderate; the insertion of the ligament is still firm. In one case the graft loosened after 1 year and the patient became unstable, necessitating a new operation (Table 4).

Follow-up took place at 6 months and at 4 years after the operation. Also, 55 patients have had a further examination. Patients were asked about their achievements, especially about sport activities before and after the operation, and about the rehabilitation which enabled them to go return to work. The same surgeon carried out all the clinical examinations. The results have been split up into four groups after Müller [11]: 16 cases were excellent, 26 satisfactory, 10 moderate and 2 poor (Table 5). A further analysis was made following the scoring scale of Lysholm and Gillquist [9] (Table 6).

Up to now six control arthroscopies have been carried out. During a further operation intra-articular results were recorded, and in two others tissue samples were

Table 7. Results of re-examination by arthroscopy or arthrotomy* ($n=8$)

Age	Time of examination (months)	Cause of examination	Anaesthetised response	State of graft	Histology
50	6	Bad range of motion	Stability	Graft viable and stable	Tense collagenous tissue with regular pattern of fibrils
	17	Pain	Stability	Graft viable and stable	Cartilage within the fibrous tissue
41	18	Pain, swelling	Lachman test (0), Jerk test (+), anterior drawer (+)	Graft viable and stable	Cartilage within the fibrous tissue
32*	18	Loosening of the graft	Instability	Graft loose at the distal insertion	Tense collagenous tissue with regular pattern of fibrils
24	8	Pain	Lachman test (0), Jerk test (0), anterior drawer (+)	Graft viable and stable	Tense collagenous tissue with regular pattern of fibrils
26*	19	Operation in another hospital	Instability	Graft ruptured because it was non-physiologically directed	Tense collagenous tissue with regular pattern of fibrils
29	14		Stability	Graft viable and stable	Tense collagenous tissue with regular pattern of fibrils
27	12	Crepitation	Lachman test (0), jerk test (0), valgus instability	Graft viable and stable	Cartilage within the fibrous tissue

taken for histological examination. Re-examination by arthroscopy was performed where there were problems with existing cartilage lesions or insufficient mobility. On one occasion, while another operation was being performed, arthroscopy was carried out on the lower extremity (Table 7).

The time between the operation and arthroscopic examination ranged from 6 to 19 months. During arthroscopic examination the grafts appeared healed and firm. The results of the histological examination differed here, which means that earlier results have had to be revised. The results are also in contrast to those of Ferkel [5] showing mainly tense collagenous connective tissue with loosened primary substances and changing density of the nucleus [4]. There was also cartilage with exposed fibres and swelling of the primary substances 17 months after operation (Fig. 3).

One patient examined by arthroscopy 6 months postoperatively showed tense collagenous connective tissue. Eleven months later, on the occasion of a second arthroscopic examination cartilaginous areas are still visible within the fibrous tis-

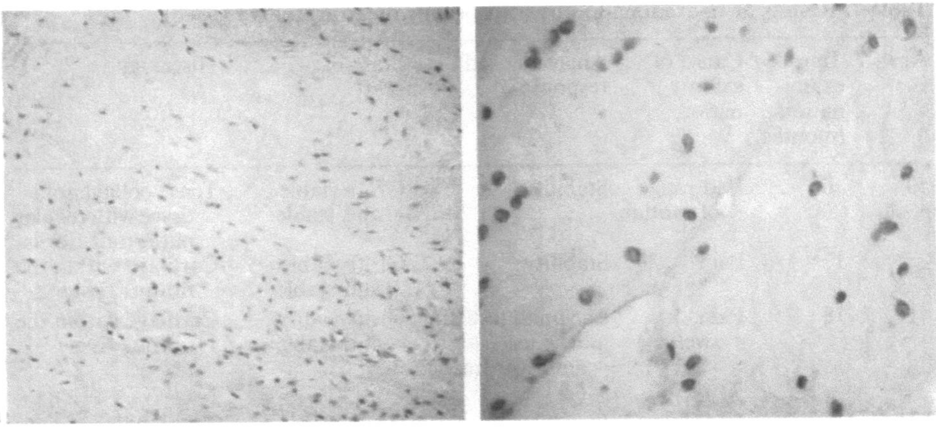

a b

Fig. 3. a Patient W., 6 months postoperatively: fibrous tissue in the graft; **b** Patient F., another patient, 17 months postoperatively: cartilage in fibrous tissue in the graft

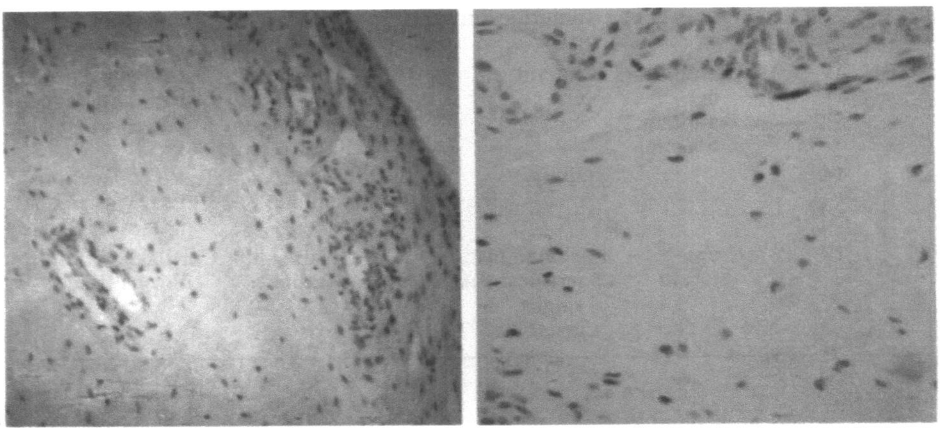

a b

Fig. 4 a, b. A rich vascular pattern under the synovial cover of the graft

sue. All histological preparations show a strong vascular profusion under the synovial cover (Fig. 4). For further clarification light microscopy, electron-microscopy and histochemical examinations were conducted. Electron micrographs indicate vacuolar degeneration and cell consumption (Fig. 5). Antigen marking for collagen type I, II and III shows an increase in types I and III in the graft but a clear decrease in type II. Immunofluorescence studies of heparan sulphate proteoglycan, a component of the basement membranes, display diffuse, faint, fibre-orientated markings in the normal meniscus tissue as well as a strong vascular increase in the graft (Fig. 6). This can be interpreted as a sign of the enormous vascularisation of the synovial cover. The results indicate that the graft is subject to a rebuilding process while strong vascularisation takes place from the surface, at the same time as degeneration. It is possible that the driven-in bucket handle which

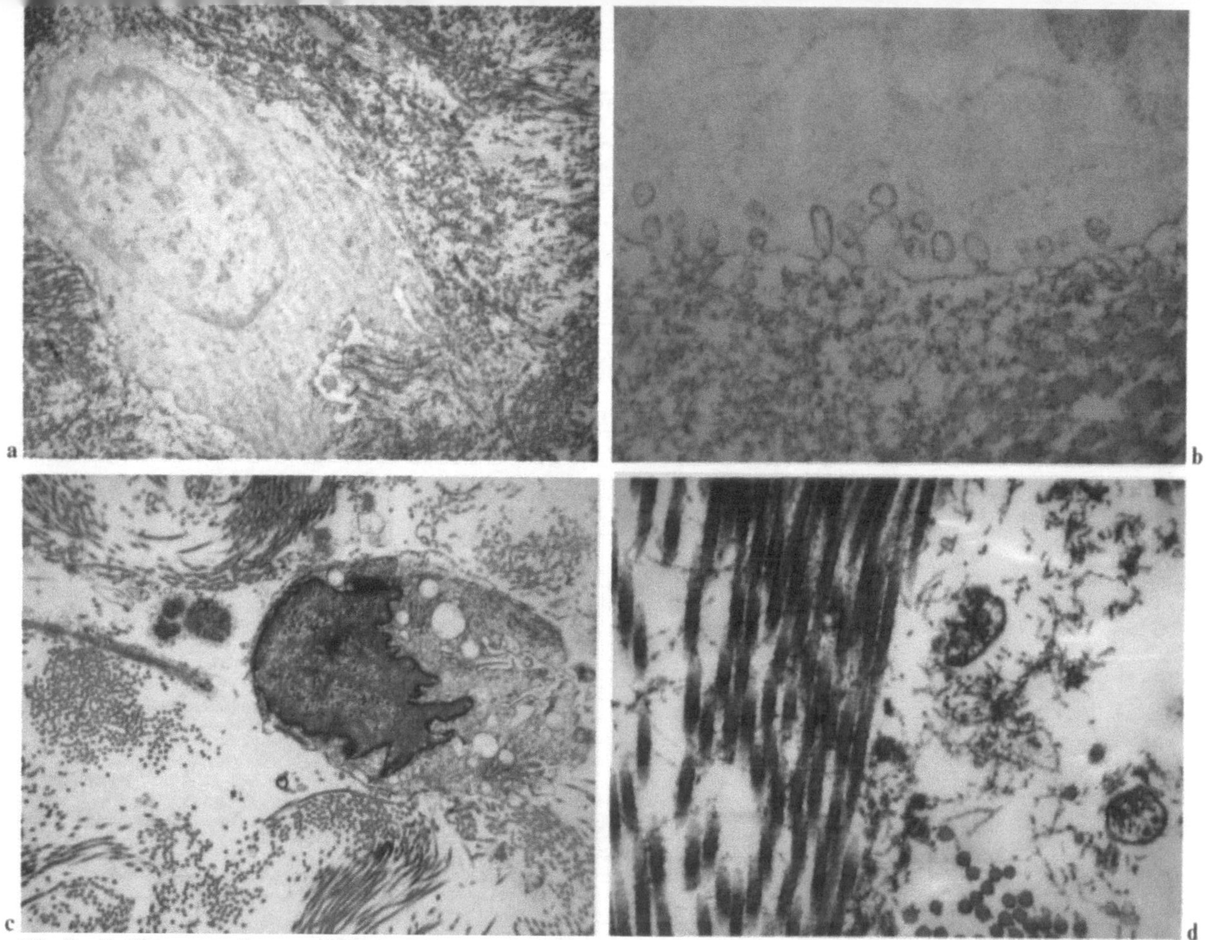

Fig. 5a–d. Electron micrograph from a 27-year-old female, 12 months postoperatively. **a, b** Normal meniscus tissue: regular arrangement of collagen fibrils, proteoglycan next to the cell and the fibrils (×6000). **c, d** The anterior cruciate ligament graft: degenerating cell, the fibrils appear more spongy; there is an increase of Heparin (×40000)

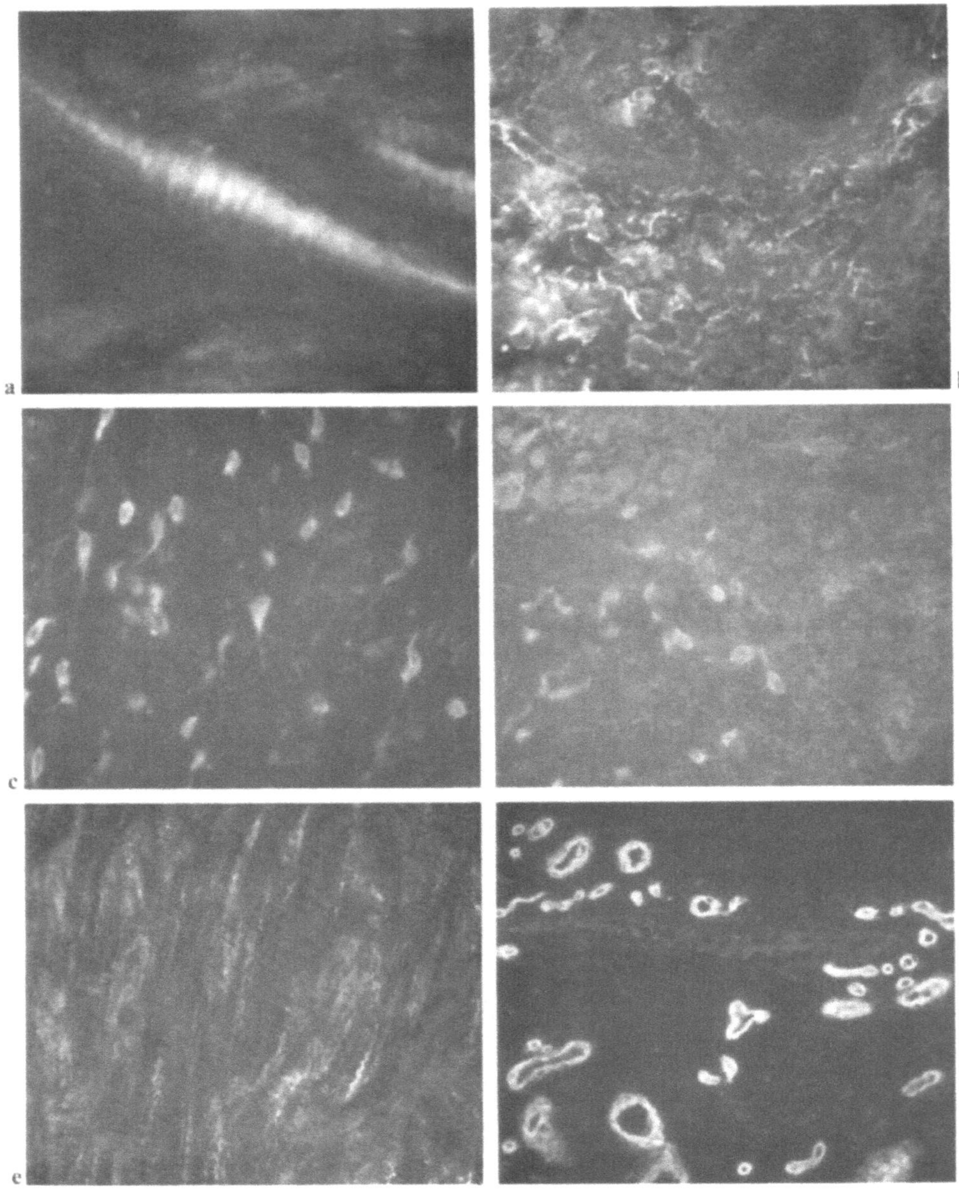

Fig. 6 a–f. 27-year-old female, arthroscopic re-examination 12 months postoperatively: **a, b** collagen type I; **c, d** collagen type II; **e, f** heparan sulphate proteoglycan (BM_1). The first picture of each pair is normal meniscus tissue; the second is the graft after 12 months is fed by diffusion has a better chance of healing or that the deterioration of the graft is less. The rebuilding processes certainly take a much longer time than originally expected. These observations also serve to explain the results of Paulos [14], Drobny [4] and Noyes [12] who found an enormous decrease in tensile strength of the tissue 1 year after transplantation. This fact should receive attention during postoperative treatment. Patients should be warned against participating in sports that risk injuries during the first year after the plasty. The frequent feature of late loosening of grafts was not observed. Patients that were followed up, especially after serious injuries, had often only been free from problems 1.5–2 years after surgery. Some of the followed-up patients began to play football again after 3 years, without any disadvantageous consequences. We had to give some patients the necessary confidence that the knee joint was resilient because fear of a new injury meant that they had reduced or even given up sports. Training of the muscles for dynamic support of the operational steps is the most important part of the postoperative treatment.

Discussion

Frequently lesions of the meniscus such as bucket handle tears or lobe ruptures are associated with long-standing instabilities of the knee. Thus it is only seldom that the tears are suitable for refixation, and they can be employed as a substitute for the cruciate ligament in ACL reconstruction. The operative procedure is simple and can be performed as for normal ACL reconstruction. Due to the stability and potentially excellent healing ability of the meniscus together with carefur stabilisation of the functionally associated knee joint structures satisfactory and even excellent results were achieved in 70%–75% of the cases in this study.

References

1. Arnoczky SP, Warren RF, Ashlock MA (1986) Replacement of the anterior cruciate ligament using a patellar tendon allograft. J Bone Joint Surg [Am] 68(3):376–385
2. Benedetto KP (1985) Der Ersatz des vorderen Kreuzbandes mit dem vasculär gestielten Drittel des Ligamentum patellae. Unfallchirurg 88:182–188, 189–197
3. Collins HR, Hughston JC, Haven KE, de Bergfeld JA, Ewarts JA (1974) The meniscus as a cruciate ligament substitute. Sports Med 2:11
4. Drobny TK, Müller W, Perren S (1984) Pediculated patellar tendon grafts for reconstruction of the anterior cruciate ligament – experimental analysis of vitality and operative technique. First European Congress of Knee Surgery and Arthroscopy, Berlin
5. Ferkel RD, Fox JM, Markolf KL, Dorey F, Del Pizzo W, Friedmann JJ, Snyder SF, Kasimian D (eds) Anterior cruciate ligament substitution using the torn meniscus. Center for Disorders of the Knee, Van Nuys, California
6. Hughston JC, Barret GR (1983) Acute anteromedial rotatory instability. J Bone Joint Surg 65:145
7. Hughston JC, Eilers AF (1973) The role of the posterior oblique ligament in repairs of acute medial (collateral) ligament tears of the knee. J Bone Joint Surg 55:923–940
8. Ivey FM, Blazina ME, Fox JM, del Pizzo W (1980) Intraarticular substitution for anterior cruciate insufficiency. Am J Sports Med 8:405
9. Lysholm J, Gillquist J (1981) The evaluation of knee ligament surgery with special emphasis to the use of a knee scoring scale. Dissertation, Linköping

10. Mathur PD, McDonald JR, Ghormley RK (1949) A study of the tensile strength of the menisci of the knee. J Bone Joint Surg 31:650
11. Müller WE (1982) Das Knie. Springer, Berlin Heidelberg New York
12. Noyes FR, Butler DL, Grood ES, Zernicke RF, Hefzy MS (1984) Biomechanical analysis of human ligament grafts used in knee ligament repairs and reconstructions. J Bone Joint Surg 66:344
13. O'Donoghue DH, Rockwood CA (1966) Repair of the anterior cruciate ligament in dogs. J Bone Joint Surg 48:503
14. Paulos LE (1986) Anterior cruciate ligament reconstruction – biologic versus synthetic replacement. Vortrag, 18. September 1986, Berlin, Rudolf-Virchow-Krankenhaus
15. Seiler H, Hager D, Kayser M, Flory PJ (1986) Ist der Meniskus am vorderen Kreuzband tatsächlich eine historische Methode? Unfallchirurg 88:315–321

Recent Rupture of the Anterior Cruciate Ligament: Clinical Results of Treatment by Reinsertion and Allogenic Augmentation

O. Kwasny [1], R. Schabus [1], M. Wagner [1], and H. Plenk [2]

[1] First University Clinic of Traumatology, Vienna, Alser Straße 4, A-1097 Vienna, Austria
– Bone and Biomaterial Research Laboratory, Institute of Histology andEmbryology, University of Vienna, Schwarzspanierstraße 17, 1090 Vienna, Austria

Introduction

Operative treatment of the acutely ruptured anterior cruciate ligament (ACL) by suturing or reinsertion has been commonly reported in the literature. As the results often appear unsatisfactory [4], alternatives have been suggested, such as the augmentation of the ACL by autologous graft or primary prosthesis [2, 3, 6]. However, reports of primary surgery of the ACL with immediate allogenic augmentation are scarce [1]. This particular procedure has been applied at the First University Clinic of Traumatology in Vienna since the end of 1983.

Patients

From November 1983 until June 1986, altogether 105 patients (60 male, 45 female; mean age: 28 years) were operated on for recent complete rupture of the ACL. Forty-eight patients had an "isolated" ACL rupture. Structures of the capsular ligament were additionally involved in 47 patients on the medial and in 10 patients on the lateral side of the knee joint. The injury was due to various kinds of accidents: during sports in 80%, in road traffic in 10%, at home and at work in 5% each. Surgery was performed after an average interval of 8 days (range 0–25 days) after the trauma.

Methods

Via a straight anterolateral skin incision, a small anteromedial arthrotomy was made, permitting inspection of the joint cavity. The following additional lesions were detected: lesions of the medial meniscus in 19 cases, of the lateral meniscus in four cases, osteochondral fractures in 11 cases.
The medial meniscus was sutured in 12 patients and had to be partially resected in seven. The lateral meniscus was sutured once and resected in three cases. In 11 patients defects of the cartilage were smoothed, glued or treated by Pridie drilling.
Reconstruction of the capsular ligaments was necessary at the medial side in 42 patients, at the lateral side in five. In four patients a concomitant rupture of the posterior cruciate ligament required reinsertion.

Surgery and Arthroscopy of the Knee
2nd Congress of the European Society
Editors: Müller/Hackenbruch

In six cases of midsubstance rupture of the ACL we used the semitendinosus tendon for augmentation. In another six cases of recent rupture of the ACL that appeared too weak for primary reconstruction, the ligament was substituted by a bone-ligament-bone graft of the patellar tendon.
When the ACL was torn out proximally or a Z-shaped rupture in the proximal third of the ACL was present, a reinsertion was carried out according to the technique described by Marshall. In addition a braid of polypropylene, 8 mm width and 13 to 16 cm long (Kennedy ligament augmentation device, LAD) was inserted. The allogenic band was passed through a tunnel and then drilled within the tibial bone and along a "through the top" tunnel within the lateral femoral

Fig. 1. Reinsertion of the anterior cruciate ligament (Marshall), ligament augmentation device through the top

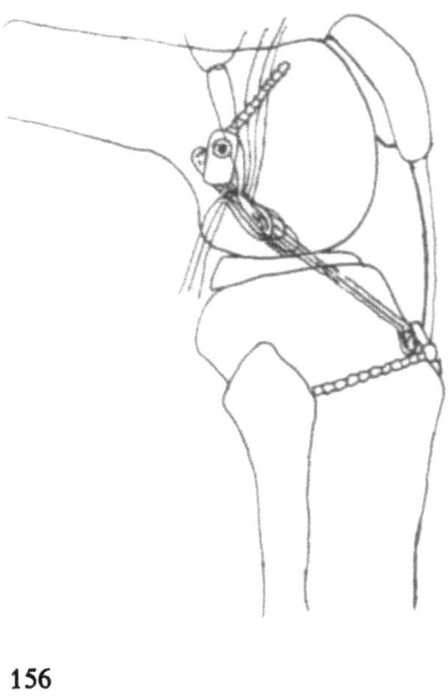

Fig. 2. Fixation of the ligament augmentation device with spike plates on both ends

condyle in such a way that it came to lie parallel to the ACL. Finally the synthetic ribbon was tightly anchored with spike plates on both ends.
In all patients except those with a posterolateral instability, an additional tractopexy was carried out with a screw and a spike washer (Figs. 1, 2).

Postoperative Management

Postoperatively the knee joint is immobilised with a flexion of 30°–40° by means of a plaster cast. For subsequent mobilisation we use a brace that limits movement. During the period of mobilisation, up to 8 weeks after surgery, weight-bearing is restricted. This restriction is prolonged to 4 months in the presence of concomitant lesions of the cartilage.

Results

Patients who had undergone reinsertion with allogenic augmentation more than 6 months previously were examined for follow-up in June 1986. Of a total of 76 patients, 72 were available for examination after an interval of 6–29 months (average 18 months) after surgery.

Clinical Stability

At the physical examination the Lachman test was negative in 33 patients, slightly positive with perceptible resistance of the ligament in another 33 patients and positive in six patients, in one of whom the ligament's resistance was uncertain (Table 1).
The pivot shift test was negative in 69 patients and positive in three. One patient with a positive pivot shift, however, had a history of a second trauma of her knee joint with an effusion 1 year after the operation. In full extension there was no loosening of the knee joints under valgus or varus stress. In 30° flexion there was a slightly positive valgus loosening in 13 patients and a slightly positive varus loosening in eight patients. The 13 menisci that had been sutured healed without problems.

Stability with KT-1000 Arthrometer

All knee joints were measured with the KT-1000 (Table 1). Compared with the contralateral side, femorotibial displacement was smaller or identical in 27 cases, between 1 and 4 mm in 42 cases, larger than 5 mm in 3 cases.

Table 1. Results: stability ($n=72$; follow-up 6–29 months, mean 18 months)

Lachman				KT-1000			Pivot shift			
–	+	++	+++	–/0 mm	1–4 mm	5 mm	–	+	++	+++
33	33	6	0	27	42	3	69	3	–	–

Mobility

The extension was unimpaired in 47 patients, restricted to 5° in 23 patients and went to 10° in two patients. The flexion was free in 31 patients, restricted at the end of the movement in 40 patients and restricted beyond 120° in one patient (Fig. 3).

Subjective Results

Altogether, 58 patients were perfectly satisfied with the result of the operation and claimed to have regained complete activity in sports. Eleven patients complained about occasional discomfort in the knee joint and considered themselves restricted in their sports activities compared with their activity before the trauma. Only three patients regarded the operative result as unsatifactory. Two of these complained of frequent pain in the knee joint, combined, in one patient who had developed Sudeck's disease after the operation, with recurrent effusions. One of the two reported a sensation of instability with attacks of "giving way" that seemed to restrict his everyday life. Another patient suffered from pain in the medial part of the joint as a consequence of a traumatic lesion of the cartilage of the medial femoral condyle (Table 2).

Removal of Metal Implants and Arthroscopy Findings

At an mean interval of 12 months postoperatively the metal implants had already been removed in 52 patients. In 45 of these patients arthroscopy was performed during this second operation. In all cases the reinserted cruciate ligament ap-

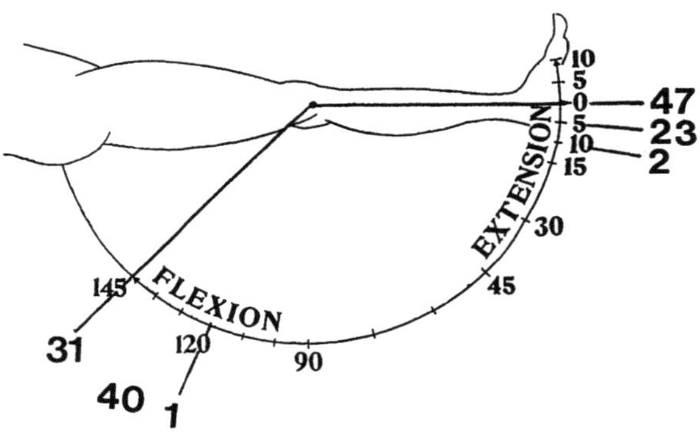

Fig. 3. Results: mobility ($n=72$; follow-up 6–29 months, mean 18 months)

Table 2. Results: patients' subjective assessment ($n=72$; follow-up 6–29 months, mean 18 months)

Result	Function	No. of patients
Very good	No pain, 100% sports activity	58
Satisfactory	Minimal complaints, sports activity restricted	11
Unsatisfactory	Major pains, no sports activity	3

peared intact and well structured and the knee joint was also tight under general anesthesia. In five patients whose knee joint had been clinically stable, arthroscopy disclosed rupture of the allogenic braid; however, an arthroscopy the ACL of these five patients was covered by a synovial sheet and the ligament appeared normally structured as demonstrated by optical control and by palpation with a hooked retractor.

Discussion

Our results suggest that the operative method analyzed in this series facilitates early mobilisation. Although the mean period of observation is relatively short for assessment of this type of reconstruction, it appears that temporary support of the reconstructed ligament by allogenic graft improves the results of reinsertion and secures the complete restoration of concomitant traumatic lesions. Furthermore, other autologous structures within the joint cavity remain unharmed by the operation and without obvious signs of foreign body reaction. Neither the clinical nor the arthroscopical follow-up could detect any disadvantages of this method, such as the "stress shielding" that one might theoretically have expected [5, 7].

References

1. Blauth W, Hassenpflug J (1985) Gedanken zur Kreuzbandrekonstruktion unter besonderer Berücksichtigung des synthetischen Bandersatzes. Unfallchirurg 88:118
2. Cho KO (1975) Reconstruction of the anterior cruciate ligament by semitendinosus tenodesis. J Bone Joint Surg [Am] 57:608
3. Feagin JA (1976) Isolated tear of the anterior cruciate ligament: five year follow-up study. Am J Sports Med 4:95
4. Donoghue DH (1963) A method for replacement of the anterior cruciate ligament of the knee. Report of twenty cases. J Bone Joint Surg [Am] 45:905
5. Kennedy JC (1974) The anatomy and function of the ACL as determined by clinical and morphological studies. J Bone Joint Surg [Am] 56:223
6. Marshall JR (1978) Ligamentous injuries of the knee in skiing. Clin Orthop 108:196
7. Peterson MU (1985) Experimental mechanical and histologic evaluation of the Kennedy ligament augmentation device. Clin Orthop Relat Res 196:185

Autogenic Versus Polylactate-Coated Carbon-Fibre Augmentation in the Anterior Cruciate Reconstruction

J. Witvoet, P. Christel, and F. Gaucher

Orthopaedic Surgery Department, Hôpital Saint-Louis,
40, rue Bichat, 75010 Paris, France

Introduction

Two major types of methods are used in the surgical treatment of chronic anterior knee laxities: the intra-articular and the extra-articular. Extra-articular techniques are designed to avoid the anterior subluxation of tibial plateaus. They eliminate pivot shift, but not entirely anterior laxity, when used alone. Intra-articular techniques reconstruct the anterior cruciate ligament (ACL) and eliminate the anterior laxity.

However, when the secondary restraints of the knee are stretched, intra-articular reconstruction alone is not sufficient to avoid the anterior tibial subluxation. Thus in the case of severe anterior laxity, we and others [4, 9, 14] consider it preferable both to reconstruct the ACL and to combine extra-articular transfers simultaneously. The Marschall/MacIntosh technique [7] has been one of the most widely used for almost 10 years. Over time several modifications have been made in the original technique. The purposes of this article are to describe our own modifications in the initial procedure and to compare the results obtained when using a pure autogenous transfer with those found when augmenting the graft with polylactate-coated carbon fibres (C-PLA, Intergraft, manufactured by Hexel).

Material and Methods

The surgical technique involved has been described previously in detail [12]. A 20–22-cm long transfer is created from the central third of the extensor mechanism. This enters the joint through a tunnel drilled in the anteromedial part of the tibia, replaces the intra-articular part of the ACL and passes over the top in a transverse groove grounded behind the lateral femoral condyle in order to position the graft anteriorly. The transfer passes deep into the fibular collateral ligament and ends on Gerdy's tubercle. In this way the ACL deficiency is treated by a combined intra- and extra-articular transfer. When a drawer sign at 90° flexion in external rotation is still present, the posteromedial part of the capsule is advanced to tighten the medial secondary restraints.

In a first series of patients the prepatellar retinaculum was elevated with a patellar bone block. However, when the patellar tendon was too short, the bone block lay in the intercondylar notch; in four cases it was necessary to elevate the tibial attachment of the transfer, creating a free transfer. With this earlier technique several ruptures of transfer occurred in surgery [10, 11] due to the weakness of junc-

Surgery and Arthroscopy of the Knee
2nd Congress of the European Society
Editors: Müller/Hackenbruch

tions of the prepatellar retinaculum with the patellar and quadriceps tendon [2, 8]. For these reasons this technique was discarded, and in a second series of patients [12, 13] the transfer was created without patellar bone block and reinforced with two strands each containing 10000 polylactate-coated parallel carbon fibres [1] woven deep within the transfer. The transfer was tubed with a non-resorbable suture to avoid spreading of carbon fibres within the joint.

On the tibial attachment the carbon fibres were woven in the graft 1 cm above its origin, preserving the physiological stiffness of the transfer. On Gerdy's tubercle the carbon fibres and the graft were pulled tight and stapled together on the tibia. In addition, the graft was sutured to the intermuscular septum and to the fibular collateral ligament.

The form of postoperative care was identical in the two groups: a posterior splint at 30° flexion was applied during 2–3 weeks, associated with isometric contractions of the quadriceps. Passive motion exercises at 15°–40° between 15 and 40 degrees of flexion began 2 days after surgery. Weight bearing was first allowed at 3 weeks, and all patients attended a rehabilitation centre.

The results were assessed retrospectively using subjective and objective criteria according to the ARPEGE rating system detailed previously [12]. For functional purposes the patients were classified into four categories of activity level (competitive, recreational, active and sedentary) and graded according to their functional stability, pain, extent of athletics and range of motion. Each of these parameters was rated on a scale from 0 to 9, with grade 9 representing normal function. Patients with "excellent" results (a total score of at least 27) were permitted to return to pre-injury athletics. "Good" results (a total score of 24–27) allowed a return to athletics but to a different type of sport. Results were considered "fair" in patients with a drop in activity level (e.g. "competitive" down to "recreational") and a total score of 21–24. All other results were rated "poor". These functional results were compared with the objective results obtained from the physical examination of laxity. Whenever possible, the anterior knee laxity was measured with the MEDmetric arthrometer model KT 1000 [3].

Results

Between January 1980 and July 1985 a total of 104 cases were operated on: 33 with a patellar bone block (group 1) and 71 augmented with C-PLA fibres (group 2). All were reviewed after at least 1 year. The average follow-up period was 3 years for group 1 (ranging up to 5.5 years) and 1 year, 9 months for group 2 (ranging up to 3 years, 2 months). The average age was virtually identical in the two groups: 25 years in group 1, 24.5 in group 2. Sex ratio and activity level were also the same. The time lapse between injury and surgery was 3.5 years in group 1 and 3 years, 4 months in group 2. All patients in group 1 were either competitive or recreational athletes; none were classified as active or sedentary. Only 1 of 71 patients in group 2 was active; the others were either competitive (47) or recreational (23) athletes. No sedentary patient was operated on. Among group 1 72.7% were competitive athletes and among group 2 patients, 66%.

All patients had a combined anteromedial and anterolateral laxity with a positive pivot shift, Lachman test and an anterior drawer at 90° flexion in neutral rotation

between 1+ and 3+. In group 1, 14 patients had previously had a meniscectomy as compared with 23 in group 2. Posteromedial reefing was associated in two group 1 cases and in 6 group 2 cases. The percentage of grade 3 (severe cartilaginous degenerative) lesions with bare subchondral bone was higher in group 2 than in group 1 (35% and 24%, respectively).
After surgery the duration of rehabilitation was identical in both groups, as well as the time preceding a return to training (7 months) and to competition (9 months in group 1, 11 months in group 2).

Functional Results

In group 1 75.5% of patients maintained or improved their activity level, while 24.5% fell by one category; corresponding figures in group 2 were 64.7% and 35.3%. Thus as many patients improved by one category as fell by one category, difference between the two groups was not statistically significant on the basis of a chi-square test (Table 1).
Functional stability was graded at 8 or 9 in 90% of cases in group 1 and in 86% of those in group 2; this also represents no statistically significant difference. As regards residual pain and athletics 84.8% of group 1 patients had a score over 8, compared with 70.4% of group 2 patients (not significant). Residual pain was the main parameter limiting functional performance. The range of motion evaluation showed that 98% of patients had a score of over 8 in each group, representing at least 130° knee flexion.
The overall functional results were rated excellent and good in 97% of cases in group 1 and in 86% of those in group 2 (without statistical difference using the chi-square test).

Objective Results

Residual pivot shift was present in only two patients in each group (6% in group 1, 2% in group 2). The Lachman test was eliminated in 66.6% in group 1 and in 63.3% in group 2 (Fig. 1). Among group 1 patients 45.4% and among group 2 patients 60% showed no drawer sign at 90° flexion (Fig. 2); this difference was not statistically significant. Measurements of anterior laxity at 30° flexion using the KT-1000 arthrometer are listed in Table 2. There were no statistically significant differences either between normal and operated knees nor between groups 1 and 2. X-ray evaluation showed no difference between groups regarding frequency of medial compartment remodeling or occurrence of osteoarthritis (Table 3).

Table 1. Pre- and postoperative activity levels

Activity level	Group 1	Group 2
Competitive	24/24	47/31
Recreational	9/9	23/32
Active	–	1/7
Sedentary	–	0/1

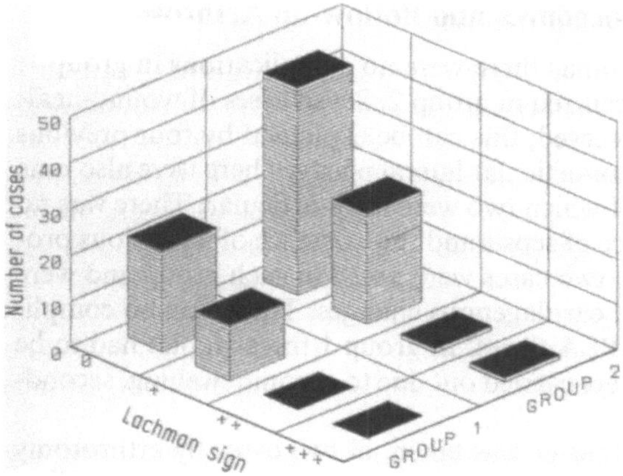

Fig. 1. Postoperative Lachman test (anterior drawer sign tested at 25° flexion)

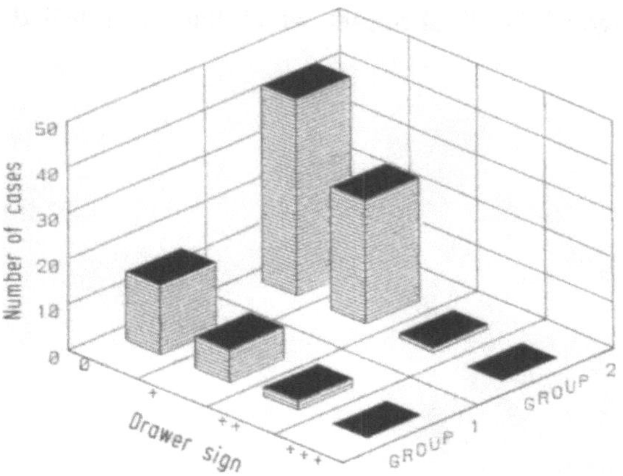

Fig. 2. Postoperative anterior drawer sign tested at 90° flexion in neutral rotation

Table 2. Post-reconstruction displacement for operated and normal knees by type of transfer. Displacement (mm) with a pull force of 89 N at 30° flexion (mean ± SEM)

	Type of transfer	
	Non-reinforced	C-PLA reinforced
Operated knees	4.7 ± 2.3	5.7 ± 2.9
Normal knees	5.1 ± 2.3	3.9 ± 2.1
	($n = 11$)	($n = 42$)

Table 3. Pre- and postoperative X-ray evaluation according to patient group. In groups 1 there was an average 3-year follow-up; in group 2, 1.75-year follow-up. In group 1, 54% of patients had a normal X-ray at the last follow-up; in group 2, 60%

	Group 1	Group 2
Normal	28/18	61/42
Medial remodelling	5/12	8/27
Arthritic changes	0/1	1/2

Complications, Subsequent Procedures and Follow-up Arthroscopy

With the exception of two haematomas there were no complications in group 1. However, several complications occurred in group 2. Seven cases of wound-healing delay or skin necrosis were observed; this can be explained by four previous lateral approaches made for an extra-articular lateral plasty. There were also nine haematomas and five infections, of which two were intra-articular. There was no relationship between the occurrence of sepsis and the existence of a previous procedure. Chronic swelling was rare; two cases were noted in each group and were related to the existence of arthritic cartilagenous changes. There was no complication related to the use of the C-PLA fibres. In group 1 three staples had to be removed, and two arthroscopies were carried out due to chronic swelling, secondary to chondritis.

In group 2 a second view of the transfer was taken, in two cases by arthrotomy and in eight by arthroscopy. One arthrotomy was conducted during a patellectomy and one for a rupture of the patellar tendon at its patellar attachment. The arthroscopies were performed while retrieving the staple on Gerdy's tubercle. An asymptomatic case of rupture was discovered during an arthroscopy; the transfer was found to be broken on the edge of the tibial tunnel, which had been drilled

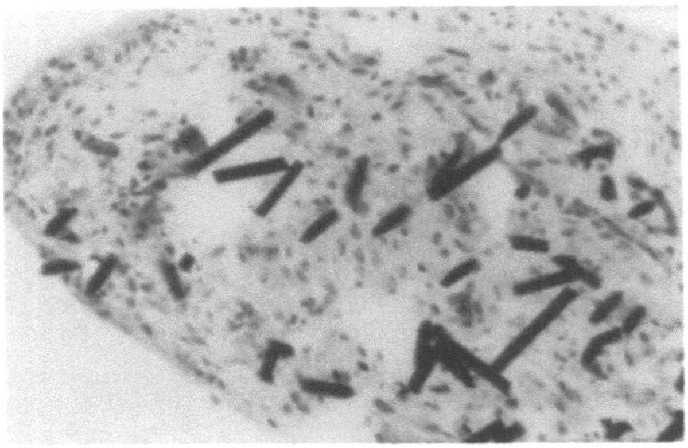

Fig. 3. Biopsy specimen taken from the synovium 1 year after surgery. Macrophages and multinucleated giant cells are in direct contact with the carbon fibres (H&E, ×25)

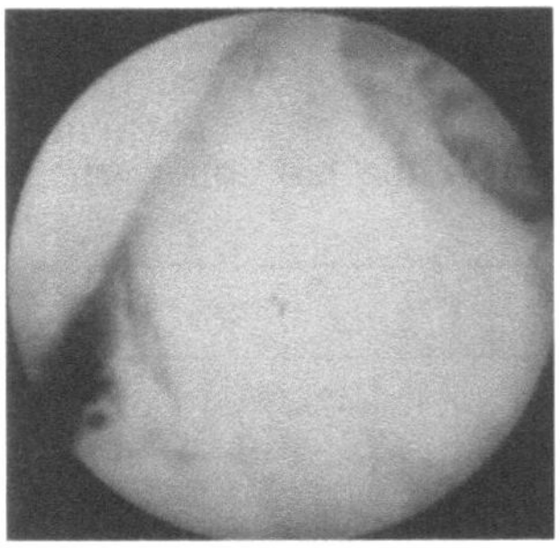

Fig. 4. Arthroscopic view of a graft reinforced with C-PLA fibres 5 months after surgery (right knee). The graft appears to be intact and covered with a thin synovium sheath and tiny capillaries

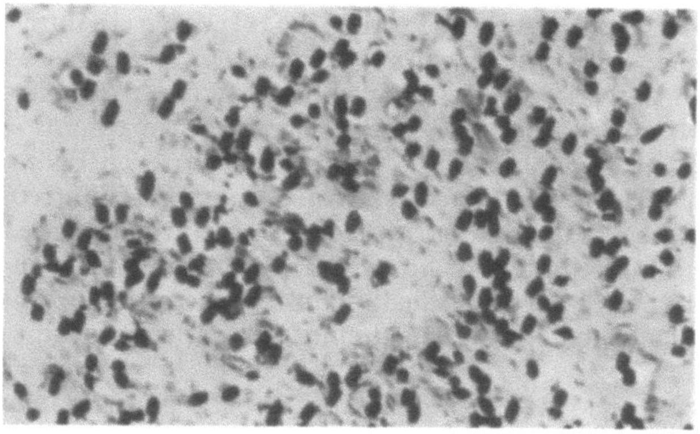

Fig. 5. Biopsy specimen taken from the extra-articular part of a C-PLA augmented graft 1 year after surgery. Collagen fibres have grown between the carbon fibres with few fibrocytes. As in Fig. 3 macrophages and giant cells are in contact with the carbon fibres (H&E, ×25)

too anteriorly and was consequently not isometric. The patient had no pivot shift, as the extra-articular part of the transfer was functioning well. Carbon fibres were found either in the synovial fluid alone (in two cases) and/or in the synovium itself (two cases) (Fig. 3). The articular cartilage was not macroscopically stained. Most of the reconstructed ligaments had moderate tension, with more tension in extension than in flexion. Capillaries were always visible on the graft surface (Fig. 4). Eight biopsies of the intra-articular part of the graft were carried out in the augmented group (group 2) and two in the non-augmented (group 1). In all these cases the transfer was found to be viable and vascularised but with a disorganised structure. Carbon fibres induced a mild foreign-body reaction, with some macrophages in direct contact and very few giant cells. Collagen fibres were always found between the carbon fibres (Fig. 5).

Discussion

No statistically significant differences were found between the two groups regarding overall subjective and objective results. Pain occurred more frequently in group 2, influencing the patients' subjective appreciation. This can be explained by a higher rate of grade 3 cartilagineous lesions in group 2 than in group 1 – 35% as compared with 24%.

The modified Marshall/MacIntosh technique presented here appears very efficient, as it led to 90.3% excellent or good results on the overall series of 104 patients after the follow-up of 1–5.5 years. Of the athletes 92.2% resumed sports after surgery and 75% returned to their preinjury athletics. The combined intra- and extra-articular technique eliminates the pivot shift in virtually all cases and anterior laxity in 65% of cases.

When compared with other reported findings these results appear quite satisfactory. The isolated reconstruction of the ACL using quadriceps tendon, iliotibial band or other tendons leads to a recurrence of the pivot shift in between 6.4%

and 45% of cases, depending upon the technique used [5, 6, 14]. When the intra-articular reconstruction is augmented with an extra-articular repair, the percentage of residual pivot shift drops. In a 2-year follow-up series of 43 cases, combining an intra-articular reconstruction with the patellar tendon and an extra-articular Ellison transfer, Fried et al. [4] reported a 15% pivot-shift recurrence. In another 2-year follow-up study [9] involving 143 cases, treated by a non-reinforced MacIntosh technique, the surgeon, Imbert, noted a 16% pivot-shift recurrence. Using the Kennedy ligament augmentation device (LAD) Roth (personal communication) observed a lower incidence of residual pivot shift and anterior laxity as compared with non-augmented grafts. This finding was not confirmed in the present report.

One must be careful regarding the validity of comparisons between these two groups: this was a retrospective study, the group 1 series being historically older than the group 2 series, with a longer follow-up and fewer patients.

Summary

This paper has compared the results of a combined intra- and extra-articular technique using the central third of the extensor mechanism to stabilise anterior knee instabilities, with or without polylactate-coated fibre augmentation. The graft is passed up through a tibial drill hole, across the intercondylar notch, and over the top of the lateral femoral condyle in a transverse groove to position the graft anteriorly and deep to the fibular collateral ligament, down to Gerdy's tubercle where it ends. Posteromedial reefing is associated when necessary. In a first series of 33 patients (group 1) the prepatellar retinaculum was freed with a patellar bone block. In a second series of 71 patients (group 2) the prepatellar retinaculum was elevated without a piece of bone and the graft augmented with two strands of 10000 polylactate-coated carbon fibres. All 104 patients were evaluated using subjective and objective criteria according to the ARPEGE rating system at 1–5.5 years after surgery. The overall functional results were rated as excellent or good for 97% in group 1 and for 86% in group 2. The pivot shift was eliminated in 94% in group 1 and in 98% in group 2. The Lachman test was reduced to 0 or 1+ in 100% of group 1 cases and in 97% of group 2 cases. There was no statistically significant difference between the groups. Functional stability of the knee was achieved efficiently in both groups. Residual pain was the main factor limiting functional performance.

Conclusion

With the operation described in this article we were able to stabilise the knee efficiently in 95% of cases. The use of polylactate-coated carbon fibres did not significantly change the overall functional and objective results when compared with the non-augmented technique. However the C-PLA fibres made the surgical technique itself easier and reduced the number of operative problems. There was no complication related to the use of C-PLA fibres as an augmentation material. Cartilagineous lesions appeared to be the most limiting factor in functional performance.

References

1. Alexander H, Weiss AB, Parsons JR (1983) Absorbable polymer filamentous carbon composites: a new class of tissue scaffolding materials. In: Burri C, Claes L (eds) Aktuelle Probleme in Chirurgie und Orthopadie. Huber, Bern, p 78
2. Butler DL, Grood ES, Noyes FR, Sodd AN (1985) On the interpretation of our cruciate ligament data. Clin Orthop 196:26–34
3. Daniel DM, Stone ML, Sachs R, Malcolm L (1985) Instruments measurement of anterior knee laxity in patients with acute anterior cruciate ligament disruption. Am J Sports Med 13:401–407
4. Fried JA, Bergfeld JA, Weiker GG, Andrish JT (1983) Two to five year evaluation of anterior cruciate reconstruction using patellar tendon and Ellison iliotibial band transfer. Read at the 50th Annual Meeting of the AAOS. 10–15 March, Anaheim
5. Insall J, Joseph DM, Aglietti D, Campbell RD (1981) Bone block iliotibial band transfer for anterior cruciate insufficiency. J Bone Joint Surg [Am] 63:560–569
6. Kornblatt IB, Warren RF (1983) Longterm follow-up of anterior cruciate ligament reconstruction using quadriceps tendon substitution for chronic anterior cruciate insufficiency. Read at the 50th Annual Meeting of AAOS, 10–15 March, Anaheim
7. Marschall JL, Warren RF, Wickiewicz TL, Reider B (1979) The anterior cruciate ligament. A technique of repair and reconstruction. Clin Orthop 143:97–106
8. Trillat A, Ficat P (1972) Symposium sur les laxités post traumatiques du genou. Rev Chir Orthop 52(1):31–116
9. Varaud MM (1983) Transplantation palliative du ligament croisé antérieur par procédé mixte intra et extra articulaire. Thèse Médicine, Université Saint-Etienne
10. Witvoet J, Faustin JM, Christel P (1984) Traitement des laxités chroniques du genou par plasties mixtes intra et extra-articulaires. Rev Chir Orthop 70(2):151–158
11. Witvoet J, Faustin JM, Christel P (1985) The preliminary results of treatment of chronic anterior cruciate laxity intra- and extra-articular plasty: a review of 53 cases. Clin Orthop 196:143–153
12. Witvoet J, Christel P (1985) Treatment of chronic anterior knee instabilities with combined intra- and extra-articular transfer augmented with C-PLA fibers. Clin Orthop 196:143–153
13. Witvoet J, Christel P (1985) Renforcement des plasties du ligament croisé antérieur par des fibres de carbone. A propos de 44 cas. Rev Chir Orthop 71(2):121–125
14. Zarins B, Rowe CR (1986) Combined anterior cruciate ligament reconstruction using semitendinous tendon and iliotibial track. J Bone Joint Surg [Am] 68:160–177

ACL Reconstruction Procedure with Autogenous Graft Under Arthroscopic Control

J.-C. Imbert

Clinique radiologique et orthopedique, 60, rue Robespierre,
42100 Saint-Etienne, France

An anterior cruciate ligament (ACL) reconstruction procedure under arthroscopic control is presented; it allows surgery without opening the joint. This procedure is based on a modified Macintosh.

Material and Methods

From June 1983 to July 1985, 213 ACL reconstructions were performed on patients mostly involved in strenuous sports. A patellar pedicled autogenous transplant is successively driven inside and outside the joint, through three bony tunnels. A skin incision is necessary for taking the graft and for the peripheral part of the procedure, but the intra-articular steps, including drilling of the tibial tunnel and positioning of the graft inside the joint, are performed under arthroscopic contol without any capsular incision. Rehabilitation includes early mobilisation in the second postoperative week. Return to sport activities is allowed after 6 months.

Results

Anatomical and functional results are reported. Anatomical data show a recurrence of an anterolateral pivot shift in 20 patients (9.6%), a persistence of a positive Lachman test in 94 patients (but the test is no more than 1+ in 79 cases). Functionally, the patients were referred to the "Arpege" evaluation system: preoperatively 35 patients were classified in high level sport activities and 53 recreational. After 6 months postoperatively 104 returned to strenuous sports and 92 to recreational.

Discussion

Although the average follow-up is shorter, anatomical results should be compared and appear to be very close to those of the first part of our experience using the same procedure with open arthrotomy in 368 patients from April 1978 to June 1983. The early functional results seem to be better, the benefit of not opening the joint appears clearly during the rehabilitation period, decreasing the number of minor complications such as swelling and haemarthosis and allowing earlier mobilisation of the joint. Technically speaking, it improves the accuracy of the intra-articular steps of the operation, allowing a better placement of the tibial tunnel and the utilisation of the remaining ACL synovial sheath.

Surgery and Arthroscopy of the Knee
2nd Congress of the European Society
Editors: Müller/Hackenbruch

Summary

A consecutive series of 213 ACL reconstructions using a combined intra-extra-joint modified Macintosh procedure were performed over a 2-year period, using arthroscopic control for the intra-articular steps of the operation. With a 6 months to 2 years follow-up, anatomical results are not consistently different from a previous series with open arthrotomy, but the improvement of short-term functional results allowed by early rehabilitation is obvious and encourages us to carry on with this procedure.

Treatment for Anterior Cruciate Ligaments Insufficiency and Combined Meniscus Injuries in Athletes

N. Kanazawa, A. Saito, T. Itagaki, Y. Tomoyasu, and K. Sato

Department of Orthopaedic Surgery, Nihon University Surugadai Hospital, Tokyo, Japan

Introduction

Anterior cruciate ligament (ACL) insufficiency of the knee is a great problem for athletes as it influences their lives, limiting or preventing participation in sport. We selected 60 patients with ACL insufficiency and performed ACL reconstruction. The cases with more than a 1-year follow-up were selected for evaluation of the pathological anatomy, treatment and prognosis of the injury, particularly combined meniscus injury. The selected subjects were athletes who belonged to sports clubs and who keenly desired to return to sports activities after surgery. The average age at time of injury in these patients was 17 years, with a range of 12–34 years: the average age at time of surgery was 22 years with a range of 16–34 years.

The cause of injury was most frequently one of three sports, basketball, American football, or volleyball. The most frequent mechanism of injury was valgus-rotation stress from the postero-external direction.

Of the 60 ACL insufficiency cases, 45 (75%) had combined meniscus injuries, which were classified into two groups: isolated injury of the medial meniscus (20 cases) and injury of the lateral meniscus (8 cases). Combined bilateral meniscus injury occurred in 17 (37%) cases.

The average times from injury to ACL reconstruction for medial, lateral and bilateral meniscus injuries were 4 years and 8 months, 4 years and 1 month, and 7 years and 6 months, respectively. This suggests an increased tendency to bilateral involvement the longer the elapsed time between injury and ACL reconstruction. On the other hand, the average time from injury to surgery was only 1 year and 2 months in cases without combined meniscus injury.

Bucket-handle tears (19 cases) and flap tears (12 cases) account for 84% of the medial meniscus injuries and for 72% of the lateral meniscus injuries.

Material and Methods

Surgical procedures of ACL reconstruction were evaluated. Extra-articular reconstruction was performed in 7 cases, Moriya's intra- and extra-articular reconstruction in 7 cases, and our modified Moriya's intra- and extra-articular reconstruction in 46 cases. In the treatment of combined meniscus injury ACL reconstruction is, as a rule, combined with arthrotomy, and partial meniscectomy is appropriately combined with posterior horn excision under arthroscopic observation.

Surgery and Arthroscopy of the Knee
2nd Congress of the European Society
Editors: Müller/Hackenbruch

The six menisci with small transverse tears (three cases) or horizontal tears (three cases) were left untreated, and ACL reconstruction alone was performed; however, no symptoms attributable to the meniscus injuries have appeared.
Our modified intra- and extra-articular reconstruction is performed as follows. A strip-shaped flap, 21 cm long by 3–3.5 cm wide, is dissected from the iliotibial tract on the proximal side and passed directly through the bone tunnel by an "over the top" route from the lateral femoral condyle. It is then inserted, from the postero-external side, through the bone tunnel in the medial side of the anterior tibial tubercle and sutured to the patellar tendon.

Results

The results were evaluated according to Nozawa's sliding ratio. According to this method of measurement, the sliding ratio of the normal knee joint was in the range between 20 and 80, the ratio of ACL tear was less than 20 and that of posterior cruciate ligament (PCL) tear was more than 90. The average sliding ratio of the 60 cases was 14 preoperatively, which improved to 28 a year after reconstruction surgery.
The three methods were evaluated by comparing their average sliding ratios 1 year after surgery. The outcome of intra- and extra-articular reconstruction was favourable; subjective evaluation indicated no giving-way episodes and negative pivot shift tests were recorded on objective examination. The two groups of patients have returned to sporting activities. The average sliding ratio of our modified method was better than that of the original.
To study the implication of medial instability, we examined 39 cases with a follow-up of more than a 1 year after reconstruction. In cases of ACL insufficiency, preoperative grade II or III medial instability was improved immediately after ACL reconstruction. However, a mild anterior drawer sign tended to appear about a year or so after the operation. These findings suggest that in cases of grade II ACL insufficiency and more medial instability, or in those necessitating total meniscectomy, ACL reconstruction needs to be combined with simultaneously performed medial reinforcement.
The results of the 2-year follow-up showed that 17 (28%) patients returned to their original sports. In these cases ACL reconstruction was conducted shortly after injury, there was no combined bilateral meniscus injury, medial instability was mostly grade II or less and our procedure of ACL reconstruction was employed every time. Of the whole study 54 (90%) patients returned to sport.
One case report presents 22-year-old female who began gymnastics in earnest during her elementary school days. She sustained an injury in the form of a violent valgus-rotation stress in one leg when she fell upon jumping down from a balance beam. Seven years after the injury she came to our hospital for treatment. Arthroscopic observation revealed that the ACL was non-functional, appearing only as a scar, and that a bilateral meniscus injury existed. A modified intra- and extra-articular reconstruction was performed with simultaneous partial excision of a flap tear of the medial meniscus and a double vertical tear of the lateral meniscus. Within 2 years after the operation the sliding ratio had improved from 9 to 34. Although arthroscopic observation revealed hypertrophy of the reconstructed li-

gament, adequate tension is kept up by the newly formed blood vessels. She has discontinued her original sport but now plays tennis and skis actively.

Conclusion

A history of haemoarthrosis, giving- way episodes, the anterior drawer sign, the Lachmann sign and the pivot shift sign were all useful indicators in diagnosing ACL insufficiency. Results were further confirmed by arthroscopy. We have observed that articular injury tends to be advanced and combined with bilateral meniscus injury when time elapses between injury and treatment, emphasising the importance of early repair and reconstruction. Of the surgical procedures evaluated, our method of intra- and extra-articular reconstruction produced the most favourable results.

Late Results After Primary Repair of Anterior Cruciate Ligament Tears

W. Kipfer [1], P. Ballmer [1], B. Grünig [1], H. U. Stäubli [1], R. Zehnder [2], and R. P. Jakob [1]

[1] Klinik für Orthopädische Chirurgie, Universität Bern, Inselspital, 3010 Bern, Switzerland
[2] AO Dokumentation, 3000 Bern, Switzerland

"What can be expected of primary repair of acute anterior cruciate ligament tears?" This question has been the subject of a number of clinical follow-up studies. The conclusions of these reports differ and thus the efficacy of primary repair of the anterior cruciate ligament (ACL) remains controversial [9, 10, 11, 13, 17, 18, 21, 22]. On the other hand, the secondary consequences of ACL insufficiency have been clearly demonstrated in several clinical and experimental studies [2, 10, 11, 15, 18, 20]. Most authors therefore advocate early operative treatment of ACL rupture, in an attempt to avoid these sequelae.
Since the results in the literature often differ, we wanted to critically evaluate our own results. The purposes of our study were to compare stability after suture of the ACL with and without augmentation and to analyse the evolution of osteoarthritic changes.

Material and Method

Between 1972 and 1982 a total of 241 primary repairs of acute ACL injuries were performed at our hospital. Out of these, 153 patients with 156 ACL ruptures were available for examination. The average follow-up time was 5.5 years, the interval ranging from 3 to 12 years.
All 156 knees followed up fulfilled the following three criteria:

1. A total rupture of the ACL was confirmed at surgery.
2. The operative treatment included primary suture of the ACL with or without augmentation.
3. No secondary reconstruction had been necessary.

The average age of the 153 patients (58 women and 95 men) at the time of injury was 32 years. 65% were sports injuries and 35% traffic accidents or work-related injuries. The right knee was involved in 66 cases and in 3 patients both knees were injured. An isolated rupture of the ACL was found in 18%, in 60% it was combined with a tear of the medial collateral ligament, and a concomitant lesion of the posterior cruciate ligament was present in 13% (Table 1). The tear of the ACL occurred in 70% in the area of the proximal third, and in 14% a tear was found in the middle of the ligament. In 5% the ACL ruptured with an osseous fragment (Table 1).
In 77% the reinsertion of the ACL was performed either through a transcondylar femoral or tibial drill hole or by the use of an intraligamentous suture. In 6% the

Surgery and Arthroscopy of the Knee
2nd Congress of the European Society
Editors: Müller/Hackenbruch

Table 1. Pathology as found at operation (%)

A Pattern of associated injuries	
ACL tear	100
Isolated	18
Associated with tear of	
Medial collateral ligament	60
Lateral collateral ligament	4
Medial and lateral ligaments	5
Posterior cruciate ligament	13
Medial meniscus	14
Lateral meniscus	8
Medial and lateral meniscus	3
B Location of the ACL tear	
Proximal	69
Middle	14
Distal	4
Proximal and distal	8
With an osseous fragment	5

stump of the ACL was reinserted proximally "over the top". In 36 cases (23%) the sutured ligament was augmented by a distally based, 20-mm-wide fascial strip of the iliotibial band in a modification of the MacIntosh procedure [14, 19]. The strip was passed underneath the lateral collateral ligament, sutured to it with the tibia in external rotation, and then passed through a drill hole or over the top of the femoral condyle into the joint and then sutured to the anteromedial part of the tibial plateau. Associated lesions of the medial and lateral capsulo-ligamentous structures or of the PCL were identified and a primary repair was performed.

Acute meniscal tears were found in 25% of patients. 40% of the torn menisci could be repaired, 16% required partial menisectomy and 44% were totally resected.

Postoperatively the knees were immobilized for an average of 7 weeks. The majority (86%) were placed in a plaster of paris cast in 30° of flexion, while in 14% a motion brace was applied. The immobilization phase was followed by a course of physiotherapy.

Complications were seen in two cases: a severe haematoma, which had to be evacuated, and a deep infection of staphylococcus epidermidis, which healed with no further sequelae. Seven knee joints had to be manipulated under anaesthesia.

Results

The follow-up evaluations were conducted according to a computerized score programme. These included subjective patient ratings and objective criteria. The latter were based on physical examinations with functional tests, radiographic examination and Cybex measurement of thigh-muscle strength. All clinical tests were performed by at least two examiners.

Table 2. Evaluation of results by patients (%)

Excellent	18
Good	71
Moderate	10
Poor	1

Table 3. Complaints by patients (%)

	None	Mild occasional	Severe constant
Pain	40	47	13
Swelling	76	24	–
Giving way	72	28	–

Table 4. Reduction in range of motion (%)

Extension	None	58
	Less than 10°	40
	More than 10°	2
Flexion	None	49
	Less than 20°	48
	More than 20°	3

Table 5. Performance in functional tests (%)

	Without discomfort	With discomfort
Hopping	80	20
Squatting	55	45
Kneeling	40	60

Subjectively 89% of the patients rated the result as good or excellent. For 10% the result was fair and for two patients it was poor (Table 2). About 40% did not suffer pain, 50% had some discomfort and 13% suffered from functionally disturbing pain. Recurrent swelling occurred in 25% of the patients and almost 30% had episodes of giving way (Table 3). 84% had returned to their sports activities. Out of these, 57% participated in the same sports and were able to achieve the same level of performance as before injury; 4% had to discontinue sports activities because of persistent discomfort (pain and/or instability) in the involved knee.

Many of the operated knees had a reduced range of motion in comparison to the non-operated knee. In 20% a lack of extension of 5°–10° was found. Flexion was decreased by 5°–20° in less than 50% of patients (Table 4). Some patients therefore had difficulties in carrying out the functional tests, especially kneeling and squatting (Table 5).

Table 6. Stability of knees (%)

	0	+	++
Anterior drawer			
In 15°–20° extension (Lachman test)	20	70	10
In 70°–90° flexion and neutral rotation	44	45	11
Posterior drawer			
In 15°–20° extension	79	20	1
In 70°–90° flexion and neutral rotation	85	14	1

		Without giving way	With giving way
Pivot shift and giving way			
Pivot shift	Negative	49	12
	Positive	23	16

In 52% retropatellar crepitation was present, but it was symptomatic in only 18%. A significant correlation between the patellofemoral complaints and the radiological findings could not be established.

The anterior drawer in 15°–20° flexion (Lachman test) was negative in 20% and in 70% the translation of the tibial plateau in the sagittal plane was 1+ and in 10% 2+. By contrast, in flexion of 70° we found a negative drawer in 44%, and in 45% it was 1+. So the stability in the tested knees is less in extension than in flexion (Table 6). In 20% a mild posterior drawer was present in extension (Table 6). A soft or altered endpoint was found in 31% of all knees.

We think the two most important criteria in evaluating knee instability are episodes of giving way and a positive pivot-shift test. In 66% of the patients the pivot-shift test was negative. 10% had a definitely positive pivot shift, i.e. the test was also positive in neutral position and in slight external rotation, according to the grading criteria proposed by Jakob et al. [16] (Table 6). We observed a positive reversed pivot-shift test in 47% of the injured knees, but this test, although more subtle, was also positive in 32% of the non-injured knees. Lateral or medial instability in extension was not seen. However, in flexion we observed in 58% a slight medial and in 25% a lateral instability.

Roentgenographic examinations showed bony alterations corresponding to osteoarthritic changes in a few cases. In none of the knees was significant narrowing of the joint space seen. According to Ahlbäck [1], narrowing by 50% of the normal width of the joint space becomes significant. An increased subchondral sclerosis occurred in the femorotibial area in 6.3% and in the patella in 5%. Subchondral cysts as a typical sign of advanced osteoarthrosis were not observed in any of the cases. Eccentric depression of the tibial plateau as described by Gerber [11], a distinct radiographic sign of chronic instability, was present in 2.8%. However, 25% of the patients had osteophytes at the tibial plateau and femoral condyle. The intercondylar notch was narrowed in 18% and 14% were found to have broadening or peak-shaped alterations of the tibial eminences (Figs. 1, 2).

On the basis of work by Arvidsson et al. [3] we also tested the strength of the quadriceps and of the hamstring muscles by Cybex dynamometer. The maximum

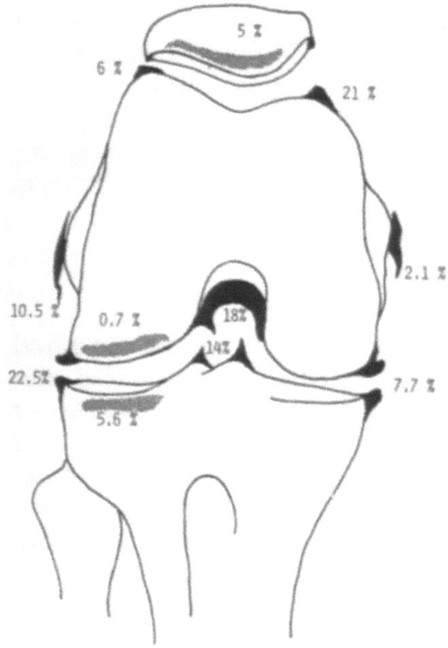

Fig. 1. Roentgenographic findings (antero-posterior view)

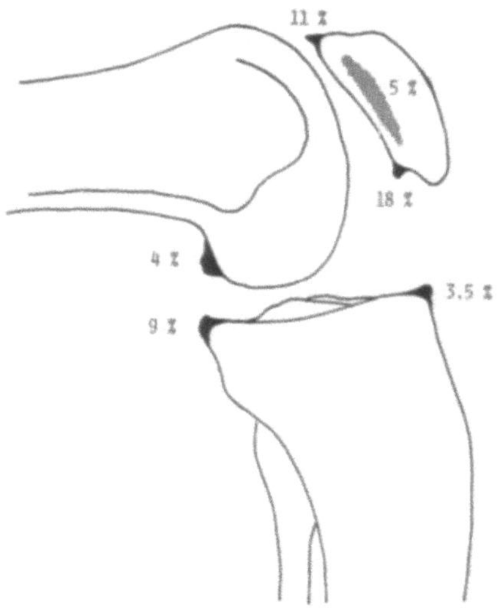

Fig. 2. Roentgenographic findings (lateral view)

peak torque in flexion and extension, the torque acceleration energy and the total work at an angular velocity of 60°/s and 240°/s were measured. From these data the quadriceps and hamstring index for the operated and non-operated knee were calculated. In 40% of the 114 patients tested the difference between the involved and the non-involved side was less than 10% for all Cybex data.
14% of all patients had some complaints when walking on stairs. These patients showed a deficit of strength in both the quadriceps and hamstrings when mea-

sured with slow angular velocity (60°/s). Patellofemoral symptoms, pain and crepitation, were compared to the peak torque data. We found a decrease in strength of the quadriceps up to 17% when crepitation and pain were present. Comparisons between strength of the hamstrings and the Lachman test confirmed the observation of Arvidsson et al. [3]. The strength of the hamstrings increases with increasing insufficiency of the ACL. However, no relation between the patient's own evaluation and the analysed Cybex data was found.

Discussion

An important finding of this study was the difference noted when we compared the two techniques of ACL repair, suture of the ACL with or without fascial reinforcement. In 50% of the cases that were treated with ligament suture alone, we found a positive pivot shift, whereas in all patients on whom an augmentation was performed the pivot-shift test proved to be negative (Table 7). This observation leads us to recommend that a primary ACL suture be combined with an augmentation procedure. This point was also stressed before by other authors [4]. In our cases the reinforcement was performed with a strip of the iliotibial band, as described above. However, this technique might lead to a weakening of the posterolateral corner, since in these patients a reversed pivot shift sign was observed more frequently than in the group with ACL repair alone (Table 7).

Other structures, e.g. the semitendinous or gracilis tendon, can be used as well for an augmentation plasty. We now recommend, based on our own experience with ACL substitution in chronically deficient knees, utilization of the middle third of patellar tendon as a free graft. This graft's qualities have proved to be almost ideal, as shown by the biomechanical studies of Butler et al. [5] and the histomicroangiographic analysis of Clancy et al. [7], Drobny et al. [8] and Shino et al. [24].

Regardless of the technique or the graft chosen, the anatomy of the ACL requires an isometric intra-articular insertion of the transplant to the femoral condyle and the tibial plateau. Odensten and Gillquist [23] have reported that anisometric reinsertion of the ACL or a graft limits the range of motion. We feel that this observation explains at least in part the loss of flexion found in some patients of our study group. Repeatedly we noted on postoperative radiographs that drill tunnels were placed too far anteriorly.

Table 7. Comparison of primary repair with and without augmentation (%)

Stability tests	Without augmentation		With augmentation	
	Negative	Positive	Negative	Positive
In 15°–20° extension	21	79	19	81
In 70°–90° flexion	39	61	61	39
Pivot shift	50	50	100	–
Reversed pivot shift				
Involved knee	62	38	22	78
Non-involved knee	(74)	(26)	(64)	(36)

Conclusion

On the basis of these results, the initial question "What can be expected of primary suture of the acute ruptured ACL?" can be answered as follows:
In the long term, early operative management of the acute ACL rupture contributes to the maintenance of secondary ligamentous restraints [6] and therefore also to the maintenance of the menisci. In this way we believe the development of degenerative changes is retarded. Our data suggest, in agreement with the opinion of the ACL study group [4], that in most cases, especially in young, active patients, a primary reconstruction and augmentation with a locally obtained autogenous transplant is to be recommended.

References

1. Ahlbäck S (1968) Osteoarthrosis of the knee. A radiographic investigation. Acta Radiologica [Suppl 277]
2. Arnold JA, Cocker TP, Heaton LM, Park JP, Harris WD (1979) Natural history of anterior cruciate tears. Am J Sports Med 7:305–313
3. Arvidsson I, Eriksson E, Häggmark T, Johnson RJ (1981) Isokinetic thigh muscle strength after ligament reconstruction in the knee joint: results from a 5–10 year follow up after reconstruction of the anterior cruciate ligament in the knee joint. Int J Sports Med 2:7–12
4. Bilko TE, Paulos LE, Feagin JA, Lambert KL, Cunningham HR (1986) Current trends in repair and rehabilitation of complete (acute) anterior cruciate ligament injuries. Analysis of 1984 questionnaire completed by the ACL Study Group. Am J Sports Med 14:143–147
5. Butler DL, Noyes FR, Grood ES, Miller EH, Malek M (1979) Mechanical properties of transplants for the anterior cruciate ligament. Orthop Trans 3:180–181
6. Butler DL, Noyes FR, Grood ES (1980) Ligamentous restraints to anterior-posterior drawer in the human knee. J Bone Joint Surg 62-A:359–270
7. Clancy WG, Narechania RG, Rosenberg TD, Gmeiner JG, Wisnefske DD, Lange TA (1981) Anterior and posterior cruciate ligament reconstruction in rhesus monkeys. A histologic, microangiographic and biochemical analysis. J Bone Joint Surg 63-A:1270–1284
8. Drobny KT, Müller W, Wentzensen A, Perren SM (1984) Das Hoffagestielte Patellarsehnentransplantat beim vorderen Kreuzbandersatz. First European Congress of Knee Surgery and Arthroscopy, Berlin, 1984
9. Feagin JA, Curl WW (1976) Isolated tear of the anterior cruciate ligament: 5-year follow-up study. Am J Sports Med 4:95–100
10. Fetto JF, Marshall JL (1980) The natural history and diagnosis of anterior cruciate ligament insufficiency. Clin Orthop 147:29–38
11. Gerber C (1983) Die Gelenkinstabilität als Grundlage chronischer Schäden am Bewegungsapparat. Schweiz Z Sportmed 31:37–44
12. Giove TP, Miller SJ, Kent BE, Sanford TL, Garrick JG (1983) Non-operative treatment of the torn anterior cruciate ligament. J Bone Joint Surg 65-A:184–191
13. Hughston JC, Barrett GR (1983) Acute anteromedial rotatory instability. Long-term results of surgical repair. J Bone Joint Surg 65-A:145–153
14. Ireland J, Trickey EL (1980) MacIntosh tenodesis for anterolateral instability of the knee. J Bone Joint Surg 62-B:340–345
15. Jacobsen K (1977) Osteoarthrosis following insufficiency of the cruciate ligaments in man. A clinical study. Acta Orthop Scand 48:520–526
16. Jakob RP, Stäubli HU, Deland J (1987) Grading the pivot shift. Objective tests with implications for treatment. J Bone Joint Surg 69-B:294–299

17. Jokl P, Kaplan N, Stoven P, Keggi K (1984) Non-operative treatment of severe injuries to the medial and anterior cruciate ligaments of the knee. J Bone Joint Surg 66-A:741–744
18. McDaniel WJ, Dameron TB (1980) Untreated ruptures of the anterior cruciate ligament. A follow-up study. J Bone Joint Surg 62-A:696–705
19. MacIntosh DL, Darby TA (1976) Lateral substitution reconstruction. J Bone Joint Surg 58-B:142
20. Marshall JL, Olsson SE (1971) Instability of the knee. A long-term experimental study in dogs. J Bone Joint Surg 53-A:1561–1570
21. Marshall JL, Warren RF, Wickiewicz TL (1982) Primary surgical treatment of anterior cruciate ligament lesions. Am J Sports Med 10:103–107
22. Odensten M, Lysholm J, Gillquist J (1984) Suture of fresh ruptures of the anterior cruciate ligament. A 5-year follow-up. Acta Orthop Scand 55:270–272
23. Odensten M, Gillquist J (1985) Functional anatomy of the anterior cruciate ligament and a rationale for reconstruction. J Bone Joint Surg 67-A:257–262
24. Shino K, Kawasaki T, Hirose H, Gotoh I, Inoue M, Ono K (1984) Replacement of the anterior cruciate ligament by an allogeneic tendon graft. An experimental study in the dog. J Bone Joint Surg 66-B:672–681

Five to Ten-Year Follow-up After Primary Suture of Anterior Cruciate Ligament Tears

T. Strand and A. Mølster

Department of Orthopaedics and Traumatology,
Haukeland Hospital, Bergen, Norway

Reasonably good results have been reported in short- and medium-term follow-ups after primary suture of ACL tears. Deteriorating knee function with recurrence of instability, however, has been reported after 5 years in a few longer-term studies.

Material and Methods

Sixty-eight patients aged 13–70 years at injury were evaluated 5–10 years after primary suture of ACL tears by the Palmer technique without augmentation. All but four patients sustained combined ligament injuries.

Results

Objectively 61% of the knees were stable by manual and instrument testing (using the KT-1000 arthrometer), whereas 21% were failures. The remaining 18% had a minor instability, with a trace pivot shift and I-N anterior laxity difference of 2–4 mm at 25 dgr/20 lb. In 80% of instrumentally tested patients the I-N laxity difference was ≦3.0 mm. Evaluation by computerized tomography indicates that most patients with minor instability had partially intact, but elongated ACLs. Only three patients had received second ACL reconstructions, and 60 patients still had both menisci intact at follow-up.
The functional results were graded as excellent to good by the Lysholm knee score in 72% of the patients. Knee scores in patients with minor instability were not significantly lower than in those who obtained objectively stable knees. A moderate or severe objective instability gave significantly lower scores in a subgroup (one-third) of our patients who were involved in vigorous turning and twisting sports before injury. In this subgroup 30% were obvious failures. Two-thirds of the highly active patients had reduced their activity level postoperatively.

Discussion

Today most authors consider the Palmer technique to be inadequate, and advocate augmented suture repairs. Although our long-term results appear to be better than those reported by others, this may partially be explained by a reduced activity level.
We consider primary operative treatment to be indicated in young active individuals with total ACL tears but are not entirely satisfied with the stability obtained by the Palmer technique in this group. Currently we are comparing the Palmer technique with two different augmentation techniques in a prospective study.

Surgery and Arthroscopy of the Knee
2nd Congress of the European Society
Editors: Müller/Hackenbruch

Anterior Cruciate Ligament Repair Combined with Tibial Valgisation Osteotomy in Medial Osteoarthritis with Insufficient Anterior Cruciate Ligament

H. Bereiter, P. Giger, and A. Gächter

Rätisches Kantons- und Regionalspital, Orthopädische Abteilung, 7000 Chur, Basel, Switzerland

Introduction

Athletic activity amongst the elderly has increased in recent years. Accordingly, combined injuries of knee ligaments with cartilage wear are also more common. The problems become compounded in an untreated, unstable knee because of the ever increasing risk of osteoarthritis. For example, many patients who previously underwent a successful medial meniscectomy may, after a new injury, experience more pain and instability than before. The main symptoms are medial compartment pain and a feeling of instability.

Clinical examination often reveals a functionally insufficient anterior cruciate ligament (ACL). Medial compartment pain is due to either a meniscus lesion or cartilage wear, or both. If a varus deformity is present, radiographic signs of medial overload must be sought while the patient is in a one-leg stance. Arthroscopic assessment plays a crucial role in planning therapy.

The question to be asked, in patients suffering from chronic ACL insufficiency combined with either early medial overload or varus gonarthrosis, is which one of the lesions should be treated. Neither a valgus osteotomy nor an ACL reconstruction alone solve the problem. Here we report the results of combined simultaneous ACL reconstruction with valgus osteotomy of the tibia.

Operative Technique

Technically, the goal is to avoid one phase of the procedure encroaching on the other. Valgus osteotomy may jeopardize ACL reconstruction and vice versa. Thus, it is absolutely necessary to perform the valgus osteotomy first, to avoid compromising the ACL reconstruction (Fig. 1); a primary transosseous fixation of the ACL at the upper-medial aspect of the tibia may be harmed (cut) if the tibial osteotomy is performed secondarily.

An infra-condylar tibial valgus osteotomy is performed in the usual manner with a proximal osteotomy of the fibula at the level of the collum. Previously, we osteotomised the fibula through the middle third. Internal fixation methods for the tibial osteotomy may be chosen individually but should not interfere with the ACL reconstruction. If one uses an ASIF-T-plate, the position of the proximally placed screws should not compromise the ACL-plasty.

Following tibial osteotomy, a modified Eriksson-Trillat ACL-plasty is carried out using the middle third of the patellar ligament. It is fixed proximally with

Surgery and Arthroscopy of the Knee
2nd Congress of the European Society
Editors: Müller/Hackenbruch

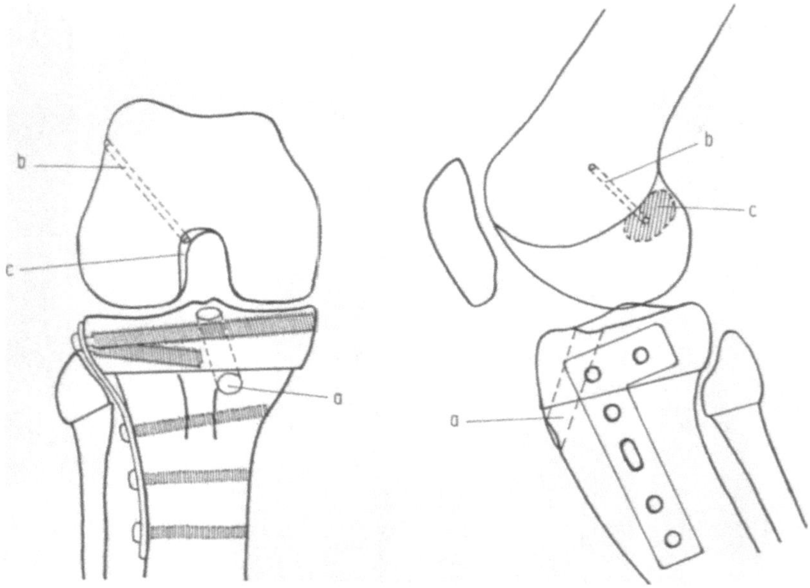

Fig. 1. The relationship of the valgus osteotomy and internal fixation to the ACL repair

transosseous sutures and distally with a screw. In order to reduce intra-articular encroachment, a transligamental (ligamentum patellae) approach is used without a medial arthrotomy. Preoperative arthroscopy demonstrating an intact meniscus facilitates this approach. The ACL reconstruction dictates immediate postoperative mobilisation: continuous passive motion and controlled active muscular rehabilitation.

Results

Seventeen patients were treated by the combined procedure with a 1.5–6-year follow up. An additional 13 patients were also treated this way, but only the long-term follow-up results will be discussed. In each case the data include case history, physical examination and a questionnaire regarding any remaining complaints. Knee stability is always described relative to the contralateral knee by the same person (patient or observer) in each instance.

In cases in which preoperative radiographs were not always available, evaluation of preoperative function and previous procedures served as a reference (Fig. 2). Stability and medial compartment pain were rated as good or excellent in most patients. Virtually all patients were able to return to the athletic activities that they had enjoyed before the initial knee injury. All patients subjectively rated their treated knee as stable and improved by the procedure.

Objective examination revealed no inflammation or swelling compared to the untreated knee. In four cases an anterior drawer sign with a tendency toward lateral subluxation was present in spite of a subjective rating of good stability.

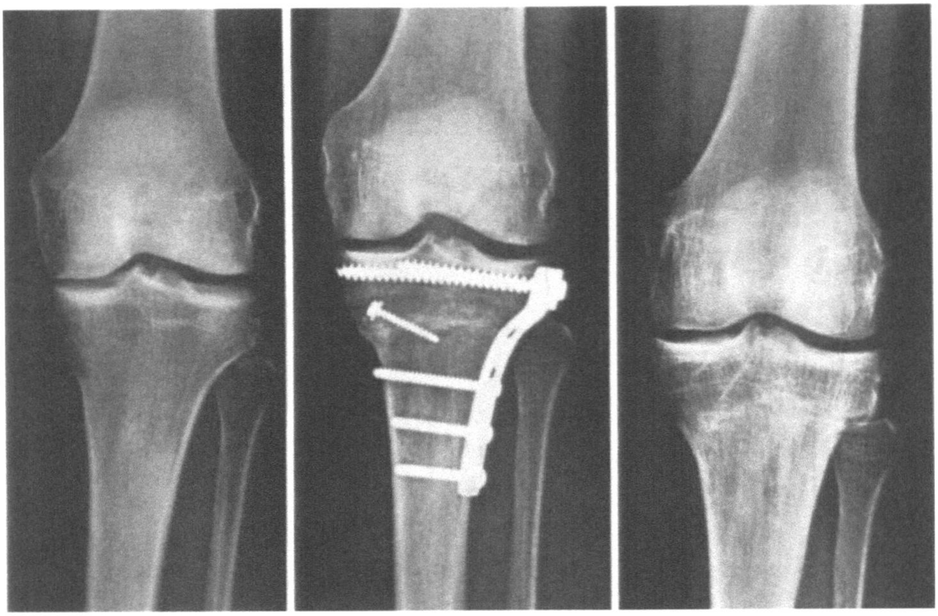

Fig. 2. Preoperative, postoperative (6 months) and 5-year follow-up radiograph studies from left to right

There were no cases of infection, and the period of rehabilitation was equal to that required for an ACL reconstruction alone. There was no apparent compromise in one procedure by virtue of the simultaneous performance of the other.

Conclusion

In the presence of functional instability and pain due to chronic ACL insufficiency and early medial gonarthrosis of the knee, a modified Eriksson-Trillat ACL reconstruction combined with an infracondylar tibial osteotomy can be therapeutically successful. Pain and instability are both relieved. Neither an ACL reconstruction nor a tibial osteotomy alone would provide adequate functional and subjective relief.

The Natural History of ACL Ruptures

J.Y. Dupont, C. Scellier, and D. Chaudières

Clinique du Marais, 75003 Paris, France

Posttraumatic osteoarthritis of the knee is the long-term result of ACL injuries. The initial lesion consists of total or partial ACL ruptures, peripheral capsular or ligamentous damage, meniscal tears, and chondral fractures. The evolution of these lesions varies greatly from one patient to another. The purpose of our study consisted in precisely assessing symptoms and lesions in patients with an abnormal ACL as seen at arthroscopy. Then the data were compared to the time spent between the initial sprain and the time of arthroscopy.

Of 54 patients having a recently torn ACL, with hemarthrosis, 34 had a total rupture and 20, a partial tear; 14 have no meniscal lesion, 11 have a torn medial meniscus, 8 a torn lateral meniscus, 21 have both menisci torn. There were 15% with chondral fractures. A total of 375 patients have a chronic laxity of the knee, with an abnormal or absent ACL at arthroscopy. We used a scale to compare functional instability, laxity and chondromalacia.

Functional instability was found to increase from the time of the initial injury for a 2-year period afterwards. Then it levels off for 8 years, and decreases thereafter. The proportion of patients who have no functional instability is 35% in the first year, decreasing to 0% after 5 years.

Laxity score is parallel to the instability score, and there is a strong correlation between them. In the first 18 months, laxity increases, because of the stretching of peripheral restraints to abnormal movements. It then settles for the next 7 years, then decreases as osteoarthritis begins. Each test (Lachman, drawer, pivot shift) follows the same evolution.

The state of the ACL changes too: complete resorption increases from the time of the accident, partial tears disappear.

In the first months, menisci remain normal in 50% of the knees. Afterwards, normal menisci decrease with time. Vertical meniscal tears increase in frequency in the first 2 years, then decrease as they are replaced by non-reparable complex tears. Chondromalacia gradually develops with time. The medial compartment is first damaged, then the patellofemoral, then the lateral.

The average chondromalacia score increases with time, as the percentage of normal surfaces decreases. At 10 years, no knee has all three compartments intact, and 30% have all three damaged.

The relations between symptoms and lesions show that there is no correlation between arthritis and pain: osteoarthritis developing in unstable knees is relatively well tolerated.

Meniscectomies, even partial, significantly increase the onset of chondral lesions.

Surgery and Arthroscopy of the Knee
2nd Congress of the European Society
Editors: Müller/Hackenbruch

Conclusions

This study confirms the spontaneons evolution of ACL-deficient knees towards osteoarthritis; it also confirms the concept of primary and secondary restraints suggested by Noyes. This evolution is caused by chronic laxity, and is independent of functional instability. Abnormal kinematics of the joint is the reason for meniscal ruptures and cartilage wearing.

It indicates that anatomical intra-articular reconstructions of the ACL are preferable to extra-articular procedures. The goal of the treatment must be to avoid any kind of abnormal movement in the knee after surgery, on a long-term basis. Incorrect reconstructions may lead to abnormal biomechanics of the knee, which increases the risk of postoperative osteoarthritis.

As long as there is no evidence that surgery is able to prevent osteoarthritis, we still believe that indications should be based upon functional instability. The best time for reconstruction should be in the first 2 years, when laxity is still limited, the menisci reparable and the cartilages normal.

A Comparative Analysis of Four Prosthetic Anterior Cruciate Ligaments in the Goat

N. P. Thomas and I. G. Turner

111, Rowland Avenue, Hatch End, GB Pinner Middlesex HA5 AW

This paper presents partial results of a two-part study on the implantation of four different prosthetic anterior cruciate ligaments. Here we shall examine the biological response to the prostheses, while a companion paper considers their comparative mechanical properties (see Thomas et al., this volume). The four ligaments studied were expanded polytetrafluoroethylene polyethylene terephthalate and two carbon-fibre/Dacron composite ligaments, the Westminster and the ABC (Active Biocomposite). The ligaments were implanted into the knee joints of 72 adult female goats – 18 goats for each of the four ligament types. Eight animals from each group were killed at 6 and 12 months. Of these, five underwent mechanical testing and three underwent histological study. A total of eight animals were kept for a longer period and were killed after 2 years.

Surgical Technique

The knee was approached via a medial parapatellar incision, and the natural anterior cruciate ligament was excised. Each prosthetic ligament passed through the femoral and tibial bone tunnels and was inserted according to the manufacturer's preferred method. The expanded PTFE (or Goretex) ligament was implanted as a true prosthesis via an "over-the-top" route, prior to fixation through each eyelet with a screw which crossed both cortices.

The Dacron ligament (Meadox) consisted of an outer knitted velour envelope with an inner high-strength core of four woven dacron tapes. This was inserted as an augmentation via the Toronto MacIntosh "over-the-top" method. The intra-articular portion of the ligament was tubed in fascia lata which had been passed behind the lateral ligament. Both proximal and distal fixation was by means of Richards' staples.

The Westminster ligament consisted of 40000 filaments of carbon fibre drawn into a polyester, or Dacron, braid. This was inserted using a principle similar to that used with the Dacron ligament. In this case the eyelet was used to secure proximal fixation with a toggle; distally, the ligament was wrapped around a screw and washer, then sutured to itself.

The ABC ligament was a novel design and comprised an "open window" of four bare strands of carbon fibre comprising 10000 filaments each and four separate polyester strands. This was positioned intra-articularly and inserted via the "over-the-top" method. A toggle was used for distal fixation through an eyelet, and for proximal fixation an AO cancellous screw and washer were used.

All the ligaments were hand-tensioned with the knee at 20° flexion.

Surgery and Arthroscopy of the Knee
2nd Congress of the European Society
Editors: Müller/Hackenbruch

Results

Synovium

In the case of the Goretex ligaments the synovium appeared either normal in colour or slightly pale yellow. Histological examination after 6 months revealed many focal granulomata. After 1 year the entire synovium was considerably thickened with multiple granulomata and with some fibres on the surface, probably-particles of PTFE. At 2 years the granulomata appeared less abundant, although the synovium was not histologically normal, being thickened and macroscopically more yellow in colour.

The synovium was "quietest" both histologically and macroscopically in the case of the Dacron ligaments. It appeared normal to the naked eye at all times, but histologically a little reactivity could be found in some sections.

Knees with the Westminster ligaments varied in terms of heavy carbon staining, with many giant cells and thickening at a few carbon fibres found in the subsynovial layer. This appearance depended upon whether or not the ligament had ruptured intra-articularly.

Those with the ABC ligament showed a mild to moderate carbon load in the synovium, with some thickening and signs of moderate synovitis. The carbon fibre was confined to the subsynovial layer.

Intra-articular and Intra-osseous Analyses

Goretex. The usual appearance of this ligament at 6, 12 and 24 months was that of the fibres predominantly put intra-articularly in soft tissue, although broken strands could also be seen. The histology of this intra-articular portion confirmed the presence of fibrous tissue around but not within the fibres. Even at 2 years there was no significant or consistent intra-articular penetration of the fibres of expanded PTFE. In the intra-osseous portion at 6 months there was consistent fibrous tissue ingrowth to the centre of the ligament but no significant penetration of the fibres at this level. At the periphery of the bone tunnel, there was after 6 months consistent penetration of the fibres lying on the edge of the tunnel. At 12 and 24 months progressive bony ingrowth was observed to the centre of the tunnels and through the expanded PTFE fibres. Haversian systems seen well with Miller's stain confirmed this in all sections.

Dacron. In its intra-articular portion the Dacron which had been tubed in fascia lata appeared to have revascularised consistently, and histologically one could observe at all time intervals an ingrowth of collagen from the fascia lata into the outer part of the velour sheath of the Dacron ligament. In the intra-osseous portion there was a progressive ingrowth of fibrous tissue into the outer part of the velour sheath; however, this did not progress after 1 year and remained fibroblastic rather than osseous in all the specimens.

Westminster. This ligament used carbon fibre tubed in Dacron and then further tubed in its intra-articular course by fascia lata. Of these ligaments 90% revealed ruptured carbon fibre within their substance, and some had ruptured actually into the joint, producing severe synovitis. There was little ingrowth from the fascia lata into the Dacron, although in one specimen at 12 months penetration had occurred. The intra-osseous aspect of this ligament at all time intervals showed some fibroblastic ingrowth into the Dacron sheath, but the central carbon-fibre core was not infiltrated in any case at any time interval.

ABC. The open window of this ligament had become completely engulfed in white soft tissue which was proven histologically to be collagen. Good ingrowth into the Dacron and carbon-fibre portions of the ligament were seen histologically in all specimens intra-articularly. It should be noted, however, that in two of the five specimens subjected to mechanical testing at 12 months, rupturing had occurred. The intra-osseous aspect of this ligament proved somewhat disappointing in that both histological examination and scanning electron microscopy (performed on all ligaments to supplement histological examination) showed very little penetration of the Dacron coating.

Discussion

The expanded PTFE ligament usually showed thickened synovium with signs of chronic inflammation marked by focal and sometimes confluent granulomata. This ligament showed no significant ingrowth in the intra-articular portion, and broken strands were seen invariably becoming more numerous in the specimens examined after 1 and 2 years. This ligament did show, however, a progressive ingrowth of bone, both between the strands of the ligament and into each strand, this being most marked at the bone/ligament interface.

Specimens with the Dacron ligament had an almost normal synovium. All ligaments were intact up to 1 year, and some ingrowth was seen in the intra-articular portion. There was good fibrous ingrowth into the outer velour sheath.

With the Westminster ligament the synovium varied between little reaction and severe synovitis associated with a heavy carbon-fibre load. There were many partial ruptures here and one complete rupture. Some intra-articular ingrowth was present, but varied in extent.

In the intra-osseous portion fibres grew into the periphery of the sheath but in no case extended to the carbon-fibre core.

The ABC ligament showed a synovitis ranging from mild to moderate. There was progressive ingrowth into the open-window intra-articular part, but there were two complete ruptures at 1 year. In the bony tunnels there was poor fibrous ingrowth into the periphery.

Conclusion

In anterior cruciate ligament replacement in the goat, a synovitis of varying intensity was present in all knees examined although less pronounced among specimens examined at 2 years than among those examined earlier. Intra-articular ingrowth varied from poor, using the expanded PTFE, to abundant, in the case of the ABC ligament. This finding coincides in part with the theory associated with each ligament, the PTFE ligament being designed as a prosthesis and the ABC relying on ingrowth for its ultimate function. In bone tunnels all Dacron ligaments showed fibroblastic ingrowth only. This was most pronounced in the ligaments with the loosest outer weave, i.e. the Dacron-only ligament; as the weave became tighter (e.g. in the ABC ligament), the amount of ingrowth declined. The expanded PTFE ligament was the only one which showed a progressive bony ingrowth, both between and into the strands.

A Comparative Analysis of the Mechanical Properties of Four Prosthetic Anterior Cruciate Ligaments

N. P. Thomas, K. Wright, I. G. Turner, and I. W. Nelson

111, Rowland Avenue, Hatch End, GB Pinner Middlesex HA5 AW

In this report we analyse the comparative mechanical properties of four different prosthetic anterior cruciate ligaments implanted into the knees of 72 goats. Details of the procedures involved and of the biological responses to the prostheses are presented in a companion paper (see Thomas and Turner, this volume). The present paper is divided into in vitro and in vivo parts.

In Vitro

The in vitro component of the study consisted of testing the ligaments to complete rupture at various strain rates, and temperatures after refluxing and after incubation in serum for 1 year. Five specimens of the same ligament were used in each test.

The two strain rates used were 150% s^{-1} to simulate injuries sustained during contact sports and 2.6% s^{-1}. At the low-strain rate and 38° C the ligaments broke at approximately the manufacturer's specified strength: the Goretex at 4900 N, the Dacron at just over 3000 N, the ABC (or Surgicraft) at 2500 N and the Westminster at 1000 N. At the high-strain rate the Goretex breaking load fell by 40% whilst the others remained virtually unchanged (Fig. 1). Further tests

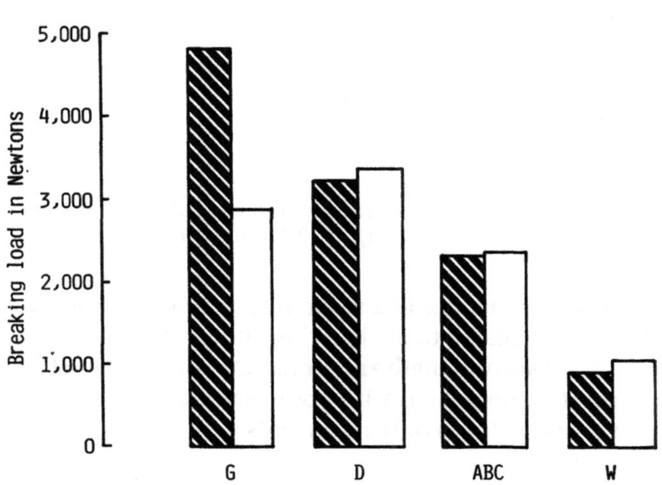

Fig. 1. Breaking loads of implanted ligaments at 38° C and at different strain rates: 2.6% (*striped bars*) and 150% s^{-1} (*solid bars*). *G*, Goretex, *D*, Dacron; *ABC*, active biocomposite (Surgicraft); *W*, Westminster

Surgery and Arthroscopy of the Knee
2nd Congress of the European Society
Editors: Müller/Hackenbruch

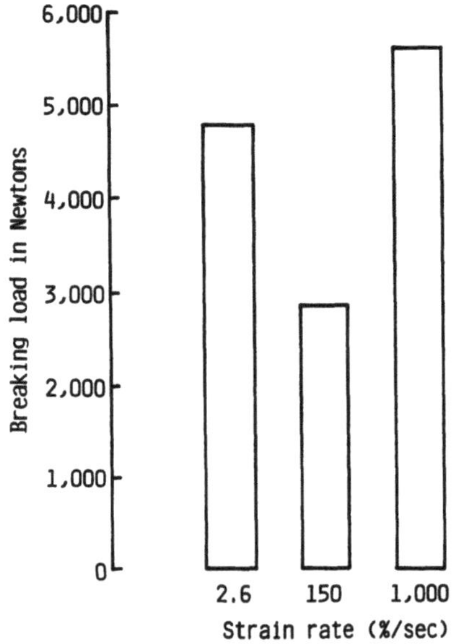

Fig. 2. Breaking load of Goretex ligaments at 38° C and varying strain rates

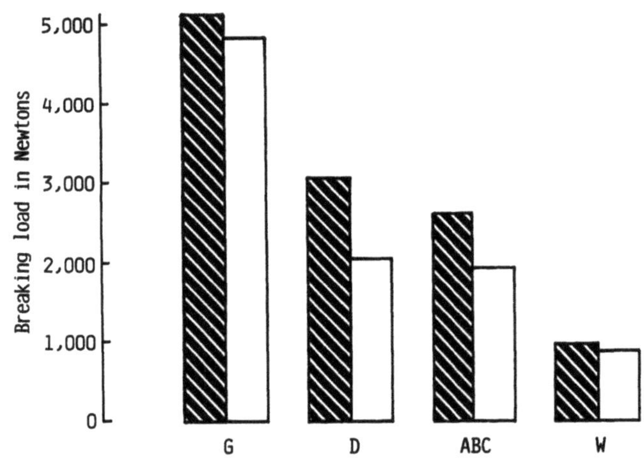

Fig. 3. Breaking loads before (*striped bars*) and after (*solid bars*) refluxing at 21° C. *G*, Goretex, *D*, Dacron; *ABC*, active biocomposite (Surgicraft); *W*, Westminster

were carried out on more Goretex ligaments at a strain rate of 1000%/s, yielding breaking loads of 5800 N. We are unable to explain this set of results which indicates a biphasic relationship between strength and strain rate (Fig. 2). The effect of temperature (room temperature compared with body temperature) made no significant difference to the breaking loads of the ligaments.

Refluxing was performed in normal saline solution at 100° C. This accelerated test sought to determine whether the mechanical properties of the material are likely to be degraded in the human environment. The test is unphysiological and has been popularised by bioengineers and manufacturers of materials. After

500 h immersion in boiling normal saline solution, the breaking loads of all ligaments declined. The ligament most sensitive to this procedure was Dacron, which lost almost 30% of its strength; this was followed (in order of descending sensitivity) by the ABC, Goretex and Westminster ligaments (Fig. 3). After incubation in serum for 12 months the breaking loads of the four ligaments remained unchanged.

In Vivo

The in vivo analyses were performed on the dissected knees of freshly killed goats. The capsule and menisci were removed, and the posterior cruciate ligament divided. The only remaining structure connecting tibia to femur was either the natural anterior cruciate ligament (ACL) (in the normal knees) or its synthetic replacement (in the test knee).

The mean ACL breaking load among the 40 animals was found to be 1600 N. The results of the tested knees were expressed as a percentage of the mean normal goat ACL. In order to evaluate the test specimens better, a baseline breaking load obtained directly after implantation was measured. These after-implantation values tested the fixation of the bone ligament to bone system. For each ligament the mode of failure was consistent. The Goretex ligament failed by the screw pulling through the tibial bone tunnel, The Dacron by the ligament pulling out from under the staple and the ABC by breaking the toggle, while the Westminster unwound itself from the screw and washer.

The breaking loads (or pull-out strength) obtained after implantation are shown in Fig. 4. Only the fixation of the Goretex and ABC exceeded 50% of the normal goat ACL. After 6 months the following changes were observed: the Goretex ligament strength had risen to greater than normal, the Dacron had risen from

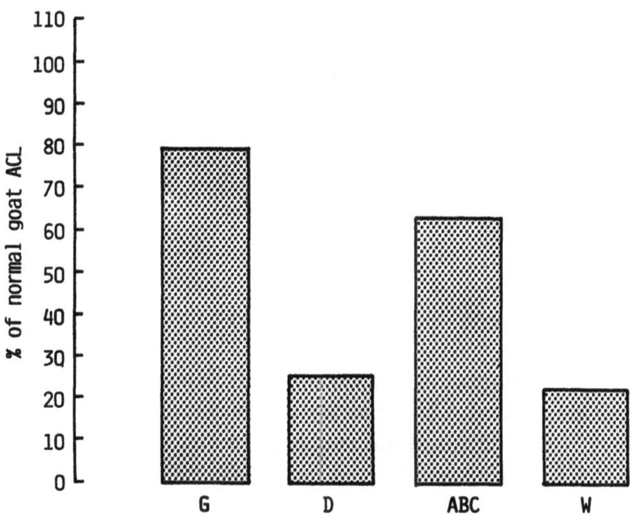

Fig. 4. Breaking loads after implantation. *G*, Goretex, *D*, Dacron; *ABC*, active biocomposite (Surgicraft); *W*, Westminster

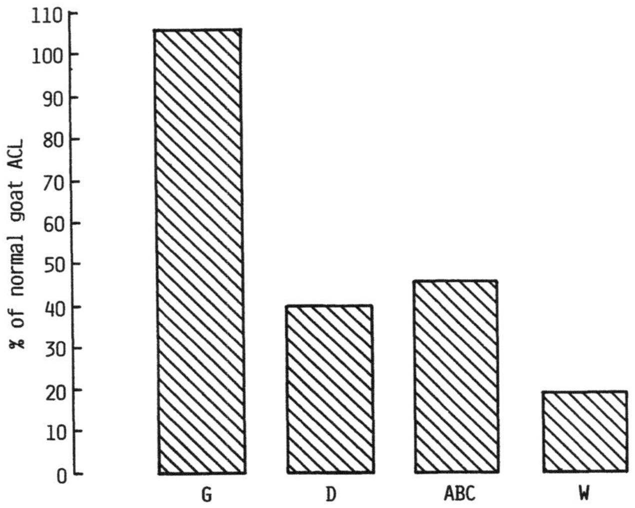

Fig. 5. Breaking loads at 6 months. *G*, Goretex, *D*, Dacron; *ABC*, active biocomposite (Surgicraft); *W*, Westminster

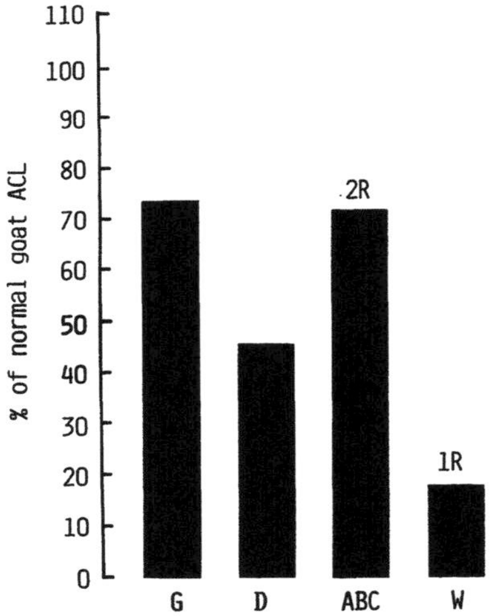

Fig. 6. Breaking loads at 12 months. *G*, Goretex, *D*, Dacron; *ABC*, active biocomposite (Surgicraft); *W*, Westminster

25% to 40%, the ABC had fallen, and the Westminster remained unchanged (Fig. 5).

The ligaments of each of the five animals killed after 12 months were tested. The Goretex had fallen to 75% of the normal ACL, and the Dacron had risen to 45%. Two of the five ABC ligaments had ruptured although the mean level of the remaining three was high. One of the Westminster ligaments had ruptured, but the mean of the remaining four was unchanged (Fig. 6).

Figure 7 represents the results obtained thus far in the form of a histogram, emphasizing the trend for each ligament over 12 months. The Goretex ligament

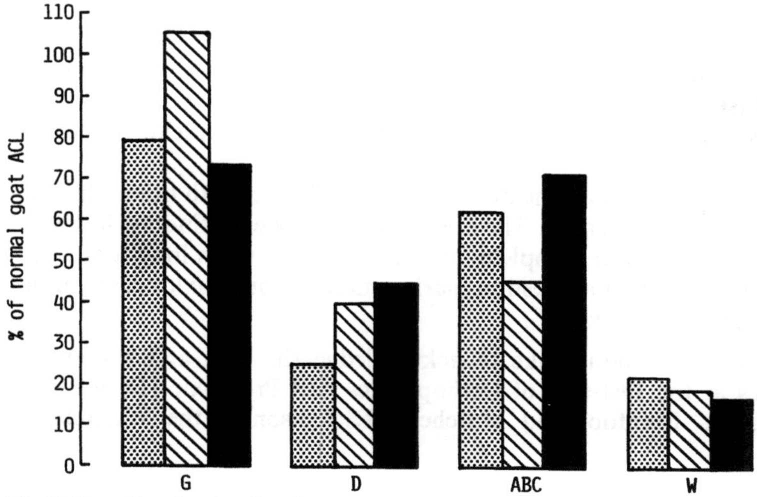

Fig. 7. Breaking loads after implantation (*shaded bars*), at 6 months (*striped bars*) and at 12 months (*solid bars*). *G*, Goretex, *D*, Dacron; *ABC*, active biocomposite (Surgicraft); *W*, Westminster

peaked at 6 months; the 1-year figure was less than the after-implantation values. The Dacron showed a slow but steady increase in breaking load. Two of the five ABC ligaments had ruptered at 1 year, and the Westminster did not change significantly.

The 2-year results could not be expressed in this way as too few animals were tested. One knee containing each type of ligament was used for histological analysis, and a pull-apart test was performed on the other. At 2 years all bone ligament to bone system failed by a midsubstance rupture. The breaking strength of the Goretex ligament was 40% of the normal goat ACL, as the majority of its fibres had already ruptured intra-articularly.

The normal mode of failure of the Dacron bone ligament to bone system was by the central core tapes pulling through the bone tunnel; this occurred in each knee within 1 year (excluding after-implantation values). However at 2 years there was a partial rupture in the specimen used for histology, and the pull-apart tests done on the side which appeared intact ruptured intra-articularly at 44% of the Normal goat ACL.

The 2-year ABC ligament ruptured at 33% of the normal ACL. This was reduced as compared with the 12-month mean of that ligament. The Westminster 2-year specimen ruptured at 20% of the normal goat ACL which was roughly the same as among the 1-year samples. These findings support the earlier histological evidence and the hypothesis that the Goretex ligament was firmly held in the bone tunnels after 12 months. Its breaking strength depended on how many strands had ruptured in the notch. Histologically and mechanically the Dacron ligament seemed to achieve a plateau at 6–12 months, but there were partial ruptures intra-articularly at 2 years. The ABC and Westminster ligaments showed problems with intra-articular and intra-osseous components, the latter probably due to the relatively tight weave of the Dacron coating.

Conclusion

The in vitro tests showed that the Goretex ligament was sensitive to strain rate in a nonlinear fashion. In the refluxing experiments the Dacron and the ABC ligaments showed a reduction in breaking load, the in vivo significance of which is unclear.

The in vivo studies showed that the mode of failure changes with time. This correlated well with histological data. The most important overall finding, however, was the increase in partial or complete intra-articular failure in long-term specimens. In this animal model a question mark remains about the longevity of all the prosthetic ligaments tested.

Acknowledgements. We should like to acknowledge the help and support of Professor Scales at the Institute of Orthopaedics and Professor Harold Ellis, Paul Aichroth, Dr. Branfoot and Marcheter Dickenson at the Westminster Hospital.

Mechanical and Biological Considerations in Prosthetic Cruciate Replacement

B. B. Seedhom [1], K. Fujikawa [2], and P. J. Atkinson [3]

[1] Rheumatism Research Unit, School of Medicine, University of Leeds, UK
[2] Dept. of Orthopaedic Surgery, Keio School of Medicine, Tokyo, Japan
[3] Dept. of Oral Biology, School of Dentistry, University of Leeds, UK

For effective and long-term success of cruciate ligament replacement, the following requirements must be satisfied: (a) placement of the ligament must be isometric; (b) anchoring must benefit from both bone and soft tissue ingrowth; (c) the ligament must be of sufficient strength; (d) the stiffness of the ligament must be matched to that of the natural ligament; and (e) the ligament must retain sufficient strength after being subjected to several millions of tensile stress cycles. Placement has been discussed elsewhere by several workers.
This paper presents the results concerning (b)–(e) obtained with a polyester prosthetic ligament made into a tubular open-weave structure.

Anchoring

The method used consisted of placing a bone plug within the tubular section of the ligament placed within bony tunnels. This was tried in five living pigs and within the first month it was shown that bone ingrowth does take place. Histological sections showed that the bone growth increases with time so that 7 months and 9 months postoperatively the ligament is firmly anchored by bone trabeculae. Further, tissue growth into the intra-articular section of the ligament was found to strengthen it. Thus 9 months postoperatively the strength of the implant ligament was 10% above that of the ligament as tested in the laboratory, and after 18 months postoperatively it was in excess of twice its original strength.

Mechanical Strength of the Ligament

The chosen structure was tested in the laboratory and was found to be in the region of 2000 N. This is above average of the breaking strength of human ligaments in young adults, which was found to be 1730 N.

Ligament Stiffness

Laboratory tests on ligaments which have been subjected to the relevant sterilisation process showed that the load extension curves of the prosthetic ligament was quite similar to that obtained with ligaments from young human adults. The strain corresponding to the maximum breaking load of the prosthesis was 30% –35%, which is quite similar to published figures.

Surgery and Arthroscopy of the Knee
2nd Congress of the European Society
Editors: Müller/Hackenbruch

Fatigue Life of the Prosthesis

The prosthetic ligament was subjected to cyclic tensile loading of 500 N amplitude at 25–30 Hz whilst kept in saline at 39 °C. Specimens were subjected thereafter to tensile loads till breaking point. The residual strength was found to depend on whether the clamp was soft or hard. With a soft clamp, after subjecting the ligament to 63 million cycles, the residual strength was 1280 N compared with a starting strength of 2000 N. This large number of cycles corresponds to walking 80 km per week for 20 consecutive years.

Conclusions

A method of fixation of a prosthetic ligament was developed and tried in animals. It was shown to be adequate and that the strength of anchoring increases with time. The mechanical strength of the ligament as well as its stiffness correlated well with the natural ligament's characteristics. On testing the prosthesis in cyclic conditions, although its strength was affected by long periods of cyclic loading, the residual strength was adequate.

Anterior Cruciate Ligament Reconstruction Using the Leeds Keio Terylene Ligament

J. B. Hamilton, G. K. Sefton, R. B. Smith, and T. J. Cain

Woodlands Orthopaedic Hospital, 5 Mornington Villas, Bradford BD87JX, UK

Introduction

Instability of the knee following rupture of the anterior cruciate ligament (ACL) has been treated by several methods in the past. This paper describes the experience of several centres in the north of England using the Leeds Keio terylene (polyethylene terephthalate) ligament as an ACL replacement. A series of 50 cases with a follow-up from 1 to 4 years is described.

Material and Methods

Preoperative Assessment. Clinical assessment is carried out by the surgeon. A history of the patients' complaints was recorded concerning: pain, swelling, giving way and their physical requirements of the knee in relation to their job and to any sporting pastimes. Full clinical examination of the knee was carried out including: flexion, extension, stability of the knee in full extension, the anterior drawer sign, the Lachmann test and the jerk test.

Radiographs of the knee were performed, including a lateral view with the knee flexed to 90° and strong forward pressure applied to the upper part of the tibia.

An arthroscopic examination of the knee was conducted, unless arthrotomy had previously been carried out to establish the integrity, or otherwise, of the ACL.

Self-Assessment. The Cincinnati Sports Medicine Self-Assessment Knee Chart was completed by all patients before and after surgery.

Stress Measurement Using the Stryker KT-1000. Some patients were subjected to stress measurement using this instrument, but due to restricted availability not all patients were tested; therefore, the results have not been included in this study.

Selection of Cases. Only those who complained of significant disability, either during professional sporting activities or in performing their jobs, were considered for surgery. No patient was considered before thorough rehabilitation after previous injury.

In general it was found that many patients who had an ACL rupture, as an isolated injury could be successfully rehabilitated to carry on with almost any but the most taxing physical occupations, and many sportsmen were able to continue with their sports without surgery. However without exception, knee joints with an ACL rupture and one other injury, either ligamentous or meniscal, required surgery. This paper deals with the patients who were thought to require only an ACL replacement; thus, patients also requiring a lateral reinforcement are not included.

Surgery and Arthroscopy of the Knee
2nd Congress of the European Society
Editors: Müller/Hackenbruch

Surgery. The Leeds Keio ligament is a partly tubular, open-weave graft of Dacron. Instrumentation allows accurate placement of the ligament at both the femoral and tibial attachments. Fixation is achieved by a bone plug technique, which results in a very firm, reliable, biological fixation.

Results

Clinical examination in 50 cases shows a marked increase in stability of the knee joint using the tests described for preoperative assessment. Patients claim a considerable reduction (or absence) of pain, swelling and giving way postoperatively. All the sportsmen in this trial returned to active sport, and all indicate an improvement of 40 to 70 points (on a scale of 100) on the self-assessment knee chart.

Discussion

From the study of these 50 cases, we are convinced that ACL reconstruction using the Leeds Keio ligament is a reliable and effective means of stabilising the knee joint, even in young athletes where instability can clearly be demonstrated to be due to the rupture of the ACL.
The postoperative behaviour of the ligament as assessed by arthroscopy and histology following repair indicates that the ACL is surrounded by firm fibrous tissues, similar to that of the original ligament, and remains stable over the period of review.

Reconstruction of the Anterior Cruciate Ligament Using a Ligament Prosthesis

J. Gillquist and M. Odensten

Department of Orthopaedic Surgery, University Hospital,
581 85 Linköping, Sweden

Patient Selection

Our previous studies have shown that muscle rehabilitation alone results in acceptable knee function for a significant number of patients suffering from anterior cruciate ligament (ACL) instability (Tegner et al., 1984). From a random sample of patients with chronic ACL instability, about 75% become satisfied with their knee function after a 3-month rehabilitation programme. Therefore, all patients in the present study had experienced a rehabilitation programme before being considered for ACL reconstruction. Patients with persistent instability symptoms who desired surgery were offered the choice between a standard patellar tendon reconstruction or implantation of a ligament prosthesis.

Material and Methods

The prosthesis selected was the Stryker-Meadox Dacron ligament (Stryker Corp., Kalamzoo). The ligament was implanted using a new guide system which increased the accuracy in placing. Prophylactic antibiotics were used. Menisci were repaired whenever possible, and 2+ medial laxity (valgus stress test) was also corrected.

The instrument system was designed for arthroscopic surgery, as well as for a miniarthrotomy. The notch width is measured with a special device, and the notch is enlarged using an arthroscopic abrader or an osteotome depending on the particular technique preferred. A guide frame allows simultaneous drilling of the tibial and femoral tunnels in a straight line with the knee held in 90° of flexion. The inner openings of the tunnels are bevelled with a router. A test ligament system is used to check isometry using a strain gauge. A maximum change in strain in the ligament of 8% over the range of motion is accepted. Strain change in the ligament during passive knee motion was identical to that described by Arms et al. (1985). The ligament position was examined on postoperative radiographs.

Results

To date 96 patients have been operated on since August 1983. The follow-up includes 86 cases, with a maximum of 3 years. Forty-five cases have been followed up for more than 2 years and only a few for more than 3 years. The mean patient age was 27 years with a range of 16–35 years.

Surgery and Arthroscopy of the Knee
2nd Congress of the European Society
Editors: Müller/Hackenbruch

Rehabilitation is started immediately postoperatively for patients without meniscal repair, with a full range of motion, weight bearing and muscle strength exercises. Patients with repaired menisci start full rehabilitation after 4 weeks.
Muscle strength was regained between 2 and 6 months postoperatively as measured by a Cybex II isokinetic dynamometer. The most rapid rehabilitation was found in patients with initially higher muscle strength (male athletes). Patellofemoral pain was very uncommon and found in only two patients. The Lysholm knee score was significantly improved from 67 to 90 after 1 year and up to 93 after 2 years. Some 84% of the patients were in the good/excellent category. The Tegner activity level rating indicated a median of 7 (range 0–9), and after 2 years 6 (range 1–10). Stability was also significantly improved. Lachman and pivot shift tests were generally normalized postoperatively. A better evaluation of stability is gained by looking at values measured with the Stryker laxity tester. Preoperatively the patients had a mean of 7-mm increased laxity as compared with the normal knee. After 1 year the difference was 0.5 mm and after two years, 0.8 mm. Looking at the absolute values for anterior and posterior drawers in both knees, there was no significant change over time in the normal knee. The surgically treated knee became tighter than the normal knee immediately postoperatively. A difference of −2 mm was recorded. This difference disappeared over the first 2 months, and then anterior laxity increased slowly. At 2 years it had almost reached the upper confidence limit for the normal knee. The overtensioning of the ligament substitute caused a slight posterior subluxation of the tibia which disappeared during the first 2 months. In the earlier part of the series we used a 60 N tension in the ligament. This was later changed so that the tension duplicates the laxity of the normal knee. Usually this means a tension of 30–40 N in the ligament. Note that these values will vary with the kind of ligament substitute used. The increase in laxity during the follow-up may be caused by slipping of the implant under the staples, remodelling of the bony tunnels or permanent creep in the ligament itself. At present the precise cause cannot be determined.

Complications

Synovitis did not occur in any of the cases. One patient had a synovial fistula which necessitated removal of the ligament. This complication was probably caused by bad positioning (free-hand drilling) and too large tunnels. In the international series a 1,5% incidence of synovitis was recorded. Graft rupture occurred in 4% of the patients, usually between 1 and 2 years after surgery, caused by improper positioning in about 1.5% and by a new trauma in the remainder. Three of our patients have had their ruptured ligaments replaced with a good result. The operation is simple when using our technique. Since the two bone channels are in line with 90° of flexion, the guide pin can be passed through the ruptured ligament and overdrilled with a 5-mm reamer without opening the knee. The new ligament is then pulled through the channels.
In summary, replacement of the ACL with a Dacron prosthesis has proved a successful operation in this 2-year perspective. It is, however, an open question as to how long this type of ligament will last before fatigue failures occur. The rupture frequency will undoubtedly become larger as time goes by, but it is possible

that the implant will suffice until many patients have passed the ACL-dependent age of demanding sports.

References

Arms S, Pope M, Johnson R et al. (1984) The biomechanics of anterior cruciate ligament rehabilitation and reconstruction. Am J Sports Med 12:8–18

Tegner Y, Lysholm J, Gillquist J et al. (1984) Two year follow up of conservative treatment of knee ligament injuries. Acta Orthop Scand 55:176

Reconstruction of the Anterior and Posterior Cruciate Ligaments of the Knee Under Arthroscopy. First Results

J. Bahuaud, J. Rebour, J. Mayer, D. Besse, F. Mautalen, and H. Baspeyre

Department of Orthopaedics and Traumatology,
Hôpital d'Instruction des Armées „Robert Picque", Bordeaux 33998, France

Introduction

It is now possible to perform knee ligamentoplasty under arthroscopy in the following two ways: by replacing the arthrotomy by arthroscopy [10] and using artificial ligaments such as Dacron (polyethylene tetraphthalate) implants inserted under arthroscopy through small incisions, resulting in minimal surgical damage [2]. Here we present the techniques we have used since 1981 [1–3] and an analysis of the first results concerning 131 operated on and re-examined patients.

Material and Methods

Between 1982 and 1986, 131 artificial ligamentoplasties of the central knee pivot, using Dacron, were performed by arthroscopy.
They can be divided into three groups: 116 anterior plasties, 13 posterior plasties and 2 double plasties (anterior and posterior). The patients, 125 males and 6 females, were all participators in sport with an average age of 25 years. The two kinds of Dacron prostheses used were the Stryker prosthesis (in 111 cases) and the Proflex (Protek) prosthesis (in 20 cases). The surgical techniques employed were as follows: MacIntosh's "over the top" technique for anterior plasty, Clancy's technique for posterior plasty [1] and combined technique for double plasty, but using only one Stryker implant, the 130–20. The *surgical techniques* employed [3] depend on the kind of artificial implant used. First, all associated intra-articular injuries are treated. All the techniques are performed under peridural anaesthesia using a tourniquet. A video camera facility is necessary during the operation.

Anterior Ligamentoplasty

The incisions made are classic arthroscopic portals, 1 cm fore – inferior and medial tibial, and a 2 cm upper and lateral femoro-supracondylar. The last incision allows insertion of the implant. A tibial tunnel is drilled towards the destroyed anterior cruciate ligament (ACL) with an aimer. The slender and bent arm of the aimer penetrates through the anteromedial portal and is settled in the centre of the stump. The final diameter of the tunnel must not be greater than 6 mm, so that the Dacron implant will be held firmly in the tunnel.

Insertion of the Stryker 130-25 Implant (Figs. 1–4). The plasticised extremity of the prosthesis is introduced into the tibial tunnel and caught with forceps on the

Surgery and Arthroscopy of the Knee
2nd Congress of the European Society
Editors: Müller/Hackenbruch

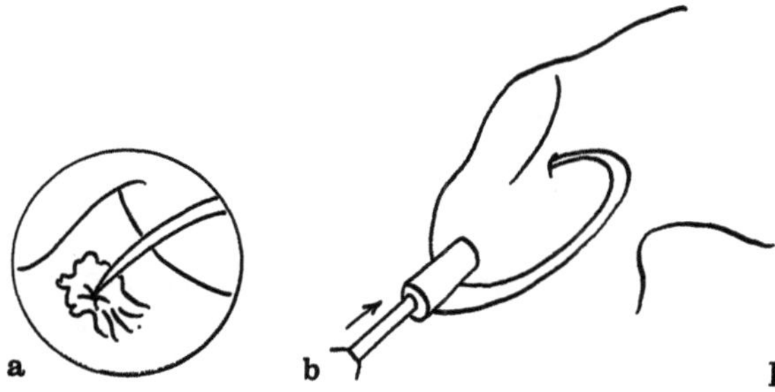

Fig. 1. **a** Aimer, **b** arthroscopic view

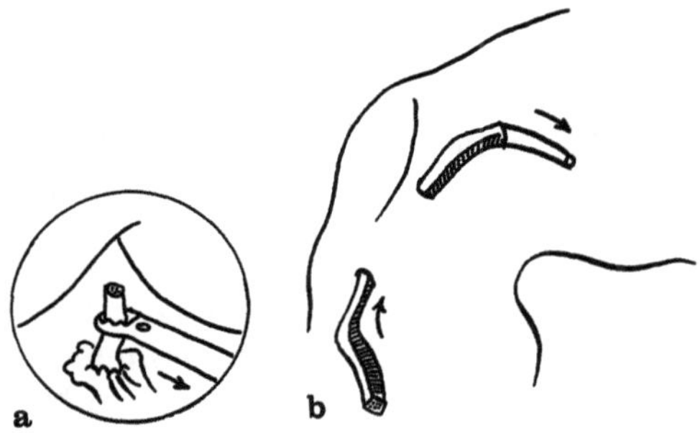

Fig. 2. **a** The plasticised extremity of the implant is caught with forceps and **b** pulled outwards

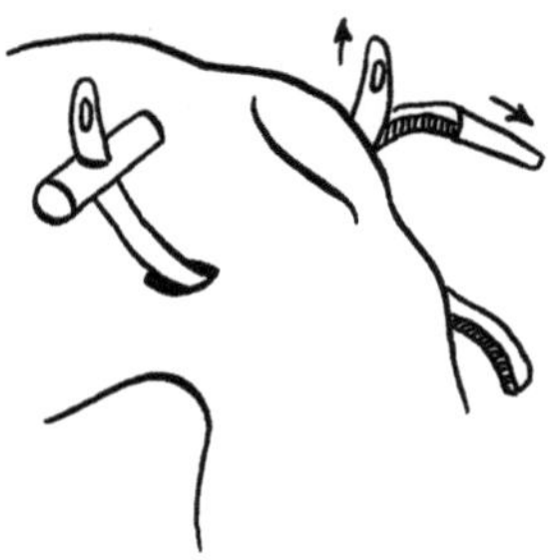

Fig. 3. Stryker implant guide

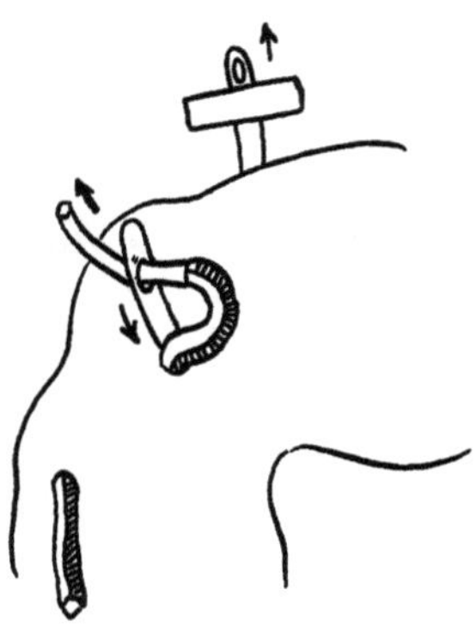

Fig. 4. Traction of the implant through the intercondylar notch

level of the inner joint space and then pulled outwards. The upper and lateral femorosupracondylar incision is made, and a Stryker guide is introduced, from top to bottom following the line of the supracondyle, through the intercondylar notch, then through the anteromedial portal incision. The guide catches the implant within the joint through its eyelet by an inverted motion in the "over the top" manner.

Insertion of the Proflex (Protek) Implant. A steel wire loop is introduced into the tibial tunnel and caught in same way as the plasticised extremity of the Stryker 130-25 implant, then laid aside. The slender extremity of the Protek implant is introduced into the intercondylar notch through the upper and lateral incision. It is caught under arthroscopy by forceps and introduced into the same steel wire loop. The loop holds the implant so that it is drawn downwards through the tibial tunnel. As soon as the implant is settled, it is secured in the "over the top" manner with a Stryker staple which is designed to smooth out the implant. While the leg is held in slight flexion the implant is tightened until the dynamometer reads between 12 and 15 kg. Finally it is fixed by another Stryker staple through the small tibial incision.

A clinical check is made to test suppression of the Lachmann's sign as well as an arthroscopic inspection for tightness and correct positioning of the implant. The skin is then closed immediately with an automatic skin staple without draining.

For this implant material, the "over the top" technique of fixation seems to be the best, i.e. the one less liable to wear. In simple cases, the operation lasts 30 min.

Usually postoperative management simply includes systematic anticoagulant treatment and very gradual rehabilitation which could start on the 1st day with passive assistance by an arthro-motor splint in directional movements which are not painful. Weight bearing can commence on the 5th day, gradually progressing to sporting activities in the 3rd month if it is a single ligament injury.

Posterior Ligamentoplasty

The incisions made are very tiny as described below: first, classic arthroscopic portals followed by an upper and lateral femoro-supra- and retrocondylar incision, which allows the penetration of the lateral gemellus muscle at three posterior levels after partial insertion, the intercondylar notch, the joint space and the retrospinal area. We would advocate using the forefinger while making this incision to shield the vulnerable anatomical connections. This incision should be 3 cm long. Next comes the lateral and inferior incision (1 cm long) on the tubercle of Gerdy, which is the lower anchoring point of the implant. Then upper and a medial incision (1 cm long) is positioned close above the femoral insertion of the medial ligament.

A condylar tunnel is drilled from top to bottom and from inside to outside with the aid of a guide. The guide's slender and bent extremity rests inside the joint at the upper insertion of the destroyed posterior cruciate ligament and is always passed through the anteromedial portal.

At the second step, a tibial tunnel is drilled exactly horizontally from front to back and from inside to outside starting at the tubercle of Gerdy and emerging in the retrospinal area. The tunnel is marked by a steel wire, the loop of which protrudes from the upper and outer side, and is laid aside.

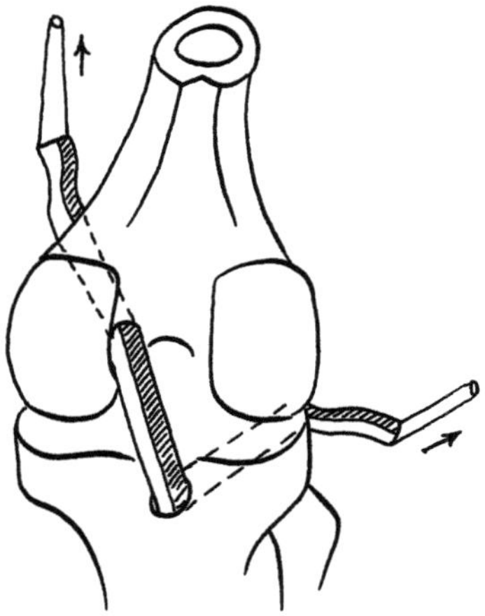

Fig. 5. Assembly of the posterior Stryker 130-25 implant

The Stryker 130-25 Technique (Fig. 5). The plasticised extremity of the implant is introduced into the condylar tunnel and caught through the anteromedial portal incision and brought to the outside. The Stryker guide is introduced through the lateral and upper incision and passed through the intercondylar notch medially, emerging again through the same incision. The implant is caught by the guide with reverse motion. The steel loop allows passage through to the lower tibial end of the tunnel. The implant is caught outside at the tubercle of Gerdy.

Protek (Proflex) Technique. This time a steel loop is introduced into the condylar tunnel, drawn outside through the anteromedial portal incision and laid outside. The Protek guide penetrates the intercondylar notch medially while the implant is introduced in a lateral and upper direction. The implant is located by arthroscopy and is caught by the steel loop and then drawn outside. The other extremity of the implant is pulled through the horizontal tibial tunnel as described above. To tighten the implant, the knee is held in 90° of flexion to cancel the posterior laxity. The upper part of the implant is fixed with Stryker staples; the bottom is tightened to 15 kg, measured with a dynamometer, then secured to the tibia in the same way. The implant positioning is checked in the same way as for anterior ligamentoplasty, and the wound is closed.

Double Ligamentoplasty

In this operation a Stryker 130-20 implant is manipulated into position through both sets of the incisions and tunnels described above. The anterior cruciate implant passes through a vertical tibial tunnel, comes out "over the top", then under the lateral knee ligament to penetrate the horizontal tibial tunnel at the tubercle of Gerdy, exiting at the retrospinal area (Fig. 6).

Results

The results were evaluated for 40 patients subjectively using the Lysholm knee scoring scale and objectively by clinical examination of the Lachmann's sign before and after surgery.

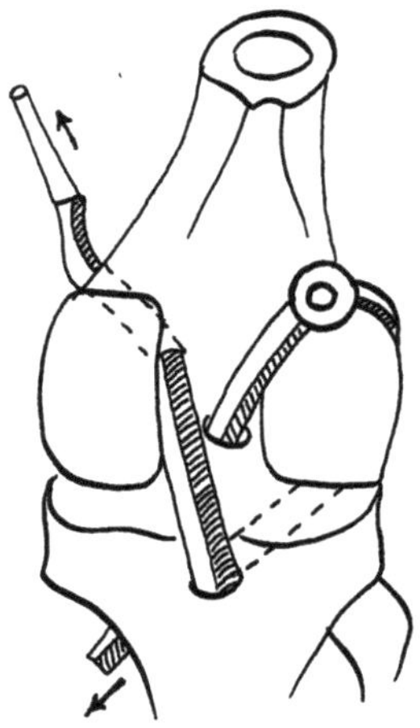

Fig. 6. Anterior and posterior plasty with the Stryker 130-20 implant

Laximetry was also conducted following Mansat's technique with a computerised accurate ligament tester (CALT).
Using this procedure, *P* represents the injured knee with an interval of 1.8–14 mm (an average of 6.8 mm for men), and *N* represents the normal knee with an interval of 0.2–9 mm (an average of 2.5 mm for men).
Before surgery the margin was: $P–N=A=4.3$ mm on average. After surgery, the margin is $P–N=B=0.4$ mm on average.
Therefore, the average postoperative improvement in laxity was:

$$A–B=D=3.99 \text{ mm}.$$

The tension was 150 N.

Follow-Up

For anterior plasties, two groups were studied: the first group consisted of 47 patients followed up between 2 and 4 years and the second of 69 patients followed up between 8 months and 2 years. It is important to include the preoperative condition, since it influences the prognosis markedly.
In all, 28 plasties were performed by broad arthrotomies (with 4 medial meniscectomies). Of these, 26 were classic plasties including 15 by Lindemann's technique, 5 using MacIntosh's technique, 3 by the Kenneth-Jones' technique and 3 using Lemaire's technique; 2 were performed with Dacron.
There were 4 straight repairs of the ACL.
Six patients underwent more than 2 plasties.
Many associated interventions were treated a the same time as the ligamentoplasties in 45 cases.
The most frequent were: partial meniscectomies [39], meniscal repairs [10], foreign bodies [3], notch cleaning [20] and surface treatments [23].
Objective and subjective functional results were analysed (Tables 1–4). For anterior plasties (group 1), results seem to be satisfactory 79% of the time with a follow-up period longer than 2 years. In 18% of the cases, the results were poor.

Table 1. Group I anterior plasties (2–4 years follow-up)

	Subjective results		Objective results	
	Percentage of cases	Lysholm scale score (points)	Percentage of cases	
Excellent	42	≧95		
Very good	16	95–90	82	small residual Lachmann's sign
Good	21	90–85		
Average	3	85–60	10	important residual Lachmann's sign
Poor	18	<60		

Table 2. Group II anterior plasties (8 months–2 years follow-up)

	Percentage of cases	Subjective results: Lysholm scale score (points)	Objective results: Lachmann's sign	CALT
Excellent Very good Good	85	85–95	95% absent or small residual	40 patients average $D=3.99$
Average Failures	12 3	<85	3% small residual	

CALT, computerised accurate ligament tester.

Table 3. Percentage of satisfactory results

	Group I	Group II
First intervention	91%	92%
Re-intervention	47% (follow-up longer than 2 years)	57% (follow-up 12 months)

Table 4. Results of double and posterior plasties (follow-up between 1 and 13 years for 16 patients)

	Subjective results		Objective results	
		No. of cases		No. of cases
Posterior plasties	Excellent	7	Small Lachmann's sign	5
	Good	4	Small Lachmann's sign	3
	Average	2	Small Lachmann's sign	2
	Failure (break)	1	Large Lachmann's sign	1
Double plasty	Average after 2 years (painful)		Tight knee	

Large Lachmann's sign, before surgery (mm); small Lachmann's sign, after surgery (mm).

These were patients who had received repeated operations but who still grew worse after 2 years. For posterior plasties, 11 of 13 patients showed successful results because, as for double plasties, the surgery is less complex.
Finally, three topics must be emphasised. For most of the patients sporting activities can be attempted in the 3rd month after the operation. Nevertheless, 4.2% of the patients suffer from implant breaks and 6.1%, from synovitis.

Discussion

Analysis of failures helps to identify the basic problems. A very important point to consider is the patient's preoperative state (surgical history including arthroscopy).
There are two types of complications, those due to the implants and those due to the arthroscopy (Table 5). Implants can result in attrition which may lead to early breakage, especially if the positioning is too angular; this is avoidable. The assembly can be examined by scanner after implantation. Each mistake can increase attrition.
Implants can also cause reactional synovitis [4], which seems to originate in two ways. One is mechanical, resulting from particle emission subsequent to a mistake in the implant's assembly. In this case the patient may be cured by removal and replacement of the implant. The other is true intolerance, perhaps due to allergic responses to the foreign implant; removal of the implant is necessary, followed by a return to classical techniques.
Wrapping the Dacron with fascia lata resulted in an intense reaction, and the technique was abandoned.
Arthroscopic complications are rarer.
Thrombophlebitis can be avoided by systematic use of prophylaxis. The risk of reflex sympathetic dystrophy has been reduced by restricting tourniquet use to 60 min. However, if the surgical procedures are numerous or elaborate, it is difficult to complete the operation within this time.
Advantages of arthroscopic techniques are that only small incisions are made, that there is a lack of intra-articular damage and that recovery time is short.

Table 5. Complications in 131 patients

	No. of cases
Complications associated with Dacron implants	
Synovitis (four Dacron implant removals)	8 (6.1%)
Breaks	6 (4.5%)
Staples	5
Stiffness	1
Complications due to arthroscopy	
Defect in the assembly (rectified by arthroscopy)	2
Phlebitis (without after-effects)	2
"Aseptic" arthritis (Dacron implant removal)	1
Reflex sympathetic dystrophy (cured)	4

Disadvantages involve the fact that it is an artificial plasty, although this seems to be quite acceptable because of the small incisions, joint washing and immediate rehabilitation [5, 8, 9, 11, 12].

There is a risk of incomplete treatment when patients have a significant laxity of the peripheral component. It would be nonsense to treat everything with arthroscopy. Decay beyond 2 years is foreseeable, as patients who have undergone repeated operations may be cured of their instability, but develop pain because of arthrosis.

Nowadays, indications are restricted to young patients disabled by laxity, who are athletes determined to return to their sport within a precise time [6].

They should by psychologically examined and warned of the limitations of this method. The future improvement of these techniques relies on the discovery of new materials and better implant designs.

Summary

A total of 131 athletes, with an average age of 25 years, underwent a central pivot plasty with Dacron implants under arthroscopy (116 anterior, 13 posterior and 2 double). The artificial implant was inserted through small incisions and fixed to the bone with staples. The postoperative rehabilitation time is very short, and return to sport can be expected after 3 months. However, patients followed up between 8 months and 4 years reveal that the preoperative state influences the postoperative result.

Thus, in surgical re-interventions, results which are encouraging at the beginning worsen after 2 years.

The type of implant is important: reactional synovitis develops in 6.1% and breaks in 4.2%. The discovery of an improved implant design will allow a wider application of this non-damaging technique.

References

1. Bahuaud J, Ferro RM, Rebour J, Richard M (1984) Ligamentoplastie antérieure du genou sous contrôle arthroscopique. Presse Méd 13(43):2641–2643
2. Bahuaud J, Ferro RM, Richard M (1984) Ligamentoplastie antérieure au Dacron type Mac Intosh à „genou fermé“ sous contrôle arthroscopique. Les Laxités chroniques du genou. Masson, Paris, pp 114–119
3. Bahuaud J, Rebour J (1985) Techniques de ligamentoplasties artificielles du pivot central du genou sous arthroscopie. J Méd (Lyon) 66:1406 1406:471–475
4. Bahuaud J, Rebour J, Mayer J, Besse D, Mautalen F, Baspeyre H (1986) Les synovites réactionnelles au Dacron implantées en intra-articulaire au genou pour ligamentoplastie. 3ème Conférence internationale sur les biomatériaux nouveaux et biocompatibles, BIOMAT 86, 15 March 1986, Bordeaux
5. Bonnel F, Petit Bon E (1984) Biométrie des ligaments croisés du genou. Les laxités chroniques du genou. Masson, Paris, pp 200–206
6. Bousquet G (1984) Indications thérapeutiques dans les laxités chroniques du genou. 5ème Journées Lyonnaises de chirurgie du genou, SIMEP, 12–15 September, pp 211–217
7. Clancy WG, Shelbourne KD, Zoellner GB, Keene JS, Reider B, Rosenberg TD (1983) Treatment of knee joint instability secondary to rupture of the posterior cruciate ligament. J Bone Joint Surg [Am] 65:310–322

8. Chambat P, Perault F (1984) Ligaments artificiels du genou, cahier des charges d'une prothèse du ligament croisé antérieur. 5ème Journées Lyonnaises de chirurgie du genou, SIMEP, 12–15 September
9. Gacon G (1986) Problèmes généraux posés par l'utilisation des ligaments artificiels en chirurgie du genou. Traumatol Sport 3:3–5
10. Imberg JC (1984) Ligamentoplastie mixte extra et intra-articulaire sous arthroscopie (laxités antérieures chroniques). 5ème Journées Lyonnaises de chirurgie du genou, SIMEP, 12–15 September, pp 117–119
11. Mansat C, Gacon G, Alain JJ (1984) Plastie antéro-externe du genou. Intérêt du DACRON à propos de 500 cas. 5ème Journées Lyonnaises de chirurgie du genou, SIMEP, 12–15 September, pp 263–269
12. Woods GW (1985) Synthetics in anterior cruciate ligament reconstruction: a review. Clin North Am 16(2):227–235

Short-Term Results in Anterior Cruciate Ligament Reconstruction with a Dacron Prosthesis

T. Wredmark

Department of Orthopaedic Surgery, Huddinge University Hospital,
14186 Huddinge, Sweden

Introduction

So far, reconstruction of chronic anterior cruciate ligament (ACL) insufficiency is an unsolved problem. In recent years several synthetic materials have been developed as ligament substitutes. One such substitute is the high-strength Dacron (polyethylene-terephthalate) ligament with an ultimate breaking strength and stiffness twice that of a human cruciate ligament. The Dacron ligament is a double-knitted velour sheet containing for high-strength cords with X-ray indicators. The envelope and knitting technique allow fibrous ingrowth, enhancing anchorage to the bone. Thus, this implant seems suitable as a substitute for the ACL. This paper presents results from the first 42 patients treated for ACL insufficiency in our hospital with the Dacron ligament. Patients referred to us with a chronic ACL rupture are examined using arthroscopy for a proper diagnosis. In addition intra-articular pathology is simultaneously done. The patients are assigned a specific training programme for 3 months. A second assessment of the patients is then performed, and those with clinical symptoms are selected for surgery. The surgical procedure we used was originally described by Mansat (1983). The Dacron ligament is reinforced with an iliotibial strip and is inserted without the help of a drill-guide through drill-holes in the femur condyle and the tibial plateau. The implant is fixed with serrated staples. Postoperatively the patients are put on a continuous passive motion machine for 1 week. Protected weight-bearing is allowed for 3 weeks; thereafter, a gradual increase of physical activity is encouraged. After 3 months the patients start jogging and running. All patients are given prophylactic antibiotics for 3 days postoperatively.

Material

A total of 42 patients (27 males and 15 females), representing 37% of the ACL insufficiencies, were selected for surgery. The mean patient age was 26 years, and the average time from injury to surgery was 4 years. Three patients were lost to follow-up: one died, and two refused to participate, claiming that they had recovered. In all, 39 patients were observed for at least 1 year and 21 of these for 2 years. The patients were assessed by radiography of the knee joint, clinical examination including the Stryker arthrometer test, muscle strength using a Cybex II isokinetic dynamometer, and finally Lysholm's functional knee score.

Surgery and Arthroscopy of the Knee
2nd Congress of the European Society
Editors: Müller/Hackenbruch

Results

There were no postoperative complications such as infection of thrombosis in this study. Radiography detected rupture of the X-ray indicators incorporated in the Dacron ligament in 14 knees within 1 year of surgery. Before surgery the mean functional score of all patients was 63 points out of a maximum of 100. The score increased to 86 point 1 year after surgery and remained unchanged after 2 years. The comparative score between patients with ruptured or non-ruptured ACL did not differ 1 or 2 years after surgery.

Subjective complaint of instability is the single most important part in the functional score. Preoperatively all 42 patients had giving-way symptoms during daily activities or frequently while playing sports. One year postoperatively 22 patients never felt giving way during daily activity, 8 patients rarely, 3 frequently and 6 patients complained of experiencing giving way during daily activity. Two years postoperatively 11 patients never had giving way, 6 rarely had symptoms, 2 frequently and another 2 complained of giving-way symptoms in daily activity. Clinical examination revealed a positive pivot shift in all patients before surgery. After 1 year a negative pivot shift was found in 26 patients, 9 patients had a trace and 4 indicated a positive pivot shift. After 2 years a negative pivot shift was found in 14 patients, 5 had a trace and 2 a positive pivot shift. However, in the group with ACL ruptures the pivot shift sign could not be used to discriminate instability. Objective measurement of instability utilising the Lachmann test with a Stryker arthrometer indicated a preoperative mean of 9-mm difference between the normal and unstable knee. The mean difference was 6 mm in the ruptured ACL patient group and 3 mm in the non-ruptured ACL group 1 and 2 years postoperatively.

There was no difference between results of the Cybex II isokinetic muscle test before or after surgery, or between the ruptured ACL group and non-ruptured ACL group.

Discussion

In this series of 42 ACL reconstructions, we found 14 ruptures by radiography, apparently despite an improvement in clinical tests. Some of the patients with ruptured ligaments did not redevelop symptoms of instability during the short observation time, probably because they still maintained a low level of physical activity. Better results using this technique have been described earlier but without utilising radiography. Radiography turns out to be of considerable value for distinguishing early ruptures of an artificial ligament. The Stryker arthrometer test, which permits objective examination of sagittal instability, revealed a difference between ruptured and non-ruptured ligaments in this material.

However, so far the athlete's sensation of giving way is the final judgement of successful surgery. In the absence of a reliable objective method we must interpret the athlete's subjective sensation and express it as a functional score. In this material, though, this score was not sensitive enough to detect early ruptures.

Some of the patients, in whom ruptured ligaments were visualized by arthroscopy, have now been reoperated on. Evidence of bone ingrowth within the drill

holes, as has been earlier described was not found in any of the cases. Fibrous tissue was present within the envelope of the ligament, but not within the high-strength cords, which are only anchored by staples. The long distance between the anchorage points may increase the risk of ligament rupture.
Another possible explanation of failure could be impingement of bone on the Dacron ligament. The ruptures observed were located approximately 10 mm above the tibial plateau where the intercondylar notch impinges on the implant when the knee is in full extension. To prevent impingement, it is necessary to create a proper notchplasty, which was not done in this material.
From the cadaver experiments of Odensten (1985) and the computerized knee models of Arnoschky et al. (1985), it is evident that correct isometric positioning of the ACL reconstruction is mandatory to avoid overload stress. Such precision can only be fulfilled using an adequate drill-guide as described by Odensten (1985).

Conclusions

High-strength Dacron augmentation using this surgical technique results in a high failure rate. It seems the implant is too dense for bone ingrowth, thus giving improper anchorage.
Objective measurements of stability such as the Stryker arthrometer test, as well as visualisation of the implant either by radiography or arthroscopy are of considerable value in identifying failures early on in ACL reconstruction.

References

Arnoschky SP (1983) Anatomy of the anterior cruciate ligament. Clin Orthop 172:19
Lysholm J, Gillquist J (1982) Evaluation of knee ligament surgery results with special emphasis on use of a scoring scale. Am J Sports Med 10:150
Odensten M (1985) Treatment of the torn anterior ligament. Linköping University Medical Dissertation, No 177

Anterior Cruciate Ligament Reconstruction in the Chronically Unstable Knee Using an Expanded Polytetrafluoroethylene (PTFE) Prosthetic Ligament

[1] Suite 560, 1815 North Capitol Ave. Indianapolis, Indiana 46202, USA

[1] Suite 560 1815 North Capitol Ave. Indianapolis, Indiana 46202, USA
[2] 820 North Chelan, Wenatchee, Washington 98104, USA

Introduction

Reconstruction of the anterior cruciate ligament (ACL) in the multiply treated, grossly unstable knee is a great challenge. Numerous different methods of autogenous reconstruction exist. The procedures commonly used include the patellar tendon [2–4, 8, 11, 12, 15, 18, 19, 24, 30–32, 34, 37, 41], the iliotibial tract [5, 13, 14, 17, 20, 23, 25, 33, 40, 43–45], the semitendinosus [10, 39, 46], and the gracilis tendon [7, 16, 37, 38, 41, 47]. Frequently, where previous autogenous procedures have failed in patients we have utilized a prosthetic ligament for further reconstruction efforts.

Concentrated efforts directed toward developing prosthetic substitutes or synthetic augmentation in the reconstruction of the ACL have developed in recent years [35]. Currently, four broad categories of implants are being tested including: carbon filament fibers for augmentation and inducement of collagen growth, polymer fibers for augmentation and replacement, porous polymers, and biological tissues (bovine or human allografts). This report is a review of our clinical experience with the use of a porous polymer, expanded polytetrafluoroethylene (PTFE), in the reconstruction of the ACL in the chronically unstable knee.

Material and Methods

The prosthesis used in this study was the PTFE prosthetic ligament (W. L. Gore and Associates, Inc.). Bolton and Bruchman reported on the mechanical properties of the device [6]. The implant is composed essentially of pure PTFE that has been expanded to form a structure consisting of solid nodes interconnected by strong, highly oriented fibrils, with loops on each end for bony attachment. The resultant microstructure contains more than 70% air by volume and is characterized by fibrils averaging 60 μm in length. The porosity of the material is sufficient to allow ingrowth of surrounding fibrous and osseous tissues. The device is considered a nonbiodegradable scaffold and is a permanent implant.

The ultimate tensile strength of the device is approximately 5300 N as shown during Instron testing. The modulus of elasticity of the device at physiologic rates is 3200 MPa.

The present study is a prospective study of 30 patients with whom previous knee operations had failed or who present with gross knee instabilities. In all cases the

Surgery and Arthroscopy of the Knee
2nd Congress of the European Society
Editors: Müller/Hackenbruch

ACL was reconstructed using the PTFE prosthetic ligament. Patients were evaluated preoperatively and at regular 3-month intervals postoperatively. The knee evaluation protocol developed by Larson and Slocum was employed, which allows comparison between the treated and untreated knees with respect to pain relief, subjective instability, and function. Objective stability was assessed by conventional clinical tests. A score of 200 points represents a normal knee, allotting 100 points each for the subjective and objective evaluations.

Operative Procedure

The procedure is performed through a medial arthrotomy incision combined with a lateral suprapatellar incision. An area 2–3 cm distal to the tibial brim and 1 cm medial to the patellar tendon is prepared for insertion of a Kirschner wire. This wire is angled so that it emerges at the anatomic center of the ACL on the tibial plateau. A 7.9-mm cannulated reamer is used to drill over the Kirschner wire. The popliteal space is approached by blunt dissection between the lateral gastrocnemius head posteriorly and the femur anteriorly. The bone tunnel in the femur begins at a point 2–3 cm proximal and anterior to the lateral collateral ligament origin on the femur. A femoral pilot hole is drilled with a Kirschner wire starting at the preselected site on the lateral femoral epicondylar area, and it is angled posteriorly, distally, and medially so that the wire exits on the popliteal surface of the distal femur. This point should be extracapsular approximately 1–2 cm proximal to the capsular attachment and 1–2 cm lateral to the midline. Extreme care is taken to protect the popliteal structures during the drilling over the Kirschner wire by a 7.9-mm cannulated reamer. The edges of the drill hole are radiused with a rasp or high-speed deburring instrument. A 36-inch length of umbilical tape is prepared as a leader for the ligament. The tape is marked in the center, followed by marks placed 16 cm and 18 cm on either side of the center. A large curved instrument is passed from the anterior joint space posteriorly through the intercondylar notch, perforating the posterior joint capsule, and exiting at the opening between the lateral head of the gastrocnemius and the femur. The length of umbilical tape is folded in half; the looped end is then caught and pulled into the anterior joint. A small tendon passer is placed through the femoral tunnel to engage the umbilical tape leader in the popliteal space, and the leader is pulled through the tibial tunnel from the anterior joint. The umbilical tape leader is utilized to determine the correct length of the ligament to implant. The marked looped end is positioned 1.5 cm distal to the edge of the exit hole in the tibia, the tape is pulled taut so that the mark (either 16 or 18 cm) which lies 1.5 cm from the exit hole determines the ligament length. The knee is placed through a full range of motion while the ligament is taut; if improver positioning of the bony tunnels has occurred, then flexion or extension are limited. The looped end of the leader is passed through the loop in the ligament device, and the entire device is then passed through the looped end of the leader. Tension is applied to the leader, and the ligament device is aligned in a distal to proximal position. The femoral end of the ligament is attached first. A specially designed step drill bit is used to drill the hole for the screw. The hole is placed 1.5 cm from the edge of the bone tunnel opening and penetrates through to the opposite cortex. A self-tapping screw is then inserted. To account for the thickness of the loop on the ligament device, one-eighth of an inch is added to the depth gauge measurement to determine the appropriate length of the screw. For attachment of the ligament at the tibial end, the knee is placed in 30° of flexion, and tension is applied so that the ligament becomes taut. A punch mark is made at least 1.5 cm from the edge of the tibial bone tunnel, and the knee is then relaxed to 90° of flexion for drilling and screw insertion. A stepped bit is used to drill a hole perpendicular to the tibial surface. Again, one-eighth of an inch is added to the depth gauge measurement and selection of the appropriate screw length is made. The screw is inserted until it comes in contact with the opposite cortex of the tibia. A full range of motion should be obtainable after correct placement and fixation of the ligament.

Postoperative Rehabilitation

Postoperative rehabilitation is divided into five phases for patients who have associated reconstructive procedures combined with the insertion of the prosthetic ligament. Phase one (0–6 weeks) consists of cast brace immobilization allowing a 40°–70° arc of motion. Crutches to limit weight bearing to 10% are used during this phase. In phase two (6–16 weeks) the patient's motion is increased to a 30°–90° arc in the cast brace. Weight bearing is increased gradually to 100% at the end of 12 weeks. Active resistance exercises in a locked knee position are started, and swimming is allowed at the end of the phase. In phase three (16–26 weeks) a custom-molded, fabricated knee brace is fitted, and use of the cast brace is discontinued. Progressive resistance exercises are started, initially using high repetitions and low weights. Short arc exercises (30°–70°) are done for the first 6 weeks of this phase. The arc of motion and resistance are gradually increased; however, no terminal resistance exercises from 30°–0° are started until approximately 6 months. Stationary bicycling at high speed but against low resistance is allowed when motion is adequate.

In phase four (26–36 weeks) the muscle strengthening program is continued and low repetition, high weight exercises are started. When the strength of the treated knee reaches 75% of the uninvolved leg on isokinetic evaluation, jogging is allowed. The patient is allowed to increase speed and distance during running in phase five which usually starts at 36 weeks. Cutting and pivoting activities are resumed when the treated knee's strength reaches approximately 90% of the uninvolved leg. Brace wearing is encouraged for protection.

Patients receiving only the PTFE prosthetic ligament with no other associated reconstructive procedures are advanced much more rapidly. Knee movement is initiated by placing the knee in an arc of continuous passive motion in the recovery room. Straight leg raising with isometric quadricep and hamstring exercises is encouraged. The patient is allowed tolerable weight bearing with the support of crutches. At 2 weeks the patient is permitted to progress to 100% weight bearing, as tolerated, and to start a vigorous physical therapy program including progressive resistance exercises and an unrestricted range of motion.

Results

The first PTFE prosthetic ligament was inserted by one of the authors (RLL) in 1982. The aim of the treatment was to provide a knee capable of enabling the patient to perform activities of daily living without pain and constant instability from both a subjective and objective basis. Such an achievement was felt to be represented by a score of 140 points out of a possible 200.

Of the 30 patients evaluated in this series, 22 were male and 8 female. The average age was 27.5 years and the average follow-up period was 24 months. Ninety-seven previous operative procedures had been performed on the 30 patients (Table 1). Twelve previous intra-articular ACL reconstructive procedures had been performed, 22 extra-articular ACL procedures of various types had been attempted, and primary ligamentous repair had been performed seven times.

The average time from the initial injury to insertion of the prosthetic ACL was 52 months. All PTFE prosthetic ligaments were inserted through an open technique and were combined with other reconstructive procedures as the instability pattern indicated (Table 2). Eight patients had had no other procedure at the time

Table 1. Previous reconstructive procedures

Procedure	Number
Intra-articular ACL reconstruction	12
Extra-articular ACL reconstruction	22
Medial collateral ligament reconstruction	7
Primary ligamentous repair	7
Capsular repair or advancement	7
Medial meniscectomy	15
Lateral meniscectomy	12
Open debridement	4
Arthroscopy	6
Other	5

ACL, anterior cruciate ligament.

Table 2. Procedures associated with polytetrafluoroethylene ligament insertion

Procedure	Number
Ellison	4
Iliotibial band tenodesis	3
Pes anserinus transfer	4
Nicholas five-in-one	1
MCL reconstruction	4
Arcuate complex advancement	2
Medial meniscectomy	5
Lateral meniscectomy	4
PCL reconstruction	1
Tibial osteotomy	1

MCL, medial collateral ligament; PCL, posterior cruciate ligament.

of the prosthetic ligament insertion and seven other patients had only meniscectomy combined with the device placement. These 7 patients raised the total number of patients who had a (partial or complete) meniscectomy prior to or during PTFE prosthetic ligament placement to 24. Among these 24, 8 had both menisci excised.

Overall, 83% of the patients achieved a satisfactory result during the review period; only five patients showed unsatisfactory results. Preoperatively, none of the patients reviewed had a total score higher than 140 points.

Objectively, knee stability was improved markedly, as graded by the Lachmann test, in 87% of the patients; 13% of the patients persisted with a Lachmann test of greater than 5 mm. The anterior drawer sign was significantly improved in 80% of the patients. The pivot shift test was improved in 87% of the cases. Only four patients (13%) continued to have a pivot shift present on examination at follow-up. Subjectively, knee stability corresponded well with the previously noted objective findings. Only 10% had actual giving way in the review period; 80% of the patients achieved subjective improvement in knee stability. Some patients noted a sensation of giving way in the initial postoperative period; however, the sensation resolved as muscle strength was regained in later stages.

Pain was improved in 67% of the patients, but only 30% had complete pain relief. Of the remaining 70%, 43% had mild pain, 20 moderate pain, and 7% severe pain. Chondromalacia or articular cartilage defects were noted in 70% at the time of PTFE prosthetic ligament insertion. Only 30% had no defects. The persistence of pain postoperatively seems to correlate with the incidence of articular cartilage changes. Of the 70% with residual pain, only those cases with persistent instability were graded as unsatisfactory results.

Functional return to activities of daily living is markedly improved. Only 10% of the patients stated they were unable to perform activities of daily life. The five patients who were graded as unsatisfactory were subjectively the only patients who felt that their knees were unimproved by the procedure.

There are two major complications. One patient developed sinus tracts postoperatively. After initial failed attempts at debridement, the prosthesis was removed after approximately 1 year. A second patient broke the prosthesis 1 year postoperatively during normal daily activities. During removal of the broken prosthesis, a prominent posterior tibial spine was noted which appeared to have been impinging on the prosthesis. Both of these patients were classified as unsatisfactory. "A second look" was taken arthroscopically in selected cases where prosthetic ligament failure was suspected.

Discussion

The lack of consistently good results of ACL reconstruction has initiated attempts to develop an acceptable design for a prosthetic ligament. A historical review reveals that early reports of synthetic ligaments in the canine stifle joint appeared in the veterinary literature. In 1960 Johnson reported on braided nylon [29]. However, Nylon (polyamide) lost its strength during in vivo testing and stretched. Butler (1964) reported on Teflon (tetrafluoroethene homopolymer) mesh as a prosthetic material substituted for the ACL in 15 dogs and cats. He found that it elicited little or no tissue reaction in the joint and maintained its strength after implantation [9]. Gupta (1969) introduced braided Dacron (polyethylene terephthalate) as a synthetic ligament in dogs and recognized that the ligament acted as a stent while repair occurred [21].

The first synthetic ligament available for commercial use in North America was an ultra-high molecular weight polyethylene ligament. Biomechanical studies warned that its tensile strength was lower than that of the ACL, and caution was recommended in its clinical use for ACL reconstruction [20]. The importance of elasticity in the ACL is emphasized by Laurin and coworkers [48]. They introduced a composite prosthesis using a silastic compressible core enveloped by a Dacron suture to provide elasticity to accommodate for changes in length during flexion and extension. Carbon-fiber implants have been evaluated by Jenkins [28]. Recently, in an attempt to improve the properties of carbon fiber, a polylactic acid polymer coating has been added [1]. Kennedy worked closely with the development of a diamond-braided polypropylene synthetic which is currently undergoing clinical trials [35].

A prosthetic stent was developed from Proplast, a porous material capable of soft tissue ingrowth. Proplast is a composite of three materials: a core of polyxaramid fibers, surrounded by a fluorinated ethylenepropylene polymer; a coating of a porous, low-modulus composite of tetrafluoroethylene; and a vitreous carbon. The preliminary results were reported by James [27], who concluded that the prosthesis provided temporary stability while allowing early joint mobility during the postreconstruction phase. The Proplast experience was recently updated in a report on 45 patients with ACL, and 65 patients with posterior cruciate ligament (PCL) reconstruction [26]. The report on 45 patients with ACL reconstruction provided a comparison group for our present study using the expanded PTFE prosthetic ligament.

Both clinical series were composed primarily of knees that had undergone multiple operations or knees with gross instability. The same 200-point knee grading scale was used to evaluate both study groups. A score of 140 points or more was considered satisfactory in both series. The comparative results of the two series are illustrated in Table 2. Subjectively, the PTFE group showed a marked in-

Table 3. Comparison of Proplast and Polytetrafluoroethylene prosthetic ligaments

	PTFE	Proplast
Number of patients	30	45
Average age	27.5 years	27.4 years
Average number of prior procedures	3	2
Percentage satisfactory results (>140 points)	83	52
Percentage with improvement in subjective instability	80	63
Percentage with improvement in functional level	83	50
Percentage with improvement in pain level	67	60
Percentage with improvement in anterior drawer test	80	55
Percentage with improvement in pivot shift	87	70
Number of prosthetic failures	1	35

PTFE, polytetrafluoroethylene.

crease in the percentage of patients who improved in the areas of instability (17% greater) and function (33% greater) as compared with the Proplast group.
In the PTFE group the number of patients showing an improvement in pain relief was only small (7% greater) compared to the Proplast group. This may be explained by the high incidence (70%) of articular cartilage changes noted at the time of prosthesis insertion. Preoperatively, patients selected for PTFE prosthetic ligament reconstruction were felt to be restricted by instability rather than by pain. This explains the fact that among the 70% of patients in which pain persisted postoperatively, only those in which instability was unimproved were graded as unsatisfactory. Objectively, the PTFE group showed a much higher percentage of patients improving in stability, for both the anterior drawer test (25% greater) and the pivot shift test (17% greater), compared to the Proplast series. The Lachmann test was not reported in the Proplast group, so no comparison can be made. In the PTFE group the improvement in both the anterior drawer test (80%) and the Lachmann test (87%) was noted. The slightly less favorable improvement in the anterior drawer test has a mechanical explanation. The PTFE device allows some laxity to be present at 90° of knee flexion due to the nonanatomic position of the ACL reconstruction using a PTFE device. The device is taut at approximately 20° of flexion as indicated by the higher improvement rate obtained in Lachmann testing compared to the anterior drawer test.
KT-1000 arthrometer testing and Cybex II isokinetic evaluations were performed postoperatively on most of the patients in the PTFE series. However, this data was not available in the Proplast group and no comparisons could be made.
Prosthetic breakage was a major problem in the Proplast group. Thirty-five ligaments broke with an average life span of 13 months. In the PTFE group only one prosthesis has broken to date, a fact which may be attributed to the ultimate tensile strength of the device (5300 N). However, care must be taken to carefully radius the tibial and femoral tunnels, or the lifetime of the device may be shortened. Also, a notchplasty usually prevents bone impinging on the device. The case that sustained a rupture of the PTFE device in our series had a prominent posterior tibial spine, noted at the time of prosthesis removal, which may have been impinging on the device. Infection is always a concern with implantation of prosthetic devices. One infection occurred in our series, which resulted in removal of the prosthesis. We encourage strict adherence to sterile surgical technique and the use of the same precautions that would be employed in performing total joint arthroplasty to avoid postoperative infection.
In conclusion, the PTFE prosthetic ligament compares quite favorably with the Proplast ligament in reconstruction of the ACL. The 83% satisfactory results obtained with the PTFE device compared to the 52% success rate reported on using the Proplast device is encouraging. However, we will continue to monitor the early results in the series and encourage future long-term studies of the PTFE prosthetic ligament in the reconstruction of the ACL.

References

1. Alexander H et al. (1981) Canine patellar tendon replacement. Orthop Rev 10:11
2. Alm A (1973) Survival of part of patellar tendon transposed for reconstruction of the anterior cruciate ligament. Acta Chir Scand 139:443–447

3. Alm A, Strombers B (1974) Transposed medial third of the patella ligament in reconstruction of the anterior cruciate ligament. Acta Chir Scand [Suppl] 445:37–49
4. Alm A, Gillquist J, Strombers B (1974) The medial third of the patella ligament in reconstruction of the anterior cruciate ligament. Acta Chir Scand [Suppl] 445:5–13
5. Arnold JA, Coker TP, Heaton LM, Park JP, Harris WD (1979) Natural history of anterior cruciate tears. Am J Sports Med 7:305–313
6. Bolton W, Bruchman B (1983) Mechanical and biological properties of GORE-TEX expanded polytetrafluoroethylene (PTFE) prosthetic ligament. Aktuel Probl Chir Orthop 26:40
7. Bousquet G (1974) Chirurgia del Ginocchio. Rome, Verduce, pp 191–194
8. Butler DL (1979) Mechanical properties of transplants for the anterior cruciate ligament. Trans Orthop Res Soc 4:81
9. Butler HC (1964) Teflon as a prosthetic ligament in repair of ruptured anterior cruciate ligaments. Am J Vet Res 25:55–59
10. Cho KO (1975) Reconstruction of the anterior cruciate ligament by semitendinosus. J Bone Joint Surg 57:608–612
11. Clancy WG, Narechania RG, Rosenberg TD, Gmeiner J, Wisnefske DD, Lange TA (1981) Anterior and posterior cruciate ligament reconstruction in Rhesus monkeys. J Bone Joint Surg [Am] 63:1270–1284
12. Drez D (1978) Modified Eriksson procedure for chronic anterior cruciate instability. Orthopedics 1:30–36
13. Ellison AE (1979) Distal iliotibial-band transfer for anterolateral rotatory instability of the knee. J Bone Joint Surg [Am] 61:330–337
14. Ellison AE (1980) Pathogenesis and treatment of anterolateral rotatory instability. Clin Orthop 147:51–55
15. Eriksson E (1976) Reconstruction of the anterior cruciate ligament. Orthop Clin North Am 7:167–179
16. Ficat P, Cuzac JP, Ricci A (1975) Chirurgie reparatrice des laxites chroniques des ligaments croises du gnou. Rev Chir Orthop 61:89–100
17. Fox JM (1980) Extra-articular stabilization of the knee joint for anterior instability. Clin Orthop 147:56–61
18. Gillquist J, Liljedahl SO, Lindvall H (1971) Reconstruction for old rupture of the anterior cruciate ligament. A follow-up study. Injury 2:271–278
19. Ginsberg JH, Whiteside LA, Piper TL (1980) Nutrient pathways in transferred patellar tendon used for anterior cruciate ligament reconstruction. Am J Sports Med 8:15–18
20. Grood ES, Noyes FR (1976) Cruciate ligament prosthesis: strength, creep, and fatigue properties. J Bone Joint Surg [Am] 58:1083–1088
21. Gupta BN, Brinker WO (1969) Anterior cruciate ligament prosthesis in the dog. J Am Vet Med Assoc 154:1057–1061
22. Hanks GA, Joyner DM, Kalenak A (1981) Anterolateral rotatory instability of the knee: an analysis of the Ellison procedure. Am J Sports Med 9:225–231
23. Insall JN, Joseph DM, Aglietti P, Campbell RD (1981) Bone block iliotibial band transfer for anterior cruciate insufficiency. J Bone Joint Surg [Am] 63:560–569
24. Ivey FM, Blazina ME, Fox JM, Del Pizzo W (1980) Intra-articular substitution for anterior cruciate insufficiency: a clinical comparison between patellar tendon and meniscus. Am J Sports Med 8:405–410
25. James SL (1981) Biomechanics of knee ligament reconstruction. Clin Orthop 146:90–100
26. James SL, Kellam JF, Slocum DB, Larson RL (1983) The proplast prosthetic ligament stent as a replacement for the cruciate ligaments of the knee. Aktuel Probl Chir Orthop 26:116–120
27. James SL, Woods GW, Homsy CA, Prewitt JM, Slocum DB (1980) Biomechanics of knee ligament reconstruction. Clin Orthop 146:90–100
28. Jenkins DHR, McKibbin B (1980) The role of flexible carbon-fibre implants as tendon and ligament substitutes in clinical practice. J Bone Joint Surg [Br] 62:497–499
29. Johnson FL (1960) Prosthetic anterior cruciate ligament. JAMA 37:846–847

30. Jones KG (1963) Reconstruction of the anterior cruciate ligament. A technique using the central one-third of the patellar ligament. J Bone Joint Surg [Am] 45:925–932
31. Jones KG (1970) Reconstruction of the anterior cruciate ligament using the central one-third of the patellar ligament, a follow-up report. J Bone Joint Surg [Am] 52:1302–1308
32. Jones KG (1980) Results of use of the central one-third of the patellar ligament to compensate for anterior cruciate ligament deficiency. Clin Orthop 147:39–44
33. Kennedy JC (1978) Anterolateral rotatory instability of the knee joint. J Bone Joint Surg [Am] 60:1031–1039
34. Kennedy JC, Roth JH, Mendenhall HV, Sanford JB (1980) Presidential address. Intraarticular replacement in the anterior cruciate ligament deficient knee. Am J Sports Med 8:1–14
35. Kennedy JC (1983) Application of prosthetics to anterior cruciate ligament reconstruction and repair. Clin Orthop 172:125–128
36. Lam SJS (1968) Reconstruction of the anterior cruciate ligament using the Jones procedure and its Guy's hospital modification. J Bone Joint Surg 50:1213–1223
37. Lange M (1962) Orthopädische-chirurgische Operation, 2nd edn. Bergmann, München, pp 692–700
38. Lipscomb AB, Johnston RK, Snyder RB (1981) The technique of cruciate ligament reconstruction. Am J Sports Med 9:77–81
39. Lipscomb AB, Johnston RK, Snyder RB, Brothers JC (1979) Secondary reconstruction of anterior cruciate ligament in athletes by using a semitendinosus tendon. Am J Sports Med 7:81–84
40. Losee RE, Johnson TR, Southwick WD (1978) Anterior subluxation of the lateral tibial plateau. J Bone Joint Surg [Am] 60:115–1030
41. Mace HB (1939) A new operative procedure for repair of ruptured cruciate ligaments of the knee joint. Surg Gynecol Obstet 69:108–109
42. Marshall JL (1979) Reconstruction of functioning anterior cruciate ligament, preliminary report using quadriceps tendon. Orthop Rev 8:49–55
43. Nicholas JA, Minkoff J (1978) Iliotibial-band transfer through the intercondylar notch for combined anterior instability (ITBT procedure). Am J Sports Med 6:341–353
44. O'Dongohue DH (1963) A method for replacement of the anterior cruciate ligament of the knee. J Bone Joint Surg [Am] 45:905–924
45. O'Dongohue DH, Frank GR, Jeter GL, Johnson W, Zeiders JW, Kenyon R (1971) Repair and reconstruction of the anterior cruciate ligament in dogs. J Bone Joint Surg [Am] 53:710–718
46. Puddu G (1980) Method for reconstruction of the anterior cruciate ligament using the semitendinosus tendon. Am J Sports Med 8:402–404
47. Ramadier JO, Benoit J (1972) Reconstruction des ligaments lateraux et croises syndesmosplasties due genou. Rev Chir Orthop 58:78–84
48. Tremblay GR, Laurin Carroll A, Drovin G (1980) The challenge of prothetic cruciate ligament replacement. Clin Orthop 147:88–92
49. Larson RL, Slocum BD (1983) Am J Sports Med II(4): 240

The Use of Proflex in the Repair of Anterior Cruciate Ligament Lesions

M. Allard, J. Bahuaud, L. Barbat, C. Chaix, G. Gacon, F. Gougeon, R. Ferro, and Ch. Mansat *

* Clinique de l'Union 31240 L'Union, France

Reconstruction of the anterior cruciate ligament (ACL) is a very controversial problem as regards both its indications and the technique used. That the functional tolerance of ACL lesions is sometimes surprising should serve as a serious caveat in assessing any findings. Some authors believe, for example, that a good muscular covering can compensate for its absence. Medium- and long-term prognoses tend to be dominated by the risk of peripheral capsuloligamentary lesions becoming aggravated or of a meniscal lesion appearing or decompensating, and by the frequency and occasional gravity of cartilaginous lesions.

An optimal technique for reconstructing the ACL must improve stability and decrease laxity, i.e. it must encourage both functional and objective improvement. Only the disappearance of laxity is capable of protecting the other anatomical structures of the knee. Most techniques in common use provide, in the short term, a not inconsiderable rate of satisfactory results, regardless of whether the technique is active, passive, mixed active/passive, extra-articular or intra-articular.

In the short term one often observes satisfactory functional results in the case of knees which, in objective terms, are disturbing. With time, however, it appears that only objectively satisfactory knees experience no deterioration in functional terms. It is highly probable that only plasties aimed at replacing the damaged structures in their form, solidity and function are capable of protecting the articular surfaces from osteoarthrosis. Since 1979 we have favoured a technique of syndesmoplasty with a prosthetic implant which permits early resumption of the articular function. This technique has developed over time, as has the implant which we currently use.

We shall examine successively the reasons which we feel justify the use of a prosthetic implant, the characteristics of this implant, the technique of inserting it, the re-education process and our initial results.

Justification for the Prosthetic Implant

Unreliability of the Various Syndesmoplastic Techniques. Most results published after a sufficient lapse of time have reflected the frequency of deterioration in the objective result and, hence, after a few years a functional deterioration and non-protection of the joint in a not inconsiderable proportion of cases. As regards the use of the articular autotransplants we consider the synovial medium to be ill-suited to ensuring the survival in mechanically satisfactory conditions of the transplanted ligament pending its possible revascularisation.

Surgery and Arthroscopy of the Knee
2nd Congress of the European Society
Editors: Müller/Hackenbruch

Absence of Iatrogenic Effect. We consider a respect for the anatomical structures of the knee to be essential. Most techniques use either a fragment of the ligamentum patellae extending over the patella and the quadriceps tendon, a more or less large fragment of the fascia lata or one or more of the pes anserinus tendons. These removals generally cause no inconvenience to the joint; this can be judged when attempts to repeat the operations fail.

Technical Simplicity. The functional consequences of ACL lesions are not dramatic. At worst a resumption of sports activity is not possible, a change of sport is desirable or certain precautions must be observed in leisure or professional activities. The surgical act, in its simplicity, its atraumatic nature and the absence of postoperative constraints, must consider such possible disablement. This is also true of techniques using prosthetic implants: these can be performed by arthrotomy or by arthroscopy. Failure brings neither problems nor consequences, for the knee is comparable before the operation and after the failure. And, further surgery begins at the same level as the first operation and poses no additional problems.

Possibility of Early Re-education. The desire to avoid postoperative immobilisation through plaster has been the decisive factor in our preference for ligamentary prosthetic implants. We no longer encounter postoperative stiffness, and algodystrophias are rare. Early re-education, i.e. active and passive mobilisation, is furthermore not without value in preventing thromboembolism.

Characteristics of the Proflex Ligament

Prosthetic replacement of the ACL is not a simple matter. Many attempts at this have been made over the past 30 years but with generally disappointing results. This is largely due to the very specific functioning of elements in the central pivot of the knee, particularly the ACL. This complex anatomical structure consists of three bundles of fibres of unequal length, poorly vascularised but well innervated, intra-articular but extra-synovial, and with broad surfaces of insertion into the tibia and the femur. In flexion-extension movements it undergoes torsion movements through rotation of the tibia in relation to the femur and, hence, through the change of orientation in its articular surfaces. The torsion of the ACL is exaggerated in the internal rotation of the tibia. Resistance and elasticity vary negatively with age.

This cruciate ligament is subjected to a number of demands of varying frequency and intensity in various types of activity (walking, going up and going down stairs, slowly running, fast running and various sports activities). These demands have been calculated by Morrison as quoted by Chen and Black, whose findings suggest that the tension demands in certain activities approach the maximum resistance of the ligament. It is protected, however, by a powerful muscular covering which is quickly triggered by a proprioceptive reflex beginning in an intra-ligamentary position. This explains the absence of rhexes in some sports activities in which the intensity of exertion sometimes far exceeds that which the cruciate ligament alone could withstand.

The design and construction of a prosthetic ligament must therefore comply with a number of technical requirements concerning form, resistance, elasticity and fatigue. As it is not innervated, its resistance should be somewhat higher than that of the ACL in order to withstand such exertion without causing serious osteoarticular deterioration when strong jarring is experienced. Its elasticity must approximate that of the cruciate ligament, so that it can fit better into the kinetics of the joint and thereby protect itself against the risk of tearing. In the sometimes frequent and intense tensional demands of various activities it must not slacken, for this would render it totally ineffective.
The Proflex ligament consists of 15 tube layers, braided around one another, and composed of Tersuisse (polyethylene terephthalate, Terylene) fibres. Each implant consists of three parts, a central functional part, surrounded by a section for fixation, and a third part for handling of the ligament, in particular for passing it through the capsule and the tibial tunnel.
On the biomechanical level the resistance of the ligament was tested in intensity and repetition cycles on an Instrom apparatus in a Ringer's solution at 37 °C. Over a million cycles at 650 Nw were conducted at the upper resistance limit of a normal ligament (corresponding to several years of use) without any sign of tearing or fatigue. Its maximum resistance is greater than 4500 Nw, with elasticity close to 11%. The only problem in repeated, intense exertion lies in the risk of slackening, which seems to be virtually non-existent for normal activity and some 2%–3% for activities involving tension over 1000 Nw. Pre-tensioning before the operation tends to reduce this risk of slackening considerably.
Because of its shape this ligament can also be correctly fixed to the articular surfaces using staples and a tight-fitting passage through the capsule and the tibial tunnel. Due to its biomechanical characteristics the implantation of this ligament is performed under some tension so as to obtain, at the end of the operation, a laxity comparable with that of a normal joint (Lachman 3–6/100 Nw).
The biocompatibility feature of the present study has been conducted by the Dental Institute Laboratory of Basle University.

Implantation Technique

The entrance to the tibial tunnel must be located at the centre of the ACL front insertion surface. This orifice is, therefore, relatively far back. We consider it important particularly to avoid conflict with the top part of the condylar fossa during extension of the knee.
The "over-the-top" passage to the back seems to give most support here, despite the criticism that this is not isometric. This simplifies the technique. More importantly, however, it avoids bending, which enables the prosthetic ligament to work in a straight line from the passage through the capsule to the exit from the tibial tunnel at the front. In time the artificial ligament adheres to the capsule, thus positioning the posterior insertion of the ligament very close to the anatomical location. We feel that the transcondylar passage exposes the implant to rhexis or to detachment from osteolysis on the front slope of the tunnel due to tension upon the ligament.

The prosthetic ligament must be located at the centre of the fossa in order to avoid all contact. The tibial orifice must be nearer still to the internal condyle, for in extension it tends to approach the external condyle as a result of automatic external rotation of the tibia.

The ligament is exposed to 100–150 Nw tension at about 60° flexion, depending on the degree of physiological laxity of the normal and on the condition of articular surfaces. Placing the ligament under too much tension can compromise the kinetics of the articular surfaces or aggravate existing cartilaginous lesions.

The Proflex ligament can be inserted either by arthrotomy or by arthroscopy. We prefer the former because of its technical simplicity, particularly considering that the arthroscopic technique eliminates only one incision of the three. Three instruments are necessary: a drill guide for the tibial tunnel, a tension instrument and an instrument for passing the ligament through the subcondylar passage. During the operation the patient lies in supine position with the knee bent at 60°, resting upon a support.

Stages of the Operation

First (Internal) Stage

Two incisions are made, an upper arthrotomous one of 4–5 cm for the meniscus and a lower arthrotomous one 4–5 cm below the interarticular space in the prolongation of the former, 2 cm in length. The drill guide is inserted, with the articular point positioned at the centre of the ACL front surface. This surface is easy to find because of virtually constant vestiges. The synovial sheath of the ACL, which is often present, must be respected as much as possible. The lower part of the drill guide is placed in the lower incision. The tunnel is made with a 7-mm drill. Articular penetration is smoothed to avoid sharp angles. A second orifice is made above the entrance to the tibial tunnel for insertion of the end of the ligament at termination of the operation. Then the first string is passed.

Second (External) Stage

An external incision of 4–5 cm is made in the axis of the fibres of Maissiat's band, starting from the posterior edge of the external condyle. Maissiat's band is incised. The lower edge of the vastus lateralis and the front part of the lateral intermuscular septum are located. Disinsertion of the femur for 1–2 cm of this septum is carried out, respecting the fibres furthest forward (Kaplan's fibres). A soft dissection opens the way to the intercondylar fossa. An instrument adapted for passing the ligament follows this path, traverses the posterior capsule and follows the axial face of the external condyle, avoiding the posterior cruciate ligament (PCL). The second string is located by the upper internal incision. Using this second string the first string can be pulled towards the back. The Proflex ligament is passed from back to front and is fixed posteriorly by staples above the external condyle. The end of the ligament is inserted into a boundary tunnel drilled to a depth of 2 cm by a 7-mm drill. External closure is performed with aspiration.

Third (Internal) Stage

The tension instrument is positioned, after checking that the ligament is correctly positioned and can slide in the tibial tunnel. Substantial pulling of the ligament is sometimes necessary as vestiges may obstruct the movement of the implant in the tibial tunnel. Fixation is performed with staples after the ligament has been placed under tension. The end of the ligament is inserted into the groove made for this purpose after truncation. Then internal closure is carried out with aspiration.

Post-operative Care and Rehabilitation

During the first 48 h following operation, the patient is confined to bed with the leg raised and is treated with a compression bandage and aspiration. Analgesic medical treatment is given on request. In addition the patient receives antibiotic therapy and platelet anti-therapy.

Rehabilitation begins on the third day, with walking authorised but with a protective apparatus. From the fourth or fifth day this apparatus may be eliminated. Passive and active rehabilitation gradually facilitates the recovery of near-normal extension and of flexion to approximately 90°.

Upon leaving the clinic, rehabilitation continues in two stages. The first stage conventionally takes place indoors. It aims at recovering mobility and muscular strength, particularly that of the quadriceps, by means of passive and active exercises. Dynamic and proprioceptive exercises are gradually integrated into the rehabilitation programme, depending upon the degree of recuperation. The second stage involves the gradual resumption of sports activity through exercises in the stadium and under the supervision of a re-educator. Exercises must put the operated knee under increasing pressure to increase muscular strength. It is sometimes necessary to use a medical treatment based on non-steroid, anti-inflammatory agents in the event of pain or swelling upon resuming sports activities.

A functional, objective and radiological balance sheet is drawn up in the 3rd month, at which time it is decided whether to resume sports training or competition.

Case Histories and Results

A total of 423 case reports from several services have been analysed by computer (Table 1). The first ligament was implanted in January 1985. From January to June the ligament was tensioned manually without instrumental laximetric control. From July 1985 tensioning was quantified and a laximetric study was made from two sides before the operation. Patients were checked after 3 months, 6 months and 1 year.

Some of the knees had already been previously operated on for meniscal, ligamentous or patellar difficulties (Table 2). The ACL repair was sometimes associated with either a meniscus operation (partial or total meniscectomy or suture of the meniscus) or a ligament operation in the external or internal compartment (Table 3).

Method of Assessing Results. Each patient's file was composed in terms of a special index including the following information: pre-operative balance sheet, operative technique, postoperative care, functional score (Lysholm score), objective clinical and instrumental score and complications.

Table 1. Patients' statistics

Average age (years)	27 (15–55)
Sex	
Male	79%
Female	21%
Side of injury	
Right	58%
Left	42%
Classification	
Acute	10%
Chronic	90%
Number of cases	423

Table 2. Causes of previous knee operations

Meniscus problems	
Meniscus medialis	28%
Meniscus lateralis	8%
Ligament problems	
Extra-articular	7%
Intra-articular	17%
Patellar problems	3%
No previous operation	37%

Table 3. Associated operative goals with anterior cruciate ligament repair

Meniscus	
Meniscus medialis	45%
Meniscus lateralis	15%
Ligament	
Medial	21%
Lateral	19%
Posterior	2%

Complications

Of the 423 knees examined 91.5% exhibited normal conditions, i.e. rapid recuperation of mobility, resumption of normal walking generally in the first month and some sports activity in the second and third months. We noted 36 complications; these were minor in 27 cases, or 6.4% of all patients and major in 9, or 2.1% (Table 4).

Synovial reactions were found to be relatively rare. They were classified into four degrees, ranging from a simple mechanical reaction, simple or more serious inflammatory reaction to articular sepsis. Swelling was observed in 24% of the 423

Table 4. Postoperative complications

	Number of cases	Percent
Minor complications		
Minor synovial reaction	21	5.0
Hemarthrosis	2	0.5
Algodystrophy	3	0.7
Skin problem	1	0.2
Major complications		
Sepsis (infection)	2	0.5
Rupture	3	0.7
Major inflammatory reaction	4	0.9

Table 5. Activity level and incidence of swelling

Activity level	At 3 months (%) ($n=330$)	At 12 months (%) ($n=51$)
Light exercise	2	2
Moderate exercise	3	0
Strenuous exercise	19	12

cases at 3 months and in 14% at 1 year (Table 5). It should be pointed out that most late synovial reactions, whether requiring puncture or not, were observed in knees with cartilaginous lesions and which had been previously operated. Almost 30% of chronic laxities had already shown the effect of mechanical hydrarthrosis prior to operation.

Functional Results

We adopted Lysholm's method of functional quantification for a scale with a maximum of 100 points. The parameters included were the following: limping, walking with cane, climbing stairs, squatting, blockage, pain, swelling and instability. Mean functional scores are presented in Table 6 and distributions by categories in Table 7. We also considered it of interest to analyse the causes for loss of points on this scale. The percentage of cases involving loss of points due to insufficiency on each of the eight composite parameters is presented in Table 8.

Table 6. Mean functional scores

Time of measurement	Mean	Number of cases
Preoperative	59	184
Postoperative		
At 3 months	91	330
At 12 months	94	51

Table 7. Distribution of functional scores

Score	Time of measurement		
	Pre-operative %	At 3 months %	At 12 months %
100–95	1	42	67
94–90	2	29	18
89–85	4	14	2
< 85	93	15	14

Table 8. Causes of scale-point loss by parameter

Insufficient parameter	Time of measurement		
	Preoperative (%)	At 3 months (%)	A 12 months (%)
Pain	91	66	39
Squatting	76	56	29
Instability	97	15	16
Swelling	77	24	14
Blockage	55	9	12
Climbing stairs	57	15	4
Limping	48	11	2
Walking with cane	13	1	0

The functional findings indicate that 81% of sportsmen at 3 months and 85% at 12 months were able to resume their sports activities, whereas 93% had been unable to do so prior to operation.
Before the operation the two symptoms most frequently motivating consultation were instability (97% of cases) and pain during exertion (91%). Swelling was present in 77% of cases, one-third of which had to be drained. At 3 and at 12 months 85% of knees were perfectly stable. Slight pain under exertion fell from 66% of cases at 3 months to 39% at 1 year. Swelling dropped from 24% at 3 months to 14% at 1 year.

Objective Results

Clinical Aspects. For each patient we noted the various frontal anteroposterior and sagittal laxities. We have analysed here only the findings obtained on two characteristic lesions of the ACL, the Trillat-Lachman test and the pivot-shift test, we present the findings in Table 9. In over 90% of cases at 3 months and at 1 year we obtained complete or almost complete disappearance of the pivot shift and of the Trillat-Lachman test. The continued presence of some instances of Lachman test (11.9% of cases at 3 months and 16.7% of cases at 1 year) and of a slight pivot shift (9.7% at 3 months and 35.6% at 1 year) may be explained by the elasticity of some implants and by their position at the top.

Table 9. Objective results: Trillat-Lachman and pivot-shift tests

	Time of measurement			
	At 3 months (%) ($n=329$)		At 12 months (%) ($n=49$)	
Trillat-Lachman				
0	85.8	97.7	75.0	91.7
+ (trace)	11.9		16.7	
++	2.3		8.3	
+++	0.0		0.0	
Pivot-shift				
0	90.0	99.7	60.0	95.6
+ (trace)	9.7		35.6	
++	0.3		4.4	
+++	0.0		0.0	

Table 10. Laximetric results (mm)

	Time of measurement		
	Preoperative ($n=175$)	At 3 months ($n=239$)	A 12 months ($n=47$)
Difference	4.5 (SD 3.4)	0.3 (SD 2.3)	1.3 (SD 2.9)
Gain		4.2	3.2

Instrumental Aspects. Several types of apparatus were used (e.g. CALT $_K$T 1000 Multistress Telos) in order to evaluate the instrumental features of recovery. In most cases (except those operated on before July 1985) we checked both knees before the operation and at each consultation. To reduce the cause of frequent errors with these techniques we have chosen here to use the laximeter. In this measurement we derive both a pre-operative and a postoperative score, each representing the difference between the measured pathological knee minus the normal knee (calibrated in millimeters). The relative "gain" represents the difference between the pre-operative score minus the postoperative score. Our results are presented in Table 10. These quantitative results must be interpreted with substantial caution, however, as errors are possible due to the technique, the apparatus, the patient and the examiner.
What needs to be noted here is that the average difference between the two knees before operation was 4.5 mm and that the average postoperative gain was 2.4 mm at the 3rd month, with a loss of 1 mm at the check after 12 months.

Conclusions

These findings and our experience from over 8 years in using artificial ligaments to replace the ACL lead us to believe that this solution certainly holds promise for the future. In the short- and medium-term it will make possible the achievement of results which are at least comparable with the techniques involving autotransplants. There are, nevertheless, certain problems peculiar to synthetic material. These techniques are simpler, recuperation is faster, and they respect optimally the anatomical structure of the knee.

References

1. Beauchamp P, Carroll Laurin A, Bailon JP (1979) Etude de propriétés mécaniques des ligaments croisés en vue du remplacement prothétique. Rev Chir Orthop 65:197
2. Bonnel F, Jeager JH, Mansat C (1984) Laxités chroniques du genou. Masson, Paris
3. Burri C, Claes L (1983) Alloplastic ligament replacement. Huber, Bern
4. Chen EH, Black J (1980) Materials design analysis of the prosthetic anterior cruciate ligament. J Biomed Mater Res 14:567
5. Contzen H (1985) Materialtechnische Voraussetzungen und biologische Grundlagen für den alloplastischen Kniebandersatz. Unfallchirurgie 2(5):242
6. Dejour H (1983) Le résultat du traitement des laxités antérieures du genou. Rev Chir Orthop 69:253
7. Insall J, Aglietti P, Campbell D (1981) Bone block iliotibial band transfer for anterior cruciate insufficiency. J Bone Joint Surg [Am] 63:560
8. Johnson RJ, Eriksson E, Haggmark T, Pope MH Five to ten years follow-up: evaluation after reconstruction of the anterior cruciate ligament. Clin Orthop
9. Kennedy JC, Hawkins JR, Willis RB, Danylchuk KD (1976) Tension studies of human knee ligaments. J Bone Joint Surg [Am] 58:350
10. Marshall JL, Warren R, Wickiewicz TL, Reider B (1979) The anterior cruciate ligament: a technique of repair and reconstruction. Clin Orthop 143:97
11. Mc Daniel W, Dameron TB (1980) Untreated ruptures of the anterior cruciate ligament. J Bone Joint Surg [Am] 62:696
12. Mc Intosh DL (1974) The anterior cruciate ligament over-the-top repair. J Bone Joint Surg [Am] 56:591
13. Morrison JB (1969) Function of the knee joint in various activity. J Biomed Eng 4:573
14. Noyes FR, Butler DL, Grood ES, Zernicke RF, Heyzy MS (1984) Biomechanical analysis of human ligament grafts used in knee ligament repairs and reconstructions. J Bone Joint Surg [Am] 66:344
15. Noyes FR, Grood ES (1976) The strength of the anterior cruciate ligament in humans and rhesus monkeys. J Bone Joint Surg [Am] 58:1074
16. Park J, Grana WA, Chitwood JS (1985) High – strength dacron augmentation for cruciate ligament reconstruction: two-year canine study. Clin Orthop 175
17. Roth JH, Kennedy JC, Lockstadt H, Mc Callum CL, Cunning LA (1985) Polypropylene braid augmented and non-augmented intra-articular anterior cruciate ligament reconstruction. Am J Sports Med 13(5):321

The Inherent Mechanical Properties of Allograft Fascia Lata

M. J. Gibbons, D. L. Butler, F. R. Noyes, and T. Schlegal

Cincinnati Sports Medicine Center and The Department of Aerospace Engineering and Engineering Mechanics, University of Cincinnati, Cincinnati, Ohio, USA

Autogenous tendon and fascia are the most commonly used grafts for anterior cruciate ligament (ACL) reconstruction. Allografts, however, are now being used in selected patients. Allografts offer the advantage of decreased surgical time and morbidity, and a wider choice of graft type, size, and shape. The purpose of this study was to determine the inherent mechanical properties of sterilely procured/freeze-dried (SP/FD), ethylene oxide/freeze-dried (ETO/FD), and irradiated/freeze-dried (IRR/FD) fascia lata specimens and to compare the results to fresh-frozen (FF) fascia data from our previous study (Noyes et al. 1984).

Material and Methods

Eighteen FF specimens from six donors, nine SP/FD specimens from three donors, eight ETO/FD specimens from three donors, and two IRR/FD specimens from two donors were dissected along the primary longitudinal fibers into 15-mm wide strips. Cross-sectional areas were measured at three different locations using an area micrometer. All tissues were gripped in sinusoidal-shaped grips and loaded to failure in an Instron materials testing machine at a displacement rate of 100% of initial length per second. The material properties, modulus (normalized stiffness), and maximum stress (maximum force per unit area) were compared among groups.

Results

Despite the similar widths for all specimens, the cross-sectional area for both the SP/FD and ETO/FD was significantly less ($P < 0.01$) than the FF. The moduli for the SP/FD (431.06 ± 24.3 MPa) and for the ETO/FD (349.6 ± 20.5 MPa) were not significantly different from the FF (397.5 ± 17.1 MPa). The modulus for the IRR/FD (284.5 ± 3.5 MPa), however, was significantly lower than for the FF. The maximum stresses for the SP/FD (112.3 ± 4.0 MPa) and for the ETO/FD (97.9 ± 4.1 MPa) were both significantly greater than for the FF (78.7 ± 4.6 MPa) while the maximum stress for the IRR/FD (74.3 ± 6.0 MPa) was lower than for the other two allograft types but not significantly different from the FF.

Discussion

This study reveals two important points regarding the inherent properties of allograft fascia lata. First, the cross-sectional area, not the width, of the graft is the

Surgery and Arthroscopy of the Knee
2nd Congress of the European Society
Editors: Müller/Hackenbruch

best indicator of the total amount of tissue present. This is important when determining the size of graft for implantation. Second, the material properties (moduli and maximum stress) for SP/FD and ETO/FD allografts were at least equal to or in some cases greater than the FF fascia results. Thus, if the same amount of tissue (based on cross-sectional area) is loaded to failure, comparable normalized stiffnesses would be obtained. This is not the case for the IRR/FD specimens. The larger maximum stress values for the SP/FD and ETO/FD allografts may indicate that more excess tissue was removed during tissue preparation. This reduces the total cross-sectional area without damaging load-bearing collagen fibers.

Reference

Noyes FR, Butler DL, Grood ES, Zernicke RF, Hefzy MS (1984) Biomechanical analysis of human ligament grafts used in knee-ligament repairs and reconstructions. J Bone Joint Surg [Am] 66(3):344–352

Alternative Operative Procedure for Anterior Cruciate Ligament Replacement with a Composite Graft for Early Functional Treatment

R. Schabus [1], O. Kwasny, M. Wagner, and H. Plenk Jr. [2]

[1] University Clinic of Traumatology, Vienna, Alser Straße 4, 1097 Vienna, Austria
[2] Bone and Biomaterial Research Laboratory, Institute of Histology and Embryology, University of Vienna, Schwarzspanierstr. 17, 1090 Vienna, Austria

Introduction

Reconstruction of damaged ligaments, especially the anterior cruciate ligament (ACL), is one of the most challenging problems in traumatologic and orthopedic surgery. The long-term success rate of primary repair is less than ideal. In recent years the emphasis has shifted towards the use of autogenous tissue grafts, because artificial total replacements failed more or less totally. Biomechanical studies on ACL biological substitutes demonstrate that none of the tissue grafts by themselves is strong enough or matches the stiffness and strength of natural ligament [5].

As a compromise, the biological graft can be augmented by an artificial graft to provide acceptable mechanical properties.

The operative therapy of chronic ligamentous instabilities of the kneejoint should be intra-articular and early load-stable reconstruction of the ACL. For this purpose part of the patellar tendon is used as the mechanically strongest autogenous substitute of the ACL.

In the University Clinic of Traumatology in Vienna we employ a modification of Clancy's operative procedure for ACL reconstruction [1]. In this procedure the central part of the patellar tendon, which is pedicled on the infrapatellar fat pad by a bony insertion, is used as a substitute for the ACL [6]. In the last reconstruction we did not use the fat pad pedicle, because experimental investigations showed that the revascularisation of the graft was the same 8–12 weeks postoperatively when using the tendon alone. We used the Kennedy LAD for augmentation of the bone-ligament-bone graft [3]. This is an 8-mm broad, braided, polypropylene band with a mean length of 12–15 cm. The advantages of this method are:

1. It provides an adequate bone-ligament-bone graft with sufficient length, bony healing abilities and satisfactory mechanical stiffness
2. A very fast revascularisation of the graft through bony healing
3. The graft is a composite consisting of the autologous patellar tendon part and the allogenic augmentation
4. Stable fixation of the graft is possible with spikes plates on both ends, which allows the implant to function as an intra-articular splint in the knee joint; the

Surgery and Arthroscopy of the Knee
2nd Congress of the European Society
Editors: Müller/Hackenbruch

autologous graft is then protected, allowing elongation during the time of healing and biological remodelling
5. The target of operative therapy is to implant the load-stable composite graft, permitting early functional physiotherapy.

Material and Methods

Since June 1983, 58 patients with combined chronic antero-lateral and antero-medial rotatory instabilities, 2°–3°, underwent the operative procedure described. ACL injury occurred between 1 month to 13 years earlier (mean 35 months). There were 44 male and 14 female patients with an mean age of 26 years (range 15–44 years). Operations on the ACL and menisci had already been performed in one-third of the patients before.
In most of the kneejoints the intra-articular lesions were cleared with diagnostic arthroscopy. When it was possible, a meniscus lesion or a cartilage lesion was maintained arthroscopically.
With a straight, antero-lateral, para-patellar skin incision the patellar tendon and supracondylar lateral femur region were prepared (in the early cases a para-patellar medial arthrotomy was made). Now the knee joint is opened through the trans-patellar tendon approach.
The central third of the patellar tendon was prepared as a bone-ligament-bone graft by cutting the tendon with an oscillating saw on the ventral side of the patella and the tibial tubercle. Before harvesting the body ends of tendon with a chisel, a polypropylene band was sutured with resorbable suture material to the ligamentous part. The LAD was fixed on the bony ends with non-resorbable sutures.
In chronic instabilities the intercondylar notch was closed through osteophytes. An enlargement of the notch (notchplasty) was necessary to avoid impingement of bone on the composite graft.
After drilling of the tibial and femoral bony tunnels to a diameter of 8–10 mm, shaving the tunnel edges and checking the isometricity, the composite graft was implanted.
After fixation of the proximal end of the LAD and the bony end of the graft with one spikes plate and a 4.5 mm AO corticalis screw, the distal end of the composite graft was fixed, while the knee joint was extended to give a tension of 8 kp. In all

Table 1. Postoperative management

Treatment	Week
Immobilisation (40° flexion cast)	2–3
Mobilisation (brace, limited motion)	4–8
No weight bearing	< 4
Half weight bearing	5–6
Full weight bearing	7–8
Crutches	<10
Physiotherapy	10 onward

cases a tenodesis of the distal part of the iliotibial tract (tractopexy) was done to act as an extra-articular protection against anterior subluxation of the lateral tibial condyle.
Additional intra-articular surgery was necessary for 16 medial and 6 lateral menisectomies. Three times the medial meniscus was refixed. In 20 knee joints with cartilage lesions a debridement and a Pridie drilling were carried out. In only one combined instability was reefing of the medial ligament necessary.

Postoperative Management

Muscle and ligament straightening, aerobic fitness, coordination and confidence are the four cornerstones of knee rehabilitation (Table 1). The operated kneejoint is immobilised in a 40° flexion cast for the first 2–3 weeks. Physiotherapy starts 2–3 days postoperatively with controlled passive motion, electric muscle stimulation and postoperative training. Then 2–3 weeks after the operation the knee is braced permitting limited motion for at least the first 8 weeks. The mobility of the brace is increased weekly by 10° flexion and extension. After removal of the brace 8 weeks postoperatively, the mobility of the knee joint ranges between 10°–90°.
Postoperatively the patients are allowed to walk with crutches without weight bearing in the first 4 weeks. This is increased to full weight bearing by 8 weeks. Ten weeks after the operation, crutches and braces are no longer required. Normally the mobility of the knee joint is then restricted to 5/–10° of extension and 30°–40° of flexion. During this time of mobilisation intensive physiotherapy is started and continued for many months.
All patients are checked clinically by the surgeon and physiotherapist every 3 months.

Complications

There are no postoperative complications. On conducting regular clinical tests during the period of postoperative management and physiotherapy, four ruptures of the alloplastic augmentation were exposed. Two LAD ruptures were exposed during arthroscopy at the time of removal of the fixation plates, although there were no symptoms, and the LADs were removed without destroying the graft. Both joints are stable, and by objective measurement there is no difference between the treated and non-involved knee joint.
One patient had feelings of instability 12 months after implantation of the composite graft, and a rupture of the LAD was exposed arthroscopically. The graft was too long, so it was necessary to make an operative tensioning. Fifteen months after the reoperation the patient is now stable, is very active in competitive sport and has no problems with his knee joint. In one knee joint a partial rupture was indicated due to symptoms of instability 15 months after the first operation. KT-1000 measurements showed that the maximal manual drawer in 30° flexion was only 2 mm, but in the active quadriceps drawer it was 6 mm.

Removal of Fixation Plates and Diagnostic Arthroscopy

Theoretical considerations lead to the conclusion that one or both ends of fixed under tension allogenic ligaments can undergo material fatique and tear. In very stiff artificial ligaments collagen structuring of the graft does not occur. So we recommended release of the LAD ends in all knee joints 6–12 months after reconstruction of the ACL, by removing the fixation plates, to allow flexibility. In the same operation we check the structure and stiffness of the graft arthroscopically and test it with a hook probe.

Clinical Results

In June 1986 36 patients underwent a clinical investigation between 12 and 30 months postoperatively (mean 16 months) [2]. In 28 of these patients removal of the metal plates was executed on average 12 months postoperatively (range 5–20 months). In 22 patients the graft was checked arthroscopically.

Stability

The results are presented in Table 2. In 20 knee joints the Lachmann test was negative. In 14 cases it was (+) positive and in 2 (++) positive. There was no anterior drawer at 90° of flexion in 27 patients but positive anterior drawer in 9 cases. The pivot shift test was negative in 35 patients and positive in 1. KT-1000 measurements, with a maximal manual drawer in 30° flexion, showed more stiffness in the treated knee than in the non-involved knee in 12 patients. In 8 knee-

Table 2. Stability of the composite graft ($n=36$)

Test	Score	No. of cases
Lachman	0	20
	+	14
	++	2
	+++	0
Anterior drawer	0	27
	+	9
	++	0
	+++	0
Pivot shift	0	35
	+	1
	++	0
	+++	0
Femorotibial translation difference (in mm) between treated and uninvolved knee joint (KT–1000) mmd/30°		
	<0 mm	11
	=0 mm	8
	1–4 mm	16
	>5 mm	1

mmd/30°, maximum manual drawer with knee in 30° of flexion.

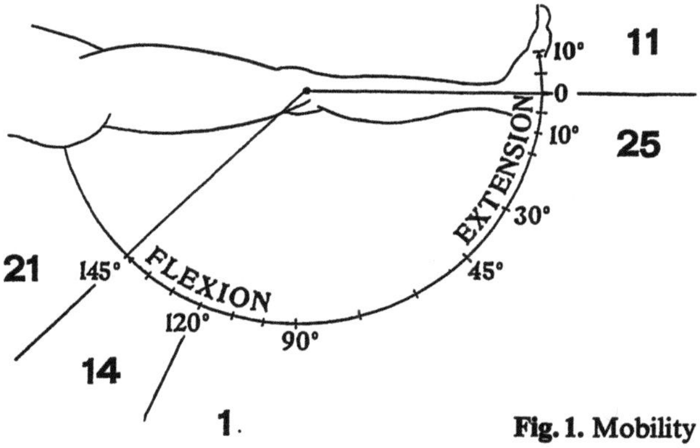

Fig. 1. Mobility of the composite-graft ($n = 36$)

joints among 16 patients there was a difference between 1 mm and 4 mm and in 1 patient the difference was greater than 5 mm.

Mobility

There was no restriction of extension in 11 knee joints but 25 patients had a lack of extension of up to 10° (Fig. 1). The flexion was free in 21 knee joints, 14 patients could flex the knee joint through more than 120° and 1 patient the flexion was restricted to less than 120° after following a patella fracture.

Sports Activities

Twenty-two patients could perform full sports activities; there were restrictions for 14 patients.

Subjective Results

Only patients with degenerative post-traumatic arthrotic joint changes had problems. Such arthrotic changes can be stabilised or stopped but joint degeneration cannot be halted.

Conclusion

The braided polypropylene band, manufactured by 3M company, was utilised for the augmentation of bone-ligament-bone grafts from the patellar tendon in chronic instabilities of the knee joint. The essential difference between the proposed Kennedy LAD procedure and our technique is the fixation of the composite graft at both ends (proximally and distally).
The clinical results of ligament repair with allogenic augmentation appear to be better than without LAD. The obvious advantages of the LAD system are better primary stability and the ability of early functional rehabilitation after the operation [4]. Clinically, there seems to be no disadvantage resulting from fixation of both ends of the LAD; however, our follow-up times are too short to speculate on the results for long-term performance.

References

1. Clancy WG (1983) Anterior cruciate ligament functional instability. Clin Orthop 172:102
2. Feagin JA, Blake WP (1983) Postoperative evaluation and result recording in the anterior cruciate ligament reconstructed knee. Clin Orthop 172:143
3. Kennedy JC (1983) Application of prosthetics to anterior cruciate ligament reconstruction and repair. Clin Orthop 172:125
4. McPherson GK, Mendenhall HV, Gibbson DF, Plenk H, Rottmann W, Sanford JB, Kennedy JC, Roth JH (1985) Experimental mechanical and histologic evaluation of the Kennedy ligament augmentation device. Clin Orthop 196:186
5. Noyes FR, Butler DL, Paulos LE, Grood ES (1985) Intra-articular cruciate reconstruction. I: Perspectives on graft strength, vascularization, and immediate motion after replacement. Clin Orthop 196:71
6. Paulos LE, Butler DL, Noyes FR, Grood ES (1983) Intra-articular cruciate reconstruction. II: Replacement with vascularized patellar tendon. Clin Orthop 172:78
7. Schabus R, Kwasny O, Wagner M, Plenk H (1986) Allogene Verstärkungsplastik bei autologer Ersatzplastik des vorderen Kreuzbandes. Hefte Unfallheilkd (in press)

Technique, Complications, and Results of Arthroscopic Anterior Cruciate Ligament Reconstructions with Bioprosthetic Ligaments

S. I. Springer

Hospital for Joint Diseases, Orthopaedic Institute, New York, USA

Introduction

Ligaments and tendons that have been compromised by injury or disease pose difficult surgical dilemmas, particularly when prior attempts at reconstructions have failed or when multiple substitutes are required. While the common approach is to use autologous structures for such reconstructions, the sacrifice of normally functioning units as donors may itself result in disability. Thus, there has been continued interest in finding suitable substitutes for these compromised structures. The alternatives to autologous structures have included: (1) freeze-dried or fresh frozen allografts, (2) carbon fiber scaffolds, either bare or with various coatings, (3) Dacron (polyethylene terephthalate)-based prosthetics, (4) polytetrafluoroethylene, and (5) bovine collagen tendons or ligaments.

The Xenotech tendon and ligament collagen prostheses are comprised of bovine tendons and ligaments which have been processed in a manner resulting in morphological preservation, conservation of biomechanical characteristics, and biocompatibility. The major component of the bioprostheses is collagen, an ubiquitous protein occurring in almost all organisms of the various animal phyla. Collagen is to a large part generic in form and function, that is, it invariably serves as the major extracellular structural protein and possesses a chemically similar composition across many distantly related species. It is also one of the unique molecules of the body in that its structure and to some degree function can be conserved long after removal from, or death of, the organism. These characteristics are significantly linked to the success of glutaraldehyde-crosslinked collagen as a prosthetic material.

Arthroscopic Replacement of the Graft

A small number of centers have participated in a subset study of replacement of cruciate ligaments with collagen prostheses under arthroscopic control. The techniques that have been developed are still being assessed and are undergoing modification as more experience is gained. It is to be emphasized that this technique is not for the inexperienced arthroscopist. The technical difficulties are great and the potential for error is high. Also it has not been determined which patients are the best candidates for this technique.

Surgery and Arthroscopy of the Knee
2nd Congress of the European Society
Editors: Müller/Hackenbruch

Material and Methods

Since October 1982, 46 patients have undergone anterior or posterior ligament reconstructions with bovine collagen prostheses under a FDA investigational protocol conducted by the author. All but one patient had the procedure done under arthroscopic control. The patient demographics include the following: 35 males and 11 females; age range 18–46 years; 43 arthroscopic ACL reconstructions, 1 open ACL reconstruction, 2 PCL reconstructions; follow-up 6–50 months.

Surgical Technique

The technique is a double tunnel procedure through the femur and tibia using standard arthroscopic portals. Since precision of placement for prosthetic devices has been shown to be critical to device longevity, clear visualization and a guide system for accuracy should be employed. The author uses a Synthes combined aiming device to pinpoint precisely the femoral tunnel posterior and superior in the notch. A generous notchplasty is made and the tunnel is drilled as vertical as possible, sized to match the diameter of the prosthesis. This is important to allow biological bony ingrowth to occur rather than the build-up of a fibrous tissue interface. The aiming device is also used for drilling the tibial tunnel, care being taken to exit anterior and medial to the original insertion site of the native anterior cruciate ligament (ACL). The tunnel edges must be chamfered to avoid contact between the bone and the prosthesis in the joint. The prosthesis is checked for isometry and adjusted if required. It is fixed by using standard, small-barbed, ligament staples, but fixation should only be completed after correct tensioning and a full range of motion have been verified. The author has added a lateral tenodesis, using a mini-incision for rotational augmentation of the intra-articular procedure, in the last ten cases of this series. Postoperatively, the patients are prevented from weight bearing in a locked hinged brace for 6 weeks. After 3 weeks additional flexion is allowed weekly until 8 weeks when a carefully regimented rehabilitation program is initiated. Patients are brought in regularly every 6 months for thorough stability evaluations.

Complications and Results

Complications and all reported incidences in this series are as follows:
- Documented traumatic ruptures: 3 (all explanted)
- Nontraumatic ruptures: 3 (2 remain stable)
- Reports of swelling during rehabilitation: 7 (all resolved)
- Synovial fluid "leaks": 2 (resolved)
- Late infection: 1 (graft explanted)
- Chondromalacia patellae: 3 (resolved)

To summarize the above, of the 43 arthroscopic ACL patients, 38 remain stable and clearly indicate successful outcomes from the procedure for periods of up to 4 years post-implant. Five patients are failures in this series as a result of either graft rupture or late infection; explant was necessitated with a resultant return almost to the preoperative instability. It should be noted that in no case did the patient's condition worsen after explant.

What has become apparent in reviewing the successful outcomes is the degree of stability which is maintained. Unlike autografts, in which a trend toward gradual loosening is seen, the review of these patients at 2, 3, and 4 years indicates a maintenance of postoperative stability.

Discussion and Summary

To date, bovine collagen prostheses appear to be a good method of providing joint stability for extended periods and are biocompatible. Attention to details of the surgical technique will further reduce complications as more experience is gained. It has already been demonstrated that there is a marked decrease in the incidence of infection, hospital stays are shorter, and there is a decrease in postoperative scarring and morbidity normally associated with the harvesting of anatomical structures. While early evaluation of this arthroscopically controlled technique is promising, the finalization of a protocol has not been completed, and this procedure should be considered developmental.

Technical problems in the clinical trial resulted in a higher complication rate early on which decreased significantly by applying principles gleaned from the early learning curve. One valuable benefit from the current clinical evaluation has been the distillation of the surgical techniques which are important when implanting any prosthetic cruciate device. Among the points crucial to a satisfactory result are: isometeric tunnel placement, proper tensioning, chamfering of the tunnel mouth, correct tunnel size, notchplasty, and appropriate postoperative rehabilitation protocols. The correction of rotational instabilities is an important adjunct to the procedure unless an isolated ACL disruption has been clearly diagnosed.

Synovitis After Anterior Cruciate Ligament Reconstruction Effusion: Confusion

W. L. Lanzer, A. Gown, and C. Piening

Department of Orthopedic Research, University of Washington,
Seattle, Washington 98195, USA

Introduction

Indications for prosthetic ligament reconstruction vary from long-standing chronic injury of the anterior cruciate ligament (ACL) to acute injuries consisting of anterior cruciate tears along with medial collateral ligament and meniscus injuries. Previous experience with anterior cruciate insufficiency and rotatory instability has demanded new concepts in applying biomechanical and kinematic research toward solving this difficult problem. A prosthetic ligament ideally would have mechanical properties very similar to the normal ligament it is meant to replace. In addition to this it must be biologically inert and provide some ability to withstand the hostile environment of the synovial cavity.

ProCol bioprostheses have a high tensile strength and excellent attachment site ingrowth in bony tunnels. However, due to the extensive cross-linking between the collagen fibrils, the flexibility or compliance is reduced as its tensile strength is increased. This becomes very important when considering the surgeons' ability to implant the graft in an isometric position. Data by Dr. John Sidles at the University of Washington show the small window of isometry that, by computer analysis, dictates placement of the ACL prosthesis in the proper position. This is difficult to achieve even in an open procedure. If a notchplasty is not done, or if there is impingement on the graft, the material which has a high tensile strength is certainly susceptible to abrasion, and therefore the intercondylar notch must be cleared of all possible structures that could impinge on the graft.

In order to investigate the synovitis associated with mechanical failure of ProCol ligaments, retrieval implants were analyzed.

Material and Methods

Xenograft prosthetic ligament and synovial fluid from 13 knees were harvested after the diagnosis of clinical rupture. All patients exhibited an effusion approximately 3–4 months postimplantation. A patient with osteoarthritis whose ACL was sacrificed during total knee arthroplasty served as a control. All ligaments and investing synovium were prepared for histologic staining with hematoxylin and eosin and special immunofluorescent monoclonal antibody stains consisting of lymphocyte, macrophage, and mesenchyme cell markers prepared in our laboratory by hybridoma technique. Synovial fluid was centrifuged and pelleted, and stained with trichrome for graft particulates. In addition, fluid was assayed for collagenase activity using a radiolabeled Type I collagen digestion assay by the

Surgery and Arthroscopy of the Knee
2nd Congress of the European Society
Editors: Müller/Hackenbruch

Table 1. Classification of responses

Type of response	Cells involved
Specific: immune-allergic	B cells, T cells
Nonspecific: foreign body	Macrophages, foreign body giant cells

method of Johnson-Wint (1980). For control histologic analyses, nonspecific synovitis patients were used and the identical stain methods were employed.

In order to understand the types of responses possible in the synovium we can classify them as specific or nonspecific depending on the cell line involved (Table 1). B cells are lymphocytes which can make antibodies, T cells are involved in cell-mediated immunity, i.e., the graft versus host reaction, macrophages are nonspecific and are involved in any inflammatory process, and foreign body giant cells are 20–30 μm in diameter, multinucleated cells with phagocytic processes.

Results

All cases exhibited a fibro-proliferative response with hypercellularity consisting of monocytes, lymphocytes, and foreign body giant cells. Common leukocyte markers failed to reveal the presence of any significant number of plasma cells or T cells within the explant itself. Particulates of graft were found embedded in synovium with particle size in the range of 20–25 μm. There were no significant differences in staining for immunoglobulins, complement, or other immune mediators between synovium from Xenograft specimens and the control groups (Tables 2, 3). No samples of synovial fluid contained active or latent collagenase enzyme, as demonstrated by no activity with aminophenyl/mercuric acetate (APMA) activation. In addition, the synovial response in patients with nonspe-

Table 2. Cell population analyses

Explants with an investing synovium					
P	FBGC	T	B	M	T+B Anti T
No	0	++	+/−	+++	+++
Yes	+	++	+/−	+++	++ → +++
Yes	+	++	+	++	+ → +++
Yes	+	+	+/−	+++	+/− → +
No	+	+++	++	+++	++ → +++
Yes	+	++	+	++	+ → ++
Yes	+	++	+	+++	++
No	+	++	+	+++	++
No	+	++	+	+++	+ → +++
Yes	+	++	+/−	++	++ → +++
Yes	+	+++	+/−	++	+ → +++
Yes	+	++	++	+++	++ → +++
No	0	0	+/−	++	+/−

Immunofluorescent monoclonal antibodies failed to reveal the presence of any significant number of plasma or T cells within any explanted graft material. P, particulates; FBGC, foreign body giant ceus; T, T cells; B, B cens; M, Macrophages; T+B Anti T, T+B Anti T-200.

Table 3. Synovitis response in patients with nonspecific synovitis

Nonspecific synovitis controls			
T	B	M	T+B Anti T
++	+	++	+++
++	+/−	++	++
++	++	++	+++
+++	+	+ → ++	+++
++	+/−	++	++
+++	+	++	+++

cific synovitis appears identical with respect to all populations and the lymphofollicular response (Table 3).

Discussion

Analyses of ruptured implants indicate that the most likely cause of synovitis is graft abrasion and the showering of the joint with small particulates. We have demonstrated that the inflammatory response to bovine cross-linked collagen Xenografts is a nonspecific foreign body reaction which does not appear to be immune mediated. The body is unable to remove the foreign synthetic material which cannot be metabolized, thus, the material remains in the body to cause synovitis inflammation. By using a bioprosthesis, the continual and chronic problems with synthetic material are obviated in that the body may break down, metabolize, and remove the cross-linked collagen, albeit over a longer time frame than that for chemically untreated collagen tissue. In addition we have seen, over a 2-year period, ingrowth of host tissue into the intrafibrillar areas of the ProCol graft with very little reaction within the graft itself. It is still not known how long total replacement takes; however, one can postulate that a composite material at 5–10 years may still present an excellent ligament as far as strength and endurance is concerned. Synovitis following implantation is becoming common with all ligament prostheses, and particulate matter can be retrieved from the joint at varying time periods postimplantion.

By antibody staining methods and cell population studies, it appears that there is not a classic allergic or immunologic basis for Xenograft failure. As with all implant materials, the grafts are sensitive to surgical technique and abrasion. In grafts that fail, particulates stimulate a nonspecific inflammatory response which results in a localized foreign body reaction. Since no enzyme activity is found, it is doubtful that articular cartilage degradation is a likely sequel of graft failure and concomitant synovitis. The use of meticulous surgical technique is strongly recommended in order to accomplish prosthetic reconstruction of the ACL successfully.

Reference

1. Johnson-Wint B (1980) Anal Biochem 104:175–181

Posterior Cruciate Ligament and Posterior Instability

Diagnosis of Posterior Cruciate Ligament Insufficiency. A Prospective Clinical and Stress Radiographical Analysis with the Knee in Extension

H. U. Stäubli and R. P. Jakob

Orthopaedic Department, University of Berne, Inselspital, 3010 Berne, Switzerland

The use of the posterior drawer test in posterior cruciate ligament insufficiency with the knee in extension is controversial. To evaluate clinical drawer tests with the knee in extension, a prospective clinical and stress radiographic study was initiated.

Material and Methods

Five clinical drawer tests were performed with the knee in extension, i.e. in 10° of flexion:

A. Without anaesthesia and the patient lying supine:
   1. The spontaneous posterior drawer (posterior sag in extension)
   2. Active reduction of the posterior sag through quadriceps innervation

B. Under anaesthesia and the patient lying supine:
   3. The passive posterior drawer test applying a manual anterior-posterior force to the proximal tibia with the distal femur resting on a thigh support
   4. The passive reduction of the posterior subluxation into the neutral position applying a posterior-anterior force to the proximal tibia
   5. The external rotation recurvatum test

C. A posterior stress radiography in 10° of knee flexion applying an anterior-posterior force of 200 N to the proximal tibia was obtained with the distal femur resting on a thigh support.

Results

A total of 24 patients were analysed prospectively. All the clinical tests were markedly positive (i.e. grade 2 or 3), with the exception of the external rotation recurvatum test. This test was found to be negative in 20 patients with proven posterior cruciate ligament insufficiency. In four patients (two with ipsilateral femoral shaft fractures and two with interposed large fragment bucket-handle tears) the clinical posterior drawer tests were not reliable. Posterior stress radiograms at 10° of flexion revealed a significant posterior tibial displacement, measuring 10.4 ± 2.4 mm in the posterior cruciate ligament deficient group ($n = 20$), in comparison with a posterior cruciate ligament intact population ($n = 114$) in which posterior tibial displacement measured 3.6 ± 2.1 mm ($P < 0.01$).

Surgery and Arthroscopy of the Knee
2nd Congress of the European Society
Editors: Müller/Hackenbruch

Isometric Points of the Posterior Cruciate Ligament

E. S. Grood, M. S. Hefzy, T. L. Lindenfeld, and F. R. Noyes

Giannestras Biomechanics Laboratory,
University of Cincinnati Medical Center, Cincinnati, Ohio 45267-0212, USA

Introduction

It is increasingly accepted that successful reconstruction of the anterior cruciate ligament requires selecting an attachment location and method of tensioning that restores normal joint laxity without subjecting the substitute to large forces during flexion-extension motions. This can be accomplished by either "over the top" placement or use of an intra-articular drill hole located to produce minimal stretching of the substitute.

The femoral attachment with this property has been referred to as the "isometric point" although no single attachment produces a purely isometric condition.

While studies have been conducted to determine isometric attachments for the anterior cruciate ligament, few data exist for the posterior cruciate ligament (PCL). This paper reports the results of a study conducted to determine the effect of tibial and femoral attachment location on length changes in the PCL during flexion and extension motions of the knee.

Methods

The full three-dimensional motion was measured in four knees obtained from different donors using a six degrees-of-freedom electrogoniometer. Whole lower limbs were mounted in the apparatus by gripping the femur in its proximal one-third while the leg hung freely under its own weight. A posterior force of 100 N (22.5 lbs) was applied to the proximal tibia to tense the PCL and the knee flexed and extended. The knee was then dissected and the location (three-dimensional coordinates) of the PCL's attachments, along with other bone landmarks, determined. The motion data and the coordinate data were mathematically combined to determine the distance between tibial and femoral attachments and how it changed during knee flexion.

Results

Similar to our prior studies of the anterior cruciate ligament, we found the tibial attachment did not affect the change in length during flexion or the flexion angle where the length was longest (most taut) and shortest (slack). The tibial attachment did, however, affect the length required for the substitute and its orientation within the joint.

Surgery and Arthroscopy of the Knee
2nd Congress of the European Society
Editors: Müller/Hackenbruch

In contrast, the femoral attachment had a strong effect on the length pattern. Attachment locations along the proximal edge of the PCL's femoral insertion were most isometric. As the attachment was moved distally on the femur the length pattern became one of progressive lengthening with flexion. The more distal the attachment, the greater was the amount of lengthening.
Along the proximal edge where the most isometric length patterns were found, a posterior attachment had a pattern that became slightly shorter and then longer with flexion from 0° to 90°. Anterior attachments had the reverse pattern, becoming slightly longer and then shorter.
A simple model was developed that explains the observed results based upon the location of the attachment relative to the average flexion-extension axis.

Posterior Instability of the Knee. Surgical Treatment, Rehabilitation, and Results

We. Müller

Kantonsspital Bruderholz, Orthopädie-Traumatologie, 4101 Bruderholz
Basel, Switzerland

Introduction

Much has been written about the anterior cruciate ligament (ACL) and anterior instability of the knee. By comparison, the posterior cruciate ligament (PCL) and posterior instability have received only scant attention in the literature. There is still diversity of opinion concerning the significance of PCL injuries. While some authors consider the PCL to be the most important of the knee ligaments, others claim that the loss of this ligament is well tolerated and not significant enough to warrant surgery. These conflicting viewpoints are clearly illustrated in the comprehensive literature survey by Barton et al. (1984).

Agonism and Antagonism

When the foot comes in contact with the ground during gait, the quadriceps (extensor-decelerator) apparatus relies on the stabilizing action of the ACL to prevent posterior displacement of the lateral femoral condyle and concomitant anterior subluxation of the tibia. This would cause unsteadiness and giving-way on descending stairs and especially during downhill walking. Symptoms in these cases tend to be localized to the patellofemoral articulation.

With an insufficiency of the ACL this femorotibial subluxation is most pronounced during "cutting," or planting the foot while abruptly changing the direction of body motion (spontaneous jerk shift–pivot shift phenomenon)

The PCL acts as a stabilizer of knee flexion during acceleration, when the hamstrings pull the tibia backward during the swing phase of gait. With insufficiency of the PCL, the tibia subluxates posteriorly during the non-weight-bearing swing phase (the "take off shoe" sign is diagnostic of posterior cruciate instability). Spontaneous reduction of the subluxation occurs while the leg is still in the swing phase and is being extended by the quadriceps, before that muscle becomes the major stabilizer in the weight-bearing stance phase.

Despite the subluxation, the femorotibial articulation and its cartilage do not undergo significant extra wear in this swing phase of gait, even though with absence of the PCL the patella becomes a prime stabilizer owing to its pulley effect in redirecting the line of action of the quadriceps. Femorotibial reduction necessarily occurs at the moment of transition from flexion to extension, and this is associated, at least in theory, with a tangential "slip" between the femoral and patellar cartilages. This explains why insufficiency of the PCL is less apt to provoke femorotibial osteoarthritis or excessive wear of the anterior horns of the menisci than it is to incite degenerative changes in the patellofemoral cartilage.

Surgery and Arthroscopy of the Knee
2nd Congress of the European Society
Editors: Müller/Hackenbruch

Yet there is also cartilage alteration on the medial femoral condyle in the weight-bearing parts that are near extension during landing and take off. The central medial facet of the condyle is no longer kept inside the notch by the PCL, an inner collateral ligament acting against varus deformation.
Thus there are more transverse translational shearing forces on the medial condyle against the tibial spine.
By nature, participation in vigorous sports activities tends to decline as the years pass, and so with reduced load posterior instability tends to cause fewer problems with passage of time. Exceptions include severe posterior instabilities that are combined with pronounced varus instability and therefore cause significant disability with the activities of daily living.

Posterior Instability

A 1+ direct PCL instability (3–5 mm) unaccompanied by other ligamentous injuries generally is not disabling enough to warrant surgery. If surgery is indicated, the techniques are the same as those described below.
The 2+ or 3+ posterior PCL instability (5–10 mm and 10–15 mm) is combined with a peripheral ligamentous instability. In rare cases the lesion is medial and leads to a valgus and posteromedial instability. However, varus and significant posterolateral instability are the rule.

Posteromedial Instability

This instability can be demonstrated radiographically (see also Fig. 1) by posterior subluxation of the medial tibial plateau on the medial femoral condyle.
A small posteromedial instability of 5 mm or less may compensate spontaneously with PCL reconstruction and posterolateral repairs.
An instability of 10–15 mm is treated by bringing up the posterior oblique ligament (POL) with its meniscofemoral component to the posterior part of the fem-

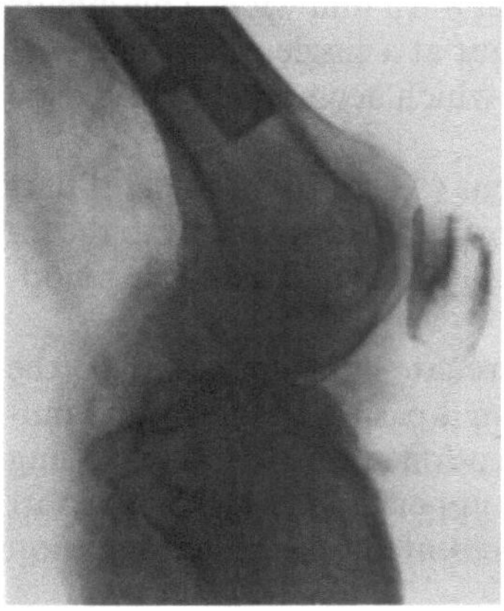

Fig. 1. Extreme global posterior instability of several months' duration

oral epicondyle to restore anatomic length to the ligament. Preferably, this is done after careful freshening of the bone at the anatomic site.
The posterior sag of the tibia removes tension from the medial collateral ligament (MCL) and produces a slight valgus laxity. With reduction of the tibia, the MCL usually regains its proper tension and does not require surgical repair.

Posterolateral and Varus Instability

Isolated posterolateral instability is most easily recognized as an increased range of lateral rotation on comparison with the healthy side with the knee flexed to 20–30°. If the range of lateral rotation is again equal on both sides with the knee flexed to 90°, it may be assumed that the lesion is confined to the popliteus corner. If lateral rotation is increased at 30° flexion as well as at 90°, then the popliteus corner *and* the PCL are insufficient, and the repair or reconstruction of both elements must be planned.
Lateral varus instability is always present in these cases but often is not apparent. It is known that the ligaments on the lateral side of the knee function as dynamic as well as static stabilizers: the arcuate ligament with the popliteus and biceps brevis tendon and muscle, the iliotibial tract with the anterolateral femorotibial ligament (ALFTL), and the lateral collateral ligament (LCL), which connects distally to the tibia by way of the fibula and receives tension from the biceps tendon.
The lateral compartment rotates more and has a greater degree of spontaneous varus laxity. Comparison with the healthy side is always important, bearing in mind that a mildly pathologic knee with a trace of effusion will show significantly greater motion than a healthy knee without effusion, i.e., will have several millimeters more of varus opening.
The three major structural elements that determine varus stability each pose different problems of reconstruction. Reconstruction of one element alone is not recommended because of the joint mode of action of the lateral stabilizers.
The LCL appears to be the easiest to reconstruct. The biceps tendon makes an excellent donor because it (a) also inserts on the fibula and (b) is sufficiently thick for graft removal. The anatomic side of attachment on the lateral femoral condyle is easily identified and is freshened by preparing a small longitudinal slot to receive the pedicled biceps tendon graft, which is about 8–10 mm wide. This avoids the less favorable attachment of the graft to the femur at a single point. The normal LCL has a wide area of proximal attachment, which accounts for its elastic reserve during function.
In the case of the anterolateral femorotibial ligament (ALFTL) an integral part of the iliotibial tract, the dynamic, isometric attachment of the ligament to the femur can restore its function with an efficiency that goes beyond a simple tenodesis.
Reconstruction is most difficult for the popliteus-arcuate ligament complex (the popliteus corner). The course of the popliteus tendon is easiest to follow and imitate. Reconstruction involves finding an isometric tendon segment and bypassing it with a graft routed through a bony tunnel in the upper tibia (Fig. 2). A report on the results of 60 popliteus bypass operations is presented following the section on posterior cruciate reconstruction.

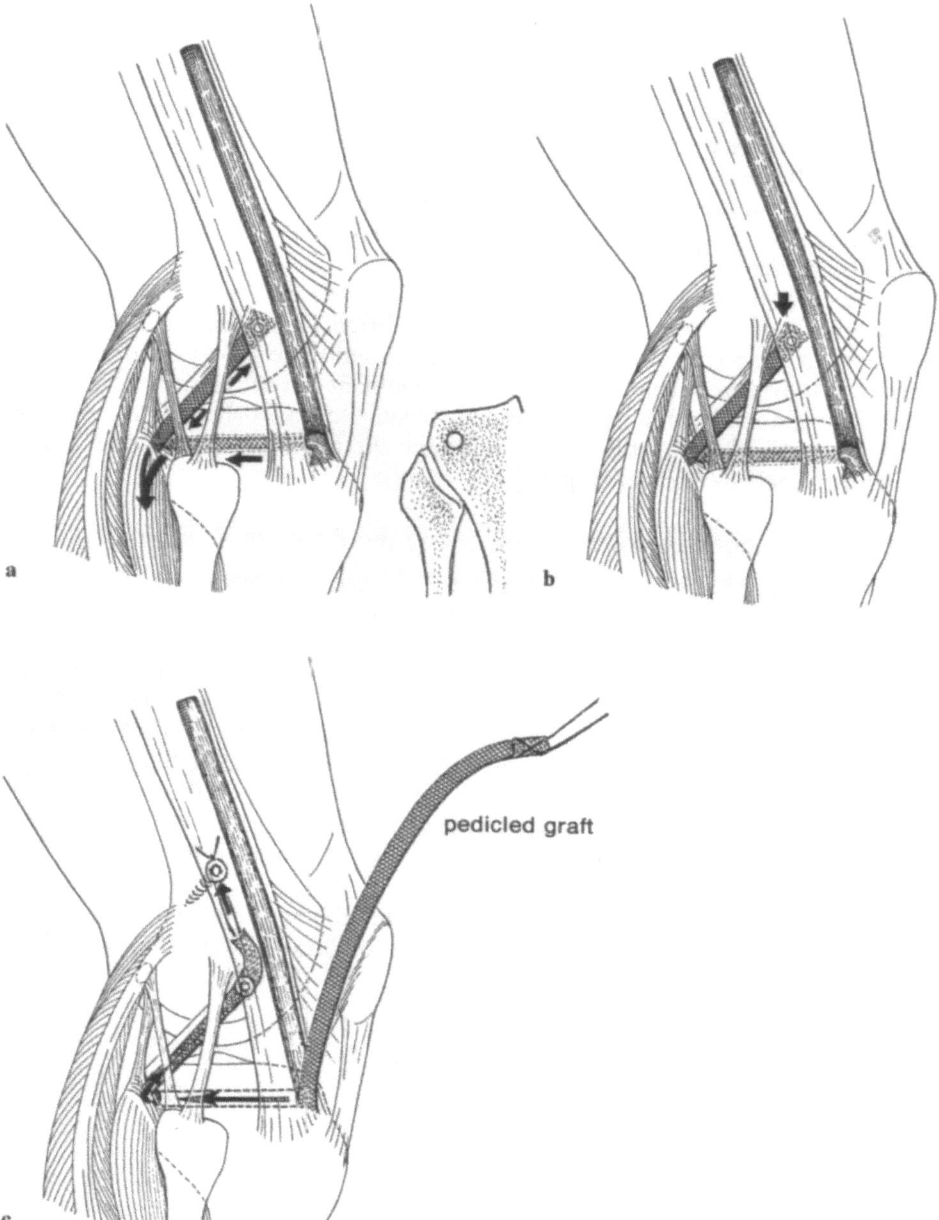

Fig. 2. a The popliteus bypass for posterolateral stabilization. The graft is dynamized by the muscle belly. **b** The pedicled graft should follow an anatomic course between its emergence from the tibia and its entry into the femur. **c** Anchoring the graft in a bony tunnel is preferable to screw fixation, which can lead to problems

Stabilizing Operations for Insufficiency of the Posterior Cruciate Ligament

Augustine's operation is a relatively minor procedure that still has value in patients with a more or less isolated, 2+ insufficiency of the PCL (approximately 5–10 mm). It preserves the normal anatomy and *does not create an irreversible state*. It increases the passive posterior stability of the knee, and its only disadvantage is an increase of tension and pressure on the patella, with a potential for increased cartilage wear.

In some cases the PCL can be repaired using the stretched posterior portion of the ligament, which may be in quite good condition. If the femoral attachment is intact, the ligament can be fixed to the correct anatomic site on the tibia. Conversely, if the distal attachment is intact, the ligament can be reattached to the femur. This is done by pulling the ligament through an oval-shaped hole (longer than it is wide) placed at the correct anatomic site in the femur and oriented at the proper angle to the roof (Fig. 4a).

Posterior Cruciate Reconstruction Using the Patellar Tendon

The techniques described by W. Clancy using a free or pedicled patellar tendon graft have significantly advanced cruciate ligament surgery. Success relies heavily on technical precision, however.

While I achieved consistently good results with ACL reconstructions, great variability was found in the results of PCL reconstructions, including a number of unexpected failures. Presumably this related to poor reproduction of the femoral attachment of the PCL, where precision is essential in order to achieve isometry. Because the round drill hole in the femur tends to force the graft into a crescent shape against the side of the tunnel, I changed to a rectangular hole into which the similarly shaped bony block at the proximal end of the graft is inserted like a drawer (Fig. 3).

In the cases with the best results, it was found that the bony callus on roentgenograms (Fig. 4b+c) followed the course of the isometric lines defined by Grood

Fig. 3. Technique of posterior cruciate reconstruction

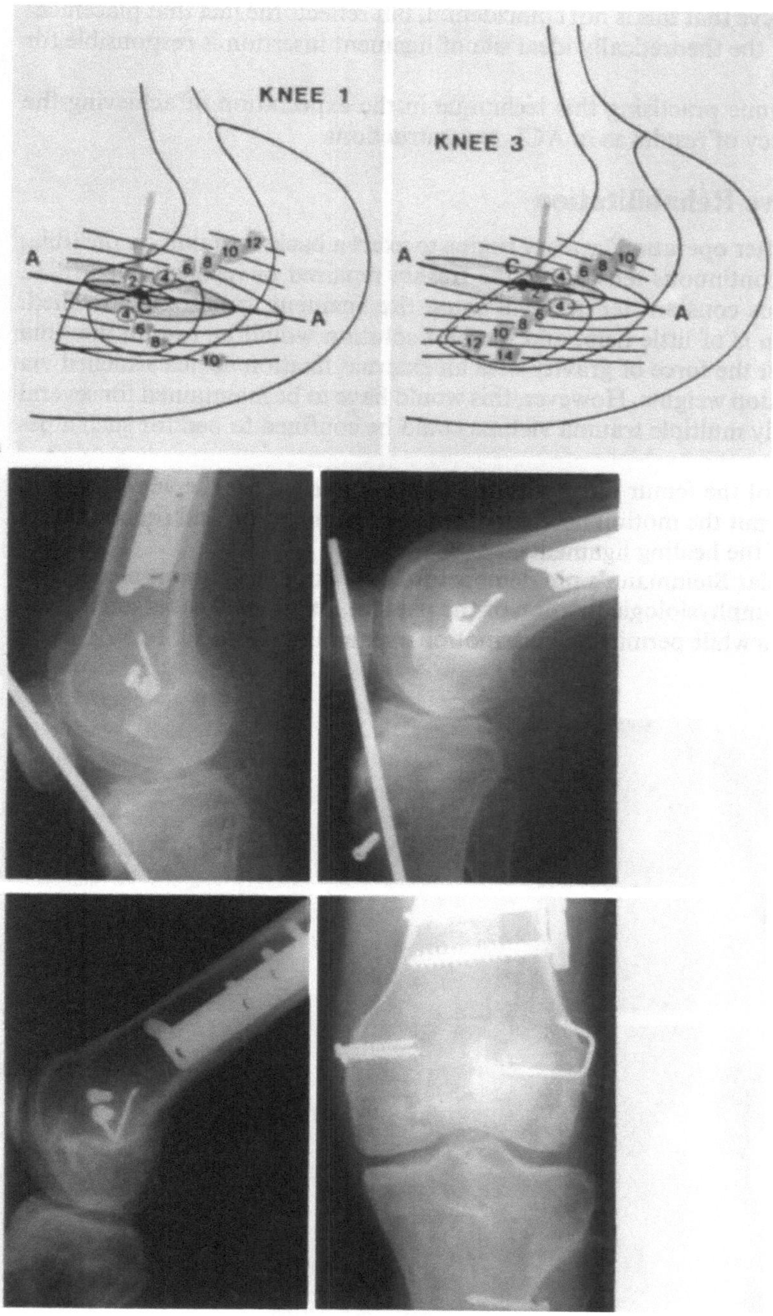

a

b

c

Fig. 4. a Grood's scheme of iometric lines. **b** Roentgenograms illustrating ideal block placement in the femur. **c** On late follow-up the bony callus from the femoral block is clearly visible as a white line, signifying correct placement. Same case as in Fig. 1, preoperative

(Fig. 4a). I believe that this is not coincidental, but reflects the fact that placement of the block at the theoretically ideal site of ligament insertion is responsible for the good result.
I plan to continue practising this technique in the expectation of achieving the same consistency of results as in ACL reconstructions.

Postoperative Rehabilitation

Immediately after operation, gravity begins to exert a backward pull on the tibia, which places continuous tension on the freshly repaired or reconstructed PCL. Unchecked, this constant tension will cause the ligament to become stretched. Plaster fixation is of little help, and the best solution would be to pull the tibia upward against the force of gravity with an external fixation device attached via pulleys to traction weights. However, this would have to be maintained for several weeks, and only multiple trauma victims could be confined to bed for such a period.
Transfixation of the femur directly to the tibia is too rigid for the whole system. It does not permit the motion necessary for cartilage nutrition and functional rehabilitation of the healing ligaments.
The transpatellar Steinmann's pin demonstrated by Paul Grammont also has its negative side (unphysiologic fixation of the patella), but it gives satisfactory support to the tibia while permitting knee motion over a range from 10° to 60° flexion (Fig. 5a+b).

a

b

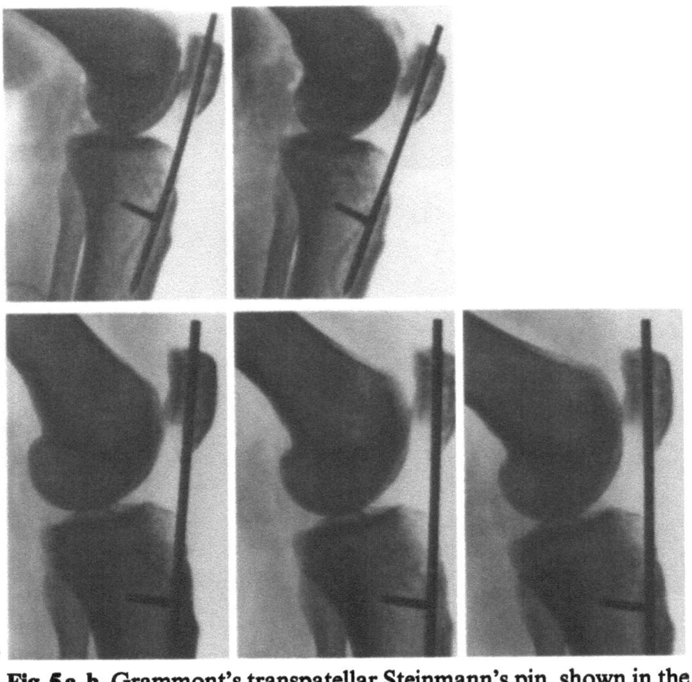

Fig. 5a, b. Grammont's transpatellar Steinmann's pin, shown in the ideal position for various degrees of flexion to 60°

In theory this fixation is bad for the patellar cartilage, so immediately after wound closure we initiate continuous passive motion (CPM) on the Kinetec or similar motorized splint between 20° and 50°, later increasing the range to between 10° and 60°/70°.

Active knee extension and leg raising with the quadriceps are permitted from the first day. Additional active, external force is generally needed to flex the knee past 60°/70°. This should be avoided, as it can cause undesired AP translation of the tibia into the posterior drawer position. Immediate CPM therapy and anatomic position of the patella is probably the main reason why this system has shown such good compatibility with the patellar cartilage in our experience.

The manner of placement of the Steinmann's pin in the patella is of critical importance. For the 3.5-mm pin, at least a 3.5-mm pilot hole should be predrilled to avoid thermal injury to the bone and cartilage. For the same reason the pin should not traverse the posterior part of the patella close to the articular surface.

The site of pin entry into the tibia must be carefully and atraumatically located after the correct position of the patella in relation to the femoral condyle has been established. For this step and for final suturing, the quadriceps apparatus is temporarily fixed with a Backhaus clamp. Usually the pin is inserted roughly parallel to the anterior border of the tibia (Fig. 5). An oblique placement is commonly shown but is kinematically unfavorable, for then the pin is not parallel to the patellar tendon, and the resulting lever-arm forces can cause early loosening and migration. In selected cases I have driven the pin in deeper as a secondary measure. Driving the tip into the posterior cortex of the tibia does not prevent migration, it may even enhance proximal migration.

If the pin has been correctly inserted and is tolerated well, we leave it in place for up to 12 weeks in difficult cases (Fig. 4). As the patient becomes more active, the pin gradually loosens in the tibial cancellous bone. This is advantageous in that it lessens the stress-protective effect of the implant on the PCL.

To date we have used this postoperative dynamic splinting technique (Grammont's "olécranisation temporaire de la rotule") in more than 100 patients and have found that it can significantly improve results in PCL reconstructions. The procedure is not entirely physiologic, however, and I shall continue to seek a principle that is more in line with the normal anatomy and physiology of the knee.

Note: Immobilization (especially in plaster) is contraindicated as an adjunct to transpatellar nail splinting.

Results

It is interesting to note in this context that despite the great number of knee surgeons and knee centers, few results have been published to date concerning the treatment of posterior instabilities.

We schedule routine follow-ups for 6, 12, 26, 38, 52, 78, and 104 weeks after surgery. So far, we have not been able to evaluate and compare all cases in numerical terms. The posterior instabilities, moreover, can involve a large combination of coexisting and superimposed problems which make evaluation difficult.

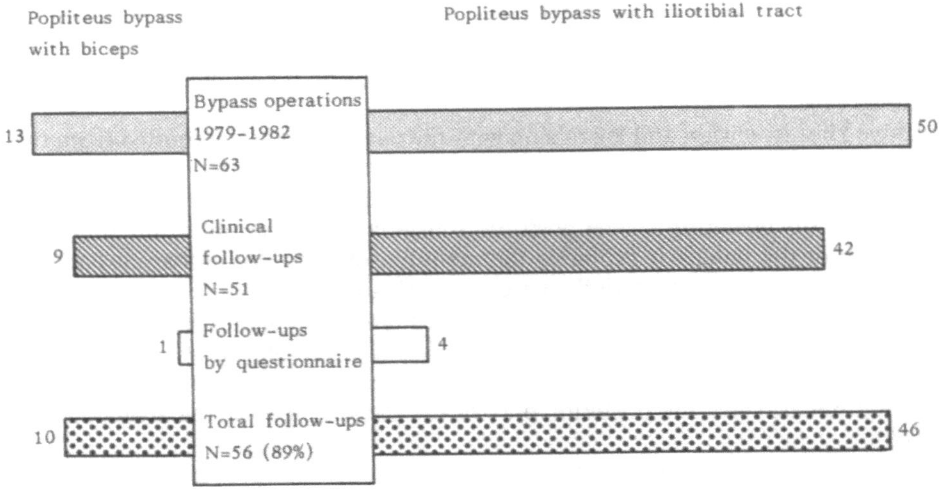

Fig. 6. Follow-ups of the first 63 patients

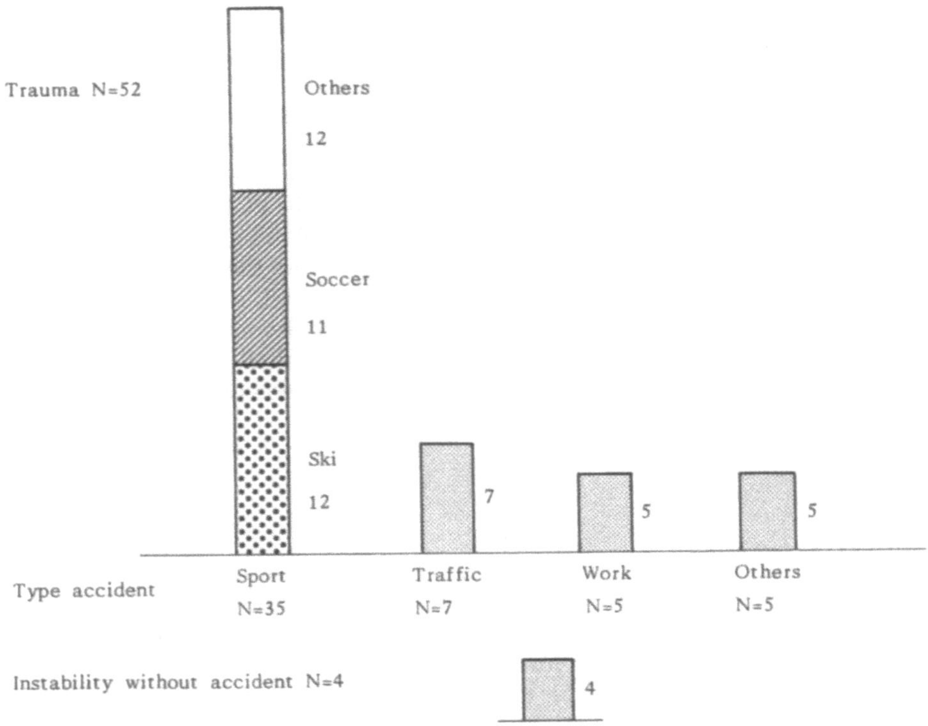

Fig. 7. Etiology of the posterolateral instabilities

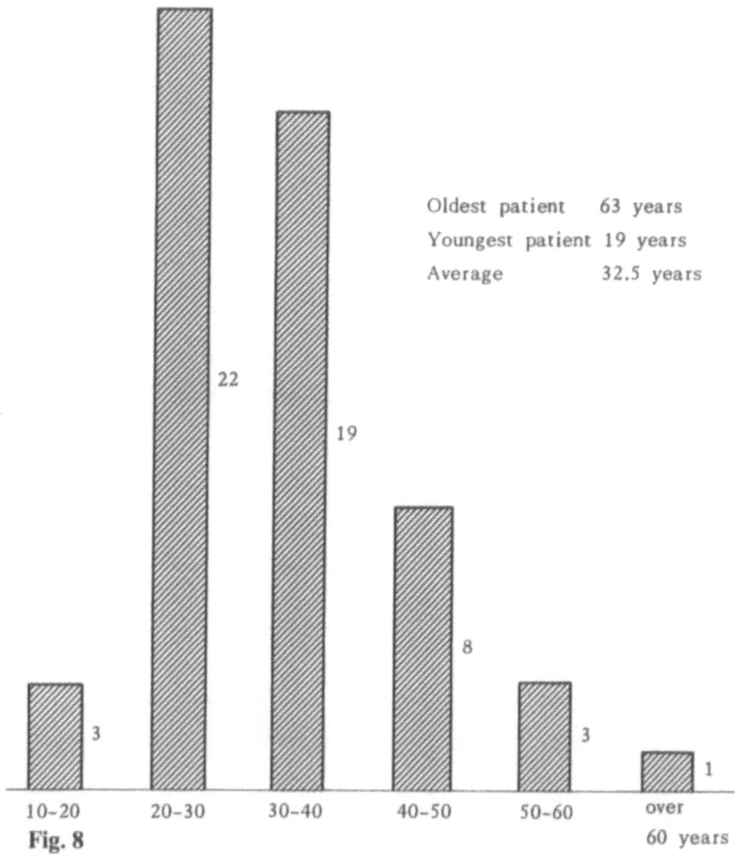

Fig. 8

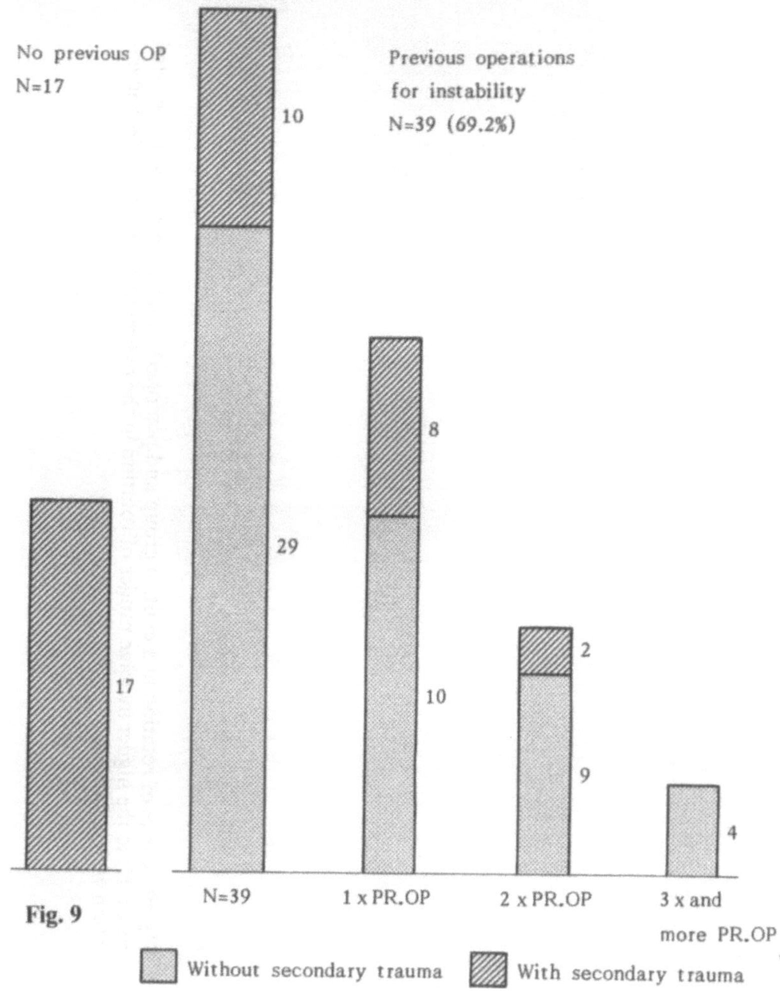

Fig. 9

Fig. 8. Age distribution of the 63 patients

Fig. 9. Number of previous operations in the 63 popliteus bypass patients

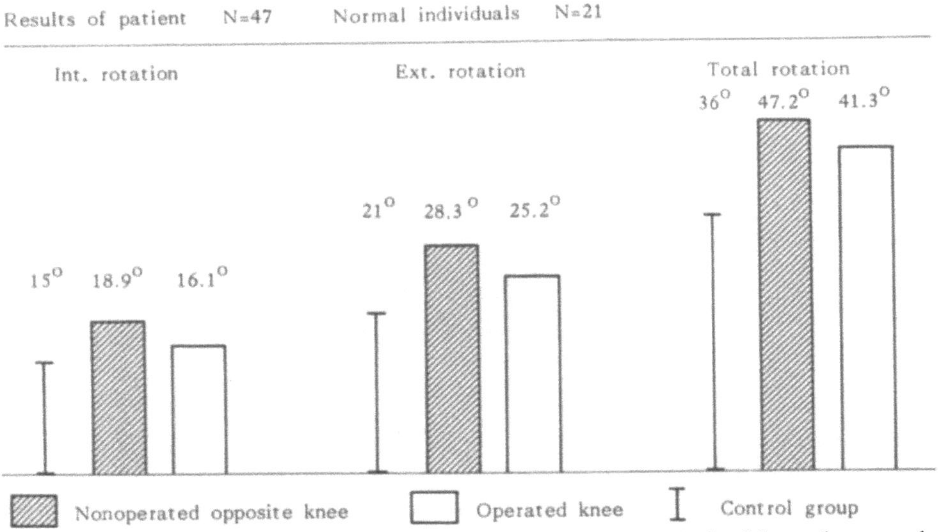

Fig. 10. Range of rotation in a control group and postoperatively in healthy and operated knees. Note the higher average ranges of rotation in the operated patients (= decompensation with instability?)

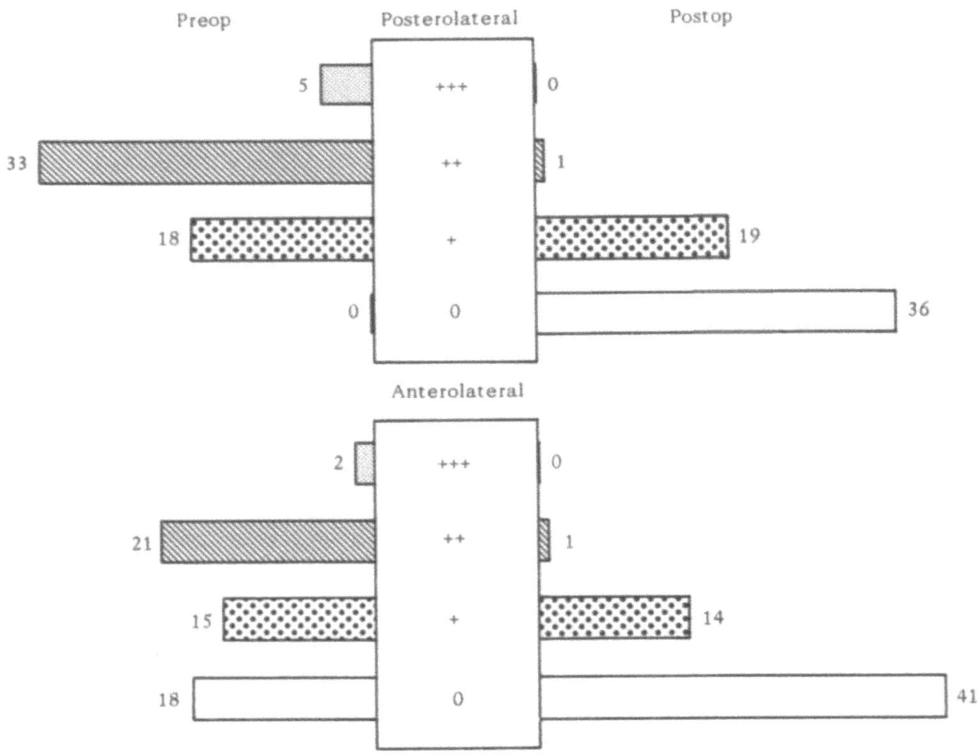

Fig. 11. Preoperative and postoperative instabilities

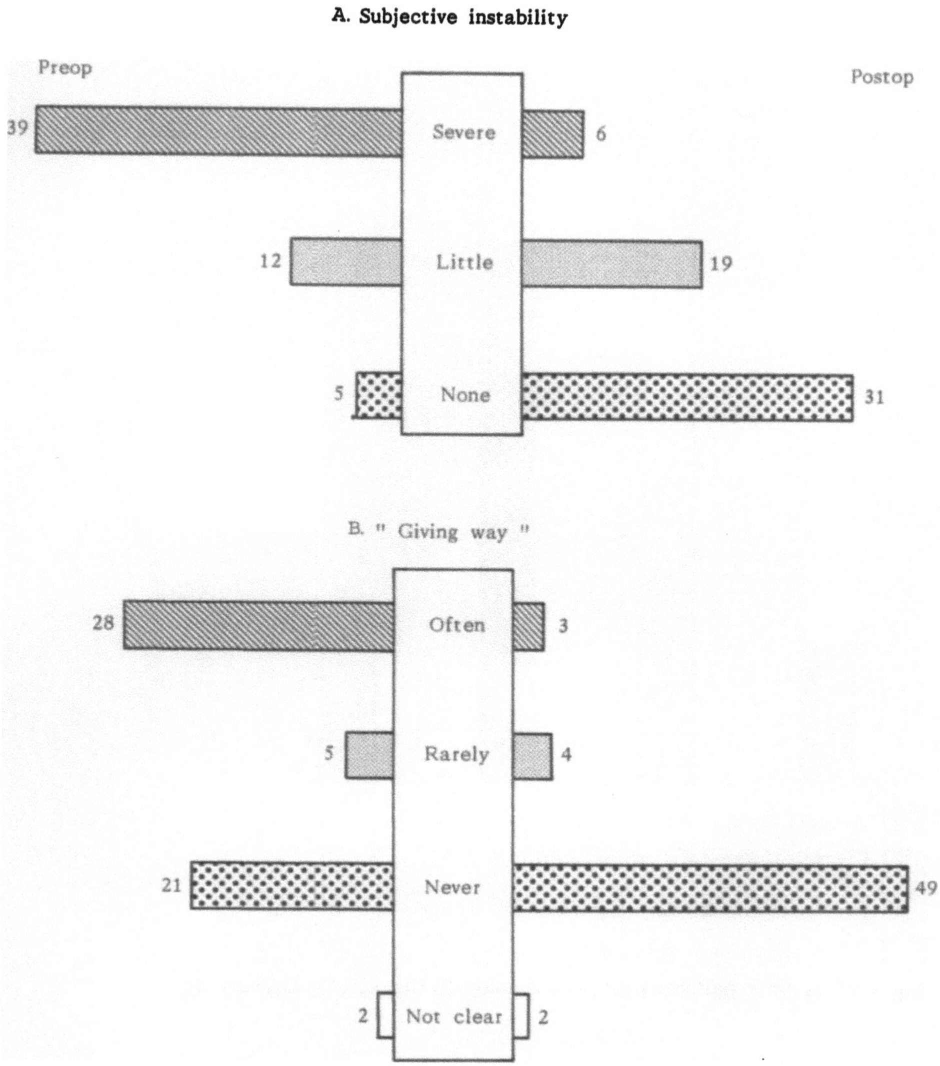

Fig. 12. Subjective instability

This proved true in our evaluation of the first 60 popliteus bypass operations performed at our center. The data presented below (Figs. 6–18) give a good appreciation of the overall quality of the results and also of the mixed problems that are involved.

Only eight patients in this initial series from 1979 to 1982 underwent a popliteus bypass alone. In five of these cases the objective and subjective results were excellent (operated side equivalent to the unaffected side). The remaining three patients had 1 + residual instability.

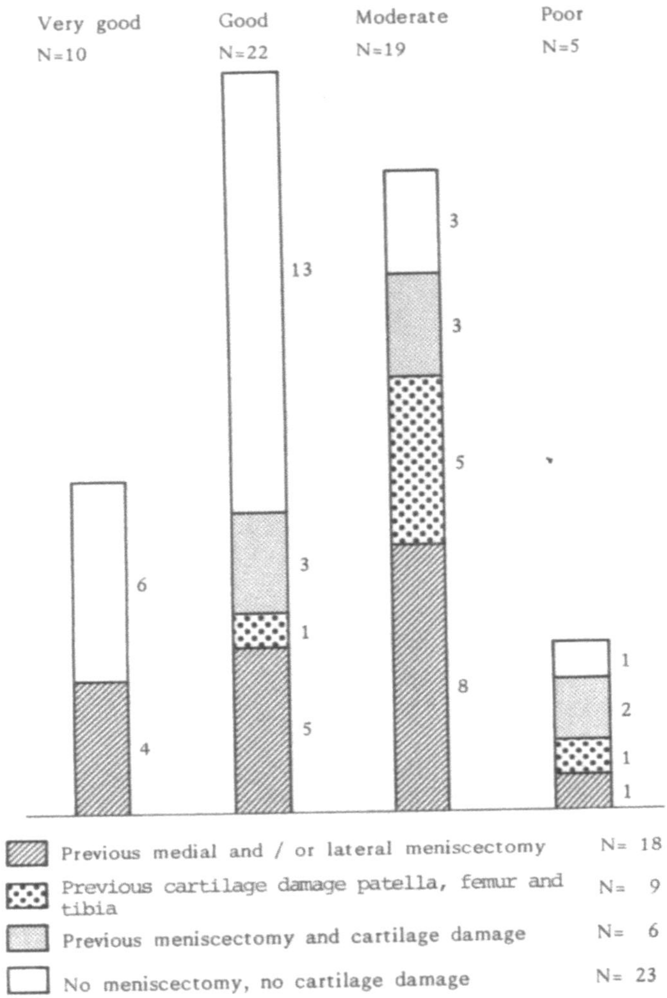

Fig. 13. Results in patients with prior damage to the menisci and cartilage

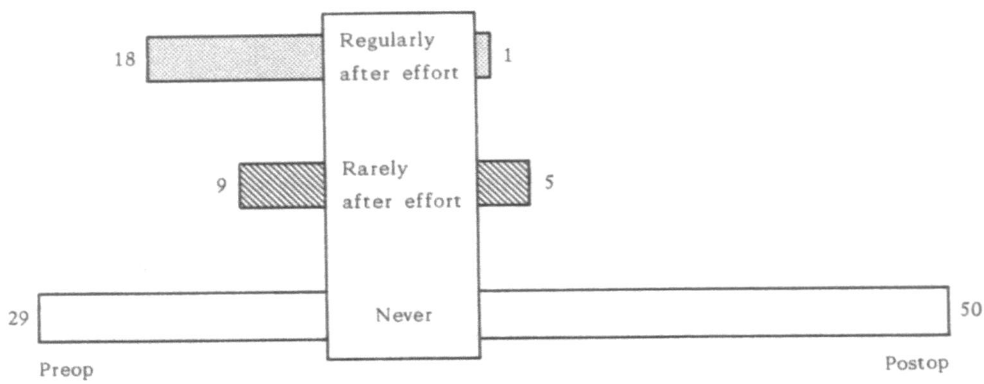

Fig. 14. Results and effusion (left preop., right postop.)

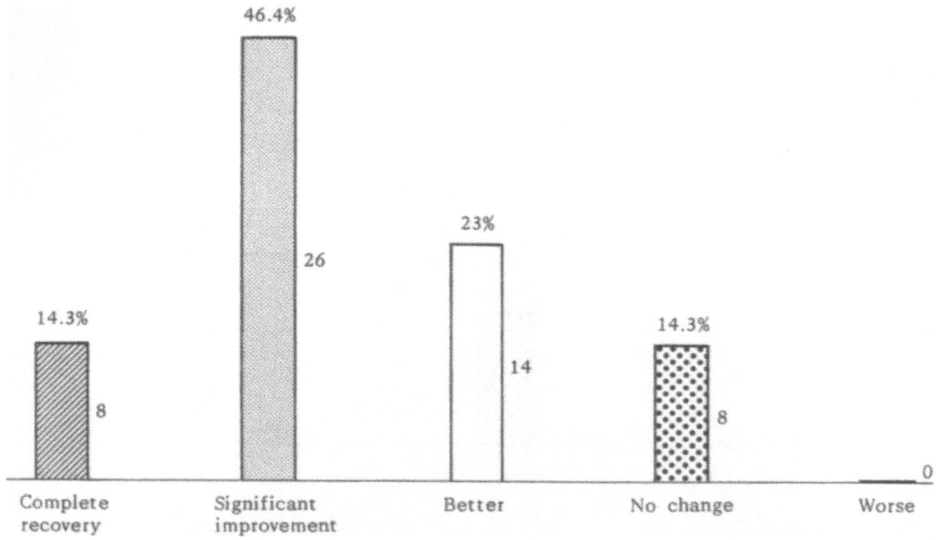

Fig. 15. Subjective results of the popliteus bypass operations

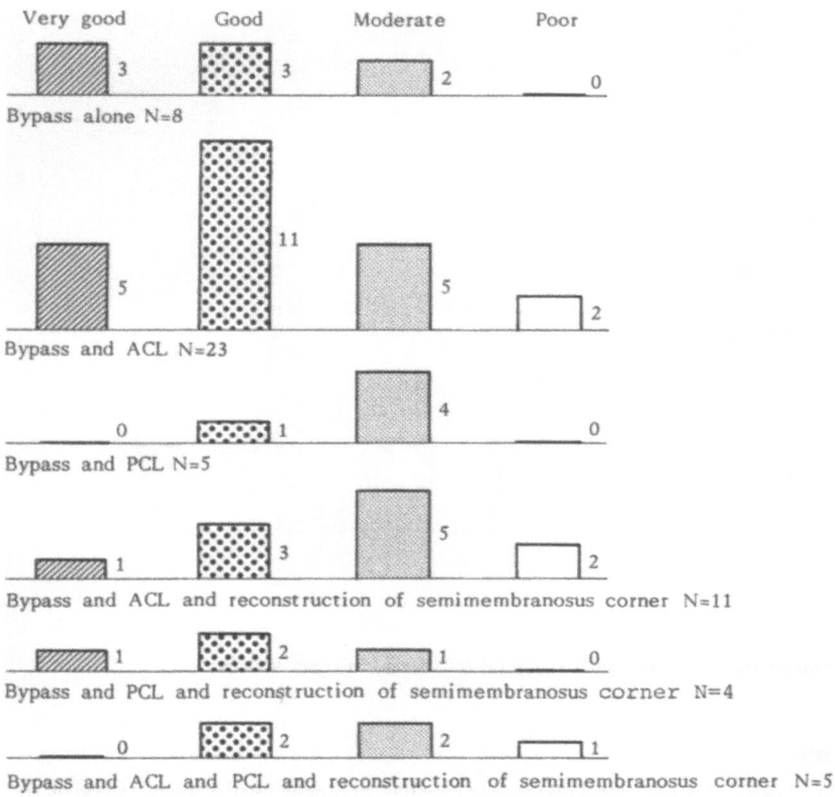

Fig. 16. Results of the popliteus bypass operations by combinations

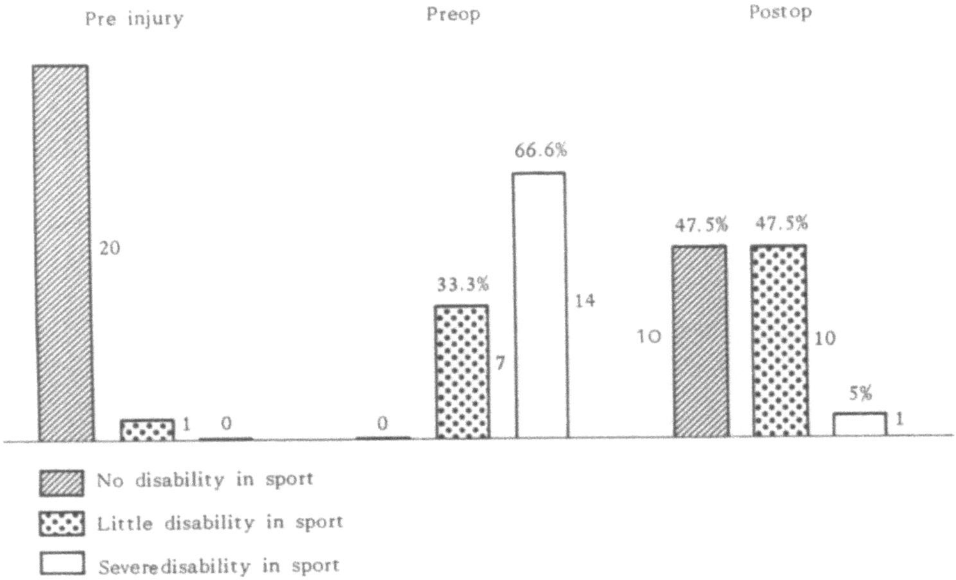

Fig. 17. Results and sports fitness

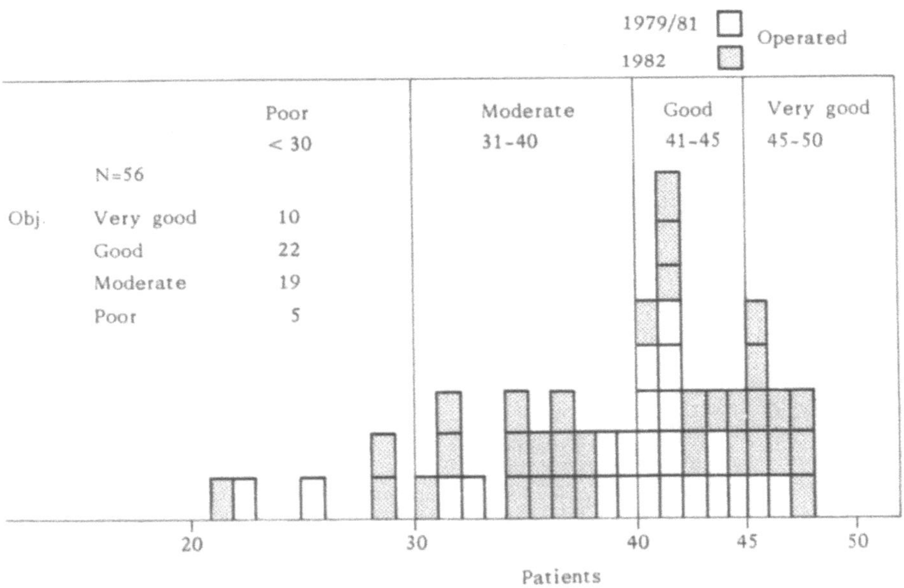

Fig. 18. Comparison of results with Hospital of Special Surgery scores

Reference

Barton TM, Torg J, Das M (1984) Posterior cruciate ligament insufficiency: a review of the literature. Sports Med 1:419–430

Treatment of Acute and Chronic Injuries of the Posterior Cruciate Ligament

H. Oshiro

Department of Orthopaedic Surgery, Nihon University, School of Medicine, Tokyo, Japan

Introduction

Injuries of the posterior cruciate ligament (PCL) frequently present diagnostic and therapeutic difficulties. This is a report of a 6-year experience with PCL injuries in the absence of avulsion fractures.

The group consisted of 50 patients (34 males and 16 females) encompassing 51 knees (28 left and 23 right). The age at the time of injury ranged from 14 to 64 years old, the mean (age) being 25 years old. The follow-up study lasted from 4 months to 6 years and 6 months; the mean period was 3 years.

The causes of injury are as follows: 20 cases from traffic accidents, 25 from sports accidents, and 5 from work accidents.

In 16 cases the mechanism of injury was anterior stress during flexion, in 12 cases an anterior abduction stress during flexion, in 6 an anterior adduction stress during flexion, in 5 hyperextension, and for 12 cases the cause was not known.

The PCL injury was acute in 25 knees and chronic in 26.

The most frequently combined ligament injuries were posteromedial injuries in which both the PCL and the medial supporting structure were affected. Only 14 cases were isolated PCL injuries.

During the diagnosis of acute cases it was evident that haemarthrosis was present in 93%. Arthrography within 2 weeks of injury revealed the ruptured site in the capsule. Arthroscopy was regarded as an essential diagnostic method.

Diagnosis of chronic cases was based on the mechanism of injury, the history of haemarthrosis and clinical examination. Recently, magnetic resonance imaging (MRI) was also used to establish the diagnosis of PCL injury. With this technique a normal PCL is indicated as a dark strip of low signal intensity, with its origin and insertion clearly visible. Normal cases demonstrate no significant differences in their coursing, thickness or shape.

Material and Methods

Of the 51 knees 37 were surgically treated.

On acute cases the surgical techniques employed were the pull out method coupled with suturing of the capsular ligament in 14 knees, the end to end suture in 2 knees, and primary reconstruction using a modification of Kennedy's or Insall's method on 7 knees in which the middle part of the PCL was injured and primary repair was impossible.

Surgery and Arthroscopy of the Knee
2nd Congress of the European Society
Editors: Müller/Hackenbruch

For the chronic cases Insall's method was performed on 13 knees, 5 of which were classified as Marshall's Grade II or above for medial instability and so concurrently received Bosworth's operation.
Two knees from the chronic cases were operated on according to our original reconstruction method. The ligament substitutes used were the iliotibial tract and the semitendinous muscle. In another case, an artificial ligament and the semitendinous muscle were utilised in combination.

Results

During the follow-up period, lasting on average for 3 years, the therapeutic result was evaluated according to Hughstone's method. Subjectively, cases in all three groups were found to be in the category of fair, except for one case which was classified as poor from the group of chronic cases. The objective and functional appraisals, however, revealed no good case from the operated on chronic group, a poorer result than for the other two groups. Conservative treatment was sufficiently effective in cases with isolated PCL injury. Reconstruction for chronic cases presenting other ligament injuries failed to achieve a satisfactory result, and the necessity of early primary repair was strongly indicated in cases with combined ligament injury.
When the reconstruction techniques were evaluated, the dynamic stabilisation, according to Insall's method, had its limitation in the stabilisation. The static stabilisation, as effected by our original method and by the use of an artificial ligament, achieved a better result than the dynamic stabilisation. These findings should be confirmed with a larger number of cases and a longer period of follow-up.
One case presents a 20-year-old male who injured the right knee in a motorcycle accident. He visited our department 2 months later. Stress radiography revealed a marked posterior drawer test, and MRI showed a blurred part more to the tibial side of the PCL. Though arthroscopy failed to detect a rupture in the ligament, the posterior drawer test indicated relaxation of the anterior cruciate ligament. Our original method was employed in the treatment. Six months of postoperative observation are not sufficient for a thorough evaluation, but the parameters have improved.

Conclusion

In view of the appraisal of treatment and prognosis, conservative treatment is effective for isolated PCL injuries. Reconstruction techniques for chronic cases associated with other combined ligament injuries prove to be less than satisfactory. It is evident that the presence of combined ligament injuries requires an early primary repair. Our policy in the future will be to employ the static approach for reconstruction cases.

Experimental Investigations of Stress Production on the Cruciate Ligaments After Transpatellar Steinmann's Nail Fixation

R. Weinstabl [1], H. Kern [2], and M. Wagner [1]

[1] The First University Clinic of Traumatology, Alser Straße 4, 1097 Vienna, Austria
[2] Wilhelminen Hospital, Department of Physiotherapy, Montlearstraße 37, 1160 Vienna, Austria

Introduction

The question arising on whether to recommend fixation or early mobilisation in postoperative treatment of posterior cruciate ligament (PCL) injuries is a very interesting one. Some authors described the technique of temporary transpatellar fixation with a Steinmann's nail to allow early mobilisation. Postoperative plaster fixation allowing movement of the knee from 0° to 90° transpatellar fixation using a Steinmann's nail has been renounced by others. This technique does seem to be non-physiological, and therefore a study was conducted to investigate the influence of the nail fixation *(Olekranisierung)* on the tension of the cruciate ligaments.

Material and Methods

In this investigation 12 knees of fresh cadaveric specimens were examined. Subjects with prepared retinacula and dissected skin and muscles could not be included in the investigation because they would present false measurements. In such cases the posterior drawer and tension on the PCL are compensated by a wide space in the patellofemoral joint.

The extremities were fixed out of the femur in the AO external fixation frame, so that free movement in the knee joint was possible. Constant tension was produced at the rectus femoris muscle.

The approach to the PCL was made through a dorsal incision orientated in a craniocaudal direction. The anterior approach was transligamentary to preserve the ligamentous and muscular structures as described by Schabus et al. (1987). Measurement of tension in the ligaments was made every 10° through an arc of 0°–90° of flexion. The system was calibrated before and after the investigation.

The Steinmann's nail, 15–20 cm long and 4.5 mm in diameter, was implanted along the lengthwise axis of the patella while the knee joint was held in 20°–30° of flexion. The nail should only perforate the first corticalis of the tibia, directly behind the patellar ligament.

Surgery and Arthroscopy of the Knee
2nd Congress of the European Society
Editors: Müller/Hackenbruch

Results

In normal flexion of the knee joint the PCL is under full tension at 0° and 90° of flexion. The PCL is relaxed from 20° to 40°, the tension rising again from 60° to 90° of flexion. A similar picture could be seen after implantation of the Steinmann's nail; much higher peaks of tension were evident, but the relief of tension persisted from 20° to 40°. For the anterior cruciate ligament (ACL) high values of tension were seen from 30° to 0°. From 40° to 60° the values descended and up to 90° remained constantly low.

Discussion

From the results it appears that during extension without patellar fixation both the ACL and PCL come under high tension, the ACL more than the PCL. From 20° to 40° both ligaments are relaxed or experience low tension. Between 40° and 90° the tension in the PCL increases while the ACL remains relaxed or under low tension. After implantation of the Steinmann's nail the distribution of tension from 0° to 90° is similar to that for the normal case with an unfixed patella, but with significantly higher tension peaks (Figs. 1, 2). The average tension of the PCL in full extension is five times as high as normal and in the ACL, five to four times as high. At a flexion of 90° the tension of the PCL increases by five to ten times the normal value to relieve the tension in the ACL. This value applies to the case in which only the first corticalis of the tibia is perforated by the nail. If the second tibial corticalis is also perforated, no patella movement is possible, and the peaks of tension are high and out of range of the measuring system. Therefore, it appears that one reason for the increase of tension in the PCL is the loss of the pivot-

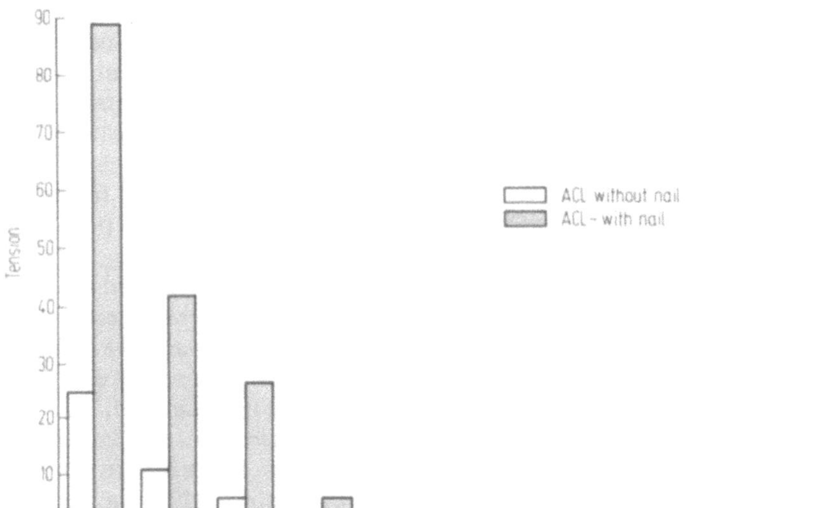

Fig. 1. ACL: Tension with and without a Steinmann's nail in 0°–90° flexion

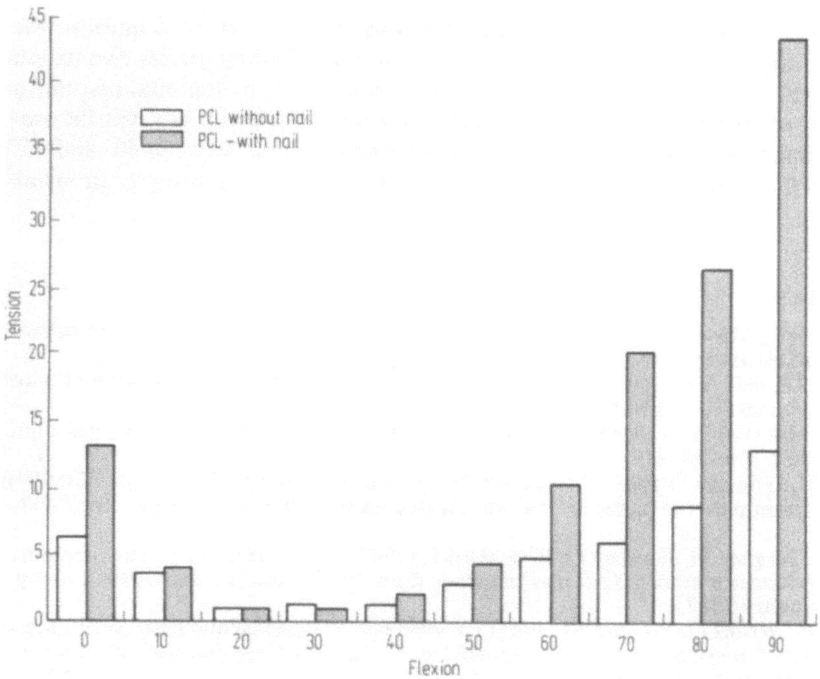

Fig. 2. PCL: Tension with and without a Steinmann's nail in 0°–90° flexion

ing action of the patella. In fact, fixation of the patella to intact ligamentous and muscular structures, results in a posterior drawer in the knee joint. In full extension this leads to high pressure and consequently damage of the patellofemoral joint cartilage.

This study indicates that functional postoperative treatment is only possible from 20° to 40° of flexion, when the PCL is relaxed. Prevention of posterior sag, which is not possible by plaster-cast immobilisation, would be advantageous in this technique. A disadvantage is the non-physiological fixation of the patella *(Olekranisierung)* preventing normal knee movement, resulting in high tension at the cruciate ligaments in certain positions of flexion. Whether reinforcement of the PCL with an elastic wire or alloplastic augmentation could assist in this kind of treatment is a point for further investigation.

Summary

The tension in human knee-joint cruciate ligaments was investigated. The system employed was an AO external fixation frame, and tension measurements were made by bucle transducers. After measurement of the normal ligament tension a transpatellar Steinmann's nail, as used in PCL reconstruction, was implanted, and then the tension measurements were repeated and compared. Without the nail, the PCL is in full tension at 0° and 90°; from 20° to 40° it is not under tension. From 0° to 30° tension in the ACL increases, but from 40° to 90° the tension is low. After implantation of the nail, the spectrum of PCL and ACL tension was similar to that for the unfixed patella, but with much higher peaks, five to ten times higher. Therefore, after transpatellar fixation with a Steinmann's nail a postoperative movement of 20°–40° can be allowed. Posterior sag can be prevented by immobilisation in a cast. Postoperative movement between 40° and 90° or 20° and 0° is very detrimental to the reconstructed PCL, leading to an insufficiency.

References

Decraemer WF, Maes MA, Vanhuyse VS (1980) An elastic stress-strain relation for soft biological tissues based on a structural model. Biomechanics 13:463–468

France PE, Daniels AV, Goble ME, Dunn HK (1983) Simultaneous quantitation of knee ligament forces. Biomechanics 16:553–564

Girgis FG, Marshall JL, Al Monasem ARS (1975) The cruciate ligaments of the knee joint. Clin Orthop 106:216–231

Monahan JJ, Grigg P, Pappas AM, Leclair WJ, Marks T, Fowler DP, Sullivan TJ (1984) In vivo strain patterns in the four major chronic knee ligaments. J Orthop Res 2:408–418

Schabus R, Wagner M, Kwasny O, Weinstabl R (1987) Preservation of normal anatomy in ACL-reconstruction. 4. Österr.-Finnischer Kurs für Traumatol. und Orthop. Ischgl, 17–24 January 1987

Weinstabl R, Wruhs O, Wagner M (1985) Experimentelle Untersuchung der Bewegungsabläufe des Kniegelenkes nach transpatellärer Steinmann-Nagelfixation. 49 Jahrestagung der Deutschen Gesellschaft für Unfallheilkunde Berlin, 13–16 November 1985

The Transpatellar Steinmann Pin as a Postoperative Safeguard After Reconstruction of the Posterior Cruciate Ligament or After Trillat's Operation

E. Orthner, O. Kwasny, R. Schabus, and M. Wagner

The First University Clinic of Traumatology, Vienna,
Alser Straße 4, 1097 Vienna, Austria

Introduction

Postoperative immobilisation, relieving tension, is compulsory after reconstructive surgery of ligamentary structures. Following reconstruction of the posterior cruciate ligament (PCL) or after Trillat's operative procedure there is a permanent risk of mechanical stress, which is apt to jeopardise the result of the reconstruction. Stress can occur, particularly when the plaster cast is applied immediately after the operation, whenever the plaster cast is changed and during the early functional treatment with a brace. During these procedures the reconstructed structures are put at risk due to the spontaneous posterior drawer in the supine position.

Since 1983, following a suggestion of Grammont and Müller, we have been immobilising the tibia in a neutral position by driving a transpatellar Steinmann pin into the tibial head after reconstruction of the PCL or after surgical anteposition of the arcuate ligament complex. The following procedure has proven the most efficient.

Material and Method

Surgical Technique

A skin incision is made across the patella laterally. Dissection and reconstruction of the knee joint's inner structures are done through a medial parapatellar arthrotomy. Before the reconstructed parts are finally tightly approximated, the tibial head is firmly held in slight external rotation and in the correct position for 20°–30° of flexion of the knee joint. In this position a canal, 3.2 mm in diameter, is drilled longitudinally through the patella into the tibia. By this canal a 4.5-mm Steinmann pin is pierced into the tibial head and fixed within the posterior corticalis of the tibial bone.

Since June 1985, owing to the experimental investigations of Weinstabl, implantation of the pin could be performed without perforating the posterior tibial corticalis. As the pin is shortened at its point of insertion, it exceeds the base of the patella by only roughly 5 mm.

Surgery and Arthroscopy of the Knee
2nd Congress of the European Society
Editors: Müller/Hackenbruch

Postoperative Treatment

A leg plaster is applied postoperatively for 4 weeks to hold the knee in slight external rotation and 20° of flexion. Provided that wound healing is free of complications, mobilisation starts with crutches and without weight bearing. After 4 weeks a brace with motion is fitted, and the patient may move his knee to an extent of S 0-10-30. After approximately 6 weeks posteroperatively, partial weight bearing is permitted, the limits of flexion and extension expanding week by week under protection of the brace. As soon as flexion is possible up to 35° (at the latest after 8 weeks) the Steinmann pin is removed under general anaesthesia. As soon as the deficit of extension is less than 5°, the patient is allowed to increase weight beaaring gradually and finally to walk without crutches.

Patients

From April 1983 until April 1986, 21 patients (16 male and 5 female) with an average age of 25.6 years (range 17–49 years) were operated on at the First University Clinic of Traumatology in Vienna for PCL insufficiency with additional securing of the reconstruction by a transpatellar Steinmann pin. The latter procedure was omitted when an osseous exfraction of the PCL was reinserted by means of a screw or when the cruciate ligament was ruptured in the presence of an open epiphyseal cartilage.

The causes of injury were: traffic accidents in 15 cases, trauma with hyperextension due to a sudden fall from a great height in 3 instances, a soccer trauma in 2 cases and a gymnastics trauma in 1 case. With the exception of 6 cases, all other patients had multiple injuries.

Twelve patients underwent surgery within the first 4 weeks after the trauma; nine received treatment later than 4 weeks post-injury.

In nine cases there was a midsubstance of the PCL, eight at the femoral origin, and one at the tibial insertion. In three cases the ligament was distended by several millimeters due to a stretch trauma. The operative procedure comprised reinsertion or suture of the PCL in 18 patients. In 13 patients the arcuate ligament complex was surgically displaced as described by Trillat, or reinforcement was provided by a polypropylene tape or the semitendinosus muscle. In five patients the PCL was sutured without any additional procedure, while in three patients the operation of Trillat was carried out exclusively.

Results

There were two postoperative complications. In one case the Steinmann pin had to be removed prematurely as it had loosened due to a second drilling during the original procedure. In the second case the pin had been inserted too deeply into the patella, and it required removing through a parapatellar incision. There were no infections of the drilling canal, and treatment was not stopped due to retropatellar pains. Moreover, ossifications within the patella ligament, which have been described by Müller and by Grammont, were never observed. There was also no sign of an osteochondritic focus at the medial femoral condyle as reported in one instance by Baumgartner.

Follow-up examinations were done in 19 of the 21 patients at an average interval of 10 months (range 4–19 months) after the operation. In 30° of flexion the dorsal Lachman test was equally stable for both sides in 2 knee joints, in 14 cases the result was slightly positive and in 3 cases moderately positive. In 90° of flexion (neutral position) the dorsal drawer test was equal for both sides in 17 cases and slightly positive in 2 cases. Ligamentary resistance was strong in 17 patients and soft in 2. Extension was uninhibited in 14 cases, restricted up to 5° in 3 cases and restricted to less than 5° in 2 patients. Those patients whose knee movement was restricted to less than 5° had also suffered rupture of the anterior cruciate ligament (ACL) during the trauma. Flexion was restricted in all patients: up to 10° in ten, up to 20° in seven and to just above 20° in two patients.
Follow-up examinations paid particular attention to any clinical signs indicating damage of the retropatellar cartilage due to the presence of the transpatellar Steinmann pin. In 17 of 19 patients the Zohlen test was negative, and in 2 patients it was positive. Originally both of these patients had suffered a typical impact trauma when riding a motorbike. In one case the trauma was combined with an open fracture of the calf and in the other case with an additional rupture of the ACL and an injury of the medial ligaments of the joint. Four patients described a sensation of retropatellar friction without pain.

Summary

Considering postoperative function and ligament stability the final result of the operation may be termed at least to be "good" in 17 of 19 patients. Eight of our patients have already returned to their previous sport, five are still in training, and four are difficult to assess as they did not participate in sport before the trauma. The two patients with a bad result had originally suffered a multiple-injury trauma followed by an extremely long period of convalescence. In both of these patients the PCL had been reconstructed by a secondary procedure in the chronic stage, when they already had considerable muscle atrophy of the thigh. The compound instability had prevented active training of the thigh musculature. In these two patients the reconstruction of the PCL proved insufficient at the follow-up examination. However our clinical results of PCL reconstruction seem to be acceptable for short-term results. The experimental results of strain measurements reported by Weinstabl show that there is a dangerous increase of strain on the ligaments when moving the knee joint through more than 40° of flexion, if the Steinmann pin is not removed.

References

1. Brantigan OC, Voshell AF (1941) The mechanics of the ligaments and menisci of the knee joint. J Bone Joint Surg 23:44–66
2. Bianchi M (1983) Acute tears of the posterior cruciate ligament: clinical study and results of operative treatment in 27 cases. Am J Sports Med 11:308–314
3. Fischer LP, Guyot J, Gonon GP, Carret JP, Courcelles P, Dahhan P (1978) The role of the muscles and ligaments in stabilization of the knee joint. Anat Clin 1:
4. Girgis F, Marshall JL, Al Monajem ARS (1975) The cruciate ligaments of the knee joint. Clin Orthop 106:216–231

5. Hughston JC, Rowden JA, Andrews JR, Norwood LA (1980) Acute tears of the posterior cruciate ligament. J Bone Joint Surg [Am] 62:438–450
6. Hughston JC, Jacobson KE (1985) Chronic posterolateral rotatory instability of the knee. J Bone Joint Surg [Am] 67:351–359
7. Schweiberer L, Hertel P (1977) Experimentelle Studie zur Biomechanik und Pathophysiologie des Kniebandapparates. Langenbecks Arch Chir 345:439–444
8. Strand T, Molster AO, Engesaeter LB, Raugstad TS, Alho A (1984) Primary repair in posterior cruciate ligament injuries. Acta Orthop Scand 55:545–547
9. Welsh RP (1980) Knee joint structure and function. Clin Orthop 147:7–14

Experiences with Reconstruction of the Posterior Cruciate Ligament Using a Free Graft of the Rectus Muscle Tendon Combined with a Patellar Bone Graft

Z. G. Harnach and V. Pokorný

Research Institute of Traumatology, Brno, Czechoslovakia

Introduction

In reconstructions of the posterior cruciate ligament (PCL) for chronic instabilities of the knee joint autogenous materials provide enough strength and firmness for ligament substitutes. The advantages of the method described are that a large enough graft of the rectus muscle tendon is obtained for ligament replacement, and the implant is firmly anchored with the help of a bone graft.

Material and Methods

This operation is performed by the Payer approach using a tourniquet. The free tendon graft is taken from the rectus muscle tendon. The complete graft is 10–15 mm wide and 11–15 cm long, terminating at the distal end in a trapezoidal bone graft. The bone graft which is, 5 mm thick, is taken from the proximal pole of the patella with the aid of an oscillating saw and thin chisel. A channel is drilled in the proximal epiphysis of the tibia, located at the original point of the ligament insertion, using a special instrument called a "cobra" chisel. The bone graft is then firmly fixed in the prepared channel. The graft thus passes through the channel in place of its anatomical origin through the femoral medial condyle and if long enough continues as a replacement of the collateral tibial ligament. After the operation a splint of plaster of Paris is applied, which is exchanged 12 days postoperatively for another plaster cast allowing 45° of flexion for a further period of 6 weeks. Cast removal is followed by rehabilitation, full weight bearing being permitted after 3 months.

Results

During the period from 1975 to 1985 this method was used for 76 reconstructions of the PCL. Over a period of 2–10 years postoperatively 52 patients were followed up. On a number of these patients a further operation was performed when the previous operation was unsuccessful. For 29 patients the results achieved were very good. In 12 patients good joint stability was obtained, although pain was still present owing to osteochondritic changes which were already present before treatment. In 2 patients full extension was reduced by 5°. In all patients flexion was more than 120°.

Surgery and Arthroscopy of the Knee
2nd Congress of the European Society
Editors: Müller/Hackenbruch

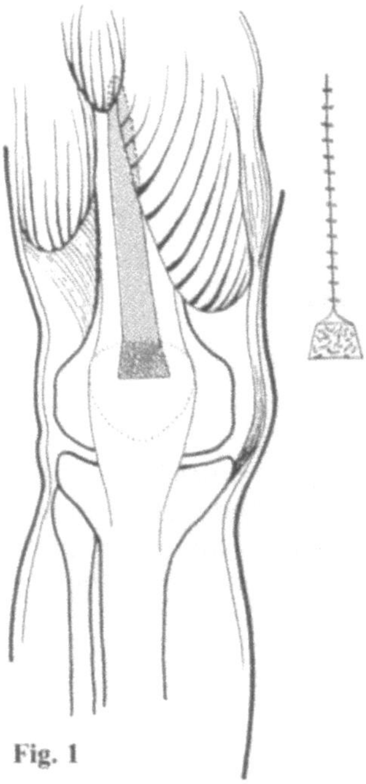

Fig. 1

Fig. 2

Fig. 1. Preparation of a free graft of the rectus tendon with a 40° patellar bone graft and suture of the tendon defect

Fig. 2. Excision of a bone graft from the patella, using an oscillating saw and thin chisel

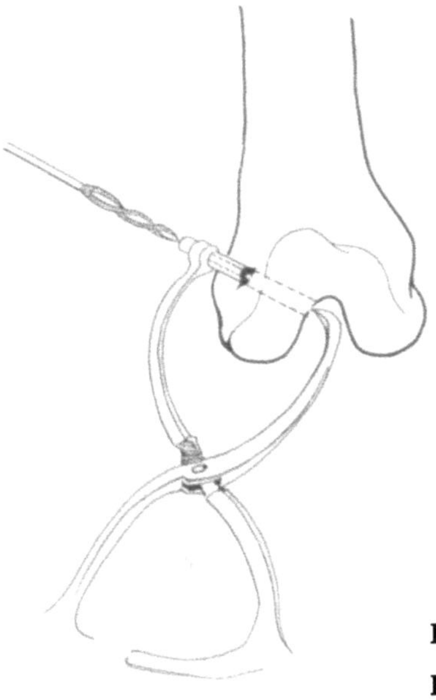

Fig. 3

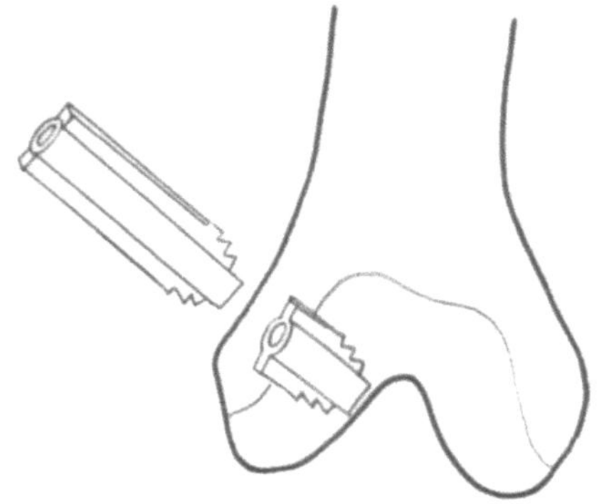

Fig. 4

Fig. 3. Guide for accurate drilling of the channel

Fig. 4. Channel for anchoring the bone block prepared with a "cobra" chisel

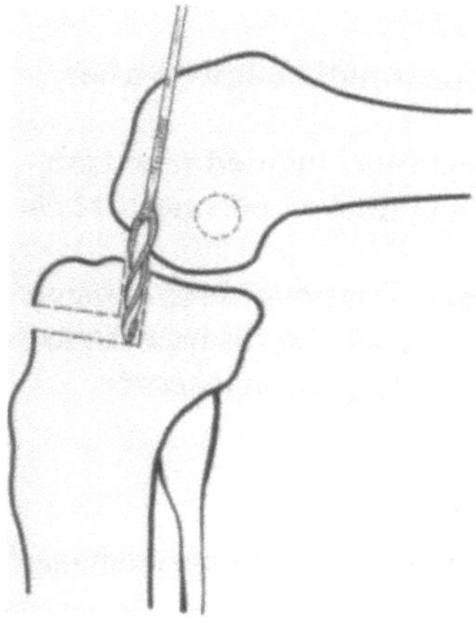

Fig. 5

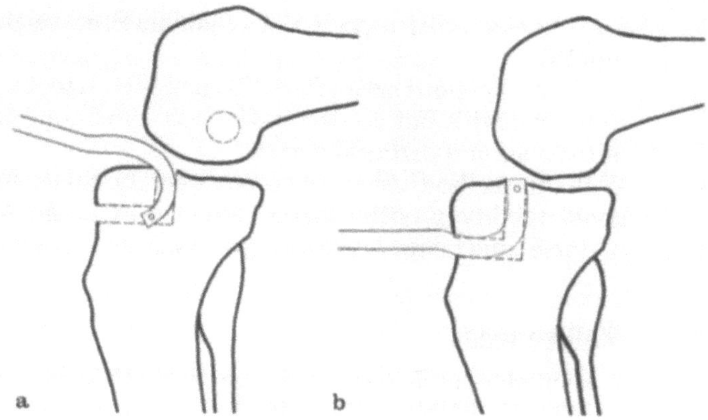

Fig. 6

Fig. 5. Drilling of the tibia for replacement of the anterior cruciate ligament

Fig. 6 a, b. Curved hook for drilling the channel for reconstruction of the anterior cruciate ligament

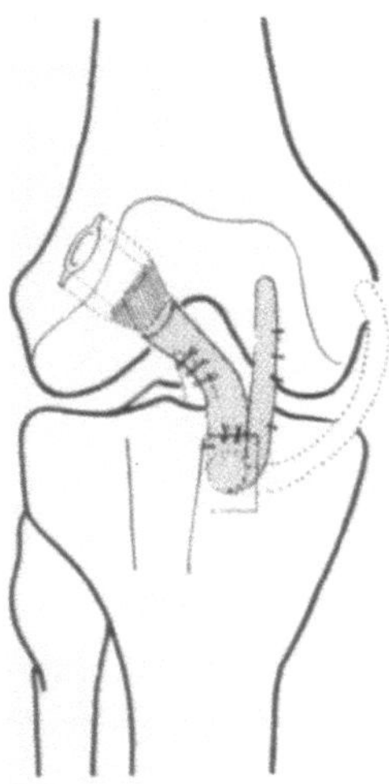

Fig. 7

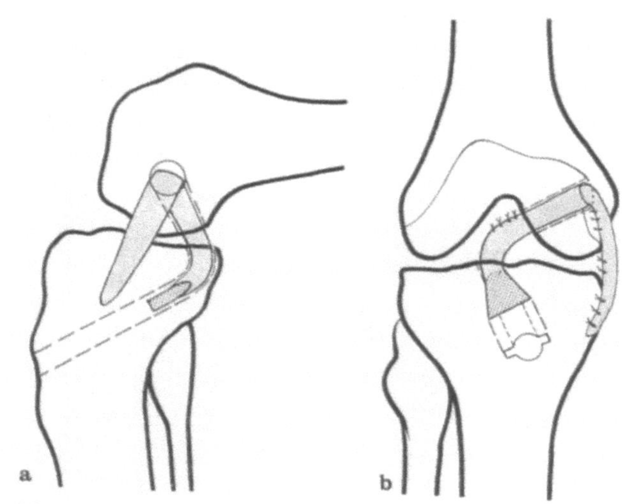

Fig. 8

Fig. 7. Reconstruction of the anterior cruciate ligament

Fig. 8 a, b. Reconstruction of the posterior cruciate ligament

This method was also used for reconstruction of the anterior cruciate ligament (ACL) with success in 280 reconstructions.

Figures 1–7 illustrate the operative technique employed in reconstruction of the ACL. The bony part of the graft is secured in a tunnel drilled in the lateral condyle of the femur. The tendon part of the graft passes through the joint cavity along the tibial channel and then serves as a reinforcement of the tibial collateral ligament. A splint of plaster of Paris is applied for 12 days, followed by a plaster cast

for 6 weeks. After a period of rehabilitation, weight bearing is allowed after 3 months.
The complications seen after PCL and ACL reconstructions included nine transitory synovitis but no deep infections, three tourniquet shocks and one paresis of the common peroneal nerve.
With this method, using the correct operational technique, it is possible to achieve good stability in both the reconstructed PCL and ACL, with the disappearance of dorsal and ventral drawer signs resulting in a very good functional score.

References

1. Harnach Z (1976) Rekonstrukce předniho zkříženého vazu kolenního kloubu modifikací Jonesovy metody. Rozhl Chir 55:8
2. Harnach Z, Rybníček K (1984) Rekonstruktionen des hinteren und vorderen Kreuzbandes. Resultate nach 120 Operationen mit eigener Methode. Hefte Unfallheilkd 167:349–357

Treatment of Posterior Cruciate Ligament Injuries: Long-Term Effects

J. Lysholm. M. Lindh, and J. Gillquist

Department of Orthopaedic Surgery, University Hospital,
58185 Linköping, Sweden

The object of this retrospective study was to determine the long-term effects after treatment for a torn posterior cruciate ligament (PCL) administered in the early phase after injury. The follow-up was conducted 3–12 years after treatment and addressed knee function, level of activity, static stability and the risk of developing secondary osteoarthrosis.

Material and Methods

The study was carried out on 55 patients – 40 males and 15 females – whose mean age was 28 years. Each had originally been treated within 4 weeks of injury. The majority of injuries had occurred in sport activities, while 25% were the results of traffic accidents. In 80% of cases diagnosis had been made on clinical examination under anaesthesia and arthroscopy (Gillquist and Hagberg 1976; Lysholm and Gillquist 1981) and in 20% on the basis of clinical examination under anaesthesia and exploratory arthrotomy. Seven patients showed a partial tear of the PCL, 12 had isolated total substance tears, 6 had isolated avulsions with a bone fragment of the tibial insertion of the PCL and another 6 had avulsions of the tibial insertion combined with a medial ligament injury. A total of 33 patients had incurred combined injuries; 19 showed a tear of the medial collateral ligament, 7 tears of both cruciates, 11, a medial meniscus tear, and 4, a lateral meniscus tear.

The seven patients with a partial tear of the PCL were treated conservatively with regard to their cruciate tear; in a few of these an associated medial ligament injury was repaired. Twenty-four of the 36 patients with a total tear of the PCL were treated with a simple suture, adapting the torn end of the ligament to the medial femoral condyle or to the tibia with 6–8 U-sutures through drill holes in the bone – largely the technique described by Palmer (1938). In the 12 cases involving avulsion the tibial fragment was reduced and fixed with nails or a screw. In five cases a primary reconstruction was carried out – in four of these with a part of the medial gastrocnemius tendon and in one with the medial meniscus. The remaining seven patients with a total tear of the PCL were treated conservatively. Associated injuries were required when neccessary; torn menisci repaired or excised depending on the site of the lesion. After surgery the leg was immobilized in a cast for 5–6 weeks, followed by rehabilitation, usually under the guidance of a physiotherapist.

The follow-up study was conducted at an average of 67 months after treatment and included the following features: (a) evaluation of knee function in a 100-point

Surgery and Arthroscopy of the Knee
2nd Congress of the European Society
Editors: Müller/Hackenbruch

knee score (Lysholm and Gillquist 1982), (b) activity grading on an 11-step activity grading scale (Tegner and Lysholm 1985), (c) traditional clinical examination of the knee, (d) measurement of thigh muscle strength in a Cybex-II dynamometer and (e) evaluation in terms of a special knee function test (Tegner et al. 1986). This knee function test involves running a figure-eight course twice for a total of 40 m; the running time is measured electronically with photocells placed to make it possible to separate running straight and running in the turns. The test, furthermore included a staircase running test (up and down a 25-step spiral staircase), a slope running test (up and down a 55-m long slope with a 180° curve at midpoint) and a one-leg hop for length.

Results

Figure 1 shows the distribution of scores for all 55 patients. One patient had been reoperated before the follow-up because of persistent instability after primary repair of the PCL. Three patients showed less than 115° of maximum knee flexion and one patient, an extension lag of 10°. There were no infections. Among all patients 74% had a knee score above 82 points, which is classified as "good" or "excellent". Symptoms of instability in this group were rare. Fourteen with a score below 82 complained of pain, a few even on activities of daily life, and 11 of these 14 complained of symptoms of instability, usually slight to moderate. Patients with partial or isolated PCL injuries had an excellent knee function with a mean score of 98 and 96 points, respectively. Patients with combined injuries had a significantly lower score, on average 81 points. Also, patients with isolated avulsions of the tibial insertion of the PCL showed a lower average score, 92 points. There was no significant difference in median level of activity (11-point scale), which varied between 4 and 7 in the different subgroups, except in patients with tears of both cruciates, who had a significantly lower level of activity (median 2).
Eighteen patients had participated in an earlier follow-up 5 years previously; among these there had been no drop in knee score over the 5 years (Fig. 2). A slight to moderate posterior drawer sign was common, at a rate fairly similar in

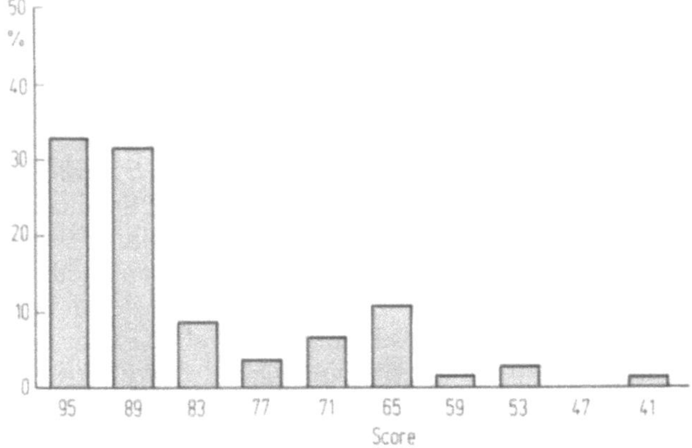

Fig. 1. Distribution of scores on the 100-point knee-grading scale ($n = 55$)

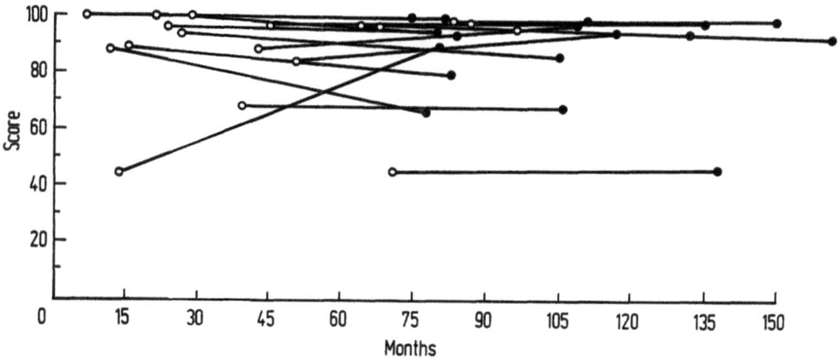

Fig. 2. Change in score on the 100-point knee-grading scale over time ($n = 18$)

all subgroups as defined by injury type. The relative isokinetic quadriceps muscle strength in comparison with the uninjured leg at 30° s^{-1} approximated 100% in all subgroups except in patients with a tear of both cruciates. This group showed a significantly lower relative strength (71%). Performance both in the hop test and at staircase running was significantly lower in patients with tears of both cruciates as compared with patients with partial or isolated PCL tears.

For 21 of the 55 patients the follow-up study took place at least 6 years after the injury. Of these, 18 were examined by X-ray in standing position and patello-femoral joint views. Ten showed no signs of osteoarthrosis (OA), two presented isolated femoro-tibial OA, one patello-femoral OA and five combined femoro-tibial and patello-femoral OA. OA was significantly more common after medial meniscectomy as compared with patients with normal menisci. Six of the seven who had had a medial meniscectomy showed radiographic signs of OA as compared with two of eleven with a normal medial meniscus.

Conclusions

We found excellent results after treatment of partial and isolated PCL tears. Patients who were treated for combined injuries, on the other hand, showed lower scores and lower levels of activity. There was no deterioration of the knee score results in the first years following the injury. A slight to moderate posterior drawer was common but did not seem to influence knee function. This coincides with our earlier observations. A 0–5 mm straight posterior drawer was associated with a high knee score, while a positive pivot shift was almost always associated with a significant decrease in knee score (Lysholm et al. 1982). A medial meniscectomy due to a medial meniscus tear in combination with a PCL tear seemed to predispose patients strongly towards the development of osteoarthritis.

References

Gillquist J, Hagberg J (1976) A new modification of the technique of arthroscopy of the knee joint. Acta Chir Scand 142:123–130

Lysholm J, Gillquist J (1981) Arthroscopic examination of the posterior cruciate ligament. J Bone Joint Surg [Am] 63:363–366

Lysholm J, Gillquist J (1982) Evaluation of knee ligament surgery results with special emphasis on use of a scoring scale. Am J Sports Med 10:150
Lysholm J, Gillquist J, Liljedahl S-O (1982) Long term results after treatment of acute knee injuries. Acta Orthop Scand 53:109–118
Palmer I (1938) Injuries to the ligaments of the knee joint. A clinical study. Acta Chir Scand [Suppl] 53
Tegner Y, Lysholm J (1985) Rating systems in the evaluation of knee surgery. Clin Orthop 198:43
Tegner Y, Lysholm J, Lysholm M, Gillquist J (1988) A performance test to monitor rehabilitation and for evaluation of anterior cruciate ligament injuries. Am J Sports Med (to be published)

Surgical Stabilisation of Serious Posterior Instability of the Knee: A Preliminary Report on Personal Techniques

J.-L. Meystre

Service Universitaire d'Orthopédie et Traumatologie,
1005 Lausanne, Switzerland

Introduction

Several recent publications seem to prove the progressive nature of posterior instabilities (Barton 1985; Dandy and Pusey 1982; Clancy et al. 1983; Kennedy et al. 1979; Hughston and Degenhardt 1982), and results of surgical stabilisation are often disappointing (Hughston and Degenhardt 1982; Dejour 1984; Müller 1985). The mechanical problem is complex, particularly because of the small articular congruence and the important physiological mobility of the lateral compartment.

Precise reduction of the posterior subluxation during an operation is often difficult to perform. Overcorrection may be due to degenerative changes and elongation of the anterior cruciate ligament (ACL).

The strains on the graft during rehabilitation are much more important than those on the ACL; also when flexion is greater than 40°, there is no longer active protection of the posterior cruciate ligament (PCL), and contraction of all the muscles of the thigh leads to strain on the ligament. The search for a surgical solution and suitable rehabilitation must take into account all the biomechanical components: morphotype, passive stability (laxity) and active control.

The chosen graft should be much stronger than the normal PCL, so that its resistance to strain remains sufficient to allow healing to take place. In any case, a mechanical protection is certainly useful during this critical period. The anatomical positioning of the graft is essential.

Material and Methods

The treatment includes:

1. Correction of any serious genu varum
2. Reconstruction of the PCL by means of a quadruple tendinous free graft using doubled tendons of the gracilis and semitendinous muscles (Fig. 1)
3. Reconstruction of the insufficient popliteus complex is achieved by transfer of the tendinous part of the lateral head of the gastrocnemius under the lateral collateral ligament (LCL) with an underlying capsule and bone block, providing an active and passive reconstruction similar to the normal popliteus (Fig. 2)
4. Postoperative stabilisation for 4 months by means of an internal guidance device fixed in the lateral condyle of the tibia and articulated with the tibia on a condylar screw

Surgery and Arthroscopy of the Knee
2nd Congress of the European Society
Editors: Müller/Hackenbruch

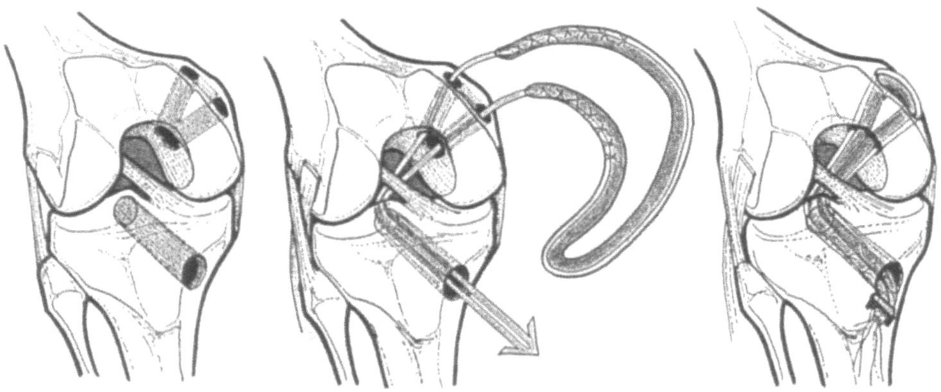

Fig. 1. Reconstruction of the posterior cruciate ligament by means of a quadruple tendinous free graft

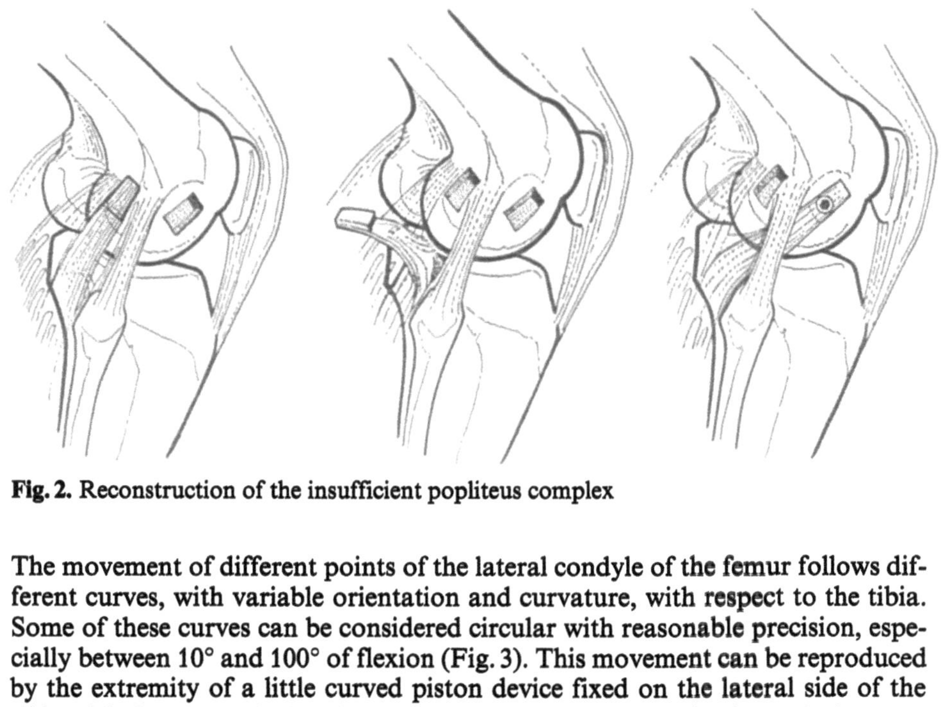

Fig. 2. Reconstruction of the insufficient popliteus complex

The movement of different points of the lateral condyle of the femur follows different curves, with variable orientation and curvature, with respect to the tibia. Some of these curves can be considered circular with reasonable precision, especially between 10° and 100° of flexion (Fig. 3). This movement can be reproduced by the extremity of a little curved piston device fixed on the lateral side of the tibia with the correct inclination (Fig. 4). The precision of this piston device must still be improved. It was first determined by means of dynamic X-ray studies and on anatomical preparations.

At the present time, studies are being conducted in collaboration with the University School of Engineering, Lausanne, by means of a computerised mathematical model of the knee.

The individual condition of the damaged knee is exactly recorded preoperatively on roentgenograms taken with the knee in different positions of flexion and extension, the knee being kept in a reduced position and slight internal rotation.

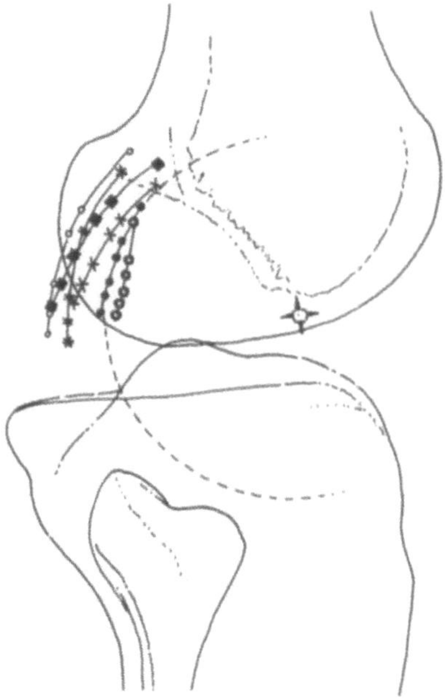

Fig. 3. Movement of the lateral condyle of the femur with respect to the tibia

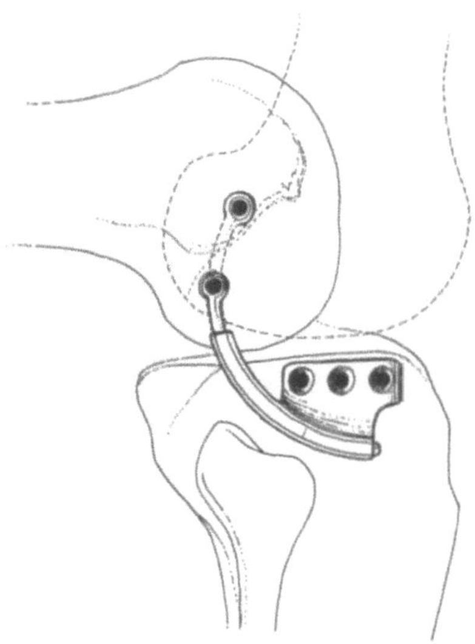

Fig. 4. Movement of the lateral condyle of the femur relative to the tibia can be reproduced by using a curved piston fixed on the lateral side of the tibia

Fivė patients have been operated on using this method since 1984. Even though a linear guidance device was fixed to a fibula which lacked insufficient biomechanical properties, the preliminary results seem to confirm our hypotheses.
The function and subjective stability of the treated knees are good. The clinical laxity after rehabilitation is indeed better than with previous techniques, the Lachmann test and the reversed pivot shift are always negative, and the posterior drawer test at 90° of flexion between + and 2+.

References

Barton TM, Torg JS (1984) Natural history of the PCL deficient knee. 4th Meeting of the international society of the knee

Clancy WG et al. (1983) Treatment of knee joint instability secondary to rupture of the PCL. J Bone Joint Surg [Am] 65:310–322

Dandy DJ, Pusey RJ (1982) The long-term results of unrepaired tears of the PCL. J Bone Joint Surg [Br] 64:92–94

Dejour H (1984) Les indications thérapeutiques dans les laxités chroniques. 5e journées lyonnaises de chirurgie du genou, Lyon

Hughston JC, Degenhardt TC (1982) Reconstruction of the PCL. Clin Orthop 164:59–77

Kennedy JC et al. (1979) Posterior cruciate ligament injuries. Orth Digest 19–31

Muller W (1984) Symposium on posterior instability of the knee. 4th Congress of the international society of the knee

Noyes FR et al. (1984) Biomechanical analysis of human ligament grafts used in knee ligament repairs and reconstruction. J Bone Joint Surg [Am] 66:344–352

Surgical Treatment of Chronic Posterolateral Laxity of the Knee: Procedure, Indications and Early Results

J. H. Jaeger and J. R. Meyer

Service de Traumatologie et d'Orthopédie, C. H. Louis Pasteur, 68021 Colmar, France

Introduction

Posterolateral instability develops as a result of a lesion in one or more structures of the knee: the posterior part of the lateral meniscus, the popliteal ligament, the lateral ligament, the arcuate ligament, and the posterolateral capsule.

The lesion leads to rotary instability which manifests itself through an abnormal posterior rotary mobility of the lateral tibial plateau [2], a posterolateral drawer, a hyperextension-varus-external rotation movement [3] and a reversed pivot shift [5]. Not all of the abnormal movements are present simultaneously, but they are more obvious if the posterior cruciate ligament (PCL) is also ruptured.

Posterolateral instability is rotatory in nature and thus is very markedly felt by the patient.

Only a few reports have been published on surgical treatment of this condition [1, 2, 4, 6]. From our own experience at least, the procedures expounded by these authors did not always lead to satisfactory results. This led to the development of a new procedure using fascia lata in two different ways.

Methods

Surgical Techniques

In both methods the thigh is supported at an angle of 80°.

Type 1. The principle aim of the surgical technique is to oppose the posterolateral laxity by building up two mattresses of support, one spreading out from the posterolateral tibial corner up to the femoral condyle at the popliteal tendon origin, and a second from the upper end of the capsule near its posterolateral femoral corner down to Gerdy's tubercle.

The operative procedure is illustrated in Figs. 1, 2. A strip of fascia lata is pulled away from Gerdy's tubercle, to which it remains attached. It should be about 15 cm long but variable in breadth, measuring about 3 cm at the upper end and 1 cm near the lower end attachment. Two tunnels are made. One passes through the tibial tuberosity, beginning near Gerdy's tubercle and exiting 0.5 cm below the plateau just medial to the peroneotibial articulation. For this a 6-mm drill is used (Fig. 1 b). The other tunnel, which is about 3 cm long, goes through the lateral condyle of the femur and underneath the attachments of the lateral and popliteal ligaments (Fig. 1 c). A curved awl is used to drill through the bone.

Surgery and Arthroscopy of the Knee
2nd Congress of the European Society
Editors: Müller/Hackenbruch

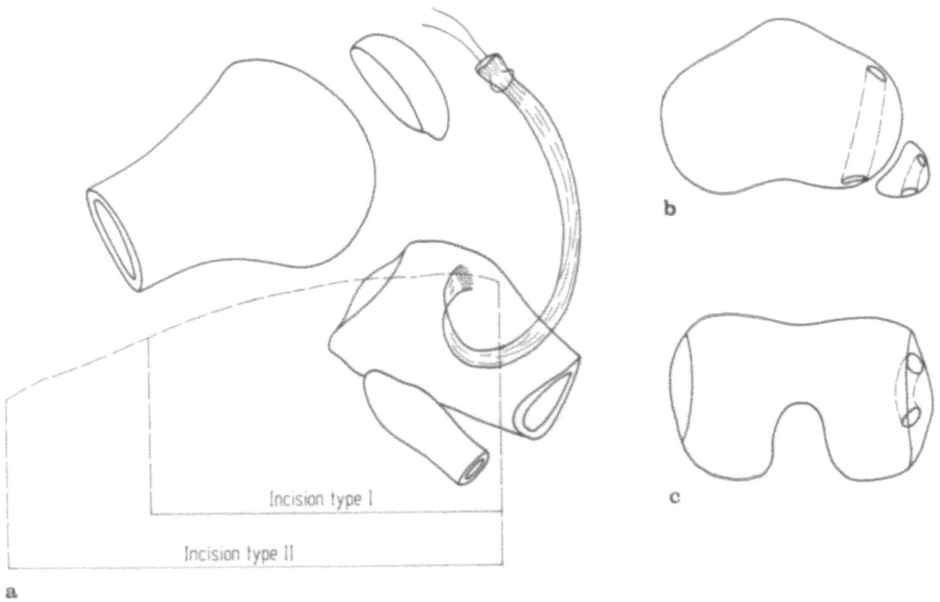

Fig. 1 a–c. The two surgical techniques: **a** incisions, **b** tunnels in the tibia and fibula, **c** femoral tunnel

The graft is passed first through the tibial tunnel, then straight on through the lateral part of the gastrocnemius. From there it is turned towards the front. Behind the lateral ligament a vertical cut opens the capsule and allows an intra-articular control. The graft is turned towards the front and back into the gastrocnemius but 1 cm higher. It is then passed through the capsule. Now located intra-articularly, the graft follows the popliteal tendon to its femoral attachment where it is inserted in the second tunnel. From here the graft is turned backwards to reappear out of the tunnel at a point 1 cm above its entrance. It is passed again through the capsule and gastrocnemius tendon before being turned down towards Gerdy's tubercle where it is attached under strong tension, while the knee is held in midrotation (Fig. 2).

Type 2. The principle of the surgical technique, as for type 1, aims at neutralising the rotatory instability but in addition obviates the incompetence of the lateral ligament to control the gaping of the joint in varus. The surgical procedure is illustrated in Fig. 3. The fascia lata strip should be about 20 cm long, and the two bony tunnels are prepared as in the first procedure but a third tunnel is drilled through the fibular head. The graft is implanted in the same way as in the first technique but on emerging from the femoral tunnel, instead of going down towards Gerdy's tubercle, it goes towards the fibular head then through the bony tunnel and is finally tied up to itself.

Postoperative Regimen

In this study immediate mobilisation seems just as effective as rigid immobilisation or articulated cast braces. During the first postoperative days we used continuous slow-moving splints.

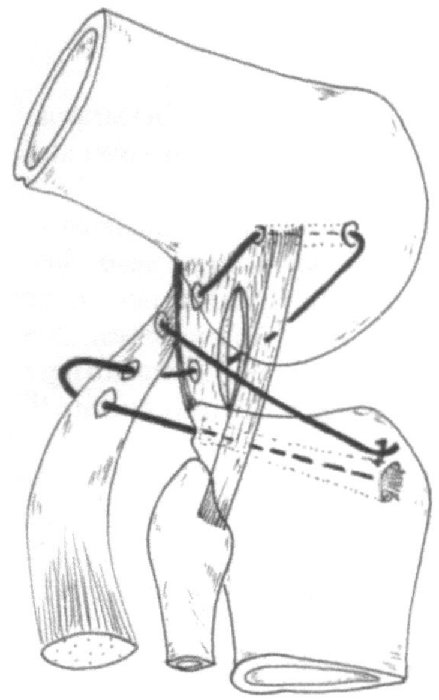

Fig. 2

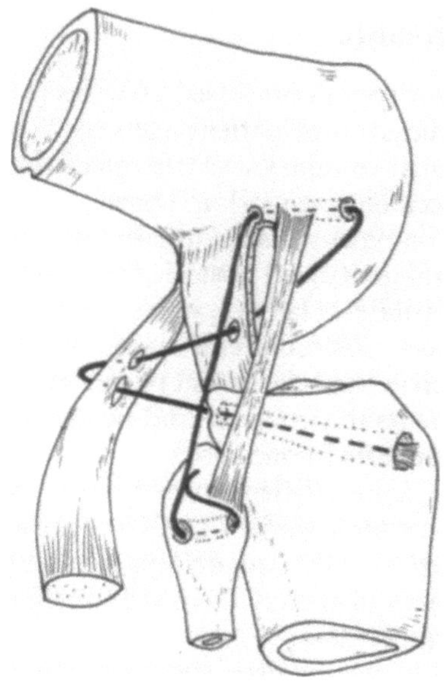

Fig. 3

Fig. 2. Passage of the graft through the prepared tunnels in technique type 1

Fig. 3. Implantation of the graft in technique type 2

In the type 2 operation, weight bearing is postponed for 6 weeks because these patients often have a varus morphology which increases the risk of stretching the fascial band.
Sports can be resumed between 3 and 6 months postoperatively depending on the kind of operation, i.e. whether it is a posterolateral replacement alone or combined with a cruciate replacement.

Indications

Posterolateral instability is of a rotatory kind. It nearly always leads to a severe disability which must be suppressed. An upper patient age limit of 50-years-old is usually imposed, and associated degenerative arthrosis should be treated by an osteotomy first. The work team ARPEGE has developed a grading system CLAS (C, competition sports; L, recreational sports; A, active life; S, sedentary life) wherein physically active persons and athletes are deemed fit to be surgically treated.
A varus deformity should be dealt with by performing an osteotomy before ligament replacement, in particular if a recurvation is associated. Such a ligament replacement should never be made in isolation unless the posterolateral laxity is pure, which is rare (only two cases were recorded at the beginning of our series).

Results

Our series comprised 16 cases (11 type 1 and 5 type 2).
The average patient age was 27 years, with a range of 20–47 years. The original cause of injury was trauma during sport in six cases, traffic accidents in seven and accidents at work in three.
The types of laxity presented were: posterolateral laxity in two cases, posterolateral laxity and posterior cruciate ligament (PCL) insufficiency in eight cases, posterolateral laxity, posteromedial laxity and posterior cruciate insufficiency in one case, posterolateral laxity and anterior cruciate insufficiency in two cases, posterolateral laxity and posteromedial laxity plus insufficiency in both cruciate ligaments in two cases and posterolateral laxity and insufficiency in both cruciate ligaments in one case.
PCL insufficiency was always treated using a quadriceps tendon graft. Anterior cruciate ligament (ACL) insufficiency was treated by a Macintosh over-the-top repair with a quadriceps tendon graft in two cases and in a third case by reattachment of the ACL. In the fourth case nothing was done as the anterior laxity was moderate.
In 1 case an osteotomy was combined with a double ligament replacement. In 12 cases, at least one surgical intervention was performed before the ligament replacement (meniscectomy of ligament repair).
Analysis of the results was divided into three groups: the patient's opinions, a score of the knee function postoperatively and clinical examination.
The patients' opinions on the success of the operation were as follows: eight patients were highly satisfied, six were satisfied, one was disappointed and one was not satisfied with the outcome. Therefore, it apears that the majority of patients were pleased with the results of the operation.
Of the two bad results one presented posterolateral laxity, posteromedial laxity and insufficiency in both cruciates with marked cartilage degeneration. This might explain the patient's pain on weight-bearing which was not relieved by a valgus osteotomy. In the second case, no explanation could be found for the poor results. The cause of injury in both cases was an accident at work; the second patient was also a chronic alcoholic.
The ARPEGE team has developed a scale to score the functional results, the score for stability is 9 points with a further 9 points for pain and endurance. The postoperative scores were as follows:

27 points – 1 case
26 points – 3 cases
25 points – 6 cases
24 points – 1 case
23 points – 3 cases
17 points – 1 case
16 points – 1 case

For the two bad results the scores were 16 and 17 points.
Measurement of laxity was difficult because of the rotatory component. This limitation throws some doubt on the accuracy of this examination method. The duration of follow-up varied from 1 year to 3 years and 6 months. A score of 3 +

was given for the most important change in laxity, i.e. on complete disappearance of posterolateral laxity. This was regarded as being a very good result. The score was reduced 2 + for disappearance of the reversed pivot shift or a disappearance of reversed pivot shift given an initial laxity of 1 +. These are classified as a good result. If the initial laxity score of 1 + was improved upon and the pivot shift lessened, the result was classed as medium. When the laxity remained unchanged the result was poor.

Of the 16 patients followed up the improvement in laxity was very good in seven, good in eight and medium in one. None of the patients was designated as poor. The medium result was achieved by a 21-year-old patient who was satisfied with the operation and scored 23 points on the CLAS grading system. He still plays European football. On clinical examination the two bad results rated as good.

Conclusion

In view of the analysis of the series of 16 cases described, the technique of posterolateral fascia lata plasty raises interest as a method for the treatment of chronic posterolateral instability. It should nearly always be seen as an additional procedure to a replacement of an incompetent cruciate ligament. However, the reconstruction cannot function and last if there is a marked varus or a recurvation. In these cases, a corrective osteotomy should be done first.

References

1. Augustine RW (1956) The unstable knee. Am J Surg 92:380–388
2. Bousquet G (1981) Les laxités du genou et l'altération du cartilage rotulien. Cahier d'Orthopédie de St. Etienne, Journées du Genou, 6–8 May 1981
3. Hughston JC, Norwood LA (1980) The posterolateral drawer test and external rotational recurvatum test for posterolateral rotatory instability of the knee. Clin Orthop 147:82–87
4. Hughston JC, Jacobson KE (1985) Chronic posterolateral rotatory instability of the knee. J Bone Joint Surg [Am] 67:351–359
5. Jakob RP, Hassler H, Staeubli HU (1981) Observations on rotatory instability of the lateral compartment of the knee. Acta Orthop Scand [Suppl] 191(52):18–27
6. Muller W (1982) Das Knie. Form, Funktion und ligamentäre Wiederherstellungschirurgie. Springer, Berlin Heidelberg New York, S. 352

Treatment of a Delayed, Nonreducible, Chronic Posterior Dislocation of the Tibia with Posterior Instability After Elsewhere Internal Fixation of a Femoral Shaft Fracture

P. Lauber, A. Kentsch, and We. Müller

Department of Orthopedic Surgery, Kantonsspital Bruderholz, 4101 Bruderholz, Switzerland

Initial Situation

The patient was referred to us from St. Elsewhere Hospital after internal fixation of the femoral shaft fracture and primary reconstruction of the ligamentum patellae 7 months ago, due to a concomitant compound knee lesion, and after secondary "lateral ligament-plasty".

Clinical Finding (Fig. 1). Nonreducible posterior dislocation of the tibia. The anterior edge of the tibial plateau was impressed and worked as a new joint surface against the posterior femoral condyles. There was excessive posterior instability

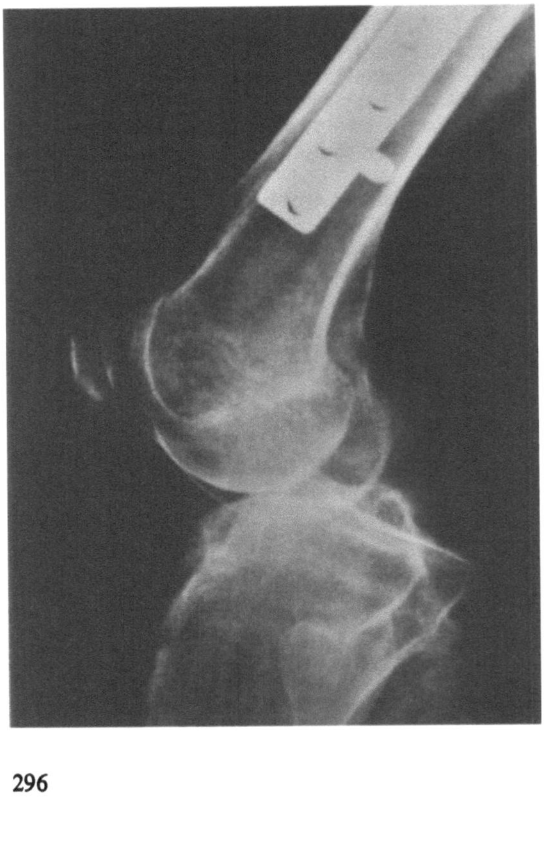

Fig. 1

Surgery and Arthroscopy of the Knee
2nd Congress of the European Society
Editors: Müller/Hackenbruch

and a range of motion between 70°-30°-0° of flexion. The patient walked on crutches with partial weight bearing.

Initial Therapy (Fig. 2)

Preoperatively

Traction for 3 weeks (anterior drawer/tuberositas tibiae 6 kg, and longitudinal traction/distal tibia 6 kg for the knee in 30° of flexion), exercises twice a day. The X-ray (Fig. 2) shows partial reduction of the posterior tibial subluxation.

Operation (Figs. 3–6)

1. Osteotomy of the medial epicondyle with the medial collateral ligament and the lateral epicondyle with the lateral collateral ligament and the popliteal tendon
2. Generalized removal of adhesions (anterior, medial, lateral, and intercondylar), reduction of the knee joint
3. Lysis and reduction of both menisci, which were dislocated and fixed behind the tibial plateau and the femoral condyles
4. The anterior edge of the tibial plateau is lifted up and supported by homologous cancellous bone

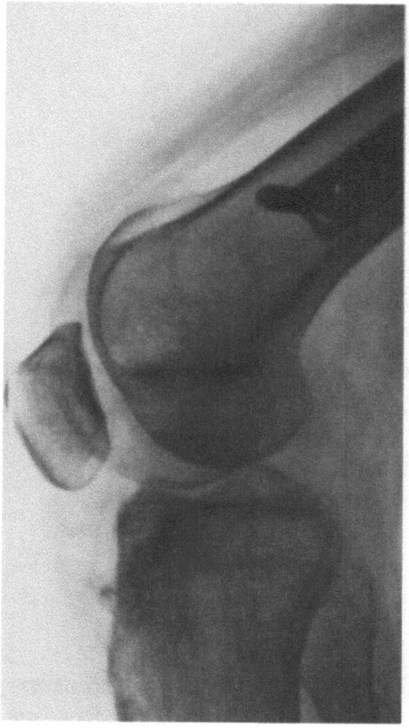

Fig. 2

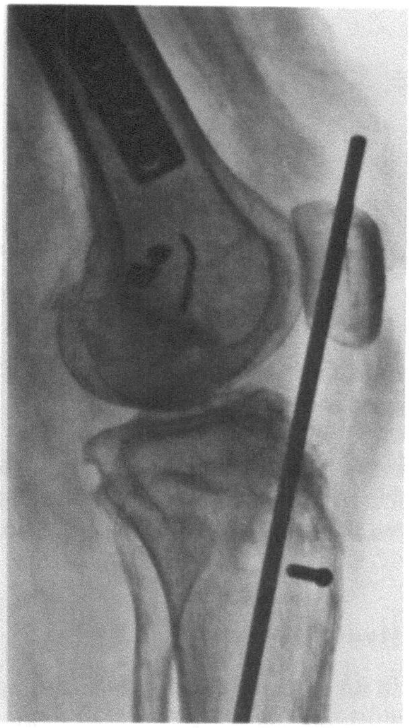

Fig. 3

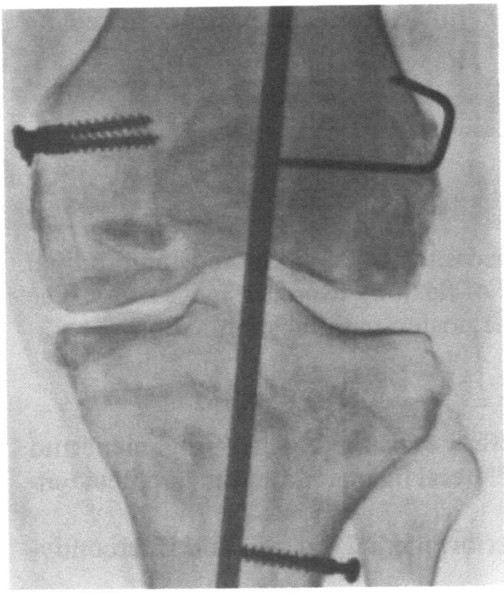

Fig. 4

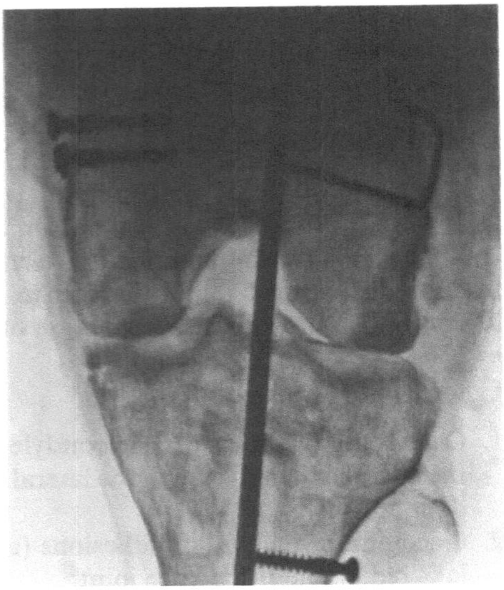

Fig. 5

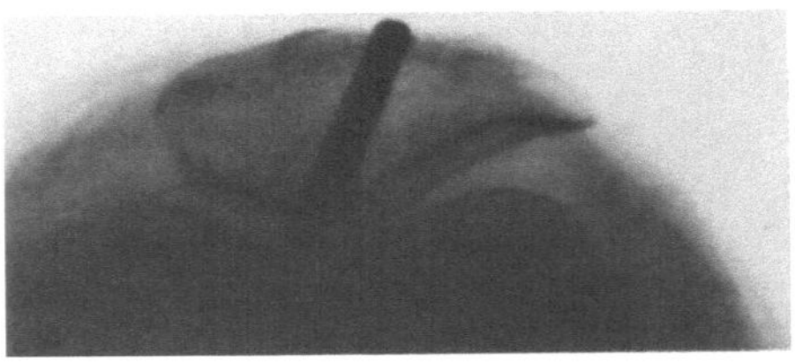

Fig. 6

5. Replacement of the LCP by a free mid-thread graft from the other knee joint (see bone blocks in Fig. 3)
6. Olecranisation with a Steinmann pin
7. Reconstruction of the medial collateral ligament
8. Stabilization of the lateral collateral system

Figure 7 shows the main intraoperative findings after removal of adhesions.

Postoperative Course

Pin traction in an anterior drawer with 4–6 kg for another 4 weeks and continuous passive motion between 20°–70° of flexion (Figs. 8–10). Nonweight bearing for 6 weeks, then partial weight bearing (15 kg) with a posterior 3/4-Scotchcast

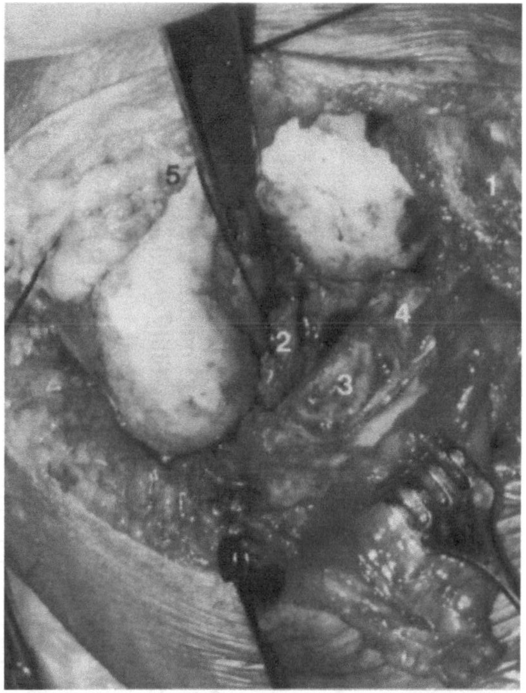

Fig. 7. Main intraoperative findings after removal of adhesions. *1*, patella; *2*, rest of the posterior cruciate ligament; *3* new joint surface at the anterior edge of the tibial plateau; *4*, dislocated medial meniscus; *5*, hook placed on the posterior border of the tibial plateau for reduction of the posterior tibial subluxation

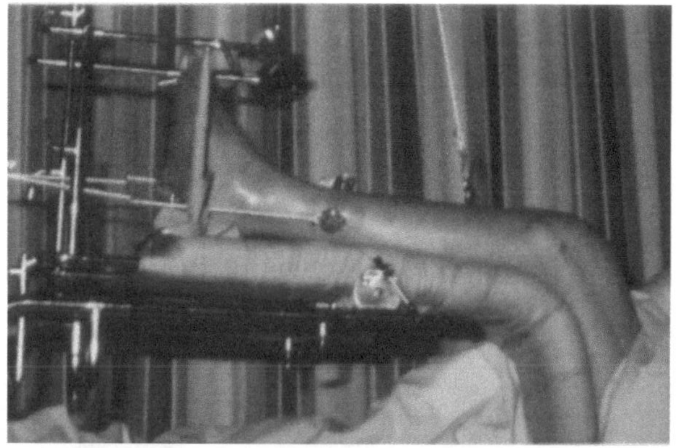

Fig. 8

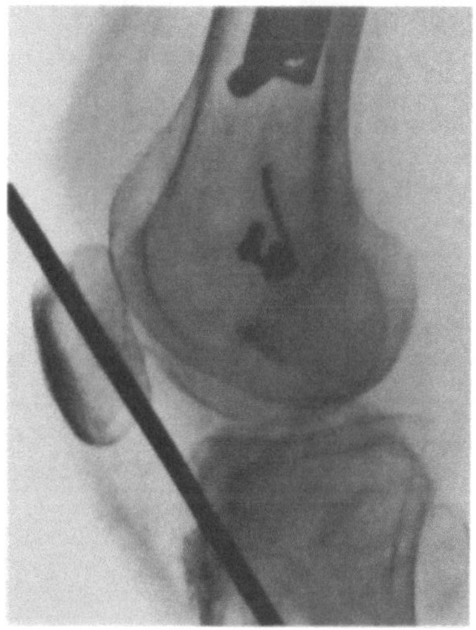

Fig. 9

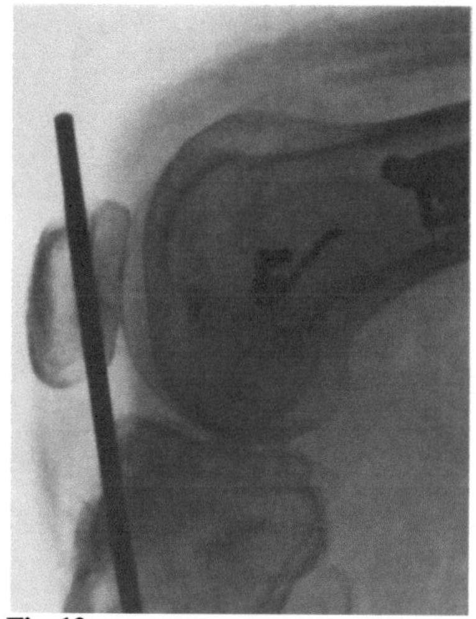

Fig. 10

Fig. 11

Fig. 12

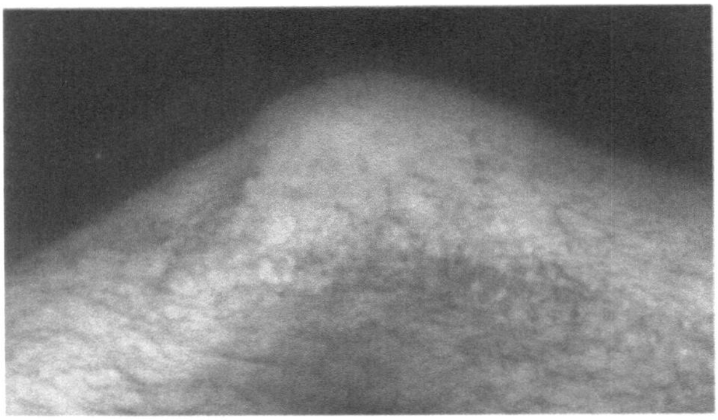

Fig. 13

splint in 10° of flexion as protection for the operated joint. Physiotherapy should be continuous.

After 10 weeks the patient had achieved a flexion between 10° and 80°; after 12 weeks the patellotibial Steinmann pin (olecranisation) was removed. At 14 weeks full weight bearing with a 3/4-Scotchcast in full extension is allowed; at 16 weeks the joint was protected by only a brace. After 15 months the plate was removed and correction of a persistent anterolateral instability could start.

Two years later the patient is pain free, has a range of motion of 125°-0°-0° for flexion/extension (Figs. 11, 12), and has no sensation of instability. For sports he uses an AOA knee brace.

The tibia shows clinically no posterior drawer (Fig. 13). Objectively, a posterior drawer remains and a lateral instability with varus stress of 3 mm (Figs. 14, 15).

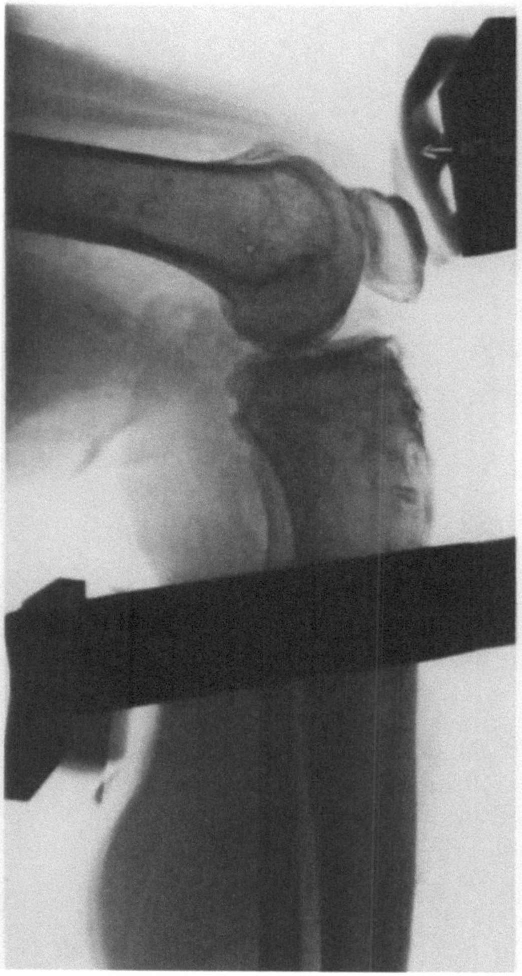

Fig. 14

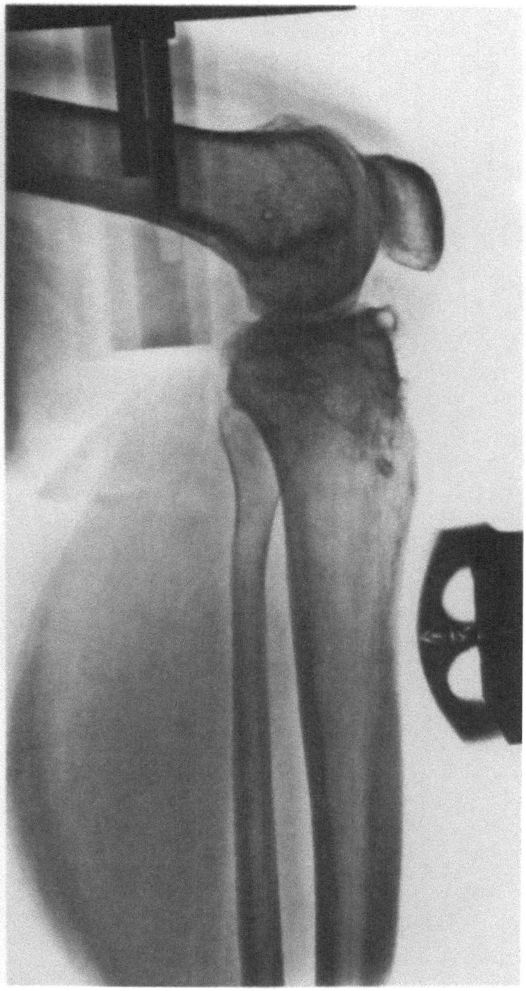

Fig. 15

Arthroscopic Operations on the Meniscus

Follow-up of More Than 150 Meniscal Repairs up to 13 Years Postoperatively

A. Schreiber and M. Rodriguez

Department of Orthopaedic Surgery, University of Zürich, Belgrist, CH-8008 Zürich, Switzerland

Total meniscectomy has a high incidence of postoperative degenerative osteoarthritis of the knee joint. Instability is also increased. Therefore, we began to perform meniscal repair and partial meniscectomy at our institution, at first in selected cases, later routinely. Close follow-up, including clinical examination, arthrography and arthroscopy, shows over 85% good results, even 13 years after the operation.

Surgery and Arthroscopy of the Knee
2nd Congress of the European Society
Editors: Müller/Hackenbruch

Meniscus Repair by Arthroscopy

J. Defrere[1], A. Franckart[1], and R. Bassleer[2]

[1] Orthopaedic and Arthroscopic Surgery, Hôpital d'Esneux, 4050 Esneux, Belgium
[2] Histology and Cytology, Faculty of Medicine, University of Liège, Liège, Belgium

Introduction

Between 1983 and 1985, 55 cases of meniscus lesions for which repair by arthroscopy was feasible were treated. The treated ruptures were located at the meniscosynovial junction of the medial meniscus posterior horn and in some cases extended to the middle horn area. Lateral menisci ruptures were located in front of the popliteal tendon and at the meniscosynovial junction, except in one case where the rupture started from the meniscus-free side, at the transition between the anterior and middle horns, and extended obliquely rearwards to the middle horn meniscosynovial junction.

This study takes into consideration 55 patients, 12 females and 43 males. Their average age was 27 years, with a range of 12 to 44 years. Meniscus repair is normally only appropriate for patients less than 35 years old; however, three patients, one patient of 38 and two of 44 years old, were also treated, one of whom had a rupture recurrence.

Two groups of lesions were treated. The first group consisted of four cases of meniscus laxity without rupture of the meniscus body. This laxity as well as meniscus subluxation in the internal femorotibial space result from alteration of the femoro- and tibiomeniscal ligaments. In the second group, 51 meniscus repairs were performed (48 on the medial menisci and 3 on the lateral menisci).

Of the 48 patients suffering from medial meniscus damage, 45 presented with a rupture at the posterior horn meniscosynovial junction, 2 at the posterior horn and middle horn, and 1 affecting the whole length of the medial meniscus. In 26 patients the complete thickness of the meniscus was ruptured, and in 22 a partial rupture affected two-thirds of the meniscal substance thickness. Regarding the three instances of lateral meniscus damage, two presented a partial rupture affecting one-half of the meniscus body thickness and the last one a complete rupture through the total thickness, affecting the anterior and middle horn.

Methods

Arthrographic Investigation

For 35 of the 48 patients presenting medial meniscus lesions, arthrographic investigations were carried out before surgical intervention. The results confirmed the presence of a lesion in 27 cases, but in the remaining 8, lesions were not detected. Amongst the cases of lateral meniscus rupture, the arthrographic investigation indicated positive lesions in two of the three cases.

Surgery and Arthroscopy of the Knee
2nd Congress of the European Society
Editors: Müller/Hackenbruch

Operative Technique

Meniscus repair always requires general anaesthesia, use of a leg holder and a tourniquet. The latter is inserted at the leg holder level and positioned at a distance of four-fingers' breadth from the patella (Fig. 1). Repair by arthroscopy has a reputation of being a dangerous procedure due to the vulnerability of neurovascular elements in the popliteal space. This seems to be an erroneous notion, if strict discipline is kept in making the appropriate knee movements during the operation. The leg holder allows three movements that are indispensable. The valgus movement opens the internal femorotibial interline, a slight flexion allows the separation of the popliteal vessels from the capsule and internal rotation facilitates the emergence of the needles from the postero-internal position.
The two paths of access used to insert the arthroscope are in an antero-external manner or via the transpatellar tendon. The antero-external path lies one-finger's breadth from the upper edge of the external tibial platform, along the patellar tendon. For the transpatellar tendon method, the knee is flexed at 90° and the trocar inserted at the patella apex level. Both methods are used for the arthroscope during the same intervention. The transpatellar location permits better visualisation of the posterior horn, but the arthroscope sometimes lies too close to the cannula; the antero-external way does not provide the best view but allows easier positioning of the cannula. The antero-internal approach is only used for the suture cannula and is located one-finger's breath from the upper edge of the internal tibial platform, along the patellar tendon. Dual suture cannulae are used to allow the passage of two needles. The major axis of the cannula must be positioned perpendicular to the knee posterior plane. In order to do this, the cannula should be placed against the patellar tendon. The design of the cannula handle allows two movements: the first makes cannula positioning along the tendon easier due to the way the handle is grasped, and the second allows the emergence of needles without butting against the femur or against the tibia if the handle and tibial platform are in the same plane. The distal end of the cannula is fitted with two blunt points to prevent it from slipping on the meniscal tissue during this movement. To make the points blunt it was necessary to sheath the cannula with a Teflon (polytetrafluoroethylene) sleeve so that the ends pass easily through the approach.

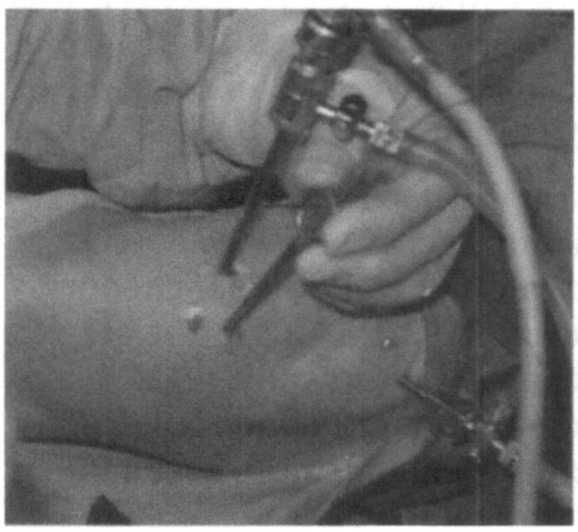

Fig. 1 Right knee with a leg holder and tourniquet positioned at four-finger's breadth from the patella. With the knee in a valgus position, slightly flexed and rotated internally, the arthroscope is inserted along the transpatellar tendon and antero-internal cannula which lies one-finger's breadth from the internal tibial platform, along the patellar tendon

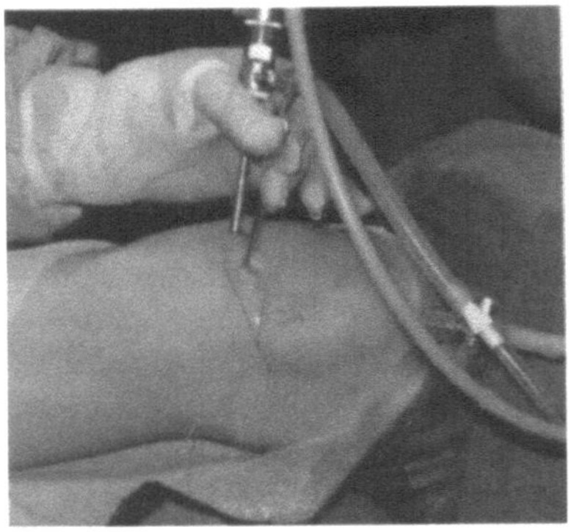

Fig. 2. Emergence of needles in the postero-internal position

The angle selected for positioning the cannula will depend on the place that has been sutured at the meniscus area level. Both suture needles must be curved in advance so that they will emerge with ease from the postero-internal position (Fig. 2). A 2/0 absorbable suture thread is used. Coordination of movements when inserting the needles facilitates their emergence from the postero-internal area. The ligature is then carried out under the skin. The meniscus rupture area must be debrided using a basket forceps or a shaver if more than 15 days have elapsed between trauma and repair. The stitches should not be too far apart. The transmeniscal canal resulting from the suture seems to be a support for the advancement of neovascularisation, allowing repair of the meniscus tissue. Immobilisation in a plaster cast for 6 weeks without traction is required for complete ruptures of the meniscus; this period decreases to 4 weeks for partial ruptures. After this period, traction is authorized and rehabilitation can start. Any sport activity that might require knee rotation will be contraindicated during a 6-month period postoperatively.

Treatment of meniscus repair for laxity follows similar techniques; however, there is a difference in preparing the suture due to resection of the meniscocapsular junction area. The junction area section is prepared by electrosurgery, and the cicatricial tissue is resected using a shaver.

The length of time between trauma and repair does not affect the type of repair operation that is indicated and seems to have no influence on the results. If this period of time is shorter than 15 days, the rupture area is not prepared: if it is longer, the wound is debrided using a basket forceps or a shaver. In the group of 48 patients with medial meniscus ruptures, the time between trauma and repair exceeded 15 days (between 15 days and 2 years) in 35 instances.

Lesions in Conjunction with the Repaired Meniscus Rupture. Amongst the 48 repaired meniscus ruptures, 14 medial meniscus lesions occurred in the absence of any other joint defects. Two cases presented an internal femorotibial chondritis lesion, 16 a rupture of the anterior cruciate ligament (ACL), 14 a rupture of both the ACL and the lateral meniscus, 1 a lateral meniscus rupture only, and 1 a rupture of the ACL of the lateral meniscus and a chondral fracture combined (Fig. 3). For the three repaired lateral meniscus ruptures, one exhibited the beginning of degeneration in the external femorotibial compartment.

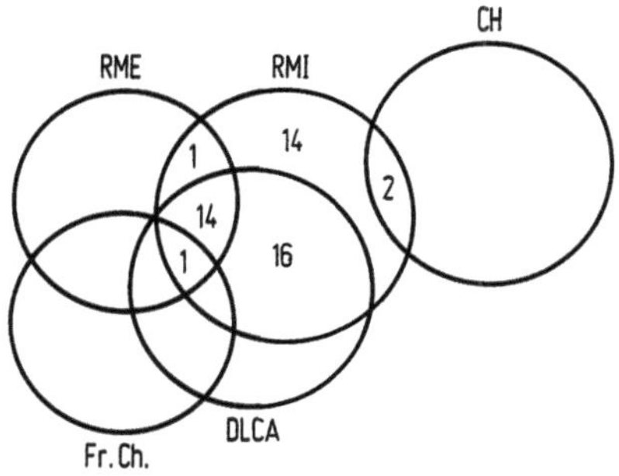

Fig. 3. Lesions associated with the medial meniscus ruptures. RME, Rupture of the lateral meniscus; RMI, rupture of the medial meniscus; CH, chondritis; Fr. C, chondral fracture; DLCA, anterior cruciate ligament rupture

Of the associated lesions, ACL rupture was the most interesting. From the 48 arthroscopic examinations performed during medial meniscus repair, 31 ACL ruptures were identified. Analysis of these cases indicated 21 of laxity without instability and 10 with laxity and instability, i.e. positive for both the jerk test and the Lachmann sign. These 10 cases require ligamentoplasty after meniscus repair. According to the observations, an unstable, repaired knee must be stabilised to avoid any recurrence of injury in the future.

Complications. There was no intraoperative incidence to be noted in this study. Regarding postoperative complications, two reactions to suture threads were observed. During the first meniscus repairs, ligature was carried out at the skin level, which resulted in the two incidents of a reaction to the suture material. Where a deeper ligature process was used under the skin, this type of reaction did not occur. As far as major complications are concerned, there were two cases of arthrolysis due to arthroscopy causing stiffening at 90° and 100° of flexion, one of phlebitis without embolic complication, one of major algodystrophy with knee blockage at 45° and one of internal saphenous nerve paraesthesia from which the patient later recovered.

Results

When the follow-up takes place during the months after the operation, the patient's subjective satisfaction does not always mean that the result is good at a clinical level. For example, a patient who was pleased with the recovery of normal activities presented, in fact, a rupture of the same meniscus at the arthroscopic control examination. Therefore, clinical results that are too subjective are preferably disregarded, and evaluation should be made from the arthroscopic control examination, which gives objective results.
The 27 postoperative arthroscopic control inspections carried out for the 51 meniscus ruptures consisted of 24 for medial meniscus injuries and 3 for lateral meniscus injuries. Of the 24 medial meniscus repairs 16 were confirmed as being successful and 8, as failures. The three lateral meniscus repair cases indicated that the intervention was successful. The macroscopic aspect of the repaired menisci presented positive results, 19 (70%) of the 27 confirming perfect cicatrisation at the meniscosynovial junction and good stability of the meniscus body. No sublax-

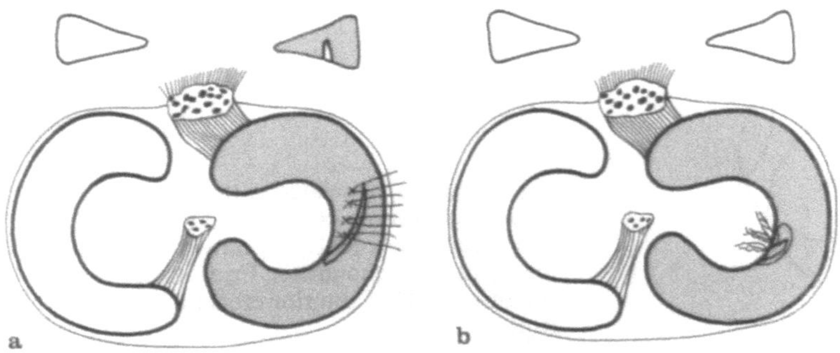

a b

Fig. 4 a and b. Efficiency of lateral meniscus repair by arthroscopy

ity appeared in the repaired area, as shown by the palpation hook. Sometimes the meniscus part near the rupture area appeared to be slightly laminated. The efficiency of lateral meniscus repair by arthroscopy is shown in Fig. 4.

Conclusions

Failure of meniscus repair by arthroscopy does not seem to result from a vascularisation deficit. In all cases (27), the meniscosynovial junction appeared to be well cicatrised, except in two cases of actual lesion recurrence. It seems that the meniscus tissue facing the suture is less resistant to mechanical strains, and in some cases this results in a new fatigue rupture of the radiated type. The answer to the question of whether the option of using the suture process should be kept open or not is yes. It is highly preferable to maintain a buffer, and so protect the cartilages from encrusting, even if it results in lower resistance, rather than perform a total meniscectomy. Taking the objective results into consideration, 70% (19) of the cases with menisci rupture escaped total meniscectomy; this inevitably means that, for these patients, the appearance of arthrosis will be postponed.

Spontaneous Meniscus Repair

Spontaneous meniscus repair is extremely rare but does exist, and four instances supporting this statement were noted. They included three ruptures at the posterior horn meniscosynovial junction, amongst which one was particularly conclusive. A patient suffering a knee trauma was diagnosed as having a sprain and the knee immobilised in a plaster cast for 21 days. Later, due to the persistence of pain, an arthrographic exploration was required, which revealed a rupture in the medial meniscus located at the posterior horn meniscosynovial junction. An arthroscopic examination carried out 4 months after the trauma reveals a cicatricial swelling at the joint junction level. The latter subluxates the posterior horn of the medial meniscus into the internal femorotibial space and is considered to be the element responsible for the painful syndrome. Meniscectomy, at least of the subluxating part, will allow sedation of the painful syndrome.

Histological Analysis After Ruptures of Different Types

The samples were fixed in Bouin fixative and prepared according to conventional methods of histology for standard microscopy. Four histochemical stains were used to differentiate tissue components: haematoxylin-eosin for nucleus, cytoplasm and fibres, light green for collagen fibres, and periodic acid-Schiff (PAS) and Alcian blue to locate polysaccharides of the extracellular matrix. The following case studies present observations recorded during the histological analysis of menisci after rupture.

A 28-year-old patient presented an internal femorotibial misalignment with a 12° valgus deformity as measured by teleradiography. Arthroscopy showed the start of cartilaginous degeneration in the condyle rest area and at the tibial platform level. The lateral meniscus presented a rupture involving two-thirds of the meniscus body thickness, starting from the anterior horn free edge and progressing obliquely rearwards to the middle horn meniscosynovial junction. Normally this type of lesion cannot be sutured: this can only be done for lesions at the meniscosynovial junction. However, a suture was carried out for this patient in order to maintain the external buffer before starting the varus osteotomy. At the end of the arthroscopic intervention, the suture aspect was not satisfactory. The patient did not mention any symptoms of gonalgia but 4 months later cicatrisation was evident, except at the free edge level. It was decided to neaten the scar tissue using basket forceps. During this intervention the section face of the meniscus tissue appeared lightly rosy, and on opening the tourniquet, bleeding occurred from small vessels at the free edge. It appears that the transmeniscal canals of the suture threads contribute to the penetration of neo-vessels from the synovial area and, consequently, allow rupture cicatrisation.

Evaluation of the macroscopic and histological observations recorded for the meniscal repairs that were classified as failures (8 cases amongst 27) provides a better understanding of how the meniscus behaves after the repair intervention. Such lesions are not usually iterative ruptures of the meniscosynovial junction but a new type of rupture. The first repaired rupture is cicatrised, and failure is characterised by a new kind of tear. Two of these eight cases indicated that recurrence of the rupture took place through a lack of cicatrisation in the repaired area (Fig. 5). One of these two cases presented a cicatricial point at the rupture's

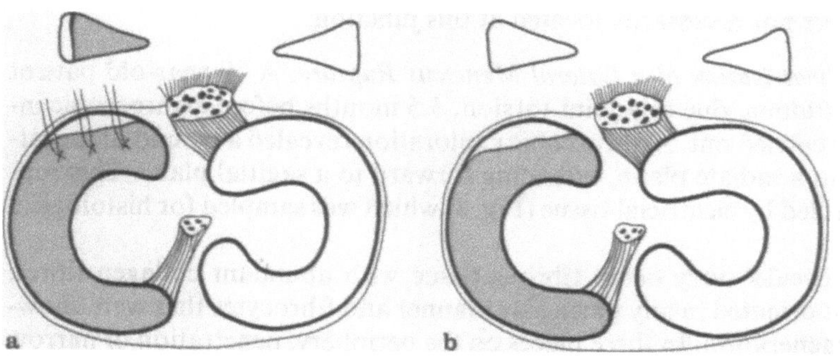

Fig. 5 a and b. Recurrence of rupture in the lateral meniscus as indicated by lack of cicatrisation in the repaired area: **a** Before repair, **b** after repair

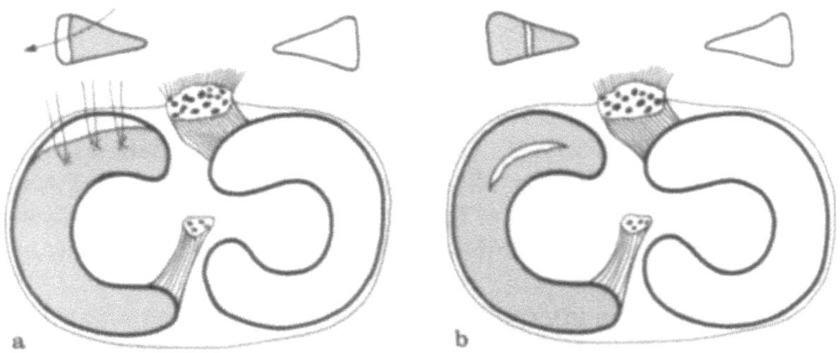

Fig. 6a and b. A sagittal rupture in the meniscus body: **a** Before repair, **b** after repair

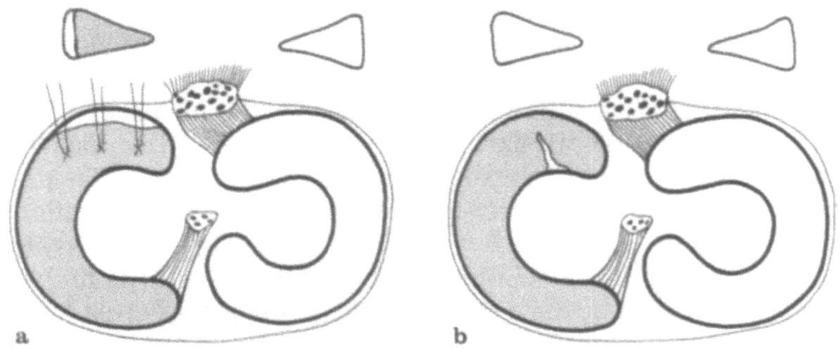

Fig. 7 a and b. Repair at the meniscosynovial junction accompanied by a new radiate type rupture : **a** Before repair, **b** after repair

middle part, which prevented posterior horn subluxation into the internal femorotibial space. One case presented a new sagittal rupture in the meniscus body parallel to the cicatrised meniscosynovial junction rupture (Fig. 6). Five cases showed perfect tear cicatrisation at the meniscosynovial junction and presented a new rupture of the radiate type (Fig. 7). The same type of rupture was found in association with internal femorotibial degeneration at the middle and posterior horn junction. However, radiate lesions, as noted during these arthroscopic examinations, were not necessarily located at this junction.

Spontaneous-Cicatrisation of a Lateral Meniscus Rupture. A 19-year-old patient experienced a trauma, due to a joint torsion, 3.5 months before arthroscopic intervention was carried out. Arthroscopic exploration revealed a wound of the lateral meniscus in a radiate plane, extending forward to a sagittal plane. This rupture area was filled by cicatricial tissue (Fig. 8) which was sampled for histological analysis.

The analysis revealed very dense fibrous tissue with abundant collagen fibres, which were not oriented in any particular manner and fibrocytes that were showing signs of degeneration. In three places on the periphery, penetration of narrow and long pieces of tissue containing many cells, associated with some blood vessels, presented narrow apertures surrounded by an abundant substance that

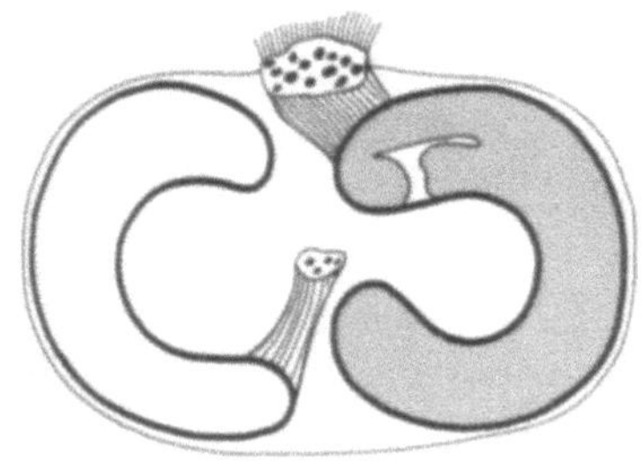

Fig. 8. A spontaneous meniscus cicatrix of a lateral meniscus rupture

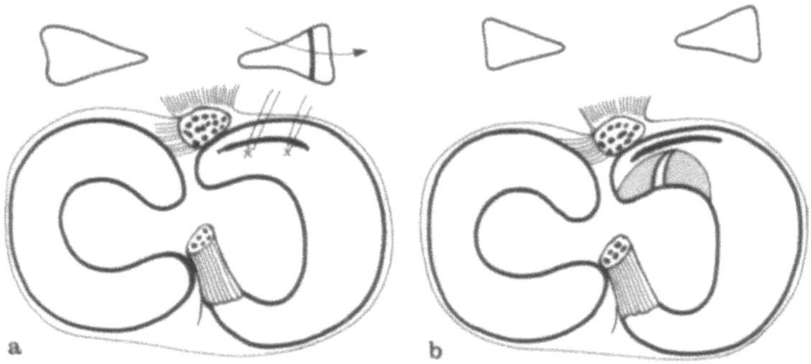

Fig. 9 a and b. Radiate degenerative rupture (posterior horn) located perpendicular to a rupture that cicatrised after surgical repair

did not stain positively for basal laminae polysaccharides. The central area was slightly necrotic.

Radiate Degenerative Rupture Located Perpendicular to a Rupture That Cicatrised After Surgical Repair. An arthroscopic examination carried out on a 29-year-old patient in February 1985 confirmed the presence of a rupture of the medial meniscus posterior horn at the meniscosynovial junction, associated with a complete rupture of the ACL (Fig. 9).

Meniscus repair was performed by arthroscopy. After treatment, instability of the joint was observed, characterised by positive jerk and Lachmann tests. In November 1985 a second arthroscopy was carried out with a view to performing a ligamentoplasty and allowing control of the meniscus suture. A new radiate rupture was detected perpendicular to the perfectly cicatrised rupture at the meniscosynovial junction.

On histological analysis, the cicatrised rupture consisted mainly of fibrous tissue bordered by a small fibrocartilage area. Extensive areas showed a high density of fibres but presented very few cells.

Iterative Meniscus Radiate Rupture. Traumatism of the left knee joint in a 33-year-old patient, resulted in an internal femorotibial conflict. Arthrography confirmed the presence of a rupture at the medial meniscus level (Fig. 10). An arthroscopic examination that was carried out in January 1985 confirmed a complete rupture of the ACL, although ligamentary fibrils persisted and there was a rupture at the posterior horn meniscosynovial junction which extended to the middle horn. Suture was carried out. The arthroscopic control examination that took

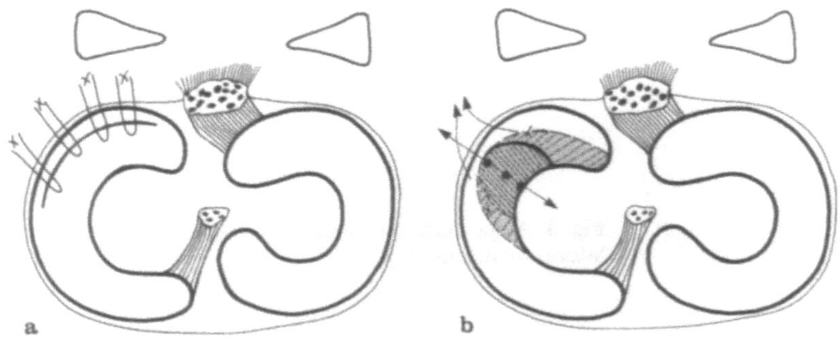

Fig. 10 a and b. Iterative meniscus radiate rupture

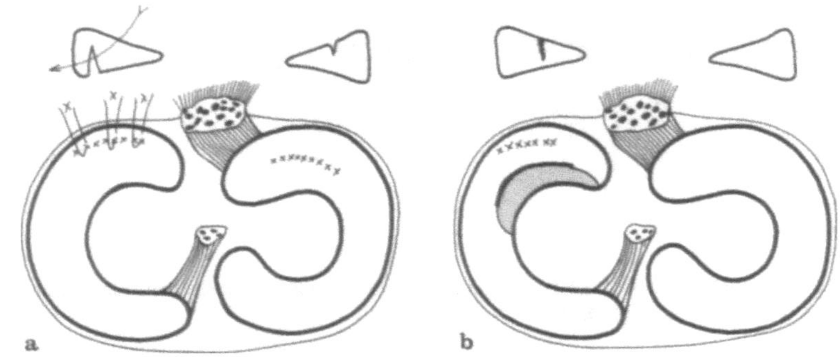

Fig. 11 a and b. A new iterative rupture of the meniscus body, lying along the same axis as the repaired and icatrised previous rupture at the meniscosynovial junction

place 4 months after the intervention confirmed cicatrisation of the repaired meniscosynovial junction but revealed a radiating rupture, perpendicular to the cicatrised area.

Samples were taken for histological analysis from the meniscosynovial junction and adjacent to the rupture extending as far as the free edge. The synovium was in excellent condition, being of the fibrous type, non-infiltrated by inflammatory cells, normal thickness and served with many blood vessels. There were indications of cellular division in the meniscosynovial tissue; the fibroblasts were active and the fibres abundant. The meniscosynovial junction was characterised by an abundant vascular plexus in excellent condition, mitotic cells in the connective tissue and very active fibroblasts.

In general the meniscus showed an abundance of collagen fibres, although connective cells were relatively rare, chondrocytes were not very abundant, isogenous groups of chondrocytes were very scarce at the free edge and dissociated collagenous fibres with very few cells but no necrotic cells were detected.

Meniscal Rupture at the Meniscosynovial Junction Adjacent to a Previous Cicatrised Rupture. A trauma of joint torsion resulting from a football accident that happened 1.5 years ago was presented by a 24-year-old patient (Fig. 11). Persistent pain in the knee incited the patient to seek medical advice. The conclusion drawn from arthrographic investigation defined a medial meniscus posterior horn

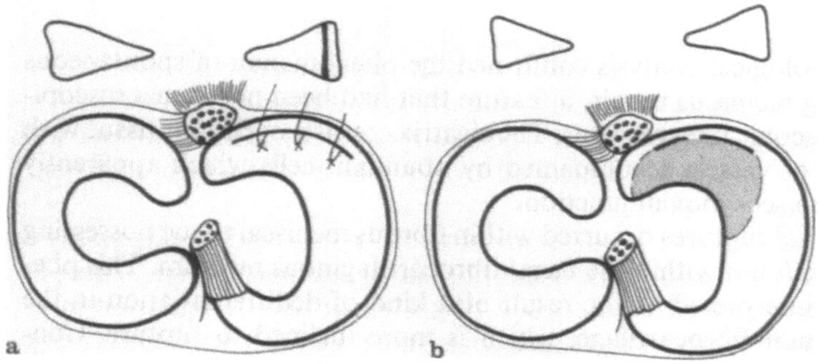

Fig. 12a and b. Recurrence on a primitive meniscus repair

rupture at the meniscosynovial junction in the posterior horn and a fractionated rupture of the lateral meniscus posterior horn, considered as a meniscus wound, but retaining the integrity of the ACL. No intervention was performed on the lateral meniscus, but there was meniscus repair of the medial meniscus posterior horn. A recent arthroscopic control examination indicated cicatrisation of the meniscosynovial junction and a new rupture at the posterior horn located half-way between the meniscosynovial junction and the meniscal free edge. Partial meniscectomy of the posterior horn was performed, and a sample was taken for histological analysis.

The triangular-shaped tissue sample consisted of fibrocartilage, with firm bundles of collagen fibres, often wavy and regularly orientated. Chondrocytes appeared normal and in good condition. The occurrence of fibrocytes was rare. There was no trace of vascularisation or inflammatory cells. In the thin region of the sample, an expansive area of necrosis was noted.

Rupture Recurrence of a Primitive Meniscus Repair. A 30-year-old patient suffered a knee injury at work, while moving from a kneeling to a standing position, resulting in intra-articular trouble at the internal femorotibial compartment level. A lesion of the medial meniscus posterior horn at the meniscosynovial junction was detected by arthrographic examination a month later. Arthroscopic examination 6 months later indicated rupture of the ACL associated with a rupture of the medial meniscus posterior horn, at the meniscosynovial junction. Meniscus repair was carried out. An arthroscopic control 5 months postoperatively confirmed a rupture recurrence, at the same level as the first; however, a small "bridge" of meniscus remained across the middle of the rupture preventing posterior horn luxation into the internal femorotibial compartment (Fig. 12). A fragment was taken from the posterior horn for histological analysis.

Areas of fibrocartilage in good condition were observed. Although large areas were acellular, they presented rich collagenous fibre content. Signs of local necrosis were associated with poor staining of highly degenerated fibres and the presence of cells with pyknotic nuclei in a few places. Areas at the periphery of the rupture were rich in cells with signs of cellular division in the chondrocytes. Inflammatory cells were present but rare; however, there was no sign of vascularisation.

Conclusions

Evidence from histological analysis confirmed the phenomenon of spontaneous cicatrisation during meniscus repair, a feature that had been noted macroscopically during arthroscopic interventions. The cicatrix consists of fibrous tissue with local penetrations of vessels accompanied by abundant cells which apparently originate at the meniscosynovial junction.

Several cases of radial ruptures occurred within fibrous meniscal tissue possessing a low cell content but not within the usual fibrocartilaginous medium. This phenomenon may be interpreted as the result of a kind of dedifferentiation in the fibres of less resistant fibrocartilage, which is more inclined to rupture. Consequently, the dedifferentiation cannot occur in previously necrotic tissue.

Cases of spontaneous repair appear after repair and cicatrisation of the initial rupture have taken place, but are located in another region of the same meniscus.

In other cases, such as a rupture parallel to a formerly repaired and cicatrised primitive rupture or simply recurrence of a primitive rupture, the histological study reveals a rather different situation. Here, in fact, fibrocartilaginous tissue presents acellular fibrous areas and necrotic areas. There is no sign of vascularisation, but in some cases signs of cellular regeneration are evident, located at the periphery of the wound.

References

1. Stone RG (1979) Peripheral detachment of menisci of the knee: A preliminary report. Orthop Clin North Am 10:643
2. Heatley FW (1980) The meniscus – can it be repaired? J Bone Joint Surg [Br] 62:397
3. Cassidy RE, Shaffer AJ (1981) Repair of peripheral meniscal tears. A preliminary report. Am J Sports Med 9:209
4. Dehaven KE (1981) Peripheral meniscus repair: Alternative to meniscectomy. J Bone Joint Surg [Am] 63:463
5. Wirth ER (1981) Meniscus repair. Clin Orthop 153:157
6. Arnoczky SP, Warren RF (1982) Microvasculature of the human meniscus. Am J Sport Med 10:90
7. Hamberg P, Gillquist J, Lysholm J (1983) Suture of new and old peripheral meniscus tears. J Bone Joint Surg 65:193
8. Clancy WG, Graf BK (1983) Arthroscopic meniscal repair. Orthopedics 6:1125–1129
9. Hendler RC (1984) Arthroscopic meniscal repair. Clin Orthop Rel Res 190
10. Rosenberg T, Scott S, Paulos L (1985) Arthroscopic surgery: repair of peripheral detachment of the meniscus. Contemp Orthoped 10:43–50
11. Barber FA, Stone RG (1985) Meniscus repair – an arthroscopic technique. J Bone Joint Surg 67:39–41
12. Warren RF (1985) Arthroscopic meniscal repair. Arthroscopy 1:170–172
13. Graig D, Casscells M, Casscells W (1986) Arthroscopic meniscus repair: A safe approach to the posterior horns. Arthroscop Rel Surg 2(1):3–12
14. Rosenberg THD, Scoot SM, Coward DB, Dunbar WH, Ewing JW, Johnson CHL, Paulos LE (1986) Arthroscopic meniscal repair evaluated with repeat arthroscopy. Arthroscop Rel Surg 2(1):14–20
15. Stone RG, Vanwincke GN (1986) Arthroscopic review of meniscal repair: Assessment of healing parameters. Arthroscop Rel Surg 2(2):77–81

Combined Arthroscopic-Open Repair of Meniscal Injuries

S. C. Marshall

University of California Medical School, San Diego,
7300 Girard Avenue, Suite 204, La Jolla, CA 92037, USA

Introduction

Recent studies have emphasized the importance of the meniscus in knee joint function and its role in weight transmission [1], stability [2], and nourishment of the articular cartilage [3, 4]. Early reports of salvage of peripheral tears of the meniscus through suture repair have been encouraging [5–8]. Preservation of at least a peripheral rim of meniscus makes the knee more stable [9] and lessens the chance of developing early degenerative changes of the knee joint [10, 11]. Therefore, every effort should be made to save the meniscus when possible.

Complete meniscectomy has been reported to result in degenerative arthritic changes [12–14]. New techniques have been developed to preserve the meniscus through repair of these tears either by open arthrotomy [5, 6, 8] or by intra-articular arthroscopic means [15–17]. Each of these methods has its drawbacks and possible complications, e.g., limited visualization and a narrow working space hindering identification of the tear when employing the direct open arthrotomy approach. Reported complications resulting from the intra-articular arthroscopic technique are caused by blind passage of needles and sutures through soft tissues of the joint which are inadequately exposed. The complications include injury to the popliteal artery, saphenous nerve and peroneal nerve, synovial outgrowth and infections secondary to placement of the skin bolsters. Published documentation of these complications is sparse and deserves more attention [17]. Inadvertent penetration of any nerve about the knee joint may cause permanent loss of function or disabling pain. This is particularly true when the saphenous nerve or its branches are injured [18].

Repair of peripheral tears of the menisci by arthroscopic techniques has been made simple and safe using an Ethibond suture (size 0) on a double-armed, 6-in. Keith needle. This is followed by a limited, transverse skin incision for an open arthrotomy to reinforce the repair firmly in the capsule which adequately exposes and protects the neurovascular structures.

A method combining the best arthroscopic and the best limited open arthrotomy techniques will be described.

Indications

This method is indicated for longitudinal tears of the medial and lateral menisci located at the synovial meniscus junction and for longitudinal tears that occur within the peripheral third of the meniscus [19, 20] (Fig. 1). It has been reported

Surgery and Arthroscopy of the Knee
2nd Congress of the European Society
Editors: Müller/Hackenbruch

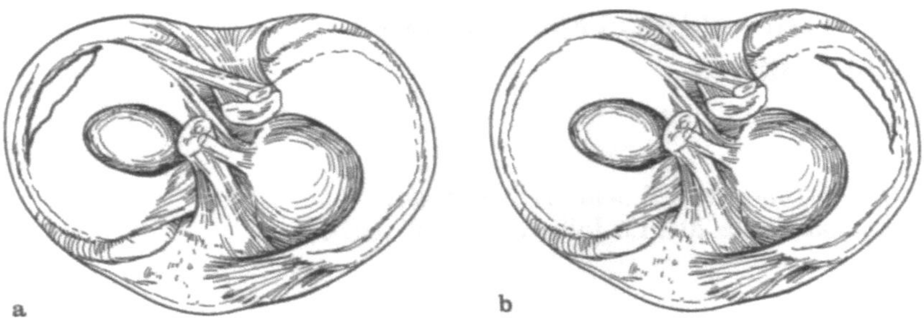

Fig. 1. a Peripheral tear of the mid and posterior lateral meniscus in the right knee. b Posterior horn tear of the medial meniscus in the right knee

that the sutures, when properly placed, act as a scaffolding for ingrowth of synovial cells which help to repair the meniscus [10]. Tears that involve the outer third of the meniscus of those that are more peripheral are amenable to repair by this method [5–8, 10]. Smillie reported that the most common longitudinal tear pattern in the meniscus, excluding complete bucket handle tears, occurred in the synovial-meniscus junction, the peripheral portion.

Contraindications

Transverse tears, flap tears, degenerative tears, or longitudinal tears where more than the peripheral third of the meniscus is intact are not amenable to this technique due to limited vascular ingrowth in order to heal the tear [10, 18]. Some surgeons have repaired radial and transverse tears as well as half-width bucket handle tears. These areas of repair are highly experimental, and it is not recommended at this time to use the combined arthroscopic-arthrotomy technique.

Anatomy

At the joint line level the most important neurovascular structures to be aware of include (Fig. 2): (1) the popliteal artery and vein, which are just posterior to the posterior horn of the lateral meniscus; (2) the medial popliteal (tibial) nerve, which lies posterior and lateral to the popliteal artery and vein; (3) the lateral popliteal (common peroneal) nerve, which lies posterior and lateral to the lateral meniscus, lateral to the gastrocnemius muscle, and slightly on the posterior lateral aspect of the biceps tendon; and (4) the saphenous nerve and its infrapatellar branch, which become superficial just above the knee joint, exiting between the musculotendinous junction of the sartorius muscle and the gracilis tendon. The infrapatellar branch of the saphenous nerve transverses the medial joint line region beneath the superficial fascia of the leg (Figs. 1, 3). The saphenous nerve is first located posterior and medial to the medial meniscus. It then becomes more superficial and anterior as it exits into the superficial fat of the medial side of the leg. Incorporating any portion of this nerve or its branches or fascia in sutures will cause deviation of the nerve and possibly severe paresthesia [18]. This will ne-

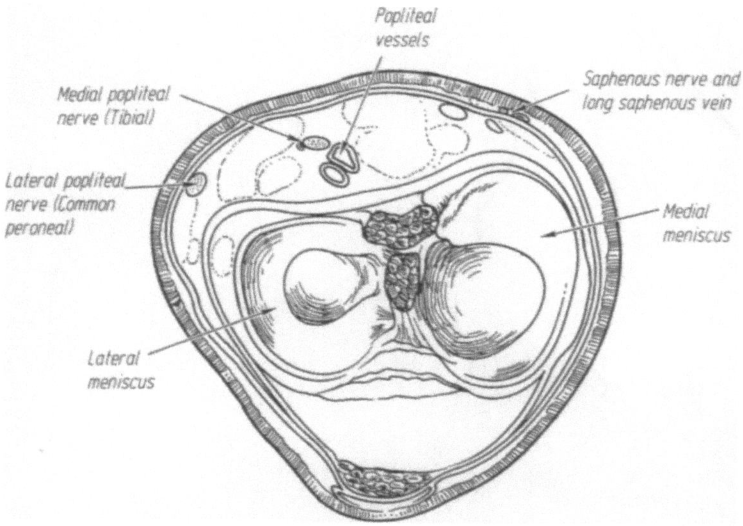

Fig. 2 Neurovascular structures in the knee joint

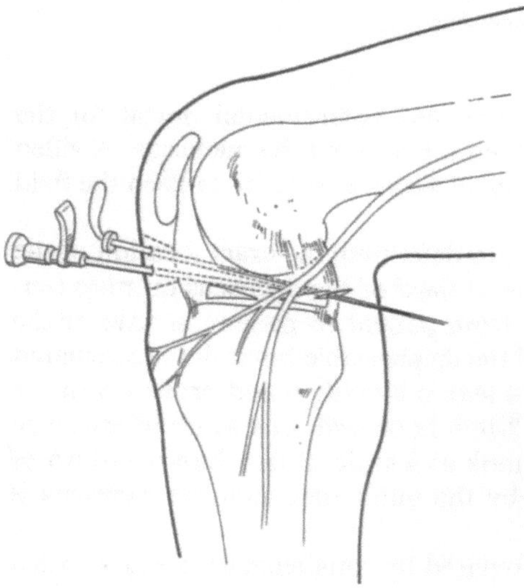

Fig. 3. Needles exiting through the skin from the posteromedial area of the knee. A transverse incision (1–2 in.) is made over the needle's exit point. Isolation and protection of the saphenous nerve is essential

cessitate a second procedure to remove the sutures. Care must be taken to avoid placing any suture material around the vessels and nerves of the knee joint region.

Operative Technique

The patient is placed in a supine position on the operating table. The leg is suspended in an adjustable, leg-holding device that can be lowered to allow easy access to the posteromedial and posterolateral areas of the knee joint when making the small transverse arthrotomy incision. Diagnostic arthroscopy is started using

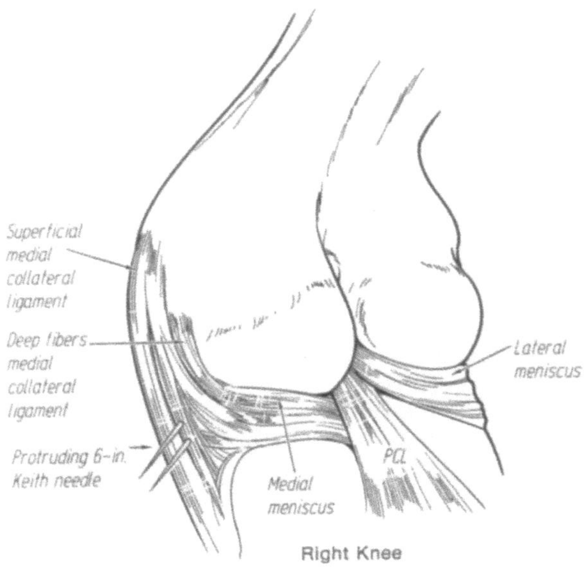

Fig. 4. Meniscus tear is not seen well from posteromedial or posterolateral incisions because of capsular and ligamentous covering of the meniscus

the anterolateral portal for the arthroscope and anteromedial portal for the Dandy nerve hook instrument, which is used for probing the meniscus. A video camera attachment is the preferred method of viewing in order to keep the field sterile during the limited arthrotomy.

The most difficult part of the procedure is to determine the exact location of the tear, i.e., whether it is in the outer peripheral third of the meniscus or more centrally located. As meniscus width varies from patient to patient, a ratio of the width of the attached outer rim to that of the displaceable inner rim is calculated to determine the location of the tear. The tear is identified and probed with the Dandy nerve hook probe that has a 90°, 3-mm hook end. The width of each side of the tear is estimated using the 3-mm hook as a scale. If less than one-third of the total meniscus tissue is represented by the outer rim, then the meniscus is judged as being reparable (Fig. 1).

The remaining intact peripheral rim is prepared by roughening the edges with a small (3-mm) basket forceps or curette to encourage bleeding; this enhances vascular ingrowth and promotes repair. If the rim cannot be seen clearly with the arthroscope it may be prepared later through the limited arthrotomy. No more than 1–2 mm of tissue should be excised to prevent buckling of the repaired meniscus.

With the knee in 80°–90° of flexion, a short 2–3-in. (5–7.5 cm) transverse incision is made over the medial joint line, starting anteriorly in the posteromedial corner and carried posteriorly parallel to the popliteal crease. The incision is carried down to the capsule, retracting all superficial structures that should not be incorporated under the suture tie, i.e., the sartorius muscle and saphenous nerve and vein distally, and the semimembranous muscle posteriorly (Fig. 4).

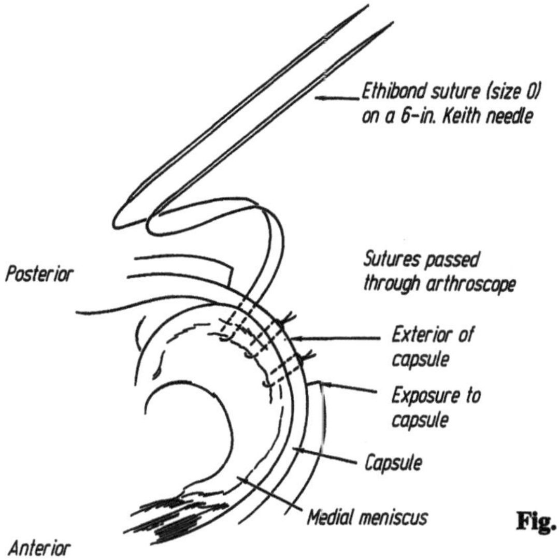

Fig. 5. Placement of the Ethibond suture

Next, the arthroscope is inserted on the side of the knee with the meniscal lesion. An Ethibond suture (size 0) (Fig. 5) on a double-armed, 6-in. Keith needle is placed in a short-barrelled, 4-mm wide, arthroscopic sheath cannula entering from the opposite portal. The needles are placed perpendicular to the meniscus tear and aimed at the anterior apex of the tear (Fig. 3). The tips may be bent manually to allow for a more anterior exit. The cannula is used to prevent snagging of the anterior fat pad. The Keith needles are gently pushed, one at a time at a distance of 2 mm apart, through the superior meniscus body and rim and finally through the capsule and skin. One or two suture passages are necessary for this anchoring stitch. A special needle holder produced by Acufex may be used to ease insertion of the needles through the meniscus. Usually, however, the tissues are compliant enough to allow the needles to be inserted by hand with minimal resistance.

This first set of sutures, placed arthroscopically, acts as a landmark for easy identification of the meniscus and the tear from the outer capsular side. This is an important step because the meniscus tear may not be very apparent through the arthrotomy incision. Thus, the suture will identify the exact location of the tear just posteriorly to the exit of the sutures from the capsule. It will also anchor the meniscus firmly to the capsule or appose the two sides of the tear well.

Anterior placement of the arthroscopic sutures avoids encroachment into regions of important neurovascular structures. These arthroscopically placed sutures usually exit in the substance of the posterior oblique ligament lying posterior to the vertical superficial medial collateral ligament. A longitudinal incision is placed in the posterior oblique ligaments and lengthened by 2 in. (5 cm) or more until adequate exposure of the posterior horn of the medial meniscus is obtained.

Loupe magnification and headlamp lighting are now used. The meniscal edges on both sides of the tear are curetted gently to promote bleeding. Then, very carefully, under direct vision, a UCL (Ethicon) needle, onto which a size 2-0 PDS (Ethicon) suture is swaged, is passed through the capsule and coronary ligament into the joint and repositioned on the needle holder. The needle is passed in a horizontal direction, through the displaced meniscal tissue, then out through the coronary ligament, capsule, and posterior oblique ligament. Two to three additional horizontal sutures are necessary to ensure a secure and stable repair.

The UCL needle is a tiny but strong needle that can be easily passed horizontally under the femoral condyles with little fear of the needle breaking or causing damage to the articular surfaces. The needle should always be in view when placing sutures. If this is not possible, a larger (size 6) Mayo needle works well, especially with more centrally located tears, allowing the needle to be viewed at all times. The landmark arthroscopic sutures may also be pulled posteriorly to draw the displaced meniscus closer in view if it tends to slide anteriorly under the femoral condyle. Knots are tied directly on the exterior of the capsule.

The posterior oblique ligaments are plicated in an overlapping fashion, causing a reefing effect, thus tightening the posteromedial ligamentous structure, which adds significantly to the stability of the knee initially [23].

Repair of lateral meniscus tears is similar to medial meniscus repair. The tear pattern seen with rupture of the lateral meniscus commonly occurs extra-peripherally through the inferior fasicular ligament which is often so thin [24].

Arthroscopic sutures are placed as far anteriorly along the tear as possible, usually in the midlateral portion of the meniscus. The needles must exit anterior to the biceps tendon in order to prevent inadvertent penetration of the peroneal nerve lying posterior to the biceps tendon. The 90° flexed knee position facilitates this passage.

A 2–3-in. (5–7.5 cm) lateral transverse incision is made over the protruding arthroscopic sutures along the joint line. The sutures identify the exact location of the tear. The iliotibial band is split parallel to its fibers. Posteriorly, near the intramuscular septum, the capsule is incised longitudinally just posterior to the landmark suture exiting from the joint. A second set of sutures are placed horizontally through the meniscus and capsule, creating a new attachment for the meniscus anterior to the popliteus tendon. Care is taken not to incorporate the popliteus tendon or lateral collateral ligament (LCL) in the sutures since the midlateral meniscus is normally not attached to these structures. A second longitudinal incision is made posterior to the LCL in the arcuate complex to insert more posterior horizontal meniscal sutures. Others have described a posterolateral removal of bone block of LCL to gain access to this area if necessary.

In normal anatomy, the lateral meniscus is attached posteriorly to the arcuate complex, capsule, and fibers of the popliteus muscle, as well as to the posterior cruciate ligament (PCL) [25]. Both menisci move slightly forward with extension of the knee and slightly backward with flexion [25]. The amount of abnormal movement can be judged arthroscopically by identifying excoriations of contiguous femoral or tibial articular surfaces.

To finish the repair on the lateral side, the capsule is reefed with Ethibond sutures (size 0) placed horizontally, the iliotibial band is reapproximated with two 0 absorbable sutures, and the traverse skin incision is closed with a subcuticular suture. The knee is then set in a cast for 6 weeks initially at 45° of flexion and neutral rotation to protect the repair. The patient is allowed partial weight bearing. At 3 weeks hinges are applied to the cast to allow motion between 30° and 90°.

Material and Methods

Twenty-six patients who underwent treatment for peripheral meniscus repair were studied 18–50 months postoperatively (average 30 months).
Preoperative symptoms included instability, giving way or locking, and recurrent effusions. Signs of joint line tenderness, crepitus, and clicking associated with meniscal tear were also present.
The average age of the 12 males and 14 females was 25 years (range 13–42 years) and the average time from initial injury to arthroscopic surgery was 8 months (range 1–24 months).
No patient had experienced prior knee surgery. Ten patients also presented complete tears of the anterior cruciate ligament (ACL). The types of meniscal tears repaired were as follows: 16 medial (14 posterior horn, 2 midmeniscus), and 10 lateral (7 inferior fascicular ligament, 2 complete longitudinal, and 1 anterior horn).
There were no complications following meniscus repair and the period of hospitalization ranged from 1 to 2 days. All patients had postoperative effusions that were relatively asymptomatic due to immobilization in the cast. Formal physical therapy was required in 15 patients to regain a normal range of motion.
The cast was removed at 6 weeks and therapy to increase the range of motion was intensified to include walking, biking, swimming, and weight lifting. Light jogging was started at 12 weeks. Interval running with sprints was started at 16 weeks. After 20 weeks, skiing, contact sports, and sprinting were allowed. Patients were evaluated through questionnaires and physical examination at an average of 21 months postoperatively.
Each patient was also asked to select the activity which was the most comfortable to execute in the presence or absence of symptoms. Activities were grouped into three classes designed to correlate with intact ACL function and agility skills as follows [26]:

Class I: Activities that involve little or no lateral movement, twisting, deceleration, or landing after a jump. These include average daily living (ADL) activities, walking, swimming, running, and biking.

Class II: Activities that require more agility and ability to cut, pivot, decelerate, or land after jumping. Examples include racquetball, tennis, baseball, and intermediate skiing.

Class III: Activities that require the most skill and agility. Football, basketball, soccer, and rugby fall into this category.

Results

Subjective evaluation revealed that all 26 patients thought that their knees functioned better after meniscal repair. None stated that the knee was worse following this surgical procedure.

Three patients with ACL deficiency suffered a further injury and were retreated by arthroscopy, although the meniscus repairs remained intact. One patient, while surfing competitively, incurred a retear of the medial meniscus adjacent to the repaired complete longitudinal tear. Two patients continue to feel tenderness over the repaired areas. Patellofemoral pain and crepitus were present in two patients. Quadriceps atrophy of the involved leg averaged 2–4 cm.

Postoperative clinical examination revealed a negative McMurray test, no effusion, and a full range of motion in all subjects. The Lachmann test, anterior drawer test, or medial lateral laxity were initially decreased due to posterior oblique ligament plications, but eventually increased after 12 months to equal the preoperative values.

Preinjury status revealed that 20 of 26 patients were participating in Class III activities. Postinjury status revealed that none of the patients was able to do Class II or III activities without symptoms, but 15 patients were capable of Class I ADL activities. Following meniscal surgery, 26 patients were able to perform Class I activities which consisted of mainly forward motion, minimal pivoting, or deceleration while 19 patients attained Class II activities and 8 were able to participate in Class III sports. Ten patients with ACL deficient knees wore braces (a graphite hinge type or neoprene with metal hinges) for sporting activities but one of them was able to do Class III activities.

Discussion

The subject of meniscal repair is in its infancy; hence long-term results are not yet available. Many techniques have been described to repair torn menisci. Reports on open repair limited to peripheral tears of the meniscus have been good with minimal complications [27]. The arthrotomy incision locating the meniscus tear is vital to expose and protect important neurovascular structures in the vicinity of the joint, including the saphenous nerve, popliteal artery, vein, and nerve, and the peroneal nerve.

Early reports describing the arthroscopic method in which needles are passed from inside the joint outward through the soft tissue have documented major damage to the neurovascular structures as well as causing deep infection [28]. With the evolution of arthroscopic meniscal repair techniques, recommendations have now been made for posteromedial or posterolateral skin incisions to expose the outer capsular area prior to needle and suture passage from the joint [29, 30].

Henning [29] advocates a combined arthroscopic and open technique to minimize the possibility of neurovascular complications. Incisions are made posteriorly to isolate and protect the neurovascular structures prior to passing arthroscopic sutures. No open suturing is performed.

Despite successful reports by many authors using open repair techniques, critics state that certain intrasubstance tears cannot be sutured by open means and must be approached arthroscopically [30, 31]. It is argued that tears located one-third of the way from the meniscocapsular junction cannot be approached or seen by open methods or that the femoral condyle blocks needle and suture passage from this direction. However, in this series, through a longitudinal incision in the posteromedial or posterolateral capsule these tears can usually be visualized well. Repair is achieved using a Mayo (6–0) needle holding the suture which is aimed anteriorly and passed horizontally from the outside towards the meniscocapsular area, through the outer meniscal rim, across the tear, and back outward through the capsule. In certain cases, this may not be possible due to the difficulty in seeing the needle point at all times. In that situation one would have to rely on the arthroscopic sutures to hold the repair.

It is important to know both the techniques (arthroscopic and open) well and apply them when indicated. When possible, a combination of the two methods gives the most secure repair. Until it can be shown by long-term studies that arthroscopically placed sutures are effective, it is best to reinforce the arthroscopic repair with an open repair.

The questions which remain unanswered are what type of suture (absorbable or nonabsorbable), what size of suture, how many sutures, and what direction of placement of sutures (vertical or horizontal) are needed to hold the repair successfully and allow vascular ingrowth.

The tear patterns seen in this study are consistent with the common types of longitudinal tears described by Smillie, being at the junction of the meniscus with the synovium or in the peripheral meniscus [21].

Morgan and Casscells [32] have described leg positions for safe passage of sutures through the joint. For lateral meniscus repairs, the knee should be flexed at 90° to allow the peroneal nerve to fall below the knee joint. Needles must exit anterior to the biceps femoris tendon. For medial repairs, the knee is placed in near or full extension which allows the saphenous nerve to be situated anterior to the suture passage.

In this combined arthroscopic-open procedure the advantages of the arthroscopic portion are as follows: (1) The meniscus tear is easily identified and sutured anteriorly using Keith needles. The external protruding sutures help locate the meniscus tear for extracapsular open repair. (2) This initial suture can be used to bring the meniscus closter to its capsular margin by placing tension on it for further repair. There are also advantages of the open repair portion: (1) The neurovascular structures can be dissected and protected so that they are not incorporated in the suture. (2) The transverse scar is cosmetic. (3) A very secure repair of the meniscus is obtained, with multiple sutures being placed in the same fashion as for a ligament under repair. (4) There are no skin problems such as fistulas or skin necrosis from a bolster pad. (5) Distraction of the joint is performed by suspending the leg over the leg-holder to obtain better exposure.

There are two disadvantages of this procedure: (1) The small open arthrotomy incision (3–4 cm in length) requires extra lighting and magnification to be seen well. The curved suture needle must be kept in view at all times. (2) It is difficult to pass the sutures longitudinally.

It is very important to have available prior to arthroscopic examination the repair kit of sutures, needles, and equipment necessary to do the meniscus repair. This will avoid a lot of confusion and save time. Included in this tray of instruments are the sutures and retracting instruments:

Leg-holder
Minor arthrotomy instruments – retractors (Z, Senn, ARMY/NAVY, Wheatlander)
Headlamp illumination
Loupe magnification
Suture and needles – Ethibond (size 0) on 6-in. Keith needles, 2-0 pds on UCL needle, Mayo needles (No. 6), small-eye needles
Short barrel arthroscopic sheath/cannula
Meniscus suture passer
Split sheets for redrape

Total meniscectomy leads to increased instability of the knee joint. This instability increases when the meniscal tears combine with rupture of the ACL. Every effort should be made to save peripheral rim tears of the meniscus in order to make the knee more stable. It is not known whether meniscal repairs are indicated in knees that are already unstable from ACL tears. Some reports have indicated that repair of the meniscus will fail in 30% of the cases unless the instability of the ACL is also repaired or reconstructed [6]. In this study, ACL deficient knees were protected with braces. These patients exhibited symptoms of instability even though they did not retear their menisci. Ideally, reconstruction of the ACL should be performed simultaneously to meniscus repair to improve knee function and allow for better results in the long-term follow-up of meniscal repair.

Summary

Arthroscopic blind passage of sutures can cause significant damage to the neurovascular structures about the knee joint. A safe technique has been devised wherein dissection of the capsular area and repair of the meniscus is done after arthroscopic passage of sutures. The repair is reinforced with sutures placed through a transverse arthrotomy incision access. To date, the results of this anatomic repair have been good for all the 26 patients studied over a follow-up of 18–50 months with only one retear reported.

References

1. Krause WR, Pope MH, Johnson RJ et al. (1976) Mechanical changes in the knee after meniscectomy. J Bone Joint Surg [Am] 58:588–604
2. Walker PS, Erkman MJ (1975) The role of the menisci in force transmission across the knee. Clin Orthop 109:184–192
3. Cox JS (1977) The degenerative effects of medial meniscus tears in dogs' knees. Clin Orthop 125:236–242
4. King D (1936) The healing of semilunar cartilages. J Bone Joint Surg 18:333–342
5. Cassidy RE, Shaffer AJ (1981) Repair of peripheral meniscus tears. A preliminary report. Am J Sports Med 9:209–214

6. DeHaven KE (1981) Peripheral meniscus repair. An alternative to meniscectomy. Orthopaedic Transactions 5:399–400
7. Hambert P, Gilquist JI, Lysholm J (1983) Source of new and old peripheral meniscus tears. J Bone Joint Surg [Am] 65:193–197
8. Wirth CR (1981) Meniscus repair. Clin Orthop 153:153–160
9. Jackson RW, Dandy DJ (1976) Partial meniscectomy. J Bone Joint Surg [Br] 58(1):142
10. Heatley FW (1980) The meniscus – can it be repaired? An experimental investigation in rabbits. J Bone Joint Surg [Br] 62(3):397–402
11. McGinty JB, Geuss LF, Marvin RA (1977) Partial or total meniscectomy. A comparative analysis. J Bone Joint Surg [Am] 59:763–766
12. Fairbank TJ (1948) Knee joint changes after meniscectomy. J Bone Joint Surg [Br] 30(4):664–670
13. Johnson RJ, Kettlekamp DB, Clark W et al. (1974) Factors affecting late results after meniscectomy. J Bone Joint Surg [Am] 56:719–729
14. Tapper EM, Hoover NW (1969) Late results after meniscectomy. J Bone Joint Surg 51:517–526
15. DiStephano V, Bizzle P (1983) A technique of arthroscopic meniscoplasty. Orthopaedics 6(9):1135–1140
16. Rosenberg T, Scott S, Paulos L (1985) Arthroscopic surgery: repair of peripheral detachment of the meniscus. Contemp Orthop 10(3):43–50
17. Stone R, Barber F (1985) Meniscus repair, an arthroscopic technique. J Bone Joint Surg [Br] 67:39–41
18. Slocum DB, Larson RL, James SL (1974) Pes anserinus transplant: impressions after a decade of experience. Am J Sports Med 2(3):123–126
19. Arnoczky SP, Warren RF (1982) Microvasculature of the human meniscus. Am J Sports Med 10(2):90–95
20. Scapinelli R (1968) Studies on the vasculature of the human knee joint. Acta Anat (Basel) 70:305–331
21. Smillie IS (1971) Injuries to the knee joint, 4th edn. Churchill Livingstone, Edinburgh, p 50
22. Williams PL, Warwick R (1978) Gray's anatomy. The knee joint. Churchill Livingstone, Edinburgh, pp 274–282
23. Hughston JC, Filers AF (1973) The role of the posterior oblique ligament in repair of acute medial (collateral) ligament tears of the knee. J Bone Joint Surg [Am] 55(5):923–940
24. Whipple TL (1982) Posterior perpheral detachment of the lateral meniscus pathogenesis with respect to the popliteus tendon. Contemp Orthop 4(5):533–538
25. Hollinshead WH (1969) Anatomy for surgeons, 3rd edn. Harper & Row, New York, pp 779–780
26. Marshall SC, Levas MG, Harrah A (1985) A simple arthroscopic partial meniscectomy associated with anterior cruciate-deficient knees. J Arthroscop Rel Surg 1(1):22–27
27. DeHaven KE (1985) Meniscus repair – open vs. arthroscopic. J Arthroscop 1(3):173–174
28. Stone RG, Miller GA (1985) A technique of arthroscopic suture of torn menisci. Arthroscopy: J Arthroscop Rel Surg 4:226–232
29. Henning CE, Lynch MA, Jelly BL, Scott GA (1984) Arthroscopic meniscal repair. Presentation. AOSS Meeting, July 1984. Anaheim, California
30. Rosenberg TD, Scott SM, Coward DB, Dunbar WH, Ewing W, Johnson CL, Paulos WE (1986) Arthroscopic meniscal repair evaluated with repeat arthroscopy. J Arthroscop 2(1):14–20
31. Clancy W, Graf B (1983) Arthroscopic meniscal repair. Orthopedies 6:1125–1129
32. Morgan CD, Casscells SW (1986) Arthroscopic meniscus repair: a safe approach to the posterior horns. J Arthroscopy 2(1):3–12

Electron Microscopical Observations on Sutured Meniscus Tears

A. Schmid, F. Schmid, and T. Tiling

Department of Surgery, University Clinic, Robert-Koch-Strasse 40, D-3400 Göttingen, FRG

Introduction

Since the number of meniscus injuries from sport and occupational hazards is steadily on the increase and diagnostic methods are improving due to the introduction of arthroscopy, treatment of meniscus ruptures is also undergoing refinement. Whereas in the past treatment was restricted to meniscus resection or meniscectomy, it has now become increasingly common to stitch longitudinal tears occurring in the well-vascularised meniscus periphery (Gillquist and Lysholm 1983; Karpf et al. 1984; Karpf et al. 1985; Rodriguez and Schreiber 1984). Using ultrastructural investigations on ruptured and sutured menisci we have attempted to establish the conditions under which a suture is indicated for treatment of meniscus rupture.

Material and Methods

From December 1983 to December 1985, 37 patients between the ages of 18 and 35 years suffering from meniscus ruptures but with no previous knee damage were treated using the technique of meniscus suture. Of the 37, 23 were suffering from isolated ruptures of the meniscus, while the other 14 patients had meniscus ruptures together with other knee damage. The time lapse between trauma and suture varied from 2 to 11 days.

A meniscus suture was only carried out under certain conditions, i.e. when the rupture was situated in the well-vascularised, peripheral third of the meniscus. Cases of capsular detachment were not included in the study. Suturing was carried out using open surgery employing ligatures 3.0 mm in diameter, made of absorbable material. In the case of five patients with isolated meniscus ruptures, the suture was inserted using the arthroscopic technique.

During the operation, tissue samples were taken from the edge of the rupture, near the capsule, and from the edge away from the capsule over the whole length and width (Fig. 1).

In cases of isolated meniscus sutures rehabilitation started with a 4-week immobilisation period; in cases of additional ligament sutures the period was increased to 6 weeks.

Up to now, 27 of the patients have been re-examined by arthroscopy at follow-up. In 11 patients a new rupture had occurred within a time interval of 3–12 months after suturing. Tissue samples were again taken from the edges of the tear next to, and on the opposite side, of the capsule. A meniscus resection ws then carried

Surgery and Arthroscopy of the Knee
2nd Congress of the European Society
Editors: Müller/Hackenbruch

Fig. 1

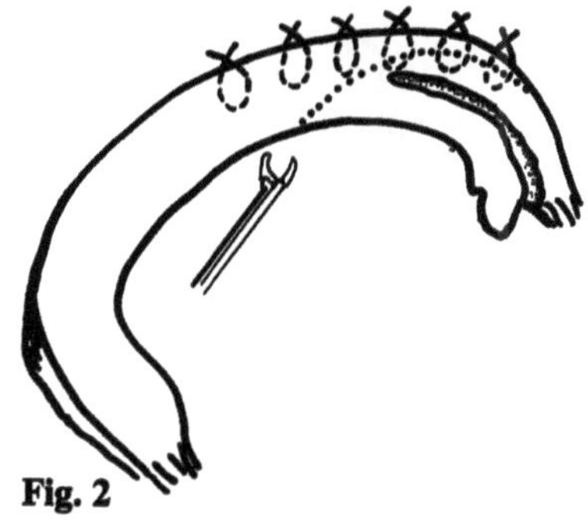

Fig. 2

Fig. 1. Tissue samples were taken from both edges of the meniscus rupture
Fig. 2. Resection line following a second rupture of the meniscus

out on all 11 patients. Parts of the stitched tissue were included in the resection; samples were also taken from this tissue (Fig. 2).
All samples were examined by light and electron microscopy. Thin sections were looked at under the light microscope to establish the condition of the fibre system and the cell density. Ultrathin sections were prepared for electron microscopy to examine cell vitality, cell mortality, morphology of individual cells, the entire fibre system and the structure of single collagen fibres.
Changes following trauma, whether involving repair or degeneration, manifest themselves in the living element of the tissue, the cell. Using electron microscopy to examine the cell morphology, it was possible to observe very early changes. In order to run morphological controls, tissue from undamaged knees was taken from organ donors, of the same age as our patients, who had died in accidents. Before each suture, an arthroscopy was carried out in order to localize the tear exactly and to determine the condition of the knee.

Results

Out of the 27 patients examined, 23 were suffering from isolated ruptures of the meniscus. In 13 of these cases the rupture was of a longitudinal-vertical type lying in the posterior horn of the medial meniscus. The tear was longitudinal, lying in the middle section of the medial meniscus in 5 cases; in another 2 the tear was longitudinal, positioned in the anterior horn of the medial meniscus; in 3 a longitudinal tear was located in the anterior horn of the laterial meniscus.
For the 14 patients with combined injuries, ten longitudinal ruptures were evident in the posterior horn of the medial meniscus and four longitudinal tears in the middle section of the medial meniscus. Light and electron microscopical examination of tissues taken during the initial arthroscopy showed that regressive and reparative changes were taking place concurrently in all cases, although with differing degrees of intensity.
Several features were in evidence.
In patients suffering from compound injuries the reparative changes clearly outweighed the regressive changes. Both light and electron micrographs reveal an

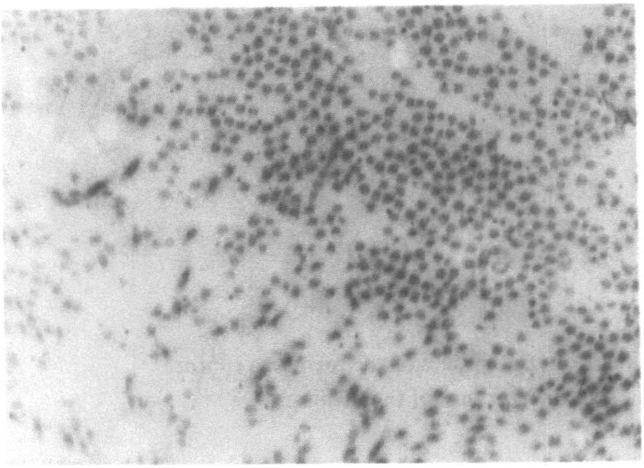

Fig. 3. Oedematous softening of the meniscal tissue, ×63000

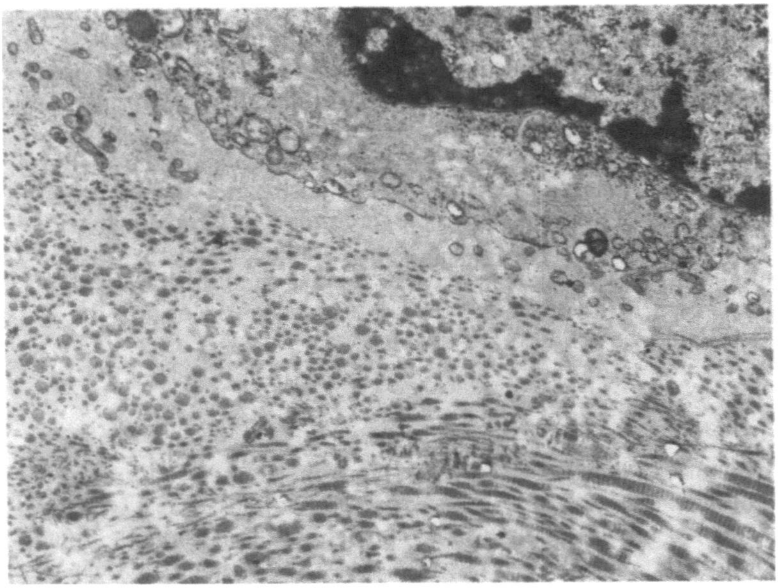

Fig. 4. Increase of intracellular filaments and pyknotic nuclei signal of degeneration in a chondrocyte, ×16500

oedematous softening of the tissue along the edges of the wound, which indicates an exudative phase at the onset of healing (Fig. 3). An increase in capillaries and a proliferation of fibroblasts implies the onset of healing, seen from the 9th day after the rupture. There is little evidence of necrosis of chondrocytes and fibrocytes.

However, in patients with isolated meniscus tears regressive changes are predominant in the overall histological picture. Chondrocytes with pyknotic nuclei are associated with chondrocyte necrosis (Fig. 4).

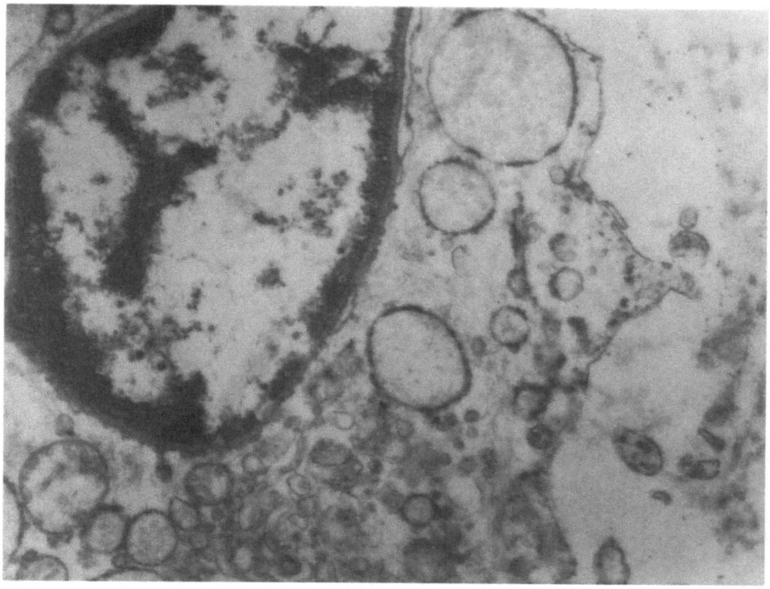

Fig. 5. A chondrocyte with swollen endoplasmic reticulum, mitochondria and nuclear membrane, ×32500

The organelles swell, and the nuclear membrane thickens (Fig. 5). An increase in the number of intracellular filaments is observed together with a greater accumulation of intracellular and extracellular fat deposits. Growth of fibroblasts is only found in one case, on the 10th day.

At the time of follow-up arthroscopy the sutured scars had, in all cases, healed smoothly. No relapse of the primary rupture was observed in the 11 patients with a second rupture. In all 11 cases, the first tear was located in the posterior horn of the medial meniscus. The second tear was found further back away from the capsule, in the region of the posterior horn. Nine of these patients had been suffering from an isolated meniscus rupture, two from compound injuries. Just once, when the patient had originally suffered a compound injury, was there adequate evidence of a trauma being the cause of the second rupture.

Histological examination of tissue taken from the sutured rupture region yielded the following results. In six cases in which the original injury was an isolated meniscus rupture and in one case of a patient suffering from compound injury, the wound appeared to be smoothly healed both from light microscopical and electron microscopical evidence. No visible scarring were present. The tissue displayed a normal, regular pattern of fibres (Fig. 6). In the sutured region fibroblast-like chondrocytes visible (Fig. 7). Therefore, tissue in the sutured region appeared to be regenerating. In the six patients with isolated tears, the follow-up arthroscopy was carried out in two cases 8 months after suture, in one case 10 months, in two cases 10.5 months and in one case 12 months. In those with compound injuries the time span was 4 months. In three cases of isolated tears and in one case of compound injury the repair process did not appear histologically complete. A large number

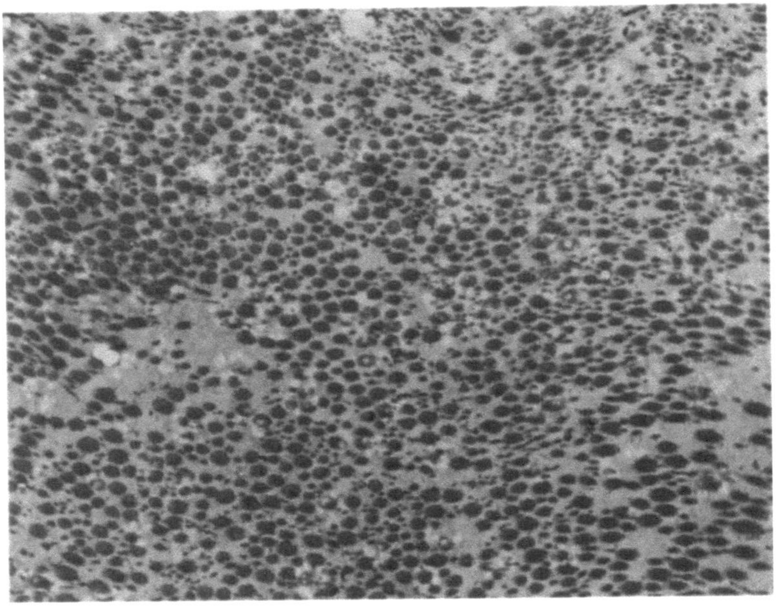

Fig. 6. The regular pattern of the collagen fibres in tissue taken from the sutured region, ×19250

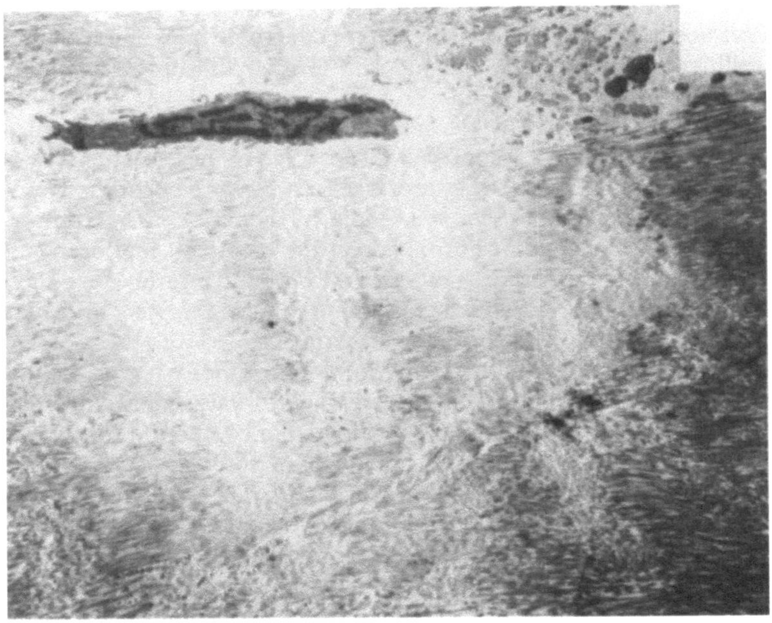

Fig. 7. Cell from the sutured region, ×8750

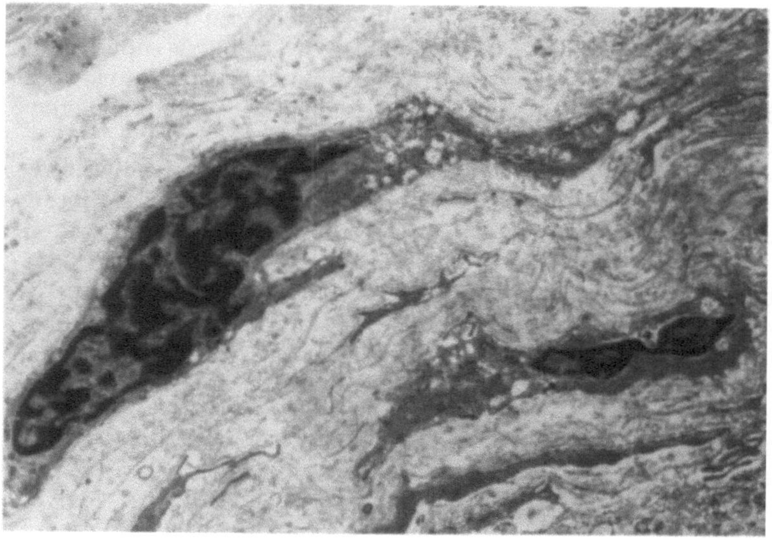

Fig. 8 Active cells from the sutured region, × 17 500

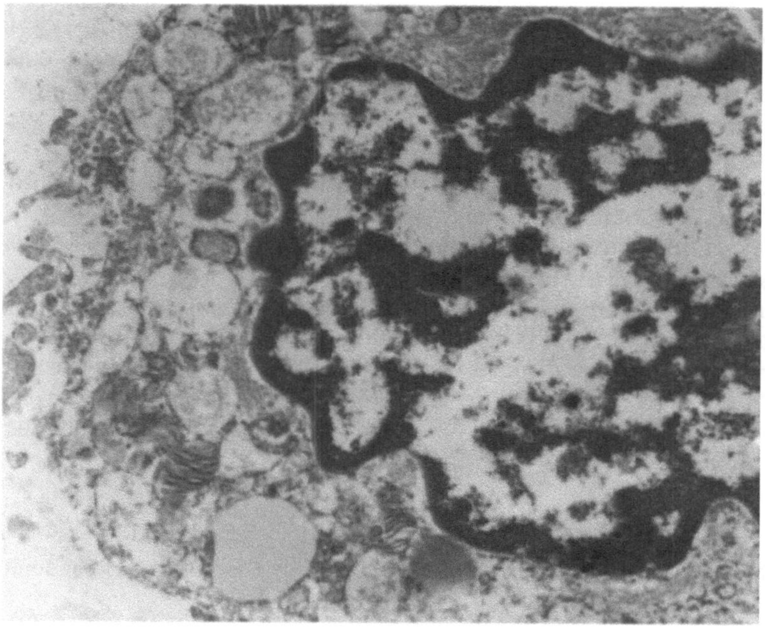

Fig. 9. Chondrocyte from the region of the second tear: the organelles are swollen and lysis of the cell membrane is evident, × 20 000

of active cells coued be seen, and thin, irregular fibres lie between normal collagen fibre bundles (Fig. 8). For the three patients with isolated meniscus ruptures, the time span between suture and follow-up arthroscopy was 5 months for two of them and 4 months for the other. In the case of the patient with compound injury the period was 3 months.

The histological examination of tissue taken from the second tear region showed a mainly degenerative pattern in patients with isolated initial meniscus rupture, with regressive changes in chondrocytes and a breakdown of the fibrous pattern (Fig. 9).

Tissues from both patients with compound injuries mainly indicated reparative changes in response to injury in the region of the second tear, displaying functional fibrocytes and capillaries in the connective tissue. In one case there was a fresh rupture caused by a trauma 4 months after suture. The scar had healed smoothly. The second patient returned to sports 2 months after the operation, although ultrastructural investigation 3 months after suture showed that repair was not yet complete.

Discussion

Evidence from light and electron microscopy studies demonstrates that meniscus ruptures in the vascularised third of the meniscus can heal after stitching. However, it does appear that regeneration advances more slowly around the scar in cases of isolated meniscus ruptures. In the case of an isolated tear, a trauma occurs in a knee which has already been damaged by some specific mechanical stress during sports or work. In all patients with isolated meniscus tears, there was greater exposure to stress. Histological results clearly showed regressive changes in the initial arthroscopy. It appears that reparative processes begin later and are therefore also completed later.

The second ruptures occurring in the nine patients with isolated meniscus ruptures appear to be due to diffuse tissue injury. However, the temporary clamp function of the meniscus suture facilitated repair of the tears by fibrocytes, thus preserving a greater portion of the meniscus.

In order for a meniscus suture to be successful, the localisation and the condition of the meniscus are important. It is also of vital importance to refrain from occupational mechanical stress and sports during the convalescent period. The optimal length of time needed, according to the results, is about 6 months following an isolated rupture and about 4 months for a compound injury without previous knee damage.

References

Gillquist J, Lysholm J (1983) Suture of new and old peripheral meniscus tears. J Bone Joint Surg [Am] 65:193

Karpf PM, Sagstetter C, Aigner R (1984) Meniscus repair. Proceedings of the First European Congress of Knee Surgery, Berlin 1984, p 78

Karpf PM, Sagstetter C, Aigner R, Rupp N (1985) Erfahrungen mit der Meniskusnaht. Unfallchirurg 88:15

Rodriguez M, Schreiber A (1984) Meniscus sutures of old tears. A report on over 100 cases. Proceedings of the First European Congress of Knee Surgery, Berlin 1984, p 77

Four Years' Experience with Arthroscopic Meniscal Suture – Where Are We?

R. P. Jakob, H. U, Stäubli, and K. Zuber

Universitätsklinik Bern, Orthopädie, Inselspital, 3010 Bern, Switzerland

Arthroscopic suture of meniscal tears has become a well-established technique in the hands of experienced arthroscopists. Our own experience over 4 years allows the following conclusions to be drawn about indications, technique, postsurgical treatment and follow-up.

The most suitable tear is an isolated fresh longitudinal rupture, 1–4 cm long, through the vascular zone or near the base of the meniscal attachment, leaving a white rim no broader than 3 mm, in a patient not over 35 years of age.

Having collected experience with all the instruments currently available, we developed our needle that allows one-handed, relatively safe, cartilage-protected insertion of nonresorbable (no. 1 or 0) sutures, perforating the upper and under surface in the middle one-third and the anterior part of the posterior horn of the meniscus and tying 7–10 sutures. The knots are tied through a skin incision over the intact posterior capsule, while protecting the neurovascular structures with an inserted finger or a special retractor. Following Henning's advice, the AO femoral distractor is used to widen the compartment and reach further posteriorly with the needle and protect the cartilage. More tears have become amenable for suture with the arthroscopic technique, mainly those underlying the collateral ligament that are hidden when using the classic open technique.

After a cast immobilisation of 5 weeks, full weight-bearing is reassumed, avoiding sports for up to 4 months postoperatively. A control air and single-contrast arthrogram after 5 months allows analysis of healing.

From October 1982 to December 1985, 45 arthroscopic sutures were performed in patients from 13 to 45 years of age. At an average follow-up of 3 years, 75% showed stable healing. Failure of healing or retears were most commonly associated with an inadequate initial technique, not roughening the meniscal surface and the synovium at the attachment site to induce pannus formation, wrong indication and concomitant ACL laxity. Although meniscal suture by open or arthroscopic technique remains a difficult intervention, the technique has become safe enough to become most gratifying for a selected group of patients.

Surgery and Arthroscopy of the Knee
2nd Congress of the European Society
Editors: Müller/Hackenbruch

Surgical Repair of Chronic Meniscal Injury: Plastic Meniscus Surgery

P. L. Melanotte and M. Baldovin

II Orthopaedic and Traumatologic Clinic, University of Padua, Italy

Reparative surgical treatment of meniscal lesions is a now well-established method of choice owing to a more complete understanding of the knee's anatomofunctional structure and advances in instrumental diagnostic and surgical techniques. The verified importance of the role of the menisci in knee physiology obliges the surgeon to respect the integrity of these structures as much as possible, since the presence of meniscal fibrocartilage is not only essential for joint function and knee stability, but also necessary for the prevention of early degenerative changes.

King (1936) studied the physiopathologic basis of operative treatment of meniscal tears in dogs, and observed that a meniscus tear never heals, a lesion of the peripheral meniscosynovial insertion tends to heal, and a meniscal tear extending peripherally into the synovium may heal. These conclusions influenced the mostly destructive tendency of meniscal surgery in succeeding years.

In 1938, Palmer showed that tears in the deep fasciae of the collateral medial ligament are associated with a concomitant meniscal lesion that must be recognized and repaired. Stone reported on the antomofunctional aspects of peripheral meniscal lesions in 1979.

Some experimental studies (Heatley 1981) have shown that acute peripheral meniscal lesions may be healed by inserting cell-rich connective fibrous tissue, which proliferates from the synovial margin, along the edge of the lesion. Suture facilitates the process of healing in that it gives a greater guarantee of stability. According to King, healing processes are advanced by the growth of blood vessels originating in the well-vascularized zone of the peripheral meniscal margin.

More recently, Hamberg et al. (1983), Bousquet et al. (1985), and Schreiber and Rodriguez (1985) demonstrated that not only meniscal lesions but also chronic peripheral tears may be sutured; however, these authors agree that such procedures do not always prevent tear recurrences. Convinced that operative treatment is also indicated in the repair of chronic meniscal lesions, in 1984 we proposed a technical modification which ensured a more lasting reinsertion. This technique consists of meniscal plastic surgery accomplished by grafting a strip of pedunculated synovial tissue which furnishes a blood supply to the appropriate meniscal area.

Surgical Technique

Using Hughston's anteromedial approach for the medial meniscus, we perform an anterior and retroligamentous arthrotomy, which is anterior and posterior

Surgery and Arthroscopy of the Knee
2nd Congress of the European Society
Editors: Müller/Hackenbruch

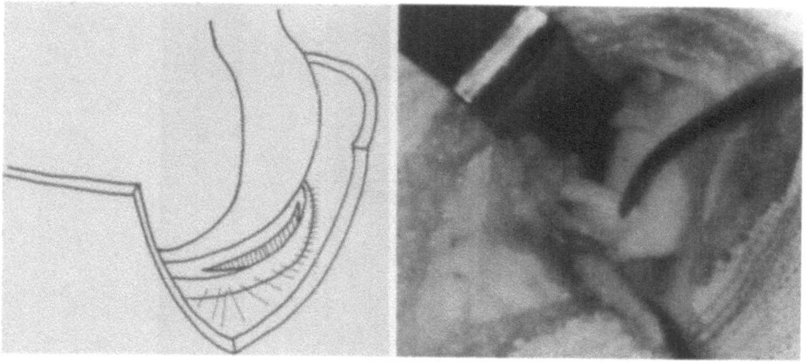

Fig. 1. Plastic meniscus surgery: arthrotomy and evaluation of the lesions

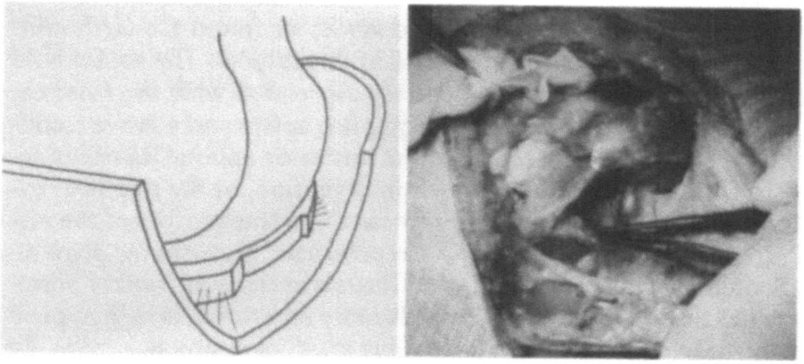

Fig. 2. Plastic meniscus surgery: removal of loose, pseudoscarred tissue

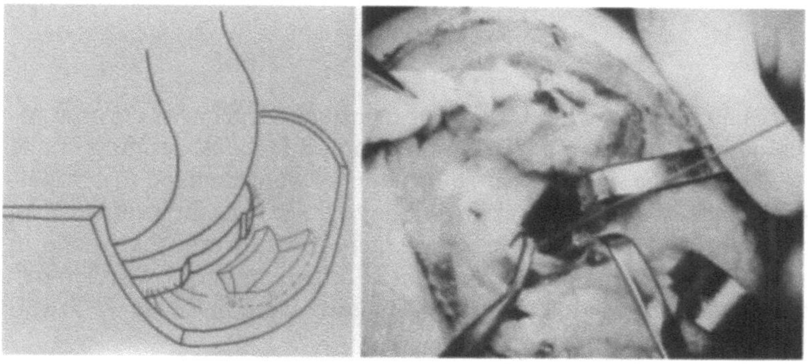

Fig. 3. Plastic meniscus surgery: preparation of the synovial flap

with regard to the posterior oblique ligament (Fig. 1). The lesions are evaluated, and the loose, pseudoscarred tissue is excised to augment revascularization (Fig. 2). A strip of pedunculated distal synovial tissue is then cut and placed within the meniscal "gap" in order to obtain an optimal blood and cell supply (Fig. 3). Lastly, an extracapsular suture of the meniscosynovial complex is made (Fig. 4) with four or five horizontal, separated stitches using Dexon 2-0 or 3-0, starting from the posterior end of the tear. To avoid causing further meniscal de-

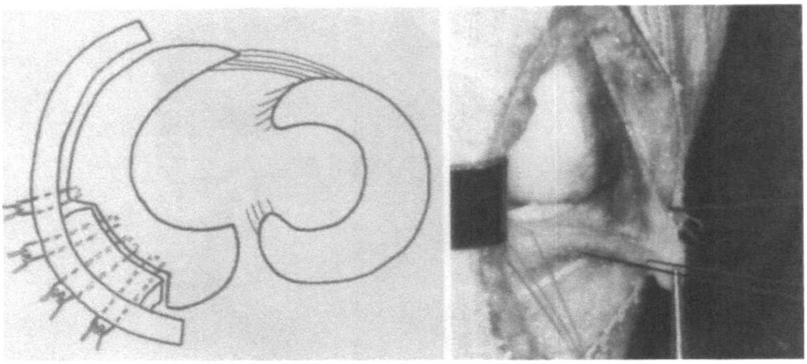

Fig. 4. Plastic meniscus surgery: extracapsular suture of the meniscosynovial complex

vitalization, as few stitches as possible should be made; we found the Orthomed instruments for meniscal sutures extremely useful in this respect. The suture hold is evaluated, and the ligamentous lesions possibly associated with the meniscal tear are treated. In the past, we preferred to accomplish active and passive reconstructive surgery by means of plastic repair of the posterior oblique ligament according to Hughston, associated with Andrew's intervention, or the popliteal bypass according to Muller, as these structures represent the true motors of the meniscal corners. Currently we prefer the Leeds-Kejo prosthetic ligament for pivot reconstruction. A Scotch-cast plaster splint in 30° flexion in cases of frankly posterior tears, or in greater flexion if the lesion extends more anteriorly, is then applied for 4 weeks; weight-bearing is allowed 10–20 days after its removal, eventually with the protection of a polycentric, derotating, regulatable device developed in our laboratory.

Case Series

The study concerns 30 patients who underwent meniscus surgery for chronic peripheral meniscal lesions, consisting of 18 males and 12 females, ranging in age from 12 to 46 years (average age, 28 years); 15 patients were engaged in competitive sports (rugby, basketball, soccer, skiing). The average time interval between the moment of injury and plastic meniscus surgery was 18 months (range, 6 months–8 years). All the lesions were documented before surgery by arthrography and/or arthroscopy; the medial and lateral menisci were involved in 20 and 10 cases, respectively.

In 18 patients the meniscal lesion was isolated, while in 12 patients capsuloligamentous lesions were associated. The types of injury consisted of damage to the posterior horn in 15 cases, posteromedial (internal meniscus) or posterolateral (external meniscus) disinsertion in 10 cases, subtotal disinsertion in 1 case, and peripheral meniscus tear in 4 cases.

Discussion

Clinical follow-up was possible in all the patients, and took place 1–3 years after surgery. In two patients it was possible to perform an arthroscopic control

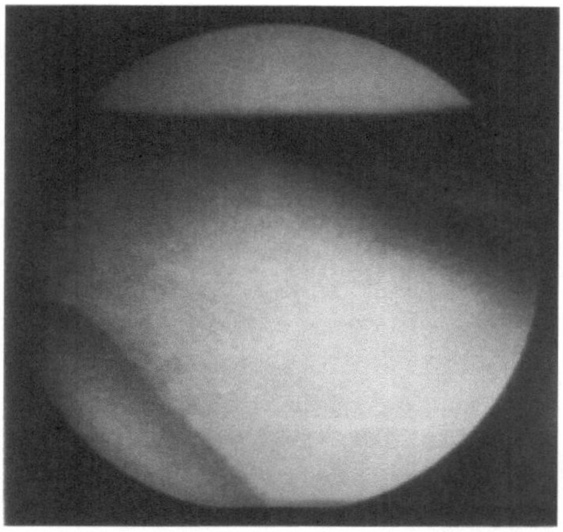

Fig. 5. Arthroscopic control following plastic meniscus surgery

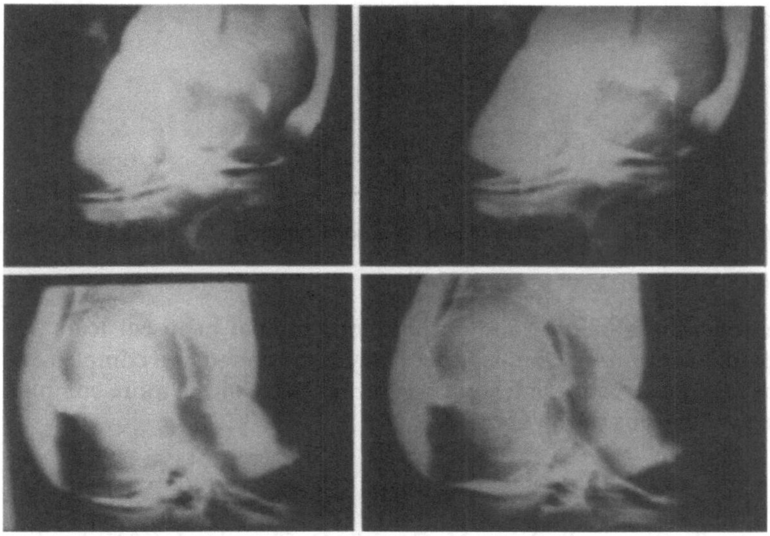

Fig. 6. Arthrographic findings following plastic meniscus surgery

(Fig. 5), in two others arthrography was carried out (Fig. 6), and in two more, CT pneumography (Fig. 7) was used. All instrumental investigations demonstrated excellent healing of the operated menisci.
In evaluating the results obtained, we followed the criteria of Lysholm and Gillquist; thus, excellent results were observed in 18 patients, good results in 10, and mediocre results in 2.

Conclusion

When a conservative surgical approach is preferred for chronic meniscal tears, we propose the synovial flap technique, which has valid anatomical and biomechanical bases.

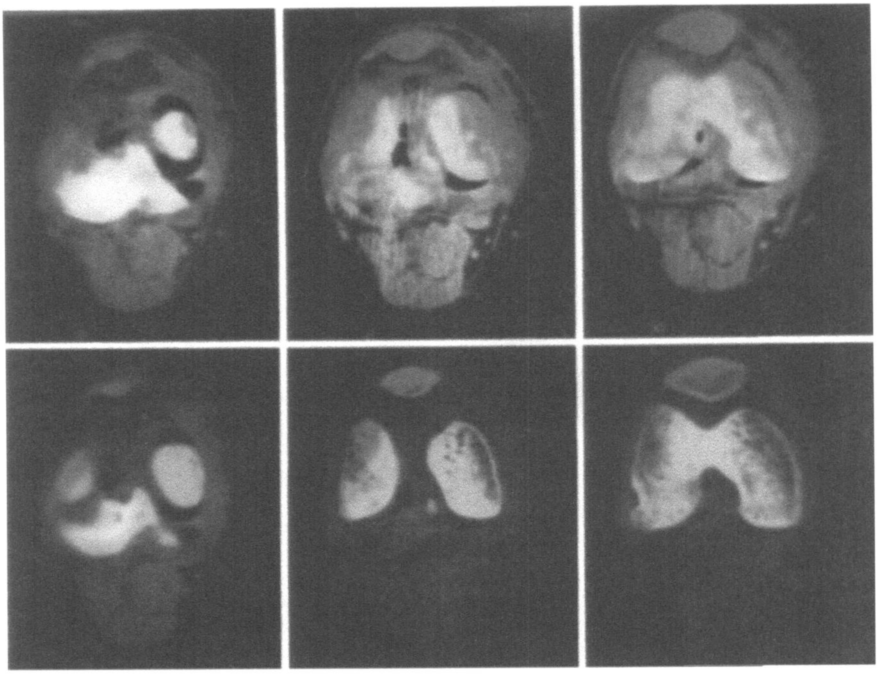

Fig. 7. CT pneumographic follow-up after plastic meniscus surgery: note the good healing of the meniscal lesion

Pheripheral lesions, disinsertions, and chronic longitudinal meniscal tears are very easily repaired with this technique, which may also be used as a complementary procedure in reconstruction interventions for sequelae of the more complex capsuloligamentous lesions of the knee.

References

Bousquet G, Passot JP, Girardin P, Relave M, Grazielly D, Charmion L (1985) La réinsertion ou suture méniscale. Rev Chir Orthop 71

Hamberg P, Gillquist J, Lysholm J (1983) Suture of new and old peripheral meniscus tears. J Bone Joint Surg [Am] 65(2):193–197

Heatley FW (1980) The meniscus: can it be repaired? An experimental investigation in rabbits. J Bone Joint Surg [Br] 62(3):397–402

King D (1936) The healing of semilunar cartilages. J Bone Joint Surg 18:333–342

Melanotte PL, Baldovin M (1984) Il trattamento chirurgico riparativo delle lesioni meniscali inveterate. Atti sertot 26(1):57–60

Schreiber A, Rodriguez M (1985) Traitement conservateur des lésions méniscales. Résultates de plus de 100 cas de sutures avec un recul jusqu' a 12 ans. Rev Chir Orthop 71

Stone RG (1979) Peripheral detachment of the menisci of the knee: a preliminary report. Orthop Clin North Am 10:643–657

Wirth CR (1981) Meniscus repair. Clin Orthop 157:153–160

Arthroscopic Management of Degenerative Meniscal Tears in Patients Over 60

G. Stroz

Clinique du Marais, 11 bis rue Barbette, 75003 Paris, France

The role of meniscal tears in painful knees of aged patients is still controversial. Recent publications [2–6] have outlined the interest in arthroscopic management of this kind of lesion. Arthroscopy allows a more precise evaluation of meniscal and chondral pathology and provides greater accuracy in the diagnosis of tears which, in many cases, were previously undiscovered or considered insignificant. Closed meniscectomy saves at least partial mechanical function of the meniscus and therefore seems less nocent in the long term than total open meniscectomy.
For this study we reviewed 43 knees in 39 patients: 23 females (average age of 73 years) and 16 males (average age of 66 years). Classification of the lesions was made according to the arthroscopic findings and recorded as follows:

- Four longitudinal traumatic tears of the medial meniscus
- One peripheral tear of the posterior horn of both menisci with concomitant tear of the anterior cruciate ligament
- Nine horizontal tears of the posterior horn of the medial meniscus
- Four horizontal tears of the lateral meniscus
- One tear of a discoid lateral meniscus
- Twelve tears of both menisci, including three bilateral cases
- Eight chondrocalcinosis arthropathies including one bilateral case, with three tears of the medial meniscus and six of both menisci

Etiology

A knee sprain or a brutal twist was reported by 9 patients, whereas 12 patients had only a history of casual traumatism, and in 18 other patients no evident etiology was found.

Clinical Data

Pain, exacerbated by walking, descending stairs or squatting, and persisting during rest in bed, is a constant factor of the clinical picture. Twelve patients complained of locking or giving way. In 18 cases intermittent swelling occurred, and was reported at examination. In 11 cases loss of extension and difficulty in completing knee flexion were noted.
Of the group in question, only three patients had a varus knee because we regarded significant malalignment as a contraindication to arthroscopic management.

Surgery and Arthroscopy of the Knee
2nd Congress of the European Society
Editors: Müller/Hackenbruch

Chondral Pathology

Seven knees were free of chondral damage, but five of these belonged to the younger patients (near 60 years old) with a history of trauma and a longitudinal tear of the medial meniscus.

A total of 32 patients had significant chondral pathology:

- Eight with femorotibial unicompartmental damage (four medial and four lateral)
- Two with patellar osteoarthritic changes
- Ten with damage to both femorotibial compartments
- Twelve with changes in the medial compartment and femoropatellar joint

More precisely, medial compartment changes were recorded in 16 knees, associated with 11 horizontal tears of the medial meniscus and 5 tears of both menisci. Ten patients had changes in two or three compartments of the knee, associated with nine bimeniscal tears and one tear of a lateral discoid meniscus. Four patients had osteoarthritic changes in the lateral compartment associated with four tears of the lateral meniscus, and two patients presented solitary damage of the patellar joint combined with two bimeniscal tears.

In summary, the association of horizontal tears of the medial meniscus with chondral damage in the medial compartment was a constant finding. Exactly the same observation applies analogously to damage in the lateral compartment and to bicompartmental lesions.

Arthroscopic Management

Arthroscopic examination was performed under general anesthesia, although other means of anesthesia are available (local, lumbar); an anterolateral approach was used routinely, and various closed procedures were executed under arthroscopic control:

- Partial meniscal resection, excising all the pathological tissue up to the vascularized portion of the meniscus, to encourage healing and reduce the risk of iterative tear
- Excision of chondral flaps
- Curettage or abrasion down to bone of necrotic cartilage
- Abrasion arthroplasty using the shaver system
- Osteophyte resection
- Extraction of loose bodies
- Lateral retinacular release

Postoperative care was eclectic, depending on the procedure which had been performed. In the case of extensive chondral procedure, weight-bearing was delayed for 4–6 weeks and walking was only allowed with crutches.

Results

Thirty patients were reviewed, with an average follow-up period of 2 years: 11 returned for examination and 19 answered a questionnaire. Results were classified subjectively by the patients as excellent (no more pain, complete return to normal

activity), good (intermittent pain and normal activity), fair (less pain than prior to the operation and better activity), and poor (no benefit after operation).
The five patients with longitudinal tears and no cartilage changes rated the results as excellent. Eighteen painful knees underwent 18 meniscal resections associated with nine chondral procedures. The results were graded excellent in eight cases, good in six, fair in three, and poor in one (concerning a severe femoropatellar osteoarthritis after prior open total meniscectomy).
More precisely, in nine patients who had a meniscal resection in association with a chondral procedure, the results were rated as five excellent, two good, and two fair. In the other nine patients, without accompanying chondral procedure, three excellent, four good, and two fair results were achieved. However, it must be pointed out that only patients with severe chondral pathology underwent a chondral procedure in addition to the meniscal resection.
In the three varus knees, a partial meniscal resection of the posterior horn led to two excellent results at follow-up after 3 years. In the third, originally rated as good, failure occurred after 18 months, and an upper tibial osteotomy was necessary.
Nine chondrocalcinosis arthropathies were debrided under arthroscopy with meniscal resection and lavage. The results were excellent in five knees and good in four knees, with a follow-up period ranging from 1 to 5 years.

Discussion

The results of arthroscopic debridement of degenerative tears of the meniscus in older patients were rated as excellent in 46%, good in 40%, and fair in only 6% of cases.
No significant difference was noted between the group of patients receiving a meniscal resection only and that undergoing an associated chondral procedure.
The overall 86% satisfying results is similar to the various percentages recently published. Salisbury et al. [4], with a 2-year follow-up, obtained 92% good results in well-aligned knees, Sprague [5], with a 1-year follow-up, reports 74% good results, MacBride et al. [3] give 65%–96%, depending on the presence or absence of osteoarthritic changes, and Jackson and Rouse [2], with a follow-up of 2½ years, announce 80% good results.
In our study, lesions of both menisci and horizontal tears of the posterior horn of the medial meniscus are predominant. A recent cadaveric study has pointed out the frequency of meniscal tears in older people but, unlike Fahmy and Noble [1], we did not find better preservation of the cartilage under torn menisci. On the contrary, our arthroscopic findings show major cartilage defects under meniscal tears and calcium pyrophosphate deposit disease of the menisci.
We think, with many others, that degenerative meniscal tears must be arthroscopically resected because they are a major cause of pain in osteoarthritic knees and because durable relief is easily obtained in the majority of cases.
In the same way, arthroscopic debridement brought complete relief of pain in the chondrocalcinosis arthropathies in our study and, like Zarins and MacInverney [7], we highly recommed this means of treatment when pain proves resistant to medical care.

Indications

The major indication for arthroscopic debridement is a painful symptomatology associated with swelling and meniscal tenderness at examination.
Anteroposterior X-rays may be negative or show various signs of osteoarthritis. Arthrography is of minor interest because of its lack of reliability in the case of horizontal tears of the menisci. On the other hand, radiographic study of leg alignment is of great interest in the determination of malalignment, which for us is a contraindication to arthroscopic treatment because of the high rate of immediate failure, although we have recently used this method on some varus knees with less than 5° malalignment with good results even after 3 years of follow-up. In the case of pseudogout, in consensus with Zarins and MacInverney [7], we highly recommend arthroscopic debridement, and at long-term follow-up all our patients are satisfied with this treatment.

Conclusion

It seems to us that meniscal pathology is an important factor in osteoarthritic knee pain and that arthroscopic debridement is a valuable alternative to medical care, with lower morbidity than other surgical procedures, and is able to provide a high rate of success.

References

1. Fahmy NRM, Noble EAN (1983) Meniscal pathology and osteoarthritis of the knee. J Bone Joint Surg [Br] 65:24–28
2. Jackson RW, Rouse DW (1982) The results of partial arthroscopic meniscectomy in patients over 40 years of age. J Bone Joint Surg [Br] 64:481–485
3. MacBride GG, Constine RM, Hoffman AA, Carson RW (1984) Arthroscopic partial medial meniscectomy in the older patients. J Bone Joint Surg [Am] 66:547–551
4. Salisbury KB, Nottage WM, Gardner V (1985) The effect of alignement on results in arthroscopic debridement of the degenerative knee. Clin Orthop 198:269–272
5. Sprague NF (1981) Arthroscopic debridement for degenerative knee joint disease. Clin Orthop 160:118–123
6. Stroz G, Chaouat Y (1984) L'arthroscopie du genou non inflammatoire. A propos de 500 observations. Rev Rhumatol 51(1):15–21
7. Zarins B, Mac Inverney VK (1985) Calcium pyrophosphate and pseudogout. Arthroscopy 1(1):8–16

Results of Posterior Horn Resection of the Medial Meniscus

P. Hertel

Department of Traumatic Surgery, Rudolf-Virchow Hospital, Berlin, FRG

The posterior horn of the medial meniscus is the most common and the surgically most difficult region for meniscal tears [1, 6]. Complex tears consisting of a combination between horizontal tears, flap tears and radial tears occur more often than bucket-handle tears, longitudinal tears and clearly defined radial tears. The use of a leg holder and angulated instruments have made the procedure safer in recent years. In the present series no open arthrotomy had to be performed, although in the early days of arthroscopic meniscus surgery an open procedure was commonly used to complete the operation in this area of meniscal pathology [3].

Material and Methods

Of 86 patients with complex posterior horn lesions 50 were reviewed (from among over 500 patients operated on arthroscopically for meniscal lesions) in the years 1981–1986. All patients were treated by the author. This follow-up was conducted between 21 and 49 months (average 36.3) after the operation and was carried out by questionnaire in 29 cases and personally in 21 cases. Evaluation was done in terms of the scoring lists of Lysholm and Gillquist [2] and Tegner and Lysholm [8]. There were 36 males and 14 females; average age at operation was 44.4 years. Excluded were patients with bucket-handle tears, longitudinal tears, simple radial tears without degeneration of adjacent meniscal tissue and patients with severe arthrosis or pseudogout. In 32 cases the central approach via the patella tendon was used for the arthroscope. The operation was done from a medial and a lateral portal, combined with a posteromedial, longitudinal, meniscal incision by a small meniscus knife. In 18 patients the anterolateral approach for the arthroscope was used, performing the operation from the anteromedial portal. This latter approach became increasingly preferred.

The technique, nevertheless, is rather uniform. Knives are normally used for the posteromedial incision and are seldom used from the anterior. Angulated scissors are normally introduced anteromedially, performing an oblique incision anterior to the lesion near the base of the meniscus. The posterior horn of the meniscus is then dissected transversely by a long curved basket. The incision is directed medially. The anterior incision is extended beneath the capsule by scissors or a basket punch and afterwards again by angulated scissors Normally the posterior horn cannot be dissolved by a homogenous incision due to the very narrow posterior joint space. The dissected part of the meniscus is grasped from the medial portal by a narrow pituitary rongeur and extracted by disrupting the last intact

Surgery and Arthroscopy of the Knee
2nd Congress of the European Society
Editors: Müller/Hackenbruch

Table 1. Complications in posterior horn resection of medial meniscus ($n=50$)

Complication	Number of cases
Cartilage damage by surgery	3–6
Saphenous nerve lesion	2
Dislocation posterior horn	1
Arthrotomy necessary	0
Infection	0
Thrombosis	0
Aspiration after dismissal	3
Arthrotomy in other hospital	2

posterior meniscal fibers. The meniscus tissue is generally fragmented, such that it must be removed piece by piece.

Severe intra-operative complications did not occur. One meniscus dislocated posteriorly, turning around posteriorly intact fibers. In two cases of posteromedial incisions the saphenous nerve was injured (causing patients to suffer from posteromedial neuroma and loss of sensitivity on the medial calf). Cartilage damage was seldom observed but may not always have been correctly recorded (Table 1).

Results

No effusion had to be aspirated during the 3 days of postoperative hospitalisation but was carried out in three cases thereafter by the general practitioner. Two patients were treated by open arthrotomy at another hospital, and their condition was classified as poor, irrespective of specific findings and procedures performed in open surgery (both had had cartilage lesions). Normally the patients were treated by their doctors at home, who were unfamiliar with arthroscopic meniscal surgery and had referred them to hospital. Sick leave averaged approximately 5 weeks (range 0.5 to 20 weeks); sports were resumed after about 12 weeks (2–26 weeks).

Table 2. Distribution of patients by functional and activity level scores

Functional score (after Lysholm)		Activity level score (after Tegner)		
		7–9	4–6	1–3
100	($n=27$)	12	9	6
90–100	($n=15$)	2	9	4
80– 90	($n=2$)		1	1
Below 80	($n=6$)	1	1	4
	$n=50$	15	20	15

Table 3. Postoperative resumption of sports ($n=50$)

Sports activity	Number of cases
Sports as before	35
Reduced sports	4
Gave up sports	
Because of meniscus	2
Independently of meniscus	6
No sports at all	3

Using the rating system of Tapper and Hoover [7], 28 patients were classified as excellent, 14 good, 5 fair and 3 poor. In terms of the rating system of Lysholm and Gillquist [2] with a maximum of 100 points the average rating was 94.2. The 3-year postoperative activity level according to the Tegner and Lysholm scale [8] was high (7–9) in 15, medium (4–6) in 20 and low (1–3) in 15 patients (Table 2). Highest scores were obtained by the patients with a high activity level. Six patients had reduced their activity level postoperatively because of the meniscus lesion and another six did so independently of the meniscus lesion (Table 3).

Discussion

In most cases posterior horn resection of the medial meniscus encompasses the entire posterior horn due to the size of degeneration and to the demands of incision. This may be considered a partial resection, but regarding only the posterior horn it is a subtotal or at times total resection. Concerning the overall meniscus, on the other hand, it is a partial resection. For safety in meniscus surgery it is mandatory to leave a stable rim at the borderline or within the vascular zone of the meniscus. It is not advisable to remove only the flap in a complex tear, for subsequent symptoms due to resulting fragmentation will arise. It has been shown that most revision arthroscopic surgery results from insufficient meniscal removal [3, 6].

Posterior horn resection of the medial meniscus does not interfere with ligament stability – contrary to what one might expect, given the narrow connection of the meniscus to the posterior oblique ligament [5]. None of the patients reported instability, and neither valgus nor rotational instability was observed. The most important cause of maintained stability after a posterior horn resection may be an intact coronary ligament, which connects the posterior insertion of the meniscus to its intact median and anterior part. It seems possible locally to remove all degenerative meniscus tissue up to the coronary ligament without hindering the aims of partial meniscectomy. Two open re-operations were carried out in this series, but these were due to other problems.

Little difference manifests itself between these results and those of others. Northmore-Ball and Dandy [4] obtained 90.5% excellent or good results in 69 patients after 2.1 years, but this included 41 bucket handles (Table 4). Lysholm and Gillquist [3] showed a score of approximately 86 (maximum 95) in 33 cases after 13 months (with all types of meniscal lesion excluding buckets).

Table 4. Results of arthroscopic meniscus resection studies

Results	Present study: posterior horn resection medial (3 years; no buckets; $n=50$)	Northmore-Ball and Dandy (1981): partial meniscectomy no cartilage lesion (2.1 years; 41 bucket-handle tears; $n=69$)
Excellent	51.5%	56%
Good	39%	28%
Fair	8%	10%
Poor	1.5%	6%

Table 5. Condition of patients with and without posterior incision ($n=50$)

Condition	With posterior incision	Without posterior incision
Excellent, good	26	16
Fair, poor	6	2

When we compare the results from our patients having an additional posteromedial incision (32 patients) with those operated only from the anterior portal (18 patients), we find only slightly better conditions among the latter group (Table 5).

Finally, this study has shown that one of the most difficult lesions for arthroscopic meniscus surgery can be treated safely without opening the knee joint in the conventional way. An additional posteromedial incision seems unnecessary and furthermore dangerous for the saphenous nerve. The results of posterior horn resections are comparable to those for bucket-handles and other types of meniscal lesions which permit easier operation. These are mid-term results, however, and must be evaluated in 5–10 years in order to ascertain long-term effects.

References

1. Löhnert J, Raunest J (1986) Arthroskopische Meniscusresektion und offene Meniscektomie – eine vergleichende Studie. Chirurg 57:723
2. Lysholm J, Gillquist J (1982) Evaluation of knee ligament surgery results with special emphasis on use of a scoring scale. Am J Sports Med 10:150
3. Lysholm J, Gillquist J (1981) Endoscopic meniscectomy – a follow up study in arthroscopy in surgery of the knee. Medical Dissertations, Linköping University, 106(4):3–19
4. Northmore-Ball MD, Dandy DJ (1982) Long term results of arthroscopic meniscectomy. Clin Orthop 167:34
5. Oretorp N, Alm A, Ekström H, Gillquist J (1978) Immediate effects of meniscectomy on the knee joint: The effects of tensile load on knee joint ligaments in dogs. Acta Orthop Scand 49:407
6. Stoffel D, Glinz W, Ricklin T, Frei E (1986) Re-Arthroskopie nach arthroskopischer Meniskusresektion. Fortschr Arthroskop 2:86–89
7. Tapper EM, Hoover NW (1969) Late results after meniscectomy. J Bone Joint Surg [Am] 51:517
8. Tegner Y, Lysholm J (1985) Rating systems in the evaluation of knee ligament injuries. Clin Orthop 198:43

CO_2 Laser Meniscectomy Under Arthroscopy

K. Inoue, K. Mikanagi, H. Morita, and T. Namiki

Department of Joint Surgery, Institute of Rheumatology,
Tokyo Women's Medical College, Tokyo, Japan

The laser is a new instrument in the field of arthroscopic surgery. There are three kinds, CO_2 laser, YAG laser and argon laser, which are applied in the medical field. In our previous study, the CO_2 laser was shown to be effective for arthroscopic surgery, that is incinerating the meniscus, etc. It is not easy, however, to employ the CO_2 laser in these circumstances. Since 1978, we have been developing and improving the CO_2 laser fibre, the arthroscope and the automatic control instrument for intra-articular pressure, with the aid of Machida Co. Ltd. and Matsushita Co. Ltd. A CO_2 laser fibre has been developed for clinical use.

Material and Methods

The knee joint is distended by CO_2 gas and the intra-articular pressure is maintained at 100 cmH_2O by the automatic control system. First, the operation is mapped out by ordinary arthroscopic observation. Second, the CO_2 laser fibre is introduced under arthroscopy, followed by CO_2 laser surgery, incineration of the meniscus and synovia, etc. During laser surgery, 3 l/min CO_2 gas is jetted out around the laser fibre to protect its lens and irrigate the joint cavity to remove the smoke, carbonate, nitrous oxide, etc. which derive from the surgery. Afterwards, the joint cavity is irrigated by 1500 ml saline solution. Laser surgery was performed in 10 cases, meniscectomy in 7 cases and synovectomy in 3 cases.

Results

Arthroscopic laser surgery was done without any serious side-effects, such as hitting the articular cartilage of the femoral condule or the tibial articular cartilage under the meniscus. Subjective, objective and X-ray findings are satisfactory in all cases in the short-term results.

Discussion

Arthroscopic surgery has not been accepted by the conservative knee surgeon, although it forms an important and necessary part. Lasers are much more expensive than ordinary arthroscopic surgical instruments. A laser is an artificial light beam and has expected and unexpected advantages. Arthroscopic laser surgery is effective and safe.

Surgery and Arthroscopy of the Knee
2nd Congress of the European Society
Editors: Müller/Hackenbruch

Results of Arthroscopic Knee Operations on Out-Patients Under Local Anaesthesia

H. Sieczewicz and H. Henkemeyer

Department of Trauma, Municipal and Teaching Hospital of the University of Freiburg, Villingen-Schwenningen, FRG

This paper deals with arthroscopic operations on out-patients at the trauma unit of the municipal hospital of Villingen-Schwenningen, affiliated with the University of Freiburg as a teaching hospital. Today the main indication for arthroscopic surgery is a meniscus lesion. Practically all meniscus lesions are manageable endoscopically. A multitude of different techniques are advocated; however, there is one common principle, i.e. minimal resection of mobile segments, preserving a stable basic meniscus rim with minimal loss of function.

Arthroscopic examination is performed on out-patients under local anaesthesia. No tourniquets are applied, the patient is in a supine position and the knee is flexed between 20° and 30° to permit free flexion and extension. Then 5 ml Xylonest (1%) is injected ventrolaterally and ventromedially of the patellar tendon. Skin and tissue are infiltrated. Additionally, 10 ml Xylonest is injected into the joint underneath the lateral patella. After 5 minutes the arthroscope can be introduced; we generally use the ventrolateral approach for the arthroscope and the ventromedial approach for the instruments. We hardly ever use two ventromedial approaches for the instruments. A strict adherence to aseptic conditions is mandatory; disinfection and covering of the operation site are performed as in all arthrotomies. The average duration of arthroscopic meniscectomies is 30 min, the longest having been 80 and the shortest 15 min.

Between 1984 and 1985, 540 out-patients were examined arthroscopically, and 246 of these underwent surgery. In the same period 609 in-patients were admitted for arthroscopic examination. The latter group consisted of patients presenting complex ligament instability, haemarthrosis or gonarthrosis, or those who refuse local anaesthesia. In eight cases loose bodies were removed, in 18 cartilage was shaved, in two plica synovialis resections were performed, and in 218 total or partial meniscectomy was necessary. In the group of 246 out-patients two-thirds were male and one-third female. The average age was 32.6 years; the youngest patient was 17 and the oldest 72 years old. Sports injuries were predominant, with football being responsible for the majority. Two-thirds of the lesions were located medially and one-third laterally.

In the follow-up we found very few complaints about intra-operative pain. Postoperatively most patients felt only a little tenderness, usually occurring on the first postoperative day, and the majority were back to pain-free weight-bearing after 2 weeks. Postoperative effusions occurred in one-third of the patients. The patients were back to work within an average of 9.1 days; there was a significant reduction in rehabilitation time in the out-patient group. Serious complications did not occur intraoperatively: damage to the cartilage is almost always

Surgery and Arthroscopy of the Knee
2nd Congress of the European Society
Editors: Müller/Hackenbruch

avoidable by careful choice of approaches and instruments. There was primary healing of the split incisions, and no infection occurred. A follow-up of 96 patients after 12–18 months showed good to excellent results: all patients were free of pain and presented no loss of function.
The obvious advantages of arthroscopic operations on out-patients under local anaesthesia are minimal traumatization of the joints, considerable saving of costs, and reduction in the postoperative rehabilitation period. Arthroscopic operative procedures avoid the typical complications of classic arthrotomies. The follow-up is in accordance with the good results in the literature. In spite of the fact that long-term results are not yet available, good to excellent early results are common.
Arthroscopic procedures under local anaesthesia have definitely improved the standard of meniscus surgery: in addition to the improvement in diagnosis, necessary measures may be taken immediately, avoiding admittance to hospital and the performance of arthrotomies. The postoperative tenderness is almost negligible, so that mobility is restored with minimum delay. In conclusion it can be expected that arthroscopic meniscectomies on out-patients under local anaesthesia will become the method of choice.

Pathological Examination of Meniscus Fragments: The Effect of Arthroscopic Excision

A. Fisseler[1], J. Witt[1], J. Krämer[2], and K.-M. Müller[1]

[1] Institute of Pathology of the Bergmannsheil Krankenhaus, Bochum University Clinic, FRG
[2] Orthopedics University Clinic, St. Joseph Hospital, Bochum, FRG

In the treatment of meniscopathy, arthroscopic methods of meniscus excision have been gaining greater importance in recent years. This method, a generally comfortable and pleasant operation for the patient, results in an altered spectrum of analysis for the pathologist. Instead of relatively large fragments or totally excised menisci, only small, fragmented specimens, taken out of their anatomical context, are available for examination (Fig. 1).

When making a morphological assessment on changes in menisci, the pathologist is concerned with three questions:

1. Are the findings normal on the whole, including physiological degenerative alterations consistent with the patient's age?
2. Are there signs of excessive nonphysiological degeneration in comparison with the relevant age group?

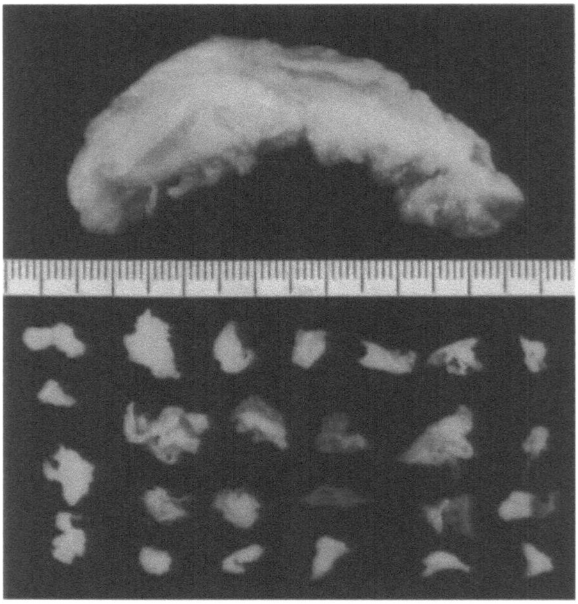

Fig. 1. Meniscus excised totally by open knee surgery and fragments of one meniscus excised arthroscopically

Surgery and Arthroscopy of the Knee
2nd Congress of the European Society
Editors: Müller/Hackenbruch

3. Are there findings present which are to be seen as the result of trauma, and are there clues which might help to determine the age of traumatic meniscus lesions?

We examined the extent to which these questions can be answered in the case of meniscus fragments excised by arthroscopy, on the basis of the material sent in for examination during a period of 6 months. The study also covered occupational disease, for example, in coal workers in Germany.

Material and Methods

Some 108 meniscus samples obtained by arthroscopy in 100 patients between 14 and 69 years of age were examined.
The findings obtained during operation were systematically documented, and the excised meniscus fragments were measured and catalogued by means of photography. The histological preparation was done by the paraffin method; sections were routinely stained with three histological stains (hematoxylin-eosin, elastica van Gieson, ferrum).
First, the histological examination is carried out without knowledge of the relevant anamnestic information. Special attention was paid to the anamnestic data, which were always obtained by the same person, dealing with the kind and course of recollectable traumatogenic events and any possible occupational strain.
In the second phase, the morphological findings of the type and age of probably traumatic meniscus lesions are added to the independently obtained anamnestic information.

Results

Our group comprised 68% men and 32% women; in 64% of the cases the patients were between 20 and 50 years of age (Fig. 2). Information about frequent participation in sports was given in 30% of cases, and 12% of the patients reported occupational strain. Regarding topographical distribution, the medial meniscus (75% of cases) was involved more often than the lateral meniscus, and the right knee (53% of cases) was involved more often than the left one (Fig. 3). The samples of menisci analyzed comprised only one fragment in 34%, up to five frag-

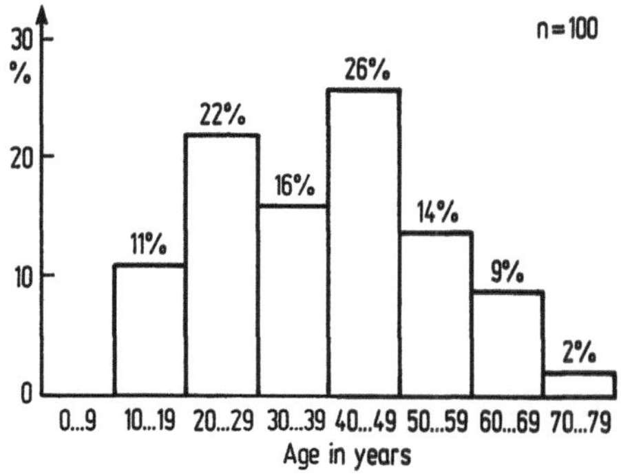

Fig. 2. Age distribution of patients arthroscopically operated upon

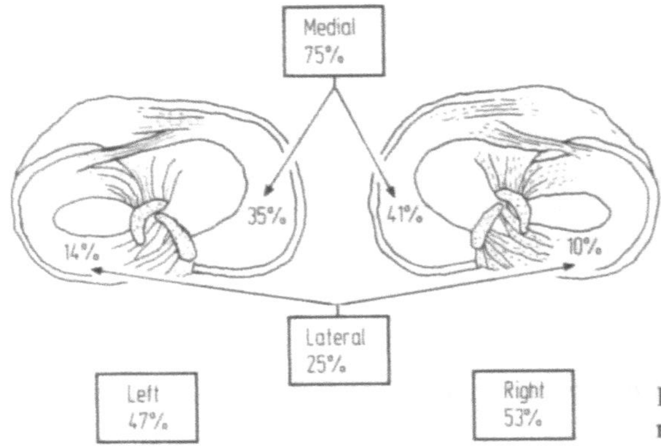

Fig. 3. Topographical distribution of menisci arthroscopically excised

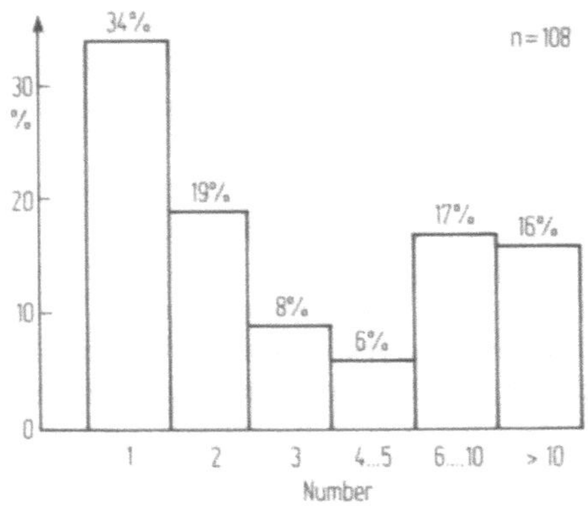

Fig. 4. Frequency distribution of the number of fragments excised per operation

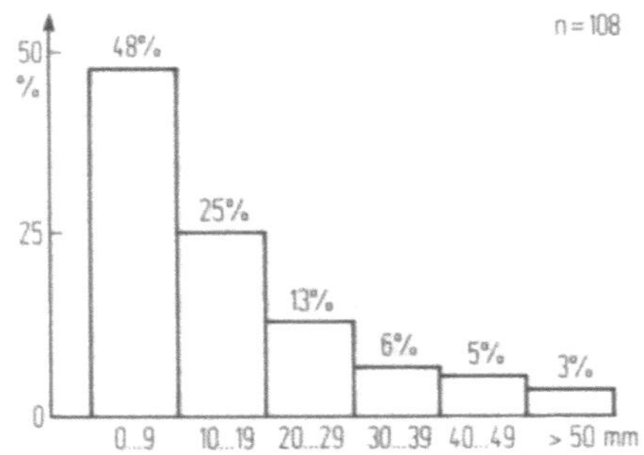

Fig. 5. Frequency distribution of fragment size (in millimeters) of arthroscopically excised meniscus fragments

ments in 33% and more than six fragments in 33% of the cases sent for examination (Fig. 4). In nearly 50% of the cases, the fragments had a size of less than 1 cm (Fig. 5). More accurate assessment was possible in the case of completely excised fragments than when shaving was involved.

In cases such as those in our study, the pathologist is confronted primarily with the question of degenerative changes and secondarily with the question of traumatic alterations in the meniscus tissue.

Degenerative alterations in menisci can be observed from the second decade of life onwards. Under normal physiological conditions, changes typical of aging and

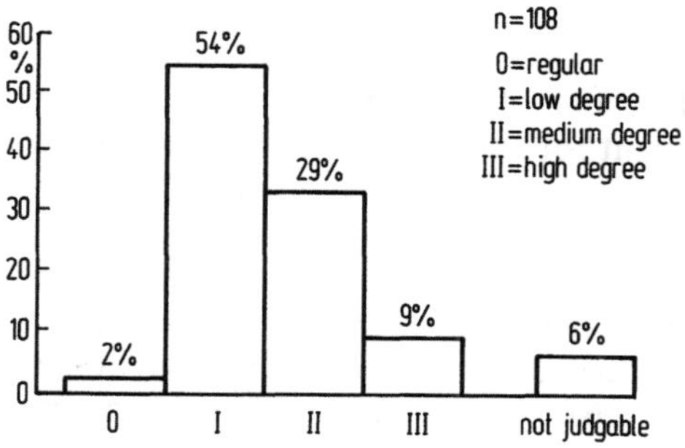

Fig. 6. Distribution of relative frequency of the degree of degenerative alteration in arthroscopically excised specimens

wear can be found; these are classified as lower grade or medium grade alterations, according to their extent. Opinion on degenerative alterations has hitherto been based on the examination of meniscus fragments which are as large as possible, because this enables comparative assessment. The meniscus fragments examined by us showed lower grade degenerative alterations in 54% of the cases, medium grade in 29% and in 9% focal high grade (Fig. 6). Degenerative changes can also be observed morphologically in arthroscopically excised meniscus fragments, and can be classified according to different stages of degeneration. In these instances, however, it is not possible to make a reliable statement about degenerative alterations in the whole meniscus; thus the value of the pathological expert's opinion is strictly limited as far as questions relating to insurance are concerned.

The consequences of a *traumatogenic* event essentially depend on the degree of pre-existing degeneration. According to studies (Könn et al. 1985; Müller and Fisseler 1985) which clarified the precise connections between adequate trauma and the lapse of time before excision of the meniscus, there is a clear correlation between the morphological findings and the temporal intervals, which can be differentiated as follows: an initial, acute phase lasts 2 weeks; the second stage of repair continues until the end of the 2nd month. Accurate predictions are still possible provided excision of the meniscus takes place within about 6 months after the trauma occurs; thereafter it is no longer possible to make a reliable assessment. Thus, clinical anamnesis is of decisive importance in the correlation of trauma to histological findings. In our study, 64% of the patients were able to recollect a definite traumatogenic event.

When the primary morphological findings concerning the type and date of a probably traumatic rupture were compared with the independently obtained anamnestic data about a trauma, a good or very good correlation resulted in 50% of cases. It must also be mentioned that 54% of the samples were not suitable for reliable correlation; the reasons for this are the removal of the meniscus fragments from their anatomical context and artifacts due to pinching and diathermy. When the interval between trauma and excision is limited to 5 months, good correlation can be found in 73% of cases.

The presented results could not have been obtained without intensive cooperation with the surgeons and orthopedists; the anamnestic data and the findings obtained during surgery were of outstanding importance.

Discussion

According to our results the question of the extent of degenerative alterations and the presence of an adequate trauma, relevant for the morphological assessment and report for insurance purposes, can only be answered incompletely by examination of arthroscopically obtained meniscus fragments. Statements about the presence of degenerative alterations are possible, even in the case of arthroscopically excised fragments less than 1 cm in size, and the graduation of degenerative alterations in our group corresponds with the results of earlier studies (Tobler 1933; Ceelen 1941; Husten 1952; Könn 1967; Aufdermaur 1971; Könn and Rüther 1976; Oellig and Rüther 1981; Könn et al. 1985; Willet-Bleich 1985). However, distinctions between primary, pre-existing degenerative alterations and secondary degenerative changes caused by trauma are not possible, because the meniscus fragments to be examined cannot be regarded as representing the whole meniscus. Thus, the expert's opinion on degenerative alterations in an excised arthroscopically meniscus has to be restricted to mere descriptive findings; as such it is considerably less useful as far as insurance questions are concerned (Müller and Fisseler 1985; Fisseler et al. 1986).

A reasonably accurate correlation between the anamnestic data and the histological findings is possible when excision takes place up to 2 months after the traumatogenic event; if no more than 5 months have elapsed, the morphological examination can result in decisive findings in three out of four cases. A reliable statement is usually not possible if the lapse exceeds 6 months. It must be mentioned, too, that 13% of the samples were in shreds or did not contain any meniscus tissue. Thus, insurance questions concerning the age of traumatic meniscus lesions can only be answered to a limited extent by examination of the tiny fragments.

Close cooperation between the pathologist, the surgeon and the orthopedics department is essential: in the absence of sufficient anamnestic and clinical data, the pathological assessment of arthroscopically obtained meniscus fragments is less likely to result in reliable findings for insurance purposes.

References

Aufdermaur M (1971) Die Bedeutung der histologischen Untersuchung des Kniegelenkmeniskus. Schweiz Med Wochenschr 101:1441–1445

Ceelen W (1941) Über histologische Meniskusbefunde nach Unfallverletzungen. Zentralbl Chir 68:1491–1509

Fisseler A, Witt J, Krämer J, Müller KM (1986) Morphologie arthroskopisch gewonnener Meniskusresektate. Versicherungsmedizinische Aspekte. Pathologe 7: (in press)

Husten K (1952) Die Auswertung des geweblichen Befundes beim Meniskusschaden des Knies. Verh Dtsch Ges Pathol 35:208–213

Könn G (1967) Zur Morphologie der Meniskusschäden. Schriftenreihe unfallmedizinischer Tagungen der Landesverbände der gewerblichen Berufsgenossenschaften 2:69–73

Könn G, Rüther M (1976) Zur pathologischen Anatomie und Beurteilung der Meniskusschäden. Hefte Unfallheilkd 128:7–13

Könn G, Oellig WP, Willet-Bleich M (1985) Möglichkeiten und Grenzen der histologischen Altersbestimmung von Zusammenhangstrennungen des Kniegelenkmeniskus. Unfallchirurgie 88:1–14

Müller KM, Fisseler A (1985) Beurteilung des Meniskusschadens aus der Sicht des Pathologen. Bericht der unfallmedizinischen Tagungen der Landesverbände der gewerblichen Berufsgenossenschaften 56:185–194
Oellig WP, Rüther M (1981) Zur Morphologie und Beurteilung der Rißbeschädigung am Kniegelenksmeniskus. Unfallheilkunde 84:295–301
Tobler T (1933) Zur normalen und pathologischen Histologie des Kniegelenksmeniskus. Arch Klin Chir 177:483–495
Willet-Bleich M (1985) Über die Möglichkeiten und Grenzen der histologischen Altersbestimmung von Zusammenhangstrennungen der Kniegelenksmenisken. Dissertation, University of Bochum

Results of Partial Arthroscopic Meniscectomies in Anterior Cruciate Ligament Deficient Knees

P. Aglietti, R. Buzzi, and A. Pisaneschi

First Orthopaedic Clinic, University of Florence, Florence, Italy

Meniscectomy in the anterior cruciate ligament (ACL) deficient knee is a controversial issue, particularly for the medial meniscus. We performed partial arthroscopic meniscectomies in a group of patients in whom the predominant symptoms were caused by mobile meniscal fragments and who were willing to reduce their activity level. This paper describes the results of a consecutive series of 100 patients.

Material and Methods

In the period 1982–1984 we operated on 100 patients with ACL-deficient knees and a torn medial (78), lateral (19) or both (3) menisci. Most of the patients were males (82%), and the average age was 28 years. The mean follow-up was 2.5 years (range 1.5 to 3.5). The results were evaluated by a numerical rating system with a maximum of 100 points divided into 40 subjective, 14 functional (including functional tests such as the hop index and the side to side jump) and 46 objective points. The three main symptoms and the level of activity were evaluated according to the level at which the patient functioned, including sports, with minimal problems. The AKT-1000 knee arthrometer was used in order to evaluate anterior knee laxity.

Results

We obtained 65 excellent or good results and 35 fair or poor results. The average score at follow-up was 83 points. The symptoms of pain, swelling and giving-way were reduced in all the excellent or good knees and allowed recreational activities in 53 and vigorous sports in 12. The remaining 35 patients continued to complain of instability as before the operation. Of them only 6 requested an ACL reconstruction, four of whom were female. No difference was found according to type or site of the lesions. The patients with two meniscal lesions did less favourably well. Chondromalacia grade 2 or greater was present in 15 knees and correlated both with symptoms and frequency of giving-way. The KT-1000 evaluation is still in progress, but our impression is that the amount of anterior laxity correlates with the results. Pivot-shift was graded clinically and when found absent or mild, in 35 cases, the results were significantly better.

Surgery and Arthroscopy of the Knee
2nd Congress of the European Society
Editors: Müller/Hackenbruch

Discussion

Arthroscopic partial meniscectomy still has a place in the "conservative" treatment of the ACL-deficient knee provided the cases are carefully selected and the indications limited to knees with symptoms of predominantly meniscal origin, in less active patients. Sex, amount of laxity, pivot-shift, and chondromalacia are important factors to be considered.

Arthroscopic Meniscectomy on Patients with Untreated Rupture of the Anterior Cruciate Ligament

A. Georgoulis, E. Lais, and P. Hertel

Trauma Section, Department of Surgery, Rudolf Virchow Hospital, 1000 Berlin 65, FRG

Introduction

The combination of a meniscal lesion and an old rupture of the anterior cruciate ligament (ACL) is a common problem. For patients with the clinical symptoms of a ligament insufficiency, e.g. subluxations, we performed a reconstructive procedure and a meniscectomy. For patients having only the clinical signs of a meniscopathy without a subjective or clinical ligament insufficiency, we carried out a partial meniscectomy under arthroscopic control. There are many surgeons who prefer to do only meniscectomy in such cases; there is some controversy about the results (Table 1). In this report we shall try to define the indications for meniscectomy on patients with an old rupture of the ACL and a meniscal lesion.

Material and Methods

Forty-six patients with an untreated rupture of the ACL and a meniscal lesion underwent a partial meniscectomy between August 1981 and December 1984 and were reviewed 1–4 years later. Forty-four patients showed a total rupture of the ACL; in only two cases was a partial tear of the ACL found arthroscopically. Six patients had already had a meniscectomy after arthrotomy, but because they continued to have symptoms, we performed partial resection on the contralateral meniscus or removed a remnant of meniscus flap.

The procedure was performed under general anaesthesia using a tourniquet and a leg-holder. As soon as possible after operation we started an intensive rehabilitation programme with muscle training. We released the patient on the 3rd postoperative day.

Of 46 patients reviewed, 36 were re-examined using the knee scoring protocol of Lysholm (Lysholm and Gillquist 1982); seven patients (six men and one woman) had been lost to the study because of change of address, and one patient because of another disease. Two patients were reoperated on.

Results

The average age of the patients was 38 years; there were 38 men and 8 women. A tear of the medial meniscus was found in 32 patients, of the lateral meniscus in 7 and of both menisci in a further 7 (Table 2). Of the 36 re-examined patients, 24 had a lesion of the medial meniscus and 6 of the lateral meniscus. Six patients had injured both menisci. Examination under general anaesthesia showed a mod-

Surgery and Arthroscopy of the Knee
2nd Congress of the European Society
Editors: Müller/Hackenbruch

Tabelle 1. Results after meniscectomy in patients with untreated rupture of the ACL

Reference	Results
Gudde and Wagenknecht (1973)	Poor, increased instability
Chick and Jackson (1978)	Good, 83% returned to sports
McDaniel and Dameron (1980)	Good, 72% returned to sports
Lysholm and Gillquist (1981)	Good for patients with anteromedial instability Poor for patients with anterolateral instability
Noyes et al. (1983)	Poor, increased pain
Paterson and Trickey (1983)	Good for patients with bucket-handle lesion Poor for patients with posterior horn lesion
Kohn et al. (1985)	Good for patients with cartilage damage

Table 2. Types of meniscal lesion: 53 lesions in the 46 reviewed patients (7 with both injured) and [42 lesions in 36 re-examined patients (6 with both injured)]

Meniscal lesion	Reviewed patients		Re-examined patients	
	Medial	Lateral	Medial	Lateral
Bucket-handle	23	2	18	2
Posterior horn	9	5	8	4
Other tears	7	7	4	6
Total	39	14	30	12

Table 3. Examination of instability under general anaesthesia

Signs of instability	Re-examined patients (%)	Reviewed patients (%)
Lachmann test positive	91.6	93.5
Jerk test positive	83.3	87.0
Valgus instability	47.2	43.5
Varus instability	8.3	6.5

There were 36 re-examined and 46 reviewed patients.

erate anterior instability. Most of the knees demonstrated positive Lachmann and jerk tests (Table 3).

The average score on the Lysholm scale was 90.57 points. A total score of 80 points was the limit between excellent–good and fair–poor results (Fig. 1). The three patients who had varus instability showed poor results, and those with valgus instability showed poorer results than those without. The results were all in all acceptable (Table 4). Other factors associated with poorer results were thigh atrophy and cartilage damage.

The postoperative results in 30 patients were good or excellent: there were no symptoms during everyday activity. Six patients had a fair to poor result; four of

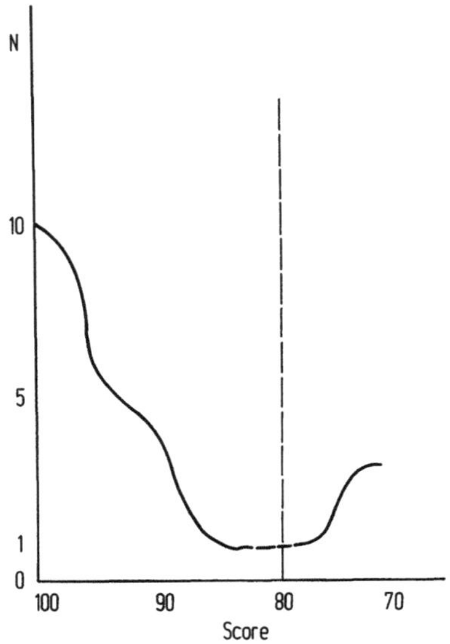

Fig. 1. Distribution of scores after arthroscopic meniscectomy using the Lysholm scale

Table 4. Mean total score of the 36 re-examined patients on the Lysholm scale

	Number of patients	Points
Only anterior instability	16	95.2
Anterior and valgus instability	17	91.2
Anterior and varus instability	3	70.0
Posterior horn tear	12	91.8
Bucket handle lesion	20	92.5
Other types of lesion	10	89.6

these had symptoms which were thought to be the result of additional cartilage damage, and two had symptoms of a ligament insufficiency. We had to perform a reconstructive procedure on two patients because of peristent instability and symptoms during sports activity. Of 28 patients who were active in sports, 12 had no symptoms and 4 had severe complaints when practising their favourite sport. Each patient was evaluated using the activity scale of Tegner (Tegner and Lysholm 1985). This scale is graded from 0–10. Activity levels 5–10 can be achieved only if the patients take part in sports (Fig. 2).

Discussion

In the case of patients suffering from an untreated rupture of the ACL, we have to differentiate between those showing symptoms of a ligament insufficiency and those showing symptoms due rather to a meniscal lesion (Fig. 3). In the latter case only meniscectomy showed satisfactory results for most of the patients.

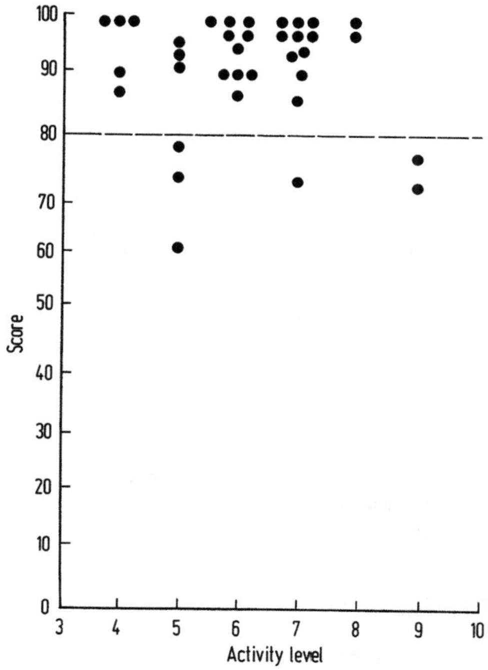

Fig. 2. Correlation between activity and score after arthroscopic partial meniscectomy according to the activity scale of Tegner and Lysholm (1985)

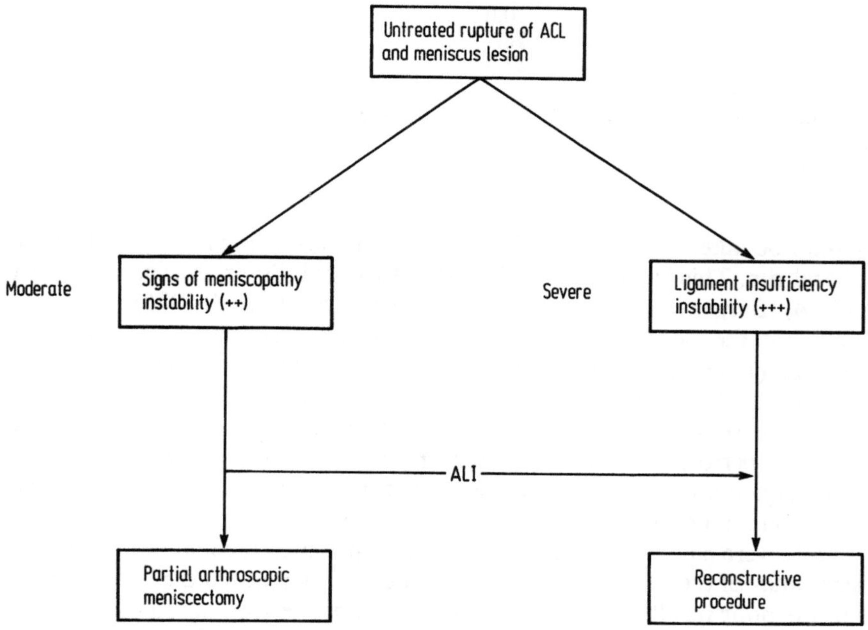

Fig. 3. Differentiation between recommended treatments according to symptoms. *ALI*, anterolateral instability; ++ anterior displacement of the tibia <1 cm; +++ anterior displacement of the tibia >1 cm

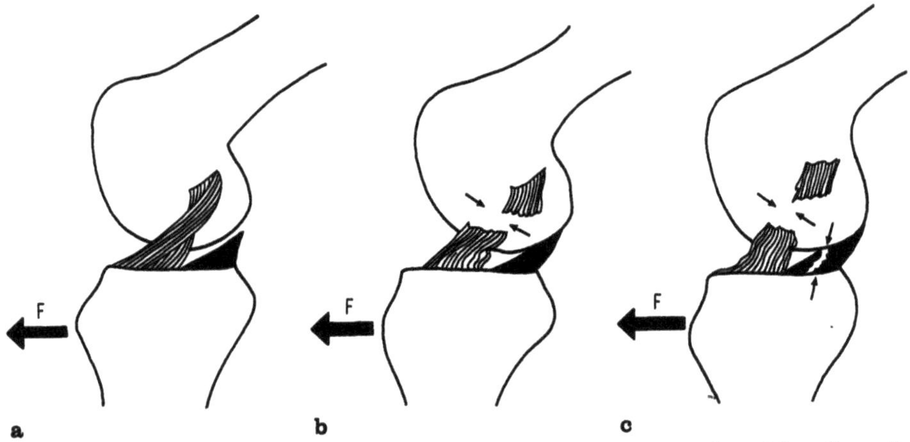

Fig. 4a–c. a and **b** The anterior cruciate ligament stops anterior motion. After disruption of the ACL, the posterior horn of the medial meniscus helps to prevent further anterior displacement of the tibia. **c** After partial meniscectomy the peripheral meniscal rim still helps to prevent further anterior displacement of the tibia. *f*, Forward direction

Table 5. Results obtained by Paterson and Trickey (1983)

Results	Bucket-handle lesion[a] (%)	Posterior horn tear[b] (%)
Good	75	20.0
Fair	15	26.7
Poor	10	53.3

[a] 50% complete meniscectomy and 50% partial meniscectomy
[b] 100% complete meniscectomy.

Care should be taken to remove only the torn portions of the meniscus, leaving a peripheral rim. This is particularly important for stability in the case of a torn ACL (Fig. 4) (Barger et al. 1980; Levy et al. 1982; Schulitz et al. 1985).

We think that the poor results after meniscectomy described by Gudde and Wagenknecht (1973) and by Noyes et al. (Table 1) could be explained by the total meniscectomies which were carried out on their patients. Noyes et al. (1983) reported an increase of pain and swelling after meniscectomy. In contrast Kohn et al. (1985), who performed a partial arthroscopic meniscectomy on 28 patients with severe cartilage damage, reported a significant improvement in the symptoms. McDaniel and Dameron (1980), who removed only the torn portions of the menisci, found severe pain in only 2% of his group of patients. It could be argued that the discrepancy of the good results after treatment of bucket-handle lesions as opposed to the poor results in the case of posterior horn lesions noted by Paterson and Trickey could be explained by the fact that a complete meniscectomy was always performed on cases with a posterior horn tear (Tables 1 and 5) (Paterson and Trickey 1983).

Anterolateral instability is associated with poor results (McDaniel and Dameron 1980; Lysholm and Gillquist 1981). It is better in these cases to perform a reconstructive procedure after the arthroscopy during which the lesion of the ACL is documented. It is our opinion that, in the case of athletes with moderate ACL instability, a reconstructive procedure is preferable. Only in cases with mild instability would we recommend partial meniscectomy (Chick and Jackson 1979).
Emphasis should be placed on the importance of rehabilitation (Giove et al. 1983; Noyes et al. 1983). Sports activity should take place only after recovery of the thigh muscle; otherwise there is a danger of a vicious circle forming: thigh atrophy–subluxation–pain and swelling–immobility–thigh atrophy.
In cases in which the result is not satisfactory in spite of postoperative treatment, plastic reconstruction should be performed.

Conclusion

Patients suffering from an untreated rupture of the ACL and a meniscus lesion benefit under certain circumstances from a partial meniscectomy with intensive muscle training.

References

Bargar WI, Moreland JR, Markolf KL, Keith L, Shoemaker SC, Amstutz HC, Grant TT (1980) In vivo stability testing of post-meniscectomy knees. Clin Orthop 150:247

Chick RR, Jackson DW (1978) Tears of anterior cruciate ligamentin young athlets. J Bone Joint Surg [Am] 60:970

Giove TP, Miller SJ, Kent BE, Sanfordant TL, Garrick JG (1983) Non-operative treatment of the torn ACL. J Bone Joint Surg [Am] 65:184

Gudde P, Wagenknecht R (1973) Untersuchungsergebnisse bei Pat. 10–12 J. nach der Innenmeniscusoperation bei gleichzeitig vorliegender Ruptur des ACL. Z Orthop 111:139

Kohn D, Lobenhoffer P, Wirth CJ (1985) Partielle arthr. Menisektomie bei AMI. Kongress der Deutschen Gesellschaft für Unfallheilkunde. November 1985, Berlin

Levy M, Torzilli PA, Waren RW (1982) The effect of medial meniscectomy on anterior posterior motion of the knee. J Bone Joint Surg [Am] 64:883

Lysholm J, Gillquist J (1981) Endoscopic meniscectomy – a follow up study. In: Lysholm J (ed) Arthroscopy on surgery of the knee. Medical Dissertation no. 106, University of Linköping

Lysholm J, Gillquist J (1982) Evaluation of knee ligament surgery results with special emphasis on use of a scoring scale. Am J Sports Med 10:150

Mc Daniel WJ, Dameron TB (1980) Untreated ruptures of the anterior cruciate ligament. J Bone Joint Surg [Am] 62:696

Noyes FR, Mooar PA, Matthews DS, Butler DL (1983) The symptomatic anterior cruciate deficient knee. Part I: The long-term functional disability in athletically active individuals. J Bone Joint Surg [Am] 65:154

Noyes FR, Mooar PA, Matthews DS, Grood ES (1983) The symptomatic anterior cruciate deficient knee. Part II: The results of rehabilitation, activity modification and counseling on functional disability. J Bone Joint Surg [AM] 65:163

Paterson FWN, Trickey EL (1983) Meniscectomy for tears of the meniscus combined with rupture of the anterior cruciate ligament. J Bone Joint Surg [Br] 65:388

Schulitz KP, Klein W, Hille E (1985) Meniscectomie – totale, partielle, offene oder geschlossene Operation? Z Orthop 123:837

Tegner Y, Lysholm J (1985) Rating systems in the evaluation of knee ligament injuries. Clin Orthop 198:43

Total Meniscectomy in Children: A Long-Term Follow-up Study

P. Abdon, A. J. G. Swanson, and M. S. Turner

Department of Orthopaedic and Traumatic Surgery, Royal Infirmary, Dundee, UK

Controversy still exists over the late effects of meniscectomy in children, and the reports available today have yielded diverse results. The satisfactory results vary as well as the incidence of secondary gonarthrosis and instability.

We therefore re-examined patients 18 years old or less at the time of operation and with a follow-up period of at least 10 years. All had had total meniscectomy according to the technique described by Smillie, and the same postoperative treatment. At the time of operation a knee card was completed, containing information regarding symptoms, signs and clinical findings. The operative findings were also noted, as was a drawing of the extirpated meniscus.

The original group comprised 227 patients, but 81 of them declined to take part, another 4 had died, and 3 were too ill to participate.

Of the 139 re-examined patients 45 had had further surgery since the original meniscectomy: 36 had had two operations, and 9 had had more than two. Only 94 patients had had a single meniscectomy. The re-examined knees were compared with those of the nonparticiparts regarding sex distribution, age at operation, meniscectomy site, follow-up time and associated ligamentous injury. Only the follow-up period differed.

We are presenting here the results of the 94 patients with a single meniscectomy. A total of 46 medial meniscectomies were performed on 38 males and 8 females, and 48 lateral meniscectomies were undertaken on 30 males and 18 females. Mean follow-up time was 16.8 years (range 10–28 years) and mean age at operation was 16.2 years (range 7–18 years). Bucket-handle tears were only observed in male meniscectomies, and there were two torn medial discs. Fourteen were either normal or hypermobile.

Ligamentous injuries at the time of operation were recorded in 14 patients. One total rupture of the anterior cruciate ligament (ACL) was seen in a male operated on for a lateral meniscectomy. At re-examination roentgenographic evaluation was performed using the technique described by Edholm et al., in which the knees are stressed into varus and valgus. Anterior drawer and rotational displacements were measured using a calliper introduced by Sylvin. Recovery was determined by the two rating systems previously used by Tapper and Hoover and by Johnson et al.

Surgery and Arthroscopy of the Knee
2nd Congress of the European Society
Editors: Müller/Hackenbruch

Results

Fifteen patients had pain (only eight different criteria), 34 experienced pain and other symptoms such as locking, giving way, catching or swelling; and only 12 had locking, giving way, catching or swelling without pair. Half of the medial meniscectomies were symptom-free whereas lateral meniscectomies were symptom-free in only one-fourth of the cases. This difference is not significant.

Range of movement was registered in 5° intervals, and 31 patients had a limited range of movement. This was more frequent with lateral meniscectomies, and those with flexion deficit or a combination of flexion and extension deficits were all lateral meniscectomies. The difference between medial and lateral meniscectomies is significant.

Five of the 14 knees with ligamentous injuries noted at the time of operation were clinically unstable, but only 2 patients had severe instability. The knee with the total ACL rupture was stable at re-examination. There were 41 knees considered stable, 40 knees showed only minor instability and 13 knees exhibited grade II–III instabilities. There were seven cases of positive pivot shift, five patients with a positive Lachmann test and one with a posterior drawer.

The mean values of anterior drawer, anteromedial and anterolateral displacements in the nonoperated knees differed between the sexes. The mean values did not differ significantly from those found in the operated knees. As there is great individual variation, the paired *t*-test was used to compare the measured difference of displacement between the operated knee and its control. The difference was significantly increased in male lateral meniscectomies, in anteromedial displacement and highly significant in the anterolateral displacement.

Adequate roentgenographic evaluation was available for 89 patients. Those missing were two male medial, one male lateral and two female lateral meniscectomies. The postmeniscectomy changes described by Fairbank were observed in 74 of the operated knees and in only 8 nonoperated knees.

Joint-line narrowing of more than 1 mm was observed in 21 medial and 20 lateral meniscectomies, and it was also noted in another 7 compartments in operated knees. Three knees which served as controls also had joint-line narrowing. The measured difference of the joint lines in operated and in nonoperated compartments was significantly decreased in all meniscectomy groups other than female lateral meniscectomies.

Arthrosis according to Ahlbäck was observed in eight patients but in no control knee. There were seven grade I and one grade II arthroses, comprising three male medial, one female medial and four male lateral meniscectomies. Two had a positive pivot shift. Half of the patients were pleased with the outcome. Six results were unsatisfactory according to the rating system of Tapper and Hoover and five according to that of Johnson et al.

Of the 71 patients who were pleased with the outcome, 10 admitted to restrictions in their activities. The remaining 23 were not pleased, and 10 of these experienced restrictions.

The results in 49 patients were satisfactory and in 45 unsatisfactory as classified by the rating system of Tapper and Hoover. The corresponding figures using Johnson's system were 56 and 38. Characteristics such as age at operation, observation time, duration before surgery, sex, meniscectomy site or trauma were not significantly different in either of the two rating systems.

Conclusion

As approximately half the patients had a decreased joint line and some 10% had a significant arthrosis, we believe that total meniscectomy in this age group is not a benign procedure.

Meniscectomy in Senior Citizens

T. Hoogland

Ignatius Hospital, Breda, The Netherlands

With the introduction of arthrography and arthroscopy under local anaesthesia, we have been able to study meniscal lesions in elderly people with seemingly arthritic knees. The results of subsequent open meniscectomy have also been studied.

Material and Methods

Between 1973 and 1982, 46 knees in 43 patients comprising 10 males and 33 females were treated by meniscectomy. The age range at the time of operation was 65–88 years with an average age of 71 years. The follow-up on these patients was 2–11 years with an average of 5.5 years. At the time of follow-up two patients had died from unrelated causes, and thus a follow-up was obtained on 43 knees (Table 1). Four patients had already had a total hip replacement, and one patient had a hemi-knee arthroplasty with a torn meniscus on the other side of the same knee.

The pre-operative studies included a positive arthrogram in 20 cases, a negative arthrogram in two cases and a positive out-patient arthroscopy in 21 cases. The follow-up included telephone interviews with the family physicians of all patients, history and clinical examination of the patient, and consultation of X-rays taken before the operation and at the time of follow-up.

The results were rated as follows. The family doctor's results were rated as good if the patient had returned to the surgery without knee complaints; as fair if the patient had noted improvement but still experienced some residual pain; and as poor if the complaints continued.

The objective results were based on the patient's history and examination. The result was rated as excellent when there was no pain during any activity; when

Table 1. Symptoms found at follow-up ($n = 43$)

	No. of patients
Pain alone	12
Pain and giving-way	6
Pain and effusion	6
Pain, giving-way and effusion	12
Pain and locking	3
Pain, locking and effusion	2
Pain and crepitation	2

Surgery and Arthroscopy of the Knee
2nd Congress of the European Society
Editors: Müller/Hackenbruch

the patient used no cane for walking and no pain medication; when there was normal knee function without effusion; and if there was no X-ray change compared with the pre-operative film. The result was rated as good when there was only occasional pain during prolonged activity; when the patient used no cane and no pain medication; when there was no loss of function and no or only little effusion and if the X-ray results had not worsened to a Fairbank rating (Fairbank 1948) of more than 3. The results were rated as fair when there was a slight improvement with regard to pain, or if the patient used a cane for the knee or occasional pain medication. The result was rated as poor when the pain had not diminished or when the X-ray showed deterioration to a Fairbank rating of 4.

Results

According to the family doctors' criteria there were 28 good results, 8 fair results, 5 poor results and 2 unknown results. The patients' subjective rating revealed that 12 felt they had an excellent result, 19 a good result, 6 a fair result and 6 a poor result. According to the objective criteria rating there were 6 excellent, 28 good, 3 fair and 6 poor results.

Pre-operative roentgenography showed six knees with a Fairbank rating of 0, seven with a rating of 1, 14 with a rating of 2, ten with a rating of 3 and six with a rating of 4. At follow-up two knees of the group rated 0 had deteriorated to Fairbank 3, two knees of group 1 had deteriorated to Fairbank 2 and two other knees to Fairbank 3. Of the group 2, one knee worsened to Fairbank 3 and one to Fairbank 4. One of the group 3 deteriorated to Fairbank 4. All knees in the group rated 4 remained unchanged, including four which were converted into a total knee arthroplasty.

When the follow-up results were correlated to the initial Fairbank classification, no significant differences were revealed in the groups rated 0–3, and the majority of the Fairbank 4 group had a poor result.

Complications

As all patients except those with hypertension received Coumadin anticoagulant for 6 weeks postoperatively, only one case of deep vein thrombosis was encountered, and no pulmonary embolism. There were no postoperative infections and no other serious complications.

Conclusion

In the elderly population with knee complaints a meniscal lesion should be suspected, particularly if the arthritic changes have not progressed beyond a Fairbank rating of 3. The diagnosis can be confirmed by out-patient arthroscopy under local anaesthesia. After an average follow-up of 5.5 years we have noticed good to excellent results in the vast majority of patients.

Reference

Fairbank TJ (1948) Knee joint changes after meniscectomy. J Bone Joint Surg [Br] 30(4):664

The Consequence of Meniscectomy in 147 Athletes: A 14.5-Year (Median) Prospective Longitudinal Study

U. Jørgensen, S. Sonne-Holm, F. Lauridsen, and A. Rosenklint

Department of Orthopaedic Surgery, Gentofte Hospital,
University of Copenhagen, Denmark

Introduction

The long-term results after meniscectomy have previously been assessed by cross-sectional studies (Appel 1970; Tapper 1969; Johnson et al. 1974; Jakson 1968). These have revealed an increased incidence of degenerative changes in the operated knee, which could be explained by biomechanical changes following meniscectomy (Krause 1976; Wang 1974; Seedhom 1974; Oretorp 1975; Walker 1975). However, the single cross-sectional design is inadequate for describing the changes in complaints and clinical and radiological findings. In order to obtain a more realistic picture of the consequences after meniscectomy we designed a prospective longitudinal study of athletes meniscectomised by arthrotomy for an isolated lesion of a meniscus in an otherwise normal knee. The primary aim was to describe the consequences with regard to:

- Time to the development of subjective complaints, objective findings and radiological changes
- The influence on working capacity and sports activity
- Identification of risk factors at the time of operation and at a short-term follow-up.

Methods

Patients

From 1964 to 1974, 439 patients had one or more meniscectomies performed in the Department of Orthopaedic Surgery T-3, Gentofte Hospital, Copenhagen. This study is based on 147 patients, who during organized athletic activity sustained an isolated meniscus injury in an otherwise normal knee. The patients were operated on according to the principles of Smillie (1970) with a postoperative regimen according to Sonne-Holm et al. (1980).

A clinical evaluation of the operated compared to the non-operated knee was performed in 131 (87%) patients at a short-term follow-up (STFU), with a median observation time of 4.5 years (range 2.5–11 years). Sixteen patients were not examined, since 11 had been reoperated for new trauma to the knee, 2 had emigrated, 2 were lost at follow-up, and 1 had died. Among the 131 athletes examined at the first follow-up, 101 (77%) were re-examined at a long-term follow-up (LTFU) 10 years later (median observation time 14.5 years; range 12.5–21 years). Thirty patients could not be re-examined, as 18 had left the area, 5 were lost at

Surgery and Arthroscopy of the Knee
2nd Congress of the European Society
Editors: Müller/Hackenbruch

Table 1. Patient description

	Short-term follow-up	Long-term follow-up
Number of patients	131	101
Males	114	91
Females	17	10
Median age (range), years	32 (17–59)	42 (27–69)
Type of meniscectomy		
Total	127	98
Partial	4	3
Medial	113	88
Lateral	18	13
Lost to follow-up	16	30
Reoperated	11	–
Emigrated	2	18
Dead	1	4
Refusal to participate	–	3

follow-up, 4 had died, and 3 did not want to participate. The patients' characteristics are given in Table 1. At both investigations changes in occupation and sports activity, subjective complaints and clinical findings were assessed.

Radiography

The radiography comprised anteroposterior views of both knees, with and without weight bearing. The focus film distance was 1 m. All radiograms were read by one of the authors (radiologist) without knowledge of the clinical data.
The degenerative changes in the femorotibial joint were described according to the principles of Ahlback (1968). Anteroposterior osteophytic formation on the femoral condyle and general flattening of the marginal half of the femoral condylar surface, as described by Fairbanks (1948), were recorded.
Comparable radiographs from the time of operation and the two examinations were available in 86 patients and were described with special regard to progression/regression of radiological changes.

Statistical Methods

Chi-square (C-s) and Fischer's exact (F-e)tests were used for impaired data and sign(s) tests for paired data. The influence of two variables on a third variable was analysed by a three-dimensional gamma test. The level of significance was defined as less than 5%.

Results

At LTFU 19% had been reoperated, 8% (11 of 142) before and 11% (11 of 101) after STFU. Problems from the operated knee had forced 7% to change occupation.

The level of sports activity decreased gradually during the 14-year follow-up period. At LTFU 34% had given up sports, while 12% (12 of 101) had reduced their activity since STFU due to complaints involving the operated knee. In the following only the complaints and findings related to the operated knee as compared with the non-operated knee are presented.

Subjective Complaints

At least one of the complaints registered was found in 67% (67 of 101) of the patients at LTFU, a progression of 14% compared with STFU assessment. Knee pain was the dominant complaint (Table 2). A significant progression (10%) was seen in the number of patients complaining of a swelling of the knee, with an increase from 19% (25 of 131) at STFU to 29% (29 of 101) at LTFU ($P<0.05$, C-s).

Objective Clinical Findings

The clinical tests were expanded at the second follow-up as shown in Table 3, especially for instability, where a significant overall increase was found. Most marked was the number of patients with anterior drawer sign; it had increased from 10% (13 of 131) at STFU to 36% (36 of 98) at the LTFU ($P<0.01$, C-s). In contrast to the great number of patients that reported a swelling of the knee, a clinically detectable effusion was only found in 5% (5 of 98) at LTFU, an in-

Table 2. Complaints concerning the meniscectomised knee

	Short-term follow-up		Long-term follow-up	
	($n=131$)	Percentage	($n=101$)	Percentage
Swelling	25	19	29	29
Sensation of instability	26	20	21	21
Pain	50	38	30	30
Analgesics	–	–	4	4
Other complaints	12	9	28	28
At least one complaint	69	53	67	67

Table 3. Objective clinical findings after meniscectomy

	Short-term follow-up		Long-term follow-up	
	($n=131$)	Percentage	($n=98$)	Percentage
Crepitation	23	18	38	38
Effusion	–	–	5	5
Limited mobility	88	6	11	11
Femoral atrophy	9	7	12	12
Collateral instability	20	15	26	26
Anterior drawer sign	13	10	36	36
Anteromedial laxity (AML)	–	–	21	21
Anterolateral laxity (ALL)	–	–	14	14
Tenderness on the joint line	13	10	12	12

Table 4. Development of radiographic changes after meniscectomy

	Length of follow-up (years, median)					
	0–4.5		4.5–14.5		>14.5	
	n	Percentage	*n*	Percentage	*n*	Percentage
Progressive changes	34	40	43	49	77	89
Fairbanks	26	30	10	12	36	42
Ahlback	4	5	3	3	7	8

crease from none at the STFU ($P<0.01$, C-s). Tenderness on the joint line was most frequent after lateral (25%, 3 of 12) as compared with medial meniscectomy ($P<0.05$, F-e).

Radiography

The radiographic changes at LTFU revealed a significant increase in the frequency and severity of pathological findings as compared with STFU (Table 4; $P<0.01$; C-s).

Progression, representing aggravation or new formation or radiographic changes, was found in 89% (77 of 86) of the operated knees at LTFU. At LTFU the most frequent radiographic changes were ridge formation (40%; 34 of 86) and narrowing of the joint line (35%; 30 of 86). Of these ridges 76% had been formed before STFU, at which time they were found in 30% (26 of 86) of the patients.

Risk Groups

The following pre-operative risk factors were identified.

Sex. At LTFU men had a higher frequency of knee instability and a higher risk of knee locking ($P<0.05$, gamma).

Age. Patients older than 36 years at operation developed a higher degree of femoral atrophy at LTFU than younger patients.

Site of Meniscectomy. Medial meniscectomy was associated with a higher frequency of pain on the joint line, effusion and joint laxity (all $P<0.05$, gamma). Lateral meniscectomy was, contrary to medial, significantly correlated to degenerative radiographic changes at LTFU ($P<0.05$, C-s).

At STFU the following risk factors were identified (Table 5):

1. Pain when climbing stairs was associated with a higher frequency of later change of occupation ($P<0.05$, gamma). Clinically, it was associated with a higher frequency of valgus instability at LTFU as compared with the STFU ($P<0.05$, gamma).
2. Feeling of instability was associated with a higher frequency of knee swelling at LTFU ($P<0.05$, gamma). Change of occupation was seen more frequently after early (STFU) reports of knee instability ($P<0.05$, gamma), and a higher degree of severe anterior drawer sign at LTFU ($P<0.05$, gamma).

Table 5. Associations between subjective complaints and objective findings at the 14.5-year median follow-up (+, $P<0.05$, gamma test)

Objective findings	Sucjective complaints						
	Analgesics	Pain at rest	Pain on weight bearing	Pain when starting	Pain when climbing stairs	Sensation of instability	Episodes of locking
Retropatellar crepitation						+	
Intercondyllar crepitation	+		+				
Femoral atropy							+
Anterior drawer sign (-slight)					+		
Anterior drawer sign (-marked)							+
Anteromedial laxity			+				
Valgus instability (ext.)				+	+	+	
Valgus instability (20°)							+
Varus instability (ext.)			+				
Tender medial joint line			+	+			
Tender lateral joint line				+	+	+	+
Decreased motility		+			+	+	

3. Early (STFU) knee swelling was a predictor for a later change of occupation ($P<0.05$, gamma).

Importance of Early Clinical Instability

Anterior drawer sign in the knee at STFU was associated with subsequent decreased or cessation of sports activity ($P<0.05$, gamma). Collateral instability was associated with a higher frequency of clinically detectable effusion at LTFU. Knee instability (drawer sign and/or collateral instability) was found in 24% (30 of 131) of the patients at STFU and was associated with a higher frequency of retropatellar crepitation ($P<0.05$, gamma) and a marginally higher frequency of intercondylar crepitation ($0.05<P<0.10$, gamma), but not with radiographic changes ($P<0.05$, gamma).

Importance of Complaints and Radiologic Changes at the STFU

The combination of complaints (at least one) and Fairbanks radiographic changes at STFU were associated with subsequent reduced or ceased sports activity and a higher frequency to change occupation ($P<0.05$, C-s) due to problems from the operated knee.

Importance of Follow-up Length

There were no significant correlations between the length of follow-up and LTFU results ($P<0.05$, C-s), but a significant overall increase in complaints and findings at the LTFU compared to the STFU was seen ($P<0.01$, C-s).

Discussion

Our LFTU results are comparable to those of others (Jakson 1968; Tapper 1969; Johnson et al. 1974; Appel 1970; Allen et al. 1984), but in contrast to these previous studies, our design allows us to describe the significance of time in the development of complaints and degenerative changes. This has been a matter of discussion, as Appel (1970) found no increase with the length of follow-up (5–43 years), whereas Allen (1984) found a slight, but not significant increase of osteoarthrosis in his 10–22-year follow-up, and Jakson (1968) found a significant increase in osteoarthrosis, when the follow-up was longer than 20 years.

We found no correlation between the length of follow-up and LTFU results ($P > 0.05$, C-s). However, observation of the individual revealed a significant increase in complaints, clinical findings and radiographic changes from STFU to LTFU ($P < 0.05$, C-s), with association of decreased or ceased sports activity and change of occupation (both $P < 0.05$, gamma).

The previous findings (Sonne-Holm et al. 1980) of an increased incidence of new knee injuries after meniscectomy was confirmed, as 11% were reoperated after STFU.

Risk factors for complaints and/or degenerative clinical or radiographic changes 14.5 (12.5–21) years after meniscectomy were found. As for Appel (1970) and Love (1923), our results revealed that an age of more than 36 years at the time of operation led to an increased incidence of femoral atrophy, indicating difficulties with the rehabilitation. Lateral meniscectomy was associated with a higher frequency of degenerative radiographic changes compared with medial meniscectomy, as found by others (Love 1923, Appel 1970; Allen 1984). This is possibly due to the greater load-bearing function of the lateral meniscus, as compared with the medial (Seedholm 1974; Walker 1975).

The presence of one of the following at STFU was associated with a higher frequency of complaints and findings and/or influenced the sports activity or occupation at LTFU. These were pain when climbing stairs, knee swelling, feeling of instability and clinical instability. Surprisingly, there was no association of clinical instability to radiographic or anamnestic signs of osteoarthrosis ($P < 0.05$).

The combination of Fairbanks changes and at least one complaint at STFU was associated with decreased sports activity and working ability at LTFU. In contrast patients with none of these findings had unchanged sports activity to years later, or had reduced their sports activity for reasons other than the operated knee.

Conclusion

Our prospective longitudinal study of the consequence of meniscectomy in athletes revealed that with time the degenerative changes increased in number and severity. After 14.5 years (mean) 67% had developed complaints, 35% knee instability and 89% radiographic changes. The degenerative radiographic changes developed later than 4.5 years (mean) after the operation in 49% of the patients. Consequently, 46% had reduced or given up their sports activity, and 7% had changed occupation.

Risk groups were male patients over 36 years old at the time of operation and patients with lateral meniscectomy.
The 4.5-year (mean) follow-up was useful as a predictor for patients at risk, as pain when climbing stairs, reports of knee swelling, clinical instability and Fairbanks radiographic changes were significantly correlated to increasing complaints, objective findings, decreased sports activity and change of occupation 10 years later.
The results indicate that a continuing effort should be given to preserve the role of the meniscus and if it has been remove then decreased knee loading might improve the late result in patient presenting risk factors. A logical first step would be to compensate for the shock absorbency provided by the meniscus by optimising the shoe/heel cushioning.

References

Ahlback S (1968) Osteoarthrosis of the knee. A radiographic investigation. Acta Radiol [Suppl] 277
Allen PR, Denham RA, Swan AV (1984) Late degenerative changes after meniscectomy. J Bone Joint Surg [Br] 66(8):666–671
Appel H (1970) Late results after meniscectomy in the knee joint: A clinical and roentgenologic follow-up investigation. Acta Orthop Scand [Suppl] 133
Fairbanks TJ (1948) Knee joint changes after meniscectomy. J Bone Joint Surg [Br] 30:664–670
Jakson JP (1968) Degenerative changes in the knee after meniscectomy. Br Med J 2:525–552
Johnson RJ, Kettelkamp DB, Clark W, Leaverton P (1974) Factors affecting late results after meniscectomy. J Bone Joint Surg [Am] 56:719–729
Krause WR, Pope MH, Johnson RJ, Wilder DG (1976) Mechanical changes in the knee after meniscectomy. J Bone Joint Surg [Am] 58:599–604
Love RMJ (1923) Prognosis after removal of semilunar cartilage. Br Med J 2:324–326
Oretorp N, Alm A, Ekstrom H, Gillquist J (1978) Immediate effects of meniscectomy on the knee joint. Acta Orthop Scand 49:407–414
Seedhom BB, Hargreaves DJ (1979) Transmission of the load in the knee joint with special reference to the role of the menisci. I Mech E 8:220–228
Smillie IS (1978) Injuries of the knee joint, 5th edn. Churchill Livingstone, Edinburgh
Sonne-Holm S, Fledelius I, Ahn NCN (1980) Results after meniscectomy in 147 athletes. Acta Orthop Scand 51:303–309
Tapper EM, Hoover NW (1969) Late results after meniscectomy. J Bone Joint Surg [Am] 51:517–526
Walker PS, Erkman MJ (1975) The role of the menisci in force transmission across the knee. Clin Orthop 109:184–192

Meniscus Reattachment or Allotransplantation? An Experimental Comparison on Sheep

K. Wulff, B. Gunterberg, G. Moberger, and P. Phillips

Surgical Department, Faculty of Medicine, University of Kuwait

The high incidence of osteoarthritic development in the meniscus-deficient knee has evoked great interest in meniscal preservation. However, as only approximately 20% of injured menisci lend themselves to repair, according to Kenneth DeHaven, our interest has been directed towards meniscal transplantation: is it technically feasible, and would any adverse reations arise? To tackle these problems experimentally we chose to work with sheep, as their knees have a meniscus of critical importance for stability and the stress load on the articulating surface.

Six knees had their medial menisci replaced with a properly fitting meniscus from another sheep. As a control, with an interval of 40 days, the contralateral knee had its medial meniscus removed and reattached again. Every other sheep had this control procedure done before transplantation. The grafted menisci were harvested from fresh cadavers and preserved in a defreezing medium at −80 °C for 2–50 days. The reattached menisci were also soaked in the same defreezing medium for 10 min before reimplantation.

In both groups the meniscus was sutured with duplication of capsule and vertical, isolated, single, nonabsorbable sutures. The horns were stitched so as to give a firm attachment to the remaining part of the removed meniscus. All sheep were kept free in a shelter and occasionally taken for a walk. Regularly, a slight to moderate limp could be observed for not more than 2 weeks. The sheep were killed 12 weeks after the first and 7 weeks after the second surgery, which means that there were three transplanted and three reattached menisci each to examine following the 7- and 12-week periods. The results were evaluated with a bone scan just before killing, together with macro- and microscopic studies following intra-arterial infusion of India ink.

Results

The bone scan did not reveal any significant difference between knees with transplanted or reattached menisci, regardless of the existence or not of chondral lesions.

Macroscopic Findings

All six transplanted menisci healed solidly to the capsules. There was macroscopically visible staining with India ink of the peripheral one-third of the menisci. Four of the six reattached menisci were solidly healed while two showed partial

Surgery and Arthroscopy of the Knee
2nd Congress of the European Society
Editors: Müller/Hackenbruch

defects between the capsule and meniscus. In two transplanted and three reattached menisci insufficient attachment of the posterior horns had created a gap filled with synovial pannus. Chondral erosions of the medial condyles were seen in three transplanted and four reattached menisci and were obviously caused by improper fixation of the posterior meniscal horn.

Microscopic Findings

Both transplanted and reattached menisci had essentially the same vascularisation as the lateral control menisci with visible staining of India ink in the peripheral third and in vessels overbridging the suture line. In the transplanted menisci complete healing between capsule and meniscal brim was confirmed. No significant degeneration was noted between lateral control menisci and the transplanted or reattached medial menisci.

Microscopic evaluation of the decalcified condyles confirmed the macroscopic findings of erosions of the articulating chondral surfaces in some cases. These chondral lesions, present to the same extent in both groups, were the only complications noted. They were obviously caused by increased surface stress load related to instability of the menisci when the posterior horn had not been sufficiently fixed, thereby allowing the meniscus to be partially extruded under joint load. Thus, this is of critical importance, and an improved fixation technique can be expected to prevent instability and the chondral lesion.

Conclusions

This study shows that in sheep the medial meniscus can be successfully transplanted with preserved mechanical function. Revascularisation without sign of rejection occurs, at least if the meniscal graft is preserved in a defreezing medium below −80 °C for 2 days or more. Postoperative immobilisation does not seem to be necessary if the meniscal horns are properly fixed.

Meniscus Transplantation – Experimental Study, Clinical Report, Arthroscopic Findings

K. A. Milachowski, K. Weismeier, C. J. Wirth, and D. Kohn

Orthopädische Klinik, Klinikum Großhadern, Marchioninistr. 15, D-8000 Munich 70, FRG

Introduction

The demand for meniscus replacement arises when the cruciate ligament substitute material does not have protection for sufficient and permanent stabilization owing to the lack of a breaking action of the inner meniscus dorsal horn in the absence of the anterior cruciate ligament and after medial meniscectomy. Consecutive loosening of the cruciate ligament replacement is the logical consequence. Preliminary animal experiments were carried out in terms of these clinical criteria.

Animal Experimental Investigations

Animal experimental investigations were carried out on thirty 1-year-old female merino sheep in two groups of equal number. Undamaged sheep menisci which had been prepared by freeze-drying and gamma-radiosterilization served as transplant material in the first group. In the second group, deep-frozen meniscus transplants were employed. The medial meniscus was replaced in each case. The medial meniscus of the same side as well as the lateral meniscus on the opposite side served as replacement material. The menisci were removed under sterile conditions from the contralateral knee joint on the occasion of prior animal experiments via meniscus refixation. The menisci were either lyophilized and gamma-radiosterilized by Braun-Melsungen, or deep-frozen at −30 °C. In the two transplantation groups (15 animals with deep-frozen meniscus transplants, 15 animals with lyophilized meniscus transplants), the postoperative follow-up of the animals was 6, 12, 24, and 48 weeks. The meniscus transplants were evaluated macroscopically with vital staining, microangiographically, histologically, by scanning electron microscopy and biomechanically. Three animals had to be killed. In two cases, there was a suture dehiscence after 6 weeks owing to the use of resorbable suture material, and in one case there was a medial knee joint necrosis with destruction of the deep-frozen meniscus transplant.

Surgical Technique

Via a skin incision over the tibial collateral ligament, the knee joint was opened in front of and behind the tibial collateral ligament, and the medial meniscus was removed. The thawed or rehydrated meniscus transplant was shaped to an appropriate form at its base with the scalpel, drawn through the tibial collateral ligament and placed in its anatomical position. The menisci were fixed with non-

Surgery and Arthroscopy of the Knee
2nd Congress of the European Society
Editors: Müller/Hackenbruch

resorbable atraumatic suture material at the base by single sutures. Under no circumstances should the medial ligament be severed for fixation of the meniscus transplant. Afterwards, Achilles tendon resection was carried out to relieve the affected hind leg, and antibiotic prophylaxis was given postoperatively up to wound healing. The animals were lame until about the 12th postoperative week and then showed an increasingly normal walking and jumping behavior after regeneration of the Achilles tendon.

Results

Transplantation of Lyophilized, Gamma-Radiosterilized Menisci

After 6 weeks, the menisci had completely healed at the marginal border. The synovium was slightly hypertrophied, and the incipient ingrowth of vessels into the transplants could be demonstrated microangiographically. Histologically, the transplant meniscus was completely covered by synovium. At the marginal border, there was a nonspecific mesenchyme reaction with numerous fibroblasts and capillaries. Occasionally, vascularized connective tissue strips penetrate into the meniscus tissue from the synovium. After 12 weeks, a broad, newly formed vascular network in the transplanted meniscus was shown microangiographically. Histologically, an increased infiltration of fibroblasts and capillaries into the meniscus was seen. After 23 weeks, the vascularization had progressed further microangiographically; this also became evident in scanning electron microscopy. An increased revascularization of the meniscus with ingrowing capillaries and fibroblasts up to the free margin of the meniscus could be discerned histologically.

After 48 weeks, the transplant had completely healed macroscopically. The synovium was still markedly hypertrophied. Microangiographically, an almost complete vascular neogenesis in the transplants was discernible (Fig. 1). Histologically, the transplanted meniscus was almost completely remodelled, progressing from the surface to the center with fibroblast proliferation and a newly formed collagen fiber structure. Over the entire experimental period, no signs of

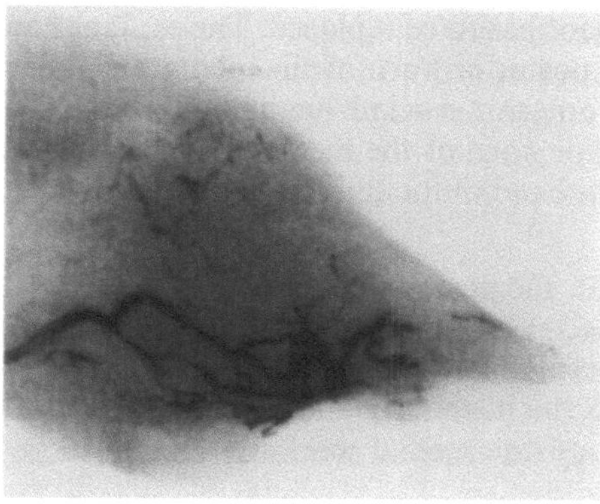

Fig. 1. Microangiography 48 weeks after transplantation of a lyophilized meniscus. Vascular neoplasia extending up to the tip to the meniscus is found

cellular inflammation or rejection reactions could be discerned. Plasma cells were only occasionally found in the region of the suture material.

Transplantation of Deep-Frozen Menisci

In general, it was shown that the deep-frozen meniscus transplant displayed an incorporation behavior analogous to the lyophilized transplant. However, the process of visible remodelling was far less pronounced. The transplants healed in all animals.

After 6 weeks, a slight synovium hypertrophy in the macroscopically intact meniscus was found. Microangiographically and with scanning electron microscopy, a vascularized synovial coat could be detected on the transplant. At the margins, there was an increased capillarization of the meniscus transplant. After 12 weeks, the synovial reaction was more pronounced, and the transplant was macroscopically unchanged. The formation of new blood vessels was increased according to the microangiogram. After 24 weeks, largely identical findings such as those after 12 weeks could be discerned macroscopically, microangiographically, and histologically. After 48 weeks, the two deep-frozen transplants had completely healed macroscopically. Histologically, only a slight vascularization of the transplant could be discerned. Remodelling of the meniscus does not occur. A broad collagenous scar plate at the margin with ingrowth of capillaries and fibroblasts along the undamaged collagen fiber bundles was uniformly shown. Cellular inflammatory reactions were not discernible. Immunocompetent cells could be detected neither in the transplants or at the base of the meniscus and in the region of the synovium.

The biomechanical investigations of the meniscus transplants were carried out on a specially constructed standardized traction device in comparison with undamaged menisci. The tensile strength of the deep-frozen meniscus transplant corresponds to that of lyophilized menisci both after 6 weeks ($P=0.27$), after 12 weeks ($P=0.09$), and after 24 weeks ($P=0.49$). However, the tensile strength of the normal meniscus was not reached even after 48 weeks ($P<0.001$).

Clinical Results

As mentioned at the beginning, we see an indication for replacement of the medial meniscus when the anterior cruciate ligament has to be replaced. The explanation for this is in the synergistic action of the posterior horn of the medial meniscus via the posterior oblique ligament on the anterior cruciate ligament [4]. Both the anterior cruciate ligament and the posterior horn of the medial meniscus act as a brake block to prevent subluxation of the head of the tibia on the extensor side. Two clinical situations are possible:

1. The anterior cruciate ligament is absent, the posterior horn of the medial meniscus has not stood up to the intensified stress and has ruptured, and the medial meniscus has moved elsewhere because of prominent signs of incarceration. The affected person now notices the instability and demands stabilization. The anterior cruciate ligament and the medial meniscus should be replaced.

2. Loss of the anterior cruciate ligament has led to a knee joint instability which can no longer be compensated by the muscles. For the purpose of anterior cruciate ligament replacement, a tattered medial meniscus is found which can no longer be preserved and which must be removed. It is necessary to replace the anterior cruciate ligament and the medial meniscus.

In a meniscectomized stable knee joint there is no indication for meniscus transplantation. Late arthrosis and possibly a secondary loosening of the capsule ligament structures are a threat, but the clinical result is very good even in the long term [6]. A lacking or destroyed lateral meniscus must not be replaced, since it does not crucially affect the stability of the knee joint.

Clinical Methods

We obtained the menisci in some cases from donors of vital organs (heart, kidneys). These were removed under sterile conditions in the operating theater and stored with the usual precautionary measures (smears, blood investigations) in accordance with the guidelines for bone transplantation in a deep-frozen state (−70 °C). We also used menisci from the knee joints of deceased persons. The menisci were lyophilized and gamma-irradiated (Braun-Melsungen) and stored in sterile packs. Rehydration of these lyophilized transplants takes about half an hour, which must be borne in mind when preparing for the operation (Fig. 2). The surgical technique corresponds to our procedure in replacement of the anterior cruciate ligament [5]. The access takes place medially via a hockey-stick incision. In anteromedial rotation instabilities grade III, the tibial collateral ligament is chiseled out from the femoral bone, so that the entire marginal border of the medial meniscus can be seen. In anteromedial instabilities grade II, the tibial collateral ligament is split longitudinally in order to be able also to fix the meniscus transplant in this region corresponding to the meniscus suture. Before carrying out the reconstructive measures on the tibial collateral ligament and to replace the anterior cruciate ligament, the medial meniscus was replaced by an allogenic, rehydrated meniscus. This was taken from the saline solution, the excess marginal border and synovium material were removed with a scalpel, and the medial marginal border in the knee joint was trimmed. The meniscus transplant was then laid on the medial tibial plateau as a trial and pressed onto the marginal border. In

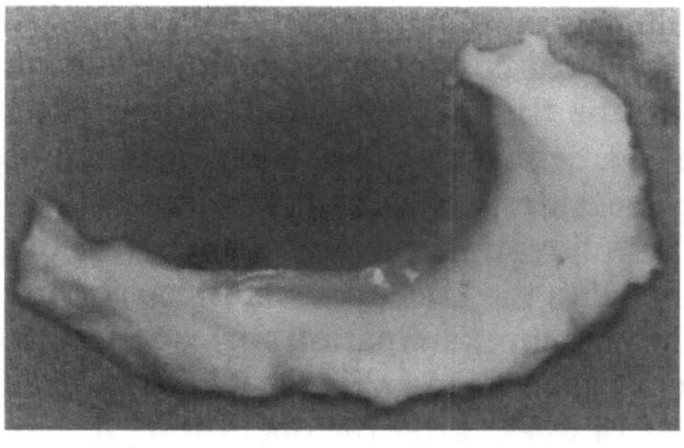

Fig. 2. Lyophilized medial meniscus before transplantation

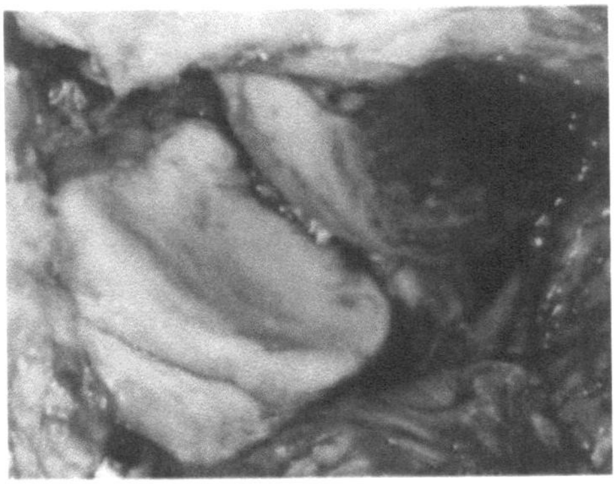

Fig. 3. Intraoperative finding for a 29-year-old patient in an old, antero-medial knee joint instability of grade III condition after medial meniscectomy. The appreciable cartilage damage at the femoral condyles and in the tibial plate are clearly discerned

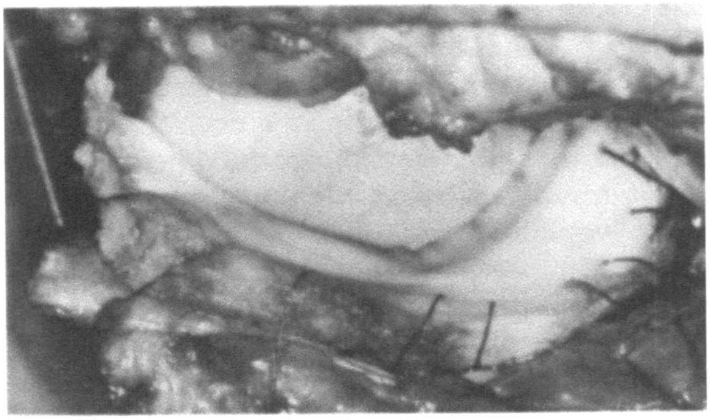

Fig. 4. Intraoperative finding of medial meniscus transplant before tying the sutures in the 29-year-old patient

the anatomically appropriate position of the joint, the knee joint was carefully moved. In this way, the meniscus transplant is given the opportunity to adapt to the required space conditions by sliding. The anterior horn and the posterior horn were reinforced. The posterior horn of the meniscus transplant was either linked with the resected meniscus still present by a suture or the suture was applied transosseously at the transition of the medial cartilage coat to the posterior intercondylar area. An additional attachment of the posterior horn at the posterior cruciate ligament secured the position of the meniscus. From dorsal to ventral, the medial meniscus was then fixed at the trimmed marginal border (Figs. 3 and 4) corresponding to the primary or secondary meniscus suture with single sutures (Vicryl USP 3-0) at an interval of 5 mm to the trimmed marginal border. After carrying out replacement of the cruciate ligament with the patellar ligament as well as concomitant, ligament-stabilizing measures, the wound was closed layer by layer after opening the tourniquet. The follow-up treatment was carried out in order to attain function at an early stage with the knee brace. Discharge took place 2 weeks after the operation. The following-up examination corresponded to that for further outpatient treatment after capsular ligament reconstructions. Full loading of the knee joint is usually allowed 14 weeks after the operation [5].

Patients

The first meniscus transplantation was carried out on 24 May 1984 at the Munich-Großhadern University Hospital. Since then, a further 14 homologous medial menisci have been transplanted (up to 11 February 1986). These involved 13 men and 2 women with an average age of 28 years. The youngest patient was 20 years old, the oldest patient, 40 years old. Eight left knee joints and seven right knee joints were affected. In 14 patients, these involved an old anteromedial knee joint instability (grade III and in one patient, an instability grade II). The injury had occurred a minimum of 11 months and a maximum of 8 years previously (on average, 3.8 years). In 14 of the 15 patients, the medial meniscus had already been removed for from 6 months to 6.5 years. Intraoperatively, only three delicate meniscus regenerates were found. Only five knee joints showed smooth cartilage surfaces on the medial femoral condyle and medial tibial plateau intraoperatively (Figs. 3 and 4). Chondromalacial areas of femoral cartilage in the loading zone were shown in three knee joints. Foci of chondromalacia in the cartilage on the medial, femoral, and tibial joint surfaces were shown in four knee joints. Two knee joints showed cartilage defect formations at the same sites. Retropatellar chondromalacia grade II–III according to Fründ was seen in 14 knee joints, and the anterior cruciate ligament was absent in 14 knee joints. In 1 knee joint, the anterior cruciate ligament had been falsely inserted in the femur too far ventrally and had appreciably loosened. The missing or resected medial meniscus was replaced in five cases with a homologous, lyophilized, medial meniscus, and in a further five cases by homologous, lyophilized, lateral meniscus from the contralateral side. Deep-frozen, homologous, medial menisci were used to replace the medial meniscus in two cases, and deep-frozen, homologous, lateral menisci in three cases. The replacement of the anterior cruciate ligament by the middle one-third of the patellar ligament was carried out following the technique which we use [5]. The postoperative complications consisted of effusions into the knee joints and febrile temperatures in a patient from the 5th postoperative day. The knee joint effusions did not require puncture and disappeared during hospitalization. The febrile temperatures of the one patient persisted for 12 days, and afterwards normal temperatures were presented. An abnormal local finding was not seen. The patient was free of symptoms on discharge. All further 12 knee joints healed without complications and tolerated the early functional follow-up treatment at the movement rail.

Results

All 15 patients remained under continuous control by our special knee joint outpatient service. The period of follow-up observation is at present from 6 months to 2 years (on average, 13 months). The subjective result was unequivocally determined by the stability result of the knee joint. Here, 13 of 15 patients were very satisfied, and 2 patients were satisfied with the result. No patient complained of knee joint effusions. One knee swelling after longer loading was reported by 6 patients. Here, the periods of follow-up observation were comparatively shorter.

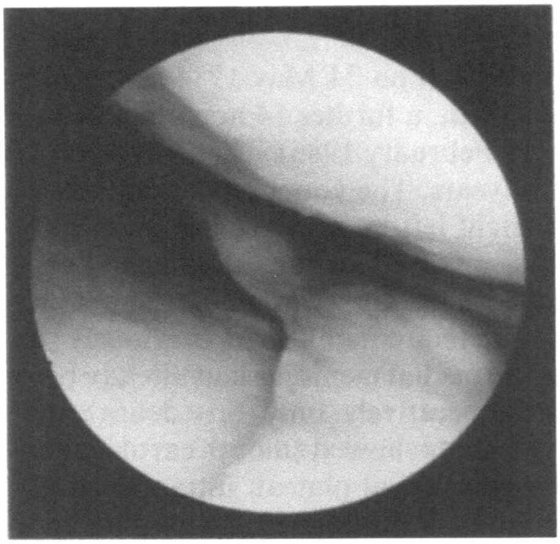

Fig. 5. Arthroscopic finding for a lyophilized, medial meniscus transplant 2 months postoperatively in a 39-year-old patient with an aged anteromedial knee joint instability of grade III

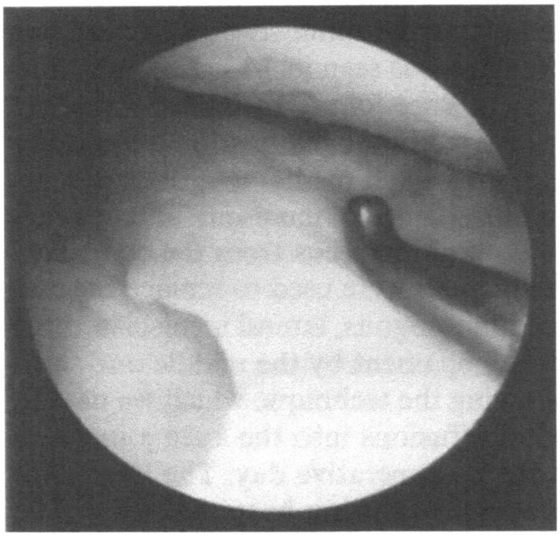

Fig. 6. Arthroscopic finding of a medial meniscus transplant in a 29-year-old patient 14 months after medial meniscus transplantation and anterior cruciate ligament replacement

Resting pain was not specified in any case, and only two patients reported loading-dependent diffuse pain in the knee joint.

The objective evaluation regularly entailed the usual stability test of the knee joints as well as the investigation for meniscus signs. Nine knee joints were checked arthroscopically on removal of the mental implants. The check was usually carried out 18 months after the operation – in one case as early as 2 months after the operation, in another case 5 months and 9 months after the operation, and in one case 8 and 14 months after the operation (Figs. 5 and 6). The following arthoscopic findings were observed.

A medial meniscus transplant had not healed at the anterior horn and the middle joint fissure and was present as a flap in the joint fissure. Macroscopically, the transplant showed degenerative alterations. In two cases, the medial meniscus transplant was completely normal 6 and 11 months after the operation and corresponded to a natural medial meniscus. A further transplant showed an irregular

surface but had a sharp margin and was of normal meniscus size. Four further meniscus transplants had reduced in size down to that of a regenerate but were likewise completely healed. Only two knee joints showed a generalized synovialitis or a synovialitis restricted to the anterior region of the joint respectively 2 and 8 months postoperatively. In a further knee joint, appreciable fusions were also found 14 months after capsular ligament reconstruction and meniscus transplantation. In one knee joint, medial arthrotomy was necessary to remove material. Inspection of the medial meniscus transplant showed a shortened anterior horn, a synovialization of the suture sites and instability of the meniscus surface at the transition to the posterior horn. The transplant had firmly grown in 11 months after the operation and was almost normal size.
Thus only 1 meniscus transplant of 15 can be designated as not completely healed and thus as a failure, since the 6 knee joints not checked arthroscopically or anatomically also did not show any signs of nonhealing of the meniscus transplant. One case deserves special mention: in this 25-year-old patient, the lacking medial meniscus had been replaced by a homologous, deep-frozen, lateral meniscus on the opposite side in the context of a grade II anteromedial knee joint instability. After an unremarkable postoperative course, the patient injured his knee joint 6 months after the operation by a rotation mechanism. There were signs of incarceration and knee joint blockades. As transplant rupture was suspected, the knee joint was arthroscopied. Here, the medial meniscus transplant was found to be reduced in size, but firmly anchored. The strong anterior cruciate ligament transplant appeared to be slightly elongated. At the lateral meniscus, a basket-handle rupture was found. This was resected arthroscopically.

Discussion

In view of possible negative consequences of meniscectomy, Lexer [1] had attempted to carry out autologous fat flap interposition arthroplasty as meniscus replacement. As communicated by his coworker Gebhard [1], corresponding animal experiments were also carried out on dogs. If one follows the biomechanical considerations of W. Müller [4], the medial meniscus posterior horn functions via the attached posterior oblique ligament as a synergist to the anterior cruciate ligament. However, replacement of the anterior cruciate ligament cannot remain stable in the long term when the medial meniscus posterior horn is lacking as a synergist. It is thus obvious that the lacking medial meniscus must be replaced in addition to the anterior cruciate ligament in such cases, in order to attain long-term stability. Prior animal experiments on 30 sheep [3] showed healing in both deep-frozen and lyophilized menisci in 27 animals over a period of 48 weeks. Immunological reactions did not occur. They are also not to be expected – at least for lyophilized tissue replacement [2]. On the basis of the positive results of meniscus transplantation in the animal experiments which have been herein described, medial meniscus transplantation in humans appears to be indicated to eliminate aged and anteromedial rotation instabilities and missing medial menisci on the basis of the early results of follow-up investigations. In the results of clinical and arthroscopic follow-ups over a period of 6 months to 2 years, only a partial healing of the transplant has so far been in one patient. Operation-specific problems of meniscus transplantation did not occur. The good early results can naturally also be explained by the regained knee joint stability after cruciate ligament replacement.

References

1. Gebhard K (1933) Der Bandschaden des Kniegelenkes. Barth, Leipzig
2. Jäger M (1970) Homologe Bindegewebstransplantation. Thieme, Stuttgart
3. Milachowki KA (1985) Meniskusrefixation – Meniskustransplantation – tierexperimentelle Untersuchung am Schaf. Habilitationsschrift, München
4. Müller W (1982) Das Knie. Springer, Berlin Heidelberg New York Tokyo
5. Wirth CJ, Jäger M, Kolb M (1984) Die komplexe vordere Knieinstabilität. Thieme, Stuttgart
6. Zippel H (1973) Meniskusverletzungen und -schäden. Barth, Leipzig

Antiphlogistics Enhance Recovery After Arthroscopic Meniscectomy

M. Bauer, D. Ogilvie-Harris, and P. Corey

Orthopedic Surgery, General Hospital, S-21420 Malmö, Sweden

An arthroscopic meniscectomy is usually followed by quicker restitution of function in the injured joint than traditional open meniscectomy. In the recovery period, some patients are, however, troubled by persistent swelling, pain and inhibition of function [2–4]. We therefore thought it was of interest to compare a group of patients who received postoperative treatment of antiphlogistics with a group who did not. The study was undertaken at Toronto Western Hospital, Toronto, Canada [6].

Patients and Methods

Patients suspected of having a torn medial meniscus at the preoperative examination were assigned to this study. All patients were fully informed of the design of the study and of possible side effects, and all signed a consent form. Patients who were younger than 16 years or over 65, with a previous history of gastrointestinal problems or hypersensitivity to NSAID, or patients with other concurrent diseases were excluded. After arthroscopy the patients who fulfilled the entrance criteria were randomised either to an active treatment group or to a group treated with placebo. Both groups obtained tablets which looked the same.

The active treatment group was treated by sodium naproxene, which is a prostaglandin inhibitor, at a dose of 550 mg twice a day for 6 weeks. The patients also received an analgesic – 500 mg of paracetamol with 30 mg of codeine to take if necessary. The use of extra analgesics was one of the evaluated parameters. Although paracetamol could have a weak antiprostaglandin effect, we did not think this to be of major clinical significance.

At follow-up 1, 3, 6 and 12 weeks following the operation the number of pills were counted. The patients were interviewed regarding subjective discomfort, and a clinical evaluation was performed.

Results

In all, 139 patients completed the study; 67 were in the active treatment group and 72 in the placebo group. There were no significant differences in the entrance criteria for the two groups except range of motion, which was slightly smaller in the active treatment group. The two groups were comparable regarding age and sex.

Pathology. The pathological abnormalities found in the two groups of patients are presented in Table 1. There were no significant differences between the two

Surgery and Arthroscopy of the Knee
2nd Congress of the European Society
Editors: Müller/Hackenbruch

Table 1. Pathological findings

	Active treatment group	Placebo group
Isolated meniscal pathology		
Bucket-handle tear	12	10
Other tear	20	19
Chondral lesions plus		
Bucket-handle tear	3	6
Other tear	14	20
Osteoarthritis plus		
Bucket-handle tear	1	2
Other tear	17	15
	67	72

groups. All patients under study had an arthroscopic meniscectomy performed by the same surgeon who used the same standardized technique. They were all treated as outpatients, and all operations were performed under general anesthesia.

Pain. The patients in the active treatment group had significantly less pain at rest up to 3 weeks following the operation (Fig. 1). The same was found for pain in connection with activities of daily living and for pain in connection with increased effort and sports. The consumption of analgesics was significantly lower in the active treatment group. These results demonstrate the analgesic property of sodium naproxene as well as its anti-inflammatory action (Fig. 2).

Synovitis and Effusion. The active treatment group had significantly fewer signs of synovitis and less effusion during the entire postoperative period. In addition to

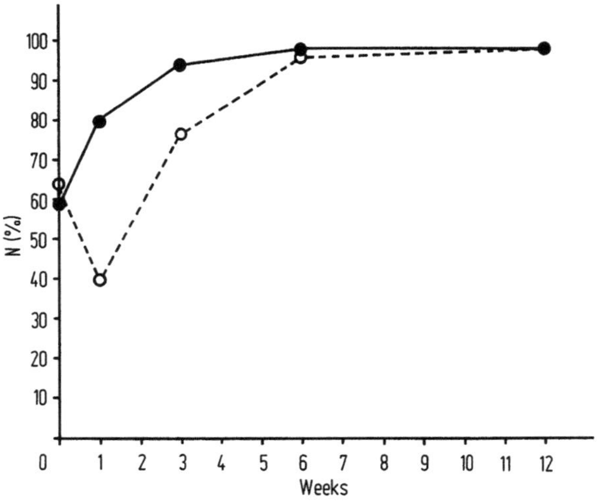

Fig. 1. Patients without pain at rest (%) in relation to treatment and time. Placebo (○), $n=72$; sodium naproxene (●), $n=67$

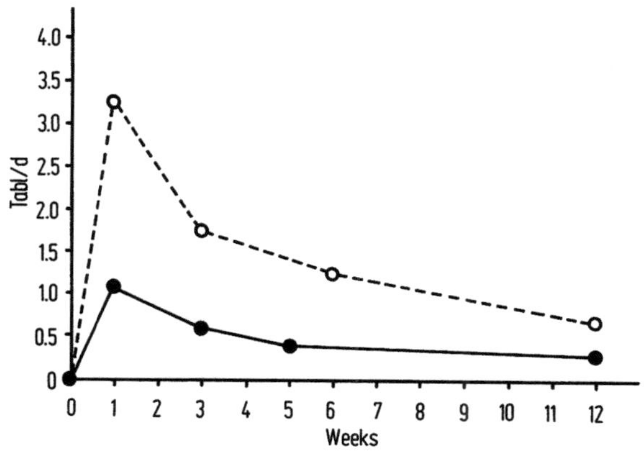

Fig. 2. The consumption of analgesics, tablets per patient per day, in relation to treatment and time. Placebo (○), $n=72$; sodium naproxene (●), $n=67$

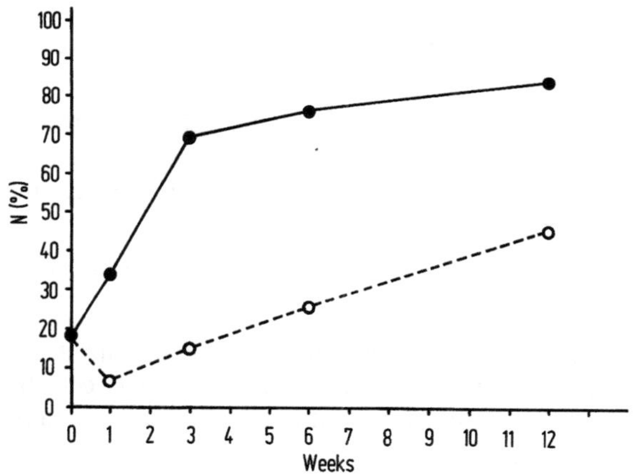

Fig. 3. Synovitis and effusion. Patients with neither (%), in relation to treatment and time. Placebo (○), $n=72$; sodium naproxene (●), $n=67$

the higher frequency of swelling, the degree of swelling was also of a greater magnitude in the placebo group (Fig. 3).

Range of Motion. The active treatment group had less range of motion at the preoperative assessment but 1 and 3 weeks following the operation the range of motion.in this group was significantly greater than that in the placebo group. After 3 weeks the range of motion was the same (Fig. 4). The earlier regaining of a full range of motion in the active treatment group is well correlated to the decrease in pain and synovitis.

Quadriceps Atrophy. After 1 postoperative week the groups were equal, but signs of atrophy started to disappear earlier in the active treatment group. At 12 weeks postoperatively there was still a significant difference in the return of quadriceps

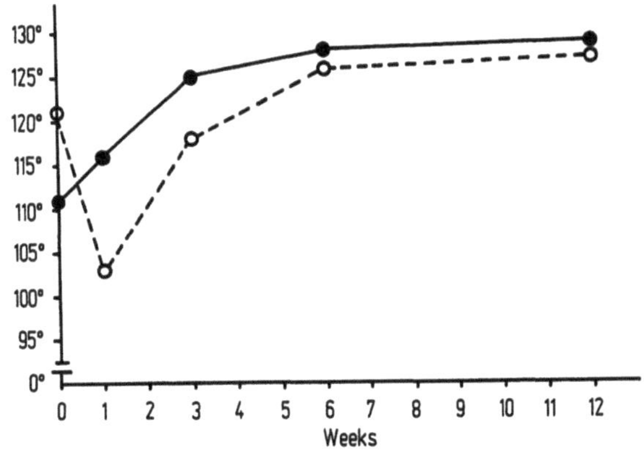

Fig. 4. Mean range of movement (degrees) in relation to treatment and time. Placebo (○), $n=72$; sodium naproxene (●), $n=67$

function: it had reached only 50% in the placebo group, but 72% in the active treatment group.

Return to Work. The mediantime for return to work in the active treatment group was 5 days; in the placebo group it was 14 days. A significantly higher percentage of the patients returned to work earlier in the active treatment group, and the same was found for return to sports.

Sports. At 12 weeks there was still a significant bias in favor of the active treatment group as regards the ability to do sports.

Side effects. These were mild. The most common was gastrointestinal upset. At week 1 there was a significant difference in the number of side effects between the two groups ($P<0.005$) but not at other times. One patient in the active group dropped out of the study due to side effects.

Discussion

Muckle [5] showed that prostaglandin inhibition enhanced the recovery after soft tissue injuries and after open meniscectomy. He could also demonstrate that results in patients who underwent open meniscectomy followed by treatment with prostaglandin inhibition were comparable to the results in patients treated by arthroscopic meniscectomy without concomitant drug treatment. Recently, Arvidsson [1] has also been able to demonstrate good results with a treatment of antiphlogistics following meniscus surgery.

In this double-blind study we have been able to show how patients treated by antiphlogistics following arthroscopic surgery had less pain, less synovitis, less effusion, faster return of the range of motion, faster regain of quadriceps function and a quicker return to work and sports. The graphs generally showed evidence of convergence, which indicates that the final restitution with or without prostaglandin inhibition would be equal.

This can be briefly expressed: active treatment leads to a faster restitution, but the final restitution is probably the same. We therefore propose that the addition of a prostaglandin inhibitor to the postoperative regime after an arthroscopic procedure might be of value.

References

1. Arvidsson I (1985) A study of rehabilitation after knee surgery with special emphasis on pain inhibition or voluntary muscle activation. Thesis, Stockholm University
2. Dandy DJ (1978) Early results of closed partial meniscectomy. Br Med J 1:1099–1100
3. Hamberg P (1983) Treatment of meniscus lesions in the knee – a selective approach. Thesis no 156, Linköping University
4. Jackson RW, Rouse DW (1982) The results of partial arthroscopic meniscectomy in patients over 40 years of age. J Bone Joint Surg [Br] 64:481–485
5. Muckle DS (1984) Open meniscectomy: enhanced recovery after synovial prostaglandin inhibition. J Bone Joint Surg [Br] 66:193–195
6. Ogilvie-Harris DJ, Bauer M, Corey P (1985) Prostaglandin inhibition and the rate of recovery after arthroscopic meniscectomy. J Bone Joint Surg [Br] 67:567–571

Problems on the Patella and Extensor Apparatus

Classification of Cartilage Damage

E. W. Morscher

Division of Orthopaedic Surgery, Department of Surgery,
University of Basel, CH-4055 Basel, Switzerland

The aim of classification of injuries or diseases is to categorise certain types of lesions according to specific characteristics such as mechanism of injury, location, type, extent and depth of lesion and to separate them according to definite criteria in order to be able to determine the clinical management. Classifications should not be made either for their own sake or to facilitate the passing of examinations.

The purpose of classification should also be to improve understanding. Thus not all classifications are suitable; incomplete and unclear classifications only add to the existing confusion. The criteria for a good classification are simplicity, completeness and general applicability.

Our knowledge of the healing processes of cartilage tissue has changed a great deal as a consequence of the studies performed by Robert Salter with "continuous passive motion" [41]. Salter showed that the propagation of a generally poor prognosis of cartilage lesions is no longer justified. Until then it was commonly accepted as impossible that a cartilage lesion could heal; progressive destruction of the respective cartilage was expected to end in osteoarthritis.

Until recent years not much attention has been paid to cartilage lesions. However, arthroscopy has made a great contribution to a better understanding of these lesions. Thus, current classifications should be based on arthroscopic examinations.

Mechanism of Injury

The location of cartilage lesions is typical for the type of accident, and therefore, as a rule, the probable type of lesion can be anticipated on the basis of the patient's description of the mechanism of injury. Cartilage lesions can be caused both by direct and by indirect forces.

The most frequent indirect mechanisms of a cartilage lesion are dislocation of the patella and ligamentous injuries such as the rupture of the anterior cruciate ligament, whereas direct trauma is caused usually by a blow on the knee cap or a fall on the flexed knee. Analogous mechanisms and lesions can be seen in joints other than the knee, i.e. in transchondral talar dome fractures in ankle sprains [2, 3, 8].

The type and location of a cartilage lesion resulting from a dislocation of the patella is highly characteristic (Fig. 1). At the moment when the patella tips over the outer edge of the lateral femoral condyle, a marked shearing action occurs between the medial circumference of the patella and the edge of the lateral femoral

Surgery and Arthroscopy of the Knee
2nd Congress of the European Society
Editors: Müller/Hackenbruch

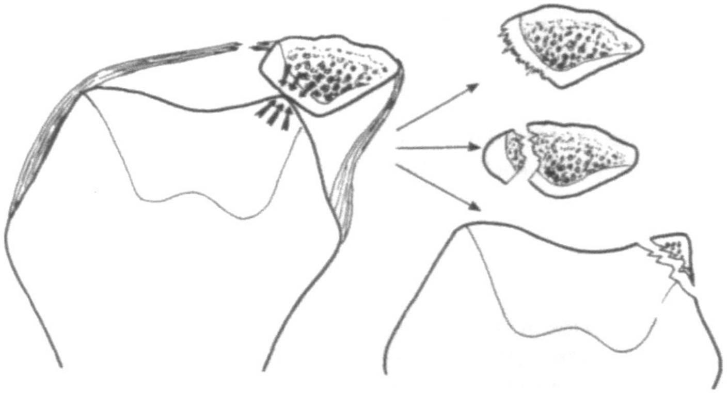

Fig. 1. Mechanisms and types of cartilage lesions caused by dislocation of the patella

condyle. The less severe the predisposing factors of a dislocation of the patella are (such as laxity of the ligaments, valgus deformation of the knee, etc.), the greater the force must be for dislocating the patella and the more probable the occurrence of a cartilage lesion. If the shearing force is sufficiently great, either at the moment of dislocation or during the following (usually spontaneous) reduction of the patella, then the cartilage is damaged either at the patella or at the lateral femoral condyle. Thus, a lesion on the medial edge of the patella can be regarded as pathognomonic for luxation of the patella.
Depending on their origin, lesions of the patella are localised on the medio-inferior circumference or on the lateral femoral condyle. As pressure and shear forces are both acting between the patella and femoral condyle during a dislocation of the patella, various types of cartilage lesions may be seen in this mechanism of injury. *Pressure forces,* not compensated by the cartilage lead to chondromalacia, osteochondritis dissecans or depressions, depending on the location and severity of the force. *Shear forces* result either in pure cartilage avulsions or – more frequently – in osteochondral fractures. Typical osteochondral fractures – so-called flakes – are seen at the medial edge of the patella, at the lateral edge of the lateral femoral condyle in dislocations of the patella or at the lateral edge of the talus in ankle sprains.

Type of Lesion

Pure cartilage lesions can be distinguished from osteochondral ones by the depth of the lesion. Pure cartilage lesions include chondromalacia and enchondral cartilage avulsions, while osteochondral lesions consist of intra-articular fractures, osteochondral fractures (flakes), traumatic osteochondritis and depressions.

Pure Cartilage Lesions

Chondromalacia

In many aspects the aetiology of chondromalacia patellae is an unsolved problem. In addition, it is very questionable as to whether every chondromalacia of the patella should be regarded as a prearthropathic condition. In a high percentage of cases, lesions of the femoropatellar joint do not produce symptoms [21, 22]. As we know, the most common form of chondromalacia, occurring during puberty and particularly in females, heals with time without treatment. By definition this is never the case in a true prearthrosis. A number of factors, alone or in combination, may predispose or cause different degrees of chondromalacia. Thus, the aetiology of chondromalacia is multifactorial [28], and it can be deduced that a uniform therapy is not tenable for all cases:

1. Trauma and mechanical overloading
2. Anatomical variations
   Wiberg types III, IV and "Hunter's hat" form
   Outerbridge ridge
   Bipartite patella, high-riding patella, etc.
3. Disturbances in alignment of the patella
   Luxations and subluxations of the patella
   Increased Q-angle (Insall)
4. Alteration of the gliding path of the patella
5. Nutritional disturbance of cartilage
6. Hormonal factors

Softening of the articular cartilage seems to be a common reaction of this tissue to different influences such as contusion and hypo- as well as hyperpression. Since articular cartilage is nourished by diffusion, squeezing during movements is mandatory. Thus, when the physiological gliding mechanism and the congruency of the joint is disturbed, it must be restored. This can be accomplished, for example, by transference of the patellar tendon in chronic dislocations or subluxations of the patella, or by sagittal osteotomy of the patella where there is a chondromalacia of the medial facet due to hypopression in severe dysplasia of the patella [28].

Chondromalacia of the patella due to direct contusion of the femoropatellar joint is very common [4, 25, 26, 31, 39]. Overloading (hyperpression) of the femoropatellar joint, which occurs in certain competitive sports and occupations, can also lead to chondromalacia [4, 37, 42, 44]. A central location of the cartilage damage is highly suggestive of such mechanisms.

Enzymatic processes play an important role in the development of post-traumatic osteoarthritis. Each time the cartilage layers are damaged, this causes destruction of cartilage cells and fissures of the cartilage surface. Enzymes, mainly cathepsines, are liberated that themselves destroy the cartilage tissue, and a vicious circle can start [9]. This pathogenetic mechanism was demonstrated in one of our follow-up studies of fractures of the patella [10]: of 44 patients (of a total of 67 patients) with a radiologically ideally reduced fracture of the patella, 32 complained of pain (73%), which was caused by a post-traumatic chondromalacia in at least 22 instances (50%). A pain-free interval of about 3 months is highly characteristic.

The most frequently used classification of chondromalacia of the patella was introduced by Outerbridge in 1961 [35, 36]. It consists of four stages according to the depth and extent of the damage:

Grade 1: softening and swelling of the cartilage
Grade 2: fragmentation and fissuring <1/2 inch (1.25 cm)
Grade 3: fragmentation and fissuring >1/2 inch (1.25 cm)
Grade 4: erosion of cartilage down to bone

With regard to the arthroscopic diagnosis of cartilage damage Lindberg and co-workers [21, 22] modified the Outerbridge classification:

First degree (C 1): the diagnosis can only be made by probing (e. g. superficial fissuring and/or softening)
Second degree (C 2): the lesion is readily seen but does not reach the subchondral bone (e. g. fragmentation, deep fissuring or "crabmeat")
Third degree (C 3): the lesion reaches the subchondral bone, which can be seen or probed

Ficat, Philippe and Hungerford [11] proposed a classification according to the location of the pathological process. On the basis of axial X-rays (in 30°, 60° and 90° of flexion) chondromalacia is divided into five groups:

1. Chondromalacia of the lateral facet
2. Chondromalacia of the medial facet
3. Central chondromalacia
4. Bipolar chondromalacia
5. Global (total) chondromalacia

Since the location of the defect reflects to a large extent the pathophysiology, the great advantage of this classification lies in the relationship between aetiology and therapy. Therefore, the location suggests a biomechanical solution. Chondromalacia of the lateral facet, for example, is mainly caused by hyperpression due to lateral subluxation of the patella. In such cases, a "lateral release" is the therapy of choice. On the other hand, a severe dysplasia of the patella (Wiberg type IV or Hunter's hat) causes hypopression and thus a chondromalacia because of a

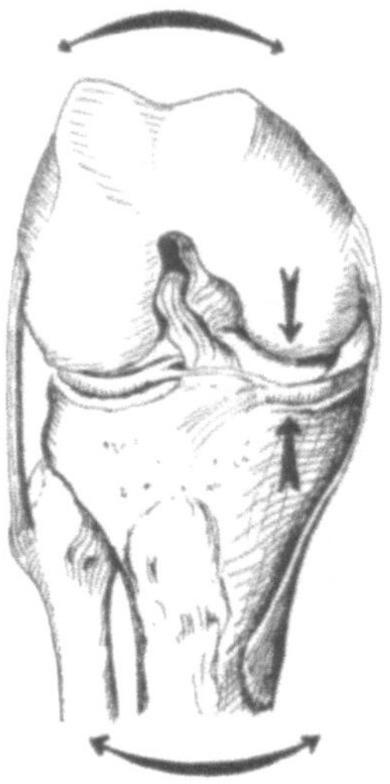

Fig. 2. Pathogenesis of a chondral avulsion. The mechanism is a combination of compression and shear

lack of physiological squeezing, and a sagittal osteotomy of the patella is to be considered.

Chondral Fractures (Enchondral Cartilage Avulsions)

Chondral fractures represent a distinct clinical entity and should be distinguished from the more frequently recognised osteochondral fractures. They occur most commonly in young adults in the medial femoral condyle and are produced by a combination of shear and compression [12–14, 16, 17, 19] (Fig. 2). Shearing forces give rise to a shift between the subchondral zone of calcified cartilage and the normal articular cartilage and produce a purely chondral fracture. Therefore, this fracture is more commonly seen in skeletally mature patients. The presence of calcified cartilage in the elderly patient allows the force of the injury to be transmitted across the "tidemark" or transition zone [17, 18, 20]. The prognosis of purely chondral fractures is poor because there is no direct bone contact and production of fibrocartilage is impossible. A flow of hyaline cartilage from the surrounding intact areas of cartilage is limited in larger defects [7, 20].

Osteochondral Lesions

Intra-Articular Fractures

Intra-articular fractures are associated – with a few exceptions (e.g. fracture of the neck of the femur) – with a fracture of the hyaline cartilage. Provided the cartilage is broken in a sharp line, and there is no additional damage due to a contusion of the cartilage – as is often the case in fractures of the patellae (see above) – there will be perfect healing, assuming that ideal reduction of the fragments can be achieved. Open reduction and proper adaptation of the fragments therefore are mandatory for prevention of post-traumatic osteoarthritis.

Osteochondral Fractures (Flakes)

Osteochondral fractures can be caused either by blunt trauma or by a twisting injury, either at the knee or an ankle sprain at the talus [35]. Osteochondral fractures predominantly occur in younger individuals because – in contrast to adults – there is no continuous subchondral calcified cartilage layer, and the shearing force is transmitted to the underlying bone. Thus, an osteochondral fracture, as opposed to a purely chondral one, is likely to result [16, 19]. Apart from dislocations of the patella, osteochondral fractures can also be caused by a sudden twist and forced valgus stress of the extended knee. The lesion is due to the presence of the lateral edge of the tibia, and it is found, therefore, at the weight-bearing area of the lateral femoral condyle [23].

The fate of an osteochondral fragment depends mainly on whether a free body remains or whether it is attached to the synovial membrane [40]. If the fragment remains free, the cartilaginous sheath continues to survive whereas the subchondral bone dies and calcifies. In contrast, fragments growing firmly on the synovial membrane become vascularised and continue to live, while the cartilagenous covering degenerates [26].

Fresh osteochondral fractures are treated as intra-articular fractures. They have a good prognosis provided they are well reduced and refixed.

Osteochondritis Dissecans

As a rule, osteochondritis dissecans is considered a disease caused by a localised, avascular necrosis of the subchondral bone in a convex part of a joint. However, there is no longer doubt that osteochondrosis dissecans can also arise as a result of a trauma [1, 5, 6, 15, 24, 26, 33, 43]. Histological proof of living bone tissue in the fragment is not compatible with the diagnosis of osteochondritis dissecans, even if the latter is the result of trauma. Bone necrosis with more or less marked degeneration of the hyaline cartilage covering is, however, characteristic of this condition.

Depressions

As reported by Wagner [45], we distinguish three types of injury of articular cartilage caused either by direct or by indirect contusion:

1. Two-dimensional depressions, depressed fractures or edge depressions (Fig. 3). These are real fractures with a fragment depressed into the underlying cancellous bone. The fragment has a characteristic jagged ridge.
2. Resilient cartilage bone depression or springy depression (Fig. 4). These occur only on osteoporotic femoral condyles or femoral heads. After the contusion causes a slight depression, the subchondral bone remains compressed while the cartilage itself regains its original shape.
3. Shallow depressions of the articular crest on the femoral condyle (Fig. 5). These were only mentioned in passing by Smillie in 1970 [43] in his classic monograph *Injuries of the Knee Joint*. He noted, especially in female patients with genu recurvatum, that not only the anterior horn is pressed between the condyles of the femur and tibia and is therefore injured, but also a depression may result in the articular cartilage of the femoral condyle.

After observing several cases in which a depression caused knee pain on the medial femoral condyle either alone or in combination with other injuries, I felt that this observation should be reported [25–27, 29, 30, 32]. Wagner reported in 1974

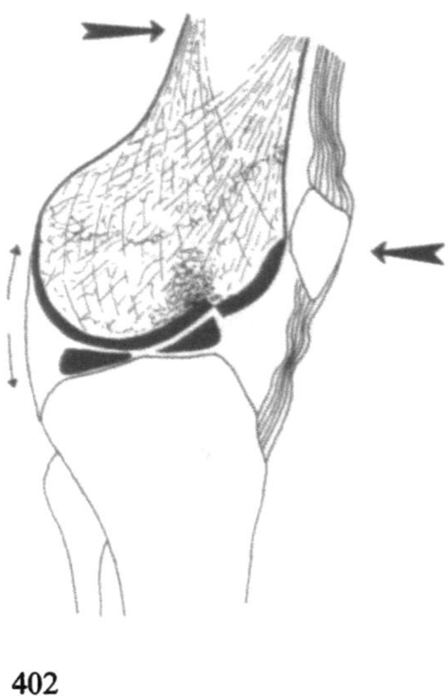

Fig. 3. Edge depression

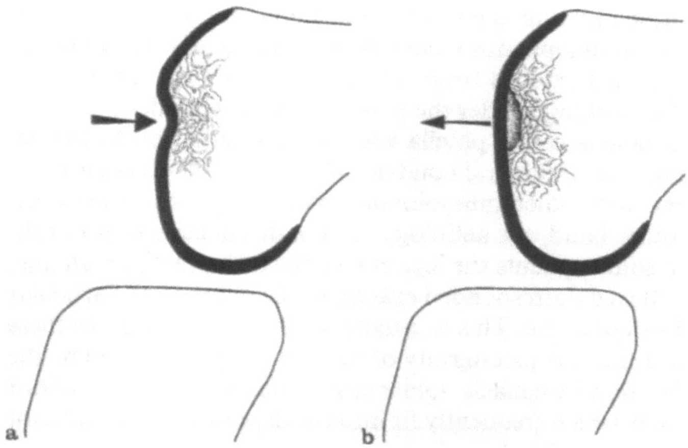

Fig. 4. a Resilent cartilage bone impression or **b** springy depression

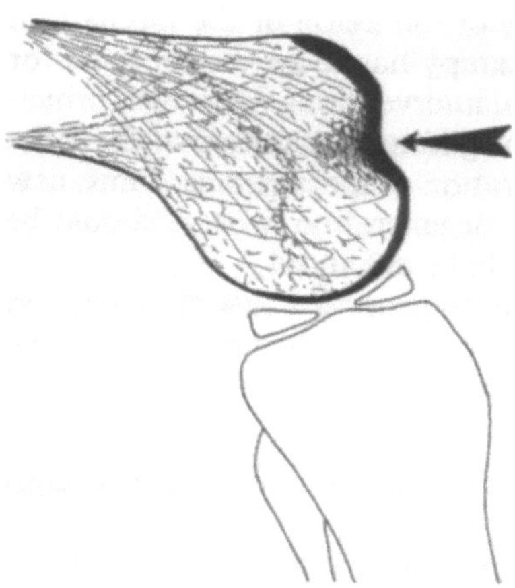

Fig. 5. Shallow depression

13 such shallow depressions [45]. In 43 depressions of the femoral condyle which we designated pathological or post-traumatic [27, 29], the most obvious mechanism of injury, forced hyperextension of the knee joint, was only found in one-tenth of the cases. In two-thirds, either direct contusion or indirect trauma to the affected knee was reported. Only 7 patients denied injury. In most cases this depression was the only pathological finding and, therefore, the only explanation for the patients' complaints. In other cases the depression was a secondary finding, but there appeared to be a pathogenic connection between the depression and the other lesions.

These depressions extend from one to several centimeters in diameter. In some, crests are visible and palpable; in others, the depression is shallow. When the knee joint is extended the depression is usually located opposite the contact point corresponding to the anterior horn of the medial meniscus. The articular cartilage above the depression exhibits clear pathological alterations in the form of soften-

ing, yellowish discoloration and dull appearance. Occasionally, genuine erosions are observed. The formation of a pannus is considered to be clearly pathological. The synovial vascular pannus extends from the cartilage synovial junction towards the depression. The cartilage under the pannus is degenerated.
In many cases chondromalacia of the patella was found in addition to the depression on the medial or lateral femoral condyle. The simultaneous occurrence of chondromalacia patella with other knee joint pathology (e. g. meniscal lesions) is well known. On the other hand, the aetiology of chondromalacia is not explicable in all cases, and in some patients the location of the retropatellar softening when the knee joint was flexed corresponded exactly to the point on the articular cartilage across from the depression. This is usually on the medial facet. In these cases it might be assumed that the incongruity of the femoral joint caused by the depression had led to the chondromalacia. Quite often alterations on the anterior horn of the meniscus occur which frequently fit into the depression. The anterior horn of the meniscus sometimes develops degeneration and in other cases a pannus develops as a sign of chronic irritation.
A classification must enable the surgeon to make the best choice of therapy. One must always consider two factors: the aetiology of the lesion or the mechanism of injury and the lesion itself. The operative therapy has to be very specific for each lesion. There are dozens of possible surgical interventions. However, principally we can distinguish measures that restore the gliding mechanism of the articular surfaces and those that enhance the regeneration of the cartilage. Thus, new damage to the cartilage should be avoided, the degeneration process should be arrested, and the regeneration of the lesion should be enhanced.
In summary, it was not the intention of this paper to propose a new classification in addition to those already existing. However, it could be helpful to have one which is generally accepted. According to my considerations such a classification should:

1. Be a description of the lesions, whether purely cartilaginous or osteocartilaginous
2. Be a description of the cartilage lesion, referring only to the cartilage itself
3. Include inspection as well as palpation
4. Be based on arthroscopic examination
5. Describe separately the extent and location of the lesion

In order to obtain such a classification I suggest that the European Society of Knee Surgery and Arthroscopy constitute a committee to develop a classification of cartilage lesions which could be called the ESKA classification.

References

1. Aichroth PM (1969) Osteochondritis dissecans of the knee. J Bone Joint Surg [Br] 51:181
2. Alexander AH, Lichtmand DM (1980) Surgical treatment of transchondral talar-dome fractures (osteochondritis dissecans); long term follow-up. J Bone Joint Surg [Am] 62:646–652
3. Baker CL, Andrews JR, Ryan JB (1986) Arthroscopic treatment of transchondral talar dome fractures. Arthroscopy 2:82–87

4. Bandi W (1977) Die retropatellaren Kniegelenkschäden. Huber, Bern
5. Baumgartl FA (1964) Das Kniegelenk. Springer, Berlin Heidelberg New York
6. Buri P, Geiser M (1962) Beitrag zum Problem der Aetiologie und Pathogenese der Osteochondritis dissecans am Kniegelenk. Z Unfallmed Berufskrank 55:95–102
7. Calandruccio RA, Gilmer WS (1962) Proliferation, regeneration, and repair of articular cartilage of immature animals. J Bone Joint Surg [Am] 44:431–455
8. Canale ST, Belding RH (1980) Osteochondral lesions of the talus. J Bone Joint Surg [Am] 62:97–102
9. Chrisman OD (1969) Biochemical aspects of degenerative joint disease. Clin Orthop 64:77–86
10. Dick W, Henche HR, Morscher E (1975) Der Knorpelschaden nach Patellafraktur. Arch Orthop Unfall-Chir 81:65–76
11. Ficat RP, Philippe J, Hungerford DS (1979) Chondromalacia patellae: a system of classification. Clin Orthop 144:55–62
12. Ganz R (1976) Isolierte Knorpelabscherungen am Kniegelenk. Hefte Unfallheilk 127:79–84
13. Ganz R (1978) The isolated shear injury to articular cartilage. In: Hastings DE (ed) The knee: Ligament and articular cartilage injuries. Springer, Berlin Heidelberg New York, pp 113–116 (Progress in orthopaedic surgery, vol 3)
14. Gilley JS, Gelman MI, Edson DM, Metcalf RW (1981) Chondral fractures of the knee. Radiology 138:51–54
15. Green JP (1966) Osteochondritis dissecans of the knee. J Bone Joint Surg [Br] 48:82
16. Hermans GPH (1982) Endogenous osteochondral fractures in the knee joint. Acta Orthop Belg 48:495–503
17. Hopkinson WJ, Mitchell WA, Curl WW (1985) Chondral fractures of the knee; cause for confusion. Am J Sports Med 13:309–312
18. Kennedy JC (1979) The injured adolescent knee. Williams and Wilkins, Baltimore
19. Kennedy JC, Grainger RW, McGraw RW (1966) Osteochondral fractures of the femoral condyles. J Bone Joint Surg [Br] 48:436–440
20. Landells JW (1957) The reactions of injured human articular cartilage. J Bone Joint Surg [Br] 39:548–562
21. Lindberg U, Hamberg P, Lysholm J, Gillquist J (1986) Arthroscopic examination of the patellofemoral joint using a central, one-portal technique. Orthop Clin North Am 17:263–268
22. Lindberg U, Lysholm J, Gillquist J (1986) The correlation between arthroscopic findings and the patellofemoral pain syndrome. Arthroscopy 2:103–107
23. Matthewson MH, Dandy DJ (1978) Osteochondral fractures of the lateral femoral condyle. J Bone Joint Surg [Br] 60:199–202
24. May E, Kuhn D, Diethelm L (1962) Knorpelläsionen an den Femurcondylen im Experiment bei Traumatisierung der Patella. Arch Orthop Unfallchir 54:301
25. Morscher E (1970) Traumatische Knorpel-Knochenläsion am Kniegelenk. Orthop Praxis 6:31–34
26. Morscher E (1971) Cartilage-bone lesions of the knee joint following injury. Reconstr Surg Traumatol 12:2–26
27. Morscher E (1976) Traumatische Knorpelimpression an den Femurcondylen. Hefte Unfallheilkd 127:71–78
28. Morscher E (1978) Osteotomy of the patella in chondromalacia. Arch Orthop Trauma Surg 92:139–147
29. Morscher E (1978) Posttraumatic cartilage impression of the femoral condyles. In: Hastings DE (ed) The knee: Ligament and articular cartilage injuries. Springer, Berlin Heidelberg New York, pp 105–111 (Progress in orthopaedic surgery, vol 3)
30. Morscher E (1979) Traumatische Knorpelläsionen am Kniegelenk. Chirurg 50:599–604
31. Morscher E (1981) Traumatische Knorpelschäden: Diagnostik, Therapie, Prognose. In: Hohman H (Hrsg) Das Knie. Vordruckverlag, Bruchsal, S. 433–442 (Praktische Orthopädie, vol 11)

32. Morscher E, Pfeiffer KM (1970) Spätschäden nach Knorpel- und Knochenläsionen am Kniegelenk. Z Unfallmed Berufskrankr 70:47–55
33. Nicolai N (1960) Die traumatische Kniescheibenverrenkung und ihre Folgen. Monatschr Unfallheilkd 63:215
34. Ogilvie-Harris DJ, Jackson RW (1984) The arthroscopic treatment of chondromalacia patellae. J Bone Joint Surg [Br] 66:660
35. Outerbridge RE (1961) The aetiology of chondromalacia patellae. J Bone Joint Surg [Br] 43-B:752–757
36. Outerbridge RE (1964) Further studies of the aetiology of chondromalacia patellae. J Bone Joint Surg [Br] 46:179–190
37. Pfeil E (1966) Chondropathia patellae bei Bergarbeitern. Zentralb Chir 91:1700
38. Pridie KH (1959) Method of resurfacing osteoarthritic knee joints. J Bone Joint Surg [Br] 41:618
39. Rohlederer O (1964) L'arthrose de la surface articulaire de la rotule. Rev Chir Orthop 50:361–368
40. Rosenberg NJ (1964) Osteochondral fractures of the lateral femoral condyle. J Bone Joint Surg [Br] 46:1013–1026
41. Salter RB, Simmonds DF, Malcolm BW et al. (1980) The biological effect of continuous passive motion on the healing of full-thickness defects in articular cartilage. An experimental investigation in the rabbit. J Bone Joint Surg [Am] 62:1232–1251
42. Schneider G (1968) Die Frühartbrose im Femoropatellargelenk des Leistungssportlers. Arch Orthop Unfall-Chir 54:401–416
43. Smillie IS (1970) Injuries of the knee joint, 4th edn. Livingstone, Edinburgh
44. Viernstein K, Weigert M (1969) Chondromalacia patellae beim Leistungssportler. Z Orthop 104:432–435
45. Wagner H (1974) Traumatische Knorpelschäden des Kniegelenkes. Orthopade 3:208

Quantitative Structural Analysis of Articular Cartilage: A Comparison of Weight-Bearing and Less-Weight-Bearing Areas in the Femoral Condyle of Young Adult Rabbits

P. S. Eggli, E. B. Hunziker, T. Naef, and R. K. Schenk

Institute of Anatomy, University of Berne, Postfach 139, Bühlstr. 26, CH-3000 Bern 9, Switzerland

Introduction

The composition of articular cartilage has previously been quantitatively evaluated under normal (Stockwell 1971; Vignon et al. 1976; Mitrovic et al. 1983; Quintero et al. 1984) and pathological conditions (Mitrovic et al. 1983). The analysis was mostly limited to a two-dimensional examination of the sections. Paukkonen and co-workers (1984) attempted to introduce stereological methods for a three-dimensional light microscopic analysis of articular cartilage. Recent advances in stereological techniques (Sterio 1984; Baddeley et al. 1986; Cruz-Orive and Hunziker 1986) together with an improved fixation method (Hunziker et al. 1982) were the basis of our studies. Our aim was to establish fundamental stereological parameters for the articular cartilage of young adult rabbits at the light and electron microscopic levels. In the second part weight-bearing and less-weight-bearing areas in the medial femoral condyle were compared. Several authors have tried to relate the structure of articular cartilage to its biomechanical properties. They focus mainly on the composition and three-dimensional arrangement of the matrix (Broom 1986), whereas information about cellular adaptation to physiological load is rare. Holmdahl and Ingelmark (1948) noticed a tendency to an increase of cell size in loaded areas. Paukkonen and co-workers (1985) observed a pronounced increase in cell volume within the middle and deep zones of young rabbit cartilage after non-strenuous physical exercise. Concerning the superficial zone of articular cartilage, higher cellularity is reported in non-weight-bearing areas (Vignon et al. 1976). As far as we know no data are available describing quantitative or qualitative changes in the cytoplasmic composition of cells under various load-bearing conditions.

Methods

Six 8-month-old NZW rabbits with a mean weight of 3.7 kg were given lethal doses of pentobarbital intravenously. The knee joints were opened and the complete distal femora were immediately immersed in a buffered prefixation solution containing 0.7% ruthenium hexamine trichloride and 2% glutaraldehyde (Hunziker et al. 1982). After 2 h the entire medial condyle was divided into four slices

Surgery and Arthroscopy of the Knee
2nd Congress of the European Society
Editors: Müller/Hackenbruch

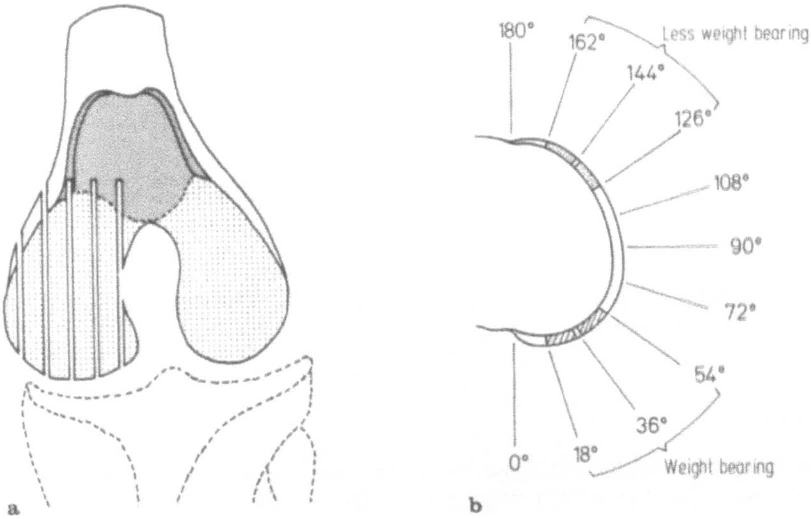

Fig. 1. a Division of the medial condyle into four sagittal slices of 1-mm thickness. **b** Subdivision of a slice into ten sectors or blocks. Posterior weight-bearing and anterior less-weight-bearing areas of the femoral condyle are indicated

of about 1-mm thickness (Fig. 1 a). Cartilage and subchondral bone were further prefixed for another 2 h, washed in buffer and postfixed in buffered 1% osmium with an additional 1% of a ruthenium hexamine trichloride analogue. The slices were then washed, dehydrated in ethanol and embedded in Epon 812. Each slice was subdivided into ten sectors or blocks by means of a template (Fig. 1 b). For examination of the total cartilage one block per slice was chosen according to a systematic sampling scheme, resulting in a final number of eight blocks per animal. For comparison of weight-bearing and less-weight-bearing areas, the posterior and anterior sectors of the two central slices were selected (Fig. 1 b). Undecalcified thick sections of 1 μm were photographed for histomorphometry (light microscopic level, magnification of ×400). Stereologic estimators were determined by means of the dissector method (Sterio 1984), point counting and intersection counts using a cycloid test system (Baddeley et al. 1986; Cruz-Orive and Hunziker 1986). Cytomorphometry (at the electron microscopic level) was carried out by point counting (Weibel 1979).

Results and Discussion

Qualitative Aspects of Articular Cartilage Morphometry

The more or less pronounced subdivision into different layers is a common light microscopic feature of mature articular cartilage, independent of location and species (Fig. 2). In sections running perpendicular to the surface the tangential or superficial zone is characterised by long and flat cell profiles orientated with their

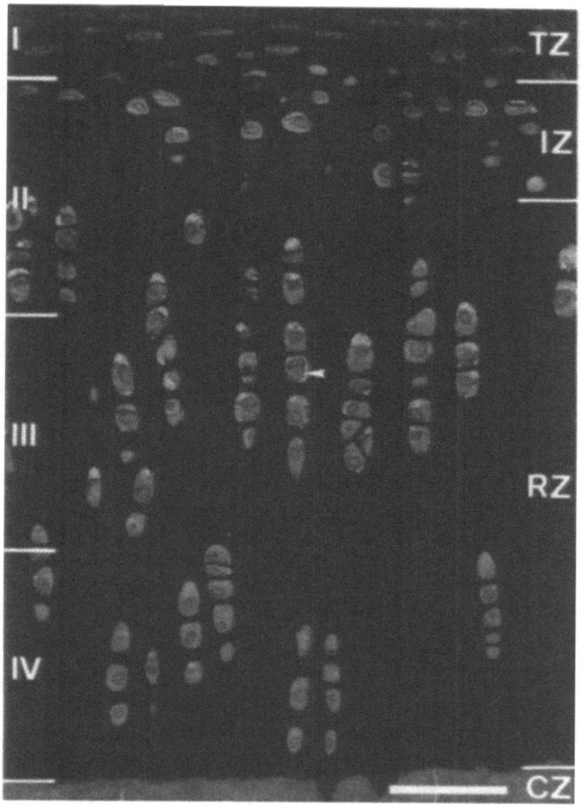

Fig. 2 a and b. Light photomicrograph of articular cartilage. Thick section (1 μm) stained with toluidine blue 0. Division into zones based on morphological (*right side*) and arithmetical (*left side*) criteria. Huge accumulations of intermediate filaments can already be identified here within the cytoplasm of a zone III chondrocyte (*arrowhead*). *TZ*, tangential zone; *IZ*, intermediate zone; *RZ*, radial zone; *CZ*, calcified zone; *bar* = 50 μm

long axis parallel to the surface. They represent discoidal cells. This layer is followed by the transitional or intermediate zone, where the chondrocytes become rounded. The most extensive part is the radial zone, in which the cells are arranged in perpendicular columns. The tidemark, a basophilic line (or multiple lines) in decalcified specimens, marks the borderline towards the heavily mineralised (calcified) zone. The rather few cells within this layer appear disintegrated or necrotic. This might be due to poor penetration of the fixative and thus represent an artefact. At its base the calcified cartilage is anchored in the subchondral bone.

The fine structure of mature articular cartilage cells shows some peculiarities which are of special interest for a quantitative analysis of the cytoplasmic components. Intermediate filaments are present in the perinuclear cytoplasm of superficial cells and near the tidemark. Huge amounts of these filaments are found within the chondrocytes of the upper two-thirds of the radial zone (Figs. 2, 3).

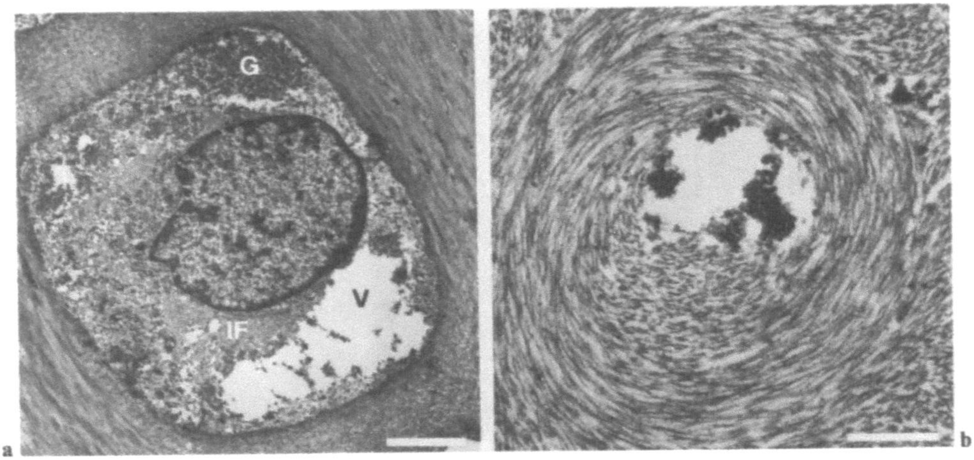

Fig. 3. a Electron micrograph of a chondrocyte of zones II and III with abundant intermediate filaments (*IF*) and glycogen (*G*). Vacuoles (*V*) are partially due to extraction of glycogen during the fixation procedure. *Bar* = 2 µm. **b** Electron micrograph of an intermediate filament whirl surrounding glycogen particles. *Bar* = 0.5 µm

Here, they are not restricted to a perinuclear region but spread out through the entire cytoplasm, often forming big whirls (Fig. 3 b). According to their diameter of 10 nm, they are classified as intermediate filaments and thought to belong to the vimentin type (Ghadially 1983). In close association and often intermingled with the intermediate filaments are large deposits of glycogen (Figs. 3 a, b) and occasionally lipid droplets. Accumulation of filaments, glycogen and lipid was often observed in older specimens and under pathologic conditions and therefore believed to indicate cell degeneration (Ghadially 1983). Though large agglomerations of filaments and glycogen are a typical feature of the cells within the upper part of the radial zone, disintegrating or necrotic cells were rarely found in this area but frequently encountered in the superficial layer just beneath the articular surface (Schenk et al. 1986).

Histomorphometry of Total Articular Cartilage

For light microscopic morphometry the noncalcified portion of the articular cartilage was subdivided into four zones (Fig. 2). Since delineation based on morphologic criteria is unreliable, we preferred an arithmetical subdivision. Zone I consists of one-tenth of the cartilage thickness and comprises the tangential zone. The rest, i.e. the intermediate and radial zones, was subdivided into three equal parts. Thus zones II, III and IV represent each three-tenths of the entire cartilage thickness.

Morphometric analysis shows that the mean cell volume is largest in zone III, confirming the visual impression (Table 1; Fig. 2). Surprisingly, the mean cell volume within zone I was found to be of the same order of magnitude as in zones II and IV, even though the vertical diameter of its cell profiles appears quite small. Because of the discoidal shape of the superficial cells, their surface is definitely

Table 1. Stereological estimators of the entire articular cartilage of the medial femoral condyles[a]

Zone	Mean cell volume (μm^3)	Mean cell surface are (μm^2)	Mean numerical density of cells (number per mm^3)	Mean matrix volume per cell (μm^3)
I	997 (15.3)	883 (10.6)	169046 (8.8)	7325 (14.0)
II	987 (7.7)	603 (10.0)	160164 (4.5)	6178 (4.2)
III	1329 (7.0)	673 (8.1)	110427 (6.0)	9728 (8.9)
IV	1092 (13.3)	660 (9.5)	89416 (8.4)	15287 (11.3)
I–IV	1115 (4.9)	670 (5.7)	126045 (3.6)	10226 (5.7)

[a] Coefficients of error are given within parentheses as percentage of SEM ($n=6$ animals).

larger than in the more spherical cells of zones II–IV. The numerical density, i.e. the mean number of cells per mm^3 was highest in zone I with a clear decrease towards zone IV. The mean matrix volume per cell was of similar size in zones I and II and markedly increased in zones III and IV.

Quantitative Comparison Between Weight-Bearing and Less-Weight-Bearing Areas

In the normal position of the rabbit the knee joint shows a high degree of flexion, and thus a somewhat posterior part of the femoral condyle contacts the tibial plateau. This area is loaded in the resting position and at the beginning of every movement. The most anterior part on the other hand has no direct relation to the tibia or to the patella even in the fully stretched position of the knee joint and has therefore been considered as the less-weight-bearing area (Fig. 1 b).

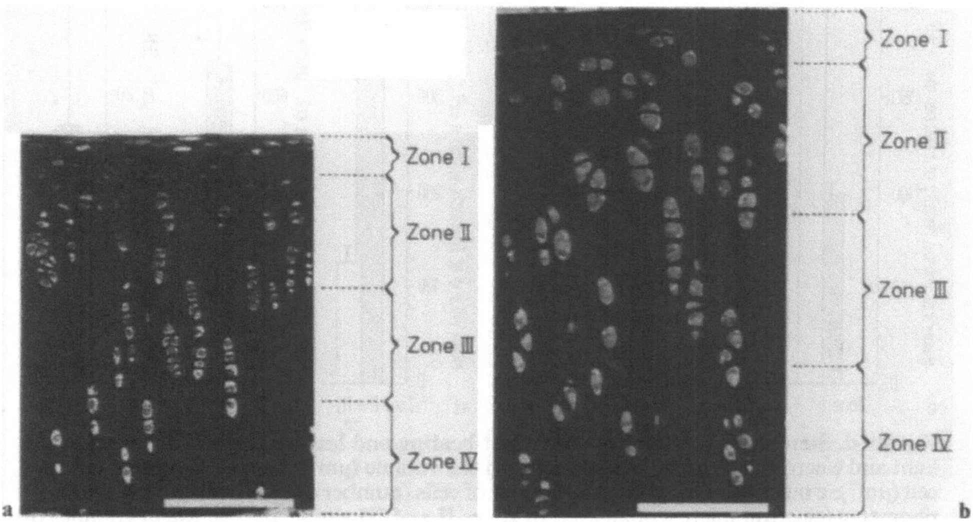

Fig. 4 a, b. Light photomicrograph of less-weight-bearing **a** and weight-bearing **b** areas at the same magnification. *Bar* = 100 µm

Light microscopical sections of these two regions show at the same magnification surprisingly clear differences (Fig. 4), which are confirmed by morphometry (Fig. 5a–c). There is a distinct tendency to thicker cartilage in the weight-bearing as compared with the less-weight-bearing areas with mean values of 415 µm and 344 µm, respectively. These differences are not significant, and they reveal that

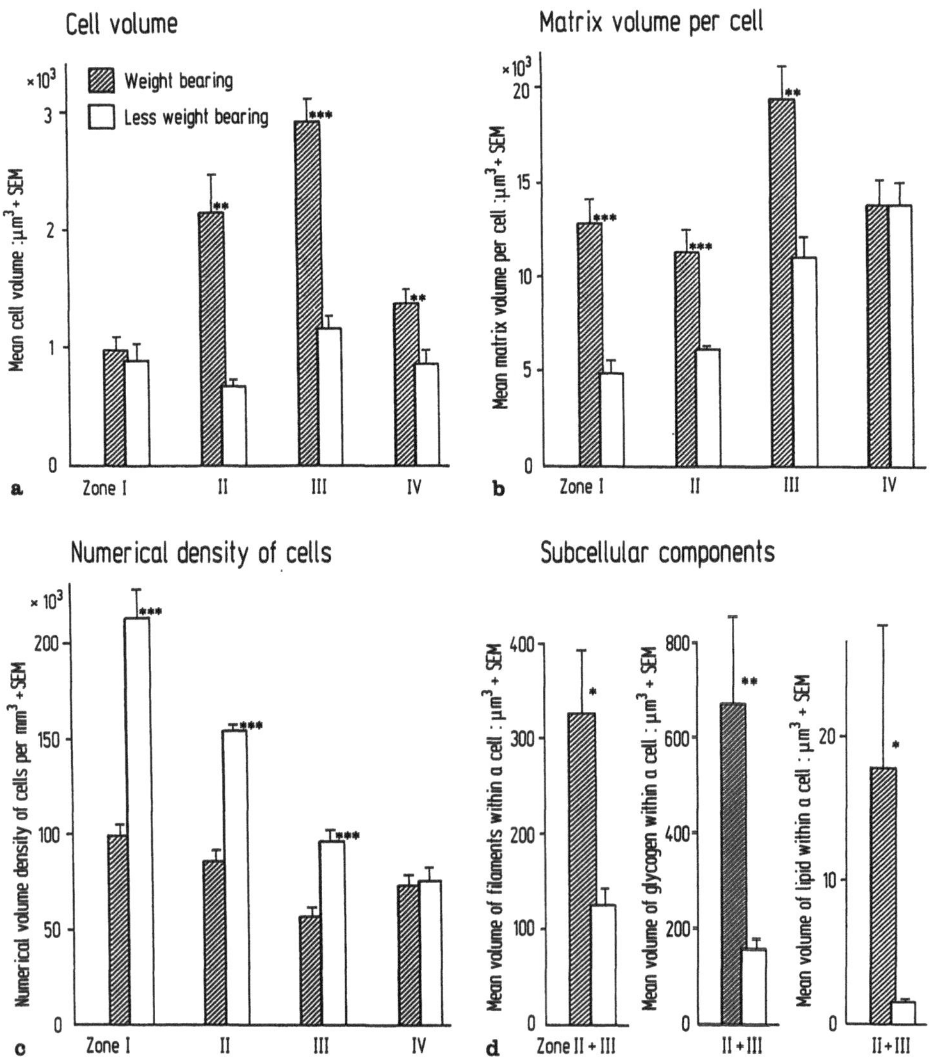

Fig. 5 a–d. Stereological estimators of weight-bearing and less-weight-bearing areas at the light and electron microscopic levels. **a** Mean cell volume (μm^3); **b** mean matrix volume per cell (μm^3); **c** mean numerical volume density of cells (number of cells per mm^3); **d** cytomorphometry (electron microscopic level) of zones II and III chondrocytes. Mean volumes of intermediate filaments, glycogen and lipid within the cytoplasm of an average cell. Mean ± SEM; * $P<0.05$; ** $P<0.01$; *** $P<0.001$

cartilage thickness is not solely influenced by the forces acting on it (Simon 1970). In the less-weight-bearing tangential zone I there are numerous, flat, spindle-shaped cells just beneath the articular surface resulting in a numerical density of over 200000 cells per mm^3 (Figs. 4a, 5c). In the weight-bearing area this value is markedly reduced, corresponding to the appearance of an almost acellular superficial layer (Fig. 4b). As mentioned before, all stages of cell degeneration and cell death were observed in the superficial zone. High stress may release or accelerate these processes which ultimately result in a 2.5 times greater mean matrix volume per cell in heavily loaded superficial regions (Fig. 5b). Towards deeper zones of the weight-bearing areas the cells show a remarkable increase in size (Fig. 4b), resulting in a 3 times larger mean cell volume within zones II and III (Fig. 5a). Simultaneously, a decrease in cell number per unit volume (Fig. 5c) and a twofold increase in mean matrix volume per cell (Fig. 5b) occur compared with less-weight-bearing areas. In zone IV near the tidemark, cell size is slightly but still significantly increased for weight-bearing areas (Fig. 5a), whereas the estimators of numerical cell density and mean matrix volume stay within the same range (Fig. 5b, c).

To identify the cellular components responsible for the enormous increase in cell size within zones II and III of the weight-bearing areas, we quantitatively analysed the chondrocytes of these layers at the ultrastructural level. Some values calculated for the average cell are more or less but not significantly higher in the weight-bearing areas, such as mean volume of nuclei (207 μm^3 versus 173 μm^3), mitochondria (13.2 μm^3 versus 11.2 μm^3), rough endoplasmatic reticulum (29.4 μm^3 versus 19.2 μm^3) and Golgi area (12.3 μm^3 versus 8.5 μm^3). The most prominent and significant increase was found for lipid droplets (by a factor of 12), for glycogen (by a factor of 4.3) and for intermediate filaments (by a factor of 2.5, Fig. 5d), whereby the accumulation of glycogen with a net increase of over 500 μm^3 was the highest.

Nearly all other parameters measured show significant differences between weight-bearing and less-weight-bearing areas. They seem to reflect an adaptation to functional requirements. Physiological loading increases the thickness and stiffness of canine articular cartilage (Jurvelin et al. 1986) and may be due to an increased production of matrix constituents such as collagen and mainly of proteoglycans. Biochemical analysis reveals a higher content of proteoglycans in habitually loaded sites (Yoshida and Azuma 1982), and an increased synthesis of proteoglycans was suggested under these conditions (Slowman et al. 1985), though direct proof is still lacking (Slowman and Brandt 1986). Within weight-bearing areas, an increased matrix production may explain the higher amount of matrix per cell and the increased cytoplasmic content of rough endoplasmatic reticulum, Golgi area and mitochondria. Intermediate filaments are part of the cytoskeleton and act as "integrators of cellular space" (Lazarides 1980). The accumulation of glycogen offers several explanations. Glycogen binds a lot of intracellular water, thus helping to withstand increased intra-articular pressure and to maintain cellular integrity especially when cartilage is deformed by shearing forces. On the other hand, high amounts of glycogen are associated with proteoglycan synthesis (Mazhuga and Cherkasov 1974) and can therefore be interpreted as an indicator of high metabolic activity of chondrocytes under high load.

References

Baddeley AJ, Gundersen HJG, Cruz-Orive LM (1986) Estimation of surface area from vertical sections. J Microsc 142(3):259–276

Broom ND (1986) The collagenous architecture of articular cartilage – a synthesis of ultrastructure and mechanical function. J Rheumatol 13(1):142–152

Cruz-Orive LM, Hunziker EB (1986) Stereology for anisotropic cells: application to growth cartilage. J Microsc 143(1):47–80

Ghadially FN (1983) Fine structures of synovial joints. Butterworths, London

Holmdahl DE, Ingelmark BE (1948) Der Bau des Gelenkknorpels unter verschiedenen funktionellen Verhältnissen. Acta Anat (Basel) 6(4):309–375

Hunziker EB, Herrmann W, Schenk RK (1982) Improved cartilage fixation by ruthenium hexamine trichloride (RHT). A prerequisite for morphometry in growth cartilage. J Ultrastruct Res 81:1–12

Jurvelin J, Kiviranta I, Tammi M, Helminen HJ (1986) Effect of physical exercise on indentation stiffness of articular cartilage in the canine knee. Int J Sports Med 7:106–110

Lazarides E (1980) Intermediate filaments as mechanical integrators of cellular space. Nature 283:249–256

Mazhuga PM, Cherkasov VV (1974) Adaptive distribution of specific biosyntheses in a homogeneous population of articular cartilage chondrocytes. Z Mikrosk Anat Forsch 88(2):364–374

Mitrovic D, Quintero M, Stankovic A, Ryckewaert A (1983) Cell density of adult femoral condylar articular cartilage. Joints with normal and fibrillated surfaces. Lab Invest 49(3):309–316

Paukkonen K, Selkäinaho K, Jurvelin J, Helminen HJ (1984) Morphometry of articular cartilage: a stereological method using light microscopy. Anat Rec 210:675–682

Paukkonen K, Selkäinaho K, Jurvelin J, Kiviranta I, Helminen HJ (1985) Cells and nuclei of articular cartilage chondrocytes in young rabbits are enlarged after non-strenuous physical exercise. J Anat 142:13–20

Quintero M, Mitrovic DR, Stankovic A, de Sèze S, Miravet L, Ryckewaert A (1984) Aspects cellulaire du vieillissement du cartilage articulaire: I. Cartilage condylien à surface normale, prélevé dans les genoux normaux. Rev Rhum Mal Osteoartic 51(7–8):375–379

Schenk RK, Eggli PS, Hunziker EB (1986) Articular cartilage morphology. In: Kuettner K et al. (eds) Articular cartilage biochemistry. Raven, New York, pp 3–22

Simon WH (1970) Scale effects in animal joints. I. Articular cartilage thickness and compressive strength. Arthritis Rheum 13(3):244–255

Slowman S, Brandt KD (1986) Composition and glycosaminoglycan metabolism of articular cartilage from habitually loaded and habitually unloaded sites. Arthritis Rheum 29(1):88–94

Slowman S, Palmoski MJ, Brandt KD (1985) Glycosaminoglycan turnover in weight-bearing and non-weight bearing canine articular cartilage in vitro. Trans Orthop Res Soc 10:64 (Abstract)

Sterio DC (1984) The unbiased estimation of number and sizes of arbitrary particles using the dissector. J Microsc 134:127–136

Stockwell RA (1971) The interrelationship of cell density and cartilage thickness in mammalian articular cartilage. J Anat 109(3):411–421

Vignon E, Arlot M, Patricot LM, Vignon G (1976) The cell density of human femoral head cartilage. Clin Orthop 121:303–308

Weibel E (1979) Stereological methods, vol I. Practical methods for biological morphometry. Academic, London

Yoshida K, Azuma H (1982) Contents and compositions of glycosaminoglycans in different sites of the human hip joint cartilage. Ann Rheum Dis 41:512–519

Comparative Morphological Examination of Experimental Patellar Dislocation

I. Földes [1], J. Gyarmati Jr. [2], I. Kiss [3], J. Csongor [4], M. Kern [1], and G. Deseö [4]

Institutes of [1]Anatomy, [2]Radiology, [3]Orthopaedics,
and [4]Central Research, Laboratory of Medicine University of Debrecen, Hungary

Numerous experiments have demonstrated that degenerative changes can result both from alterations in the mechanical, stress-dynamic features or from disturbances in the biological balance of the knee joint constituents (Barz 1982; Mohr and Kirkpatrick 1983; Stockwell 1983). Arthrosis can therefore be produced experimentally by various methods (Barnett 1956; Bentley 1975; Kalbhen et al. 1981; Dustmann et al. 1974; Simon 1978; Floman et al. 1980; Pond and Nuki 1973; Stockwell and Sprinz 1981; Kiss and Földes 1983). The aim of our experiments was to investigate the early stages in the development of arthrosis as produced specifically by the dislocation of the patella. We have examined the alterations of GAGs, collagen and cytological structures of articular cartilage in the femoropatellar and femorotibial joints, using radiochemical, histochemical and electron microscopical methods.

Material and Methods

In 60 male Wistar rats under ether anaesthesia the left patella was displaced and fixed by the lateral retinaculum to the lateral condyle of the femur. The rats were killed 3, 7, 10, 14 and 21 days after the operation. At 24 h before killing each animal received 2 μC/g body weight $^{35}SO_4$ intraperitoneally. One group of rats was used for an assay of incorporated radioactivity, while the remaining rats were used for autoradiography, testing histochemical reactions, staining the histological sections (with toluidine blue combined with hyaluronidase digestion), and electron microscopical investigations.

Results

Incorporation rates were significantly increased in the first 2 weeks by dislocation mainly in the femoropatellar and, later, in the femorotibial joints as well as generally in the operated joints, the costal cartilage and the intervertebral disc (Fig. 1). From the 2nd week the differences in incorporation values between the two arthrotic joints gradually declined.

Figures 2–4 present autoradiograms of the femoropatellar surface. At 7 days after operation a very intense concentration of grains could be seen in the cells and around the cells (Fig. 2). In the control group a mild, uniform concentration of grains was evident over the cells and matrix (Fig. 3). By 14 days after operation the concentration of grains over the cartilage had decreased (Fig. 4).

Surgery and Arthroscopy of the Knee
2nd Congress of the European Society
Editors: Müller/Hackenbruch
© Springer-Verlag Berlin Heidelberg 1988

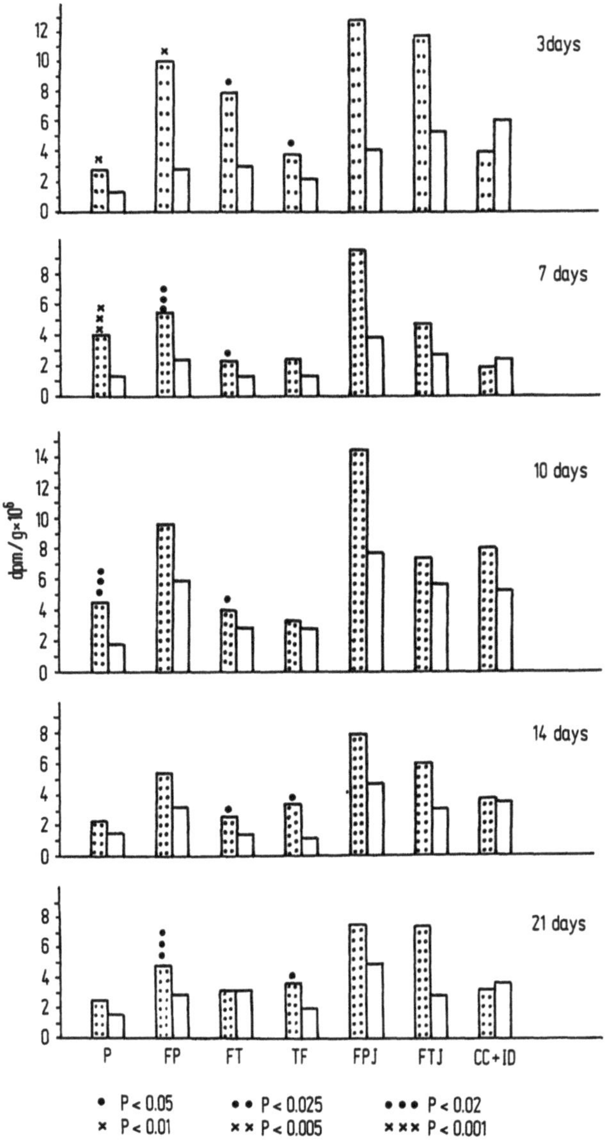

Fig. 1. Rates of $^{35}SO_4$ incorporation in femoropatellar and femorotibial joints during the first 3 weeks after operation. *P*, Patella; *FP*, patellar surface cartilage of femur; *FT*, tibial surface cartilage of femur; *TF*, femoral surface cartilage of tibia; *FPJ*, femoro-patellar joint; *FTJ*, femoro-tibial joint; *CC*, costal cartilage; *ID*, intervertebral disc; *shaded bars*, operated; *unshaded bars*, control

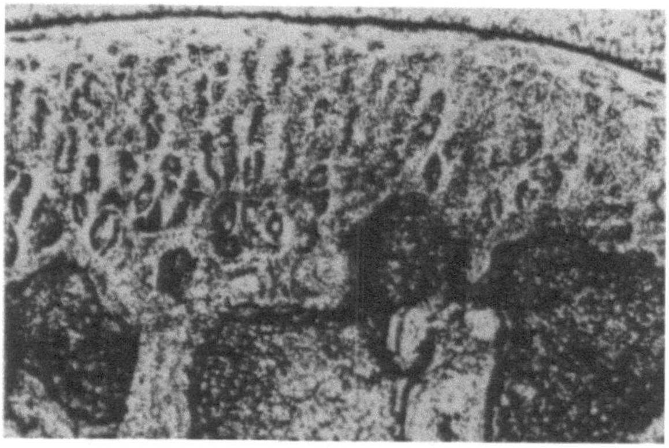

Fig. 2. Autoradiogram of femoropatellar surface, 1 week after operation (× 70)

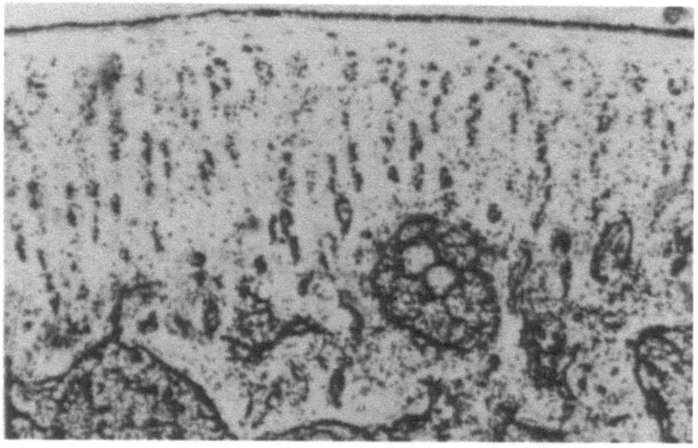

Fig. 3. Autoradiogram of femoropatellar surface, control (× 70)

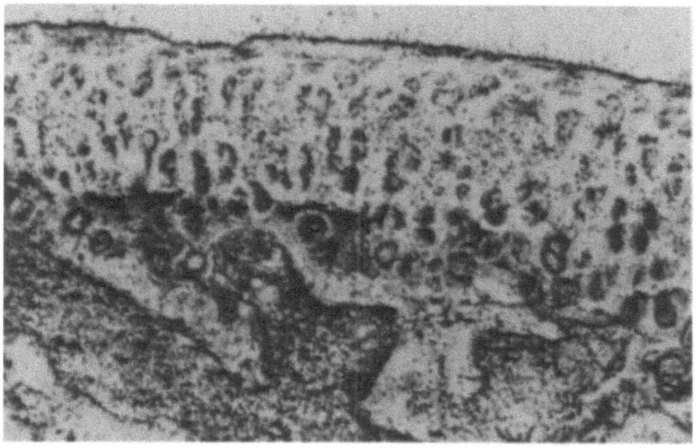

Fig. 4. Autoradiogram of femoropatellar surface, 2 weeks after operation (× 70)

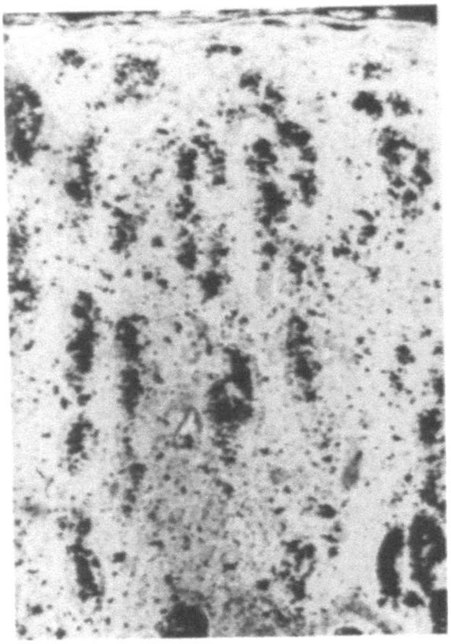

Fig. 5. Autoradiogram of tibiofemoral surface, 1 week after operation (×240)

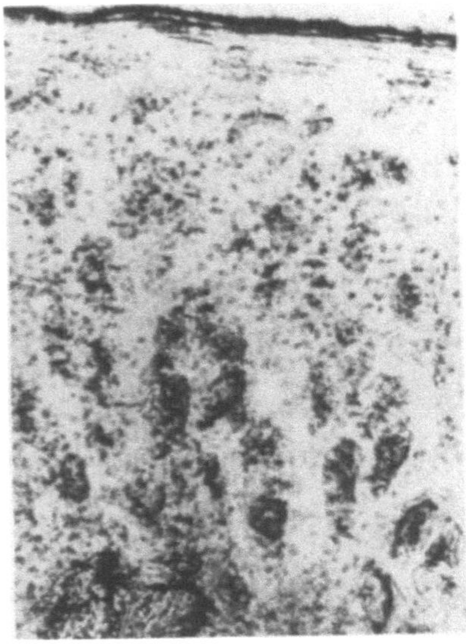

Fig. 6. Autoradiogram of tibiofemoral surface, 2 weeks after operation (×240)

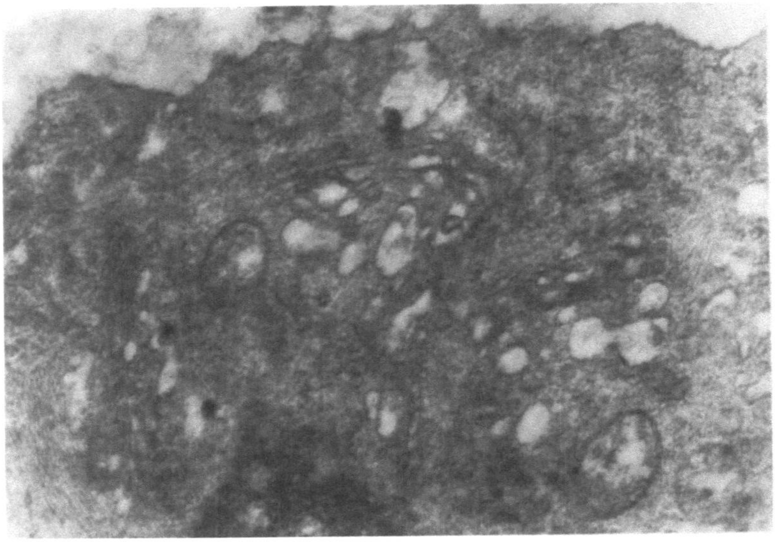

Fig. 7. Electron micrograph of cartilage cell containing well-developed Golgi complex (×66000)

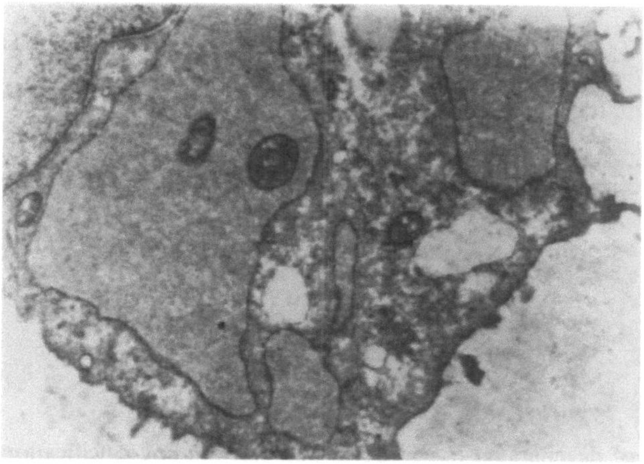

Fig. 8. Electron micrograph of cartilage cell showing the very dilated cisternae of endoplasmic reticulum (× 27 000)

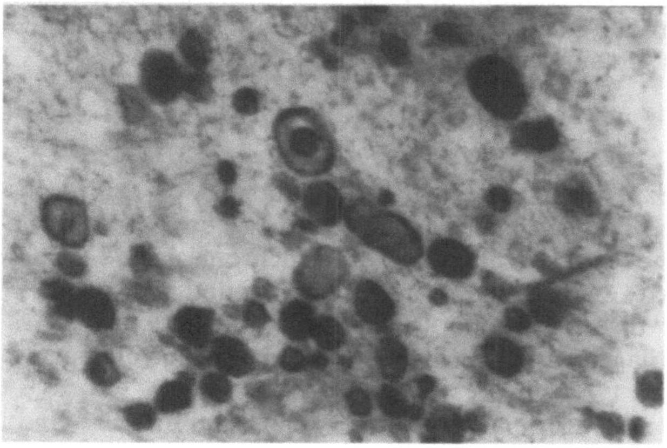

Fig. 9. Electron micrograph of extracellular matrix with lysosomes (× 57 000)

Figures 5 and 6 are autoradiograms of the tibiofemoral surface. At 7 days after the operation a mild, equal distribution of grains was found over the cartilage (Fig. 5), and at 14 days after operation a very strong concentration of grains could be seen in the cells and matrix, particularly in the middle and deep zones (Fig. 6). According to the toluidine blue staining 2 weeks after operation on the femorotibial joint, the metachromasia has decreased, compared with the control.
Figures 7–10 demonstrate the electron microscopical changes which had occurred by the stage of 10 days after the operation. The cell depicted in Fig. 7 contains a well-developed Golgi complex, which may be connected with the increased synthesis of GAGs. The very dilated cisternae of the endoplasmic reticulum seen in Fig. 8 may indicate inhibited collagen secretion. Black and grey lysosomes (Fig. 9)

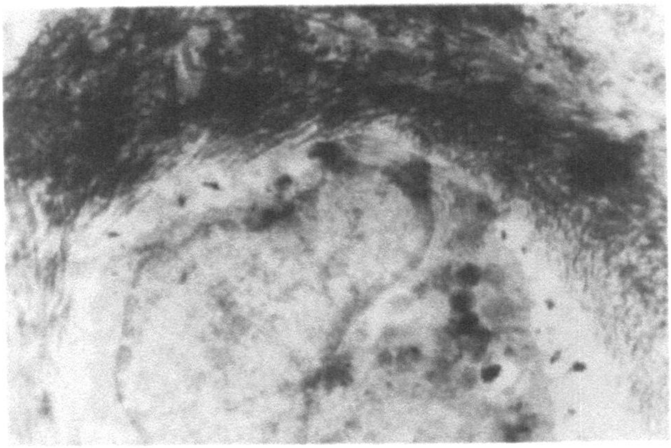

Fig. 10. Electron micrograph of degenerated chondrocyte and surrounding calcification (×18000)

may indicate the different stages of secretion of lysosomes; the lysosomic enzymes may digest the PG1s of matrix. Figure 10 demonstrates that in the deep zone the degenerated chondrocyte is surrounded by a strong calcification.

Summary

The isotopic results of our experiment indicate the stimulation of $^{35}SO_4$ incorporation by the reactive cells of damaged cartilage in the early stages of arthrotic development. By contrast, however, the alteration of metachromasia and the decrease in presence of oriented GAGs as shown by our polarization microscopical results (unpublished data) direct attention to the structural and qualitative changes of the articular cartilage matrix which are produced by dislocation of the patella.
The electron microscopical investigations provide an explanation for these findings by revealing the following:

1. The well-developed Golgi complex and the many microvesicles for the increased incorporation of radioactive sulphate and for the synthesis and release of GAGs
2. The dilated cisternae of the endoplasmic reticulum and the great amount of intracytoplasmic filaments for the inhibited secretion of collagen
3. The appearance of a great number of lysosomes in the matrix but increased degeneration of GAGs and consequently increased calcification in the structure of the matrix.

References

1. Barnett CH (1956) Wear and tear in joints. J Bone Joint Surg [Br] 28:567–575
2. Barz B (1982) Grundlegendes über Pathogenese und Therapie der Arthrose. Orthop Praxis 18:346–367

3. Bentley G (1975) Articular cartilage studies and osteoarthrosis. Ann R Coll Surg Engl 57:86–100
4. Dustmann HO, Puhl W, Martin K (1974) Der Einfluss intraartikulärer Arteparonininjektionen bei Arthrose. Z Orthop 112:1188–1196
5. Floman X, Eyre R, Glimcher MJ (1980) Induction of osteoarthrosis in the rabbit knee joint: Biochemical studies on the articular cartilage. Clin Orthop 147:278–286
6. Kalbhen DA, Scherbach E, Felten K (1981) Histologische Untersuchungen zur antiarthrotischen Wirkung von Tribenosid bei der tierexperimentellen Gonarthrose. Z Rheumatol 40:72–86
7. Kiss I, Földes I (1983) Effect of experimental patellar luxation on the knee joint in the rat. Acta Morphol Hung 34:371–385
8. Mohr W, Kirkpatrick CJ (1983) Die Relevanz von Tierversuchen für die menschliche Arthrose und die Beeinflussung der tierexperimentellen Arthrose durch Medikamente. Z Rheumatol 42:159–170
9. Pond MJ, Nuki G (1973) Experimentally induced osteoarthritis in the dog. Ann Rheum Dis 32:387
10. Simon WH (1978) Experimental degenerative joint disease: Freezing of rabbit articular cartilage in vivo. In: Simon WH (ed) The human joint in health and disease. University of Pennsylvania Press, Philadelphia, pp 122–131
11. Stockwell RA (1983) Metabolism of cartilage. In: Cartilage, vol 1. Structure, function and biochemistry. Academic, New York, pp 253–280
12. Stockwell RA, Sprinz R (1981) Glycosaminoglycan cartilage in experimental lipoarthrosis. J Anat 133:309–315

Experimental Investigations into the Reaction of Articular Cartilage and Synovial Membrane to Microtraumatical Impulsive Loading

H. Fengler, H. Schmidt, and K.-J. Schulze

Clinic for Orthopaedics, Medical Academy of Dresden, GDR

Introduction

The literature of the last four decades on orthopaedics and traumatology as well as that on sports medicine and industrial hygiene has frequently presumed that microtraumatic loading plays a major part in the aetiopathogenesis of degenerative joint diseases (Jurmain 1977; Franke 1979). This hypothesis cannot be proven using the human joint since microtraumatic loading occurs within the subclinical sphere.

In the early 1970s a number of authors, particularly Morscher (1974), continued this line of theory in articular pathology. They consider the microtrauma to be the smallest unit of an extraordinary force that achieves pathogenicity only by the summation of frequent repetition. Consequently, microtraumatisation could be possibly, particularly in activities with a repetitive and rapid load change. Such movements occur both in sports events and in industrial practice.

Late in the 1970s dynamometric investigations with the aid of force plates were made and ground reaction forces were measured on jumping and running individuals at the Institut für Angewandte Biomechanik der Eidgenössischen Technischen Hochschule Zürich. It was found that these movements result in initial peak forces of a few milliseconds duration, and therefore lie far below the latency period of musculature. They call these initial peak forces "passive loading" because, due to their short duration, they cannot be dampened by active musculoskeletal shock absorption (Nigg and Jachen 1980; Simon et al. 1981).

The impulse characteristics of these peak forces are affected by movement, footwear and type of training ground. Paul et al. (1978) published studies of animal experiments which proved that such impulses may be transmitted to bones and joints, with one portion being absorbed by the heel pad.

A number of work groups modelled passive repetitive impact loadings in animal experiments without measuring the impulse parameters. Radin et al. (1973, 1978) found a stiffening of the subchondral bone after a few week's loading of rabbit knee joints for 1 h a day with loading frequencies of 0.8 and 1.0 Hz before change in the articular cartilage occurred. That stiffening was regarded as being the consequence of the remodelling of multiple microfractures. Dekel and Weissman (1977) observed cartilage lesions after impulse loadings of rabbit knee joints over a time period of a few hours. This was associated with an enhanced synthesis of prostaglandins.

The damage pattern of the articular structure is likely to vary considerably, depending on the impulse characteristics. We carried out our own animal experi-

Surgery and Arthroscopy of the Knee
2nd Congress of the European Society
Editors: Müller/Hackenbruch

ments aimed at modelling an exactly measurable and reproducible repetitive impulse loading analogous to human impulse parameters. We assessed it by micromorphological and biochemical course checks for pathogenicity.

Methods and Results

Our test animals were young adult rabbits. The left knee joints, fixed under extension, were subjected to mechanical impulsive loading in vivo via the ankle joint and the tibia three times for 2 h each time and at 6-day intervals. The impulse measurements were taken by means of a piezoelectric measuring system with high sensitivity. The impulse parameters were made in analogy to long-distance race events. The plot of the impulse reveals an initial peak followed by a gradual rise in force with a smaller amplitude.

In order to be able to cover the course of the microtraumatic joint damage, investigations were carried out 1, 42 and 84 days after loading. This span of time appeared to be sufficiently long considering the life expectancy of the rabbit. This enabled us to comment on the prognosis of the joint.

Stereomicroscopic investigation was the first orientative method used for the assessment of the extent and the topographical localisation of the microtraumatic cartilage damage. Directly following loading, focal cartilage damage of the femoral condyle, in the form of reduced gloss, was recorded. Later, there was progressing fibrillation and a loss of proteoglycans shown here in a staining with safranin O. Safranin O binds the polyanions of the glucosaminoglycans stoichiometrically.

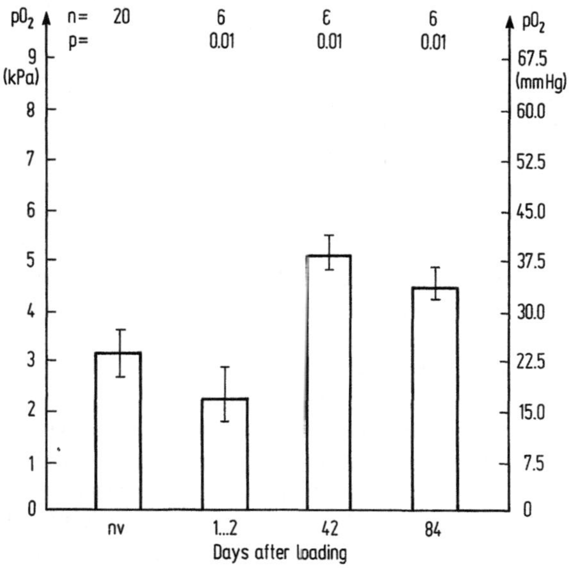

Fig. 1. Measurements of the intra-articular oxygen partial pressure in vivo after microtraumatical joint loading compared to a normal value (*nv*) with unloaded animals. *n*, number of animals; *p*, results of the *t*-test

By scanning electron microscopy the joint surfac were investigated in detail with × 300–3000 magnification. In the center of the cartilage damage irregular fibrillary structures 5–20 μm in width could be found in the absence of the normal surface relief. The stereomicroscopically intact tibial joint surfaces show fibrillary structures 2–5 μm in width when the normal surface relief is preserved. At 84 days after loading, the joint surface is subject to superficial disintegration in the centre of the cartilage damage by cell necroses and flaking of the superficial layer. The minimal lesions of the tibial joint surfaces also progress by microfissures and the exposure of superficial fibre textures. To describe the vertical distribution pattern of the cartilage damage by cell necrosis and flaking of the superficial layer. The ing surface irregularities and a reduction in the proteoglycans of the intermediate layer were detected. After 42 days a microstructural chondropathy had developed, which was distinguished by fibrillations, loss in proteoglycans and hypocellularity. Proliferative reactions can only be proven to a limited extent. The cartilage damage was evaluated numerically by using a histological-histochemical grading. The microstructural chondropathy established after 42 days showed no progression over the remaining period of observation. The tibial joint surfaces remained histologically unchanged.
For histological investigations of the synovial membrane, stepped sections were made with Goldner staining to assess the cellular alterations, with Azan staining to study the synovial fibre structure, and finally, with the van Gieson staining combined with elastica staining. By referring to 53000 morphological pieces of data, a computerised evaluation wad made. Immediately after loading there was a minor hyperplasia and hypertrophy of the synovial lining cells. Generalised exudative mechanisms were associated with a compression of the subsynovial capillaries. Further, the lining cells underwent degenerative changes. The connective tissue developed a irreversible coarse fibrosis concomitant with a concentric thickening of the walls and an obliteration of the vessels.
In order for synovial environmental changes to be measured, intra-articular pO_2 measurements in vivo were taken for the first time by the use of puncturing microelectrodes. Right after loading, a synovial hypoxia was discernible. In the morphological investigations of the synovial membrane, it was found that the hypoxia was due to the synovial exudation, which induced a generalised ischaemia. Over the later period of observation the synovial pO_2 was persistently increased (Fig. 1). Synovial hyperaemia as the cause can be excluded after histological investigations. Our findings corroborate the fact that at this stage there is an insufficiency of the synovial lining cell layer that is concomitant with a reduced oxygen consumption, thus causing the intra-articular pO_2 increase.
As a result of the animal experimental study it can be stated that episodic microtraumatic joint loading may lead to cartilage lesions, the consequence of which is the development of a microstructural chondropathy with little progression. According to our investigations even minimal lesions of the joint surface are irreversible. Microtraumatic joint loadings, in addition to cartilage damage, result in irreversible generalised lesions of the synovial membrane associated with microvascular disturbances and alterations of the synovial environment.
It is especially essential to sports medicine and industrial hygiene that methods be devised for the detection and prevention of microtraumatic loading situations.

References

1. Dekel S, Weissmann SL (1978) Joint changes after overuse and peak overloading of rabbit knees in vivo. Acta Orthop Scand 49:519–528
2. Franke K (1979) Knorpelschäden am Kniegelenk durch Fehlbelastung und Trauma. Med Sport 19:1–6
3. Jurmain RD (1977) Stress and the etiology of osteoarthritis. Am J Phys Anthropol 46:353–366
4. Morscher E (1974) Mikrotrauma und traumatische Knorpelschäden als Arthroseursache. Z Unfallmed Berufskr 4:220–231
5. Nigg BM, Jachen D (1980) Belastung des menschlichen Bewegungsapparates aus der Sicht der Biomechanik. In: Nigg BM, Jachen D (eds) Sportbeläge. Juris, Zürich, S. 31–77
6. Paul et al. (1978) Musculo-skeletal shock absorption: relative contribution of bone and soft tissues at various frequencies. J Biomech 11:237–239
7. Radin EL et al. (1973) Response of joints to impact loading. III. Relationship between trabecular microfracture and cartilage degeneration. J Biomech 6:51–57
8. Radin EL et al. (1978) Effect of repetitive impulsive loading on the knee joints of rabbits. Clin Orthop 131:288–293
9. Simon RS et al. (1981) Peak dynamic force in human gait. J Biomech 14:817–822

Electron Microscope Findings After Cartilage Shaving

A. Schmid, F. Schmid, and T. Tiling

Department of Surgery, University Clinic, Robert-Koch-Str. 40, 3400 Göttingen, FRG

Introduction

The aim of cartilage shaving is to restore a smooth articular surface after injury (Insall 1967; O'Connor 1982; Outerbridge 1964). By means of ultrastructural investigations we have attempted to find out whether this aim can be maintained over time.

Materials and Methods

Male and female patients between the ages of 18 and 42 years who had suffered traumatic damage to the cartilage of the medial femur condyle without any additional knee injury served as subjects in this study. The accidents were reported to have taken place at most 6 months beforehand. Patients with previous damage to the knee were excluded.

The cartilage damage was graded according to arthroscopic-macroscopic aspects. The selected method of grading involved examination of the cartilage by means of a special hook probe, accompanied by palpation. First degree cartilage injury was signified by small lacerations on the surface of the cartilage. These lacerations were not deep enough for the hook probe to be inserted into them. In the case of a more diffuse surface injury, the probe did not extend below the level of the cartilage surface. Second degree injury was characterised by the complete disappearance of the hook probe below the cartilage surface. There was no palpatory

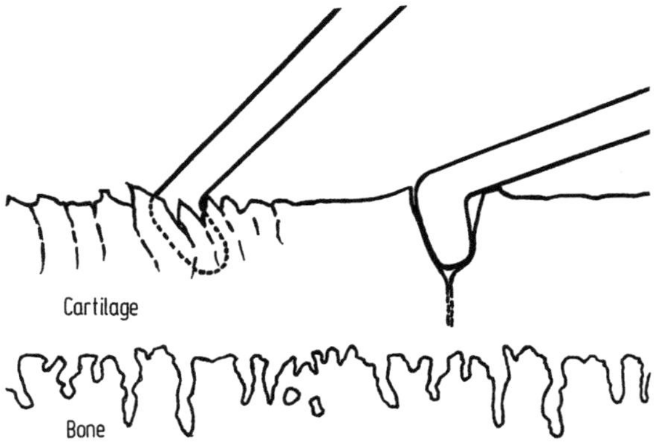

Fig. 1. Second degree cartilage injury

Surgery and Arthroscopy of the Knee
2nd Congress of the European Society
Editors: Müller/Hackenbruch

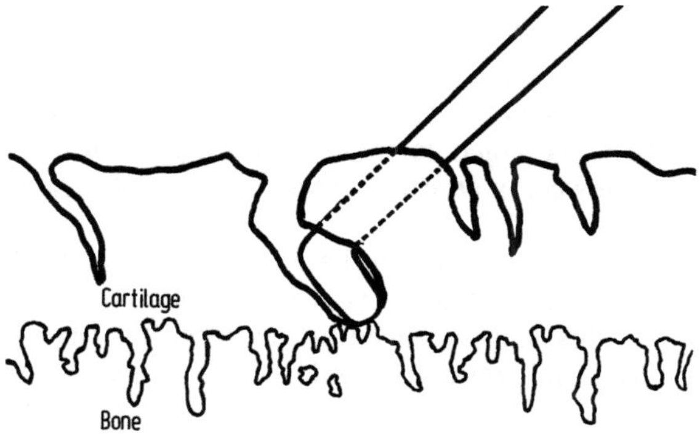

Fig. 2. Third degree cartilage injury

Table 1. Number of patients by treatment group and age

Age (years)	Second grade cartilage damage		Third grade cartilage damage	
	Shaving	Functional	Shaving	Functional
18–20	3	5	2	3
21–35	8	10	5	6
36–42	6	6	3	2

contact with the bone (Fig. 1). Third degree injury was defined by palpatory bone contact. However, at no point was more than 0.5 cm of subchondral bone visible (Fig. 2).

In patients with a second or third degree injury, cartilage shaving was performed. In another group of patients the same size as the above group and with a similar degree of injury, the cartilage damage was treated functionally (Table 1). Selection for each of the therapy groups was random.

Cartilage shaving was carried out as follows: in the case of a second degree injury, the cartilage in the vicinity of the injury was removed to such an extent that after shaving the hook probe no longer dipped below the niveau of the surrounding cartilage. In the case of a third degree injury, no bleeding from the subchondral layer should be evident. Shaving was carried out manually, without power-driven instruments, using an arthroscopic technique. Patients were allowed to put weight on the joint after the shaving.

During the first arthroscopy and during the control arthroscopy, which took place between 6 months and 2.5 years after the initial arthroscopy, samples of tissue were taken from both groups of patients with a punch needle 2.3 mm in diameter. The tissue samples were all investigated by light and electron microscopy. By means of morphometrical investigations of the cell vitality, cell mortality and pathological alterations of individual cells, quantitative results could be established.

Results

The ultrastructural investigations of samples taken within the first 4 weeks following trauma showed an increase in the rate of cell necrosis in the superficial and radial cartilage layer compared with undamaged cartilage. There was no difference in the number of cell necroses between a second degree and a third degree cartilage injury.

Investigations of samples taken between 2 and 6 months after trauma showed reactive changes in the superficial and radial cartilage layers, especially in patients between the ages of 18 and 20 years. This was evidenced by metabolically highly active cells with enlarged rough endoplasmic reticulum and nucleoli as well as an increase in the number of Golgi complexes (Fig. 3). In the case of two older patients, we found 5 months after trauma brood capsules in the radial layer as a response (Fig. 4).

For patients between 18 and 20 years old, who had not undergone cartilage shaving, the electron microscopic investigations of the control arthroscopy demonstrated a regular layering of cartilage with preservation of the superficial layer and tangential alignment of the surface fibres (Fig. 5). The fraction of degenerated chondrocytes with an increase of intracytoplasmic filaments was higher than that measured directly following trauma (Fig. 6). There was no evidence of transmission of injury to deeper layers of cartilage.

In the case of older patients without cartilage shaving we observed increased fibrillation of the surface layer of cartilage. However, the tangential layer with predominating tangential run of the fibres was still evident. The percentage and extent of degenerative changes as well as transmission into lower layers were increased as compared with initial findings. Cartilage cell clones were found in the

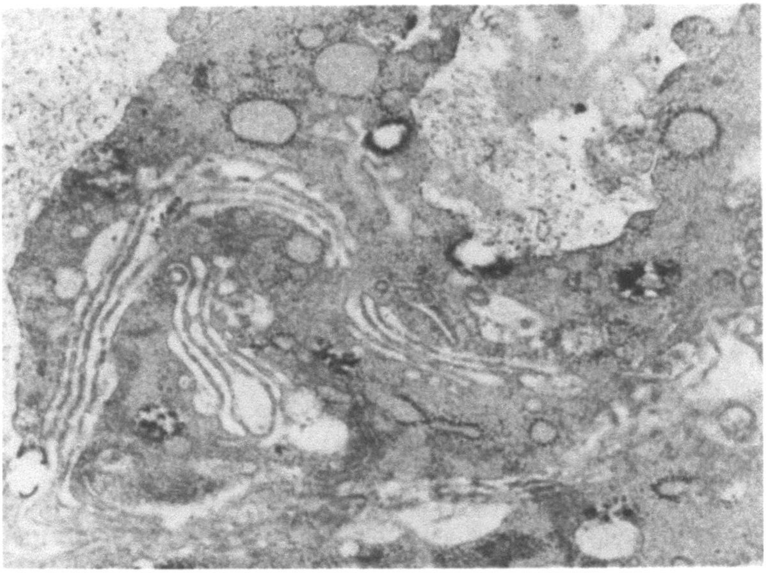

Fig. 3. Metabolically active cell from the radial cartilage layer (× 32 500)

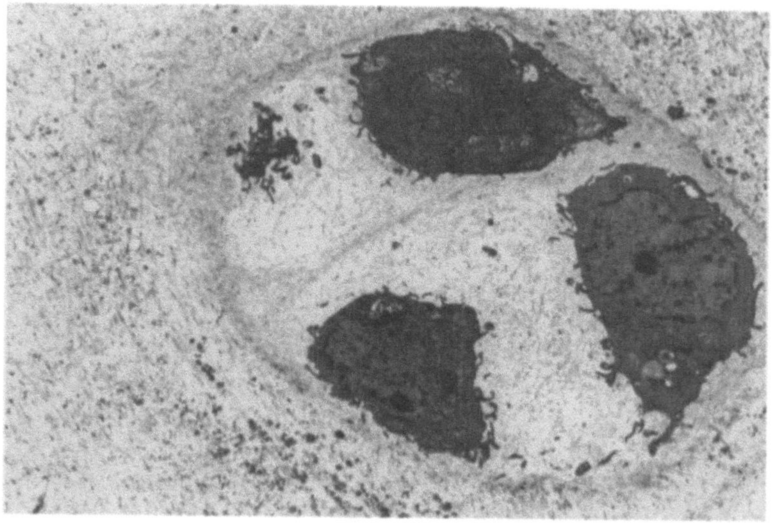

Fig. 4. Brood capsule from the radial cartilage layer (× 32 500)

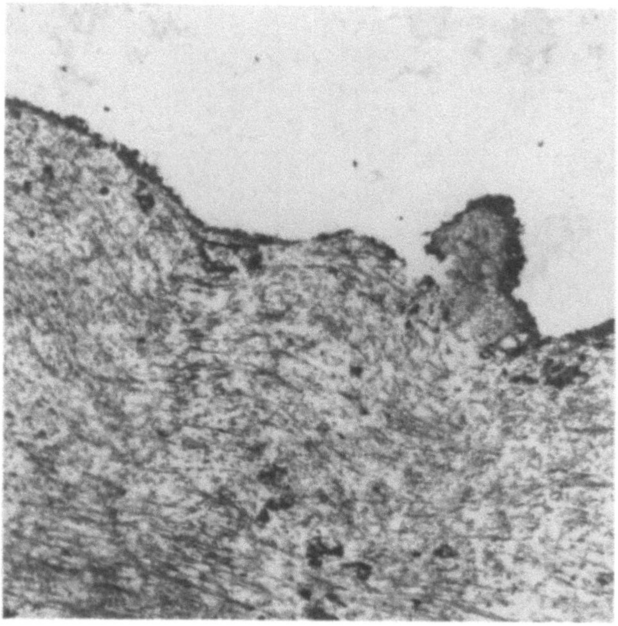

Fig. 5. Tangential alignment of the surface fibres in an 18-year-old patient who has not undergone cartilage shaving (× 12 300)

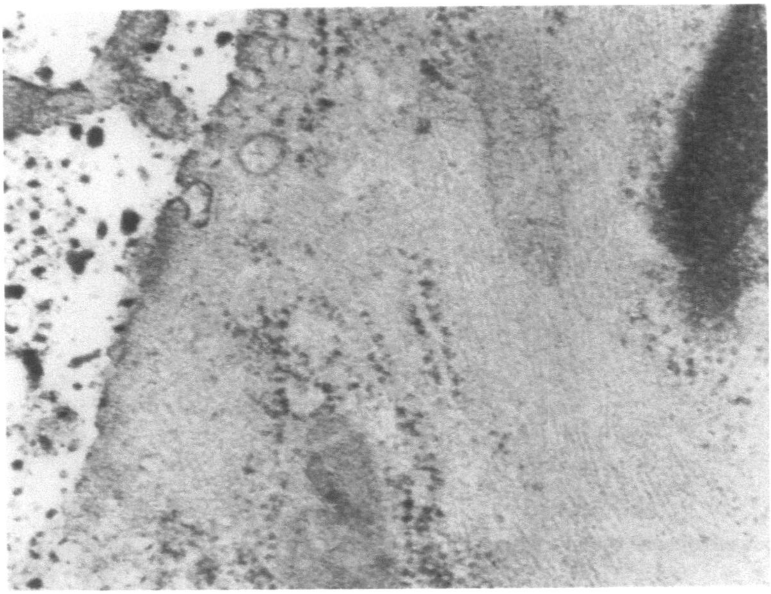

Fig. 6. Chondrocyte from a 25-year-old patient who has not undergone cartilage shaving. An increase in intracytoplasmic filaments is obvious (×60000)

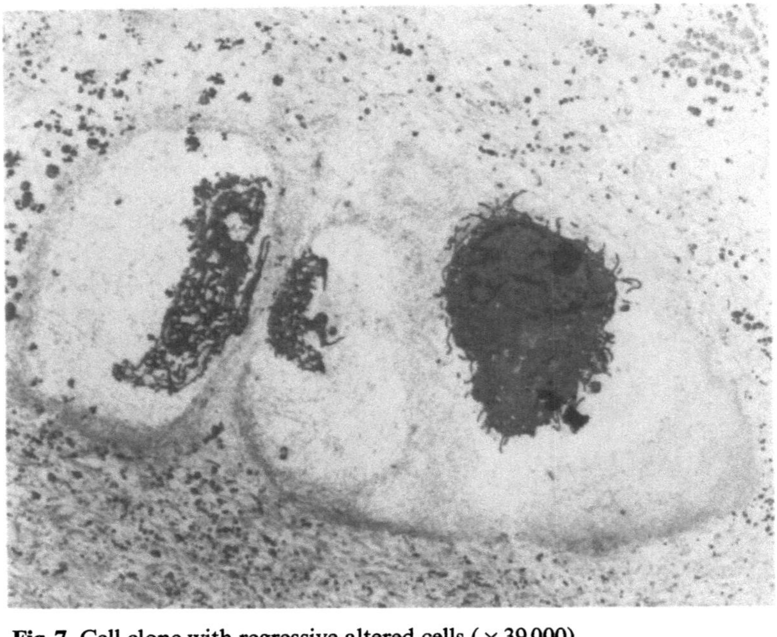

Fig. 7. Cell clone with regressive altered cells (×39000)

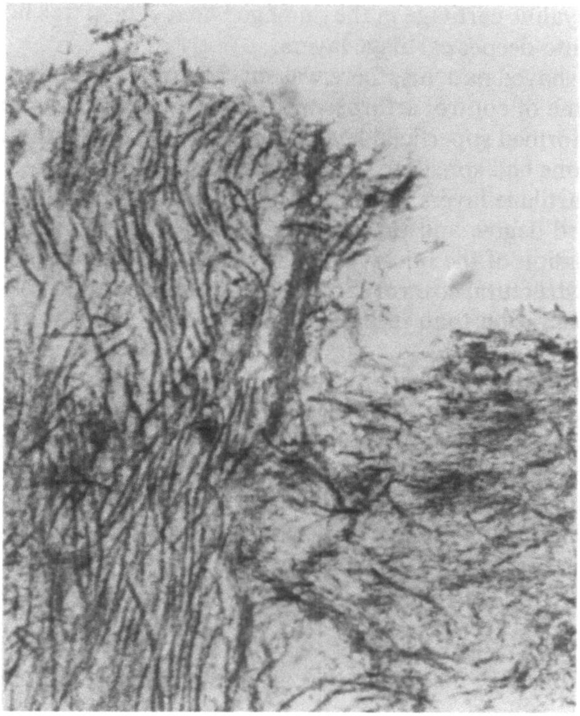

Fig. 8. Cartilage surface of a 23-year-old patient 1 year after cartilage shaving (× 32 500)

upper and middle layers. They contained mostly degeneratively altered cells (Fig. 7).

Turning to the patients who underwent cartilage shaving, there was not a single case in which a new superficial layer had been formed (Fig. 8). For the most part, the middle cartilage layer formed the articular surface. The tissue had a strikingly low cell count, with much cell necrosis. Degenerative alteration had increased in comparison with initial testing. In both younger and older patients the deeper cartilage layers had been affected also.

Discussion

How are these macroscopic and ultrastructural results to be interpreted? At the time of the initial arthroscopy, there was no ultrastructural difference at the cellular level between cartilage damage of the second and third degree. In both younger and older patients there was evidence of increased metabolic activity of the cells in the superficial layer. The cells showed age-dependent differences.

At the time of the macroscopic control arthroscopy, all functionally treated patients had kept the original degree of cartilage injury. This maintenance is probably due to metabolic performance of the active chondrocytes of the superficial layer. The younger patients with a second degree injury displayed ultrastructur-

ally a regular layering of the hyaline cartilage in the damaged area. There was no transmission of initial injury into deeper cartilage layers.
In the case of young cartilage-shaved patients, there was macroscopically no regular cartilage surface at the time of control arthroscopy. Ultrastructurally, none of these patients had a newly formed superficial layer. Compared with the initial findings, pathological alterations had spread to deeper cartilage layers.
We therefore conclude that cartilage layers are laid bare in shaving of cartilage injuries of the second and third degree and this neither effects a reparation nor prevents a downward transmission of the injury. Also, in the case of older cartilage-shaved patients, the ultrastructural control showed a greater extent of transmission of the degenerative alteration than in patients of the same age who received no cartilage shaving.
It seems that leaving the superficial layer where it is has a protective effect on the deeper layers of cartilage. Our results show that cartilage shaving in the case of second and third degree injury has no curative value. The aim of regaining a smooth articular surface is not realised in the long run. We were able to demonstrate that the spread of the injury is even accelerated. We are in favour of cartilage shaving as a palliative therapy in cartilage protrusions, in which case it prevents a further increase in protrusion of the cartilage through mechanical stress.

References

1. Insall NJ (1967) Intra-articular surgery for degenerative arthritis of the knee. J Bone Joint Surg [Br] 49:211
2. O'Connor RL, Shahriaree H (1982) Seven year follow-up on arthroscopic debridemend of the degenerative knee. Presented at the American Academy of Orthopaedic Surgeons, New Orleans, Louisiana
3. Outerbridge RE (1964) Further studies on the etiology of chondromalacia patellae. J Bone Joint Surg [Br] 46:179

A New Method of Depiciting the Relationship Between Patella and the Femur

K. W. Sloan

354 Old Hook Road, Westwood, New Jersey 07675, USA

At the present time we have developed a system using a CAD-CAM computer to predict and demonstrate the motion of the patella both actively and passively. Using this one can actually see the relationships between the patella and the femoral joint both during active and passive motion throughout the entire range of the knee.

This has produced a new method of depicting the relationship between the patella and the femur. The CAD-CAM methodology will allow the physician and surgeon to determine better correct treatment regimens for each patient. Then the multiple different types of treatment regimens and how they affect the motion of the patella and the stability of the patella femoral joint can be evaluated. The Genu-com is the integral part of this paper.

Surgery and Arthroscopy of the Knee
2nd Congress of the European Society
Editors: Müller/Hackenbruch

The Three-Dimensional Tracking Pattern of the Patella in the Human Knee Joint and the Effects of Surgical Interventions

A. van Kampen [1], R. Huiskes [2], L. Blankevoort [2], and T. J. G. Van Rens [1] †

[1] Institute of Orthopaedics, [2] Laboratory of Experimental Orthopaedics, University of Nijmegen, The Netherlands

Introduction

Precise causes for retropatellar chondromalacia are still unknown. It is generally believed that patellar tracking patterns, contact areas, pressure distribution, contact incongruence and cartilage deficiencies play a role in its aetiology [2, 3, 5, 9, 18]. Fundamental research concerning the patello-femoral joint is still incomplete and consequently so also is the scientific basis for various surgical interventions. The kinematic behaviour of the patella versus the femur has not yet been thoroughly investigated, and data from the literature, as summarized in Table 1, are inconsistent [6, 13, 16, 17].
The first objective of the present investigation was to provide an accurate data base for the normal three-dimensional tracking pattern of the patella. Operations for chondromalacia are often based on theory, intuitional convention or empirical observations. It is unknown whether and to what extent various operations cause a rerouting of the patellar tracking mechanism. The effects on the patellar tracking patterns of two operations, that of lateral retinacular release and that of tubercle elevation, both popular treatments for chondromalacia, are presented and discussed here.

Methods

Four postmortem leg specimens were fixed onto an apparatus allowing for prescribed flexion motion, free tibial rotations following prescribed tibial rates of torque and free varus-valgus rotations (Fig. 1). The four heads of the quadriceps muscle were loaded separately with static forces. Five tantalum markers (0.8–1.0 mm) each were inserted into the patella, the femur and the tibia. The knee was flexed from 0°–150° in nine steps. At each flexion position X-ray exposures were taken. The spatial coordinates of the markers as measured from the exposures were reconstructed in each position, using a roentgen stereophotogrammetric analysis (RSA) system [10, 15]. The bones were provided with coordinate systems and their relative motions expressed in terms of translation vectors along the axes and in terms of Euler rotation angles about the axes (Fig. 2). Each test series was carried out for the neutral flexion pathway and for externally and internally rotated pathways (± 3 N m tibial torque), both with total Q-forces of 110 N. These tests were repeated after lateral retinacular release and after tubercle elevations of 0.5 cm and 1.0 cm.

Surgery and Arthroscopy of the Knee
2nd Congress of the European Society
Editors: Müller/Hackenbruch

Table 1. Differences in patellar rotations and patellar translation from the literature

Reference	Number of speciemens		Q-loading	Measurements		Tibial rotation	Average patellar movements		
	In vivo	In vitro		Number	Range		Shift	Rotation	Tilt
Veress et al. (1979)	4		Isometric	4	0°– 90°	Neutral	Lateral	(Unknown)	Inconclusive
Sikorski et al. (1979)	12		Isometric	3	60°– 90°	Neutral	(Unknown)	Lateral	Medial
Reider et al. (1981)		20	92 N	7	0°– 90°	Fixed	*Type I*		
							Lateral 14 mm	Lateral 6°	Lateral 12°
							Type II		
							Medial 7 mm	Lateral 6°	Neutral 0°
Fujikawa et al. (1983)		8	20–30 N	6	25°–130°	(Unknown)	(Unknown)	Lateral 6.2°	Medial 11°

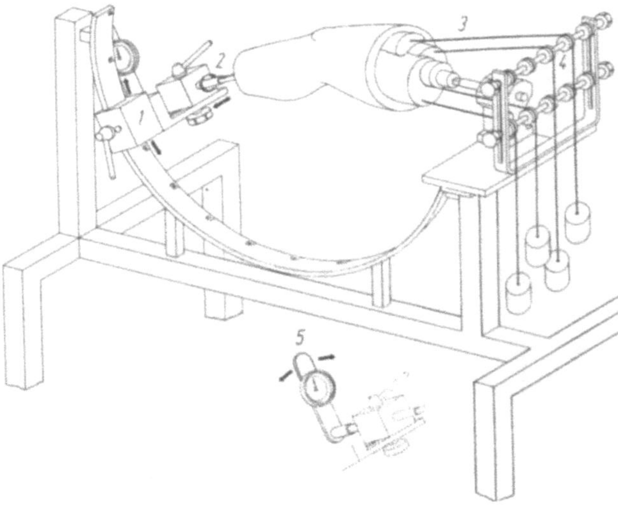

Fig. 1. Motion. *1* Clamp for flexion/extension along the rail; *2* clamp for tibial internal/external rotation and varus/valgus rotation; *3* wires, attached to the quadriceps muscle; *4* pulleys to guide the wires in the different muscle directions; *5* torque wrench

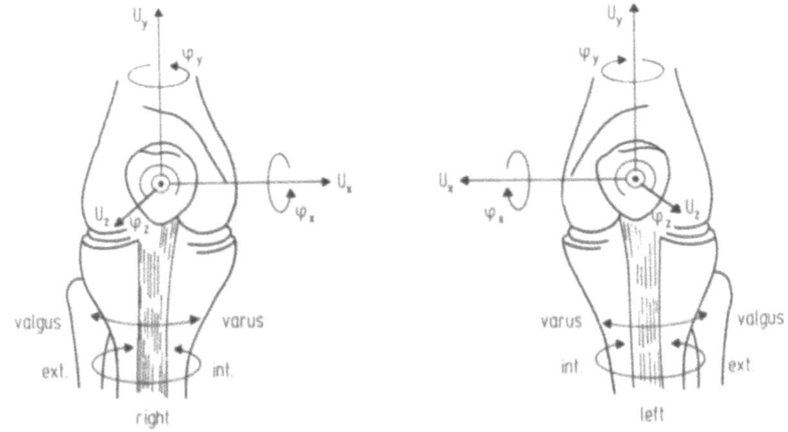

Fig. 2. Orientation and location of the coordinate axes of the patella. Subsequent rotations ($Q_{x,y,z}$) around the axes ($U_{x,y,z}$) represent patellar flexion (Qx), patellar tilt (Qy) and patellar rotation (Qz). Translation along the x-axis (Ux) represents patellar shift. For reasons of clarity, the coordinate systems of the femur and the tibia are not shown (see text)

In Fig. 3 the nomenclature of the different patellar movements is presented. Patellar rotation around the femoral x-axis describes patellar flexion (positive) or extension (negative). Rotation around the patellar y-axis describes medial tilt (positive) as the medial patellar facet rotates towards the medial femoral condyle, and lateral tilt (negative) as the lateral patellar facet rotates towards the lateral femoral condyle. Rotation around the patellar z-axis describes medial patellar rota-

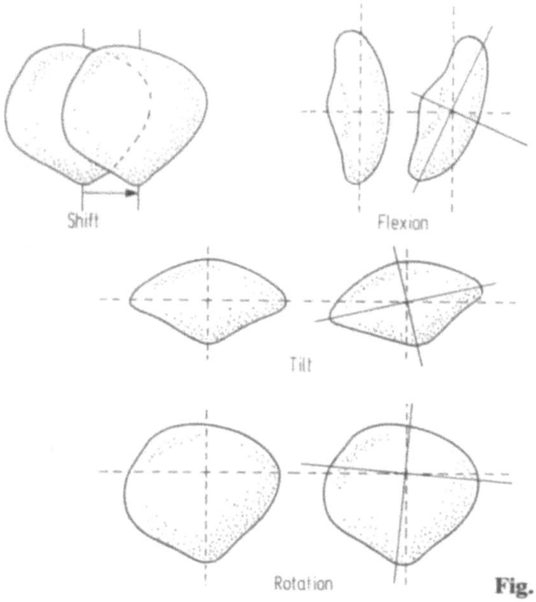

Fig. 3. Patellar rotations and patellar shift

tion (positive) as the patellar apex turns towards the medial condyle, and lateral patellar rotation (negative) as the patellar apex turns towards the lateral condyle. Translation along the patellar x-axis will be described as patellar shift: medial (positive) and lateral (negative).

Results

The precision of three-dimensional reconstruction was higher than 50 μm, resulting in rotation precisions of approximately 0.1°. Retesting one specimen resulted in maximal deviations of 5%, indicating excellent reproducibility per specimen. Effects from variation in quadriceps force magnitude were negligible. Although varying in absolute magnitudes, all specimens showed the same trends in tibial and patellar movements with increasing flexion. An example is shown in Fig. 4, in which tibial exo-endorotation, tibial varus/valgus rotation, patellar flexion, rotation, tilt and shift are shown as a function of tibial flexion, for both the internally and the externally rotated motion pathways.

Maximum tibial exo-/endorotatory laxity (approximately 35°) is reached after approximately 30° knee flexion and declines only in very high knee flexion (above 130°) due to reduced exorotation. Tibial flexion is accompanied by a slight tibial varus rotation, which mechanism is increased by external rotation of the tibia. Patellar flexion occurs in concert with knee flexion, as expected, but lags by about 20%. The flexion lag is more pronounced after approximately 105° knee flexion, as established for all specimens.

Patellar rotation is directed medially; the patellar apex moves towards the medial femoral condyle. Significant rotation occurs above 40° knee flexion. Internal ti-

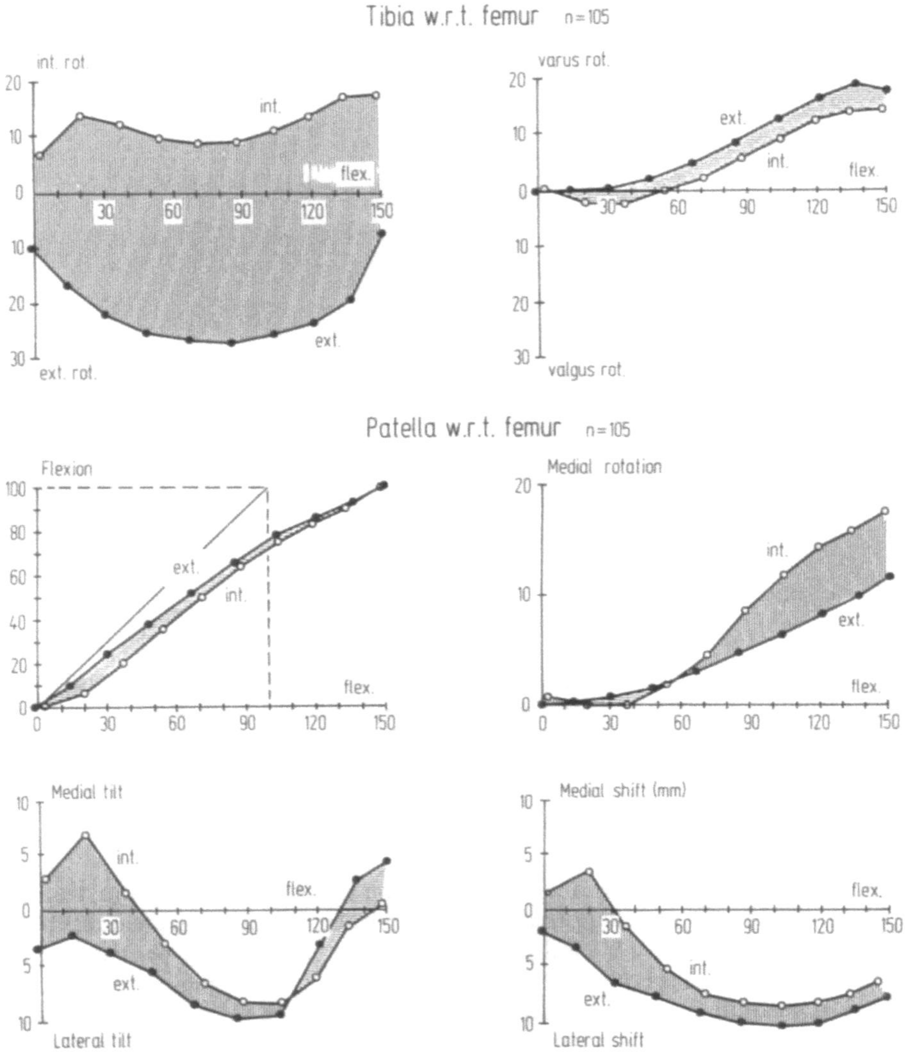

Fig. 4. An example of tibial and patellar rotations and shift versus the femur in one specimen. *Open circles,* internal tibial rotation pathways; *closed circles,* external tibial rotation pathways (±3 Nm); w.r.t., with respect to

bial rotation increases the medial patellar rotation. The tilt movement of the patella can be described as "wavering", for during flexion the patella wavers from a medially to a laterally tilted position and vice versa. The patellar tilt is highly influenced by tibial rotations, especially towards full extension. This is understandable when one considers the clinically movable patella in extension, which becomes stuck in its patellofemoral groove with further knee flexion. Along the external rotation envelope (the limit of patellar rotation induced by tibial rotation) medial tilt is converted to lateral tilt or, at least, to less pronounced medial

tilt. A possible explanation for this phenomenon is that by tibial exorotation, lateral knee structures and especially the more laterally oriented patellar ligament pull at the patella, which causes the patella to tilt laterally.
At approximately 100° knee flexion the patella exhibits a sudden jerk towards a medially tilted position, probably caused by loading of the most medial (odd) patellar facet. Translation of the patella in the frontal plane, the patellar shift, is pointed laterally and is progressive upon further knee flexion. Along the tibial exorotation motion pathway the patellar lateralisation is more pronounced.
The patellar motions relative to the femur, along the envelope of tibiofemoral flexion, did not show very significant changes in flexion, tilt, rotation or shift, after performance of a lateral retinacular release, as illustrated in Fig. 5. These trends were found in all specimens. After a tubercle elevation of 1.0 cm, according to Bandi [1], the changes in patellar flexion, tilt and shift were minimal. However, patellar rotation is evidently increased, both along the internal and the external rotated pathways (Fig. 6). Although the other specimens also showed changes in patellar tracking after the tubercle elevation, no consistent trend could be established.

Discussion

For the determination of three-dimensional tracking patterns of the patella a high precision measuring system (RSA) was used. Because of its high precision and the high reproducibility found here, the results represent true patellar motions, and general conclusions on the patellar tracking pattern may be drawn, although the number of investigated specimens was relatively small. The motions found in the four specimens varied in terms of absolute rotation and translation excursions, but the tendencies and relative motion patterns were remarkably similar. The differences that do occur can be explained by anatomical variability.
The three-dimensional patellar tracking patterns are characterised by a progressive patellar flexion, a wavering patellar tilt, a medial patellar rotation and a lateral patellar shift during knee flexion. The patellar movements can be partially explained by the bony configuration of the femoral groove. The anterior wall of the lateral femoral condyle causes in extension and during the first stage of flexion a medial patellar tilt. In the second stage of flexion a lateral patellar tilt is induced by the further caudally reaching medial condyle. The patellar tilt inversion towards a medial tilt in higher knee flexion represents the medial femoral condyle – the patellar odd-facet contact – in accordance with the results of studies on the contact patterns of the patello-femoral joint [6, 7, 14].
The medial patellar rotation and the lateral patellar shift are induced by the direction of the femoral groove, slightly laterally oriented, as seen from proximal to distal. By the pull of the patellar ligament the patella rotates medially with further knee flexion. This also explains the increment of medial patellar rotation caused by internal tibial rotation as an effect of further medialisation of the patellar tendon. In the same manner the reduced lateral patellar shift along the internal tibial rotation pathway is explained.

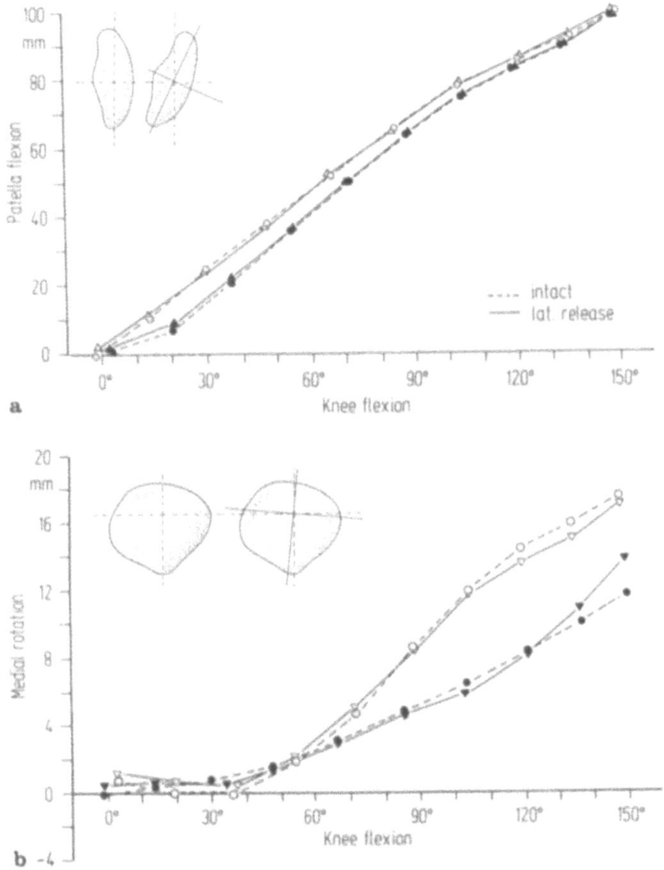

Fig. 5a–d. Effect of lateral retinacular release on patellar rotations and shift. **a** Patellar flexion, **b** patellar rotation, **c** patellar tilt, **d** patellar shift. *Open circles, triangles,* internal tibial rotation pathways; *closed circles, triangles,* external tibial rotation pathways; *broken lines,* before lateral release; *solid lines,* after lateral release

The three-dimensional tracking patterns are greatly influenced by tibial rotations: patellar tilt and shift more in the first stage of knee flexion and patellar rotation more in the second. At examination the patellar shift and tilt seem to be a coupled movement. Patellar rotation seems to follow the varus/valgus rotation of the tibia.

Lateral retinacular release for chondromalacia is intended to alter and reduce the pressure distribution in the patellofemoral joint through a medialisation of the resultant quadriceps force. This should result in a rerouting of the patellar tracking mechanism. The present results on the three-dimensional tracking pattern of the patella after lateral release, however, indicate that the patellar tracking pattern mechanism is determined mainly by the bony constraints and by the pull of the patellar ligament. It is not influenced by a lateral retinacular release, once it is

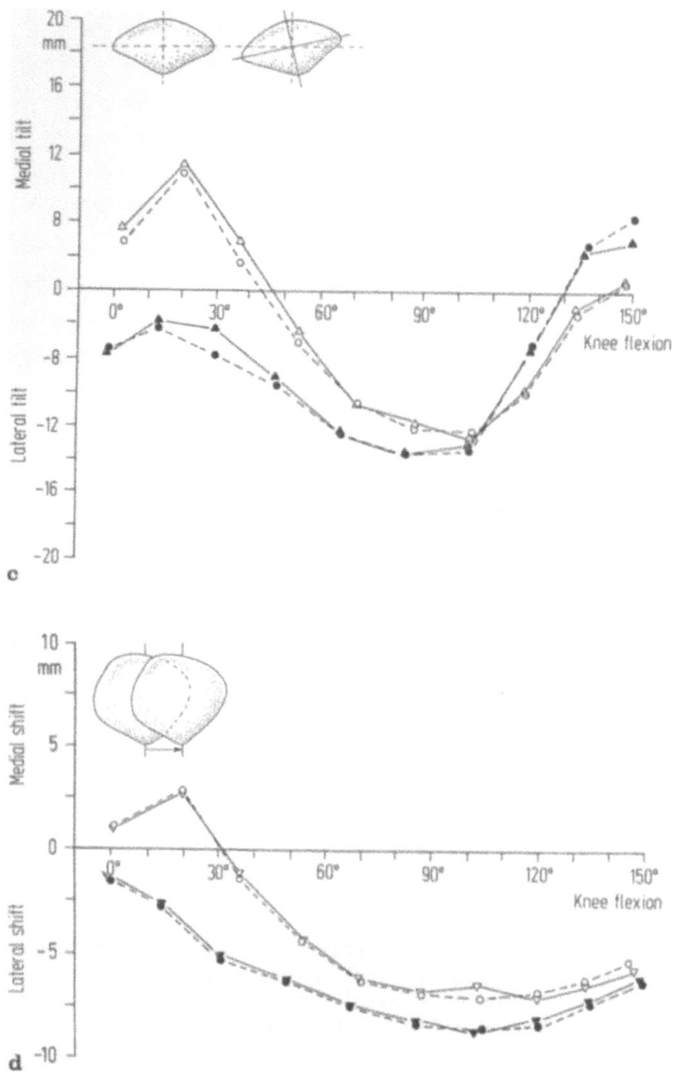

Fig. 5c, d.

pressed into the femoral groove in the greater knee flexion angles. Hence, it is likely that the shape of the contact areas, the pressure distribution and the contact area locations are not affected by this operation. This should discourage the use of this operative procedure in knees in which the lateral retinaculum is essentially normal. Loosening of the lateral retinaculum may be of value only in patients with a so-called tight retinaculum, diagnosis of which can be confirmed by axial X-ray exposures. If a beneficial effect of the lateral retinacular release in normal anatomical configuration exists, this may be explained by the cutting of the sensory nerve endings.

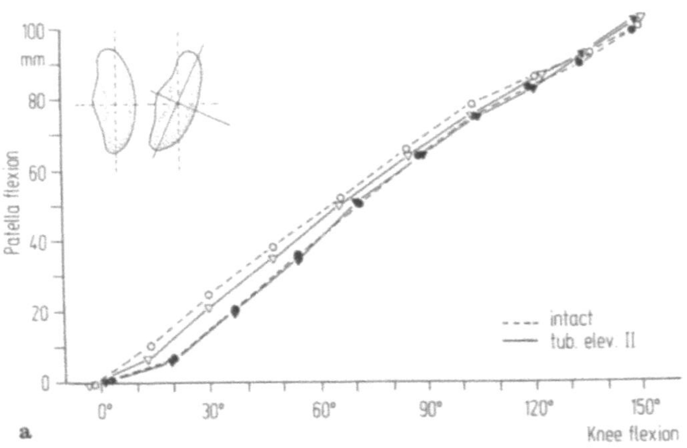

a

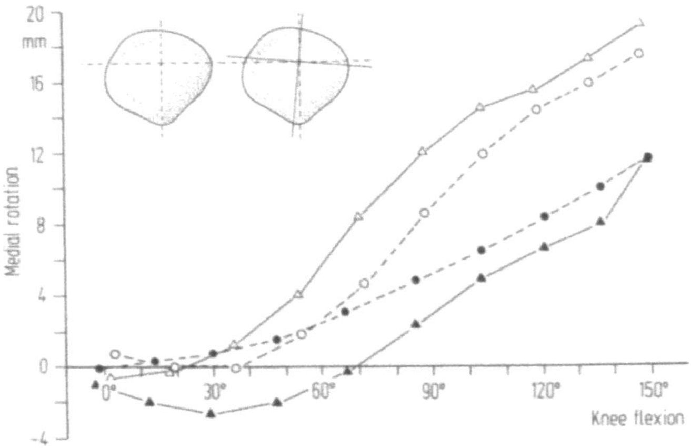

b

Fig. 6 a–d. Effect of tubercle elevation of 1.0 cm on patellar rotations and shift. **a** Patellar flexion, **b** patellar rotation, **c** patellar tilt, **d** patellar shift. *Open circles, triangles,* internal tibial rotation pathways; *closed circles, triangles,* external tibial rotation pathways; *broken lines,* before tubercle elevation; *solid lines,* after tubercle elevation

The objective of the tubercle elevation, reducing the patellofemoral contact force by increasing the moment arm of the patellar tendon, has been motivated by several studies [1, 4, 11, 12]. However, none of these studies took tibial rotations into account. Patellar rotations are greatly influenced by tibial rotations in the intact situation, and even more so after tubercle elevation. The latter is probably due to the increased medial/lateral translation of the tuberosity after elevation, as depicted in Fig. 7. The importance of a changing Q-angle on retropatellar contact areas and stresses has been emphasized by Huberti and Hayes [8], who found in an experimental study that increasing the Q-angle resulted in increased peak pressures and changes of the contact locations. Decreasing the Q-angle resulted in un-

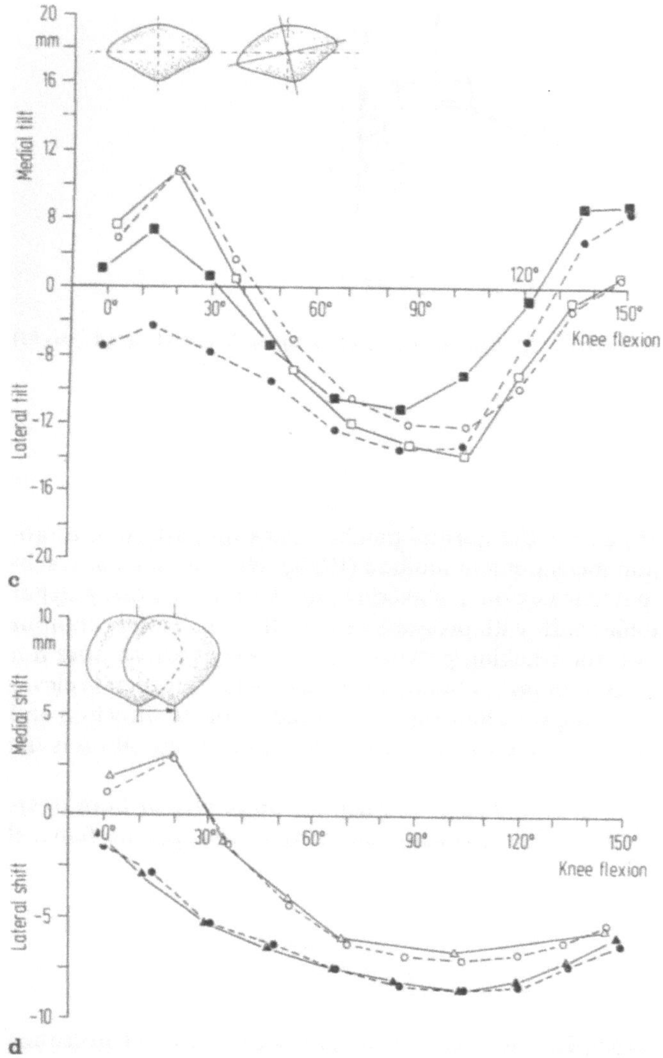

Fig. 6c, d.

loading of the vertical patellar ridge, but this was always associated with increased peak pressures at other locations. After tubercle elevation this may lead to even higher peak pressures with unpredictable contact areas due to the increased influence of the Q-angle on patellar tracking than to realising the objective of reducing the patello-femoral contact pressures.

From the present results it is evident that influences of tubercle elevation on patellar tracking pattern do exist, but their nature and extent are unpredictable. Therefore it is doubtful whether the objective of the tubercle elevation operation, the reduction of patello-femoral contact stresses, is actually accomplished.

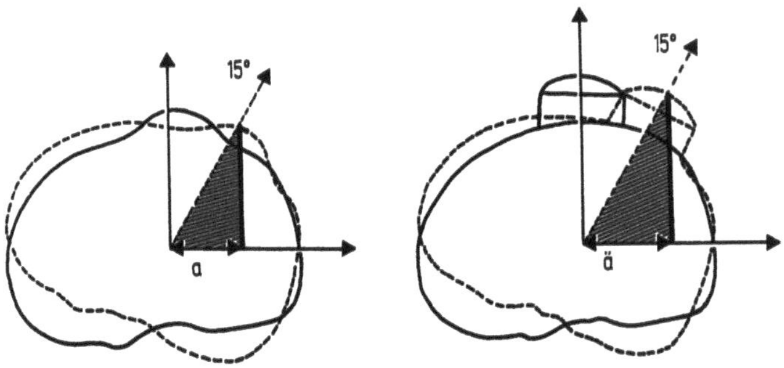

Fig. 7. Translation of tibial tuberosity (*a, left*) increases after tubercle elevation (*a″, right*) with the same tibial rotation (15°)

Summary

A data base has been offered for the normal patellar tracking pattern as established with a high precision measurement method (RSA). Normal patellar tracking is characterised by a patellar flexion, a wavering patellar tilt, a medial patellar rotation and a lateral patellar shift with progressive knee flexion. Tibial rotations have an extensive effect on the tracking pattern. The tracking pattern does not change after performance of a lateral retinacular release. After a tubercle elevation changes in patellar tracking do occur but are unpredictable in direction and magnitude. An important factor and one caused by the tubercle elevation is the change in Q-angle.

Based on the present results it is doubtful whether the objectives of both operations, rerouting the patella and/or reducing the compressive patello-femoral forces, are achieved.

References

1. Bandi W (1976) Vorlagerung der Tuberositas Tibiae bei Chondromalacia Patellae und femoropatellarer Arthrose. Hefte Unfallheilk 127:175–186
2. Bentley GCM, Dowd G, Orth MC (1984) Current concepts of etiology and treatment of chondromalacia patellae. Clin Orthop 189:209–228
3. Dandy PJ, Poirier H (1975) Chondromalacia and the unstable patella. Acta Orthop Scand 46:695–699
4. Ferguson AB, Brown TD, Fu FH, Rutkowski R (1979) Relief of patellofemoral contact stress by anterior displacement of the tibial tubercle. J Bone Joint Surg [Am] 61:159–166
5. Ficat P, Ficat C, Bailleux A (1975) Syndrome d'hyperpression externe de la rotule. Rev Chir Orthop 61:39–59
6. Fujikawa K, Seedhom BB, Wright V (1983) Biomechanics of the patellofemoral joint: part 1. Eng Med 12:3–11
7. Goodfellow J, Hungerford DS, Zindel M (1976) Patellofemoral joint mechanics and pathology: part 1. Functional anatomy of the patellofemoral joint. J Bone Joint Surg [Br] 58:287–290

8. Huberti HH, Hayes WC (1984) Patellofemoral contact pressures. The influence of Q-angle and tendofemoral contact. J Bone Joint Surg [Am] 66:715–724
9. Hughston JC, Walsh WM (1979) Proximal and distal reconstruction of the extensor mechanism for patellar subluxation. Clin Orthop 144:36–42
10. Huiskes R, Kremers J, Lange A de, Selvik G, Woltring HJ, Rens TJG van (1985) An analytical stereophotogrammetric method to determine the 3-D geometry of articular surfaces. J Biomech 18:559–570
11. Maquet P (1976) Advancement of the tibial tuberosity. Clin Orthop 115:225–230
12. Nakamura N, Ellis M, Seedhom BB (1985) Advancement of the tibial tubercle, a biomechanical study. J Bone Joint Surg [Br] 67:255–260
13. Reider B, Marshall JL, Ring B (1981) Patellar tracking. Clin Orthop 157:143–148
14. Seedhom BB, Takeda T, Tsubuku M, Wright V (1979) Mechanical factors and patellofemoral osteoarthrosis. Ann Rheum Dis 38:307–316
15. Selvik GA (1974) Roentgenstereophotogrammetric method for the study of the kinematics of the skeletal system. Dissertation, University of Lund, Sweden
16. Sikorski JM, Peters J, Watt I (1979) The importance of femoral rotation in chondromalacia patellae as shown by serial radiography. J Bone Joint Surg [Br] 61:435–442
17. Veress SA, Lippert FG, Hou MCY, Takamoto T (1979) Patellar tracking patterns measurement by analytical x-ray photogrammetry. J Biomech 12:639–650
18. Wiles PW, Andrews PS, Devos MB (1956) Chondromalacia of the patella. J Bone Joint Surg [Br] 38:95–113

Epidemiology of Chondromalacia (CHM)

K. Franke

Department of Surgery and Division of Sports Traumatology,
Municipal Hospital, GDR-1100 Berlin-Pankow

The result of knee joint surgery is influenced to a high degree by the state of the hyaline cartilage and synovium. The extent of chondromalacia (CHM) and synovitis and the possibility of their medical and surgical treatment are as decisive for the final functional loadability of the joint as exactly performed surgery and a sophisticated rehabilitation process.

There is a remarkable percentage of CHM seen in arthroscopy or open knee joint surgery. This might be a relative increase due to improved possibilities of diagnosis, but could also be an absolute increase due to a higher exposure to trauma and overloading, which are main factors in the etiology of CHM.

Autopsy findings revealed CHM in 30% of people younger than 30 years old and in 85% among those older than 50 years. Among patients undergoing arthroscopy or surgery of the knee joint, and who are in the active age range, the percentage of CHM is higher.

Not only the recognizable cartilage damage should be registered, but adequate surgical measures should be taken into consideration as part of the current procedure (arthroscopy or arthrotomy).

Percentage	*n*	Reason for surgery	Place	Year
48.0	100	Acute ligamentous lesions	Wels/A	1975–1980
51.7	122	Acute ligamentous lesions	Graz/A	1976–1980
65.0	2042	Acute and chronic ligamentous lesions	Berlin-Pankow	1971–1980
77.4	1409	Acute and chronic ligamentous lesions	Berlin-Pankow	1981–1985
42.0	500	Diagnostic arthroscopies	Groningen/NL	
60.0	470	Diagnostic arthroscopies	Münster/FRG	1973–1978
61.9	967	Diagnostic arthroscopies	Leverkusen/FRG	1975–1985
67.0	144	Control arthroscopy after meniscectomy	Zurich/CH	1985
75.8	1020	Diagnostic arthroscopies	Berlin-Pankow/DDR	1981–1985
Traumatic cartilage lesions				
13.0	139	Acute arthroscopies	Linköping/S	
14.5	200	Acute arthroscopies	Berlin-Pankow/DDR	
19.2	334	Acute arthroscopies	Zurich/CH	
21.2	1146	Acute arthroscopies	Leverkusen/FRG	
21.8	170	Acute arthroscopies	Berlin (West)	
23.5	196	Acute arthroscopies	Nuremberg/FRG	

Surgery and Arthroscopy of the Knee
2nd Congress of the European Society
Editors: Müller/Hackenbruch

The Natural History of Anterior Knee Pain in Japanese Adolescents

A. Fujimoto, Y. Mori, Y. Kuroki, R. Yamamoto, H. Hino, and H. Okumo

Department of Orthopaedic Surgery, Showa University, Fujigaoka Hospital, 1-30 Fujigaoka, Midori-Ku, Yokohama, Japan

Many reports have described that chronic pain in the front of the knee among adolescents can be caused by cartilaginous changes of the patellofemoral joint, so-called chondromalacia of the patella. These cartilaginous changes are often present in patella alta, patellar malalignment syndrome, subluxation of the patella, unstable patella, dysplasia of the patella (such as Wiberg type III), dysplasia of the femoral condyle and others. We also found cartilaginous changes of the patellofemoral joint in these diseases using arthroscopic study or during operations. However, some patients with chronic pain in the front of the knee had no instability of the patella upon physical examination, no abnormalities of the patellofemoral joint (such as incongruity) on radiological study (including skyline view) and no morbid changes on arthroscopic study. Is it reasonable to regard such cases as having a prechondromalacia condition? How should they be categorised and treated? To attempt to answer these questions, we investigated the natural history of anterior knee pain in Japanese adolescents, using questionnaires.

Diagnosis. At Showa University Fujigaoka Hospital, anterior knee pain is defined as follows: patients complain of chronic pain in the front of their knees. They have tenderness of the patellofemoral joint, mainly medial, but have no signs of meniscus or ligament lesions. Their patellas are not unstable. There is no history of dislocation or subluxation of the patella in any of them, nor is there previous trauma suggesting osteochondral lesions. Moreover, with plain radiological study there are no abnormalities in the patellofemoral joint.

Methods

A total of 220 females between 10 and 20 years of age were diagnosed according to the above criteria between July 1977 and June 1980. During 1985 all 220 pa-

Table 1. Duration of participation at hospital after initial presentation ($n = 98$)

Years	*n*	(%)
0–1	34	34.6
1–2	19	19.4
2–3	15	15.4
3–4	12	12.3
4–5	7	7.1
>5	11	11.2

Surgery and Arthroscopy of the Knee
2nd Congress of the European Society
Editors: Müller/Hackenbruch

tients were sent a questionnaire to determine the current state of their knee pain. The questionnaire elicited information as to the presence or absence of pain, its severity, its frequency and so on. We received 98 replies (44.5%). After their initial visit, we asked all patients to come every 6 months. As to the participation period, 34.5% of the patients came for less than 1 year and only 10.9% attended more than 5 years (Table 1). Arthroscopy was performed on 45 knees, in 30 of the 98 patients.

Results

The age of the patients at their initial visit is shown in Fig. 1. The mean age was 13.3 ± 1.4 years and the modal age was 13 years. The next most frequent age was 5–8 years, their average age being 6.5 years. At the time the questionnaire was administered, they ranged from 17–26 years old and the mean age was 20.1 years. Responses to the questionnaires are shown in Table 2. As to the presence of pain, 76 patients of 98 reported still having pain at present. For example, a 25-year-old housewife reported having pain for 11 years (from age 14), and its severity had not changed. Twenty-two cases had no pain at present and the mean age at which their pain diminished was 17.8 years.

The mean period of having pain was approximately 4 years. Regarding the many women who still had pain, its severity was improved on the whole. To summarise, an improvement was noted in 67.4% of 98 patients, no change in 30.6%, and

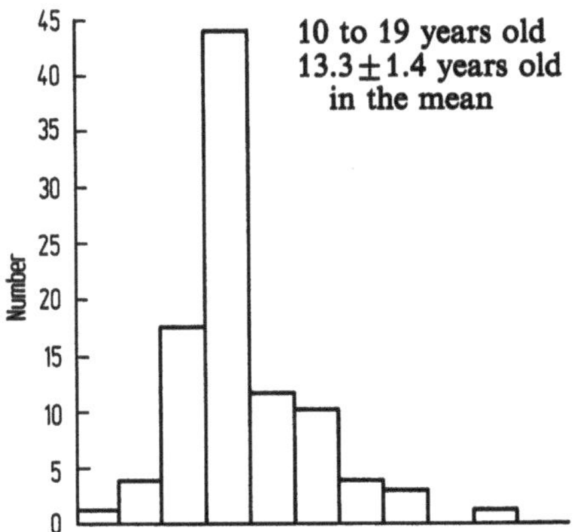

Fig. 1. Age at initial visit

Table 2. Responses to questionnaires

	n	(%)
Presence of pain		
Yes	76	77.6
No	22	22.4
Severity of pain		
Better	66	67.4
Worse	30	30.6
Same	2	2.0
Frequency of pain		
Daily	17	22.4
Weekly	22	28.9
Rarely	37	48.7
Pain-associated factors		
Chilliness	56	73.7
Sports	37	48.7
Standing	36	47.4
Sitting	32	42.1
Walking	15	19.7
Others		
Therapies by patients		
Nothing particular	48	49.0
Quadriceps drill	12	12.2
Warming	10	10.2
Give up sports	8	8.2
Rest	8	8.2

Table 3. Arthroscopic classification of the patellofemoral joint

Grade	Arthroscopic findings
I	Vascularisation and synovial infiltration on the medial facet of the patella
II	Localised fibrillation on the medial facet of the patella
III	Softening, swelling and extensive fibrillation; fragmentation and fissuring of the patellar cartilage
IV	Localized degenerative changes on both the patella and the femoral condyle
V	Extended degenerative changes to the subchondral bone

worsening in just 2%. These two aggravated cases were both beauticians, and their prolonged standing position at work may have made their knee pain worse. As to the frequency of pain, 23.0% of the 76 patients experienced pain every day, 28.4% once a week, and 48.6% rarely. The main factor causing or worsening pain was chilliness (or cold weather); next were sports activities, prolonged standing and Japanese-style sitting. Others of less frequency followed. Methods for self-treatment upon the development of pain are shown at the bottom of Table 2. Approximately half the patients did not do anything special. A small number did quadriceps drills and warming of their knees, but the effects were unclear.

Table 3 is our arthroscopic classification of cartilaginous lesions of the patellofemoral joint. Grades I and II signify slight changes of the patellar cartilage. Grade III signifies the swelling and softening of mainly the patellar cartilage. Grades IV

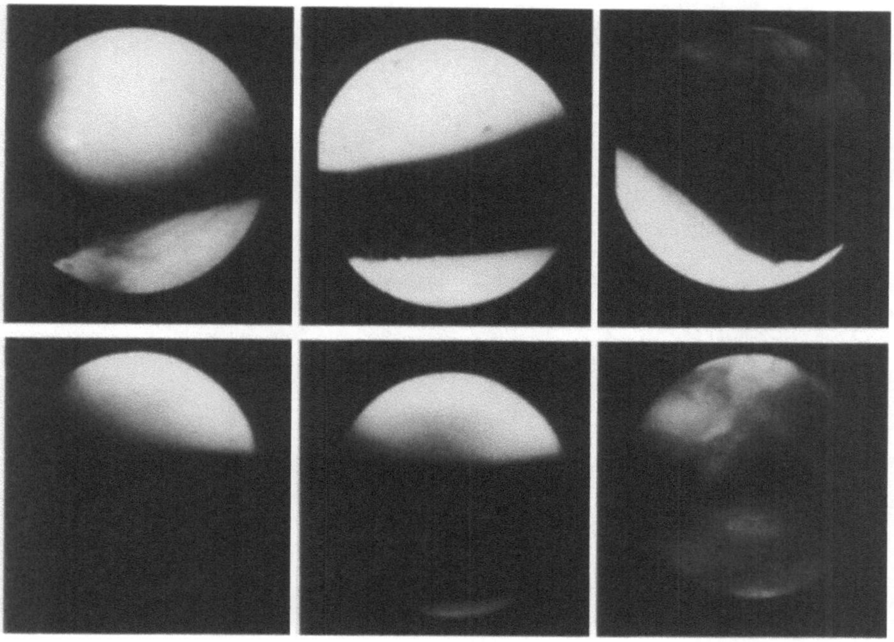

a

b

Fig. 2 a, b. Arthroscopic views in 1979 and 1986 of 20-year-old female; **b** shows much more fibrillation than **a**

and V have cartilaginous lesions in both the patella and the femoral condyle. We consider that grades I and II are usually not morbid. According to our classification, 45 knees in 30 of the 98 patients were graded as follows: grade I was noted in 11 knees, grade II in 2 knees and no abnormalities were found in 12 knees.

Case Reports

Case 1 is a 20-year-old female who is a beautician. She felt chronic knee pain at 13 years of age and visited our hospital. At present, she feels pain every day and its severity increases slightly, especially after work. Figure 2a is an arthroscopic view performed in 1979. There is vascularisation and synovial infiltration on the medial facet of the patella. Figure 2b shows the same view 7 years later; much infiltration can be seen in addition to the findings in Fig. 2a. There are no abnormalities on skyline views in 1979 and 1986, and no differences between both views.

Case 2 is a 19-year-old female who felt chronic pain in the front of her knees and visited our hospital. At present the knee pain remains, but its severity is slight, and its frequency is rare. Her response to the questionnaire indicates that, on the whole, the symptoms are improved. Figure 3a shows an arthroscopic view at 14 years of age; fibrillation can be observed on the medial facet of the patella. Figure 3b shows the same view 4 years later; fibrillation is diminished.

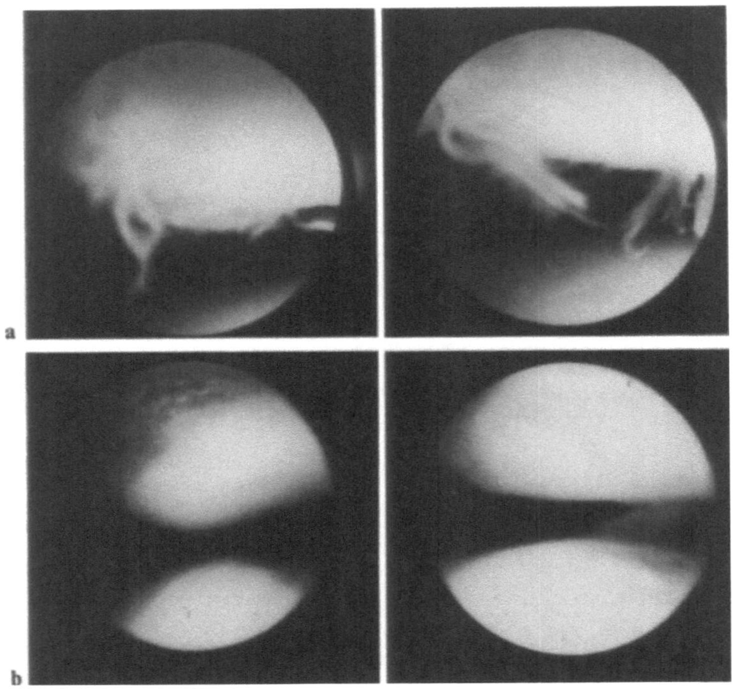

Fig. 3a, b. Arthroscopic views in 1980 and 1984 of 19-year-old female. No fibrillation can be seen in **b**

Discussion

We have not treated anterior knee pain with surgical procedures but rather have only observed. It was judged that aggravating tendencies were few in this disease. At the 6.5-year mean follow-up examination, a rather slight healing tendency was noted, so it was considered that our method was correct. Moreover, according to arthroscopic findings of 45 knees, morbid changes of the patellar cartilage were extremely slight. These results lead us to conclude that surgical treatment is not always necessary. This conclusion is nearly the same as that in Mr. Goodfellow's report of 1985.

References

Abernethy PJ, Townsend PR, Rose RM, Radin EL (1978) Is chondromalacia patellae a separate clinical entity? J Bone Joint Surg [Br] 60:205–210

Büdinger K (1908) Über traumatische Knorpelrisse im Kniegelenk. Dtsch Z Chir 92:510

Dandy DJ, Poirier H (1975) Chondromalacia and the unstable patella. Acta Orthop Scand 46:695–699

Insall J (1979) Chondromalacia patellae: patellar malalignment syndrome. Orthop Clin North Am 10:117–127

Insall J (1982) Current concepts review: patellar pain. J Bone Joint Surg [Am] 64:147–152

Insall J, Salvati E (1971) Patella position in the normal knee joint. Radiology 101:101

Insall J, Falvo KA, Wise DW (1976) Chondromalacia patellae: a prospective study. J Bone Joint Surg [Am] 58:1–8

Robinson AR, Darracott J (1970) Chondromalacia patellae: a survey conducted at the army medical rehabilitation unit, Chester. Ann Phys Med 10:286–290

Outerbridge RE (1961) The etiology of chondromalacia patellae. J Bone Joint Surg [Br] 43:752–757

Patellofemoral Dysplasia

H. A. Tomairek[1], N. A. Mahmoud[1], A. M. El-Eleimi[1], N. Galal[2], and B. El-Serwy[2]

[1] Departments of Orthopaedics and Radiology, Tanta University,
[2] Departments of Orthopaedics and Radiology, El-Azhar University,
25 Goul Gamal St. ET-Agouza, Dokki, Cairo, Egypt

Introduction

Knee complaints are very common among Egyptians in various age groups, and subsequent chondromalacia patellae and osteoarthrosis of the knee are frequent among older groups. It is well-known that the patella enjoys the thickest articular cartilage in the human body and is also the most frequent site of degenerative changes. Being subcutaneous it is frequently exposed to trauma; being a part of the skeleton it can be affected by any systemic disease; being a part of the knee joint its disorders can be mistaken for other conditions of internal damage to the knee, particularly those of the menisci. For these reasons the patella deserves substantial study.

Although a considerable amount of information is now available regarding the anatomy, biomechanics and pathology of the patellofemoral joint, there are still no precise data which contribute definitively to our understanding of the problems of degeneration and disability in the patellofemoral joint. The present study was initiated to search for any dysplasia of the patellofemoral joint that may be responsible for the frequency of knee complaints.

Material and Methods

A series of 113 patients with patellofemoral arthralgia (158 involved knees) were studied at Tanta University Hospital. Of these, 83 patients were female (73.5%) and 30 were male (26.5%). In 45 cases (40%) the condition was bilateral. There was an evident tendency of the complaint to run in families. The youngest patient was 17 years old and the oldest was 48; the average age was 37.

Orthopaedic Examination

All patients were questioned regarding previous traumas and present knee disabilities. Physical examination was carried out, for example, on such deformities as valgus, varus and rotational deformity of the foot and leg. Special attention was paid to detect degeneration in any part of the quadriceps muscle. The patella was examined regarding size, shape, mobility and particularly the course taken by the patella during knee flexion, extension and squatting.

Roentgenographic Examination

Axial View. Certain important observations are possible in axial X-ray views which need to be interpreted with reference to the normal range [17]. The func-

Surgery and Arthroscopy of the Knee
2nd Congress of the European Society
Editors: Müller/Hackenbruch

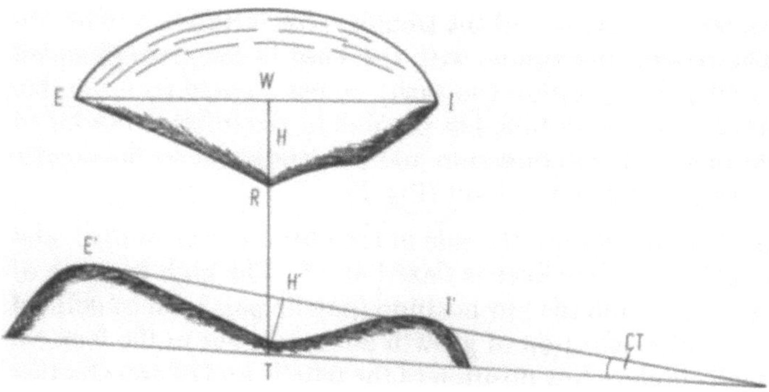

Fig. 1. Patellofemoral joint. Normal patellofemoral indices are: *I.* Patella facet ratio = RE/RI = 1.4±0.4; depth ratio = W/H = 3.9±0.3; facet angle = ERI = 130°±10°. *I'*. Trochlea depth index = E'T'/TH' = 5.3±1.2; sulcus angle = E'TI' = 140°±5°; angle of inclination = CT = 4°95'±1°45'

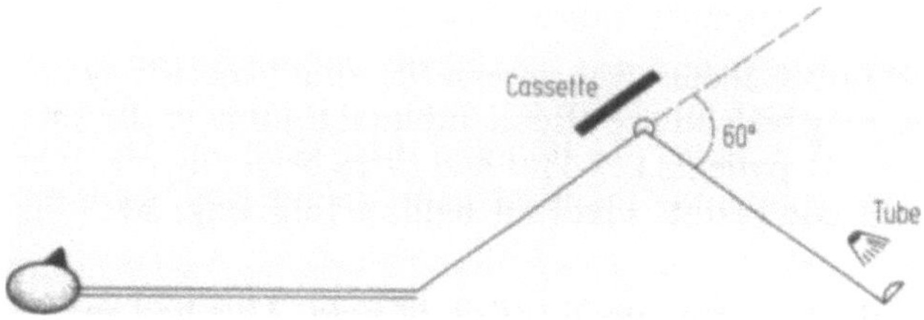

Fig. 2. Position of X-ray tube and casette

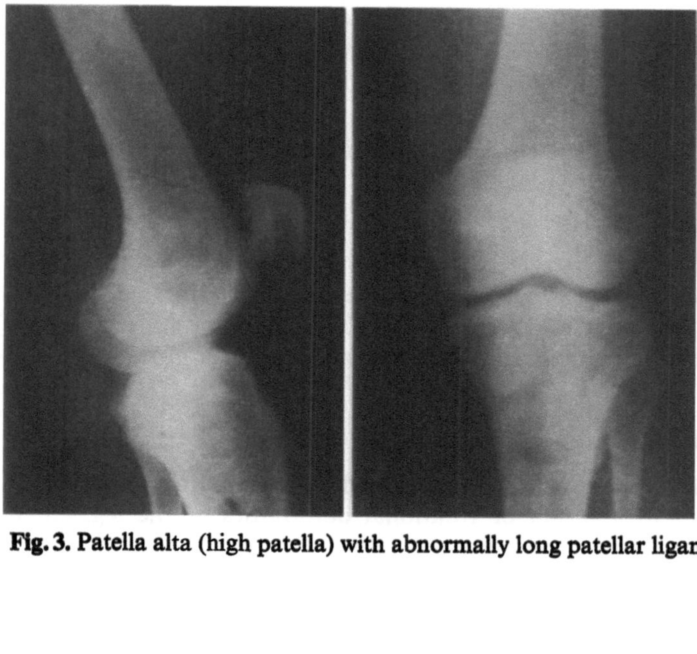

Fig. 3. Patella alta (high patella) with abnormally long patellar ligament

tional relationship between the patella and the trochlea, for instance, can be observed here (Fig. 1). The patient lies supine with the knee in the recommended angle of flexion (30° or 60°). The position can easily be maintained by a suitable cushion beneath the knee. The X-ray tube lies parallel to the inferior border of the tibia with the beam in a cephalad direction and perpendicular to the casette which lies proximal to the patellofemoral joint (Fig. 2).

Lateral View. The patient is turned onto the side of the knee to be examined, and the X-ray exposure is made while the knee is flexed at 30°. The high position of the patella (patella alta) (Fig. 3) and the low position (patella paja) can be defined [2, 3]. The lateral view permits detection of growth abnormalities in the femoral condyles, of an abnormally high or low position of the patella and of any changes in the patella or trochlea. Anteroposterior radiographs, furthermore, are sometimes recommended for viewing the position of the patella and the condition of the femoral and tibial condyles.

Results

Patellofemoral dysplasias do not in themselves necessarily entail specific symptoms. Patients suffering these may seek advice after a minimal trauma to the knee has caused discomfort. In only 31 patients (27.4%) had there been any previous complaint. The most frequent complaints involved pain, giving way, swelling, crepitus and locking.

Pain. The anterior aspect of the knee was the location of pain. This had earlier been diffuse, but medial radiation had developed which occasionally led to confusion with meniscal problems. The pain was generally dull and aching in nature, at times involving acute episodes. It was characteristically provoked by prolonged sitting, squatting or going up or down stairs.

Giving Way. Sudden relaxation of the quadriceps can occur while performing certain movements with the knee flexed or extended under load, such as in ascending stairs or walking over uneven ground. Occasionally this leads to the patient collapsing.

Swelling. Although also a frequent complaint, swelling had been only intermittent and transient.

Crepitus. Some patients reported grating sensations, sometimes audible, upon knee flexion and extension under load and especially upon squatting and rising.

Locking. A history of locking is frequently mistaken for meniscal injury but differs from this in that the patient generally refers to a temporary stiffness of the knee after prolonged sitting or squatting, which disappears after movement.

Physical Examination

Patellofemoral dysplasia is frequently accompanied by other deformities, such as genu varum, valgum, recurvatum or rotational deformities of the leg. Genu varum and abnormal position of the tibial tuberosity are the most frequent among these. The tibial tuberosity was frequently located laterally, and in some cases prominently so.

On palpating the patellofemoral joint, small patellae can be sensed in the trochlear sulcus with the trochlear condyles uncovered. On moving the patella from side to side, soft tissue dysplasia could be revealed by limitation of passive medial movement of the patella; also detectable is an abnormal attachment of thick cords passing from the iliotibial tract to the vastus lateralis, tethering the patella laterally. This was found to be associated with some degeneration of the vastus medialis. Excessive mobility of the patella in all directions due to lax attachments was observed in some cases. Tenderness was elicited in some cases on percussing the patella with a hammer while flexing the knee.

Observing the course of the patella while palpating it during active flexion and extension reveals in some cases excessive lateral and sometimes upward movement of the patella. In a squatting test while putting a load on the knee the examining hand was able to feel an audible click arising from the anterior part of the joint (which also may be mistaken for meniscal injury).

Radiological Features

Examination of the axial X-ray view while the knee is flexed at 30° or 60° can reveal the following forms of patellar dysplasia:

Hypoplastic patella (small patella or patella parva). The dimensions of the patella are small in relation to condylar dimensions (23 cases, 20.3%) (Fig. 4).

Partial hypoplasia. The patellar facet ratio shows significant variation beyond normal limits. We can distinguish three types here [19]. In type I, both facets are normal and of roughly the same size (15 patients, 13.2%). In type II, the lateral facet remains concave while the smaller medial facet is either flat or slightly convex (55 cases, 47.6%) (Fig. 5). In type III, the medial facet is narrower and markedly convex (18 cases, 16%) (Fig. 6); an occasional variant here involves marked medial facet projection.

Hyperplastic patella (patella magna). All dimensions of the patella are enlarged (7 cases, 6.1%) (Fig. 7).

Angular patellar dysplasia. Some cases show a rather more acute patellar facet angle (19 cases, 16.8%), and somewhat fewer an obtuse angle (6 cases, 5.3%). These forms of dysplasia are associated with abnormalities in the patellar facet ratio.

Patellar depth dysplasia. The ratio between the equator and the height of the facet (height/depth index) shows marked variation generally with shallow patellae (32 cases, 28.3%) (Fig. 1).

Patella externa (lateral subluxation). This is diagnosed by a line dropped at the deepest point of the trochlear sulcus (O T) [15]. It was diagnosed in 31 cases (27.4%). It is common in association with other types of dysplasia (Figs. 4, 5).

Additionally, the lateral X-ray view can reveal the following further forms of patellar dysplasia. *Patella alta* with high position was seen in nine cases (8%), excluding spastic cases (Fig. 7). *Patella paja* with low position was detected in two cases with achondroplasia, excluding poliomyelitic cases.

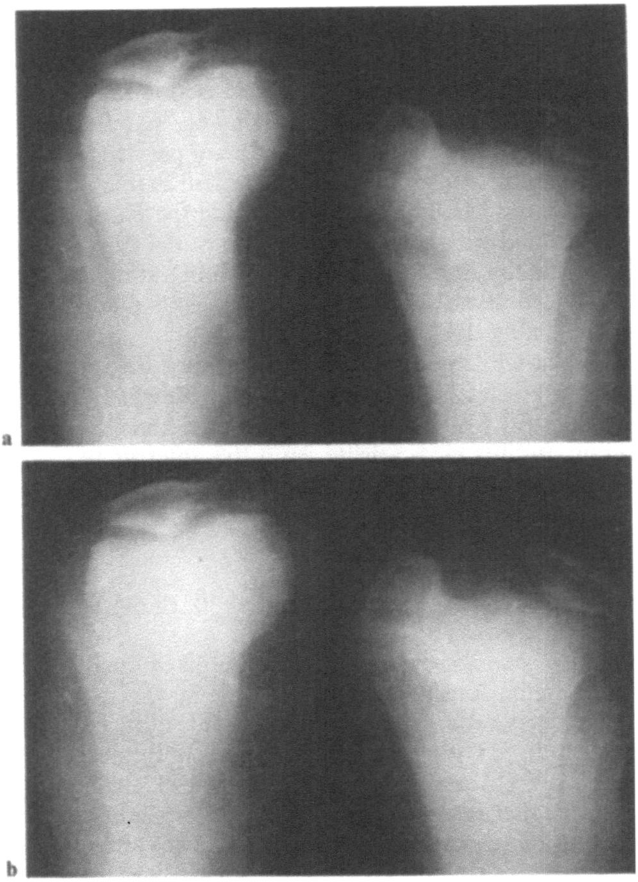

a

b

Fig. 4 a and b. Aplasia of the lateral condyle. **a** With lateral dislocation of the hypoplastic patella (note the sulcus angle of approximately 90°); **b** with marked genu valgum

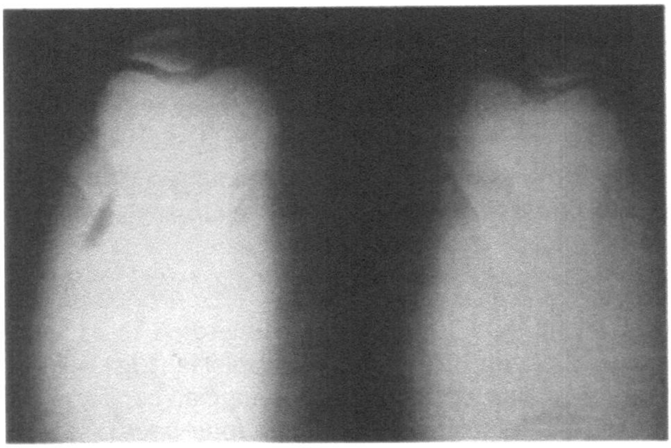

Fig. 5. Type II patella. *Right/left*, the medial facet is flat and narrow with flat sulcus; *right/left*, showing obvious subluxation

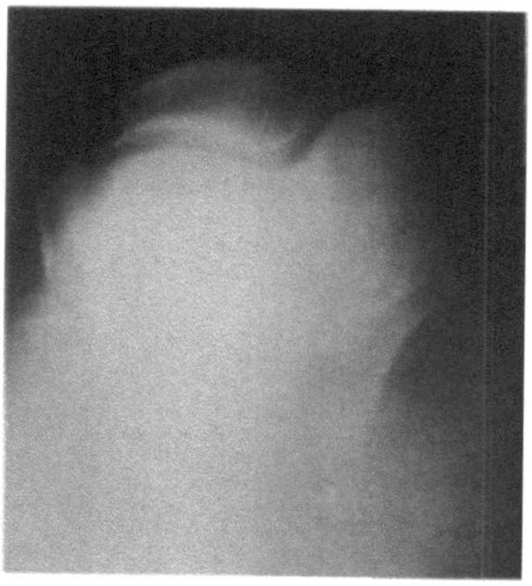

Fig. 6. Axial view of knee showing type III patella with convex and narrow medial facet. Trochlear angle and patellar angle are smaller (more acute) than normal. Hypoplasia of the lateral condyle

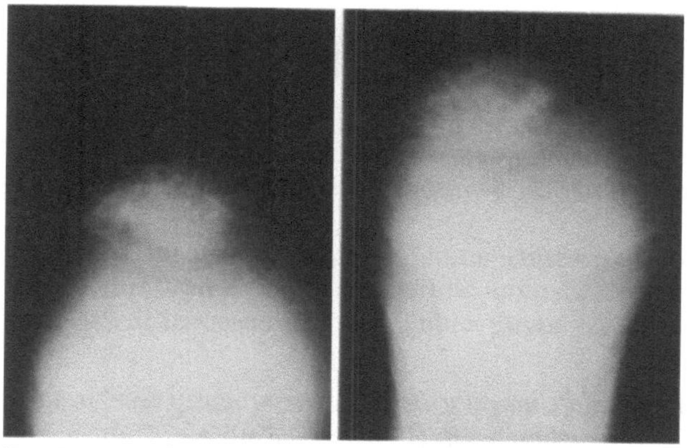

Fig. 7. Hyperplastic patella (patella magna)

Further deformities can be detected through an examination of the anteroposterior view. In addition to those of the genu valgum and genu varum (Fig. 8) these include the following.

Trochlear dysplasia is commonly associated with hypoplasia of the medial patellar facet. It may be detected by an increased angle of inclination and by a depth coefficient seen best in a 30° axial view. This was present in 49 cases (33%). It commonly leads to obvious genu varum.

Hypoplasia of the lateral condyle was not uncommon in our patients (17 cases, 15%). In these cases the angle of inclination may be reversed (Fig. 1). It usually

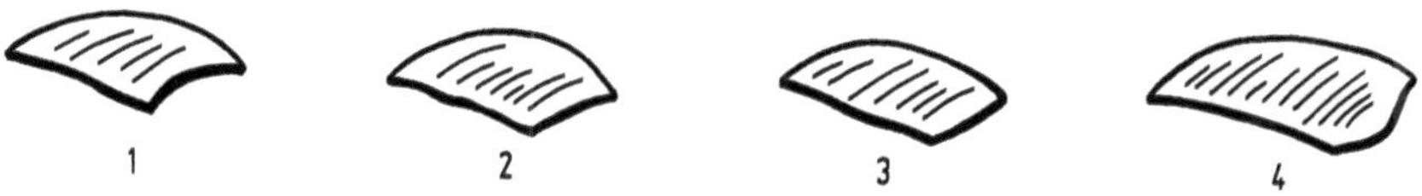

Fig. 8. Types of patellar pattern according to medial facet. *1* Concave and identical with lateral; *2* flat and narrower than lateral; *3* convex and narrower than lateral; *4* with convex projection

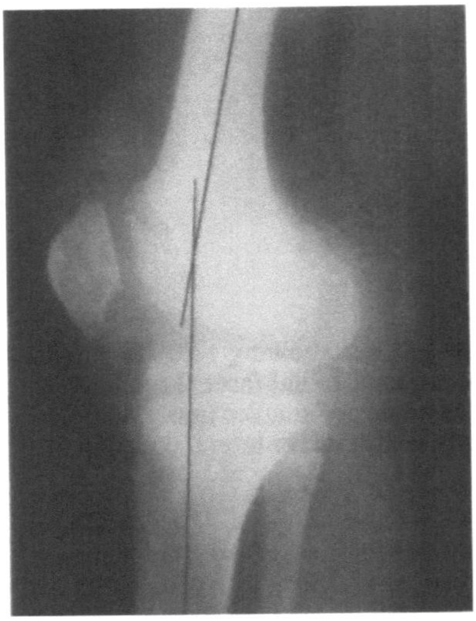

Fig. 9. Trochlear dysplasia with the formation of an anterior outer bridge

leads to genu valgum. *Trochlear angle dysplasia* is usually associated with dysplasia of the medial and lateral condyles. Two forms can be distinguished; global dystrophy and deep sulcus.
Global dystrophy of the trochlea presents with an obtuse trochlear facet angle (6 cases, 5.3%). This may sometimes show protuberance in the lateral radiograph (Fig. 9). The angle in deep sulcus is more acute. This was observed in 21 cases (18.5%).

Malposition of the patella and patellofemoral arthrosis are frequently seen as a sequel of dysplasia in patellofemoral articulation (13 cases, 11%).

Discussion

Knee complaint is a common condition in Egypt. Of 113 patients seeking our advice 93 were found to have various degrees of patellofemoral dysplasia (PFD) with patellar, trochlear and/or soft tissue dysplasia (excluding diseased cases having infections, rheumatoid or similar conditions). PFD does not produce characteristic symptoms but is probably the underlying factor for various distressing complaints and for an inability to withstand daily stresses and strains. Moreover, it may predispose one to subsequent patellar subluxation, chondromalacia and patellofemoral arthrosis. Instances of PFD, however, are frequently mistaken for other internal disability of the knee, thus leading often to unnecessary meniscectomies. We still do not know the likely effect of meniscectomy on an acutely injured meniscus with an underlying PFD.
On the basis of our study and a review of the literature, we can classify the forms of patellofemoral dysplasia.

A. Patellar dysplasia
   1. Aplasia of the patella: complete absence of the patella
   2. Hypoplasia of the patella (patella parva): entire patella small

3. Partial hypoplasia mostly involving the medial facet.
   a. Those with alterations in the patellar facet ratio [19]: in type I, both facets are gently concave and similar in size; in type II, the medial facet is small, flat or slightly convex (the most common form); in type III, the medial facet is markedly smaller and more convex; type IV resembles type III but with a more convex projection of the medial facet (Baumgartl type)
   b. Those with alterations in the patellar facet angle: flat patella (pebble-shaped patella) with an angle greater than 140°; acutely angled patella with an angle less than 120°
4. Hyperplasia (patella magna): patella large and thick
5. Patella malformation
   a. Patellar fragmentation
   b. Bipartite patella
   c. Patellar reduplication
   d. Marginal patellar deficits
6. Patellar malposition
   a. Patella alta
   b. Patella paja
   c. Patella externa
   d. Patella interna (rare)

B. Trochlear dysplasia
1. Hypoplasia of the medial condyle: medial condyle diminished in width and/or height
2. Hypoplasia of the lateral condyle: lateral condyle diminished in width or height
3. Hypoplasia of the medial or lateral condyle: an epiphyseal hyperplasia
4. Trochlear angle dysplasia
   a. Flat sulcus (obtuse angle greater than 140°)
   b. Deep sulcus (acute angle less than 120°)

C. Soft tissue dysplasia: abnormal fascia lata insertion, usually associated with some wasting of vastus medialis

D. Complicated cases
1. Subluxated or dislocated patella
2. Chondromalacia
3. Patellofemoral arthrosis

Various methods have been adopted to describe the condition of the patella. Lateral-view radiographs [2, 3, 5, 11, 13] provide information on the vertical position of the patella and on any growth disturbance of the femoral condyles. Other researchers have also noted the advantages of the axial view [12, 14]. Although the method of Bruttström [4] for studying the relationship of dysplasia in the trochlea to recurrent dislocation of the patella is accurate, it is complicated and difficult to apply clinically. That of Ficat [8] is more practical and accurate and also allows both knees to be examined simultaneously, thus controlling rotation and offering comparison on a single film. As it provides a fixed ratio, this can be applied to patients of any age with definite landmarks.

In our series the most common dysplasias were medial patella facet dysplasia (48.6% of cases) and hypoplasia of the medial condyle (43%). That these are frequently associated with one another suggests a common aetiological origin; these are, furthermore, commonly associated with genu varum. They also play a role in various forms of patellar instability and chondromalacia. Some authors have noted no relation between type III patellar dysplasia and either chondromalacia [6, 9, 16] or patellofemoral arthrosis [10]; however, it seems strongly implicated in recurrent dislocation of the patella [6, 18].
Type IV patellar dysplasia, as described by Baumgartl [1], is characterised by a convex projection of the medial facet and appears to be a variant of type III. Its importance according to Baumgartl is its frequent association with osteochondritis of the medial femoral condyles. In our cases, however, we did not find this association but rather an association with lateral subluxation.
A large-angle patella (greater than 140°) and trochlear sulcus usually lead to flat patella or flat sulcus, thus resulting in unstable patella [8, 14]. A more acute patellar angle may lead to an acute angle with a single articular facet, and this in turn may lead to lateral subluxation.
The abnormal relationship of the patella to the trochlea is sometimes due to soft tissue abnormality. The majority of such cases are due to an abnormally long patellar tendon (patella alta). Occasionally, an abnormal insertion of vastus lateralis and fascia lata in the patella causes excessive lateral tilting.
We should note that in most cases we found more than one type of dysplasia; determining the original form of dysplasia – whether in the trochlea, patella or soft tissue – is difficult and needs further study.
Excessive unit pressure may result from dysplastic deformities, and soft tissue anomalies are probably responsible for lateral pressure syndrome and chondrosis with subsequent arthrosis. These hypothetical relationships, however, also need further investigation.

Summary

Among 113 cases with 158 affected knees 93 patients were found to have various degrees of patellofemoral dysplasia. The clinical portrait of these, as presented here, emphasises that the cases should be strictly differentiated from other knee deformities. Radiological examination, including axial, lateral and sometimes anteroposterior views have been analysed. The various types of dysplasias have been classified and discussed. Patellofemoral dysplasia may cause some knee instability with subsequent subluxation, dislocation, chondromalacia and patellofemoral arthrosis.

References

1. Baumgartl F (1964) Das Kniegelenk. Springer, Berlin Göttingen Heidelberg New York
2. Blackburne JS, Peel TE (1977) A new method of measuring patellar height. J Bone Joint Surg [Br] 59(2):241–242

3. Blumensaat C (1938) Die Lageabweichungen und Verrenkungen der Kniescheibe. Ergeb Chir Orthop 31:149–223
4. Bruttström H (1964) Shape of the intercondylar groove, normally and in recurrent dislocation of the patella. Acta Orthop Scand [Suppl] 68:134–148
5. Crosby EB, Insall J (1976) Recurrent dislocation of the patella: relation of treatment to steroarthritis. J Bone Joint Surg [Am] 58(1):9–13
6. Fairbank T (1956) Dysplasia epiphysialis hemimelia. J Bone Joint Surg [Br] 38:237
7. Ficat P (1970) Pathologia femoro-patellare. Masson, Paris
8. Ficat PR, Hungerford BS (1977) Disorders of the patello-femoral joint. Williams and Wilkins, pp 36–50
9. Goodfellow JW, Hungerford M (1976) Patello-femoral mechanics and pathology. J Bone Joint Surg [Br] 58:287
10. Hungerford DS, Cockin JC (1975) Fate of the retained lower limb joints in world war II amputees. J Bone Joint Surg [Br] 57:3
11. Insall J, Goldberg V, Salvati E (1972) Recurrent dislocation and the highriding patella. Clin Orthop 88:67–69
12. Jaroschy (1924) Die Diagnostische Verwertbarkeit der Patellaraufnahmen. Fortschr Roentgemtr 31:781
13. Lancourt JE, Cristini JA (1975) Patella alta and patella infra. J Bone Joint Surg [Am] 57:1112–1115
14. Laurine CA, Levesque HP, Dussault R, Labellele H, Peides LP (1978) The abnormal lateral patello-femoral angle. J Bone Joint Surg [Am] 60(1):50–55
15. McDougall A, Brown JD (1968) Radiological sign of recurrent dislocation of the patella. J Bone Joint Surg [Br] 50(4):841–843
16. Merchant AC, Mercer RL, Jacobsen RH, Cood CR (1974) Roentgenographic analysis of patello-femoral congruence. J Bone Joint Surg [Am] 56:1391–1396
17. Outerbridge RE (1964) Further studies on the etiology of chondromalacia patellae. J Bone Joint Surg [Br] 46:179–190
18. Rohlederer O (1951) Ätiologie und Symptomatologie der praeluio patellae. Zentralbl Chir 76:103–115
19. Wiberg G (1941) Roentgenographic and anatomic studies on the femoro-patellar joint. Acta Orthop Scand 12:319–410

Chondropathia Patellae – A 5-Year Follow-Up Study After Surgical Treatment

H. Kiefer [1], W. Mutschler [1], and J. M. Friedrich [2]

[1] Department of Traumatology
[2] Department of Radiology, University of Ulm, Ulm, FRG

Abstract

Commonly used surgical techniques for the treatment of chondromalacia patellae (CP) show different levels of success. The aim of our prospective study was to obtain long-term results for operatively treated CP for a series of 100 patients within the same year. All of them had been treated conservatively over an extended period of time. Patients with intraoperatively graded CP I ($n=25$) were treated by lateral retinacular release; in those with CP II ($n=48$), additional shaving of the patellar cartilage was performed; those with CP III ($n=27$) received shaving of the cartilage and drilling of the subchondral bone combined with lateral retinacular release. Evaluation was done preoperatively, and follow-up carried out after 9 months, 2 years and 5 years. These included subjective complaints, clinical examination, radiography, arthrography ($n=36$) and arthroscopy ($n=22$). Seventy-nine patients could be followed up over the whole period. The most important clinical results, from more than 100 parameters, were examined in detail. Radiographs, arthrographic and arthroscopic results correlated well with the clinical examinations. Compared to the preoperative status, the overall result 5 years after operation showed a statistically significant ($P<0.01$) improvement both for subjective complaints and in the clinical examination.

Introduction

Today, more than 100 surgical procedures for the treatment of CP are known. The results of treatment vary between 90% success and complete failure (Burri and Rüter 1976; Jäger and Plitz 1983). A few investigators observe a positive trend after conservative treatment, particularly in young patients (Dehaven et al. 1980; Osterwalder and Zolinger 1979). Most investigators, however, consider that surgical treatment offers the best chance for long-term improvement of this aetiologically heterogeneous disease. The aims of operative therapy are preservation of the articular cartilage, removal of damaged cartilage and correction of pathogenetic factors. To improve movement of the patella on the femoral articular surface two possible corrective surgical methods can be used, namely medial displacement of the quadriceps insertion and/or a lateral retinacular release (Ficat and Hungerford 1977). Many reports deal with the open surgical and, in more recent years, with the closed arthroscopic lateral release (Bandi 1982; Bentley 1978; Baumann and Leichs 1976; Ficat and Hungerford 1977; Harwin and Stern 1981; McGinty and Mc Carthy 1981; Miller et al. 1982; Simpson and Barret

Surgery and Arthroscopy of the Knee
2nd Congress of the European Society
Editors: Müller/Hackenbruch

1984). Osteotomy of the tibial tuberosity was described by Maquet (1976) and Bandi (1982) and was believed to diminish the pressure at the patellar surface. However, experimental measurements of these pressures do not agree with that theory (Hehne 1983; Jäger and Plitz 1983).

To improve the glide path an abrasio patellae is often performed. This prevents the release of lysosomal enzymes which create pain (Bandi 1982). In cases of severe CP, additional drills are performed into the subchondral bone (Pridie 1959). This leads to an outgrowth of connective tissue and may probably reduce the pathological intraosseous pressure (Dick 1980; Pridie 1959). A comparison between the various clinical studies is difficult because differing patient groups have been examined, and differing criteria of success were used. Recent investigations by other groups have failed either to differentiate or mention the grading of cartilage damage and the long-term results. In addition, only small groups of patients, or retrospective examinations, were reported. We tried to avoid these disadvantages in the prospective randomised study presented here.

Patients and Methods

Between July 1978 and May 1979 67 male and 33 female patients were allocated to this study. The preoperative age ranged between 20 and 40 years. The patients had to fulfill the following conditions: no earlier operations because of cartilage disease; intraoperative documentation of the cartilage disease; surgery on one knee joint only; and exclusion of systemic diseases like chronic polyarthritis or gout.

In the CP grade I group (Bandi 1982) surgery consisted only of a lateral retinacular release ($n=25$). In the CP grade II group, additional shaving of the patella was performed ($n=24$). In the CP grade III group ($n=27$) lateral release and patella shaving were combined with subchondral drilling (Pridie 1959). More than 100 parameters were registered on uniform documentation sheets. The most important subjective statements that were noted and scored were: the severity and frequency of pain, giving way, painful patellar crepitus and restriction of normal activities. The most substantial clinical findings, such as joint swelling, painful patellar facets, a positive Zohlen's test, limited motion and difference in thigh circumference, were subjected to a statistical test.

Preoperative findings were compared with the results at 9 months, 2 years and 5 years after surgery. The ten most important subjective and clinical criteria given above were rated according to a modified Bentley score (Bentley 1978) which was supplemented by the clinical objective parameters. Each of the described criteria could be counted as 1 point. Addition of these points leads to a specific score for each of the patients. With a statistical sign test specific preoperative scores could be compared with those found after 9 months and 5 years. This procedure was performed for each of the 3 CP stages.

After 5 years, 79 of the 100 original patients could be followed up. Fourteen patients did not appear because they either had moved away or were suffering from poor health. However, these patients informed us that the operation was a success. The remaining 7 patients could not be found. Of the available patients 70 were examined by standard radiography (anteroposterior and lateral views) and the patellar defilee technique (Wiberg 1941). Thirty-six of the patients underwent arthrography (CP I, 10; CP II, 17; CP III, 9). Arthroscopy was performed on 27 patients between 7 and 28 months after surgery.

Results

Preoperatively, patients had suffered pain for 2 years or more. This was independent of the severeness of CP. Also, 73% of the cases had been treated conservatively over a longer period of time. Twenty-two patients had undergone other surgery previously because of ligamentous or meniscal lesions. Of the 100 patients 53 complained of severe pain, 76 of frequent pain, 63 of joint swelling, 36 of giving way and 40 of temporary disablement. Surprisingly, in CP I, many patients complained of frequent pain and swelling. Thirty-three cases were lame, and of these eight used a crutch temporarily. Clinical examination showed a typical patellary syndrome with painful patellar crepitus (77%, 26 patients) positive Zohlen's test (81%, 27 patients) and painful patellar facets (85%, 28 patients). Compared to the untreated knee, an increased circumference of the knee joint of more than 2 cm was found in 24 cases. Nine patients exhibited a decreased ability to extend the knee joint by more than 10° and 17 demonstrated a limited flexion of less than 130°. Atrophy of the thigh of more than 2 cm, measured 20 cm above the knee joint, was found in 38 cases and interpreted as an objective parameter of inactivity. Preoperative findings for each of the 3 grades are listed in detail in Table 1.

Table 1. Preoperative evaluation (in percentage; $n=79$)

Grade	I ($n=19$)	II ($n=36$)	III ($n=24$)
Pain	32	47	75
Giving way	26	44	38
Patellar crepitus (with pain)	79	86	92
Positive Zohlen's test	74	81	92
Joint swelling	16	42	71
Limitation of motion	16	28	29
Great restriction in normal activities	20	56	52

Table 2. Evaluation after 9 months (in percentage; $n=79$)

Grade	I ($n=19$)	II ($n=36$)	III ($n=24$)
Pain	5	8	4
Giving way	21	17	21
Patellar crepitus (with pain)	5	17	21
Positive Zohlen's test	26	39	75
Joint swelling	21	47	46
Limitation of motion	11	8	13
Great restriction in normal activities	11	38	30

Table 3. Evaluation after 5 years (in percentage; $n=79$)

Grade	I ($n=19$)	II ($n=36$)	III ($n=24$)
Pain	11	11	4
Giving way	21	22	9
Patellar crepitus (with pain)	11	39	33
Positive Zohlen's test	37	58	58
Joint swelling	11	17	8
Limitation of motion	0	8	4
Great restriction in normal activities	11	28	21

The findings 9 months and 5 years after surgery are shown in Tables 2 and 3. Compared to the preoperative status, the preliminary results after 9 months as well as the long-term results after 5 years showed a statistically significant improvement ($P<0.01$). Normal clinical findings after 5 years were found in 53% of the patients in CP grade I, in 33% in CP grade II and in 38% in CP grade III. The remaining patients showed more or less marked pathological clinical findings.

Figure 1 shows how the overall condition improved for each stage of the disease 5 years after surgery. The radiological stage criteria of the CP preoperatively and after 5 years are shown in Table 4. In 32 of the 36 patients who underwent ar-

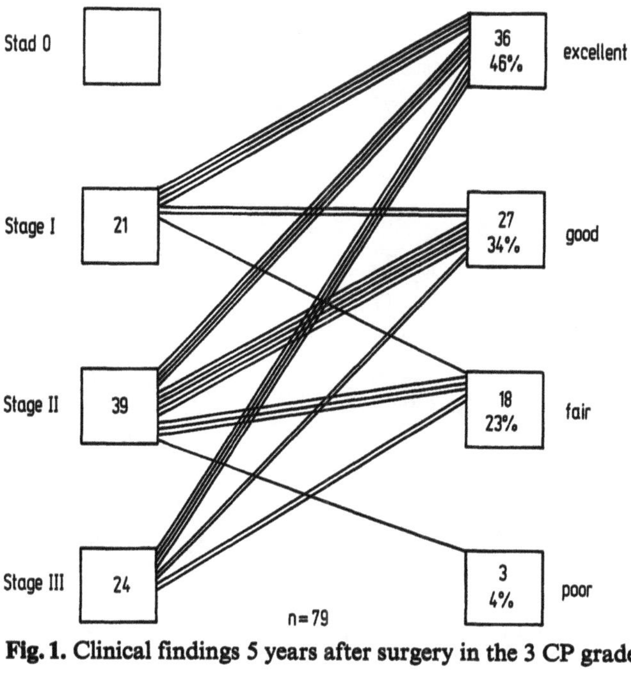

Fig. 1. Clinical findings 5 years after surgery in the 3 CP grades ($n=79$)

Table 4. Radiological findings (in percentage, $n=70$)

	Preoperatiye	5-year follow-up
Oedema	20	11
Cortical defect	47	37
Bone sclerosis	16	17
Diffuse demineralization	26	13
Localised demineralization	27	39

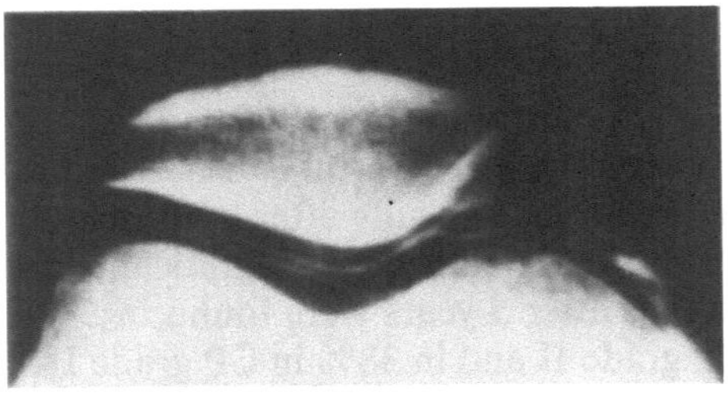

Fig. 2. Arthrograph from a 41-year-old male's knee with CP stage III, 5 years postoperatively: no complaints, positive Zohlen's test

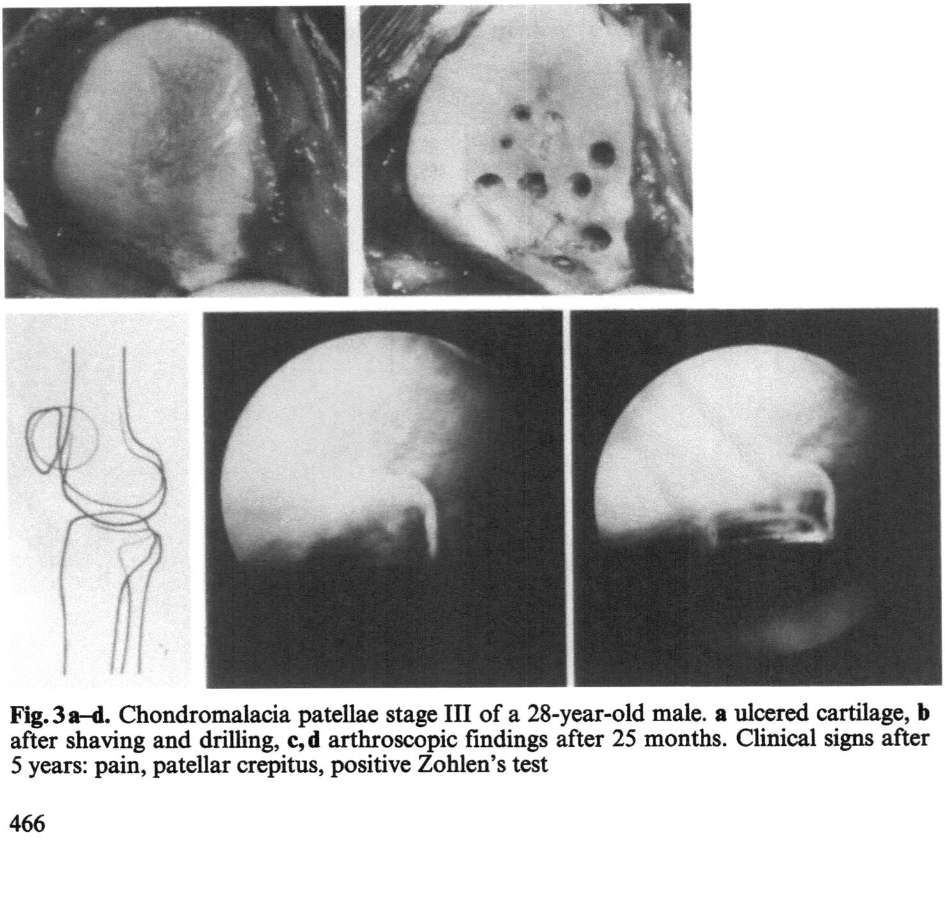

Fig. 3a–d. Chondromalacia patellae stage III of a 28-year-old male. **a** ulcered cartilage, **b** after shaving and drilling, **c, d** arthroscopic findings after 25 months. Clinical signs after 5 years: pain, patellar crepitus, positive Zohlen's test

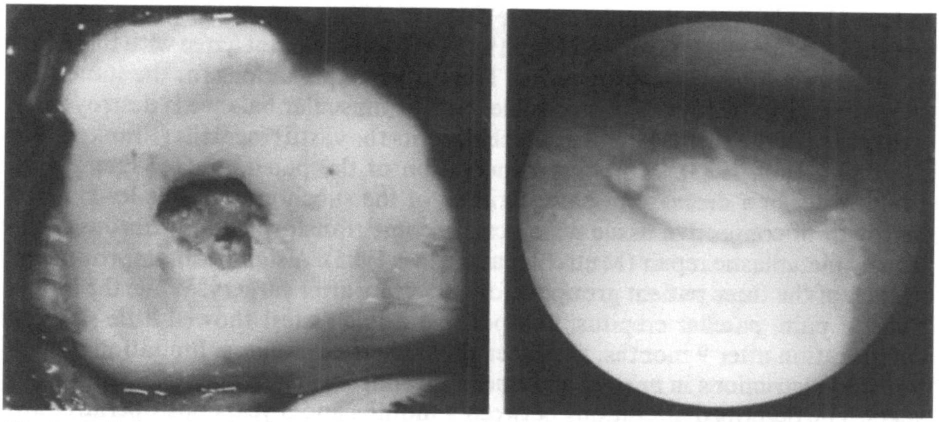

Fig. 4 a, b. Intraoperative **a** and arthroscopic (**b**, 28 months postoperative) findings of 19-year-old female with CP stage III. No complaints after 5 years

throscopy, clinical and radiological findings correlated well (Fig. 2). The remaining 4 patients showed considerably more degenerative alterations than suspected from the subjective complaints.

Arthroscopy was performed in a selected group of patients. Since only a few patients without any subjective complaints agreed to act as controls, in arthroscopical investigations, the originally intended randomisation had to be abandoned. Clinical and arthroscopical findings correlated well in the CP I ($n=3$) and in CP II ($n=14$). None of the 3 patients of group CP I experienced any softening of the remaining cartilage as determined by the instrument sensor. Those patients with CP II who had subjective complaints and pathological clinical findings also had a pitted cartilage surface as determined by arthroscopy. All 4 patients with poor results from group CP III showed a progression of CP (Fig. 3). However, in 6 patients of this group, a discrepancy between clinical and endoscopical examinations was found. While the clinical status remained constant, new areas of cartilage destruction were seen in the vicinity of the refilled drill holes (Fig. 4).

Discussion

At the present time, no generally accepted treatment exists for CP. In the literature, the described success rates vary enormously. In addition, the groups of patients and the time of follow-up differed, and the clinical and radiographical criteria of success were not uniform (Bandi 1982; Bentley 1978; Burri and Rüter 1976; Maquet 1976). The aim of our study was to treat a major group of patients, depending on the grading of the CP, in a prospective manner. Thus, the value of common surgical procedures such as lateral release and shaving and drilling of the patella (Osterwalder and Zolinger 1979) could be tested. In this study, the success of the lateral retinacular release was profound, and the difference was judged to be statistically significant. The higher pressure below the lateral half of the retropatellar plane (Ficat and Hungerford 1977; Jäger and Piltz 1983) might be

reduced by a lateral retinacular release (Ficat and Hungerford 1977). Other investigators (Hehne 1983; Jäger and Plitz 1983), however, cannot agree with this opinion after experimental investigations. Thus, other explanations for the success by surgery have to be found. It is possible that the muscular balance is destroyed by weakening of the vastus lateralis, which benefits the vastus medialis (Franke 1986; Jäger and Piltz 1983). Moreover denervation of the patella on the lateral side could result in a decrease in pain. Drilling of the subchondral bone leads to an ingrowth of connective tissue which can become transformed into fibrous cartilage by metaplastic repair (Mutschler and Claes 1985). A significant improvement in each of the three patient groups is found 5 years after surgery. While the symptoms of pain, patellar crepitus, and positive Zohlen's test showed little signs of deterioration after 9 months, the other parameters of swelling, limitation of motion and restrictions in normal activities show improvement during the following years. The described alterations between 9 months and 5 years postoperative are not statistically significant. The slight deterioration in the above-mentioned parameters can be explained by the progression of the disease, which cannot be treated causally. The longer the time between surgery and follow-up (up to 5 years), the lesser the degree of swelling, limited motion, muscular deficit and restriction in normal activities. The general decrease in symptoms can also be attributed to the patient becoming accustomed to the situation.
Although curative treatment is limited by the basic disease, compared to the non-operated knee normal clinical findings in 40% of the patients and normal radiological findings in 25% were seen 5 years postoperatively. As confirmed by other groups (Franke 1986, personal communication; Vuorinen et al. 1985), surgical procedures, depending on the grading of the disease after unsuccessful conservative treatment, offer good therapeutic results. Our study agreed with these findings. In 77% of patients there was a definite improvement. Poor results were obtained in 5%, while 18% remained unchanged.

References

Bandi W (1982) Die retropatellaren Kniegelenksschäden. Aktuel Probl Chir Orthop, vol 4. Huber, Bern

Bentley G (1978) The surgical treatment of chondromalacia patellae. J Bone Joint Surg [Br] 60:74–81

Baumann D, Leichs L (1976) Retinaculumspaltung (Indikation, Technik, Ergebnisse). In: Burri C, Rüter A (eds) Knorpelschaden am Knie. Hefte Unfallheilkd 127:168–174

Dehaven K, Dolan W, Myer P (1980) Chondromalacia patellae and the painful knee. Am Fam Physician 21:117–124

Dick W (1980) Langzeitergebnisse der Abrasio patellae in der Chondropathiebehandlung. Orthop Praxis 8:678–681

Ficat P, Hungerford DS (1977) Disorders of the patello-femoral joint. Williams and Wilkins, Baltimore

Harwin SF, Stern RE (1981) Subcutaneous lateral retinacular release for chondromalacia patellae. Clin Orthop 156:207–210

Hehne JJ (1983) Das Femoropatellargelenk. Enke, Stuttgart

Jäger M, Plitz W (1983) Die operative Therapie der Chondropathia patellae. Z Orthop 121:714–721

Maquet P (1976) Biomechanics of the knee. Springer, Berlin Heidelberg New York

McGinty JB, McCarthy JC (1981) Endoscopic lateral retinacular release. Clin Orthop 158:120–125
Miller GK, Dickason JM, Fox JM, Blazina ME, Del Pizzo M, Friedman UJ, Snyder SS (1982) The use of electrosurgery for arthroscopic subcutaneous lateral release. Orthopaedics 5:309–314
Mutschler W, Claes L (1985) Knorpeltransplantation. Thieme, Stuttgart
Osterwalder M, Zolinger H (1979) 10-Jahres-Resultate bei konservativer Behandlung der Chondropathia patellae. Orthopäde 8:110–113
Pridie K (1959) A method of resurfacing osteoarthritic knee joints. J Bone Joint Surg [Br] 41:618
Simpson LA, Barret JB (1984) Factors associated with poor results following arthroscopic subcutaneus lateral retinacular release. Clin Orthop 186:165–171
Vuorinen OP, Pakkala T et al. (1985) Chondromalacia patellae. Arch Orthop Trauma Surg 104:175–181
Wiberg G (1941) Röntgenographic and anatomic studies on the femoropatellar joint with special reference to chondromalacia patellae. Acta Orthop Scand 12:319–410

Patella Realignment Via Lateral Release: Results of Surgery and Subsequent Aetiological Investigation

W. D. Stanish, P. K. Smith, L. F. Pow, and J. Kozey

Nova Scotia Sport Medicine Clinic, Division of Orthopaedic Surgery,
Dalhousie University, Halifax, Nova Scotia, Canada

A total of 149 patients (79 athletes, 70 non-athletes) underwent patella realignment via lateral retinacular release as treatment for recalcitrant chondromalacia patella, recurrent patellar subluxation and/or patella dislocation. Pre-operatively, all patients had undergone at least a 6-month non-surgical programme as designed by DeHaven et al. and modified by us.

Materials and Methods

Each patient underwent pre-operative physical examination with gait analysis. Also static radiographs after the technique of Fichat were taken and then arthroscopic evaluation was conducted. The success of surgery was evaluated 42 months later, and the results were rated as excellent, good, fair or poor on a scoring system modified after J. Insall et al.

Results

Of the patients in this study group 82% were considered excellent or good. Biomechanical analysis revealed the following architectural abnormalities:

- 43% presented with pes planus
- 38% manifested aberration of the angle of the calcaneus
- 18% had a Morton's foot
- Standing knee angle was varus in 26% of the study group
- The problem was bilateral in 56% of patients

Conclusions

The success of surgery was a direct reflection of:

- Degree of severity of the chondromalacia patella
- Length of time from the onset of disease to surgical intervention
- Degree of the excessive lateral positioning of the patella in the trochlea
- Ability to correct the lower extremity foot and ankle malalignments while simultaneously correcting the abnormalities at the patellofemoral interface

Athletes who develop chondromalacia patella do so at a much younger age than do non-athletes (14.7 versus 23 years) but do not suffer dislocations any earlier than non-athletes (15 versus 17 years).

Surgery and Arthroscopy of the Knee
2nd Congress of the European Society
Editors: Müller/Hackenbruch

Longitudinal Osteotomy of the Patella After Morscher

M. Pećina

Department of Orthopaedic Surgery, Medical Faculty,
University of Zagreb, Yugoslavia

Introduction

It is nowadays a generally accepted idea that in view of the aetiology of patellar chondromalacia being different from other chondromalacias, its surgical treatment should also not be the same. A valuable contribution in this sense has been made by longitudinal osteotomy of the patella, introduced into clinical practice by Morscher. In 1978 he published preliminary results of the application of this technique in the treatment of patellar chondromalacia and subsequently presented the pathophysiological explanation and developed the indications and possibilities for its application (Morscher and Dick 1980; Henche et al. 1981; Morscher 1985). The pathophysiological and biomechanical bases were also investigated by other authors (Jenny 1980; Küsswetter 1980; Waisbrod and Traiman 1980; Hehne 1983; Hejgaard and Arnoldi 1984). Since 1980 we have been applying longitudinal osteotomy of the patella (following Morscher) successfully in the treatment of chondromalacia in the Department of Orthopaedic Surgery, Zagreb, and have reported on our experiences at several congresses (Pećina 1984a, b, 1985).

This paper presents clinical experiences with the application of longitudinal osteotomy of the patella over the past 5 years. With the help of the photoelasticimetric analysis of models made according to the pre- and postoperative roentgenograms of the surgically treated patients, we have attempted to show how the stresses in the patellofemoral joint tend to change after longitudinal osteotomy of the patella has been performed.

Material and Methods

We shall present the results obtained for 42 of the 54 knee joints (from 33 patients) operated on up to the end of September, 1985, with a postoperative follow-up from 1 to 5 years. There were 28 males and 5 females, with an average age of 21 years. In over one-half of the knee joints operated on (28), in addition to longitudinal osteotomy of the patella after Morscher, medialisation and ventralisation of the tuberositas tibiae were also performed because of instability of the patella. In five patients distalisation of the tuberositas was carried out as well because of the markedly high riding position of the patella. The results have been evaluated on the basis of the Bandi (1976) and Ficat (1973) criteria. The photoelasticimetric investigations were conducted at the Faculty of Mechanical Engineering and Shipbuilding, University of Zagreb, for which we owe our thanks to Professor Dr. Eng. U. Osman Muftić.

Surgery and Arthroscopy of the Knee
2nd Congress of the European Society
Editors: Müller/Hackenbruch

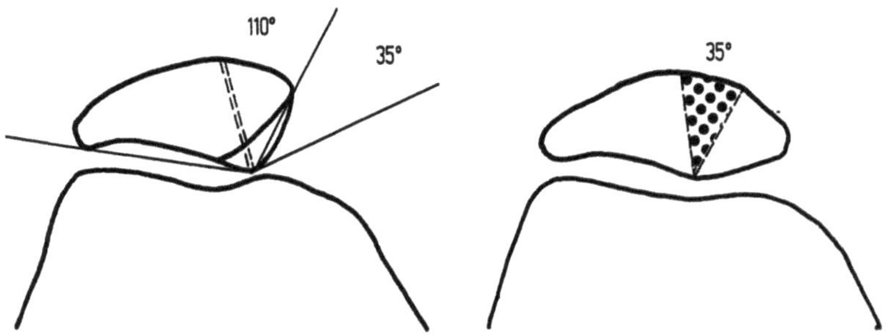

Fig. 1. The mean value of Wiberg's angle preoperatively (*left*) and postoperatively (*right*)

Table 1. Postoperative results evaluated according to the Bandi (1976) and Ficat (1973) criteria

Assessment	No. of knees	Percentage of total
Excellent	31	73.8
Good	9	21.4
Fair	2	4.8
Failure	0	0.0

Results

The mean value of the Wiberg patellar angle is 110° preoperatively and 145° postoperatively (Fig. 1). The results as judged on the basis of the criteria evaluation after Bandi (1976) and Ficat (1973), are indeed excellent (Table 1).

Figure 2 presents a roentgenogram of a female patient in whom longitudinal osteotomy of the patella was carried out in both knee joints along with simultaneous medialisation and ventralisation of the patella. The postoperative result is excellent. Successful application of longitudinal osteotomy after Morscher has been confirmed by the fact that we have operated on five top female athletes (basketball, volleyball, karate and handball players) all of whom returned to competitive sports and on three ballerinas who went back to dancing. The goal of the operative intervention is to achieve an even degree of stress in the patellofemoral joint following longitudinal osteotomy of the patella. This has been confirmed by photoelasticimetric examinations and testing of the model constructed according to the roentgenogram of one surgically treated knee. Preoperatively an unequal isochromatic distribution was visible. Following longitudinal osteotomy of the patella the stresses were uniformly distributed throughout the patellofemoral joint (Fig. 3).

Discussion

At first sight it seems improper to maintain that no poor results at all have been obtained in medicine. We had two female patients in whom only a fair result was realised. However, we consider that these were our fault and not an inadequacy

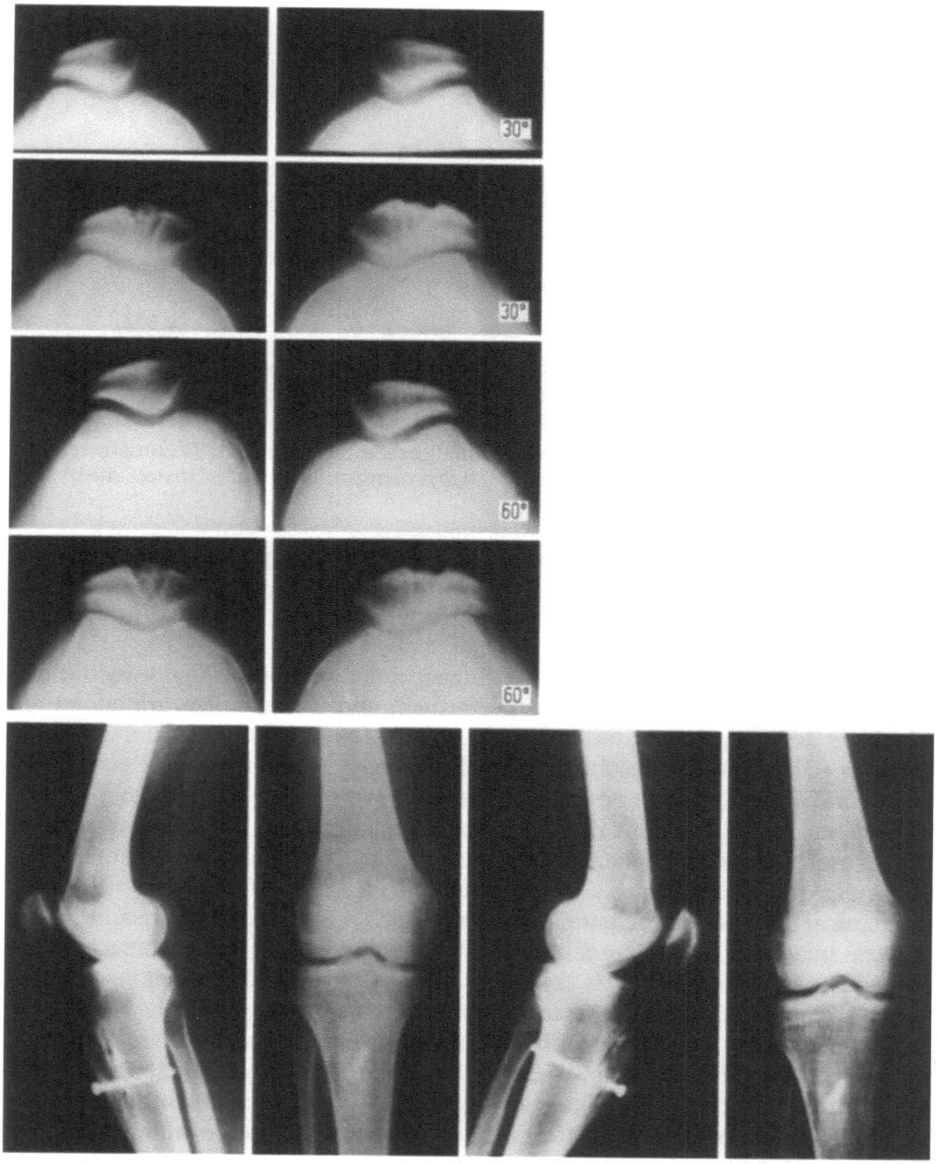

Fig. 2. a Longitudinal oesteotomy of the patella in both knee joints; **b** simultaneous medialisation and ventralisation of the tuberositas tibiae in addition

in the operative method. In one of these patients the indication for operative intervention was wrong, and in the other a technical error occurred during the operation.

The complications encountered included having to carry out manipulation of the knee under general anaesthesia some 6 weeks after the initial operative interven-

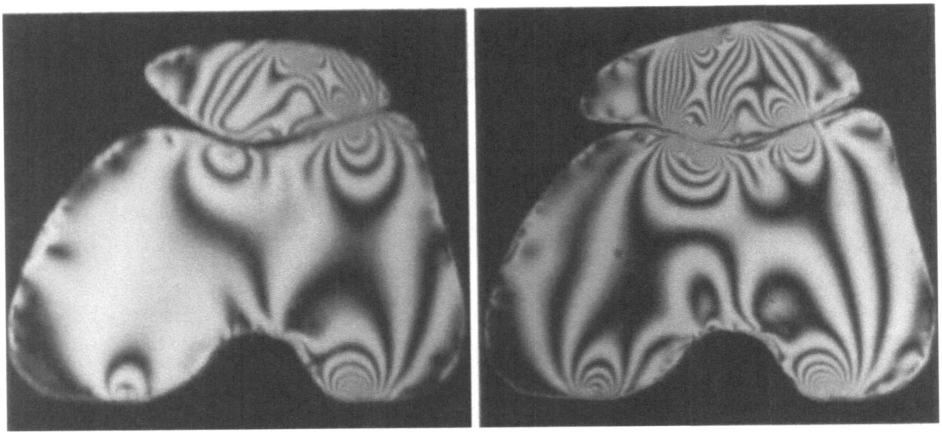

Fig. 3. Photoelasticimetric models of longitudinal osteotomy of the patella constructed according to the pre- (*left*) and postoperative (*right*) roentgenograms of a treated knee

tion in five instances. In one patient, due to a certain degree of carelessness on the part of the operating surgeon, osteosynthesis of the patella had to be carried out. The final result, however, was good, since the patient consented to undergo surgery of the other knee.

If an analysis is made of the results obtained in ten knees on which longitudinal osteotomy of the patella alone was performed, and if four knees are added on which, in addition to longitudinal osteotomy of the patella, only lateral release was carried out, the final result shows that in 12 cases excellent and in 2 cases good postoperative results were obtained. This is an even higher percentage of excellent results (85%) than that obtained when considering all the knees examined. The cause of the development of chondromalacia in the above-mentioned patients was obviously dysplasia of the patella (Wiberg IV) without any other biomechanical disorders in the patellofemoral joint. To obtain optimal results in the surgical treatment of patellar chondromalacia, it is necessary to combine surgical procedures in accordance with the aetiological factors. It would then be possible to avoid complications in the sense of lateralisation of the patella following longitudinal osteotomy as mentioned by Glymann in 1980. A question also arises in this case on the necessity of modifications in the application of patellar longitudinal osteotomy (Griso 1980). Naturally we agree with Morscher's opinion (1985) that longitudinal osteotomy of the patella is indicated in cases in which dysplasia of the patella is present in addition to incongruity in the patellofemoral joint.

In planning treatment by longitudinal osteotomy of the patella we have been following the principle that, postoperatively, Wiberg's angle of the patella and the angle of the femoral trochlea aperture should be approximately equal.

Conclusion

In concluding we should like to emphasise that we are more than satisfied with the results so far achieved by the application of longitudinal osteotomy of the patella after Morscher. However, we consider that, depending on the instability of the patella, the size of the Q-angle and the damage to the patellar cartilage, in some cases it might be necessary to carry out medialisation in addition to longitudinal osteotomy, and possibly distalisation and ventralisation of the tuberositas tibiae as well.

References

Bandi W (1976) Vorverlagerung der tuberositas tibiae bei Chondromalacia patellae und femoro-patelarer Arthrose. Hefte Unfallheilkd 127:175–186

Ficat P (1973) Les déséquilibres rotuliens: de l'hyperperssion externe à arthrose. Masson, Paris

Goymann V (1980) Die Biomechanik des patellofemoralen Gleitwegs. Orthop Praxis 16:451–461

Griss P (1980) Modifizierte Patellalängsosteotomie zur Behandlung der exzentrischen Chondromalazie und Retropatellararthrose. Z Orthop 118:822–824

Hehne HJ (1983) Das Patellofemoralgelenk. Enke, Stuttgart

Hejgaard N, Arnoldi CC (1984) Osteotomy of the patella in the patellofemoral pain syndrome. Int Orthopaed (SICOT) 8:189–194

Henche HR, Künzi HU, Morscher E (1981) The areas of contact pressure in the patellofemoral joint. Int Orthop (SICOT) 4:279–281

Jenny K (1980) Intraossäre Druckmessung an der Patella. Dissertation, University of Basel

Küsswetter W (1980) Tierexperimentelle Untersuchungen zur Keilosteotomie der Patella. Z Orthop 118:632–633

Morscher E (1978) Osteotomy of the patella in chondromalacia. Arch Orthop Trauma Surg 92:139–147

Morscher E (1985) Indikation und Moeglichkeiten der Patellakeilosteotomie. Orthopäde 14:261–265

Morscher E, Dick W (1980) Die sagittale Patellaosteotomie bei Chondromalacia patellae. Orthop Praxis 16:692–695

Pećina M (1984a) Longitudinal osteotomy of the patella and anterior displacement of the patella one stage procedure. 16th Congress SICOT, London 1984 (poster presentation)

Pećina M (1984b) L'osteotomia retulea. VI Corso di Chirurgia del Ginocchio, Firenze

Pećina M (1985) Longitudinal osteotomy of the patella. First combined meeting of the Helenic and Yugoslavian association of orthopaedy and traumatic surgery, Kavala, abstracts, no. 62

Waisbrod H, Traiman N (1980) Intra-osseous venography in patello-femoral disorders. J Bone Joint Surg [Br] 62:454–456

Patellectomy as a Salvage Procedure

G. Blatter [1], R. W. Jackson [2], and O. Bayne [2]

[1] Klinik für Orthopädische Chirurgie, Kantonsspital, 9007 St. Gallen, Switzerland
[2] Department of Orthopaedic Surgery, University of Toronto, Toronto, Canada

Patellectomy is an old surgical procedure that has been used in the treatment of such conditions afflicting the knee as comminuted fractures of the patella, chondromalacia patella, patellofemoral arthrosis, recurrent subluxation of the patella, infections and neoplasms involving the patella [1–3, 18, 29, 36, 37].
A review of the orthopaedic literature shows that patellectomy has always been the subject of much controversy. In 1909 Heineck [18] reviewed the literature and found 13 cases of total patellectomies: 5 for fractures, 4 for tuberculosis, 3 for osteomyelitis and 1 for sarcoma. He condemned patellectomy because he felt not only that the patella affords protection to the knee joint, but also that its excision reduces quadriceps power. Brooke [5] in 1937 described the patella as a mere "phylogenetic remnant" which is undergoing reduction in size and in time would disappear entirely. He reported experimental work which showed that the patella was not important in knee extension. He moreover suggested that patellectomy increases joint movement and quadriceps strength. His work was supported by Groves [19] and Watson-Jones [36]. Thereafter patellectomy became a widely accepted orthopaedic procedure.
In 1942 Bruce and Walmsley [6] disagreed with Brooks' conclusions after repeating his experiments. They also cautioned against patellectomy, suggesting that, on the basis of animal studies, the absence of the patella causes an increase in knee joint degeneration. Studies by Haxton in 1945 [17], O'Donoghue in 1952 [27], Freehaufer in 1962 [13] and more recently by Kaufer in 1971 [24] and Maquet in 1976 [25] have shown that the patella is indeed important in providing a biomechanical advantage to the quadriceps mechanism in effecting knee extension or resisting knee flexion.
Today we have a better understanding of the role of the patella in knee joint mechanics [10, 12–15, 17, 24, 25], as well as improved surgical techniques for patellar realignment [22, 25] and the use of the arthroscope in the early diagnosis of patellofemoral pathology [8, 23]. As a consequence of these developments the current indications for patellectomy have been so restricted as to limit the operation largely to a salvage procedure in the treatment of severely comminuted fractures of the patella or severe localised patellar or patellofemoral disease [3, 9, 37].
When patellectomy cannot be avoided, the object of surgery must be maximal preservation of extensor function. Several surgical techniques have been described for removing the patella and closing the defect created in the extensor tendon [4, 30, 32, 36]. However, concerns are still generally expressed as to the consequences of patellectomy, such as pain, extensor lag, limitation of flexion, recur-

Surgery and Arthroscopy of the Knee
2nd Congress of the European Society
Editors: Müller/Hackenbruch

rent effusion, giving way, limitation of activity, loss of quadriceps strength and poor cosmetic results [7, 15, 26, 32–34, 38].
In the belief that these concerns have been overstated we have undertaken a long-term follow-up study of the results following patellectomies at two teaching hospitals, the University of Toronto and the Kantonsspital of St. Gallen.

Material and Methods

We have reviewed the records of 174 patients who underwent total patellectomies between 1965 and 1983 at Toronto General Hospital, Toronto Western Hospital and Kantonsspital, St. Gallen. All patients were invited to return for a review. Of these 88 patients were seen and clinically examined. Nineteen patients were unable to return for review because of distance or other inconveniences, but responded to either a written questionnaire or a telephone interview. The follow-up therefore covered 107 patients with 113 patellectomies (6 bilateral). The average age was 42.6 years, with a range from 17 to 79. Follow-up time from surgery to review ranged from 3 years to 17.5 years (average 10.5). There were 57 females and 50 males.
The subjective and objective findings of each patient regarding pain, range of motion, effusion, giving way, quadriceps strength and limitation of activity, were recorded and graded according to our Knee Evaluation Scale (Table 1). Our grading system eliminated any knee from being pronounced satisfactory, unless it was considered so both by us and by the patient.
Patients were divided into three groups. Group 1 comprised 40 knees which underwent operations by the purse-string technique. In group 2 were 24 knees treated by the vastus medialis technique. The 49 knees in group 3 were operated on by other techniques.
For the purse-string patellectomy (Fig. 1) a midline vertical incision is made through the extensor tendon down to the bone. The patella is enucleated subperiosteally, using sharp dissection and preserving as much tendon as possible. To repair the extensor mechanism, a heavy suture of braided Dexon no. 1 is placed in purse-string fashion around the margins of the defect. The suture is then drawn up tightly, gathering the tissues together evenly. This "bunched-up" tissue adds greatly to the cosmesis of the procedure, at the same time as maintaining the original length of the patellar tendon. The knee can usually be flexed to 90° at this point without placing undue stress on the suture line. Postoperatively while the patients is still under anaesthesia, a cylinder cast is applied with the knee in 10°–15° flexion. Active isometric quadriceps-strengthening exercises are begun 1–2 days postoperatively, at which time the patient is also started on crutches with full weight-bearing. The cast is removed after 3–4 weeks, and an active range of movement (ROM) in the knee is begun. Physiotherapy commences and continues at least until maximum flexion and quadriceps strength have been achieved (average 6 months postoperatively) [20].
Using the vastus medialis technique (Fig. 2) one cuts the extensor mechanism obliquely, beginning proximal-laterally and ending distal-medially almost perpendicular to the vastus medialis. One then excises the patella subperiosteally. After closure of the synovium the vastus medialis is mobilised; it is advanced and at-

Table 1. Knee Evaluation Scale for patellectomy

Symptom				Grading
Pain				
None				*Excellent*
Mild				*Good*
Moderate				*Fair*
Marked				*Poor*
Range of motion (compared with opposite knee)				
Full flexion, extension lag 5° or less				*Excellent*
Flexion loss less than 20°, extension lag 10° or less				*Good*
Flexion loss less than 30°, extension lag 15° or less				*Fair*
Flexion less than 100°, extension lag greater than 15°				*Poor*
Giving way				
None				*Excellent*
Occasional (once monthly)				*Good*
Frequent (5–10 times) (< thrice monthly)				*Fair*
Quite frequent (> 10 times) (> thrice monthly)				*Poor*
Effusion				
None				*Excellent*
Occasional (once monthly)				*Good*
Frequent (< thrice monthly)				*Fair*
Quite frequent (> thrice monthly)				*Poor*
Quadriceps strength (as a percentage of contralateral knee)				
100%–90%				*Excellent*
90%–70%				*Good*
70%–50%				*Fair*
50% or less				*Poor*
Activities (maximum 35)				
35–30				*Excellent*
30–25				*Good*
25–20				*Fair*
< 20				*Poor*
Stairs				
Normally	With railing		Any manner	Unable
10 points	5		3	0
Distance walked				
Unlimited	5 km	1 km	Indoors	Bed/chair
10 points	8	5	2	0
Running				
Unlimited	1 km	500 m	100 m	Unable
5 points	3	2	1	0
Kneeling				
No discomfort		Minor	Major discomfort	
5 points		3	0	
Squatting				
Full		Minor	Major limitation	
5 points		3	0	
Cosmesis				
Totally acceptable				*Excellent*
Minor concern				*Good*
Major concern				*Fair*
Totally unacceptable				*Poor*

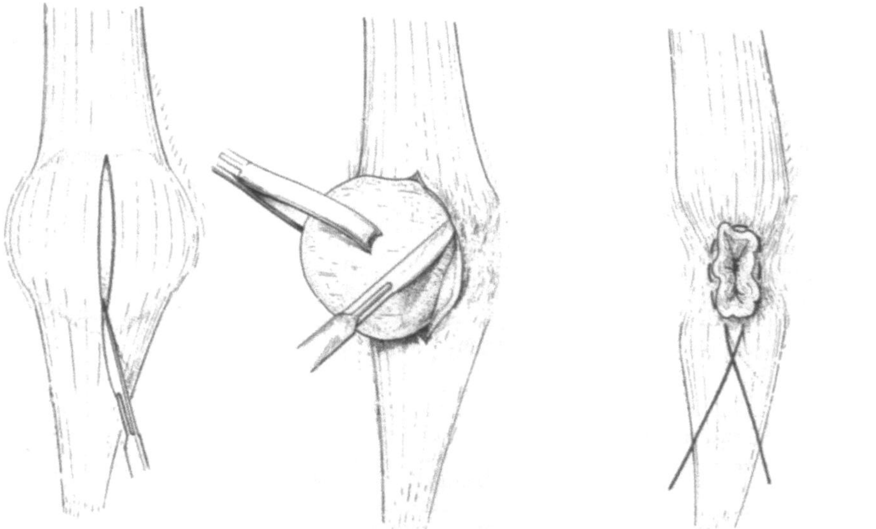

Fig. 1. Purse-string patellectomy: the extensor tendon is incised, the patella is enucleated subperiosteally, and a purse-string suture is used to dose the defect

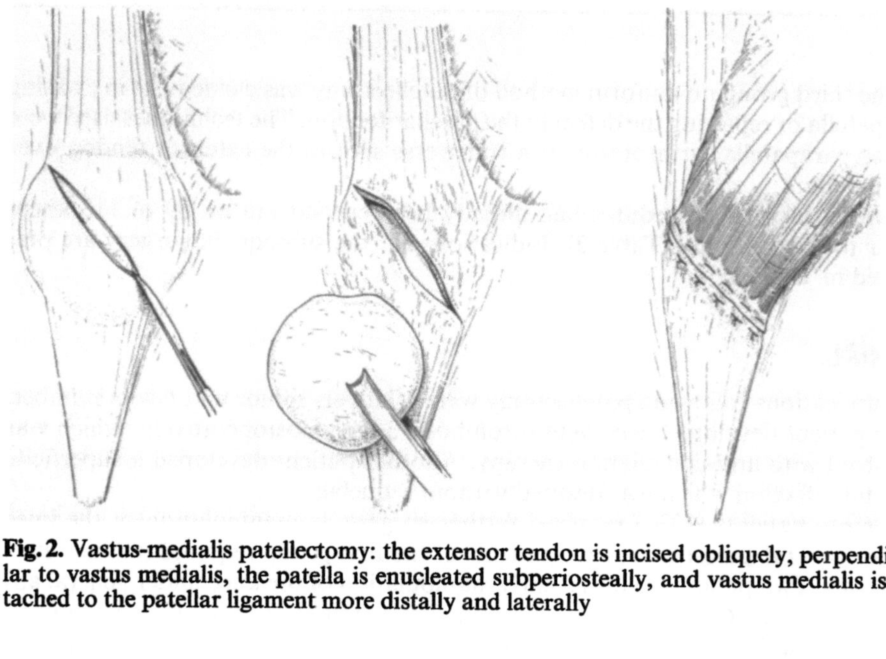

Fig. 2. Vastus-medialis patellectomy: the extensor tendon is incised obliquely, perpendicular to vastus medialis, the patella is enucleated subperiosteally, and vastus medialis is attached to the patellar ligament more distally and laterally

tached to the patellar ligament more distally and laterally. Postoperatively we apply a posterior splint with the knee in 10°–15° flexion. This splint is removed several times a day for passive ROM exercises. For the first 4 weeks we allow partial weight-bearing on two crutches. Thereafter physiotherapy progresses with active quadriceps function.

Table 2. Previous surgery

	Group 1	Group 2	Group 3	Total
Meniscectomy	8	4	10	22
Patellar shaving	9	4	15	28
Patellar tendon transfer	7	6	2	15
High tibial osteotomy	2	3	4	9
Pes anserinus transfer	0	1	2	3
Lateral reticular release	2	5	0	7
Prosthetic replacement	0	0	1	1
Previous ORIF patella	0	3	1	4

ORIF, open reduction with internal fixation.

Table 3. Surgical indications

	Group 1	Group 2	Group 3	Total
Chondromalacia patella	21	12	23	56
Comminuted fractures	12	6	14	32
Patellofemoral arthrosis	4	3	10	17
Recurrent luxating patella	3	3	2	8

In the third group no uniform method of patellectomy was carried out in excising the patella or repairing the defect in the patellar tendon. The techniques used were either parapatellar arthrotomy or a transverse split in the extensor tendon over the patella.

Previous surgical procedures had already been carried out on 89 of 113 knees prior to patellectomy (Table 2). Indications for the subsequent surgery are presented in Table 3.

Results

Complications following patellectomy were relatively minor and few in number. One patient developed deep vein thrombosis 3 days postoperatively, which was resolved with anti-coagulation therapy. Another patient developed a superficial wound infection which was drained without sequelae.

Fourteen patients (13%) required further surgery or manipulation of the knee under general anaesthesia. Five patients of the first group required further interventions. Two patients required early manipulation of the knee under general anaesthesia (5 and 6 weeks postoperatively). One required excision of an area of calcification in the patellar tendon 18 months after surgery. One had a repair of a partial tear in her quadriceps expansion following a hyperextension injury 3 years following patellectomy. The remaining patient underwent total knee arthroplasty because of progressive degenerative changes in the other compartments of her knee. In the second group two patients required further surgery for realignment of the extension tendon. Seven patients of the third group required further inter-

Table 4. Good-excellent scores on the Knee Evaluation Scale

	Group 1	Group 2	Group 3	Total
	(%)			
Pain	85	83	76	80
Range of motion	92	91	81	87
Giving	90	88	86	88
Effusion	80	83	83	81
Quadriceps strength	60	66	50	58
Activities	65	66	59	63
Cosmesis	95	91	73	85

ventions. Three needed manipulation under general anaesthesia between 1 and 6 months after patellectomy (one required manipulation of his knee three times in 1 year). Two patients had high valgus tibial osteotomies because of degenerative changes in their medial compartments. Two patients underwent total knee arthroplasty, 2 and 5 years after patellectomy.

Table 4 presents the percentage of patients in each group receiving a good or excellent score in the various parameters of the Knee Evaluation Scale. Six patients in group 1 had moderate to severe pain; three required analgesia on a regular basis for pain relief. One patient had a comminuted fractured patella, three had chondromalacia patella, and two had patellofemoral arthritis. Two of the four patients in group 2 who had occasional pain were patellectomised for chondromalacia and two for patellofemoral arthritis. No patient required analgesia on a regular basis. In group 3 pain was significant in 12 knees. Of these, 8 took analgesia on a regular basis. Four knees had chondromalacia patella, 7 had patellofemoral arthrosis, and 1 had subluxating patella prior to patellectomy. None of the patients treated for fractures had significant complaints of pain.

Only three patients in group 1 and two patients in group 2 had less than full active extension and flexion to within 0°–10° of the opposite knee. One patient was limited to 115° due to an old supracondylar fracture and was graded as fair. One patient who had a subsequent total knee arthroplasty was limited to 90° flexion and had an extensor lag of 15°. Another patient, a 42-year-old workman who had a comminuted fracture of his patella, was limited to 60° flexion. Two patients in the second group who underwent patellectomy for patellofemoral arthritis lost 30° flexion postoperatively. In group 3 were nine patients with unsatisfactory results. Seven patients had an extensor lag of 15° or greater. Of these, four had flexion losses of 10°–20° compared with the opposite knee. The two patients with total knee arthroplasties had flexion to 90° – one had an extensor lag of 20°.

Giving way occurred frequently in four patients from group 1, three from group 2 and seven from group 3. This phenomenon was noted when walking on uneven ground or climbing stairs. Five patients used canes for support. Of these, two complained of symptoms in the contralateral knee. All patients with frequent giving way had no evidence of internal derangement of the knee or ligamentous laxity, except two with total knee replacements, who had moderate anterior-posterior instability.

There were 21 knees with significant recurrent effusion. Of these, only 2 had fractures prior to patellectomy. Effusion was usually most significant at the end of the day or following severe activity. One patient in group 1 had a chronic recurrent effusion that had to be aspirated repeatedly.

Results of dynamic testing of quadriceps strength compared with the contralateral knee showed a marked reduction despite a full range of motion in most cases. Group 1 showed 12 of 30 knees and group 2 showed 7 of 20 knees to have quadriceps strength reduced by more than 30% of the opposite knee. Among group 3 patients half had greater than 30% reduction of quadriceps strength. Eight patients in group 1 and 6 patients in group 2 showed no loss of quadriceps strength.

For 68 patients with 72 knees no major problems were evident when walking, running, climbing stairs, squatting and kneeling. One patient, a 26-year-old dentist with bilateral purse-string patellectomies, jogged, skied, played squash and tennis 5 years after his patellectomies. A 60-year-old widow who lived alone climbed up and down 10 flights of stairs daily without pain or limitation 8 years after her patellectomy. Squatting was not a major problem in either group. As was expected, knee flexion correlated well with the ability to squat. As regards kneeling 33 knees in group 1 and 19 knees in group 2 had no problems. Major discomfort with kneeling was reported for 24 knees in group 3, however. Seventeen knees (5 in group 1, 3 in group 2 and 9 in group 3) were considered to have poor results in terms of limitation of activities. Of these, 10 had associated hip, back or contralateral knee pain, which they considered to contribute to their limitation in walking, running and stair climbing. Only 7 considered their limitation of activity to be on the basis of pain in the patellectomised knee.

Only four patients, two each in groups 1 and 2, were unhappy with the appearance of their knees following patellectomy. Twelve patients in group 3 were quite unhappy with the cosmetically flat-looking knee following surgery. Of these, seven were graded as having satisfactory results in regard to pain, range of motion and functional activity.

Discussion

The overall rate of satisfactory results in our series was 77%. Group 1 and group 2 knees showed better results (81% and 79%, respectively) than group 3 knees (73%) in all parameters tested, except in regard to effusion, on which there was no significant difference (80%, 83%, 83%, respectively). As regards pain relief, range of motion, quadriceps strength, giving way and cosmetic effect, groups 1 and 2 showed a major improvement over group 3 in all categories.

Our rate of satisfactory results is comparable with that of most other reported series, including those of West, 80% [38]; West and Sotohall, 75% [37]; Geckeler 74% [16]; and Peeples, 82% [28]. The satisfactory results in our series did not appear to decline with time.

Pain relief, the main indication for patellectomy (except in the fracture group), was experienced by 80% of knees. Only one patient in the fracture group complained of significant pain following patellectomy. A disproportionate number of patients with chondromalacia patella, patellofemoral arthrosis or subluxating patella still had significant pain after their patellectomies. The explanation for persistence of pain in this group was thought perhaps the result of poor patient selection – with degenerative changes in other compartments or patellofemoral degeneration resulting from repeated trauma to the condyles after patellectomy [6, 17, 18, 31]. On careful clinical assessment of these knees, however, it was noted that the pain was localised mainly over the patellar tendon, or over the lateral condyle with the knee at between 30° and 60° flexion. Subsequent X-ray evaluation and diagnostic arthroscopies carried out on these knees failed to verify significant degenerative changes to warrant the degree of pain experienced. One patient was noted to have calcification of his patellar tendon, which was then surgically removed, resulting in significant pain relief.

The reason for continuing pain in these knees may lie in the persistence or creation of patellar tendon malalignment following patellectomy, causing a distorted Q-angle and incongruence between the patella expansion and the trochlea. This maintained or produced a "lateral pressure syndrome", which caused the original pathology in most cases [21, 24, 25]. This phenomenon was most apparent in patients who had patellectomies for subluxating patella and the two patients who had high valgus tibial osteotomies following their patellectomies. The reason for the better results in groups 1 and 2 as compared with group 3 may be the arthrotomy incision which avoided unduly distorting the extensor mechanism.

Loss of knee motion was only significant in 13% of cases, while 92% of group 1 knees and 91% of group 2 knees had almost full active flexion and extension. In group 3 knees, however, persistence of an extensor lag rather than flexion loss was quite significant. Seven patients had an extensor lag greater than 15°, suggesting either poor surgical technique, which allowed undue lengthening of the extensor mechanism, or inadequate quadriceps rehabilitation.

Results of dynamic testing on quadriceps strength after patellectomy using the Cybex machine showed significant reduction in all groups and compares with that of other studies that have analysed quadriceps strength following patellectomy [33, 34]. Biomechanical studies of the quadriceps mechanism show that patellectomy alters the mechanical efficiency of the quadriceps and reduces the power of extension by 15%–30%, depending on the type of tendon repair [24, 27]. This loss, however, has been shown to be within the functional reserve of most individuals, and adequate rehabilitation therapy can return the knee to full functional strength [6, 7, 11]. This was seen in 18 of 68 knees which had no more than 10% measurable loss of their quadriceps strength. However, the fact that 26% (17/68) of patellectomised knees in our series had 50% or more reduction in their quadriceps strength, compared with 33% reported by Stougard and 24 of 26 knees reported by Sutton, serves to re-emphasise the importance of the patella in helping to maintain quadriceps strength.

Giving way was a major problem in only 14 of 113 knees (12%) in our series (4 in group 1, 3 in group 2 and 7 in group 3). Ligamentous laxity was not noted to be a major factor contributing to knee instability following patellectomy [34]. This was seen only in 2 of the 3 total knee arthroplasties following patellectomy. Both had moderate anterior-posterior instability. The patient who underwent a hinge-type total knee arthroplasty had a stable knee. This finding agrees with that of Kaufer and Sutton that a prosthesis with intrinsic stability should be selected

for total knee replacement following patellectomy. Giving way usually occurred with the knee in flexion, especially when climbing stairs or walking on uneven ground. This phenomenon may be explained on the basis of relative quadriceps insufficiency resulting from a shortened quadriceps tendon with the knee loaded and in flexion [34]. This would tend to occur more frequently in patients who had their patellar tendons unduly shortened at the time of surgery.
Cosmetic appearance of the knee following patellectomy appeared to be of concern particularly to female patients. Seven patients in group 3, all young females between the ages of 22 and 36 who had fully functional, painless knees after patellectomy, were quite conscious of the flat appearance of their knees, especially when sitting. They wore pants or long dresses to cover their perceived defects. In group 1 knees, however, 95% and in group 2 knees 91% satisfactory results were noted.

Conclusion

We may summarise our findings here in the following points:

1. In primary patellectomy for severely comminuted fractures of the patella one can expect up to 75% excellent results, with similar function and strength comparable with that on the opposite side.
2. The results of patellectomy for patellofemoral arthritis are usually less favourable.
3. The outcome of a patellectomy for chondromalacia may vary and cannot be predicted.
4. Results after patellectomy for recurrent dislocations are, in general, good.
5. Choice of a surgical technique should allow one to shorten and to medialise the extensor mechanism as well as to resume movement early; it may promise a scar which is acceptable.

References

1. Ackroyd C, Polyzoides J (1978) Patellectomy for osteoarthritis. J Bone Joint Surg [Br] 60:353–357
2. Bentley G (1970) Chondromalacia patellae. J Bone Joint Surg [Am] 52(2):221
3. Boucher HM (1958) Patellectomy in the geriatric patient. Clin Orthop 11:33
4. Bickel WH, Johnson KA (1971) Z-plasty patellectomy. Surg Gynecol Obstet 132:986
5. Brooke R (1937) Treatment of fracture of the patella by excision. Br J Surg 24:733–747
6. Bruce J, Walmsley R (1942) Excision of the patella. J Bone Joint Surg 24:311
7. Burton VW, Thomas M (1972) Results of excision of the patella. Surg Gynecol Obstet 135:753
8. Cassels WS (1979) The arthroscope in the diagnosis of disorders of the patello-femoral joint. Clin Orthop 114:45–50
9. Cave E, Rowe CR (1950) The patella – its importance in knee derangement. J Bone Joint Surg [Am] 32:542
10. DePalma A, Flynn J (1958) Joint changes following experimental partial and total patellectomy. J Bone Joint Surg [Am] 40:395
11. Duthie HL, Hutchinson JR (1958) The results of partial and total excision of the patella. J Bone Joint Surg [Br] 40:75–81

12. Frankel VH, Burnstein AH, Brooke D (1971) Biomechanics of internal derangement of the knee. J Bone Joint Surg [Am] 53:545
13. Freehafer A (1962) A study of the function of the patella. Clin Orthop 25:162
14. Garr E, Moskowitz RW, Davis W (1973) Degenerative changes following experimental patellectomy in the rabbit. Clin Orthop 97:296–304
15. Goodfellow J, Hungerford DS, Woods C (1976) Patello-femoral joint mechanics and pathology. J Bone Joint Surg [Br] 58:291–299
16. Geckeler EO, Quaranta VA (1962) Patellectomy for degenerative arthritis of the knee: late results. J Bone Joint Surg [Am] 44:1109
17. Haxton H (1945) The function of the patella and the effects of its excision. Surg Gynecol Obstet 80:389–395
18. Heineck AP (1909) The modern operative treatment of fractures of the patella. Surg Gynecol Obstet 9:177–248
19. Hey Groves EW (1937) A note on the extension apparatus of the knee joint. Br J Surg 24:747
20. Hislop HG, Perrine JJ (1967) The isokinetic concept of exercise. Phys Ther 47:114
21. Hungerford DS, Barry BS (1979) Biomechanics of the patello-femoral joint. Clin Orthop 144:9–15
22. Insall J, Bullough PG, Burnstein AH (1979) Proximal "tube" realignment of the patella for chondromalacia patella. Clin Orthop 144:63–69
23. Jackson RW, Dandy DJ (1976) Arthroscopy of the knee. Modern Ortho Monographs. Grune and Stratton, New York
24. Kaufer H (1971) Mechanical function of the patella. J Bone Joint Surg [Am] 53:1551
25. Maquet P (1976) Biomechanics of the knee with application to the pathogenesis and the surgical treatment of osteoarthritis. Springer, Berlin Heidelberg New York
26. Mishra US (1972) Late results of patellectomy in fractured patella. Acta Orthop Scand 43:256–263
27. O'Donoghue D, Tompkins F, Hays MB (1952) Strength of quadriceps function after patellectomy. West J Surg Obstet Gynecol 60:159
28. Peeples RE, Margo MK (1978) Function after patellectomy. Clin Orthop 132:180–185
29. Reilly RE, DeSouza LJ (1974) Patellectomy: an alternative technique. Clin Orthop 103:170
30. Shorbe HB, Dobson CH (1958) Patellectomy: repair of the extensor mechanism. J Bone Joint Surg [Am] 40:1281
31. Soto-Hall R (1945) Traumatic degeneration of the articular cartilage of the patella. J Bone Joint Surg 27:426–431
32. Steurer PA, Gradisar IA, Hoyt WA, Chu M (1979) Patellectomy: a clinical study and biomechanical evaluation. Clin Orthop 144:84–90
33. Stougard J (1970) Patellectomy. Acta Orthop Scand 41:110
34. Sutton FS, Thompson CH, Lipke J, Kettelkamp DB (1976) The effect of patellectomy on knee function. J Bone Joint Surg [Am] 54:537
35. Thompson WJ, Schweigel JF. Patellectomy: a 21 year follow-up. Can J Surg 11:173–175
36. Watson-Jones R (1945) Excision of patella. Br Med J 2:195
37. West FE, Soto-Hall (1958) Recurrent dislocation of the patella: end results of patellectomy with quadriceplasty. J Bone Joint Surg [Am] 40:386
38. West FE (1962) End results of patellectomy. J Bone Joint Surg [Am] 44:1689

Patellectomy: Long-Term Results

G. Brunelli, G. Stefani, F. Brunelli, and C. Milini

Brescia University Medical School, Department of Orthopaedics,
Brescia 25100, Italy

Introduction

Patellectomy is performed by different surgeons for various conditions, e.g. patellar comminuted fractures, patellar chondromalacia, patellofemoral arthrosis. Some consider patellectomy a good operation, others think poorly of it and perform it only under very restricted situations.

In 1909 Heinech published the results of 1100 patellar fractures, of which only 10 had been by patellectomy [15]. In 1925 Ludlof [18] suggested patellectomy as a treatment for patellofemoral arthrosis, and in 1937 Brooke [3] described a study on cadavers and living patients, showing that patellectomy improved the quadriceps muscle function. Hey Groves [16] and Watson Jones [21–22] agreed with this idea. In contrast Haxton [14] and Jansenius [17] thought the patella to be essential in knee biomechanics. Several authors wrote against patellectomy [4, 9, 11, 12, 20], while others were in favour [1, 2, 5–7, 10, 13, 19]. Patellectomy is therefore considered a very controversial operation in the literature. We have always considered simple patellectomy an unsatisfactory operation giving poor results (due to the weakening of the extensor apparatus), on the basis both of biomechanical research and clinical observation. We have observed severe stiffness of the knee due to patellar fractures even after careful surgery, we have seen fair results even in perfectly treated simple fractures with increasing hypertrophy and arthrosis of the patella and the patellar face of the femur.

Under particular, severe circumstances patellectomy must be performed to avoid poor results, but technical devices should be used to restore the biomechanical conditions required for good function of the extensor apparatus of the knee. To accomplish this, the correct distance of the apparatus from the centre of rotation of the joint is maintained and its angle of insertion on the tibial bone preserved. In this way a poor operation can be changed into a good one without any prosthetic replacement (which may cause either acute or late complications).

Material and Methods

In 1962, the senior author, after personally conducting biomechanical studies and a wide literature overview, developed a patellectomy technique which replaces the thickness of the patella by means of a quadriceps tendoplasty. In this technique a total thickness flap of the tendon is turned deeply under a layer of prepatellar fibrous tissue to which tiny bone chips are adherent. The tendon facing the femur is perfectly smooth and adapts to the surface of the bone, gliding without induc-

Surgery and Arthroscopy of the Knee
2nd Congress of the European Society
Editors: Müller/Hackenbruch

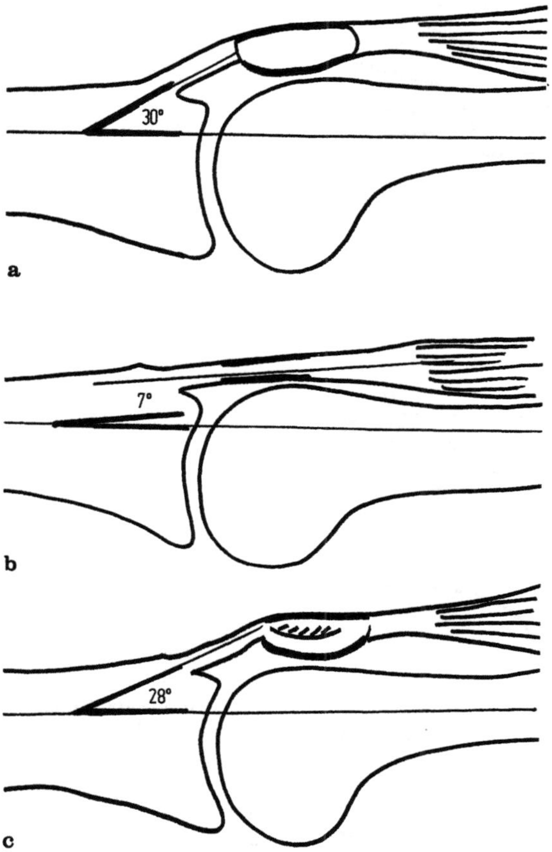

Fig. 1 a–c. In **a** the normal angle of insertion on the tibia allows the tendon of the quadriceps to exerts its force. In **b**, after patellectomy, this angle decreases considerably, thus decreasing also the extension force. In **c** the tendoplasty restores the previous thickness and a good angle of insertion

ing pain. This avoids secondary arthrosis and restores the force of the extensor apparatus, thus replacing the patella (Figs. 1–3).

Since 1962, 138 patellectomies have been performed, with ever increasingly strict requirements before operation would be considered (57 for fractures, 54 when performing total knee replacement and 27 for other indications). In the case of arthrotic changes of the patella without loss of continuity (i.e. not fractured), the operation is performed as follows: through a longitudinal anterior approach, the anterior fibrous layer of the patella is incised and detached laterally and medially with a small, very sharp chisel, together with thin lamellae of bone (Fig. 4).

In the beginning the bone was not removed.

We have been performing this technique for 10 years now, improving it continuously, e.g. obtaining thicker sections. The freed patella is removed; a pentagonal flap, about the same size of the contralateral patella (Fig. 5), is then incised in the tendon of the quadriceps, leaving untouched the distal side of the pentagon. The flap is then folded deeply under, and the apex of the proximal angle is led to the patellar ligament and secured there with two or three stitches of non-resorbable thread. Sutures are used to fix the side of the flap (under the osteofibrous

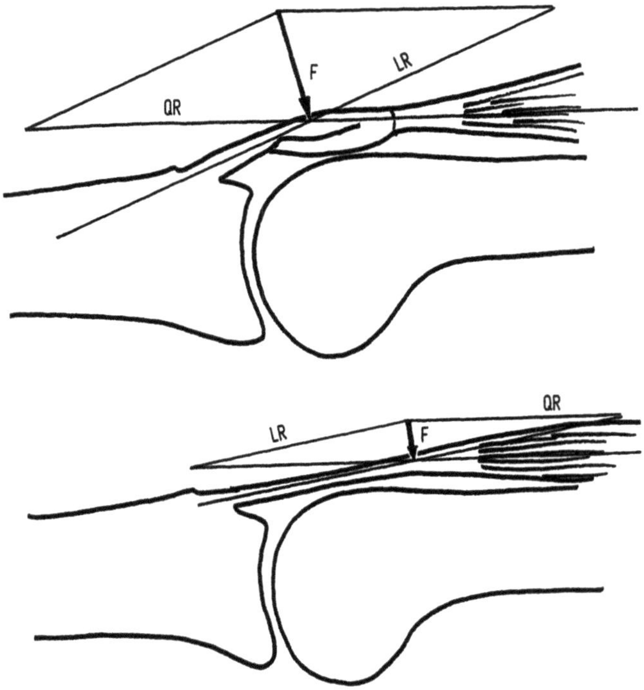

Fig. 2. The vectors of forces show a greater force *F* when the patella or the tendoplasty maintain the correct distance of the extensor apparatus from the femur. *QR*, quadriceps radius vector; *LR*, ligament radius vector

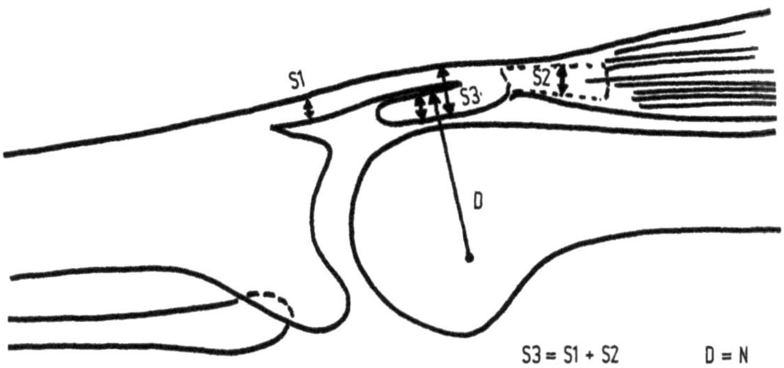

Fig. 3. With the tendoplasty we obtain *S3*, which is equal to the normal thickness of the patella. The distance *D* from the instantaneous centre of rotation is normal (*N*) again

layer) to the alar ligaments and capsule. The osteofibrous layer is closed above the flap, and the edges of the extensor apparatus at the donor site are brought together and sutured. Then by means of a thick, non-resorbable, braided thread, the tendon of the quadriceps muscle is connected with the patellar ligament (Fig. 6). The knee is flexed to judge the tension needed on this thread, which is then knotted. Finally, the skin is closed.

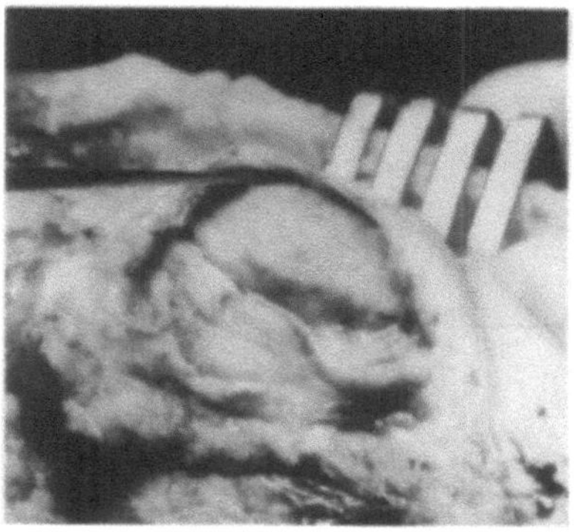

Fig. 4. The patella is excised leaving tiny lamellae of bone attached to the fibrous prepatellar tissue

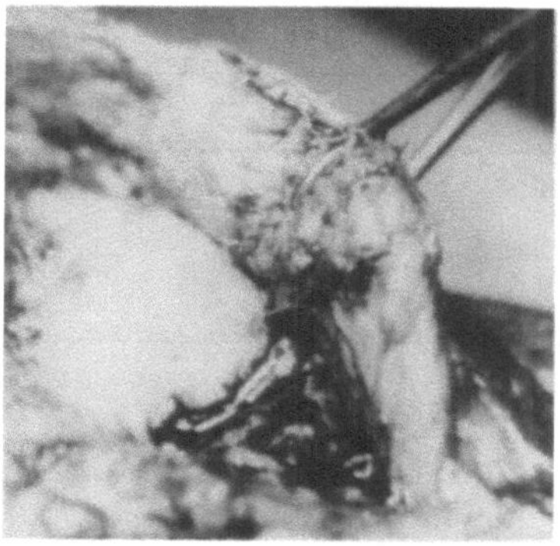

Fig. 5. The tendon flap is incised and turned under deeply to obtain a patella-like shape and thickness

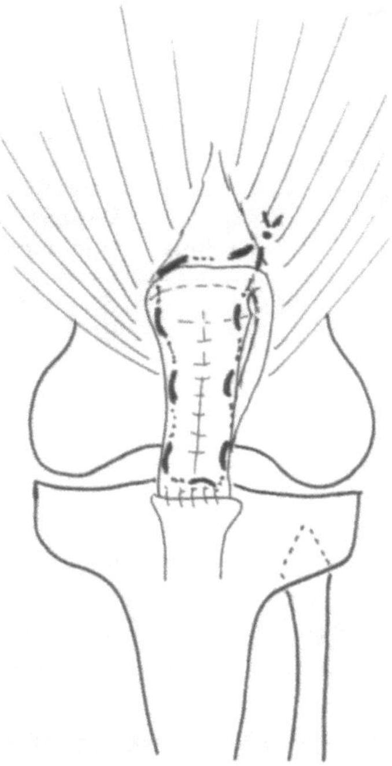

Fig. 6. A thick, non-resorbable, braided thread is used to connect the tendon of the quadriceps muscle with the patellar ligament

When dealing with a fresh patellar fracture, the technique is almost the same. The preparation of the anterior layer may be difficult, even impossible, but this does not affect the result.

Results

Of 138 patellectomies, 76 were performed on males and 62 on females; the patients age ranged from 16 to 73 years. Indications for surgery were as follows: 57

Table 1. Subjective evaluation (by number of patients)

	Pain	Stability	Force
Very good	75	78	79
Good	8	8	6
Fair	32	29	23
Poor	2	2	9

Table 2. Objective evaluation of motion

Range of motion		No. of cases
Wider than	170°–180°	55
Between	160°–190° and 170°–180°	41
Less than	160°–190°	21

The best results are often associated with total knee replacement.

Table 3. Muscular evaluation by the Highet scale

Instantaneous force	No. of cases
M5	111
M4	6
Decrease of endurance	
Less than 10%	99
Between 10% and 20%	16
More than 20%	2

Table 4. Results of patellectomies performed after 1978

	Using bone chips	Without bone chips
Very good	13	6
Good	5	10
Fair	7	2
Poor	1	0
	26	18

patellar fractures, 20 patellofemoral arthrosis (9 of which were associated with tibial osteotomy), 54 knee arthrosis in total knee replacement, 4 chondromalacia, and 3 were replacements of the tibial plate by means of the ipsilateral patella. In all 117 patients (57 male and 60 female) were followed up over a period ranging from 1 to 24 years.

Subjective and objective evaluations were made. The subjective evaluation took into consideration: pain, loss of force, difficulty in climbing and descending stairs, pain when squatting and cinema symptom (Table 1). Objective parameters were: mobility and motility range, quadriceps trophicity, muscular evaluation by the Highet scale (0, no movement, to 5, normal movement) and radiographical examination. In 11 cases we also carried out tests of maximal force by means of a Cybex 2 computerised isocinetic apparatus. Results were scored as very good, good, fair and poor (Tables 2 and 3). Measurements of quadriceps atrophy were taken 15 cm from the joint line and were always proportional to muscular force. Comparative results of patellectomies with and without bone chips are shown in Table 4.

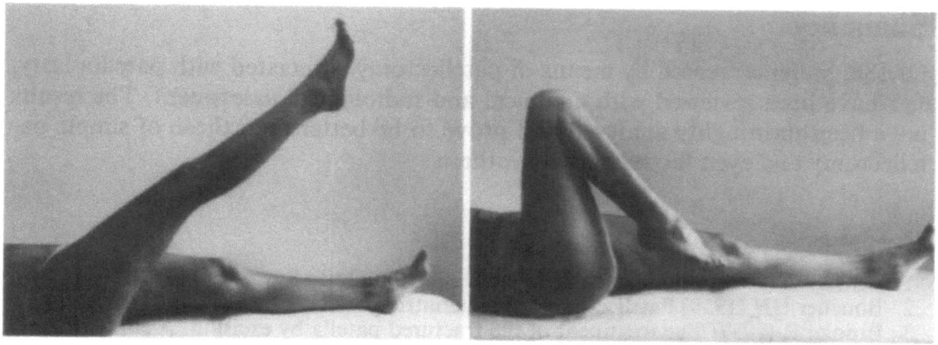

Fig. 7. Clinical result at 12 years

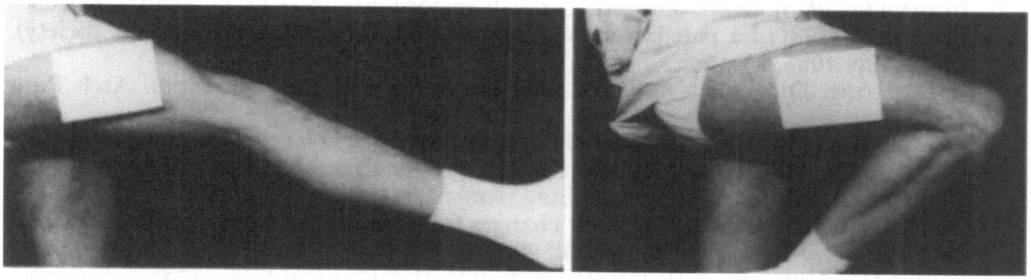

Fig. 8. Clinical result at 7 years

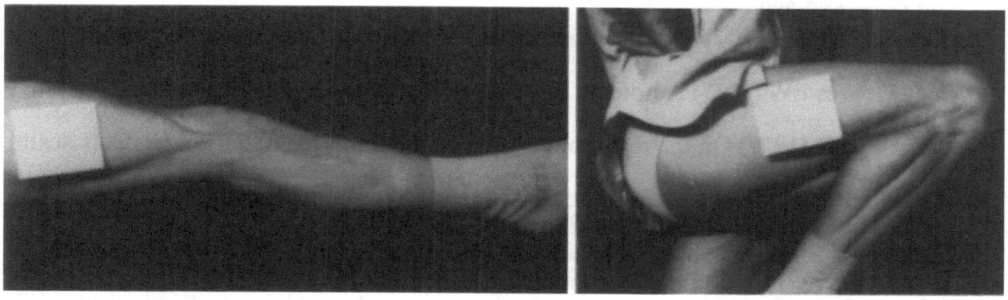

Fig. 9. Clinical result at 9 years

Conclusion

Careful, long-term, follow-up studies indicate that patellectomy with patelloplasty, achieved by turning under the quadriceps tendon, gives much better results than those obtained by means of simple patellectomy (Figs. 7–9), which is a very poor operation both from biomechanical and clinical points of view. Patelloplasty results were even better than those of good osteosynthesis.

Summary

Of 138, patients treated by means of patellectomy associated with patelloplasty, 117 have been reviewed with a clinical and radiological assessment. The results have been thoroughly analysed and prove to be better than those of simple patellectomy and even those of osteosynthesis.

References

1. Bentley G (1970) Chondromalacia patellae. J Bone Joint Surg [Am] 53:221
2. Boucher HH (1958) Patellectomy in the geriatric patient. Clin Orthop 11:33
3. Brooke R (1937) The treatment of the fractured patella by excision. A study of morphology and function. Br J Surg 24:733
4. Bruce J, Walmsley R (1942) Excision of the patella. J Bone Joint Surg 24:311
5. Brunelli G (1972) In: Borellini A (ed) Ortopedia e traumatologia, vol 1, chap 1. De Feo, Roma, p 169
6. Brunelli G (1973) Le artroprotesi. Corso di aggiornamento 1972. La Scuola, Brescia
7. Brunelli G (1981) La patellectomia. Riunione OTODI (Italian orthopaedic society), 14–15 May 1981, Mestre
8. Caruso AM, Dimicoli N (1966) La patellectomia e le sue indicazioni. Arch Putti 21:161
9. Cohn BNE (1944) Total and partial patellectomy. Surg Gynecol Obstet 79:526
10. Compere CL, Hill JA, Lewinneck et al. (1979) A new method of patellectomy for patellofemoral arthritis. J Bone Joint Surg [Am] 61:714
11. De Palma A, Flynn JJ (1958) Joint changes following experimental partial and total patellectomy. J Bone Joint Surg [Am] 40:395
12. Duthie HL, Hutchinson JR (1958) The results of partial and total excision of the patella. J Bone Joint Surg [Br] 40:75
13. Geckeler EO, Quaranta AV (1962) Patellectomy for degenerative arthritis of the knee. Late results. J Bone Joint Surg [Am] 44:1109
14. Haxton H (1945) The function of the patella and the effects of its excision. Surg Gynecol Obstet 80:389
15. Heineck AP (1909) The modern operative treatment of fractures of the patella. Surg Gynecol Obstet 9:177
16. Hey Groves EW (1937) A note on the extension apparatus of the knee joint. Br J Surg 24:747
17. Jansenius H (1951) On the result of excision of the fractured patella. Acta Chir Scand 102:275
18. Ludloff M (1925) Die Verkleinerung der Patella als funktionsverbessernde Massnahme bei bestimmten Knieaffektionen. Zentralbl Chir 52:786
19. Peeples RE, Margo MK (1978) Function after patellectomy. Clin Orthop 132:180
20. Scott JC (1949) Fractures of the patella. J Bone Joint Surg [Br] 31:76
21. Watson Jones R (1940) Fractures and other bone and joint injuries. William and Wilkins, Baltimore
22. Watson Jones R (1945) Excision of the patella. Br Med J [Clin Res] 2:195

The Biomechanical Analysis of Chronic Patellar Tendinitis and Treatment with Eccentric Loading

W. D. Stanish, H. Lamb, and S. Curwin

Nova Scotia Sport Medicine Clinic, Division of Orthopaedic Surgery, Dalhousie University Halifax, Nova Scotia, Canada

The management of chronic patellar tendinitis has long been a major frustration for the musculoskeletal doctor. Traditional programmes of treatment have included absolute, prolonged rest and surgical intervention. Our laboratory studies as well as those of Haggmark and Noyes have emphasised the deleterious effect of rest on the tensile strength of tendon, ligament and muscle. Furthermore, clinical and laboratory observations (EMG) have demonstrated that patella tendinitis is a consequence of eccentric forces that load the patellar tendon beyond its inherent tensile strength. Architecturally, eccentric loading can result in microscopic or macroscopic disruption of the patellar tendon, usually at the bone-tendon interface.

Hypothesis

Patellar tendinitis occurs most commonly as a consequence of repeated eccentric loading, which exposes the tendon to a stress 200% greater than conventional isometric concentric loading. In order for the patellar tendon to withstand eccentric loading, it must be trained and/or rehabilitated in an eccentric fashion.

Methodology

We tested our hypothesis prospectively by studying 100 patients with established, chronic patellar tendinitis. These patients had, over time, been on at least six programmes of physiotherapy, including rest, stretching, ultrasound, and in some circumstances, corticosteroid medication. They were placed on an exercise programme designed to load the muscle tendon unit in an eccentric fashion. The programme was initially conducted under the superversion of a physiotherapist and was performed within the limits of pain. Flexibility exercises and local ice were also employed. Each subject performed the exercise once daily for 20 min over a 6-week period. From week I to week III, the subjects performed three sets of 10 repetitions and from week III to week VI, they performed three sets of 20 repetitions. The patients were followed daily and kept a diary which was turned in at the completion of the study. All patients were examined and interviewed by our team at the Nova Scotia Sport Medicine Clinic.

Surgery and Arthroscopy of the Knee
2nd Congress of the European Society
Editors: Müller/Hackenbruch

Results

The overall improvement rate was excellent or good in 90% of the study population (excellent, completely pain free; good, mild discomfort, but no limitation of activity). Of the study population 45% experienced complete relief of symptoms while another 45% reported a significant decrease in the severity of their symptoms with no limitation in terms of ADL and athletic performance. Eight subjects reported no change in their symptoms and two actually felt their symptoms had worsened while on the eccentric programme. All of the subjects continued their daily recreational and athletic activities during the rehabilitation period. No other forms of treatment were employed during the study.

Conclusion

These most encouraging results have stimulated the treatment of chronic patellar tendinitis with progressive eccentric exercises. The exercise programme is biomechanically sound and respects the biology of the tissue.

Jumper's Knee Ultrasonography

D. Fritschy

Policlinique de Chirurgie, Hôpital Cantonal Universitaire,
1211 Geneva 4, Switzerland

Introduction

Jumper's knee is a disorder characterised by chronic overload of the patellar tendon and its points of insertion into the bone. Victims of this disease are usually active in sports in which the extensor mechanism of the knee is repeatedly activated. Jumping is not the only cause of the disorder, for runners, football players and skiers are also often affected. This disorder is known in international terminology by its English name, "jumper's knee". In 1973, Blazina proposed a symptomatic classification which has been adopted by most authors, some of whom added modifications [1, 4]. Some of the radiological signs accompanying this disorder have also been described, but as they vary, they are not very pathognomonic.

Using perioperative samples, the following histopathological correlations have been made: mucous degeneration and fibrinoid necrosis occur next to foci of micro-ruptures and regeneration. However in most cases, diagnosis and treatment have an empirical basis.

Ultrasonography provides us with a precise morphological description of the affected tendon and surrounding structures (skin, subcutaneous tissue, serous bursae and bone) which confirms the clinical diagnosis and allows for greater specificity when treating the patellar tendon [2, 3].

Material and Methods

Control Group

Fifteen subjects between 25 and 35 years of age, fond of sports, in good health and having no particular complaint or anomaly with respect to the patellar tendon, were examined in order to define criteria for ultrasonography of a normal patellar tendon.

The instrument used for the examinations was equipped with a 6 MHz probe, focused at a distance of 3 cm, in real time. The examination was carried out with the patient lying supine, knees in extension and then in light flexion (10°–20°). Ultrasonographs were systematically carried out across longitudinal and transverse sections of both tendons.

Ultrasonography of a Normal Patellar Tendon

Longitudinal Section. The average thickness of the sections is 4 mm. The tendinous fibres are parallel, homogeneous and smoothly linked together. Hyper- and

Surgery and Arthroscopy of the Knee
2nd Congress of the European Society
Editors: Müller/Hackenbruch

hypoechoic zones alternate regularly. The borders of the tendinous envelope are sharply and regularly outlined, slightly undulating and in connection with the volume of the underlying fibres.

Transverse Section. This section is oval in shape and has an average width of 12 mm. Here as well hyper- and hypoechoic zones alternate regularly.
There is no significant difference between left and right tendons. Note that all subjects in the control group were right-handed. The insertions of tendons into the patella and the anterior tibial tuberosity are easily identified. The subcutaneous cellular tissue and the skin are clearly visible, as is the serous pretibial bursa.

Patients with Jumper's Knee

The group consisted of 22 patients, 15 males and 7 females, with ages ranging from 15 to 45 years, the average being 27. They were active in a variety of sports (in decreasing order of frequency): running, basketball, volleyball, biking, artistic gymnastics, acrobatic skiing, athletics and tennis. In 15 of the cases, only one leg, that on the supporting side, was affected. In 7 cases, both legs were affected. Thus for 22 patients there was a total of 29 painful tendons. Ultrasonography revealed 27 pathological tendons, and in 2 patients there were no anatomical modifications.

Ultrasonography of Pathological Patellar Tendons

Three pathological characteristics are described:

1. Thickening or swelling of the tendon appears 15 times on ultrasonographs. This is usually diffuse, affecting either the entire tendon or segments of it. The more violently and acutely a tendon is affected, the more swollen it is. The degree of inflammation appears to be proportional to swelling.
2. A heterogeneous structure of the tendon appears 23 times on ultrasonographs, showing two distinct forms, hypoechoic and hyperechoic. Hypoechoic is indicated by an inflammatory or oedematous zone, fresh haematoma and segmentary micro-ruptures. Hyperechoic is indicated by a fibrous retraction zone, fibrous scarification zone and intratendinous calcification, corresponding to a scar or the final stage of a chronic inflammation.
3. Irregularities of the tendinous envelope include a hazy, poorly defined outline, corresponding to nodular zones, and modifications which take place during scarification. The tendon has a heterogeneous appearance, confirming the presence of fibrillar lesions, but is not completely swollen. These symptoms, which appear in seven patients, correspond to an orderly progression of a chronic disorder in which inflammation no longer plays a major role.

Discussion

Today, ultrasonography provides us with the means to obtain a precise evaluation of jumper's knee. The anatomical disorders of the tendon fall into three classes and, on the basis of their ultrasonographic manifestations, can be classified as follows:

1. Pure inflammatory stage, which is an initial stage, characterised by oedema of the tendon fibres. Thickening and swelling of the body of the tendon are obvious.
2. Stage with reversible anatomical lesion is characterised by tendon heterogeneity. Hypo- and hyperechoic appearance with or without oedema.
3. Final stage of lesion. The tendon envelope is irregular and the tendon fibres appear heterogeneous, but the inflammation has disappeared.

As regards treatment, the following points may be noted: patients with a jumper's knee in stage 1 should undergo conservative treatment. In stage 2 cases for which conservative treatment has been unsuccessful, surgery may be considered, but the presence of oedema would constitute a contraindication. It is in the third stage, once inflammation has disappeared, that surgery is most likely to be successful.

References

1. Blazina M (1973) Jumper's knee. Orthop Clin North Am 4:665–678
2. Fornage B (1983) L'ultrasonographie du tendon rotulien. J.E.M.U. 4(4):145–149
3. Fornage B (1984) Sonography of the patellar tendon. AJR 143:179–182
4. Roels J (1978) Patellar tendinitis. Am J Sports Med 6(6):362–368

Is Patellar Tendinitis a Surgical Affection?

G. Saillant, J. P. Benazet, G. Gagna, F. Combelles, and R. Roy-Camille

Sce d'Orthopédie, Hôpital de la Pitie, 83, Boulevard de l'Hôpital,
75013 Paris, France

Patellar tendinitis is a disabling, slowly evolving affliction which jeopardizes sport activity. It seems to evolve from a disinsertion or friction. To fight against it, we propose a resolutely surgical attitude in order to obtain functional recovery.

Material and Method

The first 76 cases were analyzed after a minimum follow-up of 1 year. Aged 17 to 45 years (average 25) half of the patients were competition sportsmen. All these patients came to surgery after the diagnosis was corroborated on xerography and echography and after correct medical treatment had been applied without success for at least 6 months.
The following surgical characteristics were constantly found: splitting of the patellar tendon, excision of the nodules, reversal of the fat pad. Two were associated: resection of the patellar apex and patellar readjustment adapted to either static or dynamic subluxation. After the operation, complete weight bearing was allowed; running was resumed 6 weeks later and competition in the 4th month.

Results

The results, evaluated by the persistence of pain, the ability to take up sport at the same grade and the morphological appearance, are arranged into two scores: excellent and good, medium and poor. Some 81% were excellent or good and 19% were medium or poor. The results improve when no patellar readjustment is done.
As for the anatomical aspect, the volume of the tendon is always increased; it has much greater proportions when carbon fibers are inserted. Five complications occurred: two reflex sympathetic dystrophy, a benign hematoma, one stress fracture, and one prepatella bursitis.

Discussion

Our study confirms the importance of surgery: ablation of calcifications and nodules, ablation of the patellar apex acting like an irritative spur, and creation of a fibrosis resulting in a continuous and homogeneous tendon.

Surgery and Arthroscopy of the Knee
2nd Congress of the European Society
Editors: Müller/Hackenbruch

The Effect of Five Different Patella Braces on Performance and Pain in an Isokinetic Quadriceps Torque Test

U. Lindberg, J. Gillquist, and J. Lysholm

Department of Orthopedic Surgery, University Hospital,
52885 Linköping, Sweden

In the treatment of patello-femoral pain syndrome (PFPS) a patella-supporting brace must be considered a potentially useful therapeutic tool. It is a matter of controversy which type of patella brace is of optimal construction. Lysholm et al. documented the effect of a brace with a simple lateral pad. Palumbo recommends a slightly different item with a more sophisticated design. A simple infrapatellar strap has been described by Levine. A simple elastic tube may be equally effective.

In 36 patients with PFPS the effect of five different patella braces on quadriceps torque and pain (visual analogue scale) was studied. The five braces were an elastic tube with a patellar aperture, three braces with different types of lateral pad systems, and an infrapatellar strap.

All the braces significantly improved the quadriceps torque and diminished the pain registration ($P < 0.01$). A ranking based on statistically significant differences between the braces was not possible. However, when all aspects are summarized, the brace with a lateral pelotte was somewhat more effective than the others.

Surgery and Arthroscopy of the Knee
2nd Congress of the European Society
Editors: Müller/Hackenbruch

Degenerative Arthritis and Osteotomy

Zed Osteotomy with Trapezoidal Resection in Internal Rotations of the Tibia and Secondary Static Arthrosis of the Knee

G. Brunelli, F. Brunelli, B. Battiston, and P. A. Guizzi

Brescia University Medical School, Spedali Civili, 25100 Brescia, Italy

Introduction

Internal rotation of the tibia is a frequent malformation. A little degree of internal rotation in newborns is considered normal but disappears in the first years of life to be replaced by a very moderate external rotation. Internal rotation at birth is often more severe than originally thought, or derotation does not take place, and in adults internal rotation can persist to varying extents (Fig. 1). Until some years ago this deformity was disregarded, notwithstanding several publications about it [1–9] and its importance in causing static and dynamic alterations of the knee [1–3]. Tibial internal rotation has to be considered as being equivalent to club foot and is often associated with varum of the tibia, especially in congenital but less often in rachitic deformities [10–16].

Flexion and extension of the knee should occur along a plane parallel to the anteroposterior (AP) plane, where it is supported by the medium and greater gluteus muscles' action. If there is internal rotation the movement of the knee in this plane provokes stumbling of the feet when walking. To avoid stumbling the patient either walks with the inferior limbs abducted at an alpha angle or automatically tries to rotate externally the tibia by means of the crural biceps (automatic contraction at each step). This compensation provokes an off-set positioning of the knee resulting in external rotation of the tibia under the femoral condyles and re-

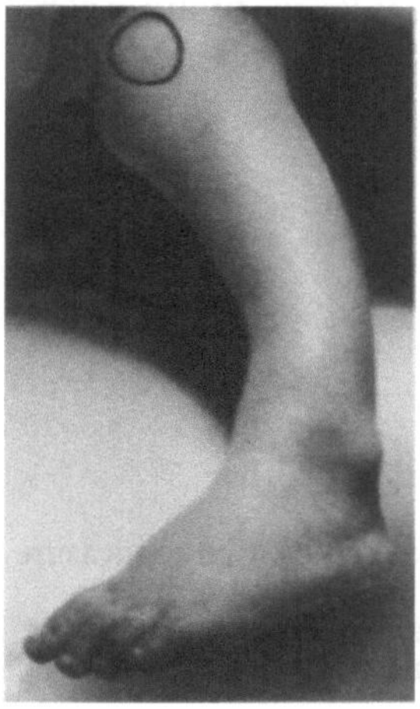

Fig. 1. Internal rotation of the tibia is frequently found in newborns; more often than previously thought, normal derotation does not take place

Surgery and Arthroscopy of the Knee
2nd Congress of the European Society
Editors: Müller/Hackenbruch

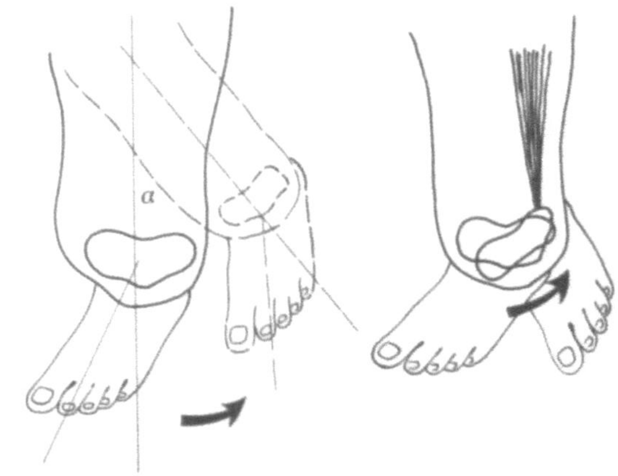

Fig. 2. Internal rotation, in order to avoid stumbling, is achieved either by an external rotation and abduction of the thigh, or (which happens more frequently) by the biceps pulling at every step to externally rotate the leg with displacement of the tibia under the femoral condyles

petitive stretch of the medial meniscus and medial collateral ligament (Fig. 2) This automatic compensatory mechanism tends to be maintained even during weight bearing (complete extension of the knee) when no physiological external rotation is possible.

Thus early arthrotic changes of the medial compartment (and eventually the whole knee) occur especially with an associated varum deformity. On the other hand, we have been interested in this lesion for years and have noticed that in static arthrosis due to varum, it is very rarely that no internal rotation occurs.

Material and Methods

Different methods can be used to assess the internal rotation of the tibia: clinical observation, tropometric devices, computerized tomography (CT) (Fig. 3). We used both a mechanical tropometric apparatus and CT although visual examination of the patient sitting on a table, with the leg hanging and the patella in front, can give a good clinical idea of the deformity; of course an internal rotation com-

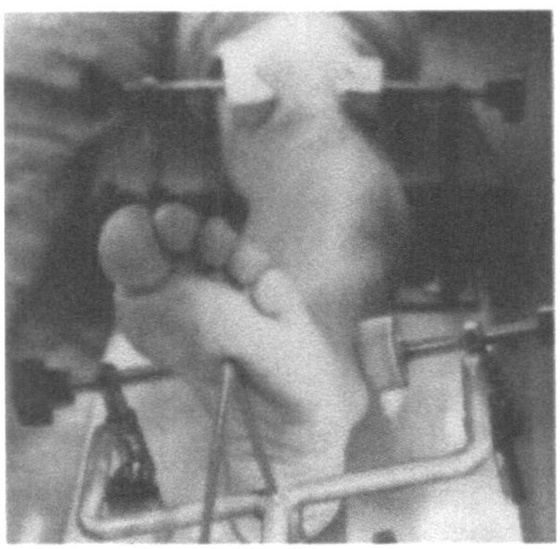

Fig. 3. Various methods can be used to measure internal rotation, e.g. axial radiography or computerised tomography. To avoid irradiation, a simple tropometric device is used as shown

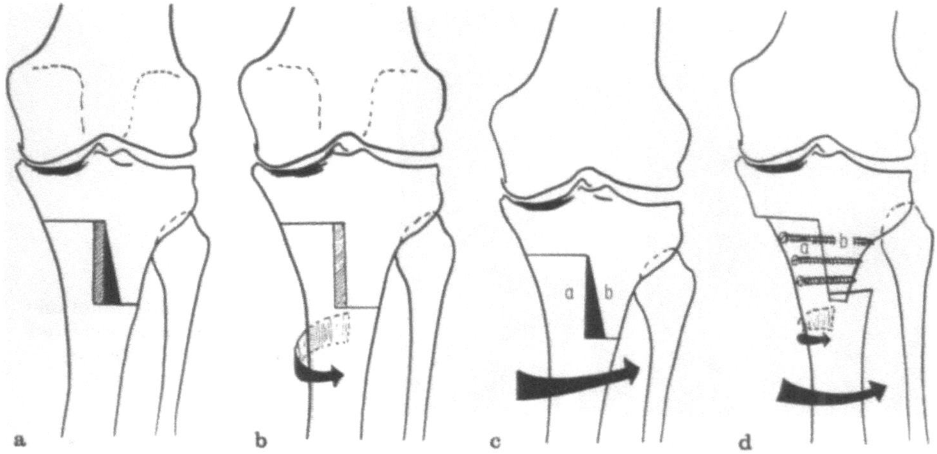

Fig. 4 a–d. Scheme of the operation. **a** The Z osteotomy is drawn, including a trapeze on the anterior aspect of the tibia. **b** The trapeze is made up of a rectangle, the removal of which will allow external rotation, and **c** of a triangle, the removal of which will give valgus correction. **d** Fixation of the bone after removal of the trapeze allows the combined correction of varus and internal rotation. *a, b* stumps of the tibia

pensatory of the femoral head and neck antiversion should be previously excluded.

Starting in 1962 to avoid or relieve arthrotic changes secondary to this deformity, we carried out derotating plane oblique osteotomies in children, and metaphyseal Z osteotomies in adults (Fig. 4). The latter technique consists of a Z osteotomy combined with removal, on the anterior vertical line of the Z, of a trapezium-shaped piece of bone formed by adding a rectangle (which determines external rotation) and a triangle (the proximal angle of which determines abduction (valgum correction). After resection of the trapezius, the proximal and distal stumps of the osteotomy are pushed together and fixed by screws. By doing this, the leg assumes

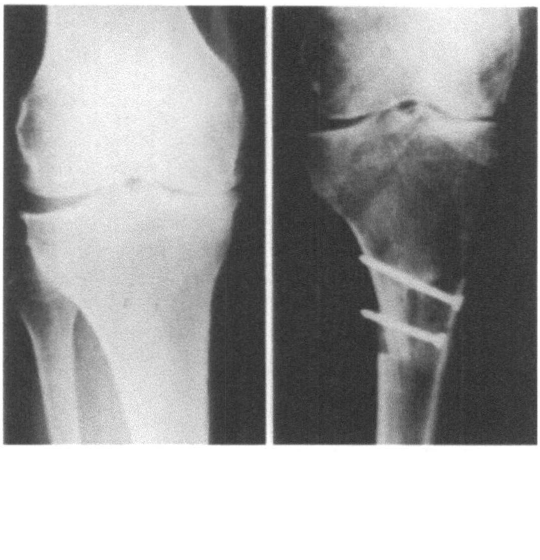

Fig. 5. At first fixation of the bone was achieved by means of two or three screws

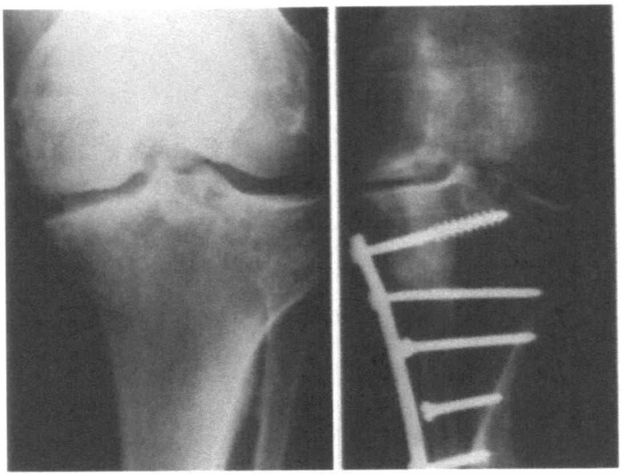

Fig. 6. Later to avoid plaster immobilisation, the bone was fixed by means of plates. This case shows the result 6 years after surgery

a new position in valgus (moderate) and external rotation. The size and shape of the trapezium have to be carefully measured before the operation. Valgus correction is evaluated by means of an anteroposterior standard radiograph and external rotation by means of a transverse CT. If the correction to be achieved is important, an oblique peroneal osteotomy has to be included.

Table 1. Scoring table according to Federici et al. [8]. Best score achievable is 28

	Score		Score
Pain		Range of motion	
No pain	6	Full motion	6
Discontinuous pain	5	60°–180°	5
Pain on walking	4	90°–180°	4
Stop walking after 1 h	3	80°–170°	3
Stop walking after 15 min	2	120°–170°	2
Walking few steps	1	150°–170°	1
Cannot walk	0	No motion	0
Swelling		Instability	
Absent	2	None	1
Discontinuous	1	Present	0
Continuous	0	Function	
Local temperature		Normal walking	6
Normal	1	Mild limping	5
Increased	0	Moderate limping	4
Axial deformity		Severe limping	3
Physiological	4	One crutch	2
0°– 5°	3	Two crutches	1
5°–10°	2	No walking	0
10°–15°	1	Venous problems	
More than 15°	0	None	1
		Present	0
		Articular blocks	
		None	1
		Present	0

When arthrosis is established, it is often also necessary to combine medial meniscectomy and removal od osteophytes or free bodies.
At first the osteotomy was secured by means of two or three screws and plaster, and weight bearing was avoided until radiological examination indicated good callus formation. For 12 years now we have been securing the correction by means of plates and screws, so leaving the knee free to move, thus avoiding secondary knee stiffness (Figs. 5, 6).

Results

In 24 years we have operated on 108 cases of internal rotation of the tibia, with more or less severe associated varum deformity, by means of this technique. Those cases treated before 1980 have already been published [3–5]. We called back all the patients from 1980 to 1985, of which 30 of them attended the follow-up. 31 knees (1 bilateral operation) were evaluated by means of the method presented in Table 1 [8].
The maximum achieveable score for a normal knee is 28. The scores achieved by our cases are given in Fig. 7. In the first years a moderate degree of stiffness of

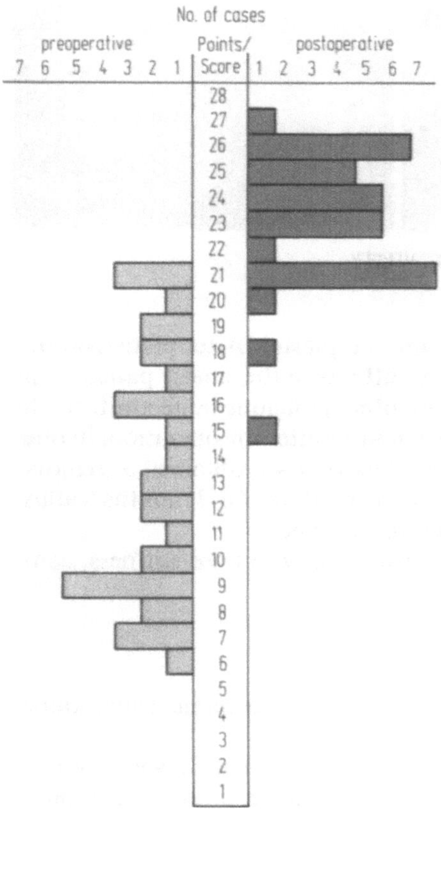

Fig. 7. Comparison between preoperative and postoperative scores

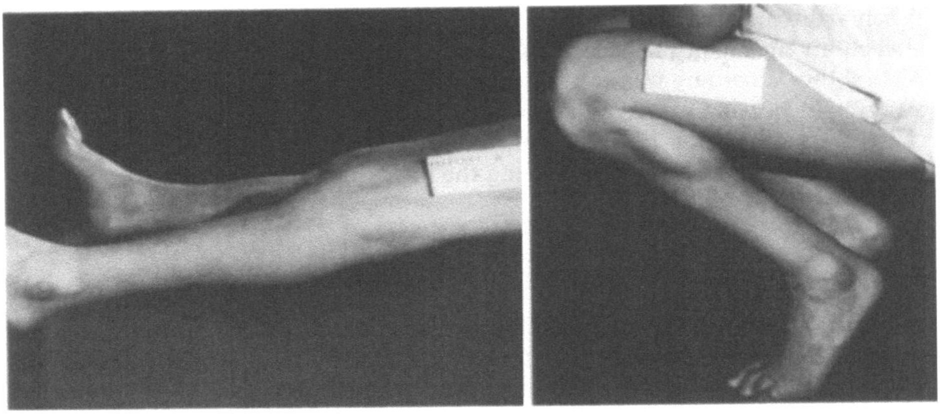

Fig. 8. Example of a late result, 4 years after surgery

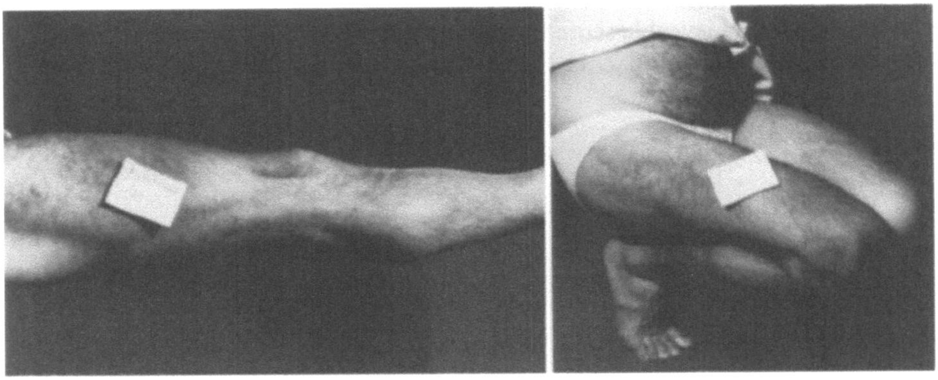

Fig. 9. Example of a very late result, 13 years after surgery

the knee was often present due to immobilisation in plaster. After plate fixation only a few cases (9) resulted in moderate knee stiffness in the elderly patients. In one case non-union occurred which recovered after spongious bone graft. In 18 cases delayed union occurred, between 4 and 9 months after the operation. In one case of early weight bearing the plate fractured. Recovery was obtained by removing the plate and screws, then supporting the knee in plaster for 2 months (callus formation was already evident when the fracture occurred).
All the results can be considered satisfying, even those with knee stiffness, considering the age of the patients (Figs. 8, 9).

Summary

Tibial internal rotation is a frequent component of the biomechanical alterations which cause arthrosis of the knee.
In the past little importance was placed on this deformity (unless it was very severe) by surgeons that look chiefly at varus and valgus deformities. Osteotomies

developed by different surgeons were mainly designed to change the point of support and friction between the femur and tibia.
In spite of this internal rotation may be the most important cause of arthrosis even in tibia vara. In fact, internal rotation compels the patient to try to obtain a forced external rotation when walking due to biceps contraction, in order to prevent the feet from stumbling over each other. Twenty-four years' experience of surgical treatment of tibial internal rotation (often combined with varum deformity) is presented. The surgery consists of a metaphyseal osteotomy with the removal of a trapezoidal anterior piece of bone which allows rotational as well as varum deformity correction. The shape and size of the trapezium determines the amount of correction.

References

1. Aherle (1933) Die Torsion des distalen Tibiaendes. Arch Orthop Unfall Chir 34:56
2. Brandt G (1928) Die Torsion der unteren Extremität und ihre Bedeutung für die Deformitätenentstehung. Z Orthop 49:481
3. Brunelli G (1968) Osteotomia a Z trapezoidale nelle artrosi statiche del ginocchio. Reumatismo 5:404
4. Brunelli G (1969) Trapezoidal Z osteotomy in static arthrosis of the knee. Societé internationale de chirurgie orthopaedique et traumatologie, 11 international congress, Mexico 6–10 October
5. Brunelli G, Brunelli F (1983) Osteotomia a Z con resezione trapezoidale nella correzione delle intratorsioni della tibia e nel trattamento delle artrosi statiche delle artrosi statiche del ginocchino. Minerva Ortop 34 (6):513
6. Cauchoix J (1968) L'osteotomie dans les gonarthrose avec deviation angulaire dans de plan frontal. Rev Chir Orthop 54(4):343
7. Dupuis PV (1951) La torsion tibiale. Masson, Paris
8. Federici A, Albanese L, Nasciuti M (1982) Osteotomie di ginocchio. Minerva Ortop 33(6):541
9. Goriepi R (1964) Genu varum treated by high tibial osteotomy. J Bone Joint Surg [Br] 46:783
10. Hohman G (1932) Ueber 2 nach Form zusammengesetzte Plattfussgelenke nebst einigen Bemerkungen über die Torsion des unteren Tibiaendes. Z Orthop 46:19
11. Jackson JP, Waugh W, Green JP (1969) High tibial osteotomy for osteoarthritis of the knee. J Bone Joint Surg [Br] 51:88
12. Le Damany L (1909) La torsion du tibie normal, pathologique experimental. J Anat Phisiol (Paris) 5:98
13. Maquet P (1977) Biomecanique du genou. Springer, Berlin Heidelberg New York
14. Sell LS (1941) Tibial torsion accompanying congenital club foot. J Bone Joint Surg 23:561
15. Smillie IS (1974) Diseases of the knee joint. Churchill Livingstone, Edinburgh
16. Wahren CPW (1930) Über die Korrektur der Tibiatorsion bei kongenitalem Klumpfuss. Acta Chir Scand 67:929

Unicompartmental Replacement of the Knee: A Long-Term Study Comparing the Results with Tibial Osteotomy *

N. S. Broughton, J. H. Newman, and R. A. J. Baily

Winford Orthopaedic Hospital, Bristol, UK

The results of 42 unicompartmental replacements and 49 tibial osteotomies, which had been performed at Winford Hospital 5–10 years previously, have been compared. As the age, sex, type of deformity, amount of deformity and radiological severity of the arthritis before surgery were similar in the two groups, a direct comparison is justified.

Of the patients treated by unicompartmental replacement, 32 results (76%) were good, 4 (10%) were fair, 3 (7%) were poor, and 3 (7%) had been revised. Of those who underwent tibial osteotomy, 21 (43%) were good, 11 (26%) were fair, 7 (17%) were poor, and 10 (24%) had been revised. Unicompartmental replacements were more effective in relieving pain, required revision less often, and showed no evidence of deterioration with time. The lateral compartment replacements had particularly good results. The prostheses were deliberately positioned to undercorrect the original deformity, so that the unreplaced compartment rarely deteriorated radiologically. It was also unusual for the patellofemoral joint to deteriorate.

We therefore conclude that the results after unicompartmental replacement of the knee were significantly better than the results after tibial osteotomy and, in view of the quicker rehabilitation, unicompartmental replacement of the knee should be strongly considered in the management of unicompartmental osteoarthritis of the knee, particularly in the elderly patient.

---

* This study has already been published in full in the *Journal of Bone and Joint Surgery* [Br] 68:447–452. This abstract is published by kind permission of the Editor.

Surgery and Arthroscopy of the Knee
2nd Congress of the European Society
Editors: Müller/Hackenbruch

Knee Prosthesis

Total Knee Replacement: The Gallannaugh Semi-Constrained Linked Prosthesis

S. C. Gallannaugh

Royal East Sussex Hospital, Bambridge Road, Hastings, UK

Introduction

This paper describes a semi-constrained linked knee prosthesis which was first used at Hastings in 1982. The early results of surgery in 126 knees followed up for between 6 months and 3 years are given.

The characteristic features of the prosthesis (Fig. 1) are a patellar flange to allow smooth tracking of the patella or patellar prosthesis in those knees in which the patella has been resurfaced. We have previously reported patella impingement on the anterior femoral shaft as a source of pain following knee replacement when a prosthesis without a patellar flange has been used (Gallannaugh et al. 1984). The stems are long to make alignment within the bone easier and to provide a secure basis for firm fixation when bone stock is deficient. The stem of the femoral component is offset by 6° to align with the anatomical axis of the femoral shaft. Flat surfaces on the proximal aspect of the femoral condyles make it possible to fit the prosthesis accurately to the lower end of the femur after precise bone cuts

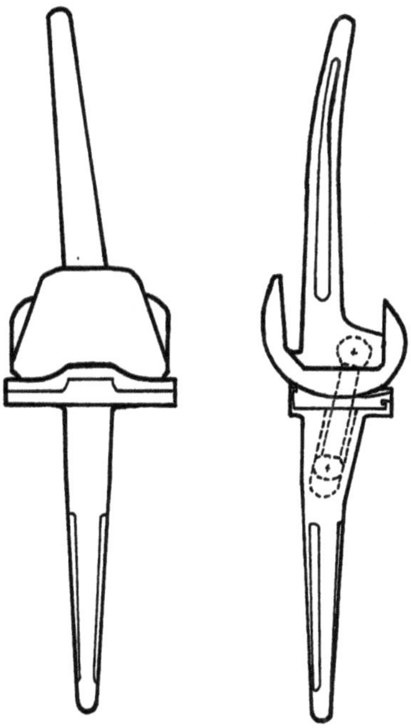

Fig. 1. Characteristic features of the semi-constrained linked prosthesis

Surgery and Arthroscopy of the Knee
2nd Congress of the European Society
Editors: Müller/Hackenbruch

have been made. Accurate alignment is thus possible and is essential if long-term stability is to be achieved. The link rod, which provides stability in the presence of pre-existing ligament or bone damage, is posteriorly positioned in the femoral component and set at 21° to the vertical within the tibial component. It maintains the normal forward location of the tibia on the femur during flexion and prevents posterior displacement of the tibia on the femoral condyles as flexion increases. Flexion of 120° can thus be achieved.
The tibial component is made of metal for strength, with high density polyethylene inserts on the bearing surface. These are available in different thicknesses to compensate for bone loss arising from disease or previous surgery.
Alignment of the prosthesis is achieved by using the intramedullary canals of the femur and tibia as guides; great care has been taken to develop instruments which enable the surgeon to fit the prosthesis very precisely to the bones. The importance of accurate positioning of any knee prosthesis within the leg cannot be overstressed.

Results

At the time of follow-up 161 knees had been replaced with this prosthesis. Three patients had died from unrelated causes, 24 had been operated on less than 6 months previously, 6 had been lost to follow-up, and 2 joints had been revised to an arthrodesis because of infection. Therefore, 126 knees were available for review.
In 97 knees the prosthesis was used with bone cement and in 15, without. In 13 it was used for a revision operation. The average age of the patients was 70 years. In 62 cases the diagnosis was osteoarthrosis and in 50, rheumatoid arthritis.
A number of complications occurred. In two knees wound dehiscence took place during the postoperative period. In one of these cases, and in one other, arthrodesis was carried out due to infection. In two others superficial wound infection occurred but was resolved with antibiotic therapy. In one knee the patella dislocated laterally and was corrected by a lateral release procedure.
The results were assessed using the British Association for Surgery of the Knee score chart. The maximum score of 100 was divided between pain, movement, function and technical merit, particular attention being paid to the absence of pain and a good range of movement. In those knees in which the prosthesis was used for a primary intervention, the results of the two groups (cemented and uncemented) are given in Table 1.

Table 1. Results assessed using the British Association for Surgery of the Knee score chart

	Cemented (97 knees)	Uncemented (15 knees)
Excellent	44	4
Good	42	8
Fair	6	1
Poor	5	2

Table 2. Scores for pain

	Cemented (97 knees)	Uncemented (15 knees)
None	63	7
Mild	29	6
Moderate	5	2
Severe	0	0

Table 3. Scores for postoperative flexion

	Cemented (97 knees)	Uncemented (15 knees)
> 100°	48	6
81°–100°	42	5
60°– 81°	5	3
< 60°	2	0

Conclusions

The prosthesis had proved to be effective in relieving pain and establishing a useful range of movement. Following the operation, in 94% of knees there was either no pain or mild pain which did not disturb sleep or reduce activity (Table 2). In 90% of knees flexion to more than 80° was achieved and in 23%, flexion of 120° or more (Table 3). Only one patient had a postoperative flexion deformity of more than 20°, but this was a reduction of the preoperative deformity.

In 86 patients the patella was resurfaced with a high density polyethylene dome. In this series there were no fractures of the patella, and to date there has been no loosening of the prosthetic components.

On 14 occasions the prosthesis was used in a revision procedure. In such cases tibial plateaux of different heights are available, and a high density polyethylene insert can be placed proximal to the femoral condyles to compensate for bone deficiency. By this means the horizontal axis of the knee can be kept in its correct position rather than being moved proximally as would be the case if all the bone deficit was made up on the tibial side. In all, up to 3 cm of bone can be replaced without the need to use excessive amounts of bone cement. In this group of revision operations, 8 results were scored as excellent, 4 as good and 1 as fair.

It is emphasised that these results are early ones, and no long-term conclusions can be drawn from them. In the short-term, however, this semi-constrained knee prosthesis has been effective in relieving pain and restoring movement in patients with osteoarthrosis and rheumatoid arthritis. It has been particularly useful for the management of revision procedures required for marked instability or loss of bone substance.

Ten Year's Experience with the Total Condylar Knee Prosthesis

M. Soudry, M. Iusim, A. Menachem, and D. G. Mendes

The Research Centre for Implant Surgery (Rothschild), Haifa, Israel

Introduction

This paper presents our 10-year experience with the total condylar knee prosthesis (TCP) beginning in March 1976. Up to September 1982, 95 patients underwent 118 total condylar replacements; of these, 15 patients died or could not be traced and thus we could only evaluate 97 knees from 80 patients. The group consisted of 66 females and 14 males (approximately five times more females), 59 unilateral and 19 bilateral replacements, representing an equal distribution between right and left knees. The average age was 58.5 years with a range of 16–79 years and the average weight was 64 kg with a range of 45–101 kg. The diagnosis was primary osteoarthritis in 74% of the patients, 15% were rheumatic and the remaining 11% included a mixed group with osteoarthritis secondary to chondrocalcinosis, osteonecrosis, psoriatic arthritis, rickets and pigmented villonodular synovitis. Ninety-three knees had a varus deformity, 19 of which had a fixed deformity of more than 10°. Twenty-five knees were in valgus deformity, and 34 knees had more than 10° of flexion contracture.

Material and Methods

Surgery was performed with the original total condylar instruments using a routine technique, starting with soft tissue release as required. Excision of the flexion gap is undertaken first, followed by excision of the extension gap.
Before 1980, an all-plastic tibial component was used, which has since been replaced by a tibial component with a metal tray.

Results

The 10-year results presented here cover an average follow-up of 6 years, with a range of 3.5–10 years and do not differ substantially from the 8-year results previously presented. Early general complications included: one myocardial infarction (MI), one case of pneumonia, one case of a pulmonary emboli, one cerebrovascular accident (CVA), one clinically proven phlebitis, and nine cases of urinary tract infection (UTI). All of them were successfully treated conservatively, except the patient with MI who died 1 month later. Early local complications included nine cases of wound drainage, three cases of superficial infection and one deep infection.
The patients were evaluated according to a slightly modified Hospital for Special Surgery scoring system. The preoperative score was 48 points with a range from

Surgery and Arthroscopy of the Knee
2nd Congress of the European Society
Editors: Müller/Hackenbruch

25 to 66. The postoperative score on follow-up remained 80 with a range of 0 to 100; 85% of the patients were still considered good and excellent, 15% fair and 4% poor. On analysing the score in relation to diagnosis we found that patients with rheumatism did worse, with an average score of 73 as compared to 83, than patients with osteoarthritis. In the mixed group the average score was 54.

Considering the range of motion, extension improved from −10° preoperatively to −1.6° postoperatively; the average flexion decreased from 98° to 92.5° and the whole range of motion increased by 3°, from 88° to 91°. An interesting finding, when displaying the postoperative flexion for all patients, was that one-third of the knees (34) achieved a flexion of more than 100° postoperatively. This is significant, since 90° was generally considered the limit of flexion for this prosthetic design. One important factor facilitating an adequate flexion range is the provision of a sufficiently large flexion gap.

Stair climbing was not optimal, 25% of the patients were unable to use the reconstructed knee correctly and chose to negotiate stairs sideways. Nevertheless pain relief was present in 90% of cases. We found quite a high proportion of patients (40%) with legs still aligned between 0° and 10° varus. This is primarily due to an error in positioning of the tibial component but also to inadequate instrumentation and insufficient medial release. To date, no negative correlation has been found between score and misalignment in varus.

The incidence of radiolucent lines was almost equal for plastic or metallic tray tibial components, and they generally appeared to be nonprogressive. They were seen mainly associated with the tibial component, under either the medial or lateral condyle but rarely around the peg. In 97% of the cases the radiolucent line thickness was <1 mm, in 10% between 1–2 mm and in 3% more than 2 mm. In 40% of the cases radiolucent lines were not seen.

The late complications included two cases of surgically treated patellar subluxation and three cases of supracondylar fracture treated conservatively with plaster of Paris cast. Infection was involved in two cases, from which Enterobacteriaceae were isolated and cultured after removal of the prosthesis resulting in arthrodesis. In a third case a Staphylococcus was found, and the infection was treated successfully with antibiotics.

Two cases of loosening occurred, one case of breakage of the metal tray and two cases of mediolateral instability. One case of intractable pain was revised to a thicker tibial component. Patellar fracture was not observed in any of the cases studied.

Of the two cases in which loosening of the prosthesis was found, the first was a symptomatic patient, who experienced pain due to a loose prosthesis as indicated by arthrography. After revision she still has moderate pain. The second case of loosening was found in a routine follow-up. Despite radiological evidence of loosening beneath the metal tray, particularly around the peg, the patient was still asymptomatic in the case follow-up.

A routine follow-up revealed a discontinuity in the metal tray in a 72-year-old patient, who was asymptomatic. A revision was done to replace and remove the broken parts, today the knee scores as excellent.

Instability was found in two cases. A 61-year-old male developed lateral instability 2 years after the joint had been implanted, but despite reconstruction of the lateral stabilisers with a carbon braid, he continued to suffer from buckling. Ultimately he was revised, 3 years later, to a TCP III. Today, 1 year after, his scoring is excellent.

Summary

In agreement with the assessment of the Hospital for Special Surgery the TCP, despite complications, is a relatively successful technique for the treatment of knee instabilities. Excellent and good results are achieved by 85% of the patients, osteoarthritic patients doing best, and pain relief was achieved in 90% of the patients. Complications included infection (3%) and revisions (6%), but no patella fractures. Revisions were performed in two cases of infection, one case of lateral instability, two cases of intractable pain and one case of mechanical failure of the implant.

Conclusion

Despite our excellent experience with cruciate retaining designs over the past 4 years, and the promising results of the newest designs, the possibility remains that for the deformed knee anatomical duplication achieved by the TCP may not be improved upon. To our knowledge no other knee implant design has yet proved to be so consistently successful over long-term follow-up.

PCA Unicompartmental Knee Endoprostheses

A. Lindstrand and A. Stenström

Department of Orthopaedics, University Hospital, 22185 Lund, Sweden

Introduction

Unicompartmental endoprostheses have been in use for more than a decade. In Europe there has been a general enthusiasm for their use, but some criticism has been raised in the United States of America. In the Swedish knee arthroplasty national survey (with approximately 10000 patients recorded over 10 years) the unicompartmental endoprosthesis is the most commonly used device for gonarthrosis; it has had the least complications and the most favourable (6-year) prosthesis life expectancy.

Material and Methods

During the past 5 years a programme of research has resulted in the development of the PCA unicompartmental endoprosthesis. The femoral component is made of Vitallium and has a porous coat. The tibial component is metal-backed polyethylene with a porous coating. Both components have a more anatomical design compared to the unidevices currently in use. New guide instruments enable precise positioning of the components and the prosthesis can be used both with and without cement. The prosthesis has been in clinical use for 2 years. One hundred unicompartmental endoprostheses have been inserted by us. In the Multicenter European Study another hundred patients have been operated on.

Results

The components were anatomically positioned with regard to both the tibia and femur. The position of the tibial component was on average 87.5° (ideal 87.5°) in anteroposterior (AP) view. The side view showed on average a position of 90.1° (ideal 90°). Early clinical results have been favourable, and 50 of the 100 patients have been followed up over 1–2 years. All patients, except one, have an improved pain-free walking distance, the majority reaching between 1 and 3 km. Of these 50 patients 49 have no pain at rest. According to the patients' own opinion 46 are much better and 3 better, compared to the preoperative situation. The flexion capacity was on average 124°, which is a gain of 4°. No major complications have been registered. The European Multicenter Study shows radiographic results similar to those mentioned for the Lund Study.

Surgery and Arthroscopy of the Knee
2nd Congress of the European Society
Editors: Müller/Hackenbruch

Discussion

The main reason for this paper is to present the biomechanical consideration behind the design of the prosthesis and to show how it is possible to position the components exactly by guide instruments, viz. radiographic results. Construction of the endoprosthesis is the result of a 5-year development programme, in which the problems of earlier prostheses have been taken into consideration in the design and instrumentation. The use of the unicompartmental endoprosthesis is probably more common in Scandinavia and Europe compared to the United States of America where total knee prostheses are widely used. Patients above the age of 65 years, with early gonarthrosis stage I–III according to Ahlbäck, should preferably be operated on with a unicompartmental endoprosthesis. It preserves both cruciates, more bone stock and gives as good results as total knee prostheses. The special advantage of the unicompartmental prosthesis is its better flexion capacity compared to total knee prostheses. The PCA unicompartmental endoprosthesis has the special advantages that it is more anatomical in its design and can be positioned very exactly by guiding instruments. Furthermore, it can be used both with and without cement.

Summary

The PCA unicompartmental endoprosthesis is a development on earlier uni-prosthesis designs. The radiographic results are very precise for 100 knees operated on in Lund and 100 knees operated on in a European Multicenter Study. The early clinical results are favourable.

Assessment of Results Following Total Knee Replacement: A Comparison of Systems

R. Binazzi, M. Soudry, L. Mestriner, and J. N. Insall

The Knee Service, The Hospital for Special Surgery, New York 10021, USA

Introduction

Total knee replacement is now widely used, but great difficulties still remain concerning the clinical evaluation of results. In fact, a large number of rating systems exist, often very different from each other. Clearly, it is extremely important to be able to compare results from different centers, using different prosthetic models, just as it is also necessary to define the severity of the arthritic condition before surgery. At the Hospital for Special Surgery (HSS) we have used a scoring system [1] that was developed in 1972: it has recently been suggested that our method may be too lenient [5].

Therefore, in 1982, we thought to examine a series of 235 knee arthroplasties, recording the results according to five commonly used rating systems: the HSS [1], the Brigham [2], the Freeman [3], the British Orthopaedic Association (BOA) [4], and the Venn diagram system [5]. Our purpose was to test the validity of our own system by comparing it with the other methods and to try to produce a "universal" rating system, which could be used for any prosthetic model.

Materials

We evaluated 235 knees (176 patients), operated on for gonarthrosis with a posterior stabilized condylar knee. There were 128 females and 48 males, the average age was 68 years, and the average follow-up was 3.6 years (range 2–6 years).

Three of the rating systems: HSS, Brigham, and Freeman are point systems. The Venn system uses three major parameters (freedom from pain, mobility, and stability), while the BOA presents the results as percentages of the group of knees.

1. HSS – 100 point total
   30 points – freedom from pain
   18 points – motion
   12 points – walking
   10 points – stair climbing and arising from a chair
   10 points – muscular strength
   10 points – stability
   10 points – absence of flexion contracture
   Deductions in points are made for the use of walking aids, extension lag, and varus or valgus deformity. Grading is as follows: Excellent (85–100 points), Good (70–84 points), Fair (60–69 points, and Poor (under 60 points).

Surgery and Arthroscopy of the Knee
2nd Congress of the European Society
Editors: Müller/Hackenbruch

2. Brigham – 100 point total
   50 points – freedom from pain
   10 points – absence of deformity
   10 points – motion
   6 points – absence of limp
   6 points – walking
   6 points – absence of walking aids
   6 points – stair climbing
   6 points – arising from a chair
   Grading is as follows: Excellent (90–100 points), Good (80–89 points), Fair (70–79 points), and Poor (under 70 points).
3. Freeman – 110 point total
   50 points – freedom from pain
   30 points – range of motion (more than 80°)
   20 points – walking outdoors (more than 30 min)
   Results are "acceptable" if pain is 40–50 points, function is 15–20 points, and motion is 30 points. If one scores "acceptable" in all above categories, 10 bonus points are allocated. A grading of overall assessment "acceptable" equals 95–110 points.
4. BOA
   The BOA system assesses: pain, ability to walk, walking aids, gait, flexion deformity, maximal flexion, extension lag, valgus angle, varus angle, arising from a chair, stair climbing; with each parameter divided into five grades. The BOA form does not directly categorize patients; some manipulation of the figures was needed to do this. We considered the percentage of knees in grades 3 and 4 (the best grades) to be equivalent to the good and excellent or acceptable categories of the other systems.
5. Venn diagram (Fig. 1)
   Three major parameters are used: freedom from pain (pain-free walking for a distance of at least 500 meters; pain on stairs or on arising from a chair are

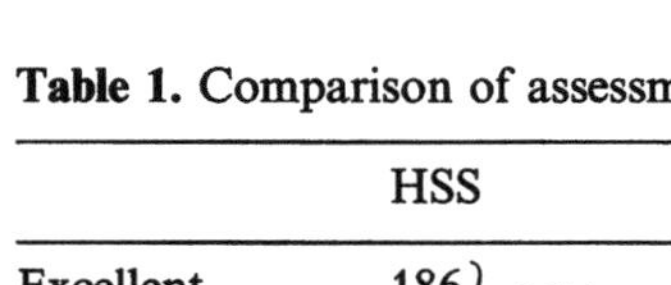

Fig. 1. The Venn rating systems

Table 1. Comparison of assessment systems for total knee replacement results

	HSS		Brigham		Freeman	BOA	Venn
Excellent	186	96%	173	92%	221 (94%)	94%	201 (85%)
Good	39		43				14 (6%)
Fair	3	4%	9	8%	14 (6%)	6%	15 (7%)
Poor	7		10				5 (2%)

not evaluated); mobility (at least 90° of flexion, with less than 5° of flexion contracture); and stability (less than 5° of varus and/or valgus instability).

Subset A is satisfactory (knees pain-free, mobile, and stable). Subsets B, C, and D are unsatisfactory but acceptable (knees that are pain-free but not mobile or stable). Subsets E, F, and G are poor results.

Results

All point systems and the BOA gave almost identical results (more than 90% of excellent and good results, with the HSS being the most lenient one, with 96%; Table 1). The Venn was the most stringent one, with only 85% of "satisfactory" knees, but by expanding the "acceptable" category to include all pain-free knees, we obtained a result that fell in line with the other systems.

Discussion

It was clear that all the rating systems had imperfections. The HSS one was designed primarily for rheumatoid arthritis, which is reflected in the 10 points given for normal quadriceps muscle strength; muscle weakness is seldom a feature of gonarthrosis. The point spread in the good category is also too wide, and some of the marginal good results could equally well be described as fair. The prime criticism of the Brigham system is that half of the points are allocated for pain, a category which is entirely subjective. The same objection can be made for the Freeman rating, which also gives nearly one-third of the points for motion, and makes no assessment of stair-climbing ability. A 110 point total also makes it difficult to calculate percentages. The BOA form, on the other hand, records the necessary information but weighs each feature of the arthroplasty equally. It is difficult to convert to a points rating or to the conventional descriptive categories of excellent, good, fair, and poor. Moreover, it does not deal very well with revision operations. The Venn diagram imparts information in the clearest manner of all and, once familiar with its conventions, it is easy to understand. However, the diagram excludes from the satisfactory category knees with minor imperfections (for example, a 10° flexion contracture). It does not take into account that old people may, voluntarily or not, limit their walking to less than 500 m for reasons entirely unconnected with the knee. The Venn system also provides no information about the stair-climbing function.

Conclusion

We maintain that a points system is preferable, being easier to handle. The HSS proved to be a good compromise between completeness and rapidity of execution in its scoring part, but it was too generous in the grading section. Thus, we have proposed a modification of the HSS system [6], which maintains the 100 point total but alters the grading: excellent (90–100 points), good (80–89 points), fair (70–79 points), and poor (under 70 points).

In this way, the results are more in harmony with our subjective clinical impressions.

References

1. Ranawat CS, Insall J, Shine J (1976) Duo-condylar knee arthroplasty. Clin Orthop 120:76–82
2. Ewald FC (1979) Surgery of the knee in rheumatoid arthritis. In: Instructional course lectures. American Academy of Orthopaedic Surgeons, vol 28, pp 285–298
3. Freeman MAR, Cundy AD, Todd RC (1977) The presentation of the results of knee surgery. Clin Orthop 128:222–227
4. Aichroth P, Freeman MAR, Smillie IS, Souter WA (1978) A knee function assessment chart. J Bone Joint Surg [Br] 60:308–309
5. Tjornstrand BAE, Egund N, Hagstedt BV (1981) High tibial osteotomy. Clin Orthop 160:124–136
6. Binazzi R, Soudry M, Insall JN (1984) Scoring systems. Scientific exhibition, Société Internationale de Chirurgie Orthopédique et Traumatologie (SICOT), London, October

Total Knee Replacement with Customized Prosthesis

R. Binazzi, M. Soudry, L. Mestriner, and J. N. Insall

The Knee Service, The Hospital for Special Surgery, New York 10021, USA

Introduction

The Insall-Burstein Total Knee System is normally able to satisfy all the necessities of a routine total knee replacement. There are, nevertheless, some cases which cannot be resolved by using a standard-type prosthesis. This happens usually when the shape of the lower third of the femur and/or of the upper third of the tibia, for congenital or acquired reasons, is abnormal. It may also occur when there is not enough bone or it is not strong enough to guarantee secure fixation of the prosthesis components.

In recent years various authors have reported their experience on the technique of total knee replacement in the presence of considerable bone stock loss. Some recommend filling the bone gap with cement, others with cement reinforced by one or more Vitallium screws or a titanium mesh, and others suggest a bone graft fixed by screws. Some authors use all these techniques according to the extent of bone deficiency. In all these cases the "recipient", that is the patient's knee, is somehow adapted to the shape of the prosthesis, yet there is no, or very little, knowledge concerning the adaptation of the prosthesis shape to the bone situation.

In this presentation we describe the results obtained at the Hospital for Special Surgery of New York, using customized total knee prostheses designed and manufactured according to the clinical and radiological features of each case.

Material and Methods

Thirty-one custom-made prostheses were implanted in 29 patients, 14 males and 15 females, the average age being 62 years (range: 37–80 years). The average weight was 86 kg (190 lb) for males and 88 kg (194 lb) for females. Nine patients (29%) were more than 95 kg (210 lb). Ten arthroplasties were primary operations and 21 were revisions, of which 14 cases were a first revision and in 7, a second. Eighteen of the revised cases showed an aseptic loose prosthesis, while three were complicated by infection. The etiology was osteoarthritis in 22 cases, Charcot's arthropathy in 5, osteonecrosis, rheumatoid arthritis, trauma and rickets in 1. The average follow-up was 2 years and 8 months (range: 1 year and 3 months – 5 years and 9 months).

All patients were evaluated using the Hospital for Special Surgery (HSS) Knee Rating Scale by the senior author (J.N.I.) or at the Knee Clinic of the HSS. The

Surgery and Arthroscopy of the Knee
2nd Congress of the European Society
Editors: Müller/Hackenbruch

necessary modifications of the prosthesis were decided on in collaboration with the Biomechanics Department of the HSS, which provided the final design and the manufacturing of the prosthesis.
All the custom augmentations were metallic and could be either in the femoral, in the tibial, or in both components. In 24 cases a posterior stabilized design, was used, in 2 a total condylar II, and in 5 a total condylar III.
The basic surgical technique was the standard one used for routine total knee replacements. In five cases a quadriceps mechanism turn-down was necessary because of stiffness of the knee, while in six cases an associated patellar lateral release was performed. Postoperatively a Robert Jones dressing was applied and kept in place for 4–5 days. The patient was able to stand on the 2nd or 3rd day and to take a few steps. Flexion was started after the removal of the dressing only if wound healing was satisfactory. Generally manipulation under anesthesia was done after 14 days if flexion was not approaching 90°. In our series this was necessary in three cases. Before being discharged from the hospital, the patient had to be able to walk with a cane, ascend and descend stairs, and arise from a chair without help.

Results

The average preoperative score was 51 points (49 for the primary cases and 52 for revisions), and scores related to pain and stability were particularly low. The average postoperative score was 86 points (89 for primary cases and 84 for revisions). Knees with only one revision showed a better result (87 postoperative score) than the ones revised twice (79 postoperative score). Twenty-two patients (71%) were rated as excellent, four (13%) as good, four (13%) as fair and one (3%) as poor. In the group of primary cases we had nine (90%) excellent and one (10%) fair result. Among the 14 first revisions there were 10 (71%) excellent results, 1 (7%) good, and 3 (21%) fair. In the group of second revisions (seven cases), 3 results (43%) were rated as excellent, 3 (43%) as good, and 1 (14%) poor.
Pain relief occurred in all patients. Only two patients 3 years after the operation were still complaining of mild to moderate pain subsequent to walking for half a mile or standing for 15–30 min, but they were two extremely obese females, weighing 123 kg and 135 kg (271 lb and 297 lb).
The average preoperative range of motion was 82° (84° for primary cases, 80° for revisions) with 6° average flexion deformity (7° for primary and 5° for revisions). Postoperatively the average range of motion was 99° (106° for primary cases, 95° for first revisions, and 99° for second revisions). Three patients had less than 90° of flexion and the minimum flexion angle was 55°. One patient, 2 years after the operation, had 10° of flexion contracture and two others had 5°.
Concerning alignment, preoperatively 20 knees were in varus alignment (13° average) and 9 in valgus (7° average). Two knees were in neutral alignment. The maximum varus was 45° in a severe case of Charcot joint. At the time of follow-up all knees were in slight valgus (6° average, range 2°–12°). Thirteen prostheses (42%), all the primary cases and three first revisions, showed no radiolucent lines about the three components at the time of the last follow-up. Eighteen knees

(58%), 11 first revisions and all the second revisions, showed radiolucency with at least one component. In these cases radiolucency was present in all tibial components having a width of 1–3 mm, usually in the medial zone. Second revisions showed a higher percentage of radiolucent lines wider than 1 mm, i.e., in 100% of cases, than first revisions (57%). The percentage of radiolucent lines wider than 2 mm was also higher in second revisions (43%), versus 21% for first revisions. Six femoral components (19%), five in first and one in second revisions, showed lucency but not more than 2 mm. Patellar button was involved in three cases (10%), two first and one second revision, the maximum width being 2 mm.
Four patients presented a positive venogram indicating blood clots a few days after the operation but showed no other clinical signs. Two patients developed a urinary tract infection and three showed some drainage from the wound, which resolved spontaneously with immobilization of the knee for a few days. One patient had a foot drop which had not completely recovered at the time of last follow-up. One year after surgery one patient showed a patellar subluxation which required proximal realignment. In one case a mid-shaft femoral fracture occurred 4.5 years after the operation on the side bearing the custom prosthesis, consequent to an accidental trauma.
Finally, an upper tibia cortical fracture, resulting from an accidental fall, was recorded in a 55-year-old, mentally handicapped, epileptic, obese female. In this case a custom prosthesis had been implanted 2 years previously, with an anteromedial wedge in the tibial component. Her last score, 1 month prior to the accident, was 77 points, satisfactory since the patient had been revised twice.

Discussion

Severe bone stock loss is a major problem for the orthopedic surgeon performing a total knee replacement; particularly involved in our series is the posterior side of the distal femur and the medial tibial plateau. In the literature, there are some descriptions of techniques to rebuild a lack of bone stock in the upper tibia. Individually, these techniques have an intrinsic validity, but they are usually restricted to cases with an inconspicuous bone stock loss. In fact, they all aim at restoring the shape of the distal femur and upper tibia such as to allow a stable fixation of standard prosthesis components. This is never easy and sometimes impossible; it is better to prepare a prosthesis which is specifically designed to fit a given bone.
The first observation which becomes evident from our series is that the results are almost the same as those obtained with a series of about 300 primary, standard, posterior stabilized condylar prostheses. In fact, custom prostheses show 71% excellent results (versus 72% for standard prostheses), 13% good (versus 23%), 13% fair (versus 2%), and 3% poor (one patient, who scored the same percentage as for standard prostheses). Thus it is clear that for the custom-made prostheses there is a shift of 10% of results from good to fair, in comparison to the standard, posterior stabilized prostheses. However, if we analyze the four fair results, we can see that at least two of them have a low score for reasons independent of the type of prosthesis. In one case the patient was wheelchair bound due to severe deformity of the contralateral knee, while the other patient was extremely obese and

had been operated on many times without considerable relief. Most of the discomfort she felt was due to her overweight. Another point to be noted is that the results become progressively worse with an increase in the number of revisions. The performance of the five Charcot joints was quite surprising. They were all primary cases and had excellent results, with an average postoperative score of 94 points.

The late complications recorded were fairly serious, one patellar subluxation, one femoral shaft fracture, and one upper tibia fracture, but the last two cases were consequent to accidents and were not related to the type of prosthesis.

We feel without doubt that the custom-made prosthesis is the best and often the only way of resolving the problem of a total knee arthroplasty in the presence of a severe bone stock loss. To date the custom-made prosthesis can only be produced in a small group of specialized centers associated with a biomechanics department, but we hope that in the future it will be possible to order them, as one does with standard prostheses, from pharmaceutical companies.

Knee Stability After Total Knee Replacement with the Biomet Prosthesis as Evaluated by Stress Radiography

B. F. Iversen, K. Jacobsen, N. Laursen, and P. Lyndrup

Department of Orthopaedic Surgery and Department of Diagnostic Radiology, Frederiksberg Hospital, University of Copenhagen, Copenhagen, Denmark

Introduction

Numerous types of total condylar knee prostheses are available. In recent years designs have emerged that allow preservation of the posterior cruciate ligament. At the Frederiksberg Hospital in Copenhagen we use the Biomet total condylar prosthesis, which differs from conventional ones in that the tibial component is contoured at its posterior aspect to allow retention of the posterior cruciate ligament. The aim of the present study was to investigate anterior, posterior, valgus and varus laxity in patients after replacement with this prosthesis.

Material and Methods

From 37 patients 41 knees were tested by means of stress radiography. The stress radiographic technique and apparatus used have been discussed previously (Jacobsen 1976, 1981; Jacobsen and Iversen 1985). The measuring technique differs somewhat from that previously described. Figure 1 shows the landmarks and lines on which the measurements were based. The mean observation time was 14 months (range 9–24). Twelve knees (12 patients) were tested with forward and backward traction of 180 N. Roentgenograms were obtained of 28 knees (25 pa-

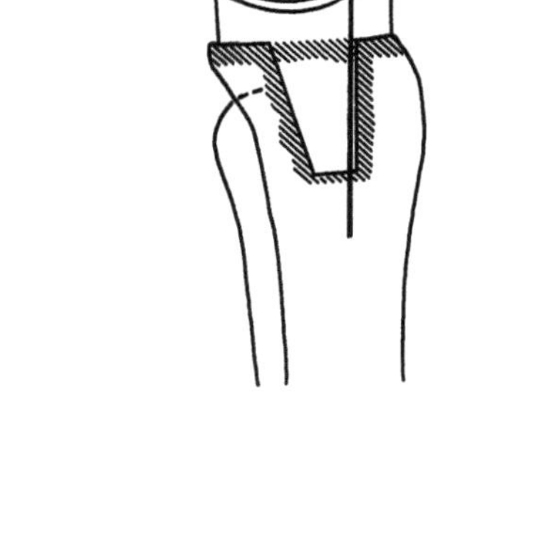

Fig. 1. As a landmark on the tibial prosthetic component we chose the anterior aspect of the small tibial stem. A tangent to this structure was drawn (*L1*). A second line was drawn perpendicular to this line through the inferior side of the pegs, at the posterior aspect of the femoral component (*L2*). AP laxity was then measured as the increase/decrease in the distance (*d*) from the articulating surface of the femoral component to the intercept of L1 and L2. L1 was also used in the measurements of valgus and varus laxity, which were calculated as the angle between L1 and a tangent to the condyles of the femoral component, minus 90°

Surgery and Arthroscopy of the Knee
2nd Congress of the European Society
Editors: Müller/Hackenbruch

tients) with 90 N, as well as 180 N of forward and backward traction. Valgus and varus testing were performed with an abducting or adducting force of 90 N. For each knee the Kettelcamp and Thompson scoring scale 1 (Kettelcamp and Thompson 1975) was completed. This scoring scale was used rather than the more commonly employed one from the New York Hospital for Special Surgery, due to its rather heavy weighting of the clinical stability, which can account for 20% of the score.
Preoperative, weight-bearing roentgenograms were obtained for both knees, and the degree of arthrosis recorded according to Ahlbaeck (1968). The radiological indication for surgery was rheumatoid arthritis (RA) in 5 cases and osteoarthritis (OA) in 38 cases. In 95% of cases of OA, the indication corresponded to Ahlbaeck grade 4 or 5 (obliterated joint space with bone attrition >5 mm and/or subluxation).

Results

On the scoring scale (maximum of 103 points) 22 knees achieved more than 90 points and were rated as excellent. Seventeen were rated as good (75–89 points), one as fair (60–74 points), and one with a score of 52 points, as poor (<60 points).

Table 1. Range and mean of the anteroposterior laxities

Force	Anterior displacement	Posterior displacement
90 N	4.2 mm (1–10 mm)	1.7 mm (0.5–5 mm)
180 N	6.6 mm (1–15.8 mm)	3.0 mm (1.0–7.5 mm)
	Valgus	Varus
90 N	4.2° (2°–12°)	6.0° (2°–14°)

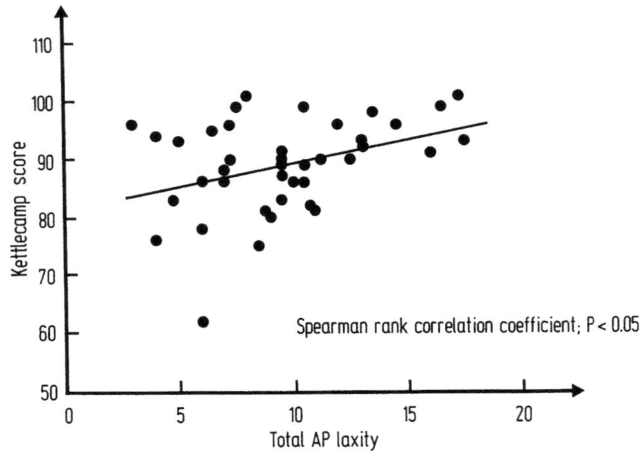

Fig. 2. Total anteroposterior laxity versus Kettelcamp score

The one rated as poor had RA. The scores of three other RA patients (representing four knees) fell within the good group.

Values for the laxities measured are given in Table 1. A positive correlation was found between increasing anteroposterior laxity and an increasing score on the Kettelcamp and Thompson scale, with a P value less than 0.05 (Fig. 2). If the points given for clinical laxity are disregarded, the level of significance increases to 99%, corresponding to $P<0.01$.

Discussion

Ninety-five percent of the knees were rated as excellent or good on the scoring scale. This is in accordance with Ritter et al. (1984) who also reported most satisfying results with this prosthesis from a 5-year follow-up study.

Knee stability following prosthetic replacement is of considerable importance in avoiding complications such as increased wear, loosening, and fractures of the prosthetic components (Minns and Hardinge 1983). The laxities measured in this study should be compared to the average anterior laxity for normal knees exposed to a traction of 180 N (3.1 mm), while the average normal posterior laxity is 3 mm (Jacobsen 1981). On the other hand, as far as we are aware, in vivo stability testing after total condylar replacement has not been performed until now. Therefore, no references exist describing what is normal after total knee replacement. In vitro studies have been carried out by Markolf et al. (1979), Shoemaker et al. (1982) and Nielsen et al. (1985). Our values for laxities correspond fairly well with those recorded by Markolf et al. (1979), for the unloaded knee.

The prosthesis investigated in this study permits preservation of the posterior cruciate ligament (PCL). Fairly modest posterior drawer symptoms which are consistent with the findings for normal knees demonstrated that the PCL remains functional (Jacobsen 1981). The moderate increase in total anteroposterior (AP) laxity after total knee replacement a. m. Biomet must consequently be attributed to the absence of the primary anterior constraint, namely the anterior cruciate ligament.

Limiting of the drawer symptoms may also be achieved by mechanical means with prosthesis designs favouring this. Such designs, however, imply generation with every step of great forces transmitted through the prosthetic stem acting as a lever arm. This results in a risk of loosening of the prosthetic components or fracturing the stems. The prosthesis investigated in this study does not have the hard end point characteristic for such a design, as obviated by the wide ranges of the laxities measured. We regard this as being important, inasmuch as the strain sustained by the secondary constraints i.e. articular capsule, collateral ligaments, etc. would be of a physiological nature and within physiological limits.

However, our results seem to indicate that a certain degree of AP laxity is indeed desirable, as a positive correlation was found between increasing total AP laxity and an increasing score on the Kettelcamp and Thompson scale, with a $P<0.05$. It should be emphasised that on this scoring scale, increasing clinical laxity involves a reduction in score, in accordance with the general opinion that stability is of importance. However, if the points given for clinical laxity are disregarded, the level of significance increases to 99%, corresponding to $P<0.01$.

Amongst our cases there were no extremely loose knees with regard to total AP laxity. Consequently, from this study, we have not been able to establish an upper limit for maximal desirable laxity (Fig. 2).
The mean values recorded for adduction and abduction are within the ranges suggested in the literature on knee biomechanics (Swanson 1976; Frankel et al. 1971; Burstein 1984), in which it is emphasised that a certain degree of varus and valgus laxity is desirable in distributing, and to a limited degree buffering, the stresses applied to the articulating surfaces.

Summary

Knee stability after total knee replacement with a posterior cruciate retaining prosthesis was investigated by stress radiography. 37 consecutive patients (41 knees) were included. On the Kettelcamp and Thompson scoring scale 95% of the knees were rated as excellent or good. Posterior laxity was found to be within the values established for normal knees, which may be attributed to the PCL being rendered functional. Anterior laxity was somewhat greater than in normal knees.
A statistically significant correlation was found between the increasing total anteroposterior laxity and an increasing score on the scoring scale.

Acknowledgements. This investigation was supported by grant No. 39/85 from the Danish Hospital Foundation for Medical Research, Region of Copenhagen, the Faroe Islands and Greenland.

References

Ahlbaeck S (1968) Osteoarthritis of the knee. A radiographic investigation. Acta Radiol [Suppl] (Stockh) 277:53–56
Burstein AH (1984) Surgery of the knee. In: Insall JH (ed) Biomechanics of the knee, chap 2. Churchill Livingstone, New York, pp 21–39
Frankel VH, Burstein AH, Brooks DB (1971) Biomechanics of internal derangement of the knee. J Bone Joint Surg 53(5):945
Jacobsen K (1976) Stress radiographical measurement of the anteroposterior, medial and lateral stability of the knee joint. Acta Orthop Scand 47:335
Jacobsen K (1981) Gonylaxometry. Stress radiographic measurement of the passive stability in the knee joint of normal subjects and patients with ligament injuries. Acta Orthop Scand [Suppl] 52:194
Jacobsen K, Iversen BF (1985) A simple rig for stress radiography of the knee. J Med Eng Technol 9:277
Kettelcamp DB, Thompson C (1975) Development of a knee scoring scale. Clin Orthop 107:95
Markolf KL, Finerman GM, Harlan CA (1979) In vitro measurements on knee stability after bicondylar replacement. J Bone Joint Surg 61(4):547
Minns RJ, Hardinge K (1983) Failure of one design of surface replacement knee arthroplasty due to loosening, deformation and wear. Biomaterials 4:147
Nielsen S, Ovesen J, Rasmussen O (1985) The posterior cruciate ligament and rotatory knee instability. An experimental study. Arch Orthop Trauma Surg 104:53
Shoemaker SC, Markolf KL, Finerman GM (1982) In vitro stability of the implanted total condylar prosthesis. J Bone Joint Surg 64(8):1201
Ritter MA, Gioe TJ, Stringer EA, Littrell D (1984) The posterior cruciate condylar total knee prosthesis. A five year follow-up study. Clin Orthop 184:264
Swanson SAV (1976) Design and testing of replacement prostheses. Proc R Soc Lond 192:173

Unicompartmental Arthroplasty with the Oxford Meniscal Knee Prosthesis

J. Goodfellow [1] and J. O'Connor [2]

[1] Nuffield Orthopaedic Centre, Windwill Road Headington Oxford
[2] Department of Engineering Science, University of Oxford, Oxford, UK

In 1978 the present authors suggested the use of free meniscal bearings in the design of a knee prosthesis [1]. From this theory we developed the Oxford knee, the first implant to provide an analogue of the natural meniscus free to glide and spin upon the flat tibial plateau (Fig. 1). In 1986, we published the clinical results of 125 knees treated since 1978 with this device in its bicompartmental form [2]. Relief of pain, restoration of mobility and recovery of function all compared well with previously reported series. In 1982, having observed a number of knees devoid of cartilage on one side and nearly normal on the other, we decided to try the device as a unicompartmental implant. Theoretically, the design would seem to be ideal, as since both the tibial and the femoral components are metal, they can both be very thin, necessitating the removal of very little bone, particularly from the tibia. The meniscal bearings come in a range of thicknesses with feeler gauges to match, so that exactly the right size can be implanted to tension the ligaments after the metal components have been fixed to the bone. The components are perfectly congruous throughout the range of movements, minimising wear and the ill effects of the products of wear upon the retained articular cartilage in the other compartment of the knee (Fig. 2).
Backwards and forwards movement of the femur and the meniscus on the tibia are obligatory if the cruciate ligaments are to be preserved. That movement is controlled mainly by the cruciate ligaments, and we believe it is essential to preserve it.
Before describing our results, we will first give, briefly, the technique of the operation. We start by removing the osteophytes from the involved compartment of the joint to allow the ligaments to move normally again. Using special saw

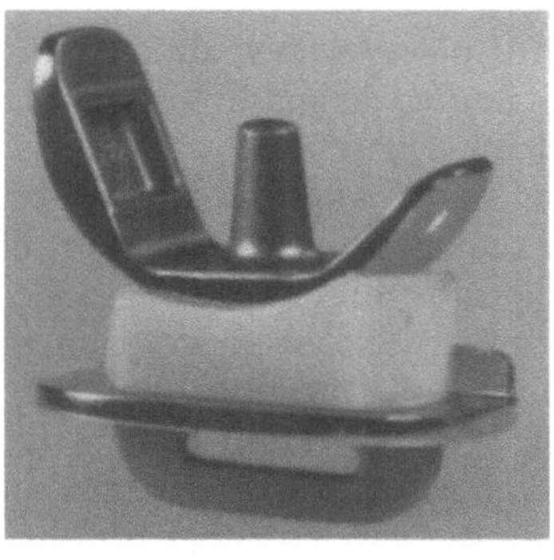

Fig. 1. Assembled components of the Oxford knee

Surgery and Arthroscopy of the Knee
2nd Congress of the European Society
Editors: Müller/Hackenbruch

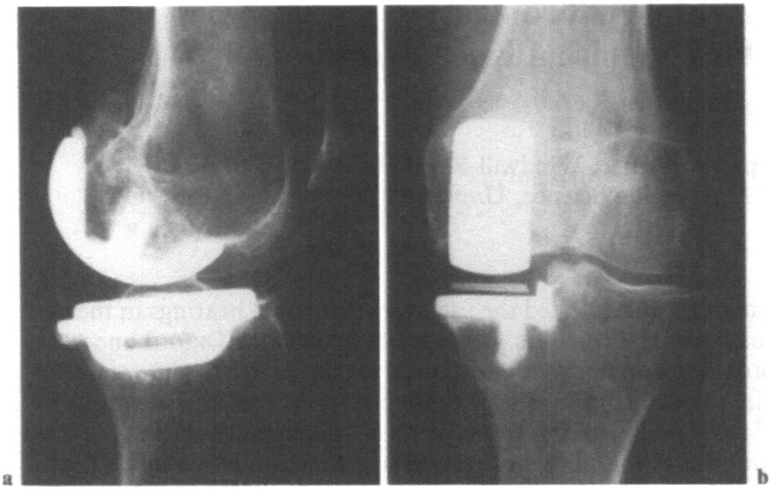

Fig. 2. a Lateral radiograph, **b** anteroposterior radiograph of medial unicompartmental Oxford arthroplasty

guides, thin slivers of bone are then removed from the damaged femoral and tibial condyles. A drill hole is made in the femoral condyle, and a groove is cut in the tibia. Trial components are then offered up, first the femoral and then the tibial. With the trial components in place, the appropriate size of bearing is chosen. The tool set contains a range of eight different thicknesses, colour-coded and varying in thickness by 1 mm. By trial and error, the size is chosen which exactly tensions the ligaments of the knee to restore stability. The joint is examined not only in flexion but in extension as well, to test for varus/valgus and anteroposterior stability. Finally the chosen bearing is snapped into place. In flexion it lies towards the back of the tibial implant and in extension moves to the front, driven backwards and forwards in flexion and extension by the intact cruciate mechanism.

In May 1986 we were able to review the early results of 55 knees which had been followed up from 1 to 4 years (mean of 2 years). There were more women than men and more medial implants than lateral. All but one were osteoarthritic, the exception being a case of primary avascular necrosis. The patients were mostly old or very old, having a mean age of 70 years, which is 5 years older than our parallel series of total joint replacements. They were also very heavy; almost one-third of them weighed more than 80 kg. The relief of pain was good; in activity, 88% of them complained of no pain or only slight pain. The relief of pain at rest was even better, with only one patient complaining of significant discomfort. The range of movements was almost the same as before the operation, an average of 102°. Varus and valgus deformities were corrected in most and improved in all. Functionally, too, the results were gratifying. Walking distance was very greatly extended; several patients could run and most could walk without limitation. Stair climbing was also dramatically improved: 80% could mount stairs normally, while only 4% could do so pre-operatively. The patients' own assessment of the operation was satisfactory or enthusiastic in all but five instances.

There have been two failures. Loosening of a medial component in a very obese woman was treated by arthrodesis, and a case of recurrent dislocation of a lateral bearing in a young man with post-traumatic osteoarthritis was also arthrodesed.

In two other cases subluxation of a bearing occurred, both times in the lateral compartment. These were replaced with thicker bearings and were looking well at the time of review. We have as yet too few unicompartmental cases to know whether this is a significant difference, but it does look as if the lateral compartment replacement poses bigger technical problems than the medial. In this series of cases, the anterior cruciate ligament was observed to be absent or severely damaged at surgery in about one-quarter of them, and the only loosening so far encountered occurred in a patient in whom that ligament was absent. It is mentioned because, in bicompartmental replacement, we have experienced much better long-term results in patients with the anterior cruciate ligament than in those without it. In 140 knees in which the anterior cruciate ligament was intact, the failure rate, 1–10 years later, was only 1.4%. This contrasts with the figures for 122 cases without a cruciate ligament in which the failure rate was ten times greater.

We conclude that the ideal subject for unicompartmental replacement is a knee in which the articular surface damage is limited to one compartment, preferably the medial, and in which the anterior cruciate ligament is intact. In these subjects, a very high success rate has already been achieved in the short term, and we may hope for a good long-term result from the experience of the Oxford knee in its bicompartmental form. In theory, this configuration of prosthesis ought to be ideal for unicompartmental replacement, and our early results suggest this is borne out in practice.

References

1. Goodfellow J, O'Connor J (1978) The mechanics of the knee and prosthesis design. J Bone Joint Surg [Br] 60:358
2. Goodfellow JW, O'Connor J (1986) Clinical results of the Oxford knee. Clin Orthop 205:21

Micromotion of Uncemented Freeman-Samuelson Knee Prostheses: A Roentgen Stereophotogrammetric Analysis

B. E. J. Albrektsson[1], L. Ryd[2], P. Herberts[3], A. Lindstrand[2], and G. Selvik[4]

[1] Department of Orthopaedics, East Hospital, Gothenburg, Sweden
[2] Department of Orthopaedics, Lund Hospital, Lund, Sweden
[3] Department of Orthopaedics, Sahlgren Hospital, Gothenburg, Sweden
[4] Institute of Anatomy, University of Lund, Sweden

Introduction

One of the most important problems in knee arthroplasty is late mechanical loosening of the tibial component. In order to avoid some of the disadvantages which are associated with using bone cement, various attempts have been made to bond prosthetic components to bone without cement. The early clinical results have been good but still very little is known about the reliability of this method over a long period of time.

Traditional clinical and radiological control methods to evaluate the quality of a new prosthetic design or a new method of fixation are not very efficient and do not give reliable answers before 5 or even 10 years have passed. Clinical follow-up is not very specific, although in the long run it is the only way to evaluate the success of arthroplasty. Ordinary roentgenograms are difficult to interpret, and bone scanning is unpredictable. None of these methods permit any conclusions to be made about the stability of the prosthetic fixation during the first 5 years; moreover, it is very difficult to perform sufficient postoperative clinical and radiological follow-ups beyond 5 years.

Radiostereophotogrammetry (RSA) was developed by Selvik in the mid-1970s and has been used in several important studies over the past 10 years. In 1986, Leif Ryd described the use of RSA for determining the micromotion of several different prostheses, with different types of fixation to bone, after knee arthroplasty (Ryd 1986). RSA has an accuracy of 0.2 mm and can be used to detect migration of prosthetic components with time as well as inducible displacement when a physiological load is applied to the knee in different directions. During surgery, spherical markers made of tantalum are inserted into the tibial metaphysis and tibial prosthesis (Fig. 1). After the operation stereoradiographs are taken with the knee placed inside a calibration cage with positional markers indicated on each of the four walls (Fig. 2). First, the three-dimensional coordinates of the markers are measured and processed by computer. The two-dimensional coordinates are then calculated, and the movements of the prosthesis relative to the bone can be determined.

Freeman, in 1977, introduced a conceptually new technique for component fixation employing two flanged pegs of polyethylene, inserted by a precise surgical technique which provide an immediate interlock between the prosthesis and bone. In the latest design, all components of the Freeman-Samuelson knee can

Surgery and Arthroscopy of the Knee
2nd Congress of the European Society
Editors: Müller/Hackenbruch

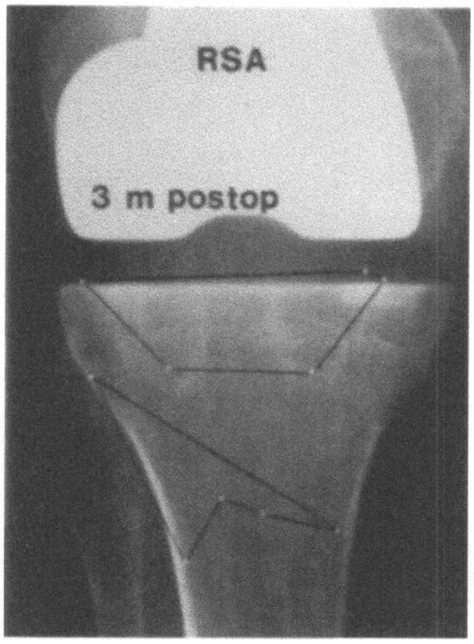

Fig. 1. A radiograph of a knee reconstructed with a Freeman-Samuelson prosthesis: markers are inserted in the tibial component and the tibial metaphysis

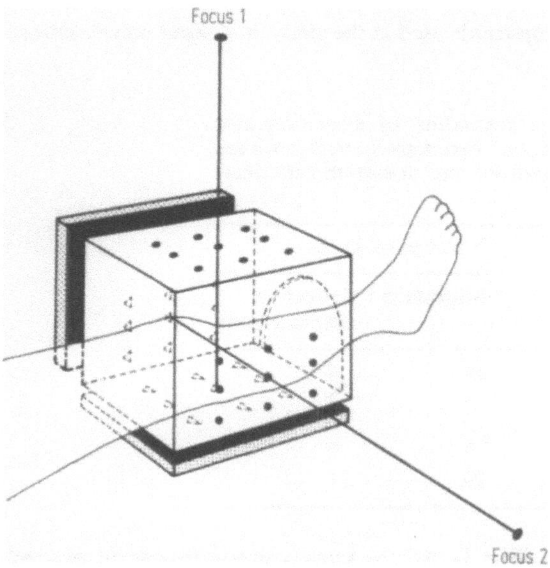

Fig. 2. Stereoradiographs are taken with the marked knee inside a calibration cage

be fixed without cement. The quality of tibial fixation using the uncemented Freeman-Samuelson knee prosthesis, with two different types of tibial component design, are presented as evaluated by RSA.

Material and Methods

The Freeman-Samuelson prosthesis was used in all of the knee reconstructions performed in this study. Two types of tibial components were used: in 23 cases a high density polyethylene tibial component was implanted and fixed with two pegs, and a further 6 knees were reconstructed with the same component rein-

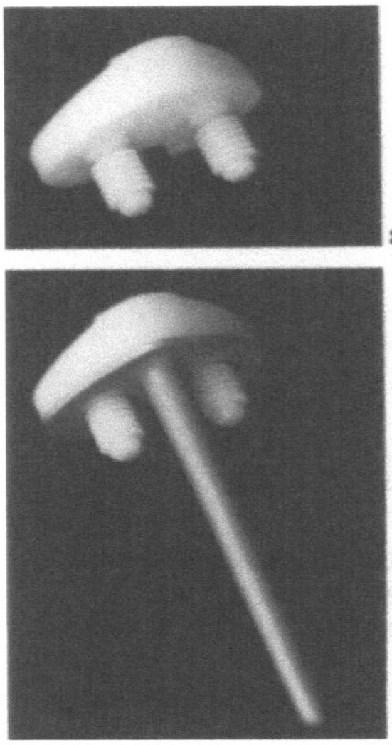

a

b

Fig. 3 a, b. The two different tibial components used in the study: **a** without metal reinforcement **b** with a metal back and stem

Table 1. Radiostereophotogrammetric evaluation of migration and inducible displacement for components of Freeman-Samuelson knee prostheses (the first three groups involved the non-stemmed tibial component)

Component	Number of knees	
	Migration	Inducible displacement
RA Metal femur	14	4
RA Delrin (polyacetate) femur	4	5
OA Metal femur	5	6
RA+OA Metal back and stem	6	6
	29	21

forced with a metal back and stem (Fig. 3). All the knee components were secured without cement; the femoral and patellar components were the same in both series.

The study is prospective but not consecutive. The cases were not specifically chosen but were representative of patients requiring ordinary knee arthroplasty within a population. The patients were followed up clinically, according to a special protocol, at 3 months and 6 months, then after 1 year and 2 years. All reconstructions appeared to be clinically successful at the last follow-up (good or excellent according to the Hospital for Special Surgery Scoring Scale). Migration of prosthetic components was studied in 29 knees 1–2 years postoperatively, and inducible displacement was evaluated 1 or 2 years after surgery in 21 knees (Table 1). Induced motion was measured from several pairs of radiographs taken while reproducible external forces, within a physiological range, were presented to the knee from different directions. The patient stood on a special platform, and the examined knee was encased within a special calibration cage. Standing and weight bearing as well as inward and outward rotation positions were studied.

Results

All tibial components showed migration as well as inducible displacement. Most of the prosthesis components migrated predominantly during the first 3–6 months and then seemed to stabilise (Fig. 4). The components in patients with rheumatoid arthritis (RA) migrated more than those in the osteoarthritic (OA)-knees group. The greatest difference in migration was found between the two different types of tibial component, regardless of diagnosis. The metal back and stemmed tibial component migrated only by about 1 mm, while the others (polyethylene only) sunk into the bone by about 1.5–2 mm. The only cemented knee studied using this component was the most stable which corresponds well with the results indicated for other cemented prostheses (Ryd 1986).
Four RA-knees failed to stabilise during the follow-up period (Fig. 5). Cases 1 to 3 all involved polyethylene tibial components, while case 4 had a reinforced component with a metal back and stem as well. Case 4 showed only migration and a very slight rotation around the stem but did not tilt or rotate in any other direction as did the other three.
The results regarding inducible displacement are presented in Fig. 6. The magnitude of motion varied between 0.5 and 1.9 mm, with the values for each group shown in the diagram. All knees exhibited rotational as well as translational displacements. The rotation in one OA-knee was exceptional, with a maximum total point motion of 5 mm. This was a combination of motion between the bone-prosthesis interface and a reversible, and totally painless, distortion of the prosthesis. On further examination only one of the pegs was found to be fixed. This particular knee was not considered to be representative.

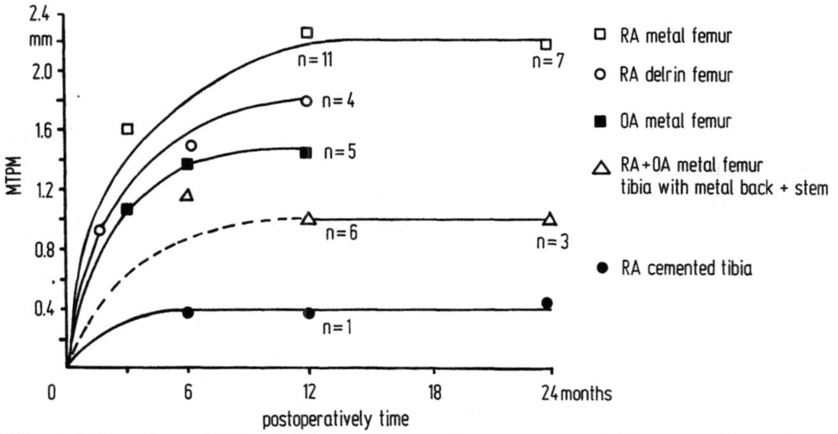

Fig. 4. Migration of the tibial component in uncemented Freeman-Samuelson knee arthroplasties: stable fixation (*MTPM*, maximum total point motion)

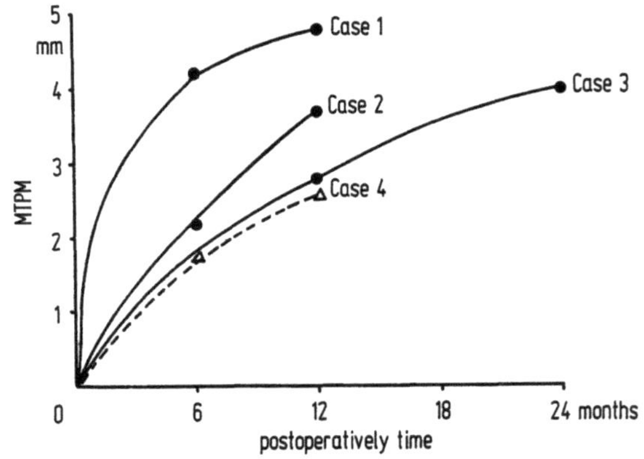

Fig. 5. Migration curves for four RA-knees considered to have unstable tibial component fixation: cases 1–3 represent components without a metal back and stem, *case 4* represents a component with a metal back and stem

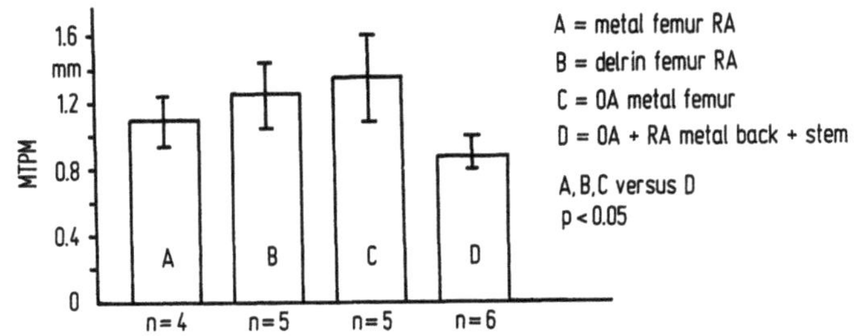

Fig. 6. Stress-induced displacement in uncemented Freeman-Samuelson knee prostheses (*MTPM*, maximum total point motion

For the majority of knees, the rotatory stress positions were the most revealing. The six knees with metal backed and stemmed tibial components were significantly less sensitive to external forces and the variation between individuals was smaller in this group than for the others.

Discussion

The aim of the immediate interlocking type of fixation is to obtain a strong bond between the prosthesis and the bone. The clinical results reported using this method have been good (Freeman et al. 1986). Indeed, all our reconstructions have been clinically successful. Nevertheless, all these implants show some movement, and, except for the prosthesis components reinforced with metal, the magnitude of the movements is two to three times greater than for conventionally cemented prostheses and approximately equal to the noncemented concepts studied by Ryd (1986). The stemmed prostheses, however, show migration and inducible displacement which is only slightly greater than for cemented knees.
Most of the components migrated during the first 6 months postoperatively and then stabilised. These cases represent the knees which indicate clinical and radiological success during the first 5 to 10 years after reconstruction. The smaller group of knees with an apparent unstable fixation, indicated by continuous component

migration, could represent early failures, although none of them has failed clinically so far. Experiences at Gothenburg using the ICLH-prosthesis with cementless tibial component fixation show a failure rate of approximately 10% during the first 5–6 years, when all the tibial components are uncemented irrespective of diagnosis, age and grade of joint destruction (Albrektsson and Herberts 1987). The metal-stemmed prostheses tended to migrate less and showed a significantly smaller inducible displacement than the all-polyethylene ones. RSA offers a very accurate and useful technique for early detection of instability differences between prostheses of different design and with different types of fixation. It is possible, even for a small study group, to collect very valuable information using this method within a reasonable postoperative follow-up period. This is obviously in the interest both of the patients and the surgeons.

References

Albrektsson BEJ, Herberts P (1988) ICLH knee arthroplsty in Gothenburg 1978–1982. J Arthroplasty (to be published)

Freeman MAR, Samuelson KM, Levack B, Alencar PGC (1986) Knee arthroplasty at the London Hospital 1975–1984. Clin Orthop Rel Res 205:12–20

Ryd L (1986) Micromotion in knee arthroplasty: a roentgen stereophotogrammetric analysis of tibial component fixation. Acta Orthop Scand [Suppl] 57:43

Application of Bench-Mounted Saws for Precision Replacement Arthroplasty of the Arthritic Knee – The Questor * Systems

T.D.V. Cooke [1], Y. Harada [1], G. Saunders [2], D. Siu [1], H. Wevers [3], and Y. Yoshioka [1,4]

[1] Clinical Mechanics Group, Division of Orthopaedics, Department of Surgery, Queen's University at Kingston, Ontario K7L 3N6, Canada
[2] Department of Surgery, Queen's University at Kingston, Ontario K7L 3N6, Canada
[3] Department of Mechanical Engineering, Queen's University at Kingston, Ontario K7L 3N6, Canada
[4] Department of Orthopaedic Surgery, Kobe University School of Medicine, Kobe, Japan

Introduction

Prosthetic replacement arthroplasty has revolutionised surgical treatment for arthritis of hip and knee joints, but as time passes, the number of failures, (mainly) due to loosening of the implant, increases. Although poor design and material features have a recognised importance in loosening, a malpositioned prosthesis must carry the greatest responsibility [1–4]. The precision involved in the manufacturing process of the implant is set to a level of tolerance ± 0.2 mm. This is, in all likelihood, a factor of over ten times greater precision than current bone cutting/placement techniques. The geometric accuracy employed in the construction of the implants sets a standard to be emulated in the orientation and accurate placement of the prosthesis; it recommends a comparably high level of placement-alignment accuracy for the bone cuts.

The first major attempt to provide accurate placement and alignment of knee implants was made by Freeman and colleagues [5]. He and co-workers [4] and subsequently others [6] pointed out that the release of soft tissue contractures, prior to making any cuts in bone, preserved bone stock. These authors indicated that the end-bearing cuts of the femur and tibia had to be made horizontal to the load-bearing axis (LBA) [5, 6]. The LBA is a physiologically ideal line of force which passes through the hip and knee centres to the middle of the ankle. To provide balance of the collateral soft tissues and lessen the amount of bone resected, Freeman employed a joint distractor or ligament-tensioning device inserted between the cut ends of the tibia and the femur. This reconstructed the joint space, where cartilage and bone had been lost, and realigned the leg. Insall et al. [6] and Townley [7], among others, have modified these ideas to permit the optimal alignment and location of condylar and Townley prostheses respectively. Cloutier's [8] approach to knee replacement added a feature of knee distraction as a preliminary

* Questor is the copyright name given to orthopaedic devices developed by the Clinical Mechanics Group at Queen's University, Kingston, Ontario, Canada

Surgery and Arthroscopy of the Knee
2nd Congress of the European Society
Editors: Müller/Hackenbruch

step to bone cutting in both the flexed and extended positions. This technique resects 2 cm of bone in flexion and in extension which matches the thickness of the implant, thereby providing balanced collateral ligaments in both positions. More recently, Hungerford et al. [9–12] have brought forward a complex but interactive system of cutting jigs and alignment devices by which the bone is marked by jigs and cuts made in a systematic way to fit the implant specifications. Hungerford and Krackow have emphasised reconstitution of the LBA with the joint line anatomically located in 3° valgus of the distal femur to the hip and varus of the proximal tibia to the ankle [11, 13].

It is now apparent that the number of cutting devices and insertional tools increases in proportion to the number of new knee implant designs. This multiplicity of design and techniques portrays a confusing picture for the orthopaedic surgeon interested in this field. In spite of the many attempts to improve the precision and reproducibility of the bone-cutting process, the actual cuts are still being made under the hand/eye control of the surgeon manipulating a hand-held saw. The attendant problems of object surface concavity and convexity with slanting planes on curved long bones contribute to the imprecision. Furthermore, saw blades are easily damaged by direct contact with metallic jigs and guides with a consequent loss of smoothly sectioned bone surfaces and a concomitant increase in tissue debris [14].

Added interest in bone-cutting accuracy attends the development and use of porous coated implants. These necessitate a precise, contoured resection of the bone ends (± 1 mm) to give a good fit; indeed, this fit may be required for successful (uniform and extensive) ingrowth of bone [15, 16]. Our experience, shared with that of many other groups, indicates that a high degree of bone-cutting accuracy cannot reliably be obtained by the use of jigs that are applied to deranged and deformed arthritic joint surfaces and/or in combination with the use of hand-held saws or routing tools.

The objective of our research has been to provide a method for cutting bone in a measured and reproducible manner. The method should be of use in the preparation of bone for many different implant styles and multiple sizes; of importance may be the applicability of the technique to corrective osteotomy. On this account, we concentrate on descriptions of the devices developed for knee arthroplasty and clinically tested with six different condylar-shaped resurfacing designs (DePuy, Townley; Howmedica, Kinematic and PCA, Zimmer, Cloutier; Ceraver, Hermes; Richards, Tricon-P; and Johnson and Johnson, Brigham unicompartment). These designs have an absolute requirement for retained collateral and posteriorly strengthened soft tissues (an intact posterior capsule and cruciate ligament) for the maintenance of functional knee stability. The principles involved do not in any way limit the use of the Questor cutting devices to unconstrained knee designs; rather, the designs tested exemplify the maximal (and probably optimal) requirements for minimal bone resection, soft tissue balance and axial alignment. Our method reconstitutes the physiologically optimal LBA of the patient with the joint line alignment at a right angle in both planes [1–4].

Orthopaedic Aims

The aims and limitations of replacement arthroplasty of the knee embodied in our developments may be summarised as follows:

1. Resurfacing of deranged articular surfaces with prosthetic implants that are better than the arthritic surfaces but less than anatomically perfect in form and function
2. Minimal bone resection (especially the bone of the proximal tibia) with re-establishment of the joint line in flexion and extension
3. Component placement about the posterior cruciate ligament (PCL) femoral origin (representing a constraint of major biomechanical significance and a "centre" of knee kinematics)
4. Balanced collateral ligaments
5. Restoration of the LBA with neutral tibial and femoral rotation
6. Placement of the joint line at 90° to the LBA in the coronal and sagittal planes
7. Components closely matched to articular dimensions

The Concept. In a joinery workshop, a piece of wood to be cut is held in a vice to a reference frame, the workbench, which in turn is the base on which a saw (such as a radial arm saw) is mounted. Square, precision cuts are easily made at exactly prescribed distances along the piece. To extend the analogy to the operating room (OR), either the bone (and the attached patient) must be taken to the saw bench or the bench on which the saw is mounted must be applied to the bone. We explored the latter approach initially, but now work with a saw bench jig mounted on the OR table to which the knee is fitted.

Designs

Questor Mark I Saw Jig

The saw jig used in our initial work is shown in Fig. 1. The design and application have been described in detail elsewhere [17]. In essence a reciprocating saw (Black & Decker or 3M) is mounted on a three-plane slide system attached to a base plate, aligned in three planes to the tibial shaft and fixed to the tibia by contractable arms proximally and a distal fixation pin. A detachable distractor/alignment device (B in Fig. 1), mounted on the base plate, gains a purchase on the intercondylar notch of the femur via a 0.25-cm drill hole precisely located at the most anterior and lateral point of the PCL insertion. This is the intra-articular kinematic landmark about which all of the measurements and alignments are based for the knee implant positioning [18]. Central distraction is obtained in either extension or flexion. Extendable locking arms (A in Fig. 1) are used to link onto a transcondylar femoral pin. When locked in place, the femur is now held distracted in either full extension or in 90° of flexion to the tibia.

Alignment of the knee/jig assembly is obtained by using long rods that extend from the distractor system to the hip in extension and then in flexion. With the knee fixed at 90°, precise, debris-free bone cuts are made at a right angle to the

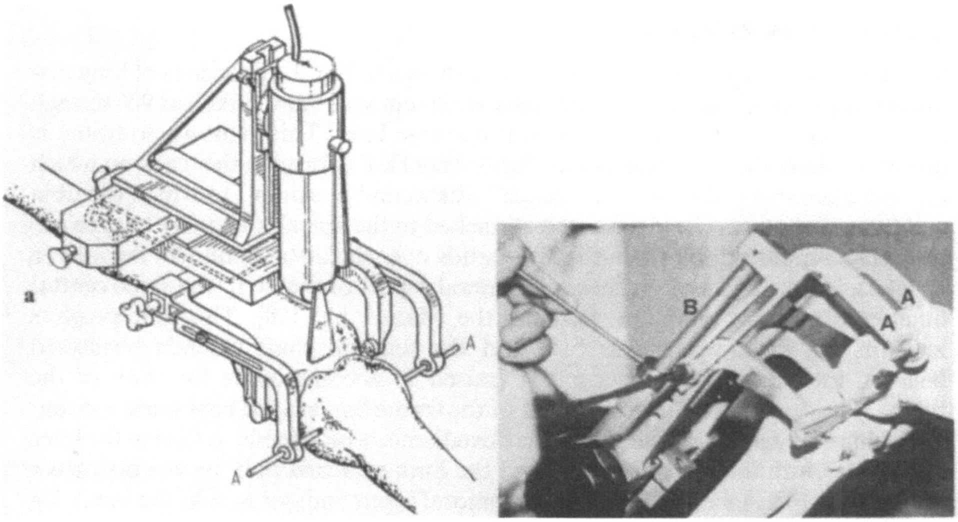

a

b

Fig. 1. a Saw jig mounted on the tibia, carrying an oscillating type of saw. The femoral condyles, having been distracted, are fixed to the jig via side arms *A* and a transosseous pin. **b** Close-up view of the saw jig applied to a laboratory bone preparation of the knee. The collateral ligaments have been simulated by elastic bands. The knee is held exactly at 90° while it is being distracted via the central distraction system *B*. Once distraction has been accomplished, the femur is locked to the jig in the flexed position via side arms *A* and the locking nut at each side

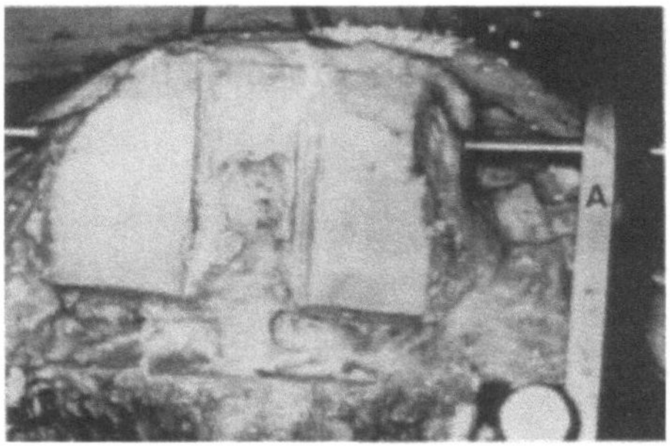

Fig. 2. Intraoperative view during knee arthroplasty, face on to the femur which was held at 90° to the tibia via the femoral fixation pin and side arm *A*. Note the smoothly cut bone surfaces and parallel spaces between the posterior femoral condyles and the upper tibia

base plate to accommodate placement of the femoral component and tibial base plate. The knee is then extended, redistracted, realigned and refixed in the extended line of position. The distal femoral bone cut is then made (Fig. 2).

Comments. The Mark I system has been used in 12 instances. In each, it provided smooth, easy bone cutting with excellent component fits. However, the neccesity to change position from a flexed to an extended position means that errors in alignment of the hip could occur more easily. Further, the realignment/distraction procedure adds to the OR time required (seldom completed within 2 h). Finally, the base plate attachment to the proximal tibia necessitates a tibial tubercle osteotomy (adding to OR time and potentiating problems of non-union).

Questor Mark II Saw Jig

The bench frame system currently in use is shown in Fig. 3. It consists of long distal and short proximal parallel steel bars set 20 cm apart and linked at 90° to each other by apical, distal and proximal transverse bars. This right-angle frame is mounted adjustably to a base plate (Fig. 3, bars D, P), forming the track on which the saw assembly slides and the "bench" (distal and proximal) to which the tibia and femur are successively mounted. Attached to the apical transverse bar, an adjustable distractor hook (A in Fig. 3) extends over the distal femur (in flexion) to engage a 0.3-cm drill hole at the anterolateral aspect of the PCL (the same central kinematic marking point as used for the Mark I jig) [18]. The foot projects beneath the distal transverse bar, and the femoral shaft extends headward beneath the proximal transverse bar (raised to accommodate the bulk of the thigh). The "top-heavy" arrangement of the frame beyond the base plate is counterbalanced in use by the weight of the flexed limb, which tends to flatten the knee out. Thus, with the frame in place and the limb attached only by the distractor hook at the PCL location point in the femoral intercondylar notch, the lower leg hangs free.

The frame-limb complex is pointed towards the hip (marked by a towel clip or an alignment device, see Fig. 4) and positioned so that the knee is roughly at 90° of flexion. The tibia is then suspended via an ankle gauntlet from the distal transverse bar of the frame by an adjustable pin in such a way that the lateral ankle centre coincides with a guide rod mounted in the vertical support of the distal frame (B in Fig. 3). This is at an equal distance from the distal frame bars as is the PCL distractor hook (Figs. 3 and 5).

The tibia is now rotated to neutral (second toe ahead) and held centrally opposite the distal anterior marker guide (C in Fig. 5). It is fixed via a proximal, threaded,

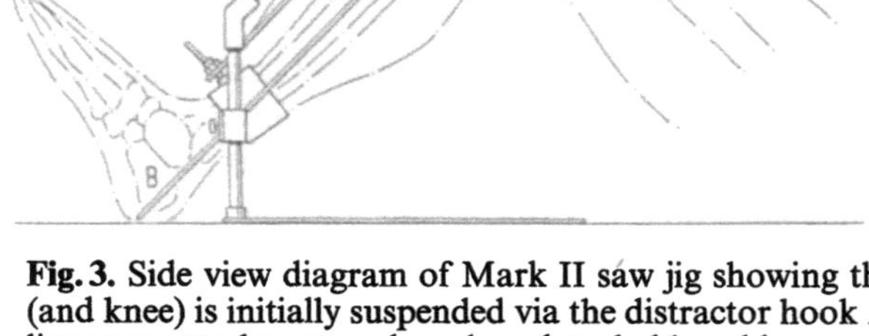

Fig. 3. Side view diagram of Mark II saw jig showing the bench frame to which the tibia (and knee) is initially suspended via the distractor hook *A* located at the posterior cruciate ligament attachment and a velcro detachable ankle gauntlet fixed to the centre of the distal transverse bar of the frame. The ankle is raised (or lowered) until the lateral alignment rod (*B*) is at its centre point (just ahead of the fibula). This insures that the ankle is equidistant from the distal frame bars as the PCL – thereby aligning the tibia (via the pull of ligaments) parallel with the frame. *D*, distal, *P*, proximal bars of frame

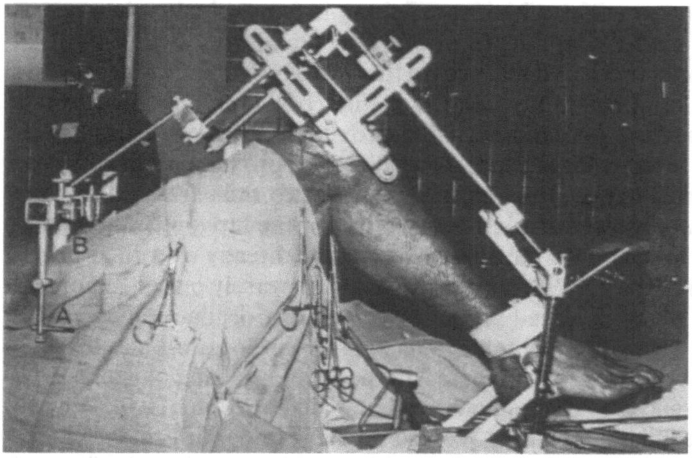

Fig. 4. Operating room (OR) picture of the hip alignment device used to provide an external reference system for location of the hip. In essence, the *transverse bar* (attached to an adjustable verticle mount on the OR table side bar) is positioned on the anterior superior iliac spine (ASIS) on plastic pods. The ipsilateral pod is used to define the hip depth (for lateral alignment) by an adjustable gauge *A* positioned about 1.5 in. (2 cm) below and distal to the ASIS. The anteroposterior hip location is marked on a line at the midpoint between the ASIS and pubic tubercle using a moveable rod holder (*B*). This position is checked by an intra-operative radiograph once the jig is set up

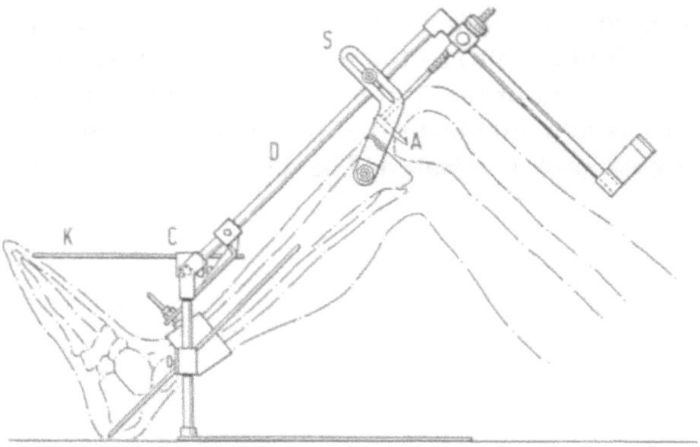

Fig. 5. Knee attached to the frame in distraction. The tibia, rotated to neutral in three planes, has been fixed to the distal bars (*D*) by attaching a proximal threaded transfixion pin to the bars by an adjustable side arm (*S*). The central location of the ankle is monitored by the guide rod on the distal transverse bar (*C*) while proximally the posterior cruciate ligament (suspended by the distractor hook *A*) is referenced to the distal frame by a short guide rod (*K*) in an adjustable rotating mount on the bar

transfixion pin, placed roughly at right angles and parallel to the distal frame bars (D in Figs. 3 and 5) to which the pin is attached by adjustable angled side arms (S in Fig. 5). This positioning fixed the tibia while perfectly aligned in three planes to the distal frame. The knee is now distracted until the PCL is under tension and the diseased joint compartment(s) open; firm distraction is confirmed by the bend in the tibial transfixion pin. This tibial transfixion pin is now locked in place to the side arm holders via locking nuts (S in Fig. 5). The kinematic knee center – PCL origin – is now referenced and marked on the side bar with a short transverse 0.4-cm rod mounted on a rotatable base (K in Fig. 7). A long alignment rod is then exchanged for a short one and extended proximally beside the thigh. Another alignment rod, mounted in a vertical swivel on the proximal transverse bar above the thigh, is extended over the hip area. These two rods are used to align for the lateral and anteroposterior (AP) perspectives of the hip joint. The frame-knee complex is now moved until the alignment bars lie directly over their proximal reference points for the hip. We currently use a transversely placed alignment hip reference system which is placed on the anterior superior iliac spines (ASIS) and held to the side bar of the OR table (Fig. 4). The hip location may be estimated from this location by measurement. Alternatively, the long axis of the femur may also be referenced to the frame via the extended portion of a 0.5-cm rod placed in an intramedullary location. We use an AP radiograph of the hip to check the accurate positioning of the anterior hip alignment rod to the femoral head. Any malposition is corrected by rotating the frame, and the alignment is then checked again by radiograph. The base plate of the frame is now firmly locked onto the OR table with a modified G clamp.

A transfemoral threaded pin is placed some 2 cm below and proximal to the anterior distal surfaces of the femur. It is attached loosely via adjustable mounts to the proximal side bars of the frame. The femur is now rotated to neutral via the transfixion pin such that the medial and lateral epicondyles are equidistant to the apical transverse bar. Maintaining neutral rotation, the femur is now *manually* distracted (pushing up on the transfemoral pin) to the point that the distractor hook is unloaded. The pin is then clamped firmly to the proximal frame via side arm mounts (S in Fig. 6). Missing this step out means that the femur will descend towards the tibia when the distractor hook is released. The distractor is now removed.

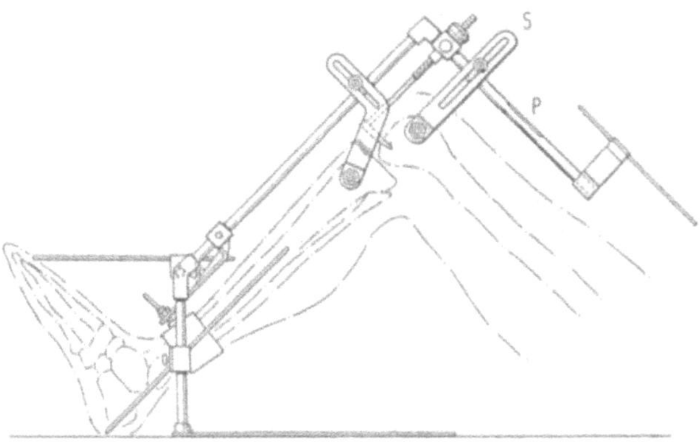

Fig. 6. Knee distracted in the frame with the femur neutrally rotated in three planes (epicondyles equidistant from the apical transverse bar) and held to the proximal frame bars (*P*) by adjustable side arm mounts (*S*) attached to a threaded transcondylar pin placed 2–3 cm away from the distal and anterior condylar surfaces

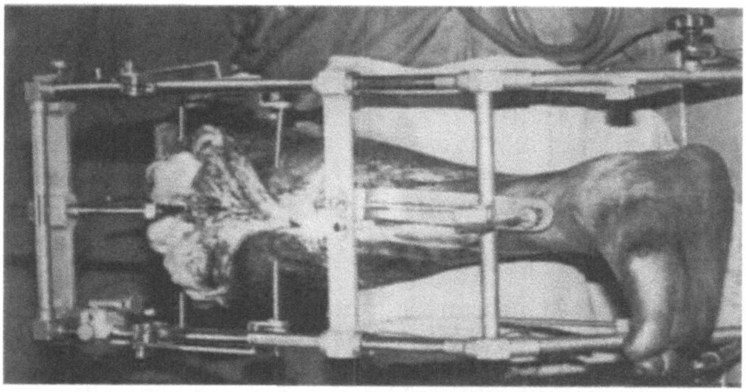

Fig. 7. Operating room photograph of the knee distracted and aligned to the hip and ankle respectively. Femur and tibia neutrally rotated and firmly fixed to the table

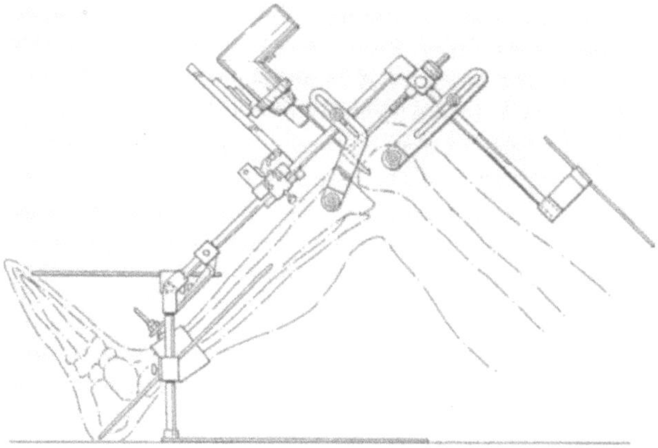

Fig. 8. The saw assembly, zeroed to the posterior cruciate ligament, is mounted on the distal frame. The template (referencing the interior dimensions of the desired prosthesis) is mounted at the zero position below the saw

This setting-up procedure has now fixed the knee to the frame at 90° flexion, with the tibia and femur neutrally rotated and each bone respectively parallel to the distal and proximal bars of the frame; the frame itself is firmly fixed to the OR table, pointing precisely at the hip joint (Fig. 7, OR setup at surgery).
The saw assembly is mounted on the frame bars (rails), and bone cuts are made according to prefabricated templates mounted on the distal limb of the frame below the saw. The templates reproduce the interior dimensions of the chosen prosthesis, basing the anterior and posterior femoral cuts equidistant to the PCL as a zero reference point (Fig. 8).

Saw Assembly. The system designed in our laboratories consists of a cylinder spring mount for the saw attachment (vertical axis of motion), fixed to another horizontally oriented cylinder mounted at 90° (horizontal axis of motion), firmly suspended between lockable slide mounts that fit on the frame rails (Fig. 8). The horizontal mount has a central axial adjustment which allows rotation of the ver-

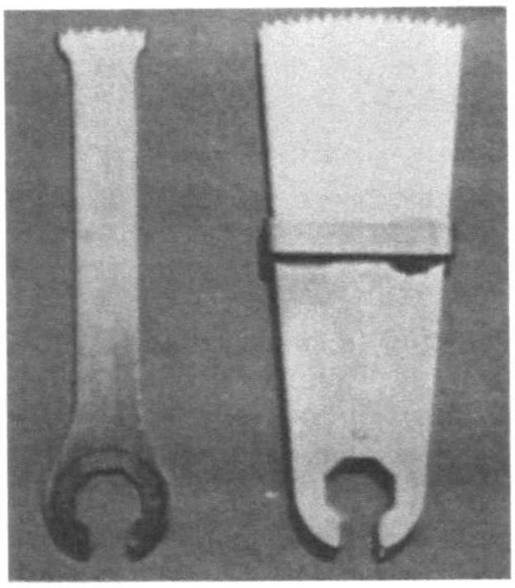

Fig. 9. Two blade modifications are illustrated: on the left is a narrowed blade with side cutting teeth, and on the right is the standard osteotomy blade with the stiffener routinely used for beginning the bone cuts

tical mount, and thus of the saw, up to 50° away from the right angle. This feature is especially useful for accurate anterior and/or posterior femoral and tibial cuts that are designed to be angled away from either the long axis of the femur or the 90° transverse axis of the tibia. Thus, for exact PCA femoral component fit, the anterior femoral cut is made at precisely 6° to the femoral long axis; for the Townley tibial component, an 8° posterior angle cut may be made for the tibia. The saw assembly has been designed to accommodate either the 3M or Black & Decker (battery-operated) reciprocating saws. When attached, the blade must be accurately checked to make sure it is at precisely a 90° angle with the frame. The problem of bone skiving has been considerably lessened by the use of a detachable blade stiffener (Fig. 9).

Comments. This system, with only minor modifications, has been used by the senior author over the past 3 years (some 50 cases). The system has distinct advantages over the Mark I jig:

1. Once positioned, the knee is never moved; all bone cuts are made with the knee in 90° of flexion. This procedure reinforces joinery and machine engineering principles and much improves the bone-cutting precision and alignment.
2. The frame itself has many positive features. Paralleling the limb elements, alignment of each to the frame is easily made by guide rods attached to transverse and vertical parts of the frame. Further, the knee centres and faces (femoral or tibial, anterior, distal femur) are easily referenced, and femoral and tibial component orientation to a central location (and correct AP placement for the tibial component) are readily checked.
3. Controlled distraction is possible with the potential to calibrate this force with a torque wrench. This may become important when considering differences in ligament qualities in rheumatoid and osteoarthritic cases.
4. Bone cuts in flexion lessen the risk of posterior soft tissue damage.
5. Minimal extra equipment is required to undertake the arthroplasty for a number of different condylar knee designs. Essentially, the additional equipment relates to templates to locate positional pegs or chamfering cuts. Specific templates of femoral interior dimensions are made for each prosthesis. The central axis of the frame may again be used for positioning and location of these templates.

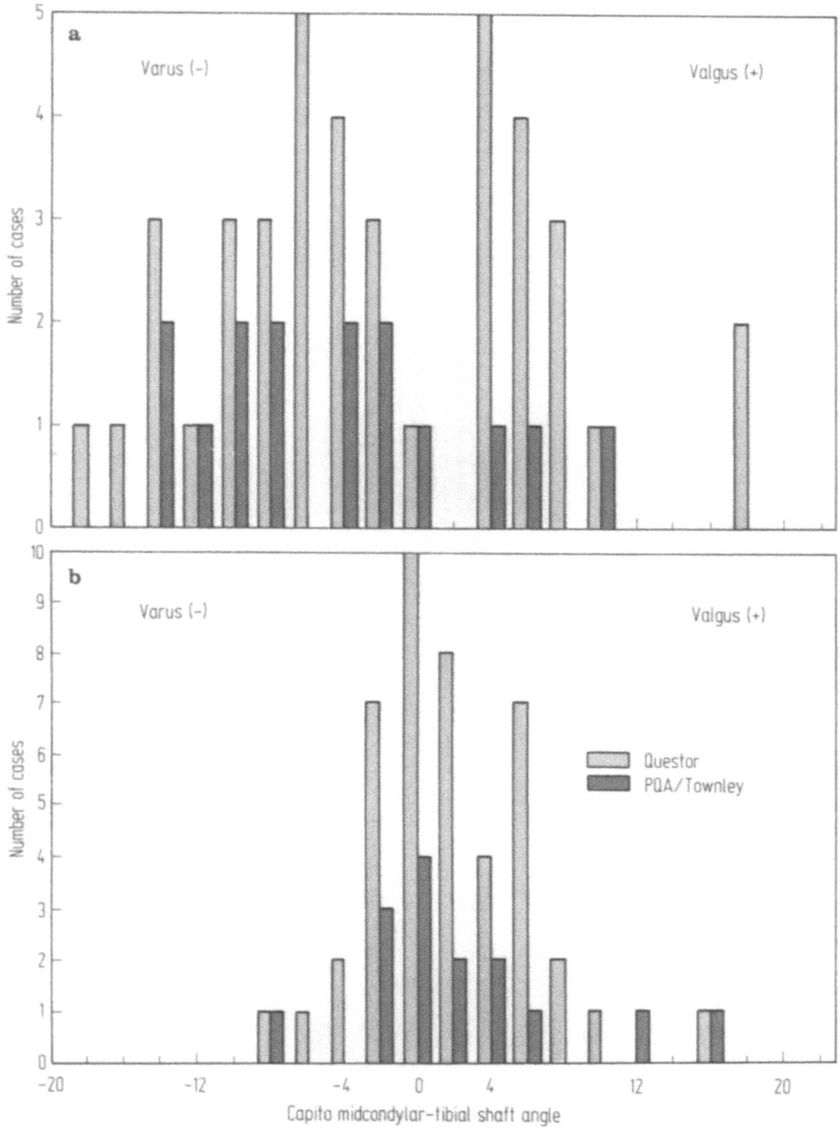

Fig. 10a,b. Pre- (**a**) and post- (**b**) operative alignment of the knees (capito midcondylar-tibial shaft angle) handled by the three instrumentation systems tested

The major disadvantages are:

1. Set-up time for the frame (20 min non-tourniquet time)
2. Set-up time in attaching the limb to the frame (20 min tourniquet time)
3. Experience and training needed for use due to the complexity of the attachments (user-unfriendly)

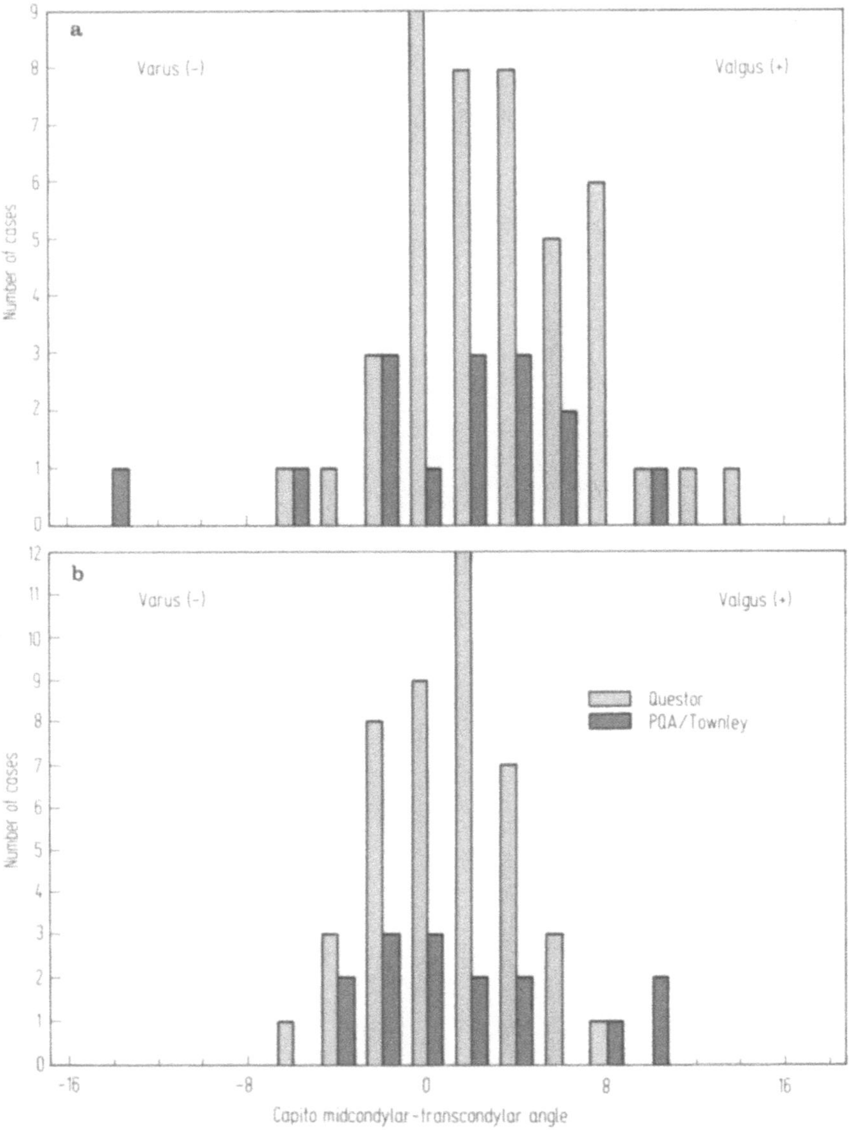

Fig. 11 a, b. Pre- (**a**) and post- (**b**) operative results of implant positioning based on the capito midcondylar-transcondylar angle of the femoral component to the hip

Prospective Comparison of Implant Component Position. The most important feature of this system is its accuracy of implant positioning and alignment. In a prospective study at Kingston General Hospital comparing Questor to the Howmedica Universal and DePuy "first instruments" applied to cases of equal or more deformity, the Questor saw-bench system has been shown to be more accurate in all respects.

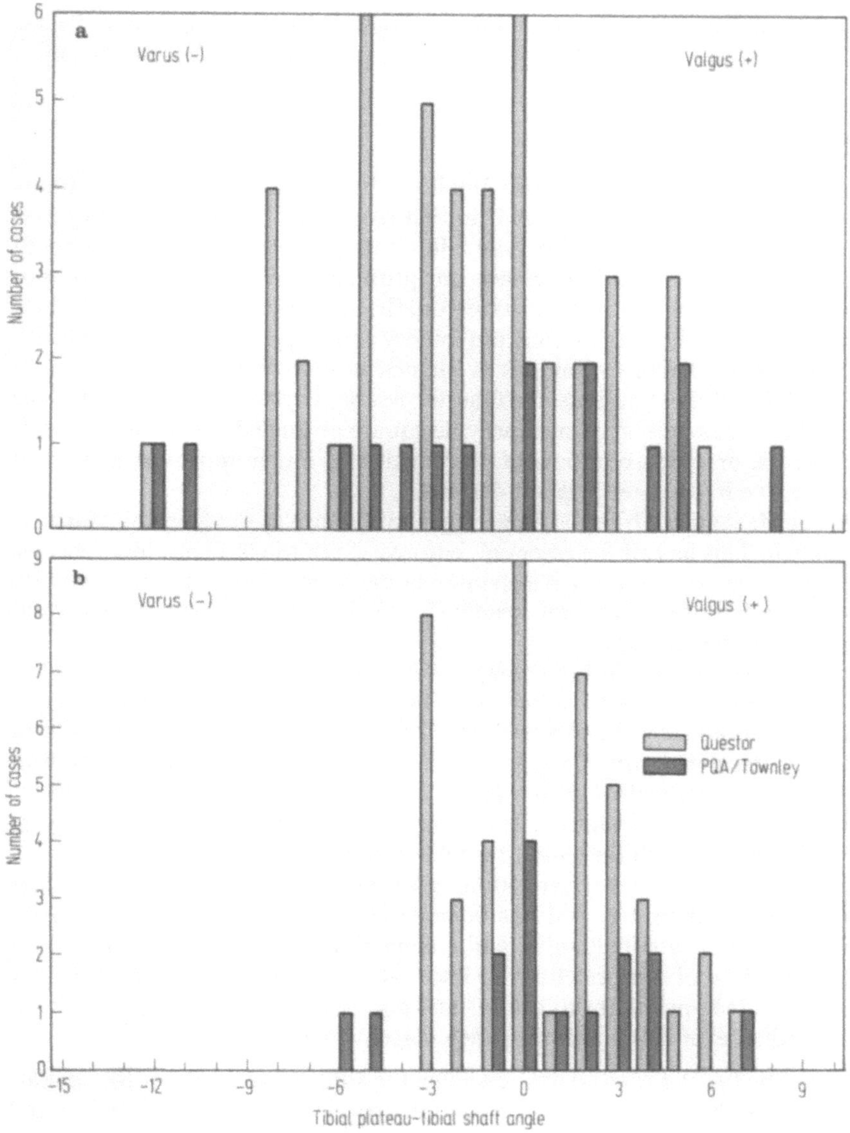

Fig. 12a, b. Pre- (**a**) and post- (**b**) operative results of implant positioning/alignment based on the transtibial ankle alignment (tibial plateau-tibial shaft angle) for the three systems tested

Shown in Fig. 10 are the pre- and postoperative alignments of the knees handled by the three instrumentation systems tested. In Fig. 11 are the pre- and postoperative results of implant positioning based on the capitomidcondylar-transcondylar position of the component to the hip. Finally, in Fig. 12, the pre- and postoperative transtibial ankle alignment for the three systems is exhibited. Note that, in Fig. 11, the desired ideal for Howmedica Universal Instrument alignment would be 3° valgus. The data reveal considerable variation in the desired alignment in spite of claims for accuracy within 2° by the manufacturers [9, 10, 18].

Discussion

Despite major gains in pain relief and function obtained by knee replacement arthroplasty, component loosening and limited recovery of knee kinematics mar these results. The major reason for these failures may be attributed to component malplacement [1–4]. This outline reviews one group's attempt to improve component placement. Our ultimate aim is to restore the arthritic knee to an ideal alignment, close to normality. Machining and joinery concepts have been applied with up-to-date bioengineering techniques in the design of saw-bench systems applicable in the OR. Of these designs, the Questor Mark II system holds considerable promise, since it has been shown to provide square or angled bone-cuts, press fit of components, precision component placement and much improved alignment for a variety of current knee implant designs.

The Clinical Mechanics Group is finalising manufacture of the Questor Mark III saw jig system. This has all the essential features of the Mark II but incorporates a much simpler means for limb attachment to the frame (less time for assembly) and a versatile femoral alignment system. Clinical trials will begin shortly, and our progress will be reported.

In this day and age of various knee component designs, many with similarities but subtle differences, a means by which one may test the kinematic advantages of one model versus another is not available. The Questor system, which permits an accurate or better placement of any knee placement/bone cutting instrumentation available (but is independent of the particular features embodied in the instrumentation linked to that design) offers a possibility by which each design may be compared in kinematic studies and prospective clinical trials.

We wish to indicate that this system, although a significant improvement on other methods used in our centre, still has deficiencies. Yet the approach developed here, aiming to replicate machining and engineering principles, offers a method by which placement of components may approach the precision employed in their manufacture. It is hoped that this outline and discussion will encourage the development of other, even more accurate knee placement systems.

Acknowledgements. We wish to acknowledge the support and help of the operating room staff at Kingston General Hospital, especially Mrs. Rose Landon and Ms. Judy Cambell, and the Orthopaedic Resident Staff at Queen's during the development of the saw jig systems, the support of all members of the Clinical Mechanics Group and the diligent typing skills of Ms. Francine Pelletier, Mrs. Clare Grant and Lee Watkins in the preparation of this manuscript.

References

1. Lotke PA, Ecker ML (1977) Influence of positioning of prosthesis in total knee replacement. J Bone Joint Surg [Am] 59:77–79
2. Hvid I, Nielsen S (1984) Total condylar knee arthroplasty. Prosthetic component positioning and radiolucent lines. Acta Orthop Scand 55:160–165

3. Kagan A II (1977) Mechanical causes of loosening in knee joint replacement. J Biomech 10:387–391
4. Bargren JH, Blaka JD, Freeman MAR (1986) Alignment in total knee arthroplasty – correlated biomechanical and clinical observations. Clin Orthop 173:178–183
5. Bargren JH, Freeman MAR, Swanson SAV, Todd RC (1976) ICLH (Freeman/Swanson) arthroplasty in the treatment of arthritis knee, a two to four-year review. Clin Orthop 120:65
6. Insall J et al. (1979) The total condylar knee prosthesis: the first 5 years. Clin Orthop 68:77
7. Townley CO (1985) The anatomic total knee resurfacing arthroplasty. Clin Orthop 192:82–96
8. Cloutier JM (1983) Results of total knee arthroplasty with a nonconstrained prosthesis. J Bone Joint Surg [Am] 65(7):906–919
9. Hungerford DS, Krackow KA, Kenna RV (1980) The Howmedica universal total knee instrument system. Howmedica, Rutherford, NJ
10. Hungerford DS, Krackow KA, Kenna RV (1982) The porous coated anatomic total knee. Orthop Clin North Am 13(1):103
11. Hungerford DS, Krackow KA (1985) Total joint arthroplasty of the knee. Clin Orthop 192:23,33
12. Greenberg RL, Kenna RV, Hungerford DS, Krackow KA (1984) Instrumentation for total knee arthroplasty. In: Hungerford DS, Krackow KA, Kenna RV (eds) Total knee arthroplasty: a comprehensive approach. Williams and Wilkins, Baltimore, pp 36–40
13. Krackow KA (1983) Approaches to planning lower extremity alignment for total knee arthroplasty and osteotomy about the knee. Adv Orthop Surg 7:69
14. Wevers HW, Espin E, Cooke TDV (1987) Orthopaedic sawblades. A case study. J Arthroplasty 2:43–46
15. Cameron HU, Pileiar RM, Macnab I (1976) The rate of bone ingrowth into porous metal. J Biomed Mater Res 10:195–302
16. Bobyn JD, Pilliar RM, Cameron HU, Weatherley GC (1981) Osteogenic phenomena across endosteal bone-implant spaces with porous surfaced intramedullary implants. Acta Orthop Scand 52:145–153
17. Cooke TDV, Saunders G, Siu D, Yoshioka Y, Wevers H (1985) Universal bone cutting device for precision knee replacement arthroplasty and osteotomy. J Biomed Eng 7:45–50
18. Yoshioka Y, Siu DW, Cooke TDV (1987) The anatomy and functional axes of the femur. J Bone Joint Surg [Am] 69:873–880

Incidence and Form of Knee Deformity Associated Osteoarthritis

T. D. V. Cooke, D. Siu, B. Fisher, and J. Rudan

The Clinical Mechanics Group, Division of Orthopaedics,
Departments of Surgery and Mechanical Engineering,
Queen's University at Kingston, Ontario, K7L 3N6, Canada

Introduction

The association between abnormalities of joint surfaces and osteoarthritis is well known; for instance in the hip, dysplastic configuration of the joint is associated with a well-defined incidence of superolateral osteoarthritis [1, 2]. Likewise, in the knee, Blount's disease in children results in a significant incidence of medial compartment osteoarthritis among adults. Nonetheless, a large number of varus or valgus knee alignments do not result in osteoarthritis. While the correlation between osteoarthritis and excessive mechanical stress is very well-defined [3], it is not absolute. Other factors, including systemic disposition towards polyarthritis [4], inflammatory or immune changes [5], and crystal deposition disease [6] may all lead to an increased risk or act as accelerating factors promoting OA. In fact, it has been suggested that most cases of osteoarthritis represent the interaction of at least two or more such factors in their genesis [7].

The observation of progressive medial compartment osteoarthritis in patients with an inward sloping joint in the coronal plane (excessive distal femoral valgus and tibia vara) was first reported in 1984 [8]. These cases, called knee dysplasia, possessed a medialized, load-bearing axis (LBA) and marked lateral tibial subluxation. Although initially thought to be relative uncommon, the incidence of this problem in the population of cases coming for arthritic assessment was actually unknown. Clinical observations in cases of dysplasia suggested that the bony configuration of the tibia and femur might play a major role in subjecting the knee to excessive medial compressive stress and shear forces [9]. This study was conducted to define carefully the incidence of dysplastic knees amongst the random consecutive population of knee arthritic cases referred to our orthopaedic clinics. The preliminary data, shown here, indicate a much higher incidence of dysplastic configurations of the knee than hitherto suspected [9]. Further, these data raise questions on the role of dysplastic malformations as one of the mechanical risk factors involved in the genesis of osteoarthritis.

Material and Methods

Patient Population

The results are derived from analysis of preoperative radiographs for 220 of 256 consecutive osteoarthritic knee cases which were referred to three orthopaedic surgeons at Kingston General Hospital between 1981 and 1985 (Fig. 1) (36 cases

Surgery and Arthroscopy of the Knee
2nd Congress of the European Society
Editors: Müller/Hackenbruch

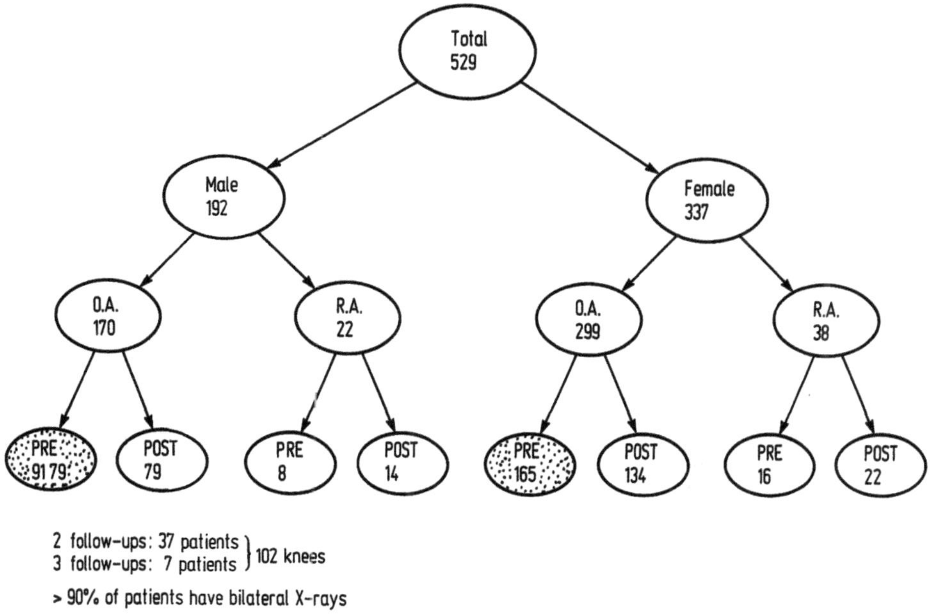

Fig. 1. Details of outpatient referrals for knee arthritis: O.A, osteoarthritis; R.A., rheumatoid arthritis; PRE, preoperative; POST, postoperative

have no standardized radiographs). The group comprised 82 males (37%) and the age range for the cases presenting osteoarthritis was from 29 to 88 years old. All cases were subject to a standardized radiograph assessment of both knees. In 18% of the cases, knee surgery in the form of an osteotomy or a replacement was performed. This surgical population was then entered into a prospective study for future data analysis.

Questor Precision Radiographs

A technique has been devised by the Clinical Mechanics Group (CMG) of Queen's University to assess knee alignment accurately while standing relative to the hip using orthogonal radiographs in the anteroposterior (AP) and lateral mode [10, 11]. The patient stands in a frame on a rotating platform, with the legs spread in an even stance and the ankles set against markers on the platform (Fig. 2). The patient's hips are aligned in the frame by adjustable pads; both knees are aligned forwards with the flexion-extension plane directly anterior; tibial rotation is referenced by the position of the second toe on radians marked on the rotating platform. AP radiographs of both knees and hips are obtained using cassettes held in the frame behind the patient. These radiographs are carefully assessed for alignment errors and retaken if the limbs are not positioned correctly. The patient is rotated through precisely 90°, to the left and to the right, without any change in knee position, and a lateral orthogonal view of each knee is obtained with the cassette held in a magnetic holder between the knees. Two plexi-

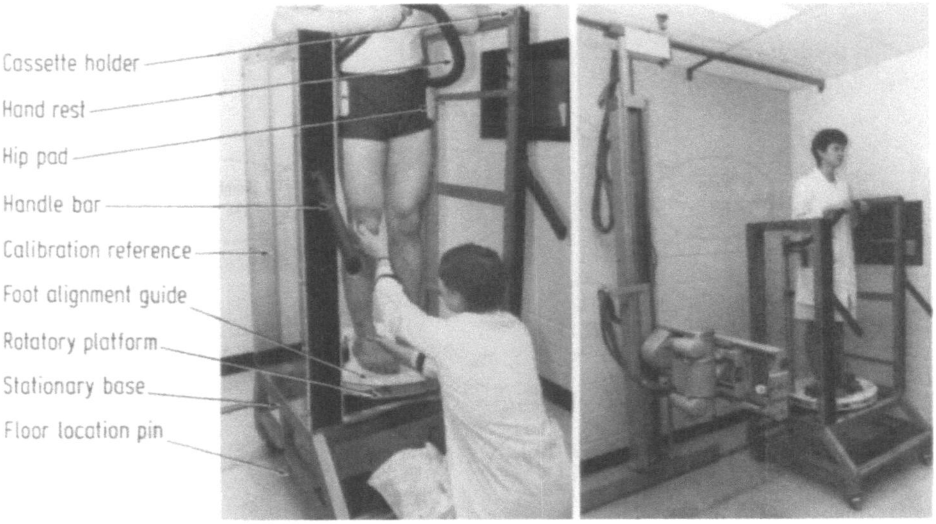

Fig. 2. Questor precision radiographs: the frame setup and patient stance

glass screens in front of the frame contain marker dots and wires whose images, along with those of the patient's knees and hips, are also transferred to the radiographic plates.

Analysis of Radiographs. Landmarks of the femur, proximal tibia and hip are identified on each film. Their relative positions, plus those of radiographic markers, are digitized using a digitizing tablet (Science Assessories Corporation).

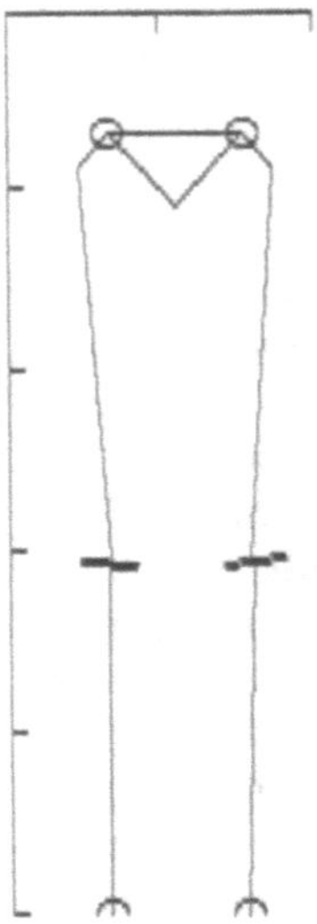

		KNEES RIGHT	KNEES LEFT
ANGLE	Capito Midcondylar-Tibial Shaft	-0.3	2.2
	Femoral Shaft-Tibial Shaft	-5.2	-2.6
	Capito Midcondylar-Capito Midmalleolar	-0.4	0.6
	Capito Midcondylar-Femoral Shaft	4.9	4.9
	Femoral Neck-Femoral Shaft	133.8	133.0
	Femoral Shaft-Transcondylar	9.1	10.4
	Capito Midcondylar-Transcondylar	4.1	5.6
	Tibial Plateau-Tibial Shaft	-3.6	-6.9
	Femoral Shaft-Tibial Shaft Flexion	7.9	9.8
	Foot Rotation	102.0	100.0
DISTANCE	Femoral Length	472.7	469.0
	Tibial Length	385.2	386.2
	Standing Leg Length	860.8	859.0
	Transcondylar Width	75.2	75.9
	Lateral Condylar A/P Depth	72.6	70.2
	Tibial Plateau Width	74.2	78.6
	Medial Tibial Plateau A/P Depth	42.6	45.7
	Lateral Joint Space	4.3	7.9
	Medial Joint Space	5.1	6.7
	Coronal Subluxation	3.3	6.6

Fig. 3. Sample of a computerised output of angular and linear dimensional parameters together with a diagrammatic outline (stick diagram) of the lower limbs

The recorded data are stored in a personal computer (Zenith Data Systems Corporation) and can be displayed as a series of numerical values expressing the dimensional and angular data of each knee together with a diagrammatic representation (stick diagram) of the hip, knee, and ankle alignment (Fig. 3). In the laboratory this technique has proven to give accurate measurements of alignments down to less than 1 mm and 1° [10]. In the clinic, however, standard errors of the mean up to 2 mm in geometric assessment and 2° in alignment have been observed using this technique.

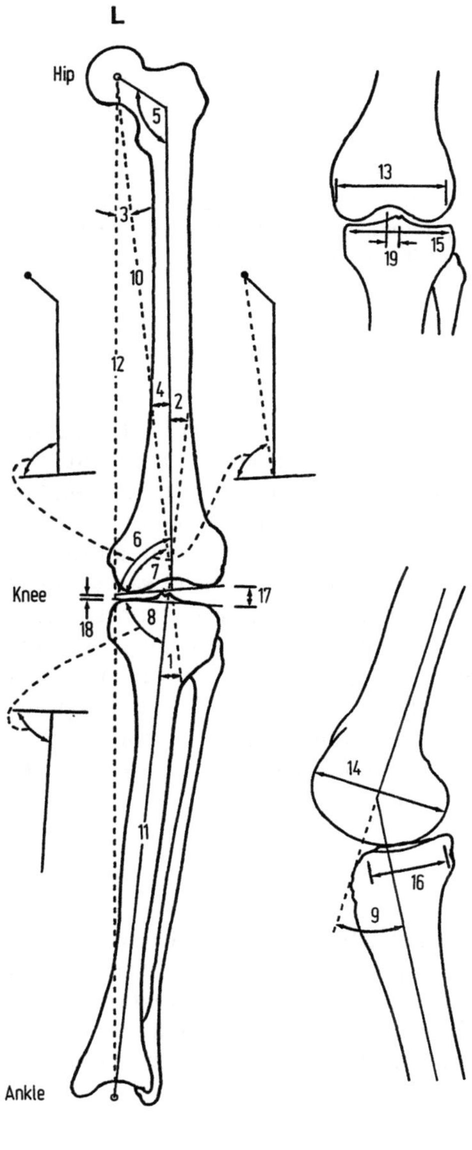

Fig. 4. Definition of the Questor nomenclature for standardised knee radiographs of articular geometric relationships

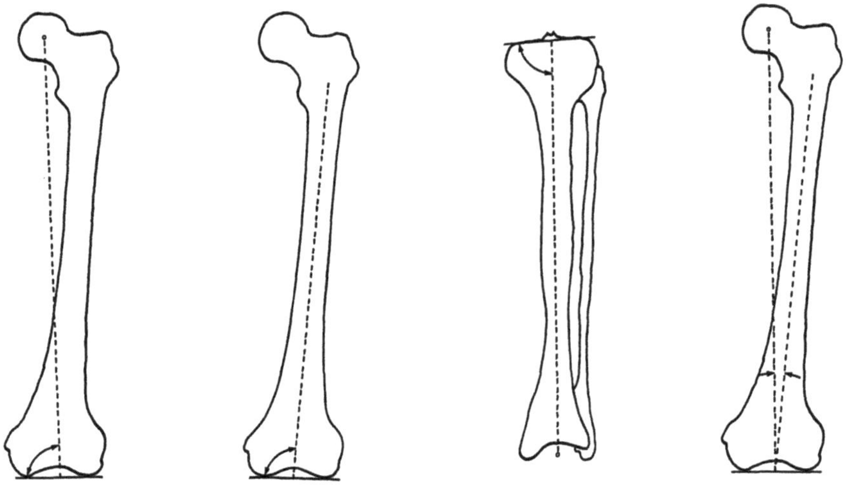

Fig. 5 a–d. Details of the angular orientation made by the tangents of the knee articular surfaces to the hip, femoral shaft and ankle. **a** Capitomidcondylar-transcondylar angle, **b** femoral shaft transcondylar angle, **c** tibial plateau tibial shaft angle, **d** capitomidcondylar-femoral shaft angle

Data Analyis and Nomenclature

The angular orientation between the bony parts of the knee relative to one another varies according to the description of each clinical group. Therefore, the CMG have developed a standardized nomenclature to describe these essential features of the knee (Fig. 4) [11]. Of special note are the angular orientations made by the tangents of the knee's articular surface to the hip, femoral shaft, and ankle. The capitomidcondylar-transcondylar angle represents the angle of varus (−) or valgus (+) when the tangent of the knee's articular surface is more than at a right angle to the tangent to the hip. The femoral shaft-transcondylar angle presents a similar relationship between the condylar tangents to the femoral shaft (this angle is commonly termed distal femoral valgus). The tibilar plateau margin-tibial shaft angle represents the degree of varus (−) or valgus (+) greater than a right angle to the ankle (Fig. 5 A–C). Statistical analysis of the data was conducted using a software program known as SYSTAT.

Results

The frequency of cases presenting rheumatoid arthritis (RA) and osteoarthritis OA is shown in Fig. 1. As can be noted, the preponderance of the arthritic patients were female with a ratio of about 2:1 females to males. The demographic data for osteoarthritic patients are given in Table 1. The age of the patients ranged from 29 to 88 years old, the mean ages of males and females were 64 and 68 (not significantly different), respectively.

Table 1. Demographic data of the osteoarthritic patients

Sex	Knees	Age (years)		Incidence of bilaterality	
		Range	Mean	No. of patients	Percentage
Male	82	41–83	64	23	21
Female	138	29–88	68	35	32

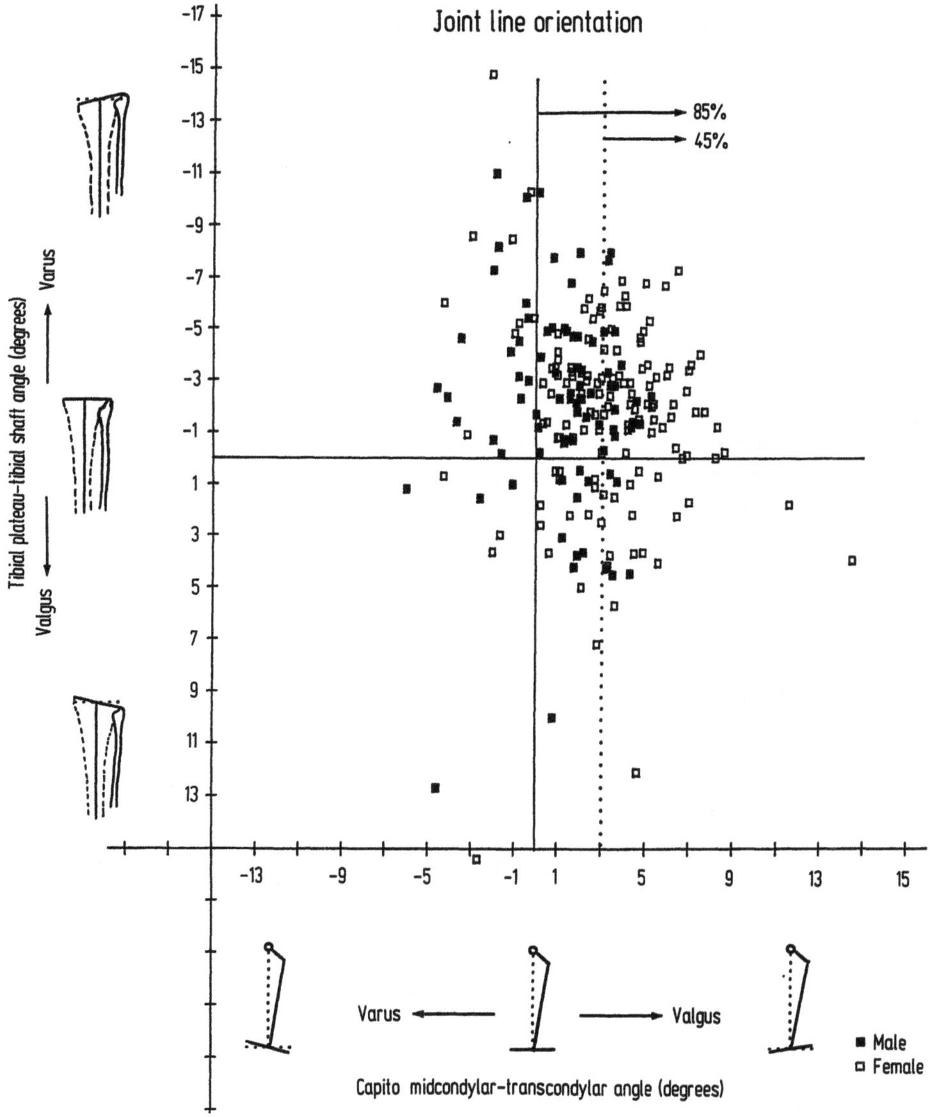

Fig. 6. Graphic representation of knees indicating osteoarthritis: a plot of tibial plateau-tibial shaft angle versus capitomidcondylar-transcondylar angle (varus −, valgus +)

Incidence of Coronal Misalignment and the Associated Geometry of Femur and Tibia in Osteoarthritis

The overall pattern in OA showed deviation of the LBA medially in two-thirds (varus knees) of the cases, and laterally in one-third (valgus knees). Of the total, 21% were within 2° of neutral alignment, i.e. showing no significant deformity. The data for all the knees are collated in Fig. 6; males are represented by single solid squares and females as open squares. The variation in varus displacement (−) or valgus displacement (+), from the right angle, of the tibial surface to the ankle (zero) is plotted on the vertical axis. The variation of the transcondylar tangent beyond the right angle to the hip (− varus, + valgus) is presented on the horizontal axis. The zero intercepts represent a limb in which the tibial margins are at a right angle to the ankle with the femoral surfaces at a right angle to the hip.

Femoral Alignment. There was a vast preponderance of cases with tibia vara and a valgus capitomidcondylar-transcondylar angle (a valgus orientation of the distal femur to the hip). The majority of cases (85%) had a valgus femoral orientation to the hip, the maximal deviation being +14°. The incidence (45%) of dysplastic femurs (with capitomidcondylar-transcondylar valgus of more than 3°) was surprisingly high. It was noted that excessive valgus femur dysplasia was associated with tibia vara in 34% of cases. The incidence of valgus dysplasia was 12%.

Tibial Alignment. A range of deformity from 15° varus to 13° valgus was seen in the tibial geometry. The mean was 2° varus, correlating with the overall medialisation of knee alignment in the majority. However, where a neutral tibia was associated with excessive valgus of the femur relative to the femoral head, an overall valgus deformity resulted. These 47 cases exemplify the "valgus form" of knee dysplasia. In 22 of these cases the lateral compartment rather than the medial compartment showed predominant arthritis.

Relationship Between Differences in Joint Space (Medial-Lateral Compartment) to Load-Bearing Axis

The differences in joint space in the medial or lateral compartment, as measured from the lateral projection of the standard radiograph, have been plotted on the vertical axis in Fig. 7. The knee alignment is shown on the horizontal axis. The data show a broad correlation between relatively greater medial or lateral joint space loss and varus or valgus knee alignment, respectively. However, an unexpected incidence of mismatched data was found where lateral joint space loss occurred in varus knees and medial joint space loss in valgus knees. In the seven varus mismatched cases, an unexpected incidence of femoral dysplasia of more than 3° was found in six knees. Likewise, in the 47 valgus knees with medial joint space loss, the incidence of excessive valgus dysplasia of the femur to the hip was also found in 45 cases.

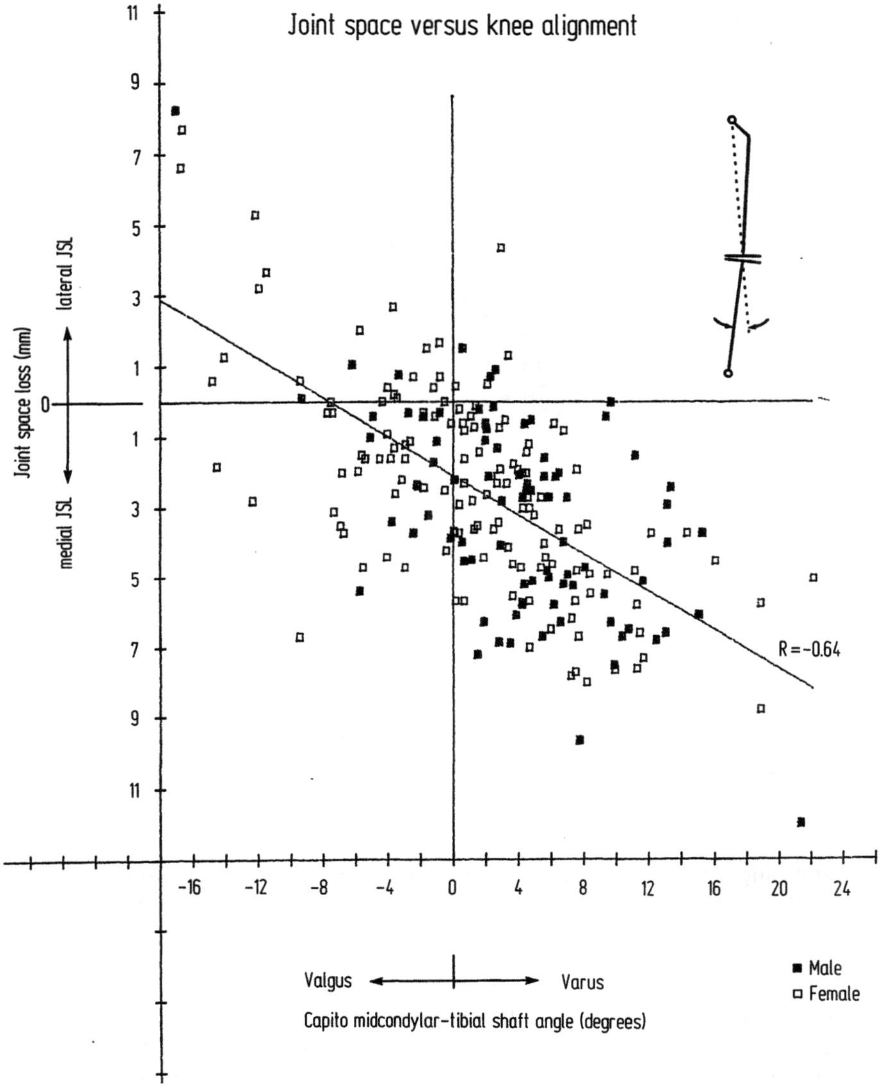

Fig. 7. Relationship between differences in joint space (medial and lateral) and knee alignment (capitomidcondylar-tibial shaft angle): varus, +; valgus, −; *JSL*, joint space loss

Relationship Between Capitomidcondylar-Transcondylar Valgus and Femoral Shaft Transcondylar Valgus

The graphic relationship between the femoral tangent to the hip and the femoral shaft, respectively, are shown in Fig. 8. The excellent correlation obtained indicates that in 92% of cases the angular orientation of the femoral-bearing surface to the hip varied in a direct and proportional relationship to the shaft. In this arthritic population, the capitomidcondylar-transcondylar valgus mean was 2.5°.

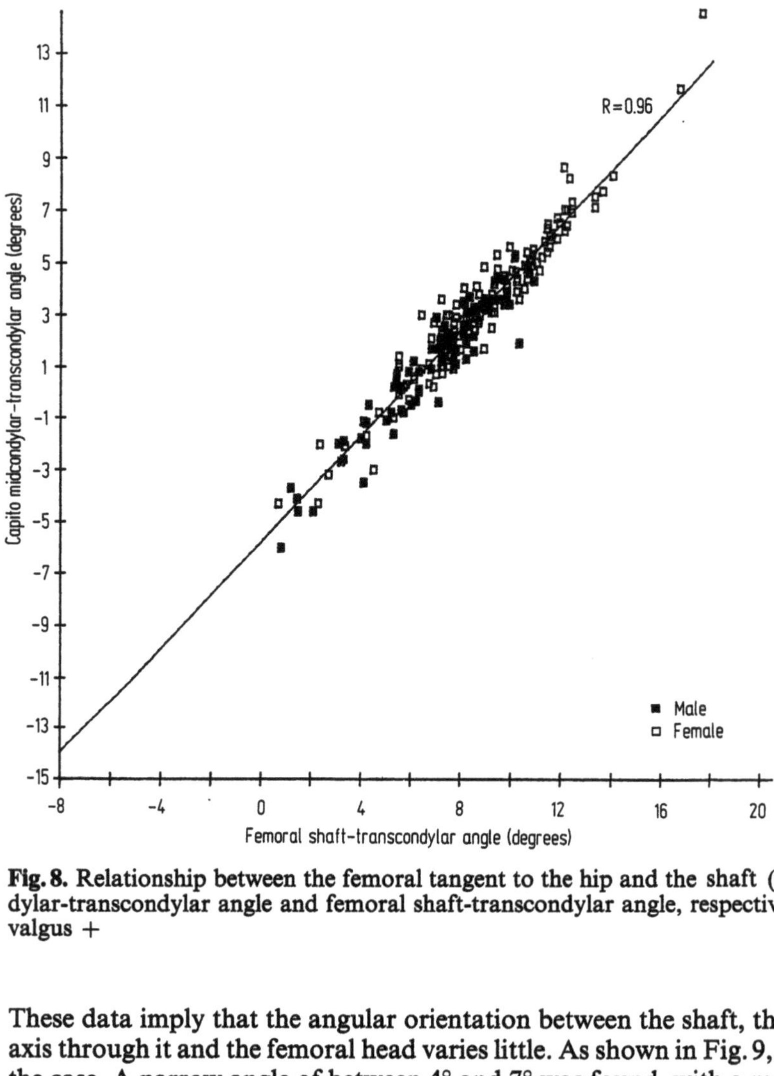

Fig. 8. Relationship between the femoral tangent to the hip and the shaft (capitomidcondylar-transcondylar angle and femoral shaft-transcondylar angle, respectively): varus −, valgus +

These data imply that the angular orientation between the shaft, the mechanical axis through it and the femoral head varies little. As shown in Fig. 9, this is indeed the case. A narrow angle of between 4° and 7° was found, with a mean of 5.6°.

Discussion

This study outlines the preliminary analysis of osteoarthritic knees presented at a Canadian teaching hospital. The cases were not selected and were all assessed with standardised radiographs using the QPR technique. While there may be geographical selection factors influencing the referral patterns, nothing especially obvious is known in this regard. Confirmation of this will require an epidemiological study of arthritic cases in this area of Ontario and a comparison to similar data obtained at other centres. However, in general there is agreement between the proportion of osteoarthritic versus rheumatoid patients referred to Kingston

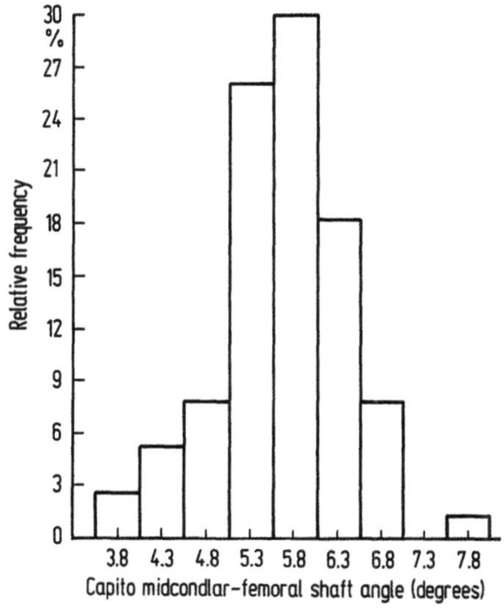

Fig. 9. Histogram of the capitomidcondylar-femoral shaft angle

General Hospital as compared with other hospitals in Canada and the United Kingdom. The male/female ratio represented here correlates with the distribution observed by others [12].

One purpose of this study was to evaluate the incidence of dysplasia of the knee associated with osteoarthritis. The findings revealed an unexpectedly high incidence of dysplasia which, in retrospect, clearly relates to the precision of the radiographic technique. Without a way to define and measure the femoral condylar tangent orientation to the hip joint, the excessively valgus dysplastic femur can easily be missed. The generally accepted range of angular orientation of the distal femur to the hip varies between 5° and 10° [13]. As noted, 45% of these cases were in excess of this amount (Fig. 8), with a valgus orientation of greater than 3° to the hip. In this analysis of OA knees, and in our studies of femoral anatomy in non-arthritic cadavers [14], as well as preliminary studies of knee alignment in normal control cases (Chow, Siu, Cooke; unpublished observation), no evidence was found for any relationship between distal femoral orientation and the hip. Rather, the case of the distal femur valgus to the hip carries an extremely high correlation to the valgus orientation to the shaft (Fig. 8). This means that in the general population, the variation of shaft to knee to hip alignment is small. In this osteoarthritic population a range of 4°–7°, with a mean of 5.6°, was found.

In the clinical assessment of dysplasia in OA knees, we initially screened patients with a distal femoral valgus of greater than 10° [9]. However, in the context of the data shown here, any angular orientation over 7° may be considered abnormal: The varus dysplastic knee, with a capitomidcondylar-transcondylar valgus of more than 3°, was found in 45% of patients examined. A medialised LBA, causing varus deformity, was the result of a combination of tibial vara and medial joint space collapse. Among one-third of the total cases represented by valgus OA knees, a dysplastic femur was associated with arthritis in 63%. Here, the neutral tibial orientation allowed for an overall lateralisation of the LBA. In the majority of these cases, lateral joint compartment OA predominated over medial.

Fig. 10. Dysplastic angular aberration of distal femoral surfaces

The implications of the dysplastic knee are not dealt with in detail in this study. However, the analysis for specific cases suggests that these deformed knees carry a high risk of lateral tibial and patella subluxation and rapidly progressive arthritis [9]. The analysis further suggests that dysplastic angular aberration of distal femoral and proximal tibial surfaces is often in more than one plane, i.e. rotation about two axes (Fig. 10). An intriguing feature illustrated in a knee showing varus dysplasia is that the rotational aberrations tend to vary in opposite directions for the distal femur and proximal tibia, respectively. Thus, as shown in Fig. 11, the distal femur tends to be twisted inwards about its long axis and outwards in the coronal plane. Complementing this is an outward twist of the upper tibia about

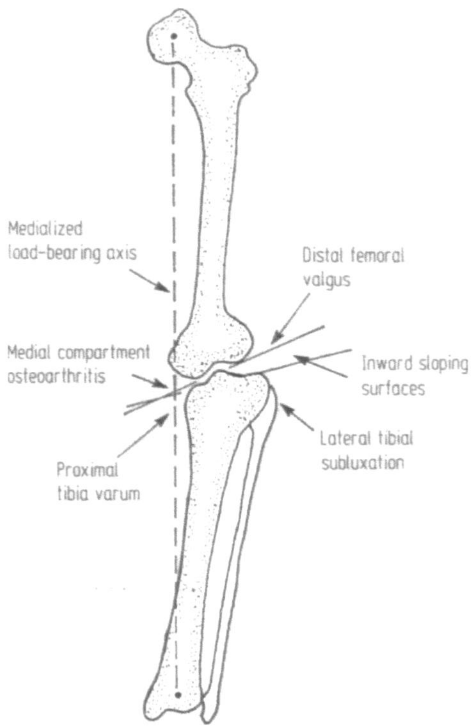

Fig. 11. Varus form of knee dysplasia with condylar asymmetry

its long axis and the varus alignment of surfaces to the ankle. The net result is an inward sloping joint line with proximal distal axial rotations in the opposite plane. When considering surgery of the knee, these data have important implications. Cases suitable for osteotomy, with mainly single joint compartment changes, must be carefully assessed for the extent of femoral dysplasia. In the varus knee, where the femoral condylar tangent is within a few degrees of the right angle to the hip, an upper tibial osteotomy is highly logical. However, in the varus dysplastic knee, medial compartment OA is inadequately treated by upper tibial osteotomy alone. In this case a varus distal femoral osteotomy is performed to correct excessive femoral valgus and then a valgus osteotomy of the upper tibia to bring the LBA lateral to the knee. The aim, in these cases, is to bring the inwardly slanting knee joint closer to a horizontal (right angle) relationship to the LBA. In this context, osteotomy for lateral compartment osteoarthritis in the valgus dysplastic knee requires a varus osteotomy of the femur alone since the tibia is usually in the neutral position [15, 16].

For total knee replacement, arrangement of the components must *not* be carried out with reference to the excessively valgus and inwardly rotated medial condyles. The transverse axis of the epicondyles should be used to define neutral rotation of the femur, placement of the femoral component is made with reference to this. Lack of recognition of the dysplastic configuration, in the varus or valgus knee, easily leads to misplaced components (usually excess valgus and internal rotation) with a continuing tendency for lateral tibial subluxation and lateral patellar dislocation.

These data also provide information on ideal knee alignments, a field which is incompletely defined. Preliminary data collected from 50 young adults without arthritis show some agreement with the results of Hungerford [17] and Krackow [18]. These volunteers usually had a neutral LBA with a slight inward slant of the joint line (2°–3°), but wide variations in angular geometry were observed. In considering the analysis of the dysplastic non-arthritic knee cases and the overall very high incidence of dysplasia in the arthritic population shown here, it is suggested that excessive obliquity of the joint line may predispose patients to arthritis. Any inward obliquity of the arthritic knee surfaces must increase the tendency for instability (shear forces) in the coronal plane. We believe, in fact, that obliquity is a potent agent for tibial subluxation in the varus knee. Thus, our recommendation, along with others, is for component placement at a right angle to the LBA in total knee replacement [13, 19, 20].

This study leaves many questions unanswered. The relevance of discrepancies in data between medial joint space loss and valgus knee alignment and lateral joint space loss and varus alignment, as well as characterisation of rotary abnormalities of the femur and tibia, await further study.

The recommendations of this preliminary account are that a standardised method for assessing alignment of knee components in a stance is of great value, both to allow accurate identification of the dysplastic configuration and to make a better assessment of the variations in geometry of the femur and tibia. The data suggest that knee malformations may indeed be a factor contributing to the development of osteoarthritis and that the dysplastic excessively valgus distal femur is commonly implicated.

Summary

A novel observation of progressive medial compartment osteoarthritis (OA) was reported in patients with an inwardly sloping knee joint (distal femoral valgus, >10° and tibia vara <90°), i.e. knee dysplasia. These cases had a medialised load-bearing axis (LBA) and significant lateral tibial subluxation. In this report, the incidence of dysplasia and other associated forms of knee misalignment were assessed in 220 consecutive arthritic knees by standardised radiographic techniques. In addition to controlled positioning, including an anteroposterior view of the hip, a 90° lateral view of each knee is taken without changing the knee position. Such radiographs are obtained by using a rotating platform on which the patient stands. This precise technique provides data on the alignment of the geometry of the knee articular surface relative to the hip (capitocondylar alignment). The development of a standardised format for defining angular and linear dimensions has resulted in the Questor nomenclature and stick diagram.

Preliminary analysis revealed an overall incidence of 67% varus and 33% valgus deformities. In 25%, unexpected medial or lateral joint space loss was found: 7 varus knees had lateral joint space loss, all had excessive distal femoral valgus dysplasia, and of the 47 valgus knees with medial joint space loss, 26 had dysplasia. Ideally, alignment of the knee surfaces in a neutral LBA (line from hip to ankle passing through the knee centre) might be considered at 90°, but verification of this point in the literature is limited. Analysis of all the knees revealed valgus capitocondylar alignment (more than 90° relative to the hip) in 85% of the cases; in 45% it was greater than 3°. Only 33 knees (15%) had a varus capitocondylar alignment. An excessive valgus orientation of the dysplastic femur predisposes the knee to instability. These data indicate that bony aberrations of the knee (dysplasia), which have hitherto been poorly defined, are commonly associated with arthritis. Such a high incidence of dysplasia suggests their implication in the induction of OA.

Acknowledgments. The authors wish to acknowledge the support and help of other members of the CMG, the radiology staff at Kingston General Hospital, especially Mrs. Linda McKinven, and the diligent typing skills of Lee Watkins.

References

1. Lloyd-Roberts GC (1955) Osteoarthritis: a study of the clinical pathology. J Bone Joint Surg [Br] 37:8
2. Hely DP, Salvati EA, Pellicci PM (1984) The hip. In: Cruess RL, Rennie WRJ (eds) Adult orthopaedics, vol 2. Churchill Livingstone, New York, pp 1209–1274
3. Radin EL (1976) Mechanical aspects of osteoarthritis. Bull Rheum Dis 26:862–865
4. Kellgren JH, Lawrence JS, Bier F (1963) Genetic factors in generalized osteo-arthrosis. Ann Rheum Dis 22:237–255
5. Cooke TDV (1986) Immune pathology in polyarticular osteoarthritis. Clin Orthop 213:41–49
6. Dieppe PA, Doyle DV, Huskisson EC, Willoughby OA, Crocker PR (1978) Mixed crystal deposition disease and osteoarthritis. Br Med J 1:150
7. Cooke TDV (1985) Pathogenetic mechanisms in polyarticular osteoarthritis. In: Solokoff L (ed) Clinics in rheumatic diseases, vol 11(2). Saunders, Eastbourne, pp 203–238

8. Cooke TDV, Pichora D (1985) Knee dysplasia: an unusual but important problem associated with progressive arthritis. J Bone Joint Surg [Br] 67(2):332
9. Cooke TDV, Pichora D (submitted) Abnormalities of articular surface alignment and their relationship to arthritis of the knee joint. I. The varus forms of knee dysplasia. J Bone Joint Surg
10. Wevers HW, Siu DW, Cooke TDV (1982) A quantitative method of assessing malalignment and joint space loss of the human knee. J Biomed Eng 4:319–324
11. Cooke TDV, Siu DW, Wevers HW, Sorbie C, Bryant JT, Wyss U, Fisher B, Saunders G, Chow D (submitted) Standardized radiographic technique and proposed terminology in the analysis of the arthritic knee – QUESTOR precision radiographics and nomenclature
12. Johnson F, Leith S, Waugh S (1980) Distribution of load across the knee. J Bone Joint Surg [Br] 62:346–349
13. Hood RW, Insall JN (1984) The knee. In: Cruess RL, Rennie WRJ (eds) Adult orthopaedics, vol 2. Churchill Livingstone, New York, pp 1275–1406
14. Yoshioka Y, Siu DW, Cooke TDV (1987) The anatomy and mechanical axes of the femur. J Bone Joint Surg [Am] 69:873–880
15. Shoji H, Insall J (1973) High tibial osteotomy for osteoarthritis of the knee with valgus deformity. J Bone Joint Surg [Am] 55:963–973
16. Coventry MB (1973) Osteotomy about the knee for degenerative and rheumatoid arthritis. J Bone Joint Surg [Am] 55:23–48
17. Hungerford DS, Krackow KA (1985) Total joint arthroplasty of the knee. Clin Orthop 192
18. Krackow KA (1983) Approaches to planning lower extremity alignment for total knee arthroplasty and osteotomy about the knee. Adv Orthop Surg 7:69–88
19. Laskin RS (1985) Alignment in total knee replacement. Clin Orthop 192
20. Clayton M, Thompson TR, Mack RP (1986) Correction of alignment deformities during total knee arthroplasties: staged soft-tissue release. Clin Orthop 202:117–124

Arthroplasty of the Knee Joint with the PCA (Porous Coated Anatomic) Endoprosthesis

H. Jenny

Orthopaedics Division, Department of Surgery, University of Basle, Switzerland

Introduction

The objective of knee joint arthroplasty is to eliminate pain and to maintain or restore knee stability and the best possible joint mobility. Until recently, it has often proved difficult to attain these goals, above all to retain them over a long period of time. Experiences reported the last 10 years describing a large number of knee joint arthroplasties [2, 3, 6–8, 10–12, 14, 21, 22, 26, 29, 41, 42, 50, 51, 56, 57, 59, 60, 61, 63, 68, 72] indicate that the complicated structure of the knee joint requires the application of appropriately complex surgical techniques with a high possibility of failure [15, 20, 32, 67, 73, 74, 77, 79, 82, 84]. In particular, as in the hip joint, it has been shown that loosening of the implant constitutes a major problem in knee joint arthroplasty. Loosening of the implant is a very frequent feature occurring in long-term follow-ups, especially for hinge prostheses with long anchoring stems [4, 13, 25, 28, 30, 31, 48, 49, 64, 69, 75]. Knee joint arthroplasty has clearly advanced sufficiently to allow more physiologically related movement through the preservation of the natural stabilising elements, i.e. muscles, tendons and ligaments, especially the collateral ligaments and, independent of the corresponding prosthesis model, the cruciate ligaments, in particular the posterior cruciate ligament (PCL). It has also been possible to preserve or largely restore normal kinematics in the knee joint [27, 38, 44, 46, 66, 76, 81], which also provide good preconditions for long-term success of the knee reconstruction [39, 40, 78]. However, to preserve or restore normal kinematics the tensioning of the ligaments should be coordinated, and as little bone as possible should be resected; thus, knee joint arthroplasty has become ever more elaborate in terms of surgical techniques and requires increasingly sophisticated instruments [24, 43].

Furthermore, today an artificial joint should be so constructed that under given circumstances, i.e. above all in young individuals or patients with a correspondingly good bone quality, it can be anchored to the bone without methylmethacrylate [5, 9, 16–19, 23, 33–35, 52, 54, 65, 70, 71, 83, 85]. The porous coated anatomic (PCA) total knee system was developed with these aims in mind. A special PCA prosthesis model was also produced for revision arthroplasty [36]. A unicompartmental prosthesis based on the PCA principle has now also been available for a short time. It would exceed the scope of this paper to discuss the kinematics of the knee joint, which are very important in understanding the design both of the prosthesis and that of the instruments. However, the importance of correct physiological positioning of the axes cannot be emphasised sufficiently [58, 62, 80].

Surgery and Arthroscopy of the Knee
2nd Congress of the European Society
Editors: Müller/Hackenbruch

The axis of alignment is also crucial during load bearing, since that is the time at which the knee joint is under the greatest strain. While in a one-leg stance, the plane of the knee joint should be horizontal to the standing surface.

Material and Methods

The technique of operation and the instruments are logically constructed but complicated and sophisticated for the unicompartmental prosthesis, the primary PCA knee joint total prosthesis as well as the revision prosthesis [see 24, 36, 53 for further details]. In principle, a painstakingly exact incorporation technique is indispensable, irrespective of whether the prosthesis fixation is carried out with or without bone cement, even with optimally designed prostheses.
For PCA prosthesis models, all surface components which are in contact with the bone are coated with a porous coating, i.e. a 1.5-mm thick double layer of sintered-on chromium-cobalt beads, with a mean pore size of 425 μm. Autologous tissue such as bone or healthy connective tissue can grow into this sintered-on layer; the coat also provides a very firm bonding with the application of bone cement.
According to the manufacturer's specification, the porous coating has a very high adhesivity. However, recently there has been an increasing number of reports on beads becoming detached from the coat, in particular in cement-free fixation. We have also observed this phenomenon occasionally. Although most patients experience absolute freedom from symptoms in subjective terms, when individual beads become detatched or even migrate the greatest attention must be paid in order to recognise possible future damage in time and correct the problem appropriately.
Our experience with the PCA knee joint endoprosthesis is based on 300 primary arthroplasties carried out between 1981 and 1986 at the Orthopaedics Division

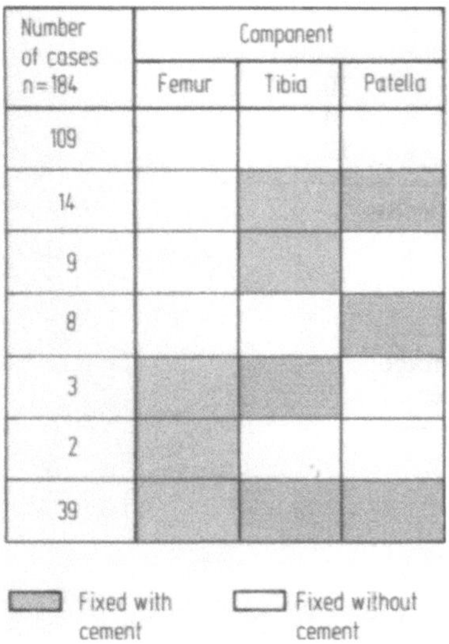

Fig. 1. Comparison of the porous coated anatomic prosthesis fixed with or without cement ($n = 184$)

in Basle. In addition, 40 unicompartmental PCA knee joint prostheses have been implanted since November 1985, and around 20 operations exchanging old prostheses for the PCA revision system performed. Clinical and radiological follow-up studies were carried out for 184 primary PCA knee joint arthroplasties implanted between October 1981 and October 1984. Of these 184 prostheses, 109 (60%) were fixed without the use of cement. In 39 cases (21%), the entire prosthesis was anchored with cement, and in 36 knee joints (19%), only individual components were fixed with cement (Fig. 1). The average age of patients was 74 years (minimum 22 years, maximum 87 years). This paper only discusses the short-term data; nothing can be stated on the effect of prosthesis fixation with or without cement on the long-term results.

In 106 cases joint prostheses were implanted in one knee only; for 29 patients both knee joints were replaced in one operation, and in a further 10 patients both knees were operated on in two stages (Table 1). The nature and number of the prior operations are shown in Table 2, and the indications for arthroplasty are given (Table 3). In the cases of idiopathic gonarthroses, there was a varus deformity in

Table 1. Knee arthroplasties employing porous coated anatomic prostheses ($n = 184$)

Cases	Replacement		Percentage of cases
106	Unilateral	60 right	32.6
		46 left	25.0
29	Bilateral	1 stage (operation)	31.5
10	Bilateral	2 stage (operation)	10.9

Table 2. The number and type of previous interventions ($n = 74$)

Number of operations	Intervention	Percentage
37	Osteotomy of the tibia	50
3	Supracondylar osteotomy of the femur	4
9	Meniscectomy	12

In total 74 previous operations were conducted on 53 knees.

Table 3. Indications for arthroplasty ($n = 184$)

Indication	Number of cases	Percentage of cases
Osteoarthritis	133	72.3
Rheumatoid arthritis	23	12.5
Post-traumatic osteoarthritis	21	11.4
Post-infectious osteoarthritis	4	2.2
Avascular necrosis (femur)	1	0.5
Others	2	1.1

90 cases (68%), a valgus deformity in 39 cases (29%) and a zero axis in only 4 cases (3%). Of the total 184 knee joints treated, 11 prostheses were implanted in 7 patients who later died. However, none of their deaths were connected with the surgical operation.

All operations were carried out in a conventional operating theater. Patients received antibiotic prophylaxis, in the form of cephalosporin of the second generation (cefazolin) administered intravenously, preoperatively and for 48 h following treatment. The operation was performed under pneumatic tourniquet. In 96% of the cases, the knee joint was opened via a median access. In 28% of the cases (51 knees), corrections were carried out on the stabilising structures. The medial capsular ligament apparatus and the pes anserinus were severed 44 times, in some cases with displacement and refixation of these structures. In two cases, the iliotibial tract was completely severed; in another two cases the dorsolateral capsule together with the popliteal tendon were cut; and in one case the lateral ligament also had to be divided in order to carry out the corresponding axis correction. It was necessary to lengthen the quadriceps tendon in only one instance.

Autologous corticospongiosa chip grafts were necessary on the femur in 4% of the cases and on the tibia in 10% because of inadequate bone support for the prosthesis components. In all cases, the bone material used for grafts was derived from bone fragments resulting from trimming the bones to shape.

During postoperative rehabilitation it was judged important that the patients left their beds on the day of the operation, or at the latest on the 1st postoperative day. Actively supported movements were commenced in all patients from the 3rd postoperative day. Since the beginning of 1984, all treated knee joints were mobilished 48 h postoperatively, after removal of the redone drainages, on a motor movement rail in the form of continuous passive motion. This method results in a crucial improvement of knee function and a reduction in the number of mobilisations performed under anaesthesia which are otherwise necessary in occasional cases. Walking on two crutches under partial weight bearing (10–15 kg) was allowed as a rule after 3–5 days. Full loading should be applied at the earliest after 6 weeks if the prosthesis has been implanted without using cement. In elderly patients who experience difficulties in taking weight off the leg, which involves use of the hip joints and shoulder joints, full loading is allowed immediately after the operation with a cemented, anchored prosthesis.

Thrombosis prophylaxis employing acenocoumarol (Sintrom) was administered from the 2nd postoperative day in all cases.

Results

The results presented below are not treated separately as regards the aetiology of the gonarthrosis or the indication, although there are appreciable differences with regard to the quality of the bone, the condition of the ligaments and the degree of severity of the deformity in relation to the different aetiologies. This is also shown in the different distribution of cement application depending on the quality of the bone bed (sclerosis, osteoporosis, extensive defects, cysts, etc.). At the last check-up, the postoperative mobility of the treated knee joints was on average

Table 4. Walking ability and use of walking aids after implantation of porous coated anatomic prostheses

	Percentage of cases	
	Pre-operatively	At last follow-up
Ability		
Unable	5	–
Few meters	7	–
Up to 100 m	39	9
100 m–1 km	41	21
More than 1 km	8	70
Aids		
Unable	5	–
With two crutches	35	12
With one crutch	41	36
Without crutches	19	52

95° (±16.5°) for flexion, and there was a deficit averaging only 2° (±4°) for extension.

The ability to walk reported by the patients at the last check-up and the use of walking supports are shown in Table 4. It is noteworthy that nine patients who were initially unable to walk after the operation regained their ability to walk well and that 70% of the patients could walk more than 1 km afterwards. This was possible in only 8% of the patients prior to the operation.

Complications

In three cases (1.5%) infection occurred which was treated as follows. In one instance the prosthesis had to be removed, and an arthrodesis was carried out. In another the function of the joint could be preserved after healing the infection by means of a suction-flushing drainage and revision of the PCA knee joint. In a third patient, the prosthesis had to be removed altogether because of severe, uncontrollable infection. Further complications occurred postoperatively in 25 prostheses (Table 5).

Postoperative complications have markedly decreased with increasing experience and modified surgical techniques. In an early series of 100 patients, 17 incidences of complications occurred; in the present series only 8 of 84 treated knees presented complications. Mobilisation of treated joints under anaesthesia had to be performed on 23 joints (12.5%) because of inadequate mobility expressed, as a rule, by markedly restricted flexion.

The results obtained up to now with the PCA knee joint are superior to those with other knee joint prostheses implanted in comparable numbers. An observation period of 5 years is too short; consequently enthusiastic recommendations may not be made despite good early results. However, other groups have already obtained good follow-ups 10 years postoperatively, with PCA prosthesis models which are similar in terms of design and anchorage [1, 45, 47, 55].

Table 5. Complications associated with porous coated anatomic prostheses

Complication	Number of cases	Percentage of cases
Patella and extensor apparatus	17	9.5
Loosening of tibial component	3	1.5
Loosening of femoral component	1	0.5
Ligament instability	2	1.0
Peroneal nerve palsy (reversible)	1	0.5
Infection (deep)	3	1.5
Heterotopic ossification	1	0.5

On the basis of this study, it is also not possible to make a definitive appraisal yet on the advantage of cementless anchorage of PCA prostheses as compared with the technique using cement. However, it appears that the PCA system provides an important model which leaves open the possibility of cementless incorporation for suitable cases, i.e. young patients and those with a vital bone bed [37].

A very crucial point in favour of the PCA knee joint system is the universal incorporation instrument set. Of further importance is the fact that the PCA knee joint system, including the instrumentation, has been developed both for the unicompartmental prosthesis and for revision prosthetics.

References

1. Aglietti P, Rinonapoli E (1984) Total condylar knee arthroplasty; a five-year follow-up study of 33 knees. Clin Orthop 186:104–111
2. Arden GP (1973) Total knee replacement. Clin Orthop 94:92–103
3. Aubriot JH (1984) Réinterventions pour complications de prothèses totales du genou. Rev Chir Orthop 67:189–194
4. Bain AM (1973) Replacement of the knee joint with the Walldius prosthesis using cement fixation. Clin Orthop 94:65–71
5. Bayne O, Cameron HU (1984) Total knee arthroplasty following patellectomy. Clin Orthop 186:112–114
6. Bryan RS, Peterson LFA (1979) Polycentric total knee arthroplasty; a prognostic assessment. Clin Orthop 145:23–28
7. Bryan RS, Peterson LFA, Combs JJ Jr (1973) Polycentric total knee arthroplasty; a review of 84 patients with more than one year follow-up. Clin Orthop 94:136–139
8. Convery FR, Beber CA (1973) Total knee arthroplasty. Clin Orthop 94:42–49
9. Cook SD, Walsh KA, Haddad RJ Jr (1985) Interface mechanics and bone growth into porous Co-Cr-Mo alloy implants. Clin Orthop 193:271–280
10. Coventry MB (1979) Two-part total knee arthroplasty: evolution and present status. Clin Orthop 145:29–36
11. Coventry MB, Upshaw JE, Riley LH, Finerman GAM, Turner RH (1973) The geometric total knee arthroplasty. Clin Orthop 94:171–184
12. Cracchiolo III, Benson AM, Finerman GAM, Horacek K, Amstutz HA (1979) A prospective comparative clinical analysis of the first-generation knee replacements: polycentric vs. geometric knee arthroplasty. Clin Orthop 145:37–46
13. Deburge A, Aubriot JH, Genet JP, Guepar Group (1979) Current status of a hinge prosthesis. Clin Orthop 145:91–93

14. Deburge A, Morin C, Neukirch F (1984) Complications des arthroplasties totales du genou. Rev Chir Orthop 70:168–170
15. Dorr LD (1984) Revision of total hip and knee. University Park Press, Baltimore
16. Ducheyne P, Aernoudt E, de Meester P (1978) The mechanical behaviour of porous austenitic stainless steel fibre structures. J Mater Sci 13:2650–2658
17. Ducheyne P, Heymans L, Martens M, Aernoudt E, de Meester P, Mulier JC (1977) The mechanical behaviour of intracondylar cancellous bone of the femur at different loading rates. J Biomech 10:747–762
18. Ducheyne P, Martens M, Aernoudt E, Mulier J, de Meester P (1974) Skeletal fixation by metal fiber coating of the implant. Acta Orthop Belg 40:799–805
19. Ducheyne P, Martens M, de Meester P, Mulier JC (1983) Titan- und Titanlegierungsprothesen mit poröser Drahtdeckschicht. In: Morscher E (ed) Die zementlose Fixation von Hüftendoprothesen. Springer, Berlin Heidelberg New York Tokyo, pp 111–120
20. Fidler MW (1983) Knee arthrodesis following prosthesis removal. J Bone Joint Surg [Br] 65:29–31
21. Freeman MAR, Samuelson KM, Bertin KC (1985) Freeman-Samuelson total arthroplasty of the knee. Clin Orthop 192:46–58
22. Freeman PA (1973) Walldius arthroplasty: a review of 80 cases. Clin Orthop 94:85–91
23. Galante J, Rostoker W, Lueck R, Ray RD (1971) Sintered fiber metal composite as a basis for attachment of implants to bone. J Bone Joint Surg [Am] 53:101–114
24. Greenberg RL, Kenna RV, Hungerford DS, Krackow KA (1984) Instrumentation for total knee arthroplasty. In: Hungerford DS, Krackow KA, Kenna RV (eds) Total knee arthroplasty. Williams and Wilkins, Baltimore, pp 36–40
25. Grimer RJ, Karpinski MRK, Edwards AN (1984) The long-term results of Stanmore total knee replacements. J Bone Joint Surg [Br] 66:55–62
26. Gschwend N, Müller H (1984) Arthroplasties au genou G.S.B. à plus de 5 ans. Rev Chir Orthop 70:182–185
27. Gunston FH (1973) Polycentric knee arthroplasty: prosthetic simulation of normal knee movement. Clin Orthop 94:128–135
28. Habermann ET, Deutsch SD, Rovere GD (1973) Knee arthroplasty with the use of the Walldius total knee prosthesis. Clin Orthop 94:72–84
29. Hagena FW, Hofmann GO (1984) Lang- und mittelfristige Ergebnisse nach Implantation der GSB-Kniegelenks-Endoprothese. Unfallheilkunde 87:133–143, 298–308
30. Hanslik L (1973) First experience on knee joint replacement using the young hinged prosthesis combined with a modification on the McKeever patella prosthesis. Clin Orthop 94:115–121
31. Herbert JJ, Herbert A (1973) A new total knee prosthesis. Clin Orthop 94:202–210
32. Hirsh DM, Bhalla S, Roffman M (1981) Supracondylar fracture of the femur following total knee replacement. J Bone Joint Surg [Am] 63:162–163
33. Homsy CA (1973) Implant stabilization, chemical and biomechanical considerations. Orthop Clin North Am 4:295–311
34. Homsy CA, Cain TE, Kessler FB, Anderson MA, King JM (1972) Porous implant systems for prosthesis stabilization. Clin Orthop 89:220–235
35. Homsy CA, Tullos HS, Anderson MS, Di Ferrante NM, King JW (1972) Some physiological aspects of prosthesis stabilization with acrylic polymer. Clin Orthop 83:317–328
36. Hungerford DS, Kenna RV (1983) The PCA revision total knee system: surgical technique. Howmedica, Rutherford
37. Hungerford DS, Kenna RV (1983) Preliminary experience with a total knee prosthesis with porous coating used without cement. Clin Orthop 176:95–107
38. Hungerford DS, Krackow KA, Kenna RV (1982) The porous coated anatomic total knee. Orthop Clin North Am 13:103–122
39. Hungerford DS, Krackow KA, Kenna RV (1984) Total knee arthroplasty. A comprehensive approach. Williams and Wilkins, Baltimore

40. Hungerford DS, Krackow KA (1985) Total joint arthroplasty of the knee. Clin Orthop 192:23–33
41. Huten D, Duparc J (1984) Résultates à plus de 5 ans des prothèses de Freeman. Rev Chir Orthop 70:177–179
42. Imbert JC (1984) Résultats fonctionnels à plus de 5 ans de la prothèse géométric. Rev Chir Orthop 70:180–181
43. Insall JN (1981) Principles of surgical technique in total knee replacement. In: AAOS Instructional Course Lectures. Mosby, St. Louis, pp 324–334
44. Insall JN, Binazzi R, Soudry M, Mestriner LA (1985) Total knee arthroplasty.Clin Orthop 192:13–22
45. Insall JN, Hood RW, Flawn LB, Sullivan DJ (1983) The total condylar knee prosthesis in gonarthrosis. J Bone Joint Surg [Am] 65:619–628
46. Insall JN, Lachiewicz PF, Burstein AH (1982) The posterior stabilized condylar prosthesis. A modification of the total condylar design. J Bone Joint Surg [Am] 64:1317–1323
47. Insall JN, Tria AJ, Scott WN (1979) The total condylar knee prosthesis: the first 5 years. Clin Orthop 145:68–77
48. Jackson JP, Elson RA (1973) Evaluation of the Walldius and other prostheses for knee arthroplasty. Clin Orthop 94:104–114
49. Jones GB (1973) Total knee replacement – the Walldius hinge. Clin Orthop 94:50–57
50. Kaufer H, Matthews LS (1979) Spherocentric knee arthroplasty. Clin Orthop 145:110–116
51. Kaufer H, Matthews LS (1981) Spherocentric arthroplasty of the knee. J Bone Joint Surg [Am] 63:545–559
52. Kenesi C (1979) The "Caviar" madreporic knee prosthesis. Clin Orthop 145:94–100
53. Kenna RV, Hungerford DS, Krackow KA (1983) The PCA primary total knee system: surgical technique. Howmedica, Rutherford
54. Landon GC, Galante JO, Casini J (1985) Essay on total knee arthroplasty. Clin Orthop 192:69–74
55. Laskin RS (1981) Total condylar knee replacement in rheumatoid arthritis. J Bone Joint Surg 63:29–35
56. Letournel E, Lagrange J (1973) Total knee replacement with the "LL" type prosthesis. Clin Orthop 94:249–256
57. Lettin AWF, Kavanagh TG, Scales JT (1984) The long-term results of Stanmore total knee replacement. J Bone Joint Surg [Br] 66:349–354
58. Lotke PA, Ecker ML (1977) Influence of positioning of prosthesis in total knee replacement. J Bone Joint Surg [Am] 59:77–79
59. McKeever DC (1985) Tibial plateau prosthesis. Clin Orthop 192:3–12
60. Marmor L (1973) The modular knee. Clin Orthop 94:242–248
61. Marmor L (1985) Unicompartmental and total knee arthroplasty. Clin Orthop 192:75–81
62. Matthews LS, Goldstein SA, Kaufer H (1985) Experiences with three distinct types of total knee joint arthroplasty. Clin Orthop 192:97–107
63. Matthews LS, Sonstegard DA, Kaufer H (1973) The spherocentric knee. Clin Orthop 94:234–241
64. Mazas FB, Guepar Group (1973) Guepar total knee prosthesis. Clin Orthop 94:211–221
65. Mittelmeier H (1983) Keramikhüftgelenkprothesen mit zementfreier Verankerung (Autophor) In: Morscher E (ed) Die zementlose Fixation von Hüftendoprothesen. Springer, Berlin Heidelberg New York Tokyo, pp 231–249
66. Moreland JR, Thomas RJ, Freeman MAR (1979) ICLH replacement of the knee: 1977 and 1978. Clin Orthop 145:47–59
67. Murray DG (1985) Total knee arthroplasty. Clin Orthop 192:59–68
68. Oglesby JW, Wilson FC (1984) The evolution of knee arthroplasty: results with three generations of prostheses. Clin Orthop 186:96–103
69. Phillips RS (1973) Shiers' alloplasty of the knee. Clin Orthop 94:122–127

70. Pilliar RM, Cameron HU, Binnington AG, Szivek J, Macnab I (1979) Bone ingrowth and stress shielding with a porous surface coated fracture fixation plate. J Biomed Mater Res 13:799–810
71. Pilliar RM, Cameron HU, Macnab I (1975) Porous surface layered prosthetic devices. J Biomed Eng 10:126–131
72. Riley LH Jr (1985) Total knee arthroplasty. Clin Orthop 192:34–39
73. Ritter MA, Stiver P (1985) Supracondylar fracture in a patient with total knee arthroplasty. A case report. Clin Orthop 192:168–170
74. Roffman M, Hirsh DM, Mendes DG (1980) Fracture of the resurfaced patella in total knee replacement. Clin Orthop 148:112–116
75. Sheehan JM (1979) Arthroplasty of the knee. Clin Orthop 145:101–109
76. Sledge CB, Ewald FC (1979) Total knee arthroplasty experience at the Robert Breck Brigham Hospital. Clin Orthop 145:78–84
77. Thompson FM, Hood RW, Insall J (1981) Patellar fractures in total knee arthroplasty. Orthop Trans 5:490
78. Townley CO (1985) The anatomic total knee resurfacing arthroplasty. Clin Orthop 192:82–96
79. Wade PJF, Denham RA (1984) Arthrodesis of the knee after failed knee replacement. J Bone Joint Surg [Br] 66:362–366
80. Walker PS, Greene D, Reilly D, Thatcher J, Ben-Dov D, Ewald FC (1981) Fixation of tibial components of knee prostheses. J Bone Joint Surg [Am] 63:258–267
81. Walker PS, Shoji H (1973) Development of a stabilizing knee prosthesis employing physiological principles. Clin Orthop 94:222–233
82. Walker RH, Schurman DJ (1984) Management of infected total knee arthroplasties. Clin Orthop 186:81–89
83. Waugh TR (1985) Total knee arthroplasty in 1984. Clin Orthop 192:40–45
84. Webster DA, Murray DG (1985) Complications of variable axis total knee arthroplasty. Clin Orthop 193:160–167
85. Yamamoto S (1979) Total knee replacement with the Kodama-Yamamoto knee prosthesis. Clin Orthop 145:60–67

The Anatomical Endoprosthesis System for the Knee Joint with Metal Spongiosa Surface Structures Allowing Biological Fixations

W. Thomas

Allgem. Krankenhaus Barmbek, I. Orthopäd. Abteilung, Rubenkamp 148, 2000 Hamburg 60, FRG

An improvement of the results obtained in the construction and application of knee joint endoprostheses is only possible if three main problems are solved: material, shape and fixation.

As to the material, the already proven association of metal and polyethylene is still considered to be the most favourable combination from the point of view of tribology. Our attempts to use ceramics for the manufacture of endoprostheses for the knee joint have not succeeded well enough, because the ceramic material will break under high stress due to its low impact strength.

The shape is easy to design in the case of the hip joint, which has the characteristics of a ball joint, since such a joint can very simply be rebuilt with technical materials. However, the extremely complicated anatomical and functional conditions of the knee joint make it much more difficult to form an endoprosthesis. A great number of knee joint endoprostheses used hitherto have a technically simplified, only functional shape in the form of a sliding endoprosthesis as well as in the form of a hinged endoprosthesis. This will inevitably lead to difficulties because normal physiological movements are impeded, especially if rigid hinge joints are used. The regrettable consequences are restrictions of mobility, an abrasion of the material, and, last but not least, a partial loosening of the endoprosthesis. This is why, during the years from 1972 to 1977, we developed the anatomical GT sliding endoprosthesis. It consists of a bicondylar femur part made of the nickel-free endoprostheses metal "Endocast". This femur part, corresponding to normal anatomical conditions, possesses anatomically shaped femur condyles with different radii of curvature in the medial, lateral, frontal and sagittal planes. At the same time, the condyles diverge from the ventral to the dorsal direction. The slide path for the patella is incorporated into the upper part with a lateral superelevation. The endoprosthesis is further provided with two polyethylene plateaux which are fixed like an inlay within the head of the tibia. The polyethylene plateaux are differently shaped on the medial and lateral sides. Corresponding to normal anatomy, they have an elevation in the direction of the cruciate ligament attachments, and their surface slopes by 5° to the dorsal end. This construction guarantees congruence between femoral and tibial endoprosthesis parts in all phases of motion, allowing rolling and sliding movements, rotation of the tibia around the femur at increasing flexion and a stabilisation of the horizontal joint plane. The fourth part of the endoprosthesis is an articular surface substitute for the patella which is made of polyethylene. This part, too, is implanted in the way of an inlay into the prepared bed for the patella. The patella substitute has a

Surgery and Arthroscopy of the Knee
2nd Congress of the European Society
Editors: Müller/Hackenbruch

gable-shaped surface with different medial and lateral facets, corresponding to the normal type Wiberg II. This patella slides congruently within the metallic femur bed in every phase of motion.

In our opinion, endoprosthetic material should be substituted for all articular surfaces in order to maintain the functional unity of the knee joint. This will result above all in a pain-free action of the extensor muscle of the knee and a firm walk (going upstairs, rising from a chair, squatting). This endoprosthesis has been constructed for fixing with bone cement. The treatment results achieved with 660 grafts have been compiled in a multicentre study. With respect to motion, the results were still satisfactory after 6 years. The flexion was not essentially improved as compared with the pre-operative situation (the starting situation was already very good). The number of flexion contractures was considerably reduced. The axis correction and the ligament stability were maintained to a high degree. At the check-up examination, more than 90% of the patients had a "very good" or "good" result; the same percentage of patients felt subjective satisfaction with the result of their operation. In 16 cases, the parts of the endoprosthesis showed an aseptic loosening. The remaining complications were insignificant with regard to the large number of grafts. The radiological evaluation of the radiolucent lines at the bone cement borders had an interesting result. At the femoral and patellar parts, there was a very low rate of appearance, amounting to 5%; the rate at the lateral tibia plateau was about 10%; at the medial tibia plateau, however, a rate of about 25% was found. Although the radiological findings with respect to radiolucent lines did not show a strong correlation with loosened parts (only 16 cases of loosened parts) in our series, we later thought about finding a way to improve the fixation. This type of sliding endoprosthesis can only be applied in cases of limited destruction grades (valgus/varus less than 20%, flexion contracture less than 25°, stability of capsular ligament). In cases of more extensive destruction, especially if there is an instability of the capsular ligament, a coupled endoprosthesis has to be applied. According to the present state of knowledge as to the kinetic and mechanical conditions of the knee joint, it is absolutely necessary to dispense with a rigid axial fixation when implanting such an endoprosthesis; indeed, a rolling and sliding movement must be possible with this sort of endoprosthesis. We have, therefore, developed a sliding axis endoprosthesis, making possible physiological motion after the individual parts of the endoprosthesis have been joined. The sliding axis, which is covered with polyethylene, guides the movements without bearing any load. When the flexion becomes stronger, a normal rolling and sliding process with an AP thrust motion takes place, and the possibility of rotation between tibia and femur increases. In extension, the endoprosthesis is stabilised by its closing shape. The shape of the patella substitute is identical with that of the sliding endoprosthesis patella.

We have been using this endoprosthesis since 1978. Despite the considerably worse starting situation of these patients with high grade indications, the results of our multicentre study have been favourable, too. A perfect correction of severe axial deformities and postoperative stability, together with excellent mobility of the knee joint, were achieved.

The long-term results obtained with this type of endoprosthesis depend above all on the long-lasting stability of its fixation. Until now, we have implanted endoprostheses into the prepared osseous bed by means of polymethyl methacrylate. This bone cement, however, is subject to decomposition and therefore does not provide a durable fixation. Parts will also be loosened by the often very significant formation of granulations. Knowing this, we have been searching for a possibility to dispense completely with bone cement fixation. Two conditions have to be fulfilled for such a cement-free fixation: immediate mechanical primary stability and long-term stability by vital healing up into the osseous tissue.

In my opinion, the first condition is fulfilled by most of the endoprostheses of the knee joint fixable without cement. Mechanical stability is achieved by different surface shapes (pellets, threads). The studies conducted by Galante in 1971 show, however, that an invasion by osteocytes inducing real neoformation of osseous tissue can only take place through a capillary structure. By means of a special procedure, we have succeeded in producing a metallic structure showing the desired properties. In experiments with sheep, we first implanted these metal spongiosa blocks into the sacro-iliac joints and then observed the healing process. The healing took place without irritation in all cases, and in most a gap-free healing up into the metal spongiosa capillaries was found. Recent experiments on dogs performed by Ascherl with hip endoprostheses consisting of these metal spongiosa structures also showed a gap-free and well-vascularised healing. These favourable results gave occasion for us to provide our endoprostheses for the knee joint with these structures. The cement-free anatomical sliding and endoprosthesis consists of a metallic socket with two distal stabilisation tendons. All fixation surfaces are made of metal spongiosa. The two polyethylene inlays, having the same shape as for the cement-applying method, are inserted into the metallic socket. The shape of the femoral slide is identical to that of the cement-applying method, and at the bone surface side are two stabilisation tendons. The patella is firmly pressed into the inlay inserted in the bony bed by means of a laminated tendon.

This endoprosthesis has been used since 1983. It is especially suitable for cases of repeated surgery. After the residual bone cement has been removed, the large destruction cavities can then be lined with bone grafts. The immediate mechanical fixation allows an early mobilisation in the follow-up period, with an increasing physical load after the 10th postoperative day of 10 kg every second day.

Results

GT-Sliding Endoprosthesis

In all, 56 surgical interventions were carried out without cement from 1983–1986, and 51 followed up (Tables 1–3).

The follow-up period ranged from 1 to 4 years, with an average of 3.1 years. Of these patients, 36 (71%) were female with an average age of 62 years (range 39–76 years) and 15 (29%) were male with an average age of 54 years (range 35–72 years). There were 22 right knees involved and 29 left.

As for results, 39 patients found it simple to go up stairs. General complications included Sudeck (1 patient) and effusion (1). Special complications involved anaesthesia mobilisation (2), extension deficiency (3), patellar lateralisation (2),

Table 1. Diagnoses leading to treatment

	n
Arthrosis	22 (43.1%)
Prearthrotic deformity	10 (19.5%)
Inflammation (CP)	7 (13.7%)
Endoloosening	6 (11.7%)
Inveterate trauma	3 (5.9%)
Recent trauma	1 (1.9%)

Table 2. Results according to Merle d'Aubigne

Points	*n*
18	4 (7.8%)
17	7 (13.8%)
16	8 (15.7%)
15	9 (17.6%)
14	8 (15.7%)
13	7 (13.8%)
12	3 (5.9%)
11	2 (3.9%)
10	2 (3.9%)
9	1 (1.9%)

Very goog 19 (37.2%); good 24 (47.1%); satisfactory 7 (13.7%); poor 1 (1.9%).

Table 3. Mobility

Flexion[a]		Extension		Use of stick[b]	
Degrees	*n*	Degrees	*n*	Months	*n*
70	1	5	2	1	2
50	1	0	25	2	4
65	2	5	5	3	7
70	1	10	11	4	8
75	4	15	5	5	10
80	6	20	1	6	4
85	6			7	2
90	6			8	1
95	11			9	4
100	4			10	2
105	2			12	2
110	5			13	1
120	2			14	2
130	1			15	1
135	1			16	1

[a] 85°–100° degrees = 27 (53%); more than 80 degrees = 39 (76.5%).
[b] 6 months on an average.

patellar pain (2), infrapatellar nerve (4), loosened collateral ligament, medial (3) and loosened collateral ligament, lateral (1). Radiological analysis showed a radiolucent line on tibia part (1) and valgus position of tibia part (1). Periarticular ossification was found in varying degrees: degree 0 = 41 patients (84%); degree I = 9 [16%; patellar tendon 4 (7.8%) and superior recessus 5 (8.2%)]; degrees II and III none.

GT-Sliding Axis Endoprosthesis

In order to treat higher grades of destruction with cement-free knee endoprostheses also, we created our anatomical GT sliding axis endoprosthesis with metal

Table 4. Diagnoses leading to treatment

	n
Arthrosis	12 (40%)
Prearthrotic deformity	6 (20%)
Inflammation	6 (21%)
Psoriasis	1 (3%)
Rheumatoid arthritis	5 (18%)
Endoloosening	5 (18%)
Infection	1 (3%)

Table 5. Results according to Merle d'Aubigne

Points	*n*
18	2 (6.6%)
17	3 (10%)
16	3 (10%)
15	9 (30%)
14	1 (3.3%)
13	8 (26.7%)
12	2 (6.67%)
11	2 (6.67%)
10	0

Very good 8 (27%); good 18 (60%); satisfactory 4 (13%).

Table 6. Mobility

Flexion[a]		Extension		Use of stick[b]	
Degrees	*n*	Degrees	*n*	Months	*n*
65	1	10	1	1	1
75	1	5	3	2	2
80	2	0	19	3	3
85	3	5	3	4	5
90	3	10	3	5	3
95	7	15	1	6	2
100	1			7	4
110	6			8	1
115	2			10	4
120	3			11	1
125	1			12	1
				13	1
				15	1
				19	1
				29	1
				30	1

[a] More than 80 degrees = 26 (87%).
[b] 9 months on an average.

spongiosa surface structures. The pedicles of these endoprostheses are firmly anchored in the osseous beds of femur and tibia. The other features of shape and function correspond to those of the cement-applying method. Thus, a normal functional mobility is possible with this type of endoprosthesis. Besides its primary application in patients suffering from high-grade instability of the capsular ligament, this endoprosthesis is used mainly in cases of repeated surgery. The often enormous destruction cysts are then filled with spongiosa. In special cases showing extreme destruction and extensive losses of bony substance, we implant specially designed endoprostheses. Particular designs are also necessary in the treatment of tumour patients.

In children and still-growing adolescents who have undergone leg-preserving tumour extirpation for osteosarcoma, the implantation of an endoprosthesis involves the problem of precalculating the increase in length to be expected for the leg. By means of a special Cardan technique, we have succeeded in developing an endoprosthesis for the knee joint which can be elongated by moving the Cardan gear with a little stab incision.

The number of surgical interventions without cement totalled 37 between 1983 and 1986, and 30 were followed up (2 died, 1 changed address and 2 moved far away). The follow-up period ranged from 1 to 4 years, with an average of 3.3 years (Table 4). Of these, 25 (87%) were female, with an average age of 70 years (range 56–81 years), and 5 (13%) were male, with an average age of 64 years (range 41–78 years). There were 15 right knees involved and 15 left.

As for results, 20 patients found it simple to go up stairs. General complications included thrombosis (1 patient), embolism (1), wound dehiscence (1), decubitus ulcer of the knee (1) and temporary fistula (1). Special complications involved extension deficiency (1), patellar lateralisation (2), fracture of the patella (1), patellar pain (1) and luxation (repositionable) (2). Radiological analysis showed a valgus position of the tibia (3) or femur (1) and extensional position of the femur (1). Periarticular ossification was found in varying degrees: degree 0 = 24 patients (80%); degree I = 6 [20.2%; patellar tendon 2 (6.67%), superior recessus 2 (6.67%) and posterior capsule 2 (6.67%)]; degrees II and III have none.

The Alloarthroplasty of the Knee Combined with Bone Reconstruction

R. Hoos [1], H. J. Refior [2], and H. Stürz [2]

[1] Abteilung für operative Rheumatologie des Roten-Kreuz-Krankenhauses, St. Pauli-Deich 24, 2800 Bremen 1, FRG
[2] Medizinische Hochschule, 3000 Hannover, FRG

The alloarthroplasty of the knee joint seems nowadays to be an established operation (Hackenbroch et al. 1987). In our opinion, the indication for implantation is given if there is no other chance to make the patients own knee joint painless, and the patient would otherwise become an invalid. We use the GSB alloplasty mainly in cases of bone destruction and joint instability.
Just as many other authors (Gschwend et al. 1981; Hagena and Jäger 1981; Hagena and Hofmann 1984; Kerschbaumer 1981; Refior and Hagena 1979) we saw good results after 2 years in 85% of our cases [Hoos and Refior (in press)]. Problems seem to arise in the patella, even after cementless implantation. Some 80% of these cases were patients with rheumatoid arthritis and severe bone destruction in one or both knee joints.
While resection at the tibial and femoral part of the joint is limited (Gschwend et al. 1981; Hagena and Hoffmann 1984; Jäger 1981), we have had to do bone grafting in the same operation. For this, we used autologous stock of the resected bone, based on our experience with bone reconstruction in alloplasty of the hip [Hoos and Refior (in press)] (Fig. 1). We used the graft primarily for reconstruc-

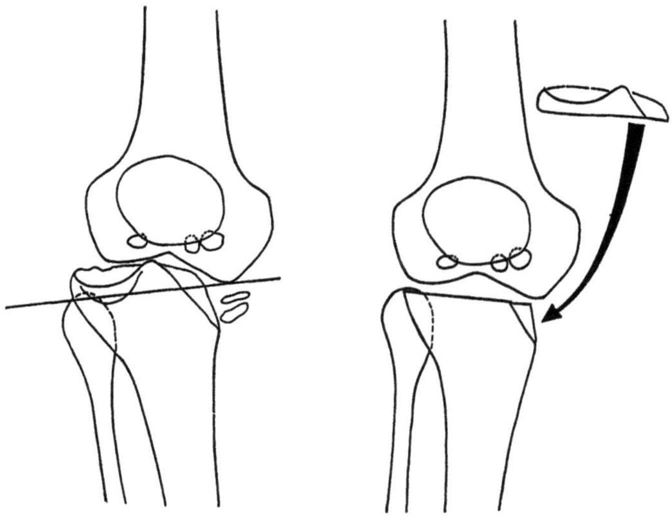

Fig. 1. Possibility for bone reconstruction in the medial part of the tibia

Surgery and Arthroscopy of the Knee
2nd Congress of the European Society
Editors: Müller/Hackenbruch

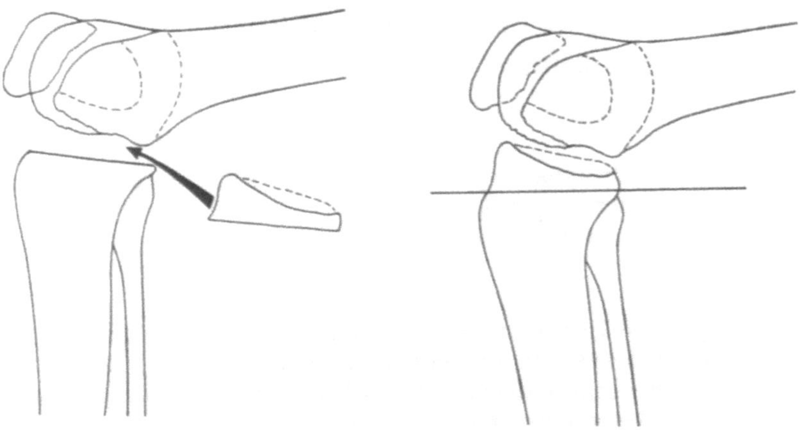

Fig. 2. Possibility of bone reconstruction in the femoral part

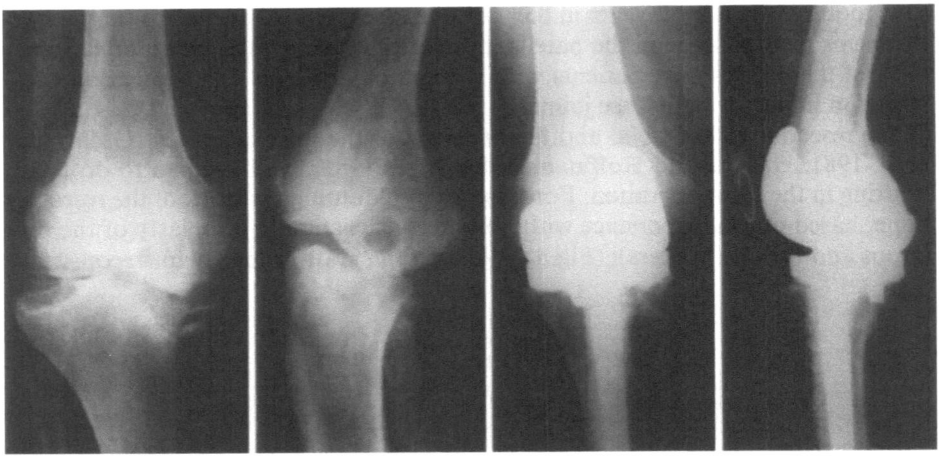

Fig. 3. Development of alloplasty of the knee joint with autologous medial tibial bone grafting: before operation and 3 years after operation

tion of the tibial and femoral part of the joint but also to fill up bigger cysts. It is possible, if necessary, to make a homologous graft also (Fig. 2).

Figure 3 shows a destruction mainly to the medial part of the tibia. Even after resection, as is necessary for implantation, there remains a great bone defect in the medial part of the tibia. With autologous grafting, it was possible to build a real tibial stock under the implant, as is shown by the X-ray taken 3 years after operation.

Figure 4 shows advanced destructions with large cysts in the femoral part of the joint, which were caused by rheumatoid arthritis. In this case, it was necessary to make an autologous and homologous bone graft, which stabilised the implants, as shown by the X-ray taken 2 years after operation.

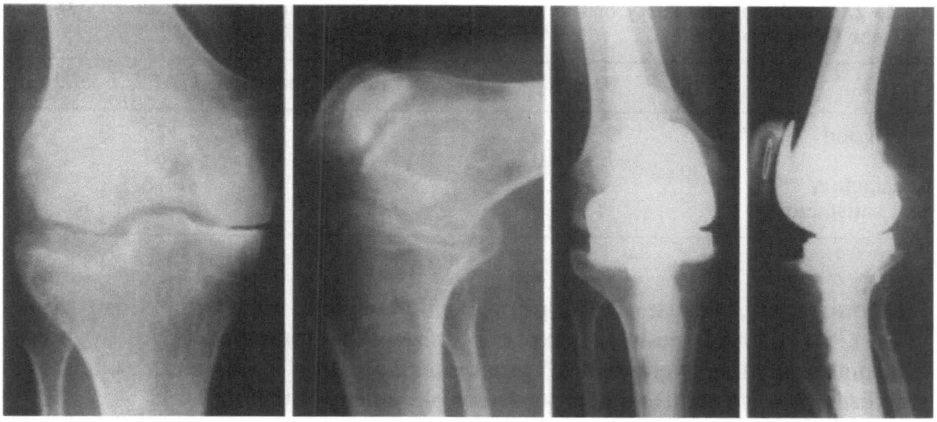

Fig. 4. Development of alloplasty of the knee joint with autologous and homologous bone grafting of the femoral part of the knee joint: before operation and 2 years after operation

In the time from August 1980 to August 1985, we had to operate (on 30 knee joints) in the described way of bone grafting, and we implanted 150 GSB endoprostheses. The first 15 patients were available to be examined clinically and by X-ray. The follow-up took place 1–4 years, with an average of 2.5 years, after operation. There were 14 women and 1 man. The age at the time of implantation differs from 56 to 78 years, with an average of 65 years. Of these, 14 patients showed disease of the knee joint due to rheumatoid arthritis and only one due to osteoarthritis. In 2 cases, it was necessary to make the implantation in both knee joints: in all 10 in the right and 8 in the left knee. Bone reconstruction was necessary primarily in 14 cases and in 4 cases during a replacement operation of implants. The grafting was done 4 times in the tibial, 4 times in the femoral and 10 times in both parts of the joints.
To examine the patients, we used the examination sheet that was developed by N. Gschwend. The results not only showed that the patients were free from pain and swelling but also that their function was better than in the whole group of patients with implanted GSB prostheses without bone grafting. Ten joints were able to move in flexion of more than 110° (Table 1). All examined joints with a follow-up of 2.5 years showed from good to very good results (Table 2). It seems to be difficult to evaluate the healing of a bone graft with X-rays, especially of

Table 1. Range of flexion at 2.5-year follow-up of alloplasty of the knee combined with bone reconstruction ($n = 18$)

Degrees	Before operation	After operation
> 90°	3	18
70°–89°	11	–
50°–69°	3	–
< 50°	1	–

Table 2. Results of alloplasty of the knee combined with bonereconstruction ($n=18$) at 2.5-year follow-up

	Points	Before operation	After operation
Very good	9.5–10.0	–	12
Good	7.5– 8.0	–	6
Satisfactory	5.0– 6.5	–	–
Not satisfactory	< 5.0	18	–

the femoral part, because of the high proportion of metal used by the construction of the GSB endoprosthesis. Our experience shows that tomography can give more information. On an average of 2.5 years after operation we saw with each X-ray stabilised bone structures and a persistent placement of the implants.

In conclusion, we would like to improve the use of bone reconstruction in alloplasty of the knee. With this technique it is also possible to make implantations in severely damaged knee joints, paying attention as well to the biomechanical necessities (Hagena and Jäger 1981).

References

Gschwend N (1981) Die GSB-Knieprothese. Aktuel Probl Chir Orthop 15:24

Gschwend N, Scheier HG (1980) Die GSB-Knieprothese. Med Orthop Techn 100–128

Gschwend N, Scheier HG, Bähler H (1981) Die GSB-Knieprothese. Prakt Orthop 10:439

Hackenbroch et al. (1987) Symposium Kniegelenksendoprothetik, Orthop Uni Köln, 1985. Roland Berger Studie BMFT

Hagena FW, Refior HJ (1981) Die GSB-Kniegelenksendoprothese bei chronischer Polyarthritis. Aktuel Probl Chir Orthop 15:132

Hagena FW, Jäger M (1981) Das femoropatellare Gleitlager nach totalem Kniegelenksersatz bei chron. Polyarthritis. Verh Dtsch Ges Rheumatol 7:594

Hagena FW, Hofmann GO (1984) Lang- und mittelfristige Ergebnisse nach Implantation einer GSB-Kniegelenksendoprothese, I. Unfallheilkunde 87:133

Hagena FW, Hofmann GO (1984) Lang- und mittelfristige Ergebnisse nach Implantation einer GSB-Kniegelenksendoprothese, II. Unfallheilkunde 87:298

Hoos R, Refior HJ (in press) Der alloarthroplastische Kniegelenksersatz bei chronischer Polyarthritis unter Verwendung der GSB-Endoprothese mit und ohne Patellagleitflächenersatz

Kerschbaumer F (1981) Vergleichende Nachuntersuchung von 74 Kniegelenksendoprothesen bei chronischer Polyarthritis und Arthrosen. Aktuel Probl Chir Orthop 15:47

Kerschbaumer F, Bauer R (1976) Problematik und Ergebnisse der Alloarthroplastik des Kniegelenkes. Arch Orthop Unfallchir 86:15

Refior HJ, Hagena FW (1979) Die GSB-Kniegelenksendoprothese bei chronischer Polyarthritis. Eine Analyse mittelfristiger Ergebnisse unter Berücksichtigung des femoropatellaren Gleitlagers. European congress of rheumatology 96, Vortrag 402

Refior HJ, Hoos R (1986) Pfannenrekonstruktive Maßnahmen für den alloplastischen Hüftgelenksersatz bei chronischer Polyarthritis. Aktuel Rheumatol 4(9):127–130

Patellar Complications with the Total Condylar Posteriorly Stabilized Knee Prosthesis

P. Aglietti, R. Buzzi, and A. Pisaneschi

First Orthopaedic Clinic, University of Florence, Florence, Italy

Since 1974 we have completely replaced the patellofemoral (PF) joint in all our total knee replacements and now we use a domical patella. This provided very satisfactory results with the total condylar (TC) mark I prosthesis, but with the posteriorly stabilized (PS) version fatigue fractures and impingement symptoms of the patella were reported to be more frequent. This paper correlates patellar problems with a clinico-radiographic evaluation of the PF joint in a consecutive series of TC-PS prostheses.

Material and Methods

We considered the first 100 knees and excluded those with tibiofemoral problems or infection. This left us with 90 knees with an average follow-up of 3.5 years (range 1–7 years). The knees were studied clinically with specific attention to patellofemoral function and problems and radiographically with Merchant axial views and lateral views to evaluate the vertical patellar position.

Results

Patellar complications included ten impingement syndromes, one fatigue and one traumatic fracture, and one subluxation. Of the ten cases with impingement, six had mild pain, six had problems with stairs, three had moderate pain, but only one was reoperated to remove a peripatellar soft tissue nodule. Radiographically a medial patellar tilt was frequent (46%), but it was asymptomatic. It was allowed by the shape of the patella and caused by the lateral retinacular release or in nine cases by a slope in the patellar cut or eccentric implant positioning. In 1982 a modified femoral component was made available with a smoother anterior intercondylar edge. This significantly reduced the impingement symptoms from 16% to 6%. When patellar height was less than 25 mm above the joint line impingement symptoms were 19%; with a height of 26–35 mm they were 6%; and with more than 35 mm they were absent. Maximum flexion also correlated with the vertical position of the patella. The average flexion in the three mentioned patella height categories was 94°, 97°, and 105°, respectively.

Surgery and Arthroscopy of the Knee
2nd Congress of the European Society
Editors: Müller/Hackenbruch

Discussion

A complete PF resurfacing improves pain relief and function. Patellar complications are the single most frequent problem with the TC-PS prosthesis. Soft tissue impingement and other patellar complications have been substantially reduced with a modification of the femoral component and specific attention to patella tracking. With the accepted technique of ligament release and an economical bone cut sequence starting from the tibia, it is difficult to avoid patella baja.

The Failure of a Semi-Constrained Total Knee Prosthesis

M. I. B. Besser

Knee Unit, Department of Orthopaedics "B", Rambam Medical Center, Haifa, Israel

The stabilized gliding prosthesis – the Attenborough total knee replacement – is a very carefully conceived implant. It was designed to ensure stability of the knee, at the same time allowing for a controlled laxity to avoid loosening of the components. The actual design was described by Attenborough in 1974, 1976 and 1978.

The early results of this prosthesis were very promising (Attenborough 1978; Vanhegan et al. 1979; Stahl and Besser 1982; Besser 1983). However, as time went on, the complication and failure rate increased (Simison et al. 1986). This, as will be shown, may be attributed to a design failure and to the rigidity of the system.

Materials and Methods

Between 1978 and 1982, 76 total knee operations were performed on 46 patients; the average follow-up was 7 years. Twenty-one patients underwent bilateral replacement at the same session, three (five replacements) had adult rheumatoid arthritis, three (six replacements) had Stills disease, and two had a conversion from a Marmor prosthesis and one from a McKintosh.

The technique of surgery was that employed and recommended by Attenborough, including primary patella resurfacing; however, in later cases, the lateral incision was moved to a mid-line one.

Results

It was found that there was an increasing development of pain as the years progressed (Fig. 1), which naturally affected the walking ability and the range of movement of the knee. This effect may be summarized as patient satisfaction (Fig. 2); dissatisfaction increased to 20% over a 4-year period. The chief causes of failure will be analysed.

Loosening. This is usually the main cause of pain and is found in repeated follow-ups of patients to be a recurring problem. X-rays usually appear normal; however, in this series 15% were definitely loose on both X-rays and Tc bone scan, and about 4% were shown only on bone scan to be loose. In one case, there was pain; we considered it to be solely patella-femoral pain since the patella had not been resurfaced. During the operation for resurfacing, the femoral component was found to be loose within the cement and was recemented successfully. In another patient the loosening resulted in a large amount of bone destruction, and the only alternative was to use a Guepar prosthesis.

Surgery and Arthroscopy of the Knee
2nd Congress of the European Society
Editors: Müller/Hackenbruch

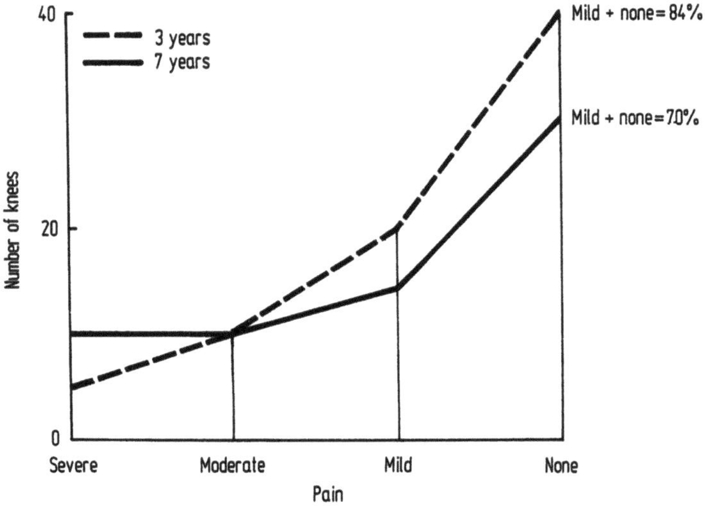

Fig. 1. Increase in pain

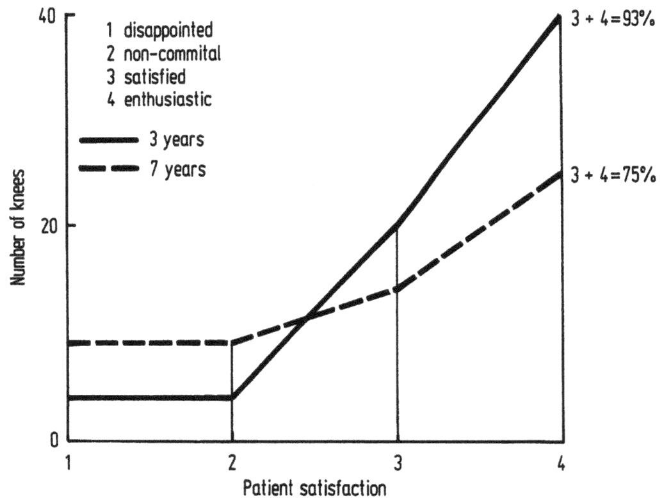

Fig. 2. Decrease in patient satisfaction

Patellar Fracture. All patients except for one had undergone a resurfacing of the patella. In five cases (7%) a fracture of the patella developed at a late stage; two of these underwent patellectomy, and the others were able to tolerate the minor discomfort.

Infection. There were five cases of infection (7%). Two were controlled with long-term antibiotics, and three underwent removal of the prosthesis and cement and debridement. In two cases, it took up to 9 months for bony union, and the third patient is walking with a brace with a painful nonunion after 2 years.

Wound. The original lateral skin incision was found to lead to early problems such as wound break-down, especially when there was early mobilization. This incision was later replaced by a mid-line skin incision, which healed more readily.

Disintegration of the Tibial Plateau. This occurred in three cases and will be discussed later.

Discussion

The increasing number of cases of loosening determined clinically and roentgenologically, especially of the femoral component, can be related to the rigidity of the system because motion and rotation, although intended, especially in flexion, are about a fixed axis (Fig. 3). Furthermore, it can be seen during the trial reduction that there is a "pistoning-out" effect of the femoral component (Fig. 4) on flexion due to the fixed axis caused by the stabilizing rod; this puts much stress on the short femoral stem and leads to loosening. From the above, it may be concluded that the increasing incidence of pain is probably due to a progressive loosening, which initially may not be seen roentgenologically.

Another reason for loosening may be surgical technique. Freeman et al. (1978) stressed the importance of alignment and accurate cutting of the bone and positioning of the components. Because the stems are short, there is no automatic self-alignment as with long stems. The instrumentation is very simple and requires an "eye-balling" approach by the surgeon. This often results in a mal-position of the prosthesis; the leg itself may be aligned perfectly, but the prosthesis may be inserted obliquely thus adding further stress at the bone-cement interface. In some cases severe destruction of tibial plateau was seen, and in these patients the pain was often of a burning nature, probably due to inflammation caused by polyethylene debris.

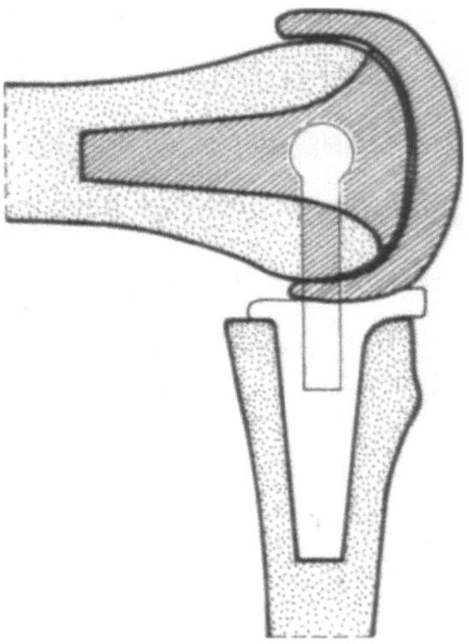

Fig. 3. Attenborough prosthesis

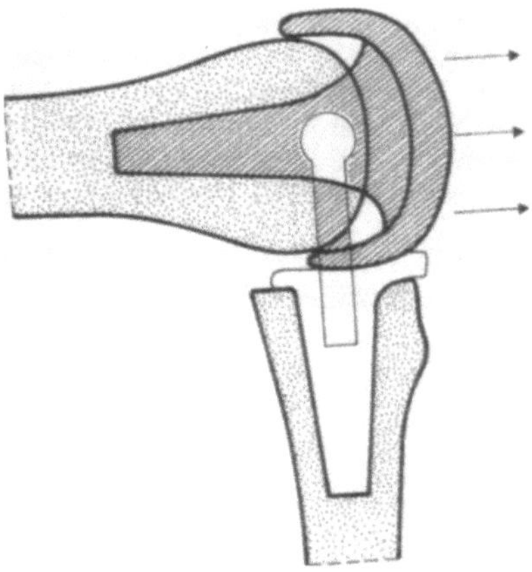

Fig. 4. "Pistoning-out" effect on flexing knee

Patella fracture may be attributed to the large stem of the patella button. The hole made in the patella and the under-cutting for cement can lead to considerable weakness of the patella and thus to its fracture.

Although a total of only about 1 cm of bone is excised from the femoral and tibial surfaces, there is a large resection of cancellous bone to take the implant and the stems. This situation may lead to two problems.

1. In the case of severe loosening, there may be further destruction of bone, leading to loss of bone stock. The only solution then is the insertion of a stemmed hinge prosthesis. One may even have to consider a custom-made prosthesis.
2. In the case of infection, arthrodesis may be very difficult. In our series, it took nearly a year to effect union; in one case, there is painful nonunion.

It can be seen from the CT scan that there is a thin cortex with minimal bone stock, which makes it very difficult, if not impossible, to effect a union. It is therefore wise to consider the addition of a bone graft at the time of arthrodesis.

Conclusion

The attempt to produce a prosthesis with a "built-in" stabilizing mechanism with short stems is shown to result in failure with time. It causes complications which make salvage procedures very difficult. The nonconstrained "total condylar" type or related prosthesis and their instrumentation, which have stood the test of time, apparently do not lead to failure or the problems described here.

References

Attenborough CG (1974) Total knee replacement using a stabilised gliding prosthesis. In Total Knee Replacement. IME Conference CP16 Paper C220/74, pp 92–95. London: Institution of Mechanical Engineers

Attenborough CG (1976) Total knee replacement using the stabilized gliding prosthesis. Annals of the Royal College of Surgeons of England 58:4–14

Attenborough CG (1978) The Attenborough total knee replacement. J Bone Joint Surg [Br] 60:320–326

Besser MIB (1983) Bilateral Attenborough total knee replacement as a single procedure. Arch Orthop Traum Surg 101:271–272

Freeman MAR, Todd RC, Bamert P, Day WH (1978) ICLH arthroplasty of the knee: 1968–1977. J Bone Joint Surg [Br] 60:339–344

Simison AJM, Noble J, Hardinge K (1986) Complications of the Attenborough knee replacement. J Bone Joint Surg [Br] 68:100–105

Stahl S, Besser MIB (1982) The Attenborough total knee replacement. Harefuah 53:191–193

Vanhegan JAD, Dabrowski W, Arden GP (1979) A review of 100 Attenborough stabilised gliding knee prosthesis. J Bone Joint Surg [Br] 61:445–450

Revision Arthroplasty Total-Knee Implants in Infected PCA

M. Rodriguez and A. Papandreou

Department of Orthopaedic Surgery, University of Zürich, Balgrist, 8008 Zürich, Switzerland

Treatment of infected or loosened total-knee implants usually requires performing an arthrodesis. We report our experiences with kinematic and PCA-type implants with special emphasis on revision arthroplasty.

Material and Methods

Since 1982 a total of 36 kinematic and 65 PCA total-knee arthroplasties were performed at our institution. In two of these, infection with progressive loosening of the femoral component occurred a few months postoperation. Revision arthroplasty had to be performed. Meticulous surgical joint debridement was followed by continuous joint lavage. Histopathologic and bacteriologic examinations were done. Two weeks after debridement, having achieved a locally infection-free situation, a revision arthroplasty total-knee implant was inserted. Methylmethacrylate with gentamicin was employed. Perioperative antibiotic therapy was initiated according to the antibiogram. The postoperative management consisted of early mobilization with continuous passive motion. Antibiotic therapy continued for 3 months.

Results

The postoperative course was uneventful in both cases. One and a half years after surgery the knees show good and stable motion and are ambulant, pain free without crutches. Up to now, no infection has occurred.

Discussion

Most total-knee implants on the market today require intramedullary fixation or methylmethacrylate. Due to major bony defects, arthrodesis is usually the only means of treatment, if infection occurs. PCA prosthesis implantation does not require major bony resection. Revision arthroplasty may be done without loss of limb length or of stability of the knee joint.

Surgery and Arthroscopy of the Knee
2nd Congress of the European Society
Editors: Müller/Hackenbruch

Arthrodesis of the Knee After Infected Total Knee Prosthesis

K. H. Müller

Klinik "Bergmannsheil", Bochum, FRG

The prognosis of osteomyelitic infection is usually more serious after prosthetic replacement of the knee joint than after replacement of a hip joint. The successful removal of the total knee is, even with orthotic devices, hardly ever possible. Even a stiffening of the former knee joint is a therapeutically satisfying result. Osseus fusion is difficult because of lack of support of the contact surfaces, reduced transfer activities in aged people, cement and infection devitalisation of bone structures and the lack of a spongiosa reservoir for the replacement of defects. The arthrodesis after removal of infected knee prostheses is very similar to the treatment of an osteomyelitic infection pseudoarthrosis under unfavorable conditions. A modified form of external fixation is the treatment chosen. This report concerns about 17 infected knee prostheses implanted elsewhere between 1974 and 1984 that were then treated here. The average age of the 11 women and 6 men was 67 years. Fourteen prostheses were removed and arthrodesed. Two patients were treated because of osteomyelitic complications after removal of their prostheses elsewhere. In one case the infected implant was left intact. Nine hinged prostheses and five sledge prostheses were removed. A specific technique of external fixation-arthrodesis was used. In 14 cases a stable bony fusion between femur and tibia could be achieved. One above-the-knee amputation had to be performed.

Surgery and Arthroscopy of the Knee
2nd Congress of the European Society
Editors: Müller/Hackenbruch

The Value of Cruciate Ligaments in an Unconstrained Knee Prosthesis

J. L. Lerat, B. Moyen, L. P. Jacomy, G. Py, and J. J. Lalain

Service de chirurgie orthopédique et traumatologique et de médecine du sport, I. Hôpital Edouard Herriot, 69374 Lyon, France

Introduction

Conservation of cruciate ligaments where possible has been the preoccupation of J. M. Cloutier since 1972. He was convinced that a knee prosthesis with femoral and tibial components simulating the normal knee anatomy could allow unconstrained rotational and sliding motion as dictated by the retained cruciate ligaments [2, 3]. Analysis of the results of a series of 67 consecutive knee arthroplasties employing the Cloutier knee, conducted between 1979 and 1985, convincingly demonstrates the value of the cruciate ligaments.

Material and Method

The series was comprised of 45 females (12 bilateral cases) and 9 males (1 bilateral), average age 67 years. In 18 cases the preoperative diagnosis was rheumatoid arthritis and osteoarthritis in 49 cases. All together, 33 left and 34 right knees (two bilateral arthroplasties were done in the same procedure in 2 cases) were treated. The average follow-up was 1–6 years. Of the 67 arthroplasties, 19 knees were operated on anteriorly, and osteotomy was also performed in 6 cases.

Operative Technique. It is a very difficult operative technique because the femoral and tibial components must be implanted exactly perpendicular to the axes of the bone, the ligaments must be symmetrically taut and the limb in physiological alignment. The tibial tuberosity was removed in 21 cases to allow a better approach to the joint, but it is not always necessary and can be a bad thing for rehabilitation because of the fragility of the tibial tubercle after screwing and during active extensive motion. This measure is perhaps necessary only after an osteotomy and for the revision of a knee prosthesis. If the cruciate ligaments are not destroyed they do not present an obstacle to the correction of valgus or varus and flexion deformity with a capsular release and removal of the osteophytes. Patellar surfaces were replaced in 44 cases (38 polyethylene buttons and 6 metal-backed buttons, uncemented). In 4 cases a bone graft was positioned under the prosthesis on the medial plateau. Physiological alignment was obtained in 59 knees (0°–5° of varus or valgus); the remaining 8 knees indicated between 5° and 10° of deformity.

Results

In 87% of the knees, pre-operative pain was adequately relieved and none of the cases were complicated with infection.

Surgery and Arthroscopy of the Knee
2nd Congress of the European Society
Editors: Müller/Hackenbruch

Postoperatively, 50% of the patients were not limited in their walking ability: 85% did not use a support while walking. Also, 92% could rise normally from a chair using the involved knee while 87% could descend stairs in a normal manner but usually with the aid of the hand rail. Ascending stairs was normal but they preferred to use the hand rail in 60% of cases.

Gait studies performed, using the method of Cloutier and Andriacchi [1], with an opto-electron digitizer, a piezoelectric force plate and a microcomputer showed that all the patients, with five different types of prosthesis, walked on level ground with a shorter than normal stride length for a given speed and used a smaller than normal range of knee flexion during the stance phase. Knee flexion while ascending and descending stairs was greater in the patients with the Cloutier prosthesis than in those with other types of prosthesis tested. While ascending stairs flexion was 84° in knees reconstructed with the Cloutier prosthesis versus 73° for the others; while descending stairs the flexion angle was 90° versus 74° for the others.

Postoperatively the average maximum angle of flexion in our series was 105° (range 70° to 140°): 32 cases had more than 110°, 16 more than 120°, and 19 cases had limitation in full extension (more than 10° in 4 cases). There was no difference in flexion between the two series with or without a patellar button.

Even though the flexion was more than 120°, only one patellar fracture occurred 1 year after arthroplasty. The fracture was successfully treated by a cast, and the result still remains very good after several years. This fracture occurred after damage to the polyethylene button which has a hole for anchorage. The hole for this type of button is too large, and it is preferable to use a metal backed polyethylene button with only two little peg insertions into the bone.

Three knees presented loosening of the femoral and tibial components without any evidence of an infection. They were revised using a Guepar prosthesis.

A radiolucent line at the cement-bone interface of the tibial component was seen on the roentgenograms of 50% of knees. The line was less than 1 mm in width and was non-progressive.

Postoperatively, the average internal tibial rotation measured at 90° of knee flexion was 16° (range 5°–25°) and the average external rotation was 20° (range 15°–30°). External rotation of the tibia during extension has been shown to increase the efficiency of the extensor mechanism. This rotation also allows the normal physiological screw home mechanism to occur there by increasing stability.

Postoperative testing for medial-lateral instability showed that at 20° of flexion there was an instability ranging from 5° to 10° in 15% of the knees. Instability developed in both knees in one patient who had rheumatoid arthritis where the posterior cruciate ligament (PCL) was inadequate to support an unconstrained prosthesis. Both cases were revised with a Guepar prosthesis.

Anterior-posterior stability was evaluated by the anterior and posterior drawer test and by stress radiography at 20° of flexion. This objective and precise method of measurement demonstrates the ability of cruciate ligaments to maintain good stability several years after an unconstrained arthroplasty.

Anterior Drawer Stress Radiography. The patient is positioned dorsally on the X-ray table: the hips on the edge of the table, the lower limbs off the table.

The tested knee is placed on a rigid support and flexed to 20°; the leg is fixed by straps. A load of 9 kg is applied to the thigh above the patella, weighting down the femoral condyles.

Posterior Drawer Stress Radiography. The patient is positioned in the same way as for the anterior drawer test with the support placed under the thigh and foot. The knee is flexed to 20° and a load of 9 kg is applied by a strap to the tibial tubercule. The measurement of the global drawer is achieved by comparison of the two roentgenograms. It is possible to measure the true anterior and posterior drawers by comparison with the preoperative roentgenograms. Stress radiography was performed for 15 knees (among the most recent patients) and for 5 knees 6 years after arthroplasty.

The average global drawer was 11 mm when the anterior cruciate ligament (ACL) was absent and 7.4 mm when the ACL was still present.

In 5 cases after a follow-up of 6 years the average global drawer was still 5.2 mm when both cruciates were present at the time of surgery and 10 mm when the ACL was absent.

In another study [4, 5] the physiological anterior drawer was demonstrated to be 3.5 ±1.6 mm in 180 knees. Where the ACL was torn, the anterior drawer was 11.6 ±3 mm in 125 knees. These results show that stability is still good during follow-up, and these reconstructed knees can be compared with normal knees with or without ACL.

The ACL was present in 39 cases, absent in 25, and in 3 cases it was present but in poor condition.

Conclusion

In arthroplasty employing the Cloutier prosthesis the geometry of the femoral component duplicates normal anatomy, and the articular surfaces of the tibial component allow unconstrained rotatory and anterior-posterior displacement as dictated by the ligaments. This type of arthroplasty presents several advantages. The PCL can control posterior roll back of the femur during flexion as it does in the normal knee (it was always present only in one case complicated by secondary luxation). Unconstrained axial tibial rotation, which is permitted by this prosthesis, may also enhance the prosthesis' compatibility with ligament function.

The good stability results achieved in this series suggest that if the ligaments are adequate and if their physiological tension is restored, they can provide stability without the need for any constraint by the prosthesis components. The PCL can absorb large forces that would otherwise be transmitted to the component cement-bone interfaces when a cruciate-substituting design is used. Normal kinematics of the knee can be preserved, and shear stress at the component cement-bone interface can be minimised. Unconstrained rotation may also increase stability since it will enhance ligament function. External rotation of the tibia during extension of the knee has been shown, experimentally, to increase the efficiency of the extensor mechanism. This rotation also allows the screw home mechanism to occur there by improving stability.

A unconstrained, cruciate-preserving prosthesis can provide adequate relief of pain and good stability, but it is necessary to have several components to restore the complete knee anatomy and preserve a good tension in the ligaments.

References

1. Andriacchi TP, Galante JO, Fermier RW (1982) The influence of total knee replacement design on walking and stair climbing. J Bone Joint Surg [Am] 64(9):1328–1335
2. Cloutier JM (1979) Communication, 1st meeting of International Society of the knee, Lyon, 1979
3. Cloutier JM (1983) Results of total knee arthroplasty with a non constrained prothesis. J Bone Joint Surg [Am] 65(7):906–919
4. Lerat JL, Moyen B, Dupre La Tour L (1986) Revision de 100 patients opérés d'une laxité anterieure chronique par une methode dérivée de celles de K. Jones et D. Mac Intosh. Intérêt d'un examen objectif de la laxité par des radiographies dynamiques. Congress of SOFCOT, 11 Nov 1986, Paris
5. Lerat JL, Moyen B, Mainetti E, Dupre La Tour L, Brunet-Gued JE (1986) Measure of laxities of the knee with stress radiographies and with the KT 1000 arthrometer. Communication, 2nd meeting of ESKA, Basel

Endoprosthetic Replacement of the Knee in Rheumatic Diseases

K. Tillmann

Rheumaklinik, Bad Bramstedt, FRG

From 1971 through 1984 in the Orthopaedic Department of the Rheumaklinik, Bad Bramstedt, 977 knee endoprostheses were implanted – the great majority in rheumatoid arthritis patients. Surface replacement was used as well as the hinge of the "St. Georg" type, and after 1978 a constrained total joint with migrating axis.

A survey will be given of the indication and differential indication of the results and of the problems.

Additionally, the special situation of endoprosthetic replacement in juvenile rheumatoid arthritis (after growth has finished) is mentioned. Possibilities and limits of remobilisation of stiff knees is discussed.

Surgery and Arthroscopy of the Knee
2nd Congress of the European Society
Editors: Müller/Hackenbruch

Prosthetic Replacements of Joint Surfaces and the Influence on Activity and Evolution of Rheumatoid Disease

R. Neumann

Rheuma-Orthopedic Department, Rheuma-Hospital,
8202 Bad Aibling-Harthausen, FRG

Between 1976 and 1985, 574 knee alloplasties were performed. Of these 350 were done for chronic polyarthritis. Since the patients were relatively young and had bilateral trouble, we decided to use double sledge prostheses even in cases of advanced destruction of the joints (62 unilateral sledge, 164 double sledge, and 128 total knee replacements). Those double sledge prostheses that were inserted in border-line cases where a total knee replacement was considered showed poorer results at a 5-year follow-up and a higher complication rate than the total knee prostheses in chronic polyarthritis. This was especially true in comparison with patients operated for degenerative arthritis (arthrosis). The follow-up shows that the result of the double sledge prostheses is dependent on the course of the rheumatic disease.

The deterioration of the patients is characterized by recurrence of synovitis, increasing instability, attenuation of capsule, popliteal cysts, and periarticular pain of the type seen in highly active knee synovitis. When these cases were reoperated it was not possible to find any loosening of the implants. The reason for changing their implant to total prostheses was the ligamentous instability. In most cases, however, it was possible to improve their status by a second synovectomy without changing the implant. It is important to note that all these patients suffered from highly active systematic disease with SR values of over 100 and that the basic therapy had failed or for different reasons had been discontinued.

The correlation between inflammatory activity and surgical end result is so significant in our material that it will influence our decision concerning surgery in these types of cases.

Surgery and Arthroscopy of the Knee
2nd Congress of the European Society
Editors: Müller/Hackenbruch

Total Knee Arthroplasty in the Treatment of Rheumatoid Arthritis and Other Forms of Osteoarthritis

F.-W. Hagena and G. O. Hofmann

Staatliche Orthopädische Klinik Harlaching, Orthopädische Klinik
und Polikliniken, Chirurgische Klinik und Poliklinik
der Ludwig-Maximilians-Universität,
8000 München, FRG

Introduction

Many papers concerning arthroplasty of the knee joint have been published. Nevertheless, comparative investigations of the different results of knee arthroplasty according to various diagnoses are rare (Gschwend and Loehr 1981). This paper examines the specific differences between degenerative and inflammatory diseases of the knee as regards the results after endoprosthetic replacement of the joint. Between 1970 and 1981 about 27 500 total knee prostheses were implanted in the Federal Republic of Germany. Of these arthroplasties 3 879 were followed up by clinical and X-ray examinations through support by the Federal Department of Scientific Research and Technology (Weber 1985). Some 30.1% of the knee arthroplasties had been performed in cases of severe rheumatoid destruction of the knee joint (Fig. 1). In 64.3% various forms of osteoarthritis (OA) led to the in-

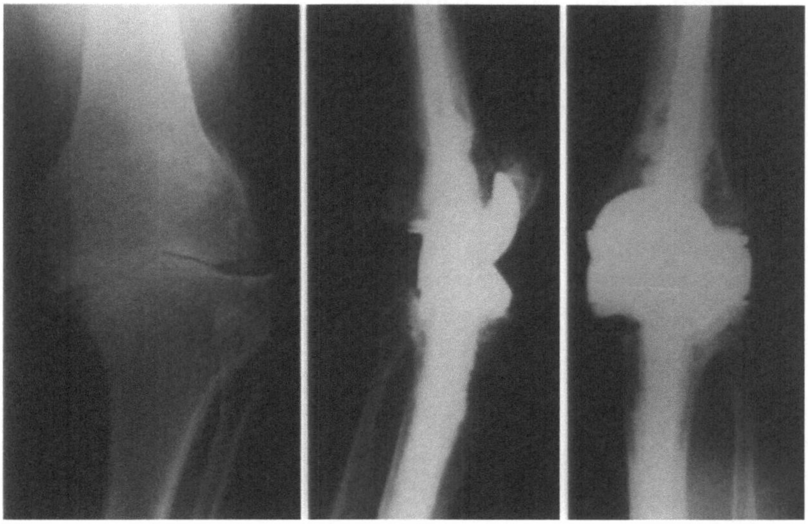

Fig. 1. Osteoarthritis of the left knee with varus instability in 63-year-old female patient. Walldius knee prosthesis was implanted; X-ray taken 16 years after operation; joint is painfree

Surgery and Arthroscopy of the Knee
2nd Congress of the European Society
Editors: Müller/Hackenbruch

dication for total knee arthroplasty (Weber 1985). In the United Kingdom 1.5 million people suffer from rheumatoid arthritis (RA); in 70% of them one or both knees are involved (Bain 1973).
Accordingly Gschwend and Loehr (1981) reviewed 300 patients with RA with an involvement of the knee joint in 74% of these cases. This analysis of three large investigations demonstrates the significance of RA for indication of total knee arthroplasty and shows the importance of knee replacement for the rheumatoid patient (Bain 1973).
The knee has a central function in the chain of joints of the lower extremity. Therefore, damage to the knee joint should be an indication for operative revision. The problems involved in reconstruction of the knee joint in rheumatoid arthritis differ from those in OA. In this study the indications, the complications and the results of total knee arthroplasty are compared and discussed for RA and other forms of OA (Hagena et al. 1981; Hagena and Hofmann 1984).

Material and Methods

Between 1967 and 1984 there were 273 total knee arthroplasties performed at the Staatliche Orthopädische Klinik in Munich. Between 1967 and 1975 a total of 58 Walldius knee prostheses were implanted in 52 patients, and from 1975 to 1984 there were 215 GSB knee prostheses implanted in 191 patients (Fig. 2). Compared with this, 411 malalignments of the knee joint in 343 patients were treated with femoral or tibial osteotomies in the period from 1965 to 1976 (Wasmer et al. 1985). This numerical example demonstrates the restrictive indication for knee replacement in this clinic.

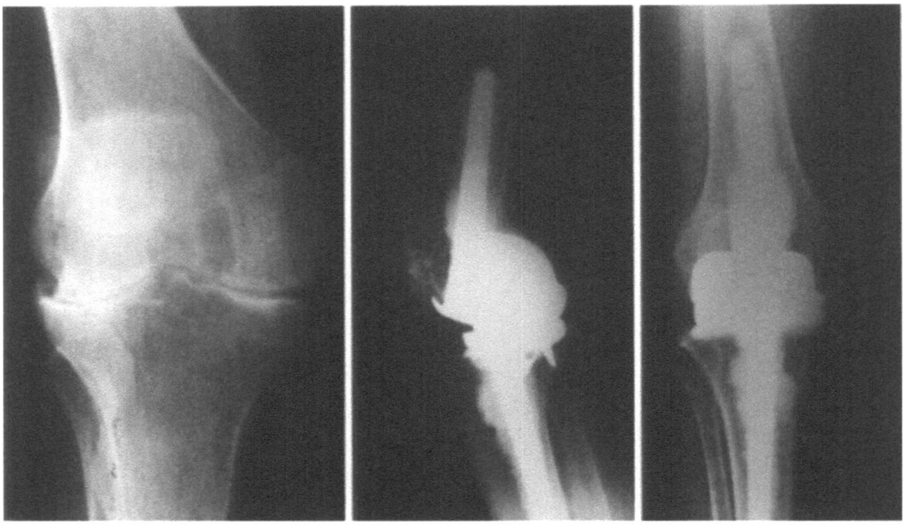

Fig. 2. Rheumatoid arthritis of the right knee in 52-year-old female patient. GSB knee prosthesis (MKI; used until 1982 without patella resurfacing) was implanted; X-ray taken 11 years after operation; joint is pain-free

Table 1. Diagnosis leading to the implantation of 58 Walldius knee endoprostheses

	Patients	Prostheses
Rheumatoid arthritis	33 (63.5%)	39 (67.2%)
Osteoarthritis	18 (34.6%)	18 (31.0%)
Other	1 (1.9%)	1 (1.8%)
	52	58

Table 2. Diagnosis leading to the implantation of 157 GSB knee endoprostheses

	Patients	Prostheses
Rheumatoid arthritis	53 (39.8%)	73 (46.5%)
Osteoarthritis	63 (47.4%)	67 (42.7%)
Other	17 (12.8%)	17 (10.8%)
	133	157

In the group of the 58 Walldius prostheses, severe rheumatoid destruction of the knee joint led to total knee arthroplasty in 63.5% of patients and 67.2% of the prostheses. The same diagnosis indicated GSB knee arthroplasty in 39.8% of the patients and 46.5% of the prostheses (Tables 1 and 2). Figures 3 and 4 show the distribution of the 58 Walldius and 157 GSB prostheses versus different age groups at the time of implantation. The climax of the age distribution for both types of prostheses is significantly earlier for patients with RA than for those with other forms of OA. The mean age of the RA patients at the time of the first implantation was 61.5 years, the mean age of the patients with OA, 72.2 years. The ratio between male and female in all the patients was 1:6; inside the RA group it even reached 1:7.

Rheumatoid patients in a destructive stage of their disease very often present more or less generalised rheumatoid manifestations and involvement of several other joints. In addition to a knee prosthesis 29.1% of the patients had a total hip prosthesis, compared with only 16.4% in cases of OA. As expected there is a striking difference in the involvement of other joints at the time of the first operation. Patients with RA have significantly more joints involved in the disease

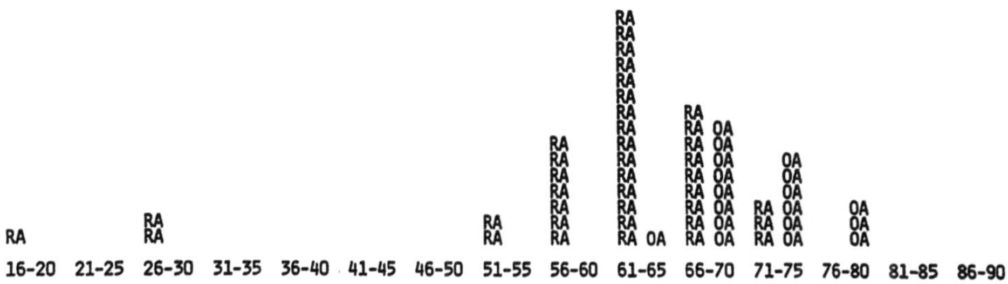

Fig. 3. Age of patients at time of implantation of Walldius knee prosthesis. *RA*, rheumatoid arthritis; *OA*, osteoarthritis

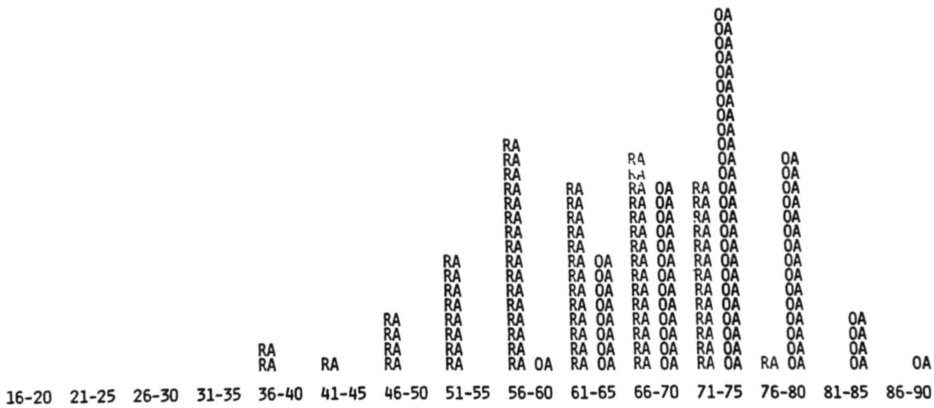

Fig. 4. Age of patients at time of implantation of GSB knee prosthesis. *RA*, rheumatoid arthritis; *OA*, osteoarthritis

than those with other forms of OA (Table 3). Eight patients receiving bilateral Walldius arthroplasty and 20 of 24 patients receiving bilateral GSB arthroplasty showed severe knee joint destruction as a consequence of long-standing RA.

It is remarkable that patients with the diagnosis OA had previous knee operations often much more before total knee arthroplasty than those with RA. While synovectomy is the most frequent previous procedure in RA, tibial osteotomy is most frequent in OA (Table 4).

Table 3. Involvement of joints other than knee at the time of the first operation

	Rheumatoid arthritis	Osteo-arthritis
One more joint	2	38
Two more joints	9	12
Three more joints	4	9
Four more joints	11	4
Five more joints	15	3
Six more joints	27	1
Seven more joints	5	–

Table 4. Previous operations before total knee arthroplasty

	Rheumatoid arthritis	Osteo-arthritis
Synovectomy	10	–
Meniscectomy	2	9
Femoral osteotomy	3	–
Tibial osteotomy	2	10
Patella operations	–	2
	17 (15.2%)	21 (24.7%)

Results

The main diagnosis for the indication of total knee arthroplasty in our patients was RA. The other big group involved patients with degenerative forms of joint disease. Aseptic forms of osteonecrosis, such as M. Blount found, fell in the rarely diagnosed category. Therefore, a greater proportion of patients with RA are represented in our clientele than in other centres (Hagena et al. 1981; Weber 1985).

A review examination was performed for 118 of the 215 GSB prostheses after a mean time of 47.2 months. A painful knee had been the indication for operation in the majority of cases (only 1.7% of the knee joints were free of pain pre-operatively). Postoperatively 59.3% of the operated knees were completely free of pain, and in 20.3% of the prostheses only incidental pain occurred. In 89% there was no extension deficit postoperatively, while only 26.8% had none before the operation. In 72.9% flexion in the endoprosthetically replaced knees was possible to 90° or more. While only 37.6% of the operated knees had a fine function of the quadriceps before the operation, this function was good in 70.3% postoperatively. Also unlimited mobility was improved from 1.7% to 44.2% by the endoprosthetic replacement. All together, 88.1% of the patients declared themselves satisfied with the results of the operation, while retrospectively 11.9% of the patients would not undergo the operation once more (Hagena et al. 1981).

Concerning the results of total knee arthroplasty, the range of movement postoperatively is the same in both groups (RA and OA) in flexion ability as well as in the extension deficit (Tables 5 and 6). It has to be noted that a rather large number of patients from both groups show a remaining flexion deformity (mainly due to the Walldius knee prosthesis).

Postoperative pain does not differ significantly in the two groups (Table 7).

In the OA group more intra-operative complications occurred such as femoral fractures than in the group with RA (Table 8). In contrast there were more early postoperative complications in the cases with RA.

A great number of superficial wound infections and wound dehiscences in the RA group led to the fact that the complication rate during the first weeks after implantation is about twice as large in the RA group as it is in the OA group (Table 9). This is in spite of the fact that patients with RA usually show involvement of several other joints. Late complications such as loosening or deep infection demanding re-operation are not the same in both groups (Table 10). Aseptic

Table 5. Range of movement (flexion ability) results

	Rheumatoid arthritis	Osteo-arthritis
90° and more	54	42
70°–89°	38	36
50°–69°	18	7
30°–49°	2	–
	112	85

Table 6. Range of movement (extension deficit) results

	Rheumatoid arthritis	Osteo-arthritis
0°	44	31
1°– 5°	15	10
6°–10°	30	29
11° and more	23	15
	112	85

Table 7. Postoperative pain results

	Rheumatoid arthritis	Osteo-arthritis
No pain, full activity	49	41
Little pain	38	22
Pain after short distances (maximum 400 m)	19	20
Analgesics necessary	5	2
Pain at rest, no activity	1	–
	112	85

Table 8. Intra-operative complications

	Rheumatoid arthritis	Osteo-arthritis
Femoral fracture	1	4
Stem perforation (femoral)	2	–
Rupture of patella tendon	–	1
Change of prosthesis	2	–
Death of patient	–	1
	5 (4.5%)	6 (7.1%)

Table 9. Early postoperative complications (within 4 weeks after the implantation)

	Rheumatoid arthritis	Osteo-arthritis
Delayed wound healing	21	9
Cardiopulmonary	3	6
Deep infection	4	1
Vein thrombosis	–	1
Joint luxation	1	1
Patella luxation	1	–
Tendon ruptures	2	1
Haemarthrosis	1	–
	33 (29.5%)	19 (22.4%)

Table 10. Number of re-operations necessary after first implantation

	Rheumatoid arthritis	Osteo-arthritis	M. Blount[a]
One	17	12	1
Two	2	2	2
Three	2	–	1
Four	1	–	–
Five	1	–	–

[a] Aseptic osteonecrosis of the medial tibial plateau.

loosening of an endoprosthesis is the main problem for patients with OA: the femoral stem was certainly loose in 3 of 118, the tibial stem also in 3.
Suspicious radiological signs of loosening were found in 4 femoral and 7 tibial stems of 118 cases (Hagena and Hofmann 1984b). The patient with RA is, however, much more threatened by the dangerous possibility of late joint infection. Patients with the diagnosis "osteonecrosis M. Blount" (as a special form of aseptic osteonecrosis of the medial tibial plateau) are much more likely to be re-operated. The re-operation frequency does not differ very much between RA and OA.

Discussion

The necessity for reconstructive knee surgery in patients with rheumatoid lesions is large for the maintenance of walking ability and the relief of weight-bearing on the upper extremities, which otherwise run the risk of accelerated joint destruction. Total knee arthroplasty had been performed in 215 knees of 185 patients. In all cases the indication to total knee arthroplasty was due to the exhaustion of all other conservative or operative techniques, as a last resort.
Patients with RA undergo operations at an earlier age than those with OA; this is found also by other groups (Ahlberg and Lindén 1977; Bain 1973; Hassenpflug 1985; Larsson 1979). Hassenpflug (1985) reports that 54% of RA patients receive knee prostheses when less than 60 years old while only 19% of those older than 60 years have the diagnosis RA. Ahlberg and Lindén (1977) give a mean age at operation of 69 years in OA and 62 years in RA.
Bilateral knee replacement in RA occurs frequently, and this is confirmed also in the literature (Gschwend 1977; Gschwend and Loehr 1981; Larsson 1979). Habermann et al. (1974) and Larsson (1979) point out that the total functional result after knee arthroplasty has to be discussed with respect to the varying extent of additional affection of all other joints in the rheumatoid patient. Early postoperative complications appear twice as often in patients with RA than in those with other forms of OA. In 1963 Young reported that all wound dehiscences and necrosis of skin after knee arthroplasty occurred in patients with RA. Four of five patients had received corticoids pre-operatively (Young 1963). The reason for this high incidence of skin necroses and wound dehiscences may be the systemic character of RA, with a disturbed microcirculation, in contrast to the local problem represented by OA, as well as corticoid-induced skin alterations (Bain 1973; Blauth 1971). There is also a significant coincidence between previous operations of the knee and disturbed wound healing after implantation. Thornhill et al. (1982) also emphasise the necessity of special surgical care of the specific skin situation in rheumatoid patients, for example, to avoid an approach like the Textor one. Intra- and peri-operative complications are the same in RA and in OA, as found also in the literature (Gschwend and Loehr 1981; Larsson 1979).
As already reported, persistent pain due to motion and weight-bearing postoperatively are nearly the same in OA and RA. The range of movement after knee endoprosthesis is worse (10% difference) in OA. Some 50% of all patients with RA and a replaced knee joint are limited to movement inside their homes (Hagena 1985). Ranawat et al. (1976) report better results in patients with

RA than in those with OA: the failure rate of replaced knees is 8.6% in RA and 12.5% in OA (Hassenpflug 1985) (Fig. 3). Hassenpflug explained this difference by a lower stress of the prosthesis in patients with RA due to the lower level of activity of these patients. The connexion between the higher level of activity of the OA patient and the higher frequency of prosthetic loosening is common sense according to the literature (Bain 1973; Cracchiolo 1975; Cracchiolo et al. 1979; Schmitz et al. 1985; Thornhill et al. 1982). From a total of 113 Guepar knee prostheses in the 97 patients of le Nobel and Patterson (1981), 7 showed loosening, and all were OA patients. Ahlberg and Lindén (1981) had to re-operate on 31 of 267 Guepar and Shiers prostheses (12%), 8 RA patients (6%) and 23 OA patients (18%). These reported results concerning the problem of loosening are identical to those in our patients.
Total knee arthroplasty opens a new dimension of therapy in patients with severe rheumatoid destruction of the knee, as a release from pain and to preserve function of the joint. Mobility and stability without pain is a tempting alternative for patients with polyarticular joint destruction but should be carefully prescribed (Baumann and Jäger 1977; Tillmann 1981). Arthrodesis should only be the last line of defence after failed total knee arthroplasty.

Summary

The intermediate results after 8 years of total knee arthroplasties with Walldius and GSB knee replacements show different problems in patients with the diagnosis RA or OA. In this comparative study the clinical results do not show obvious differences. Whereas in the OA group more intra-operative complications occurred, patients with RA had significantly more early postoperative complications, especially delayed wound healing (within the first weeks). The rate of septic loosening was higher in RA, while aseptic loosening was found more often in patients with OA. The most striking differences between knee replacement in rheumatoid- and osteoarthritis depend on operation technique. These have to take into account the special problems for different diseases, for instance by regarding the skin and bones. The operation techniques have developed during the past few years with respect to special instrumentation for correct implantation, by resurfacing the patella and through different knee exposures.

References

Ahlberg A, Lindén B (1977) Arthroplasty of the knee in osteoarthritis and rheumatoid arthritis. Acta Orthop Scand 48:99–104

Ahlberg A, Lindén A (1981) Secondary operations after knee joint replacement. Clin Orthop 156:170–174

Bain AM (1973) Replacement of the knee joint with the Walldius prosthesis using cement fixation. Clin Orthop 94:65–71

Baumann D, Jäger M (1977) Indikationen zu operativen Maßnahmen bei progredient chronischer Polyarthritis. Arztl Prax 66:2814–2816

Blauth W (1971) Zur Kniegelenk-Totalendoprothese nach Walldius. Z Orthop 109:417–435

Cracchiolo A III (1976) Statistics of total knee replacement (editorial comment). Clin Orthop 120:2–3

Cracchiolo A III, Benson M, Finerman GAM, Horacek K, Amstutz HC (1979) A prospective comparative clinical analysis of the first generation knee replacements. Clin Orthop 145:37–46
Gschwend N (1977) Die operative Behandlung der chronischen Polyarthritis, 2nd edn. Thieme, Stuttgart
Gschwend N, Loehr J (1981) The Gschwend-Scheier-Bähler (GSB) replacement of the rheumatoid knee joint. Reconstr Surg Traumatol 18:174–194
Habermann ET, Deutsch SD, Rovere GD (1974) Knee arthroplasty with the use of the Walldius total knee prosthesis. Clin Orthop 94:72–84
Hagena F-W (1985) Klinische Kniegelenksendoprothetik. Ergebnisanalyse: Schmerz und Gehvermögen. In: Weber U, Hackenbroch MH (Hrsg) Endoprothetik am Kniegelenk. Thieme, Stuttgart
Hagena F-W, Hofmann GO (1984a) Lang- und mittelfristige Ergebnisse nach Implantation der GSB-Kniegelenksendoprothese. Teil 1: klinische Ergebnisse. Unfallheilkunde 87:133–143
Hagena F-W, Hofmann GO (1984b) Lang- und mittelfristige Ergebnisse nach Implantation der GSB-Kniegelenksendoprothese. Teil 2: radiologische Ergebnisse – radiologische und klinische Korrelationen. Unfallheilkunde 87:298–308
Hagena F-W, Refior HJ, Jäger M (1981) Die GSB-Kniegelenksendoprothese bei chronischer Polyarthritis – mittelfristige Ergebnisse. Aktuel Probl Chir Orthop 15:32–34
Hassenpflug J (1985) Klinische Kniegelenksendoprothetik. Ergebnisanalyse: Verweildauer. In: Weber U, Hackenbroch MH (Hrsg) Endoprothetik am Kniegelenk. Thieme, Stuttgart
Larsson S-E (1979) Reconstruction of the knee with endoprosthesis in rheumatoid arthritis. Clin Orthop 145:117–125
le Nobel J, Patterson FP (1981) Guepar total knee prosthesis. J Bone Joint Surg [Br] 63:257–260
Ranawat CS, Insall J, Shine J (1976) Duo-condylar knee arthroplasty. Clin Orthop 120:76–82
Schmitz B, Engelhardt P, Menke W (1985) Erfahrungen mit achslosen Knieendoprothesen vom Typ Geomedic und Total Condylar 1975–1984. Komplikationen und Fehlschläge. Med Orthop Techn 105:54–57
Thornhill TS, Dalziel RW, Sledge CB (1982) Alternative to arthrodesis for the failed total knee arthroplasty. Clin Orthop 170:131–140
Tillmann K (1981) Operative Probleme des Kniegelenkes bei chronischer Polyarthritis. Möglichkeiten der operativen Behandlung. Verh Dtsch Ges Rheumatol 7:137–138
Wasmer G, Milachowski KA, Rosemeyer B (1987) Langzeitergebnisse der Pendelosteotomien. Orthop Praxis 6:453–458
Weber U (1985) Klinische Kniegelenksendoprothetik – Ergebnisanalyse: Komplikationen. In: Weber U, Hackenbroch MH (Hrsg) Endoprothetik am Kniegelenk. Thieme, Stuttgart
Young HH (1963) Use of a hinged Vitallium prosthesis for arthroplasty of the knee. J Bone Joint Surg [Am] 45:1627–1642

Unconstrained Knee Replacement in Patients with Rheumatoid Arthritis

C. T. Trepte, W. Noack, and B. Gondolph-Zink

Orthopädische Klinik im RKU, Forschungs- und Lehrbereich der Universität, Oberer Eselsberg 45, 7900 Ulm, FRG

Deformity (axial deviations) and instability of the knee of patients with rheumatoid arthritis (RA) are caused by: ligamentous insufficiency, osseous defects or a combination of both. While the joint capsule and the collateral ligaments become distended by the intra-articular effusion, the cruciate ligaments are affected by direct assault of the overgrowing pannus, leading to damage, particularly of the anterior cruciate ligament (ACL). Quite often there exist osseous defects due to necrosis or partial necrosis of the tibial head or the femoral condyles.

According to many authors (Arden 1973; Lütschig 1979; Thomas and Grundei 1982) deformity and instability are indications for constrained knee endoprostheses. In recent years unconstrained knee endoprostheses were favoured even in cases of gross deformity, flexion contracture and loss of bone stock. Normal axial alignment and physiological guidance can be accomplished by restoration of a proper tension of the collateral ligaments. Therefore a tibial prosthesis of an appropriate thickness has to be inserted (Fig. 1). In cases of ligamentous imbalance, medial or lateral soft tissue release has to be done to achieve a correct alignment. Only in rare cases does transposition of collateral ligaments have to be carried out. It is particularly significant to tighten the medial collateral ligament, whereas a slight lateral instability is of minor importance in a physiological valgus position of the knee joint.

The ACL, if not previously destroyed by the inflammation, has to be resected. The posterior cruciate ligament (PCL) should be preserved to prevent a roll-back

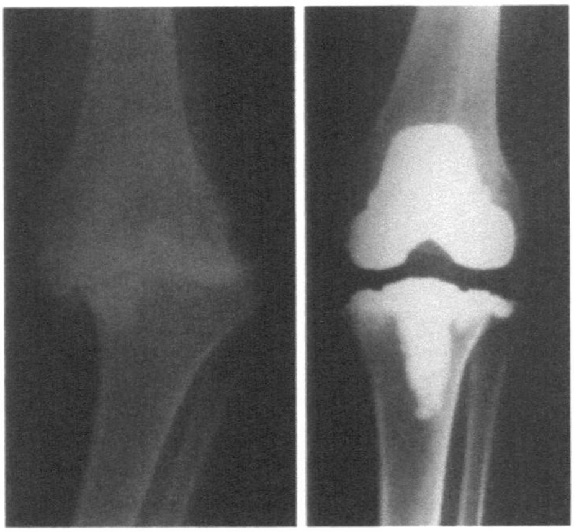

Fig. 1. Correction of varus deformity with unconstrained RMC knee – restoring of stability by insertion of a tibial component of appropriate thickness

Surgery and Arthroscopy of the Knee
2nd Congress of the European Society
Editors: Müller/Hackenbruch

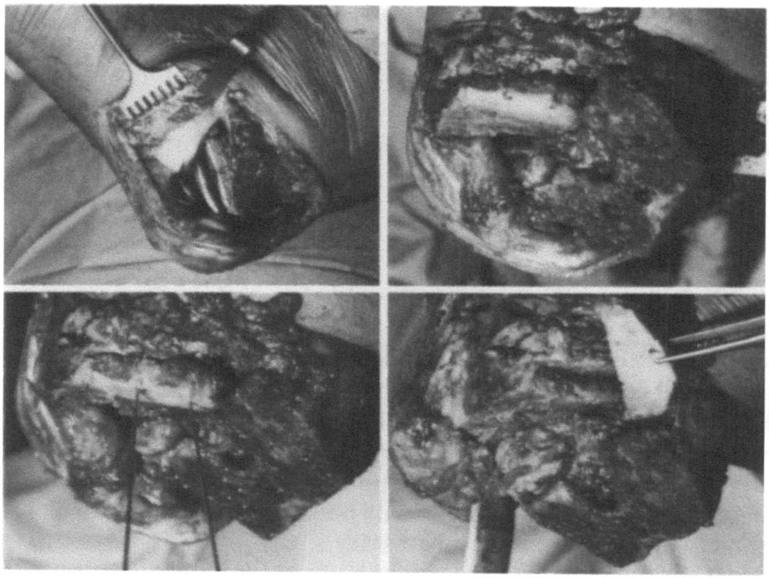

Fig. 2. Bone grafting to restore bone stock for component seating

of the tibial head in flexion of the knee. In cases with active synovitis, the operation is completed with a partial synovectomy. Bone defects of the implant site are substituted by grafting of autologous or homologous spongy bone. Major grafts are fixed by K-wire or AO bone screws (Fig. 2). Our experience shows that grafted bone will be completely incorporated in a short period of time.

Material and Methods

From October 1984 to January 1986 a total of 83 unconstrained knee endoprostheses were inserted. The primary diagnosis was rheumatoid arthritis in 31 patients. Joint replacement was performed: 25 RMC (cemented) and 6 PCA (uncemented). The mean age of patients was 67 years (range 45–80 years); 25 were female and 6 were male. Twenty patients showed valgus or varus deformity of more than 10°, 7 of more than 20°; 23 patients had a flexion contracture, 3 of more than 30°.

Postoperatively the knee was treated with passive continuous motion with a fixed range of movement between full extension and 60° of flexion. Physical therapy started after removal of the suction drains. It was very convenient to do the operative procedure and the subsequent physical therapy under continuous peridural anaesthesia.

Results

All patients had an ideal alignment between 4°–10° of valgus angulation. The range of flexion achieved in 31 successful arthroplasties varied between 90° and

130°, 24 came to 0°, while 7 knees had mild flexion contractures of less than 10°. Three knees with complete destruction of the PCL showed postoperatively a roll-back of the tibial head.
One patient had a thromboembolic complication but recovered completely after embolectomy. No patient in the series showed evidence of loosening of the prostheses or superficial or deep infection.

Discussion

Joint resurfacing in patients with RA is an accredited procedure (Bargren et al. 1976; Laskin 1981).
Deformity and instability of the knee joint, which is quite common in this group of patients, can be well corrected and stabilised with an unconstrained knee replacement. The grade of deformity and instability that still allows arthroplasty with unconstrained prostheses is ultimately dependent on the experience of the surgeon. Resurfacing reveals a number of advantages in comparison with constrained prostheses – the implantation needs only minor bone resection which allows revision surgery or an arthrodesis, if necessary.
Axial alignment and ligamentous balance reduce stress at the implant-bone interface, whereas constrained prostheses create unequal stress concentration there. The high stress concentration in the axis of a hinged prosthesis leads to substantial wear (metalosis). Metalosis can result in a local, reduced, immunological response (Rae 1975). This could explain the increased incidence of infections in patients with constrained knee endoprostheses. According to Gschwend (1983) constrained knee endoprostheses show the highest incidence of infection. This fact should be considered particularly in patients with RA, who have a long course of consuming disease and an increased risk of infection because of the long-term cortisone medication. Unconstrained prostheses in our opinion are the knee replacements of first choice in patients with RA.

References

Arden GP (1973) Total knee replacement. Clin Orthop 94:92–103
Bargen JH, Freemann MAR, Swanson SAV, Todd RC (1976) ICLH (Freemann/Swanson) Arthroplasty in the treatment of the arthritic knee. Clin Orthop 120:65–75
Gschwend N (1983) Replacement arthroplasty of the knee joint. Trends and treatment after failure. Can J Surg 26:407–409
Laskin R (1981) Total condylar knee replacement in rheumatoid arthritis. J Bone Joint Surg [Am] 63:29–35
Lütschg J (1979) Klinische Ergebnisse der Knie-Totalendoprothese. Nachkontrolle von 51 Fällen. Z Orthop 117:911–921
Rae T (1975) A study on the effects of particulate metals of orthopedic interest on murine macrophages in vitro. J Bone Joint Surg [Br] 57:444–450
Thomas W, Grundei H (1982) Die anatomische G.T. Gleitachsenprothese des Kniegelenkes. Z Orthop 120:22–28

Our Experiences with Kinematic Total Knee Replacements

F. Kerschbaumer, R. Siorpaes, R. Hingshammer, M. Eichenauer, and M. Kofler

Department of Orthopaedics, University of Innsbruck, 6020 Innsbruck, Anichstr. 35, Austria

Introduction

The results so far obtained from the implantation of non-constrained knee prostheses, in particular using the kinematic total knee replacement system [1], outdo earlier results of implantations of constrained knee prostheses. A special series using a non-constrained total knee prosthesis system was begun to check whether the significant superiority of this system as opposed to models used earlier would apply to our group of patients as well.

Material and Methods

From November 1981 to November 1985 a total of 146 kinematic prostheses were implanted. With 110 patients (86.9%) representing 127 operations we were able to carry out follow-up examinations. The average age of patients at operation was 68.5 years (range 40–86 years), and the mean follow-up period was 2.4 years (range 0.5–5 years). The indication for operation was arthrosis in 87 patients (68.5%) and rheumatoid arthritis (RA) in 40 (31.5%). In 68 cases (53.5%), i.e. primarily with RA patients, the articular surface of the patella was replaced. The posterior cruciate ligament was left in place. Cases of severe defective valgus position or of false varus position called for soft tissue release; 47 patients (37%) required a lateral release.

Evaluation was effected by means of a special ARO knee curve [2, 4] modified at two points [4]. The total assessment was based on a scheme of points, taking into consideration the stability and axial alignment of the leg. The stability of all four quadrants was clinically tested and divided into + to + + + positive. The axis of the leg and the position of the prosthetic parts were measured on X-rays taken of the whole leg in a standing position. So as to be able to eliminate early signs of loosening, the X-rays were searched for radiolucent lines at the cement-bone interface.

Results

Radiological Findings. In the mean the mechanical axis showed a varus deviation of 1.5°. The femoral part of the prosthesis showed an average inclination of 2.2° flexion to the longitudinal axis of the femur in the sagittal plane, and an average varus of 1.3° in the frontal plane. With eight patients thin, regular, non-progressive, radiolucent lines at the cement-bone interface (six tibia, two femur) were observed.

Surgery and Arthroscopy of the Knee
2nd Congress of the European Society
Editors: Müller/Hackenbruch

Clinical Findings. Symptomatic pain may be studied from Table 1. Pre-operatively, the severe (pathological) changes rendered 11 patients (8.9%) unable to walk; postoperatively that number was reduced to 3 (2.4%). At the time of the follow-up examination all patients were able to walk.
Table 2 shows the improvement in walking. The active mobility as regards flexion remained nearly uninfluenced by the operation. The functionally important active deficit of extension, however, was corrected well (Table 3). Also, the stability of the knee joint was – to a moderate degree – positively affected by the operation (Table 4). At the time of the follow-up examination, five of the unstable joints showed an anterior instability, three a posterior one, and the remaining ones showed a mostly medial instability. Table 5 displays the total results judged by subjective and objective points of view. A comparison between the results pre-operatively and at follow-up for RA patients (3.2 and 8.3, respectively) and patients

Table 1. Symptomatic pain findings (by number of knees)

Degree of pain	Preoperative	Postoperative	Follow-up
Severe	62 (49%)	0	1 (0.9%)
Medium	62 (49%)	9 (7.2%)	10 (9.9%)
Slight	2 (1.6%)	57 (45.6%)	29 (28.7%)
0	0	59 (47.2%)	61 (60.4%)

Table 2. Improvement in walking results

Walking over a distance	Pre-operatively	Post-operatively	Follow-up
100 m	61 (65.6%)	5 (6.3%)	7 (7.1%)
1 km	27 (29%)	41 (51.9%)	25 (25.5%)
10 km	5 (5.4%)	33 (41.8%)	66 (67.4%)

Table 3. Range of movement results

Range of movement	Pre-operatively	Post-operatively	Follow-up
Flexion	93.3°	94.6°	97.7°
Extension deficit	−10.5°	− 3.1°	− 1.3°

Table 4. Stability of knee joint

	Preoperatively	Follow-up
0, +	72 (64.3%)	77 (77%)
2+, 3+	40 (35.7%)	23 (23%)

Table 5. Overall results of operation

	Subjective	Objective
Excellent	66 (65.3%)	35 (36.5%)
Good	28 (27.7%)	41 (42.7%)
Moderate	5 (4.9%)	13 (13.5%)
Poor	2 (1.9%)	7 (7.3%)

with gonarthrosis (3.4 and 8.1, respectively) makes it apparent that no significant differences were produced.

Complications. Six patients incurred postoperative complications. One developed a sizeable haematoma which had to be discharged, in three cutaneous necrosis occurred and had to be treated surgically, while two developed postoperative saphenal thromboses. Three years after the operation one female patient presented with a deep infection which has been cured by antibiotics therapy for the time being.

Statistical Analysis. By means of contingency tables and the chi-square test (after Pearson), further factors possibly affecting the results of the operation were checked. Thus, it has become apparent that the position of the tibial part of the prosthesis may significantly ($P=0.01$) affect the results of the operation. Other factors, such as the anatomical position of the femoral part of the prosthesis, diagnosis or age at the time of operation did not influence the results in any significant way.

Discussion

Today's expectations of a total knee replacement system are undoubtedly higher than they were 5 years ago. Ewald and co-workers [1] name three specific aims of the kinematic prosthesis: improvement of the range of movement, improvement of the fixation of the tibial part of the prosthesis and retention of the excellent clinical results already achieved with similar condylar prostheses [3, 5]. Judging from our own observations we seem to have attained two of these aims. So far, the fixation of the metal-backed tibial part of the prostheses presents no problem, and radiological examinations did not provide any indications of prospective loosenings of these prostheses either. Anatomical implantation of tibial prostheses is, however, highly important since its orientation in a varus position or in too strong a flexion position will render bad results. The total results are comparable to those of other authors [1, 3, 5] and may be regarded as positive. Also, the patellofemoral joint did not give rise to any problem. On the other side, however, the range of movement of the knee joint, in particular the active flexion attained by our patients, was rather a disappointment: in that respect the non-constrained knee prosthesis system at issue did not prove superior to earlier constrained prostheses.

References

1. Ewald FC, Jacobs MA, Miegel RF, Walker PS, Poss R, Sledge CB (1984) Kinematic total knee replacement. J Bone Joint Surg [Am] 66:1032–1040
2. Hagena F-W, Hofmann GO (1984) Lang- und mittelfristige Ergebnisse nach Implantation der GSB-Kniegelenksendoprothese. Teil 1: klinische Ergebnisse. Unfallheilkunde 87:133–143
3. Insall J, Scott WN, Ranawatt CS (1979) The total condylar knee prosthesis. A report of two hundred and twenty cases. J Bone Joint Surg [Am] 61:173–180
4. Kerschbaumer F (1981) Vergleichende Nachuntersuchung von 74 Kniegelenksendoprothesen bei c.P und Arthrose. In: Jäger M, Hofer H, Häckel H (eds) Aktuelle Probleme in Chirurgie und Orthopädie, vol 15. Huber, Bern, pp 47–50
5. Scott RD (1982) Duopatellar total knee replacement. The Brigham experience. Orthop Clin North Am 13:89–102

Autologous and Homologous Bone Grafts in Fixation of Knee-Joint Prostheses

G. Waertel, D. Wessinghage, and E. Kißlinger

Orthopädische Klinik des BRK-Rheumazentrums, 8403 Bad Abbach, FRG

Introduction

Osseous defects which endanger the stable anchorage of the knee prosthesis are encountered particularly with inflammatory rheumatic diseases. Cystic destructions can lead to extensive defects, primarily on the femoral condyles. On the tibial plateau we also encounter anchorage problems with varus or valgus deviations; in the case of patients suffering from rheumatoid arthritis, as a result of inferior bone quality, these frequently attain a degree of severity not met in cases of osteoarthritis. Bone defects constitute a particular problem with revision operations. Earlier total prostheses of the constrained hinge type involved extensive bone resection at the time of the implantation (Baumann et al. 1978). Through replacement of such prostheses with the GSB models which we implant, stable anchorage can only be realized by building up the condyles with blocks of homologous bone.

In the case of primary operations, osseous defects can under certain circumstances be overcome by filling with bone cement. However large amounts of cement are considered unfavorable – an observation which has been substained by the experience gained with hip-joint prostheses. Alternatively, we must consider

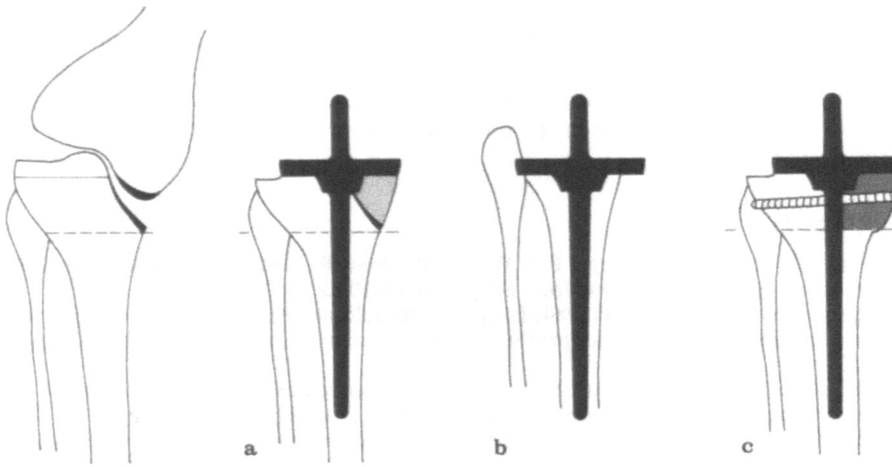

Fig. 1 a–c. Schematic illustration of bone defect due to varus deformity; alternative treatments: **a** filling with bone cement; **b** bone resection up to the height of the defect; **c** bone grafting

Surgery and Arthroscopy of the Knee
2nd Congress of the European Society
Editors: Müller/Hackenbruch

the resection of bone up to the height of the defect (Fig. 1). These procedures also have their limitations, for they disturb the locomotor system considerably. For example, extensive bone resection on the tibial head leads to a patella alta with extensive retropatellar problems, such as were reported by Hagena and Hofmann (1984) in their follow-up study of GSB knee prostheses. In addition extensive resection of the cancellous bone bed will endanger the stable anchorage of the implant. In keeping with the opinions of other authors, we therefore believe that resection beyond a height of 1 cm should be avoided (Dorr et al. 1986). If a defect remains, it can be remedied by means of bone graft.

Method

In the period from January 1984 to June 1986 bone transplantations were undertaken in the case of 44 knee-joint replacements in order to stabilize the anchorage of the implant (Table 1). Forty cases were associated with rheumatoid arthritis and four with osteoarthritis of the knee. In 28 cases insufficient autologous cancellous bone was available, and homologous bone had to be obtained from the bone bank. In 14 cases each, homologous cancellous bone chips and solid homologous bone blocks were used (Table 2). Because of our good experience with homologous bone transplantations, autologous bone grafts from another position, e.g., from the crest of the ilium, were not used with any of the patients. The autologous bone grafts were only obtained from the material that had to be resected as part of the implantation. All the homologous bone material originated from femoral heads that were resected for replacement of hip joints. Except in the case of four operations (in each of which a loose tibial plateau belonging to a sledge-type prosthesis was refixed) replacement of the knee joint was made with a GSB prosthesis supplied by Allo Pro of Baar (Switzerland) (Gschwend 1977;

Table 1. Indications for bone transplantation

	Rheumatoid arthritis	Osteo-arthritis	Total
Cystic destruction	18	–	18
Varus or valgus deformity	12	2	14
Loose implant	10	2	12
Total	40	4	44

Table 2. Bone transplantations for total knee-joint prostheses

	Cancellous bone chips	Solid bone chips	Total
Autologous	9	7	16
Homologous	14	14	28
Total	23	21	44

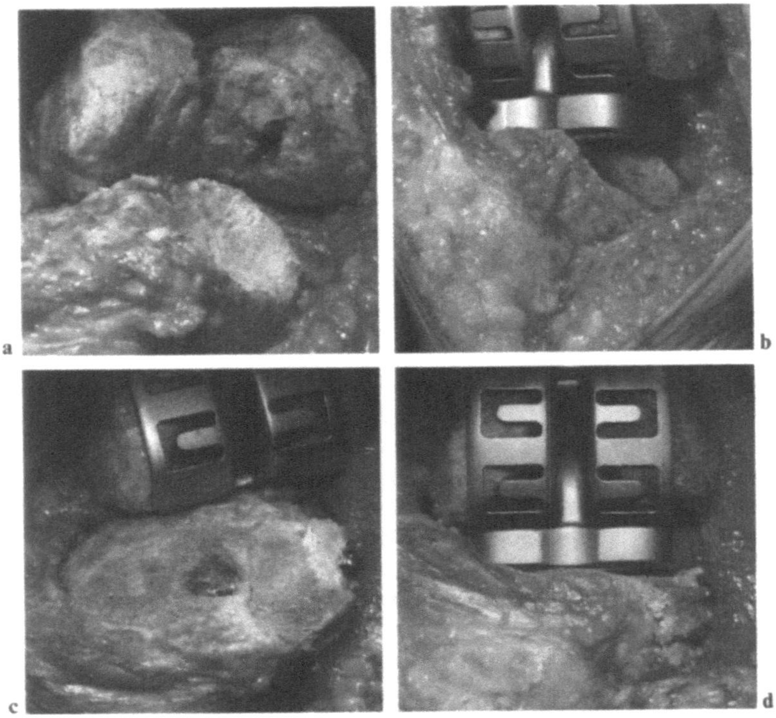

Fig. 2 a–d. A 58-year-old woman with rheumatoid arthritis, unable to walk for 7 years. **a** Intraoperative findings: deep bone defect of medial tibia plateau, cystic destruction of medial condyle, extreme osteoporosis. **b** After correct bone cut the test prosthesis is inserted, the dimension of bone defect is demonstrated. **c, d** The bone defect is filled up with a homologous bone graft

Gschwend et al. 1981). This model features a mobile axis which permits an almost natural rolling and gliding motion of the knee joint. The osseous transplant bed was cleaned carefully in each case. The cystic defects were filled with cancellous bone chips. In the case of large cysts and defects resulting from varus or valgus malposition (Fig. 2), as well as defects subsequent to loosening of the implant, the solid bone blocks were generally fixed with a screw. For postoperative treatment weight relief was recommended for 6 weeks when autologous bone blocks had been used, and for 12 weeks in the case of homologous material.

Results

We were able to conduct follow-up examinations with all the patients who had undergone bone transplantation for the purpose of stabilizing the anchorage of the knee joint prosthesis. No evidence of loosening was established. Radiological follow-ups were conducted after periods of 3, 6, 12, and 24 months. In many cases there was radiological evidence of osseous ingrowth of the autologous bone graft after only 3 months. The 6-month follow-up indicated unison between the autol-

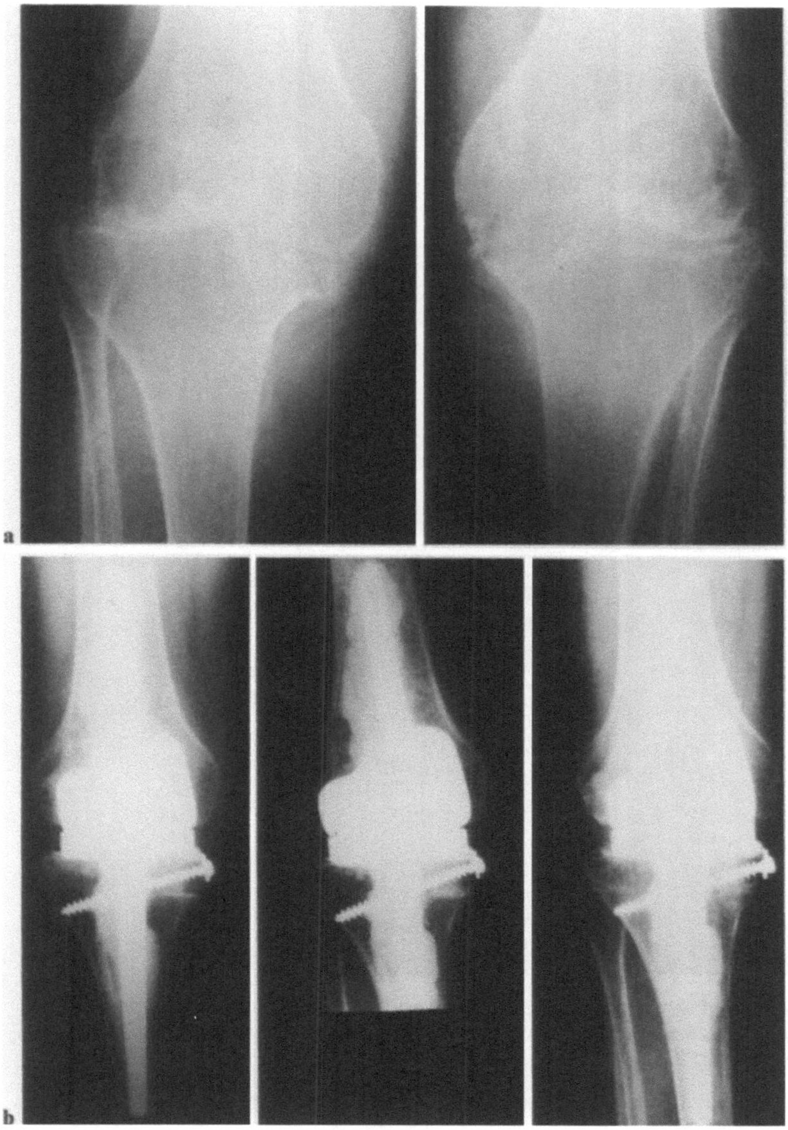

Fig. 3 a, b. Same patient as in Fig. 2. **a** Preoperative radiographs; **b** 12 weeks, 6 months, and 1 year after knee joint replacement with homologous bone graft in the right medial tibia plateau

ogous cancellous bone and the osseous bed in every case. Particular attention was given to the solid homologous bone blocks. Marked osteolytic changes, which are sometimes experienced with other homologous bone transplants, were not observed here (Fig. 3). Six homologous bone blocks, implanted to remedy the defects resulting from a varus or valgus malposition, exhibited radiological evidence

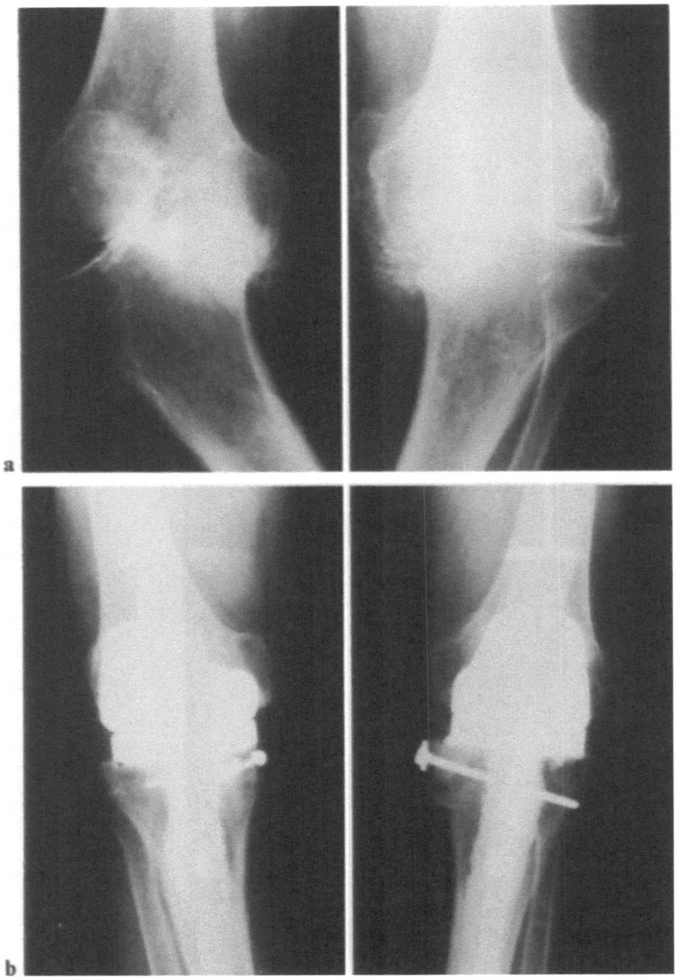

Fig. 4 a, b. A 70-year-old woman with osteoarthritis and varus deformity. **a** Preoperative radiographs; **b** 1 year after knee joint replacement with autologous bone graft (*right*) and homologous bone graft (*left*)

of osseous ingrowth after a period of about 12 months (Fig. 4). Optimal control could not be effected in the region of the femur, however, because of the radiological superposition of the implant.

Discussion

Osseous defects, encountered especially with patients suffering from rheumatoid arthritis, commonly take the form of cyst destruction and result in severe varus or valgus deformities. They can considerably impair the stable anchorage of total knee-joint prostheses. Bone defects that occur after loosening or fracture of total knee-joint prostheses create particular problems, especially in the case of implantations necessitating extensive bone resection. The filling of bone defects with cement is not possible in every case. Furthermore, there is a higher risk here of loosening in the presence of large amounts of cement. Extensive bone resection up to the level of the defect both weakens the osseous bed and leads to a disturbance of the articular mechanics and thus to pain for the patient. It therefore appears desirable to replace the missing bone with a bone graft. The latest prosthetic models, with which only slight bone resection is needed for implantation, yield only a small amount of autologous bone material that can be used elsewhere to remedy bone defects. The good results obtained with homologous bone grafts have encouraged us to dispense with the possibility of using such autologous bone material, e.g., from the crest of the ilium. For cases in which the available autologous bone material proves insufficient, we use homologous bone from the bone bank. The favorable experience gained with bone transplantations used to stabilize the floors of defective acetabula, namely in the case of patients suffering from rheumatoid arthritis or following the loosening of the socket (Fig. 5), has induced us to secure the anchorage of the implant by means of bone grafts for total knee-joint prostheses, as well. Within the past 7 years 54 homologous and 420 autologous bone transplantations in total hip replacements have been performed at our clinic (Holzhauser et al. 1984; Waertel et al. 1985; Wessinghage 1983, 1984; Wessinghage et al. 1984; Wetzel et al. 1986). Also Heywood (1980), McCollum (1978), McCollum et al. (1980), Ranawat et al. (1980), and Udvarhelyi et al. (1985) report good results after bone grafting for protrusio acetabuli and acetabular deficiency (Harris 1977, 1982).

To the present not a single case of loosening due to failure of the bone graft has been observed. In the case of stabilization of the acetabular floor with homologous bone, the follow-up conducted over a period of up to 6 years allows us to conclude that permanent new formation of the acetabular floor by means of homologous bone graft is possible. Our experience with bone transplantations associated with the implantation of total knee-joint prostheses, however, is limited by the comparatively short follow-up period. Nevertheless, we have observed no significant resorption of a bone graft in conjunction with knee joints, particularly with homologous transplants in combination with prosthetic implantations. In

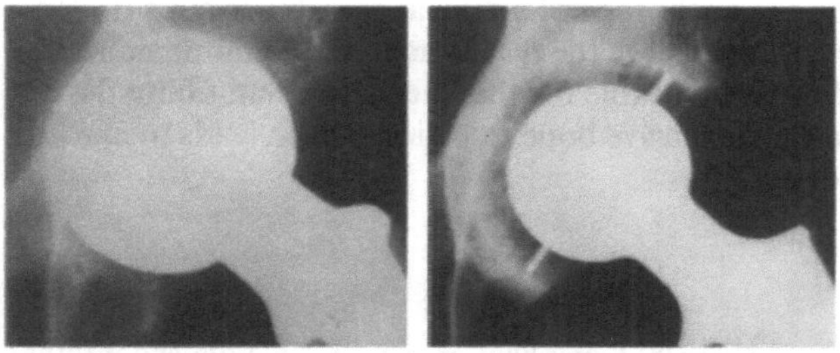

a b

Fig. 5a, b. An 86-year-old woman, treated with a femoral head prosthesis after femoral neck fracture; reoperation was performed because of penetration of the prosthesis into the acetabulum. **a** Penetration of the prosthesis into the acetabulum 18 months after arthroplasty; **b** newly formed acetabulum floor by homologous bone graft, 4 years after reoperation other cases extensive resorption of the transplants is frequently observed e.g. in treatment of traumatic bone defects without prosthetic implantation. All six homologous bone blocks affixed to the head of the tibia exhibited radiological evidence of osseous ingrowth after 1 year. In the region of the femur, however, we were unable to assess the reaction of the transplants sufficiently because of the radiological superposition of the prosthesis.

A great many questions remain open with regard to the behavior and reaction of large homologous bone blocks. Clinical results often differ substantially and moreover cannot be compared in many cases because of great variation in the area of application, operation localization, method of preservation for the homologous bone grafts and the surgical technique used. The results of experiments with animals may be related only conditionally to situations involving human beings. We can safely assume that autologous bone transplants will be completely integrated into the transplant bed within 6 months, although there have been no definitive investigations with regard to solid homologous bone transplants. Despite such problems the experience we have gained with homologous bone transplantations on the acetabulum and with total knee-joint prostheses, suggests the autologous, and in problematic cases the homologous, bone transplantation to be a suitable method for stabilizing the anchorage of an implant.

Summary

Twenty-three autologous and 21 homologous bone transplantations for the purpose of stabilizing the anchorage of a knee prosthesis were performed at the Orthopedic Clinic of the BRK-Rheumazentrum, Bad Abbach, between January 1984 and June 1986. The indications for these transplantations were cystic destructions, pronounced valgus or varus malposition, and bone defects in the case of loose total prostheses. However, loosening of the implant was not observed in any of the knee joints treated in this manner. Moreover, solid homologous bone blocks exhibited radiological evidence of ingrowth into the transplant bed within a period of 6–12 months. It must be admitted, nevertheless, that the follow-up period is still very short, and the behavior of the bone transplant could not, therefore, be followed to a sufficient extent in the region of the femur because of the radiological superposition of the prosthesis. In light of the wide experience gained with bone grafts for securing the anchorage of artificial acetabular cups (follow-up period of more than 6 years), we believe the results achieved so far demonstrate bone transplantations to be more favorable for the knee joint than filling the defects with cement or undertaking extensive bone resection, which leads to marked disturbance of the locomotor system.

References

Baumann D, Küsswetter W, Witt AN (1978) Komplikationen nach Knietotalendoprothesen vom Scharniertyp. Unfallheilkunde 81:145

Dorr LD, Ranawat ChS, Sculco TA, McKaskill B, Orisek BS (1986) Bone graft for tibial defects in total knee arthroplasty. Clin Orthop 205:153

Gschwend N, Scheier HG, Bähler A (1981) Die GSB-Knieprothese. Prakt Orthop 10:439

Gschwend N (1977) Die operative Behandlung der chronischen Polyarthritis, 2nd edn. Thieme, Stuttgart
Hagena FW, Hofmann GO (1984) Lang- und mittelfristige Ergebnisse nach Implantation der GSB-Kniegelenks-Endoprothese. Unfallheilkunde 87:133
Harris WH (1982) Allografting in total hip arthroplasty. In: Adults with severe acetabular deficiency including a surgical technique for bolting the graft to the ilium. Clin Orthop 162:150
Harris WH, Crothers O, Oh I (1977) Total hip replacement and femoral head bone grafting for severe acetabular deficiency in adults. J Bone Joint Surg [Am] 29:752
Heywood AW (1980) Arthroplasty with a solid bone graft for protrusio acetabuli. J Bone Joint Surg [Br] 62(3):332
Holzhauser P, Wessinghage D (1984) Die aktive Pfannenbodenstabilisierung als Ergänzung zur Endoprothese bei polyarthritisch bedinger Protrusio acetabuli. Z Orthop 122:777
McCollum DE, Nunley JA, Harrelson JM (1980) Bone grafting in total hip replacement for acetabular protrusion. J Bone Joint Surg [Am] 62:7
McCollum DC (1978) Bone grafting in acetabular protrusion. Orthop Trans 2:262
Ranawat CS, Dorr LD, Inglis AE (1980) Total hip arthroplasty in protrusio acetabuli of rheumatoid arthritis. J Bone Joint Surg [Am] 62:1059
Udvarhelyi I, Risko T, Lakatos J, Zsengeller J (1985) Pfannenbodenplastik wegen Protrusio acetabuli bei Totalendoprothesen des Hüftgelenks. Z Orthop 123:973
Waertel G, Wessinghage D (1985) Die aktive Pfannenbodenstabilisierung bei totalendoprothetischem Hüftgelenkersatz. In: Draenert K (ed) Symposion in Orthopädie und Chirurgie des Bewegungsapparates – Die Implantatverankerung. Art and Science, Munich
Wessinghage D, Zacher J, Holzhauser P (1984) Die Protrusio acetabuli bei rheumatisch bedingter Coxitis. Aktuel Rheumatol 9:113
Wessinghage D (1984) Polyarthritisch bedingte Veränderungen des Hüftgelenks und ihre Behandlung durch prothetischen Gelenkersatz. In: Rahmanzadeh R, Faensen M (Hrsg) Hüftgelenksendoprothetik. Springer, Berlin Heidelberg New York Tokyo
Wessinghage D (1983) Operative Therapie. In: Mathies H (Hrsg) Handbuch der Inneren Medizin, Rheumatologie A. Springer, Berlin Heidelberg New York Tokyo
Wessinghage D (1983) Operative Behandlung bei Polyarthritiden. Orthop Praxis 3:189
Wetzel R, Wessinghage D (1986) Knochentransplantation als aktive Pfannenbodenstabilisierung beim künstlichen Hüftgelenkersatz der Protrusionshüfte. Orthop Praxis 6:454

Knee Arthroplasty with GSB and Engelbrecht's and Tönnis' Sledge Prosthesis in Rheumatoid Arthritis and Osteoarthritis

E. Kißlinger, D. Wessinghage, and J. Zacher

Orthopädische Klinik des BRK-Rheumazentrums, 8403 Bad Abbach, FRG

Introduction

Among the surgical procedures employed for severe deformities of the knee associated with rheumatoid arthritis and osteoarthritis, joint-preserving interventions are preferred at our clinic. These include articular synovectomies, arthrolyses, osteotomies of the articular surfaces and osteotomies (varus/valgus) in the vicinity of the knee joint, supplemented at times by necessary soft-tissue release, such as the elimination of contractures and the extirpation of Baker's cysts (Wessinghage 1983a, b; Wessinghage et al. 1982). In view of the possibilities available today for artificial joint replacement, it is regretful that more patients are not subjected to such promising joint-preserving therapy in proper time.

We believe the implantation of the "sledge" prosthesis to be indicated in cases involving the destruction of articular surfaces. This phenomenon is largely limited to the medial or lateral compartment, where the axial deviation is less than 30° and the ligament apparatus is stable following correction of malposition. Here we implant the ENDO model (manufactured by Link in Hamburg, FRG); this represents a modification of the previously used St. Georg prosthesis (Engelbrecht 1975; Dreyer et al. 1984), and the earlier prosthesis of Tönnis (Tönnis 1979). Below, we shall describe the sledge prosthesis which has been developed at our clinic. Generally we use the sledge prosthesis exclusively unicompartmentally. However, in cases in which after the implantation of a sledge prosthesis and with increasing endurance and range of motion there is an impression on the opposite compartment, we sometimes find it necessary to undertake a bicondylar replacement with a second sledge prosthesis. With total instability of the capsule-ligament apparatus, subluxation or luxation tendency (radiograph under load), and pronounced malposition, we feel that the implantation of a total knee-joint prosthesis is also indicated in the interest of realizing a correction in stress-dynamic disproportions, stability of the joint, and an improvement in the range of motion.

For total knee-joint replacement we use only the GSB model as described by Gschwend, Scheier, and Bähler (manufactured by Allo Pro in Baar, Switzerland); this is a physiological semi-constrained joint with mobile axis. This prosthesis is characterized by its minimal bone resection, wide condylar supporting surfaces, physiological kinematics with mobile axis, stability through intramedullary anchorage, and extensive range of motion (Gschwend 1981; Gschwend et al. 1973). With the integrated femoral shield as patella gliding tract and with the partial replacement of the articular surface of the patella, the present version attempts to counteract earlier disadvantages, such as erosion of the patella, with attendant painful limitation of function.

Surgery and Arthroscopy of the Knee
2nd Congress of the European Society
Editors: Müller/Hackenbruch

Method

In all, 582 total knee-joint prostheses were implanted in 401 patients (354 females, 47 males) in the period from 1 January 1977 to 30 June 1986. Predominant among inflammatory joint diseases were cases of rheumatoid arthritis (259 of 266 patients) and among osteoarthritis cases those of primary arthrosis (127 of 135 patients). Osteonecrosis, posttraumatic arthrosis, tumors in the region of the knee joint, and revision surgery figured relatively seldom among the indications (Table 1). A 23-year-old female patient with mutilating juvenile arthritis offered an exceptional indication: due to a pronounced valgus deviation, a lateral sledge prosthesis was implanted, as well as a total hip prosthesis on both sides and, furthermore, the contralateral knee joint was synovectomized. At the time of operation the average age of those suffering rheumatoid arthritis was 61.1 years; this was almost 10 years younger than that for patients with osteoarthritis who averaged 70.8 years of age (Fig. 1). Sledge prostheses had been implanted 3.3 years earlier.

Of the 528 implants 368 (69.7%) were for patients suffering rheumatoid arthritis and 160 (30.3%) for those with osteoarthritis. Due to pronounced changes the majority were total knee-joint prostheses in the case of rheumatoid arthritis (202 GSB and 165 sledge-type). The unicompartmental replacement was considered the most suitable for osteoarthritic patients (69 GSB, 91 sledge type). It is also interesting to note that in cases of rheumatoid arthritis 62.2% of prostheses were implanted laterally, whereas among patients suffering osteoarthritis 75.8% were implanted medially. The prominence of (secondary) valgus deformity is probably attributable to damage to the hip joint through contractures, the role of the tractus iliotibialis in cases of instability through synovitis, and flexion contracture, and a rheumatoid deformity of the feet, evident in the majority of cases (Table 2).

Due to multiple diseases in joints with rheumatoid arthritis knee prostheses were implanted on both sides of 85 patients (32%); this compares to only 22 patients

Table 1. Indications for knee arthroplasty

	Number of patients
Rheumatoid arthritis	235
With fracture in vicinity of knee	2
With change of prosthesis	22
Juvenile arthritis	4
Psoriasis arthritis	3
Osteoarthritis (genuine)	122
After trauma	3
With change of prosthesis	7
Osteonecrosis	2
Giant cell tumor distal femur	1
Total	401

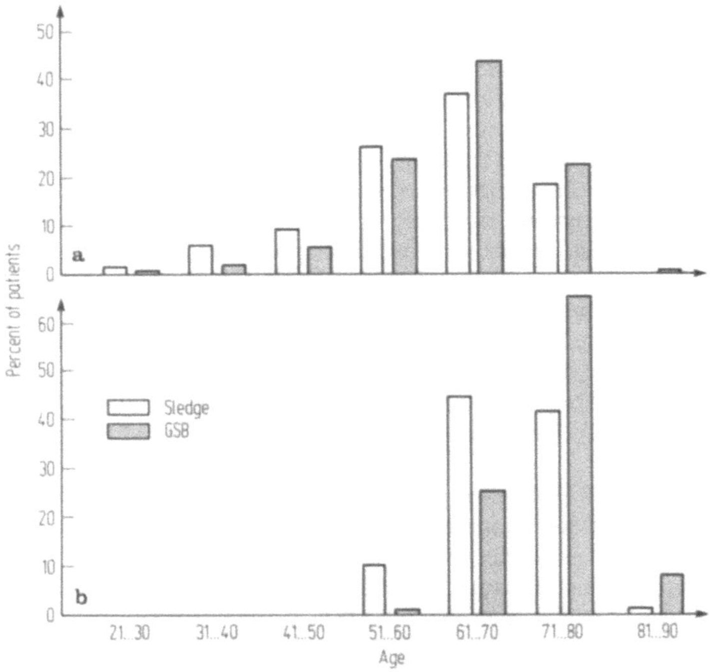

Fig. 1 a, b. Distribution of patients by age groups. **a** Patients with rheumatoid arthritis; **b** patients with osteoarthritis. *Striped bars*, those receiving sledge prostheses; *solid bars*, those receiving GSB prostheses

Table 2. Type of implant by type of arthritis

Implant	Rheumatoid arthritis		Osteo-arthritis		Total	
	n	%	*n*	%	*n*	%
GSB total prosthesis	203	55.5	69	43.1	272	51.5
Sledge prosthesis	165	44.8	91	56.9	256	48.5
Medial compartment	62	37.6	69	75.8	131	51.2
Lateral compartment	103	62.4	22	24.2	125	48.8
Total	368	69.7	160	30.3	528	100.0

with osteoarthritis (16.2%). For 49 patients suffering rheumatoid arthritis an endoprosthetic replacement was provided for three or all four large joints of the lower extremity. A synovectomy was performed in 61 rheumatoid arthritis patients (26 at another clinic), before the increasing destruction made it necessary to undertake an alloarthroplasty after an average period of 4.7 years (0.2–13 years).

All surgical interventions were carried out in an Allo Pro ultrasterile box (laminar flow), usually with a tourniquet. The less traumatizing medial longitudi-

nal incision on the extensor side with staggered offset severance of the lower layers (Scheier 1977) has proven successful. This was followed by synovectomy, including debridement, removal of the menisci and Hoffa's fatty pad with GSB resection of the ligament, and – in cases of flexion contracture – detachment of the dorsal portions of the capsule. A lateral release was adopted as routine in the event of valgus stress: severance of the contracted fascia lata as well as of the septum intermusculare between vastus lateralis and the biceps femoris. A set of instruments is available to ensure a standard implantation procedure. After introduction of a medullary plug consisting of autologous cancellous bone in the femoral and tibial cavities, the implantation was effected with Refobacin-Palacos. Extreme deformities, especially in rheumatoid arthritis and in revision cases, made it necessary to build up femoral condyles and/or the tibial plateau with autologous/homologous corticocancellous bone grafts.
Intensive physiotherapy is started on the first postoperative day, and additional treatment provided in a kinotherapeutic bath 14 days after surgery. Partial weight bearing begins after 3 weeks. Zacher et al. (1986) report a good range of motion following such postoperative stationary treatment. The routine provision of motor splints therefore appears neither necessary nor indicated.

Complications

The fractures that occurred intraoperatively healed without problems. Four cases of thrombosis were experienced. Three patients with poor general condition who underwent surgery due to their total inability to walk, died in the 1st postoperative week. The surgical restoration produced good functional results in three cases

Table 3. Complications

Implants	Rheumatoid arthritis		Osteo-arthritis		Total
	GSB	Sledge	GSB	Sledge	
Intraoperative bone fractures	3	2	2	1	8
Wound-healing disorder (operative treatment)	15	9	6	3	33
Embolism		1	2	1	4
Death	2	1			3
Mobilization under anesthesia		1	2	1	4
Transient paresis of nervus fibularis		2	1		3
Rupture of tendon (musculus quadriceps), fracture of patella	2		1		3
(Sub)luxation of patella (operative treatment)	2		1	1	4
Deep infection	11	7	4	1	23
Change of prosthesis					
Aseptic loosening		5		2	
Instability		15		3	25
Orthesis due to instability	1				1
Total	203	165	69	91	528

of rupture of the quadriceps tendon or patellar fractures (in two cases as a result of a fall). In four cases, the tibial tubercle was medialized because of luxation or subluxation of the patella.

With the previously employed incision according to Mori (two parapatellar incisions) and the subsequently used, mostly medial Payr incision the percentage of superficial wound-healing disorders (10.7% and 13.4%, respectively) and deep infections (9.3%, 8.9%) was clearly higher than that with the median longitudinal incision used since the beginning of 1983; here the superficial wound-healing disorders were found among 4.1%, deep infections among 2.7% of patients. The corresponding figure for the rheumatoid arthritis group (3.0%) was slightly higher than that for the osteoarthritis group (1.7%).

Except for a single case deep infections could be healed with retention of the prosthesis. This was accomplished by means of a timely and extensive revision with resynovectomy and suction-irrigation drainage. Otherwise an arthrodesis with fixateur externe or orthesis was performed after removal of the prosthesis. In the instance of sledge prostheses that had to be removed and replaced (because of aseptic loosening) the tibial component was loose in each case and also the femoral sledge in one. The medially implanted prostheses were found to be loose in six instances, but only in one case involving lateral implantation. The second intervention was performed on average 3.6 years after the first. In those cases in which surgery had been performed because of instabilities or malpositions, preference was given to the lateral side (6 medial, 12 lateral) (Table 3).

Results

It was possible to conduct a follow-up examination of 222 patients with 270 implants after an average of 2.5 years (Table 4). The evaluation was made according to the ARO system (Kniespezialbogen der Klinik Wilhelm Schulthess, CH-Zürich) which considers the clinical parameters of pain, swelling, inflammation, walking aids, walking distance, and flexion and extension deficiencies. To facilitate a comparison of the sledge prostheses instabilities were also rated, as in the

Table 4. Follow-up statistics

	Rheumatoid arthritis		Osteoarthritis	
	GSB	Sledge	GSB	Sledge
Sex				
Male	6	8	3	6
Female	69	66	27	37
Implants	99	89	36	46
Age at operation (years)	61.3	57.5	71.3	69.0
(range)	(34.8–73.3)	(23.1–76.1)	(56.3–83.0)	(54.1–80.5)
Follow-up (years)	2.1	2.8	2.2	2.6
(range)	(0.5–7.7)	(0.5–6.1)	(0.5–7.6)	(0.5–8.5)
Total number of patients	75	74	30	43

Table 5. Follow-up criteria: ARO rating scale

	Scoring range (points)
Criterion	
Pain	0 – 3
Swelling	0 – 0.5
Inflammation	0 – 0.5
Walking aids	0 – 2
Walking distance	0 – 1
Flexion	0 – 2
Extension	0 – 1
Instability	−2 – 0
Results	
Very good	8.5–10
Good	7 – 8
Satisfactory	5 – 6.5
Poor	<5

Table 6. Patients reporting pain (percentages)

Pain	Rheumatoid arthritis				Osteoarthritis			
	GSB		Sledge		GSB		Sledge	
	Pre-operative	Follow-up	Pre-operative	Follow-up	Pre-operative	Follow-up	Pre-operative	Follow-up
None	–	17.4	–	14.3	–	18.8	–	25.0
Slight	3.2	72.8	7.8	66.2	3.1	75.0	12.3	52.3
Moderate	44.6	9.8	59.7	10.4	59.3	6.2	68.2	22.7
Severe	52.2	–	32.5	9.1	37.6	–	29.5	–
Less pain	91.9		85.4		86.1		84.8	
Same pain	2.0		7.8		2.8		8.7	
More pain	6.1		6.8		10.1		6.5	

case of Kerschbaumer (1981), who introduced the additional criterion, ranging over a points score of −2–0 (Table 5). A reduction in subjective pain was realized with 84.8%–91.9% of patients, whereby the rheumatoid arthritis and GSB groups exhibited the better results in comparison with their respective counterparts. Unsatisfactory results with a sledge prosthesis may be attributed to pain associated with instability or to the increasing destruction of the nonreplaced compartment among the GSB group, but above all to the pain caused by the retropatellar imbalance (Table 6). For example, in the rheumatoid arthritis group with GSB knee prosthesis without femoropatellar joint replacement, 14.5% (8 cases of 55), as opposed to 31.5% (6 of 19) in the osteoarthritis group, complained of severe retropatellar pain; for those receiving the GSB knee prosthesis with femoropatellar joint replacement the proportion was only 9.1% (4 of 44) among those with rheumatoid arthritis and 5.8% (1 of 17) in the osteoarthritis group

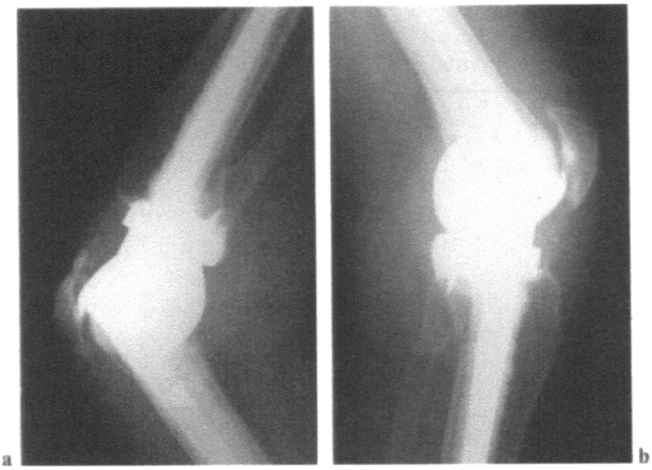

Fig. 2 a, b. 58-year-old female with rheumatoid arthritis. **a** Postoperative; **b** erosion of patella 2 years after implantation of GSB knee prosthesis without femoral shield and patella inlet

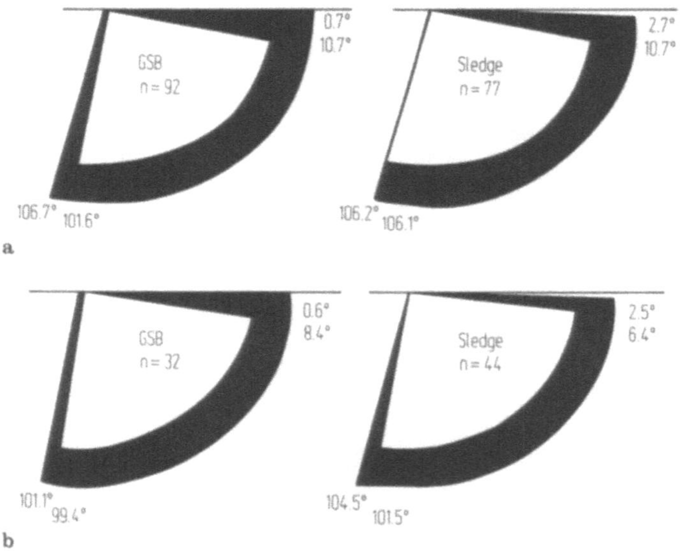

Fig. 3 a, b. Average range of motion (flexion and extension). **a** Patients with rheumatoid arthritis; **b** patients with osteoarthritis. *Inner unshaded area,* preoperative; *outer shaded area,* improvement at time of follow-up

(Fig. 2). The preoperative extension deficiency of 6.4°–10.7° was more or less equal with total prostheses in all four groups (0.6°–0.7°); with the sledge prostheses, deficiencies of 2.5°–2.7° still remained. An average degree of flexion in excess of 100° (allowing the patient to climb stairs or rise from a sitting position) was realized in all groups (Fig. 3).

Table 7. Patient mobility (percentages)

Patient mobility	Rheumatoid arthritis				Osteoarthritis			
	GSB		Sledge		GSB		Sledge	
	Preoperative	Follow-up	Preoperative	Follow-up	Preoperative	Follow-up	Preoperative	Follow-up
No walking	8.6	–	2.6	1.3	6.3	–	–	–
< 100 m	48.9	18.5	51.9	22.0	43.7	3.1	36.3	11.4
<1000 m	35.9	44.6	37.7	32.1	40.6	37.5	54.5	25.0
>1000 m	6.6	36.9	7.8	44.6	9.4	59.4	9.2	63.6
Wheel chair	8.6	–	1.3	1.3	6.3	–	–	–
2 Canes	53.3	50.0	42.9	31.2	28.1	28.1	27.3	13.6
1 Cane	33.7	34.8	48.1	41.5	59.4	59.4	56.8	38.6
Without cane	4.4	15.2	7.7	26.0	6.2	12.5	15.9	47.8

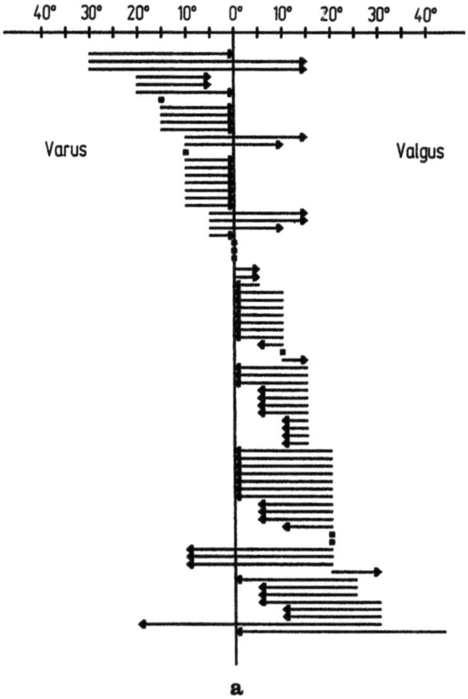

Fig. 4 a, b. Correction of deviation of axis with unicompartmental replacement. **a** Patients with rheumatoid arthritis ($n=77$); **b** patients with osteoarthritis ($n=42$). *Right*, valgus; *left*, varus. *Arrows* indicate change from preoperative to follow-up observations

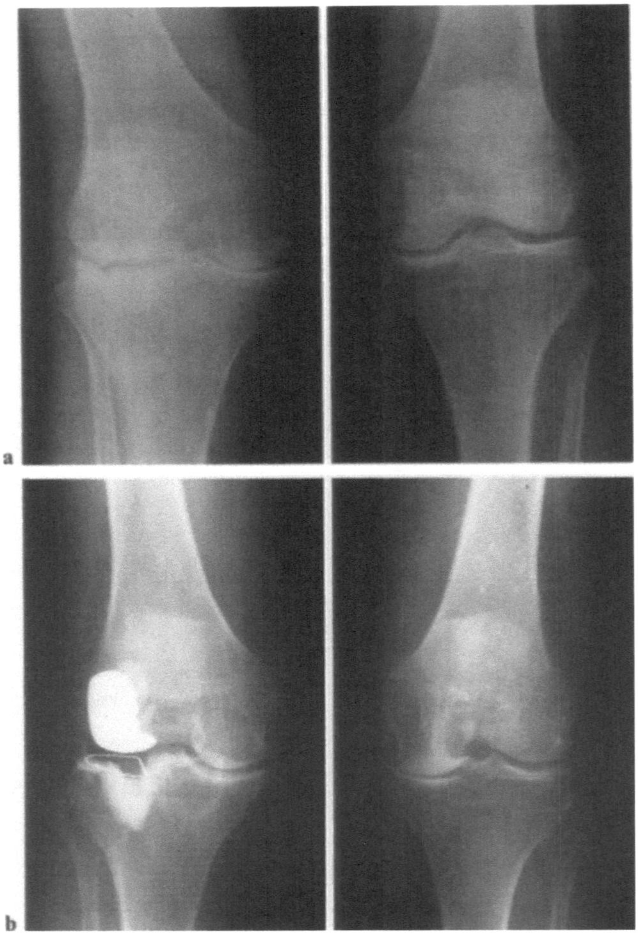

Fig. 5 a, b. A 69-year-old patient with rheumatoid arthritis. **a** Preoperative; **b** good result 5 years after implantation of lateral sledge prosthesis

Rating of walking distance and use of walking aids is problematic because of the commonly multiple diseases in a joint. However, as is shown in Table 7, it was possible to make those patients confined to bed or wheel chair mobile again in virtually every case. With sledge prostheses and in cases of osteoarthritis the axial deviation could be corrected and stabilized. On the other hand, follow-ups on rheumatoid arthritis patients revealed an irregular picture with overcorrections and malpositions, irrespective of the original indication (Fig. 4).
The overall assessment of very good or good results ranged from 65.0% (sledge prostheses and rheumatoid arthritis) to 85.1% (GSB and osteoarthrosis). Those with rheumatoid arthritis and receiving sledge prostheses showed the least satisfactory results; this can probably be attributed to their developing instabilities and to the progressive destruction in nonreplaced compartments. This overall as-

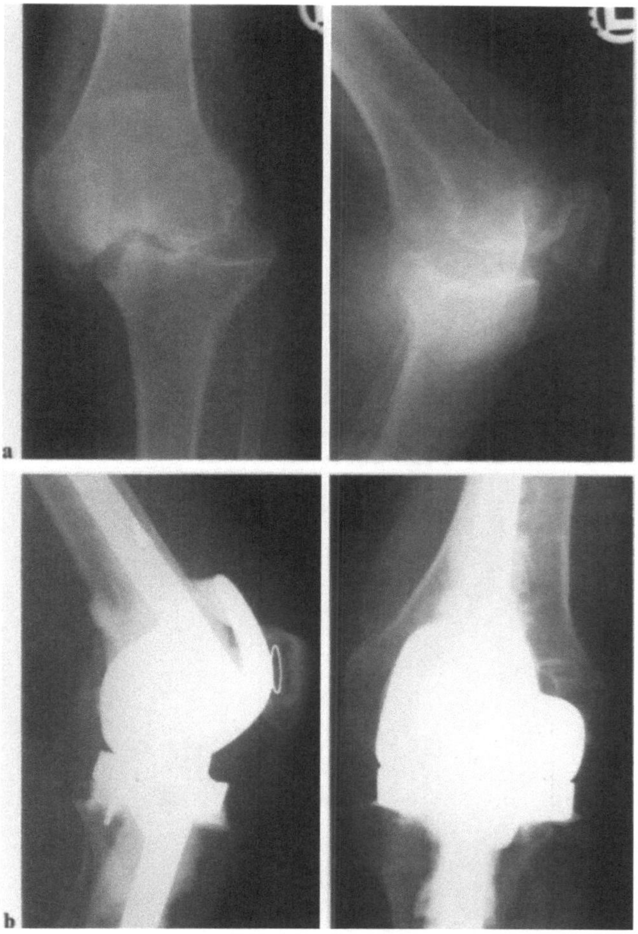

Fig. 6 a, b. A 61-year-old patient with rheumatoid arthritis. **a** Preoperative; **b** implanted GSB prosthesis after total destruction of knee with subluxation

sessment correlates also with the subjective reports by the patients themselves, whereby the reduction in pain – a decisive subjective criterion for them – actually led to a better rating (Table 8).

Discussion

Despite the high standards that have been achieved with knee-joint prosthetics, the higher complication rate in comparison with hip-joint replacements admonishes caution with regard to indications. Post-operative results in terms of objective criteria and patients' assessments appear, however, to indicate that this intervention can be a valuable tool in therapy; over 85% of our patients were satisfied with the result and would be willing to undergo the operation again. This figure

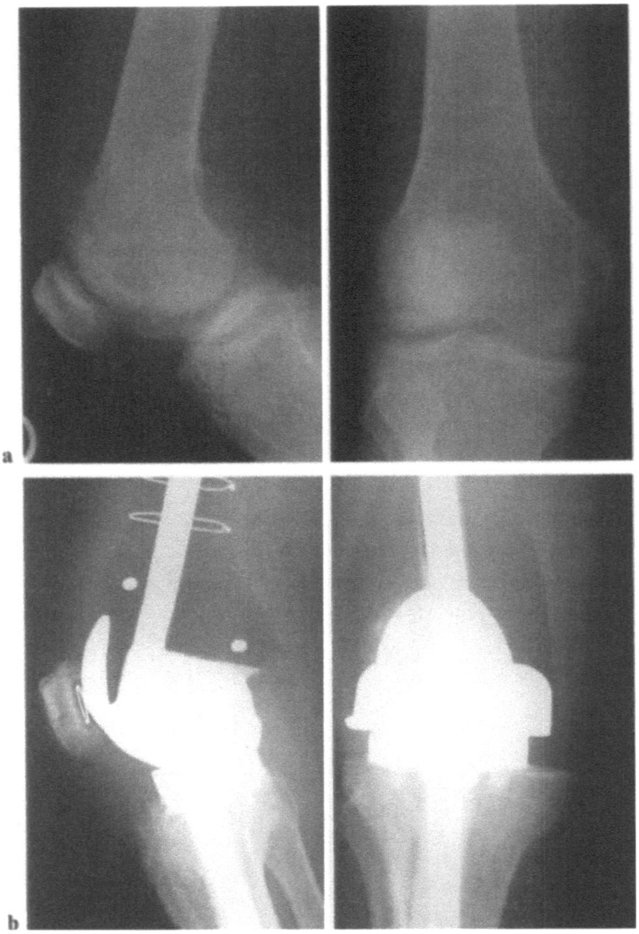

Fig. 7 a, b. Patient with giant cell tumor in distal femur region with invasion of the knee joint. **a** Preoperative; **b** implanted special custom-made "tumor" GSB knee prosthesis

corresponds to that reported by other authors (Table 9). After the implantation of the sledge prosthesis as a gliding-surface replacement, we must, nonetheless, contend not only with loosening but also particularly with corresponding painful malposition and instability. Among patients with rheumatoid arthritis 12.1% (20 of 165) and among those, with osteoarthritis 5.5% (5 of 91) of sledge prostheses had to be changed, implanted bicondylarly or replaced by a total prosthesis. This rate of failure conforms with the findings of Röttger and Heinert (1984) in a follow-up of 945 sledge prostheses. On the other hand, the minimal bone resection associated with the sledge prosthesis allows for good salvage procedures. Consequently we consider this to be an indication for gliding-surface replacement especially in the case of young patients suffering from rheumatic diseases. In keeping with the opinions expressed by other authors (Gschwend 1981; Hagena

Table 8. Results in follow-up

	Rheumatoid arthritis		Osteoarthritis	
	GSB	Sledge	GSB	Sledge
Results				
Very good	24.2%	24.7%	41.7%	45.5%
Good	59.6%	40.3%	44.4%	34.1%
Satisfactory	8.1%	16.9%	2.8%	6.8%
Poor	8.1%	18.1%	11.1%	13.6%
Patient's assessment				
Very content	88.0%	74.3%	83.3%	80.4%
Fair	2.7%	1.4%	5.5%	2.2%
Not content	9.3%	24.3%	11.2%	17.4%
ARO point average	8.5	6.9	8.1	7.6

Table 9. Follow-up results with GSB and sledge prostheses: this and other studies

<table>
<tr><th rowspan="3">Source of results</th><th rowspan="3">Number of implants</th><th rowspan="3">Follow-up (months)</th><th colspan="2" rowspan="2">ARO points</th><th colspan="4">Very good/good results (%)</th></tr>
<tr><th colspan="2">Objective</th><th colspan="2">Subjective</th></tr>
<tr><th>R.A.</th><th>O.A.</th><th>R.A.</th><th>O.A.</th><th>R.A.</th><th>O.A.</th></tr>
<tr><td>Geschwend (1981)</td><td>285</td><td>6–78</td><td>7.8</td><td>8.4</td><td>79</td><td>82</td><td></td><td></td></tr>
<tr><td>Schwägerl (1981)</td><td>80</td><td>24</td><td colspan="2">8.4</td><td colspan="2">87</td><td></td><td></td></tr>
<tr><td>Vogt (1981)</td><td>79</td><td>28</td><td colspan="2">7.7</td><td colspan="2">77</td><td colspan="2">67</td></tr>
<tr><td>Reichelt (1984)</td><td>41</td><td>37</td><td>8.1</td><td>7.0</td><td>84</td><td>63</td><td>94</td><td>82</td></tr>
<tr><td>Hagena (1984)</td><td>118</td><td>47</td><td></td><td></td><td colspan="2">74</td><td colspan="2">88</td></tr>
<tr><td>Own results</td><td></td><td></td><td></td><td></td><td></td><td></td><td></td><td></td></tr>
<tr><td>GSB</td><td>135</td><td>27</td><td>8.4</td><td>8.1</td><td>84</td><td>86</td><td>88</td><td>83</td></tr>
<tr><td>Sledge</td><td>135</td><td>33</td><td>6.9</td><td>7.5</td><td>65</td><td>75</td><td>74</td><td>80</td></tr>
</table>

R.A. = rheumatoid arthritis; O.A. = osteoarthritis.

and Hofmann 1984; Reichelt and Trümper 1984), retropatellar pain certainly constitutes the most specific problem with the GSB prosthesis.

We have not been compelled hitherto to undertake a number of the operations performed by other authors, such as a patellectomy and dura sheathing of the patella. Nevertheless, we have been concerned with medialization of the tuberositas with medial stabilization and lateral retinaculotomy on three occasions. Although the results obtained using the GSB prosthesis with femoral shield and patellar posterior surface replacement cover only a short follow-up period, they allow us to expect good results in future.

In contrast to Gschwend (1981) who does not report high infection rates among patients suffering rheumatic diseases the rate of deep infection in this group of our series (4.9%) was higher than that of patients suffering from osteoarthritis of the knee (3.1%). However, as Reichelt and Trümper (1984) and Scheier (1977) we, too, were able considerably to reduce the rate of superficial wound-healing

disorders and deep infections following the introduction of the median longitudinal incision.
Our experience with the implantation of these prostheses and with conservative and joint-preserving procedures (especially in the case of patients suffering from rheumatoid arthritis) has enabled us to attain the primary objectives of endoprosthetics: arthrogenous freedom from pain, sufficient range of motion for daily activities, and good salvage possibilities should the knee-joint replacement fail.

Summary

A total of 528 knee-joint prostheses were implanted in 401 patients at the Orthopedic Clinic of the BRK-Rheumazentrum in Bad Abbach over the period from 1 January 1977 to 30 June 1986. GSB total prostheses and the unicondylar sledge prosthesis (ENDO model) were employed throughout, these having replaced the earlier St. Georg prosthesis and that of Tönnis. The 266 patients suffering from rheumatic diseases were provided with 368 prostheses (203 GSB and 165 sledge) and the 135 patients with degenerative diseases were given 160 implants (69 GSB and 91 sledge).
A follow-up involving 222 patients with 270 implants was carried out, using the ARO rating system. Unsatisfactory results attained with GSB prostheses are attributable to retropatellar imbalance and those in the case of the sledge prostheses to the destruction of the nonreplaced compartment with axial malposition or loosening of the implant. Revision surgery has therefore been necessary in 9.8% of cases (25 of 256). The deep infection rate exhibited a correlation with the type of skin incision employed. The figure has fallen to 2.7% since the introduction of the median longitudinal incision, a rate that is significantly lower than that attained with the incisions according to Mori and Payr.
In overall terms 83.8% of GSB and 65.0% of sledge prostheses were rated very good or good in the rheumatoid arthritis group, as with 86.1% of GSB and 79.6% of sledge prostheses in the osteoarthritis group. The subjective assessment made by the patients showed that between 74.3% and 88.0% were very satisfied.

References

Dreyer J, Späh HJ, Teicher A (1984) Längerfristige Ergebnisse mit Schlittenprothesen St. Georg. Z Orthop 122:72–77
Engelbrecht E (1971) Die Schlittenprothese, eine Teilprothese bei Zerstörungen im Kniegelenk. Chirurg 42:510
Gschwend N (1981) The Gschwend-Scheier-Bähler-(GSB)-replacement of the rheumatoid knee joint. Reconstr Surg Traumatol 18:174–194
Gschwend N, Scheier H, Bähler A (1974) The GSB knee prosthesis. The Knee Joint Int Congr Rotterdam, 1973. Excerpta Medica, Amsterdam, p 264
Hagena FW, Hofmann GO (1984) Lang- und mittelfristige Ergebnisse nach Implantation der GSB-Kniegelenks-Endoprothese. Unfallheilkunde 87:133–143
Kerschbaumer F (1981) Vergleichende Nachuntersuchung von 74 Kniegelenksendoprothesen bei c.P. und Arthrose. Aktuel Probl Chir Orthop 15:47–50
Reichelt A, Trümper L (1984) Ergebnisse mit der GSB-Kniegelenksprothese bei Arthrose und rheumatoider Arthritis. Orthop Praxis 20:472–479
Röttger J, Heinert K (1984) Die Knieendoprothese St. Georg. Beobachtungen und Ergebnisse nach 10 Jahren Erfahrung mit über 3700 Operationen. Z Orthop 122:818–826

Scheier HG (1977) Spezifische Komplikationen des Totalersatzes im Kniegelenk. Orthopade 6:233–234

Schwägerl W, Böhler N, Czurda R (1981) Erfahrungen mit der endoprothetischen Versorgung des Kniegelenkes: eine mittelfristige Vergleichsstudie von verschiedenen Prothesentypen. Aktuel Probl Chir Orthop 15:51–58

Tönnis D (1979) Eine abgeänderte Schlittenprothese für den Aufsitz auf Kortikalisflächen. Z Orthop 117:509–511

Vogt KH, Mintrop J (1981) Mittelfristige Ergebnisse nach Kniegelenkstotalprothese vom Typ Guepar und GSB. Aktuel Probl Chir Orthop 15:37–42

Wessinghage D (1983a) Operative Therapie. In: Mathies H (ed) Handbuch der Inneren Medizin. Rheumatologie A. Springer, Berlin Heidelberg New York Tokyo

Wessinghage D (1983b) Operative Behandlung bei Polyarthritiden. Orthop Praxis 3:189–206

Wessinghage D, Zacher J, Mohr W (1982) Artikulo- und Tenosynovektomie bei chronischen Polyarthritiden. Int Welt 5:280–289

Zacher J, Wessinghage D, Rieger W, Wessinghage-Schäfler D (1986) Behandlungskonzept und Ergebnis der aktiven krankengymnastischen Übungsbehandlung nach Kniegelenkseingriffen beim Polyarthritiker und Arthrotiker. Z Phys Med Baln Med Klin 15:292

Cementless Knee Arthroplasty in Rheumatoid Arthritis

G. Bontemps

Fabricius-Klinik Remscheid, Brüderstr. 65, 5630 Remscheid, FRG

Considering the often young age and weak bones of rheumatoid arthritis patients, the long-term fixation of a prosthesis to the bone is still the central problem. The durability of the fixation depends on the interface stresses, which are related to the constraint of the prosthesis. To minimise these stresses the least constrained prosthesis providing sufficient stability is favoured. Even in newer cementing techniques the ageing of the acrylic cement cannot be avoided. These techniques also have the disadvantage of sacrificing more bone stock in cases of revision because of the greater depth of cement penetration.

Therefore, cementless fixation promising a durable biological fixation would be advantageous. In 1983 I decided to use the PCA system. It is a semi-constrained surface replacement prosthesis for all three compartments. It provides the physiological 6° of freedom of the natural knee joint and relies on the elastic capsulo-ligamentous structures for stress attenuation.

The design of the PCA prosthesis aims at reproducing the physiological knee-joint kinematics. The resulting rotational freedom reduces considerably the implant-bone interface stresses. The posterior cruciate ligament should be preserved. This is still a controversial matter. However, I do not see any sense in sacrificing this important stabilising agent (especially in flexion activities like walking down stairs), when afterwards one is forced to implant a more constrained prosthesis. The posterior cruciate ligament controls roll-back, thus maximising flexion and quadriceps power. This is of special importance for severely incapacitated patients with rheumatoid arthritis.

The fixation interface of the prosthesis has even planes to allow a precise fit of the bone resection surface. With irregularly arched or undercut contact surfaces this cannot be achieved satisfactorily or is at least difficult. All fixation interfaces are coated with a double layer of chrome cobalt beads. The resulting porosity offers the possibility of tissue ingrowth.

The instrumentation is also matched to the needs of cementless implantation. Jigs allow the precise fit of the bone to the prosthesis. They also grant restoration of the alignment as well as the adequate and symmetrical tension of the capsulo-ligamentous system. Apart from the release techniques for contracted structures described by Freeman [1] and Insall [3], the techniques of advancement for elongated capsulo-ligamentous structures can be used [2, 4].

It is difficult to judge the bone quality and its dynamic responses. According to my experience at least 75% of the supporting bone should have the vitality to provide fixation to the porous structure of the prosthesis. The bone stability must provide a solid fixation of the implants after press-fit. While flexing the knee

Surgery and Arthroscopy of the Knee
2nd Congress of the European Society
Editors: Müller/Hackenbruch

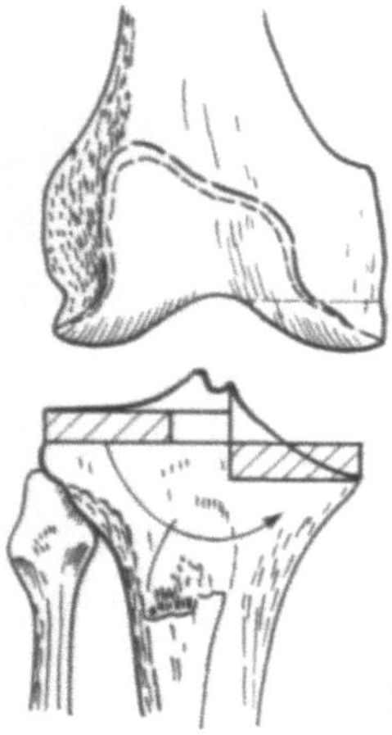

Fig. 1. Bone grafting sources

through the range of motion, no component tilting should occur. Initial stability of the prosthesis is required if ingrowth fixation is to be successful. Small cystic areas can be filled with spongiosa. Larger tibial defects have to be bone grafted with material from the lower end of the femoral condyle or from the counter lateral tibial plateau (Fig. 1). All these measures could also correct extensive deformities, offering a wide range of indication for the use of the PCA prosthesis. Non-reconstructable ligament laxity and extreme bone destruction are limiting factors.

My experience is based on 38 PCA prostheses implanted during the period from April 1983 to March 1986 in cases resulting from rheumatoid inflammatory diseases. At least two of the three components have always been implanted without cement; in four cases the tibial component was cemented, in two the femoral and in three the patella. All cases have been reviewed and checked according to a slightly modified BOA system. The average follow-up was 22 months. In 17 cases additional bone grafting was needed. Age distribution showed most patients to be in the sixth and seventh decades; mean age was 58 years.

The main reason for surgery was pain, which was totally or almost totally eliminated by this procedure in 95% of the cases. Walking ability was definitely improved. Postoperatively 83% could walk more than 0.5 h. Only three patients incapacitated by other rheumatic problems were limited to 10 min walking ability. Ten patients used walking aids. Average flexion was improved from 86° to 103° (58% were able to flex more than 100°; 35%, 80°–100°), flexion contracture decreased from 17° to 3°. In almost all cases (92%) a good varus-valgus stability was noted. Only in three cases did a slight instability remain.

A standing X-ray revealed the exact leg alignment with physiological conditions in 92% (Fig. 2). In two cases a varus deformity was not corrected sufficiently, because the tibial component was implanted in slight varus position. One knee remained in valgus, because a femoral component was malpositioned. Checking the bone/prosthesis interface showed in eight cases radiolucency in the outer medial and lateral area of the tibia plateau. Only in one case was a demarcation found on the dorsal femoral condyles and in two cases at the patella.

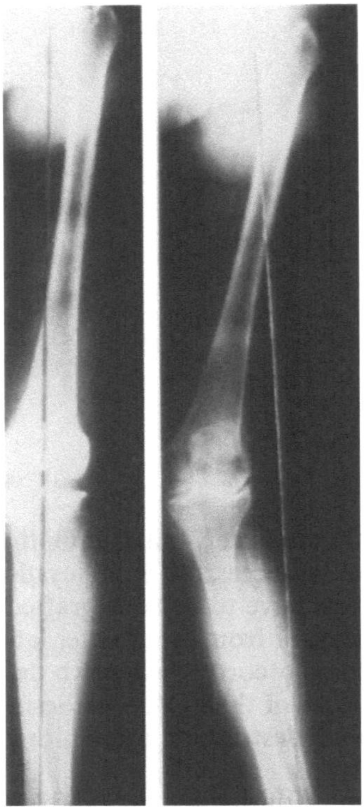

Fig. 2. Restoration of alignment

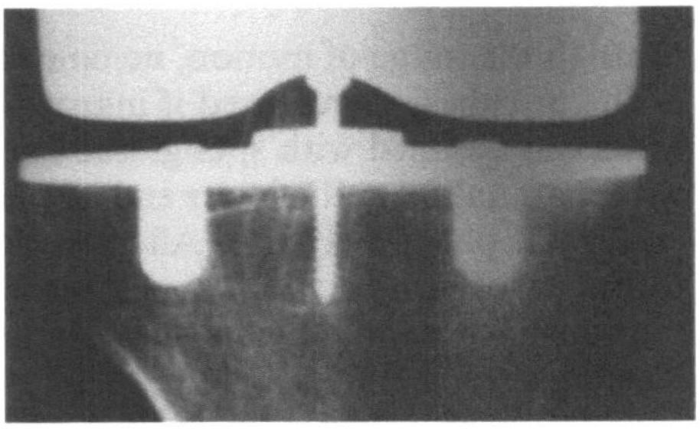

Fig. 3. Postoperative X-ray from 3-year follow-up

There were 13 cases of loose beads on the tibial side, less frequently on the femoral (5 ×) and only twice in the vicinity of the patella. If the beads were not knocked off during impaction of the prosthesis, they are, in my opinion, evidence of micro-motion at the interface. As long as their number stays small (not exceeding five), I consider them harmless, as in this instance (Fig. 3) 3 years postoperative with an excellent result. Sclerotic bone is in intimate contact with the prosthesis, and the trabecular pattern suggests stress transfer.

Overall classification with reference to the Hungerford score yields 94% good and excellent results One case was judged fair, because its range of motion was below 80° and had a flexion contracture of 15°. Another case shows patellar problems with effusions. This is also the only patient not satisfied with the result of the operation. The reason is loosening of the patella component (Fig. 4). No other serious complications like infections have occurred. One delayed healing and thrombosis left no permanent damage.

At this point I would like to mention that within the group of osteoarthritis patients also provided with a cementless PCA prosthesis I had to revise two cases because of tibial component loosening. Both were very active overweight women.

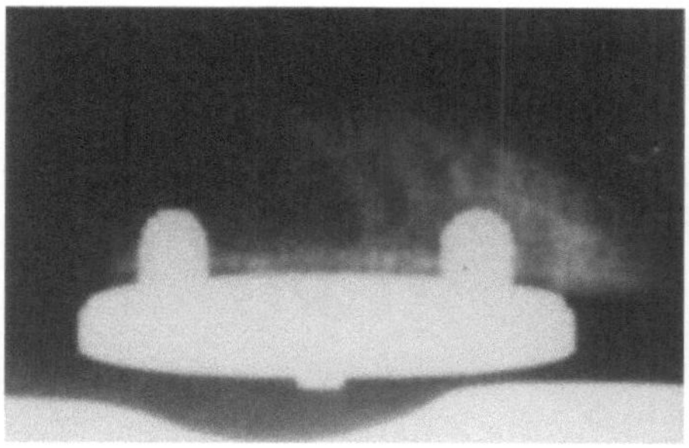

Fig. 4. Patella loosening

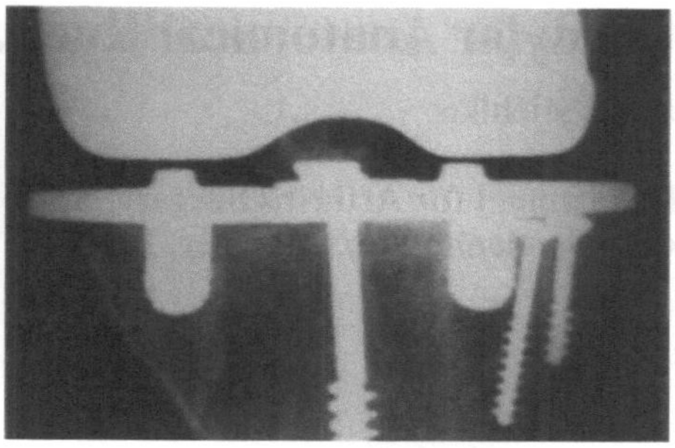

Fig. 5. Bone grafting of tibial defect

Finally, there is an illustration of bone grafting a tibial deficit with fixation by counter-sunk screws (Fig. 5). These techniques entail the necessity of prolonged protected weight-bearing.

Conclusion

In 3.5 years follow-up the cementless implantation of the PCA prosthesis in rheumatoid destruction of the knee joint was extremely successful. So far, only the loosening of one patella component has occurred. For the result there was no loosening of any biological fixation. A continuation of these good results would justify these more demanding operative techniques.

References

1. Freeman MAR (1980) Arthritis of the knee. Springer, Berlin Heidelberg New York
2. Hungerford DA, Kenna RV (1983) Preliminary experience with a total knee prosthesis with porous coating used without cement. Clin Orthop 176:95
3. Insall JN (1984) Surgery of the knee. Churchill-Livingstone, Edinburgh
4. Laskin RS, Denham RA, Apley AG (1984) Replacement of the knee. Springer, Berlin Heidelberg New York

Condylar Anatomical Knee Joint Implants

R. K. Miehlke

Department for Arthritis Surgery, St. Josef Stift,
4415 Sendenhorst/b. Münster, FRG

The SKI knee joint is a four-piece tricompartmental prosthesis which has been in use in this department since 1982. Besides the separate polyethylene tibial components there are metal-backed tibial components, a one-piece tibial component preserving only the posterior cruciate ligament, and a stabilising tibial component with intramedullary stem available. In this report the results of the first 100 insertions of the SKI knee prosthesis with separate polyethylene tibial plateaux are assessed.

The femoral component resurfaces the condyles and the trochlear groove; the tibial components and the patellar implant are used in an onlay technique. As a mere resurfacing and fully unconstrained arthroplasty of the knee, the SKI prosthesis depends largely upon the surrounding soft tissues for knee stability.

In the series of the first 100 cases there were 55 patients with 72 knees available for review. The follow-up period covered a minimum of 21.5 months and a maximum of 45.5 months, with an average of 30 months. Indications for arthroplasty were rheumatoid diseases in 74% and osteoarthritis in 26%. The overall results were 62% excellent, 24% good, 8% fair, and 6% poor, using a rating system including the factors of pain, range of motion, stability, prosthetic alignment, strength of the quadriceps mechanism, walking ability, ability to climb stairs and use of walking aids. Pre-operatively, the range of movement was between 104.6° of flexion and 16.6° lack of extension. Postoperatively, there was on average a practically identical flexion of 104.2°. The extension increased to an average of 4.9°.

Pre-existing instabilities could be, generally speaking, corrected in a satisfactory way. Clinical stability was gained in 78% of all cases. Slight insufficiency of the medial collateral ligament was found in 11%, and more severe combined instabilities of either the medial or lateral collateral ligaments and the anterior cruciate ligament were seen in another 11%. There was no insufficiency of the posterior cruciate ligament, and it has to be pointed out that the remaining instabilities postoperatively were by far less marked in comparison with the pre-operative findings.

Pain was differentiated into initial pain, additional pain on load and additional pain at rest. Pre-operatively, more than half of the patients (55.1%) showed initial pain, pain on load and pain at rest. Another 37.7% presented with pain on load and initial pain, and thus only 7.5% were practically pain-free before surgery. In these cases arthroplasty was indicated for other reasons, such as gross deformity or marked instability. Postoperatively, 58.4% of the knee joints were absolutely pain-free. Rheumatoid knees and osteoarthritic knees did not show significant

Surgery and Arthroscopy of the Knee
2nd Congress of the European Society
Editors: Müller/Hackenbruch

differences. Initial pain alone was registered in 18.1%. The osteoarthritis patients were affected in only 5.6%, whereas rheumatoid patients showed initial pain in 22.2%. Although the extensor mechanism in rheumatoid patients can be well trained following knee arthroplasty, some muscular pain in the initial phase of walking may exist. It does not seem to be capsular pain in these cases. Additional pain on load was still present in 13.8% postoperatively. Both the arthritic and the osteoarthritic groups showed similar percentages. In another 9.7% pain at rest was registered. Osteoarthritic knees were considerably more affected, with a percentage of 27.7% versus 3.7% in the arthritic knees. The reason for this might be that capsular pain is empirically more frequent in osteoarthritis. Assessing patellar symptoms critically, only 23.6% of all cases did not show any. Capsular peripatellar crepitation or slightly reduced patella mobility, and also slight tilt as shown in the skyline view, were seen in 44.4%. However, these minor symptoms were not responsible for any complaint, so that altogether 68% could be rated as very satisfactory results. In 20.8% capsular crepitation or reduced patella mobility was accompanied by mild peripatellar pain. Osteoarthritic knees were affected roughly twice as often as the arthritic knees (33.2% against 16.6%). Marked tilt or lateralisation of the patella combined with patellar pain was seen in 9.7%; patellar dislocation was registered in one case. As patellar tilt or lateralisation in the SKI knee can lead to bony contact between the remainder of the patella and the lateral condylar flange of the femoral component, capsular pain in this area can be induced. For this reason primary lateral capsular release is more frequently carried out after the first series. In this series, varus deformities up to 15° and valgus deformities up to 20° were corrected. The femorotibial axis was pre- and postoperatively assessed by radiographs taken in the one-leg stance position: 87.5% of the knees showed a femorotibial angle of 4°–10° postoperatively which represents correct alignment, 2.8% had a femorotibial angle of 1°–3°, 6.9% had a valgus position of 11°–15°, and in 2.8%, that is to say two cases, varus position below 0° was found. Those two knees belonged to one patient. She showed a varus position of 10° postoperatively on the right side and a varus position of 4° on the left side. On the right side the medial tibial component was loosened, and on the left side partial radiolucency could be seen around the medial tibial plateau.

Radiolucent lines were found in 7% of the femoral components; however, none of the components showed clinical signs of loosening. In 9.8% of the medial tibial components and in 7% of the lateral tibial components radiolucent lines appeared. Around the patellar components radiolucency could be seen in only 1.4%. Radiolucency did not exceed 1 mm in 30%, except in one medial tibial plateau.

The following complications refer to the series of the first 250 cases. Deep infection was seen in two cases. Both underwent arthrodeses, and one died due to pulmonary embolism. Aseptic loosening was seen in only three medial tibial components. In all three cases the varus deformity was not corrected sufficiently, which was obviously the reason for the loosening. One knee was plated and grafted; in the two other cases the tibial plateaux were replaced by a stabilising tibial component.

In two of the early cases displacement of the marking wires in one of the tibial components was noted. After the first case the manufacturer was informed, and subsequently a change was made in the mode of fixation of the wires in the tibial plateaux.

Patellar-related problems make up the majority of the complications so far found. Lateral displacement of the patella was seen in ten cases, dislocation of the patella in three. Loosening of the patellar component occurred once. Patellar fracture was found in two cases: one occurred through the distal part of the central fixation hole and was treated by wiring, and one fracture of the inferior patellar pole healed uneventfully. Patella tendon rupture occurred in one case, and after a traffic accident there was once a combined quadriceps and patellar tendon rupture.

Mobilisation under anaesthesia was necessary in 13 cases. Superficial delayed wound healing occurred in 14 cases, 12 of which had been on long-term corticoid therapy or on immunosuppressive drugs. In 4 cases secondary wound closure had to be carried out. Deep venous thrombosis occurred in 3 cases. The results presented here compare favourably with the data published by other authors.

It is obvious that patellar problems cannot be denied, and discussion is still open whether patellar replacement is absolutely necessary or not. In the SKI knee joint one of the reasons might be that the prosthetic trochlear groove is fairly narrow and rather prominent on the lateral side in order to give some resistance against lateral dislocation. Since lateral capsular release was carried out in all cases in which patellar tracking was not absolutely ideal during surgery, patellar-related problems have clearly diminished. As the frequency of radiolucency is still low after an average follow-up of 30 months, and loosening did not occur to a higher degree, the SKI tricompartmental arthroplasty with separate tibial plateaux is still being employed as a routine procedure at the present time.

Clinical results of noncemented knee joint prostheses seem to be comparable to those with cement fixation. For that reason indication for the SKI knee is limited nowadays. Patients must be over the age of 55, valgus deformity should not exceed 25°, and varus deformity, 10°. In varus deformities of more than 10°, especially if fixed varus deformity is present, adequate realignment may not be possible. In these cases preference should be given to stabilising the tibial component. Correction of flexion contractures up to 40° is possible.

However, one of the primary objectives for the use of an unconstrained knee joint implant must be meticulous surgical technique, especially adequate release of the surrounding soft tissues, in order to eliminate deforming forces prior to replacement and correct realignment of the knee axes. These factors, more than others, will guarantee long-lasting favourable results, and only a long-term follow-up series will answer the question whether the non-cemented or cemented technique should be preferred.

Perhaps the femoral and patellar components can more easily be left uncemented, and the tibial components preferably cemented in the majority of cases.

Innovations and Free Communications

Magnetic Resonance Imaging of the Knee: Normal and Pathological Imaging and Correlations with Arthroscopy

M. Denti, C. Bonizzoni, V. Ramondetta, and G. Peretti

The Third Orthopaedic Department, University of Milan, Milan, Italy

Introduction

Since May 1985, magnetic resonance imaging (MRI) has been performed on 35 normal and 53 pathological knees at the Third Orthopaedic Department of the University of Milan.

The knee joint is the site of a wide variety of diseases owing to the complexity of its components and the object of examination by means of diagnostic techniques that are in some way invasive.

In the light of initial studies on MRI (Moon et al. 1983a, b; Turner et al. 1985; McConkey et al. 1985; Reicher et al. 1985) this paper sets out to examine the actual possibilities of using MRI as a new tool in the diagnosis of knee conditions.

Material and Methods

All subjects in the series were examined with a superconductive apparatus with 0.5 T field strength. At first, a cylindrical coil designed for investigation of the brain was used. Subsequently, this was replaced by a surface coil of appropriate size and shape with a field of vision (FOV) of 13 cm, since this was better suited to the anatomical features of the relatively small region to be examined.

Images were taken in accordance with the multilayer technique in the three cartesian planes (sagittal, coronal and axial) and on oblique planes. It should be stressed that oblique sections could only be taken at an angle to one axis. In its present stage of development the technique does not allow visualisation of sections which are oblique to two or more axes. Images were obtained with the spin echo sequence: the pulse repetition time ranged from 1000 to 1400 ms with an echo time of 50 ms.

Depending on the size of the joint between 8 and 14, 5–7 mm thick layers, were taken. Acquisition matrices measuring 128 × 128 were constructed with 256 × 256 matrices on a 130–140 mm FOV. Axially positioned image sections were found useful for the menisci, transverse ligament, popliteus tendon and femoral insertion of the cruciates. Oblique (particularly parasagittal) sections were necessary when examining the major section of the anterior cruciate ligament (ACL).

The joint must be kept absolutely still during this period. The only real contraindication and obstacle to the use of MRI is the presence of metal components. Employment of a surface sound meant that the knee pathology could only be studied with the joint extended. The interpretation of surface sound MRI is still at the experimental stage. Surface sounds cannot provide the multiecho charac-

Surgery and Arthroscopy of the Knee
2nd Congress of the European Society
Editors: Müller/Hackenbruch

teristics that are useful, for example, in the differentiation of haemarthrosis and hydrarthrosis.

Subjects. MRI was performed on 35 normal knees (19 right, 16 left) from a group of patients consisting of 22 males and 13 females and on 53 knees with clinical evidence of both acute (16) and chronic (37) injury. The injured group was composed of 44 males and 9 females aged between 15–73 years including 31 right knees and 22 left knees. In all there were 35 athletes among the injured. Surgery or more frequently arthroscopy was performed in 36 cases with an indication for surgical management and served as a check on the effectiveness of MRI.
The method of MRI of knee components was described in our previous paper (Denti et al. 1986).

Results

MRI was not used for the study of bone fractures but solely for examination of cartilage components. An assessment was therefore made of the imaging of osteochondritis dissecans and patellar and femoral chondromalacia. Lesions of capsule and collateral ligaments are clearly indicated. The collaterals appear as a thin black band; total or partial discontinuity of this band can be seen (Fig. 1).
Isolated and combined acute and chronic lesions of ACL were studied. Their presence or absence can readily be diagnosed in coronal, sagittal and parasagittal projections. In acute injuries, various degrees of interruption can be detected (Figs. 2, 3). In chronic lesions the intercondylar notch, usually occupied by the ligament, is invaded by hypertrophic reactive synovial tissue and the remaining ligament fibres.
The low-signal band typical of the cruciate is replaced by a higher signal representing less structured tissue (Fig. 4).
Complete reconstruction of the ACL has not yet been fully studied.
The posterior cruciate ligament (PCL) can be readily identified, and its pathology can be precisely and systematically assessed.

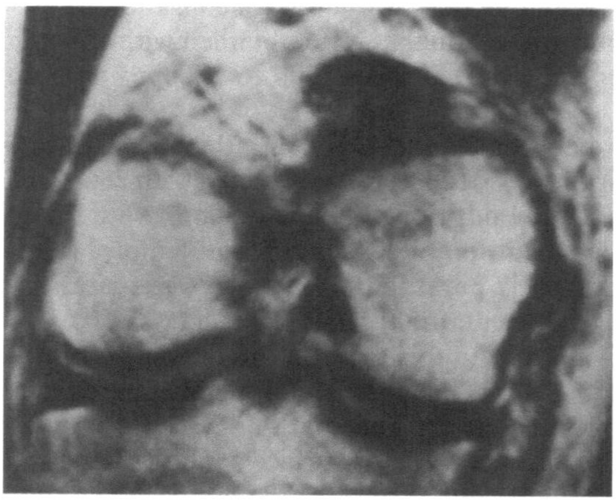

Fig. 1. Medial collateral ligament lesion with disinsertion of medial meniscus and capsule lesion

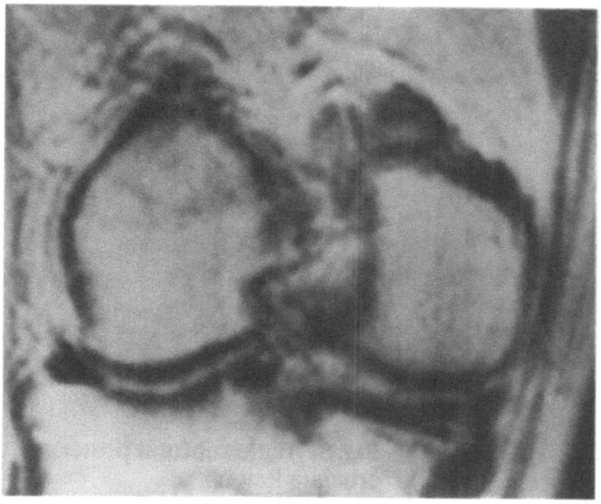

Fig. 2. Acute cruciate ligament lesion: the ligamentum mucosum is clearly visible

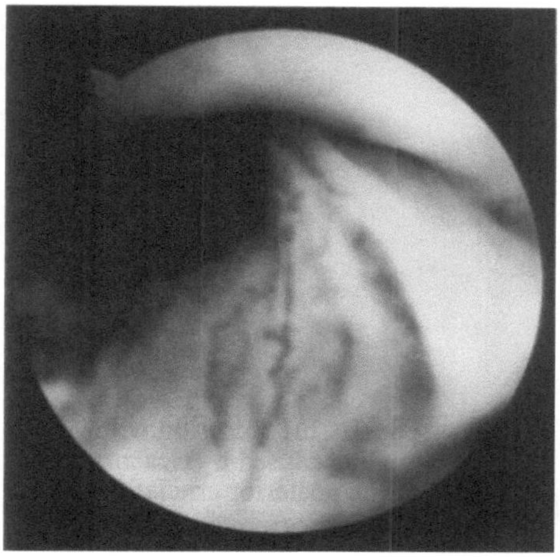

Fig. 3. Arthroscopy of the same knee as in Fig. 2

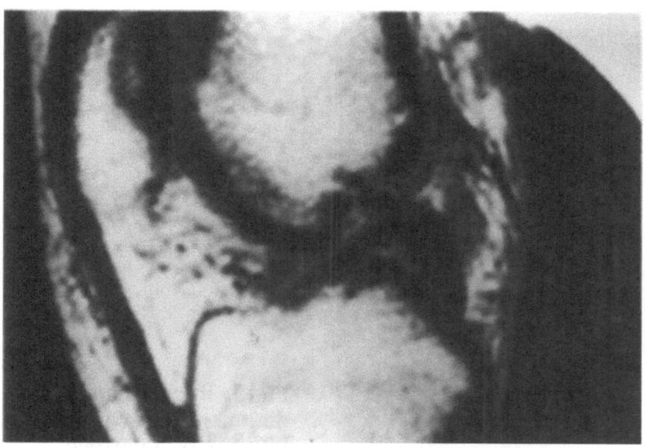

Fig. 4. Chronic anterior cruciate lesion: the intercondylar notch is filled with synovial tissue

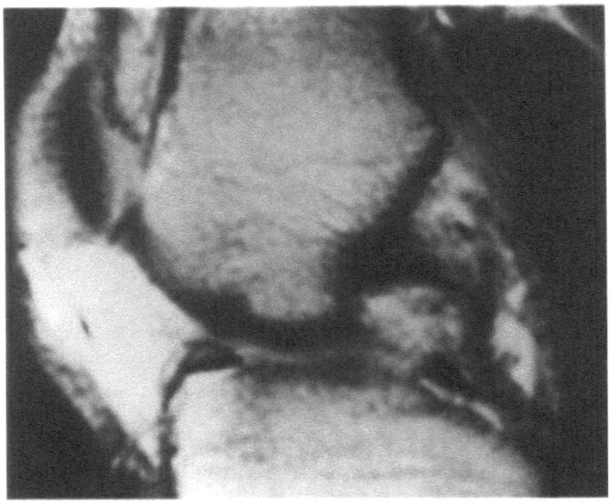

Fig. 5. Acute-subtotal posterior cruciate lesion

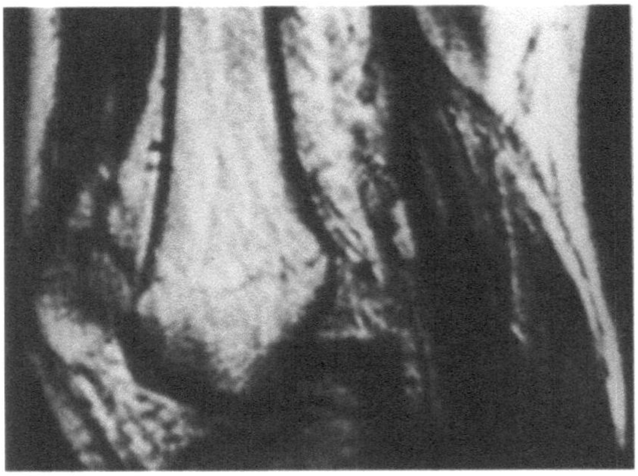

Fig. 6. Chronic lesion with angling caused by cicatrisation of the posterior cruciate stumps

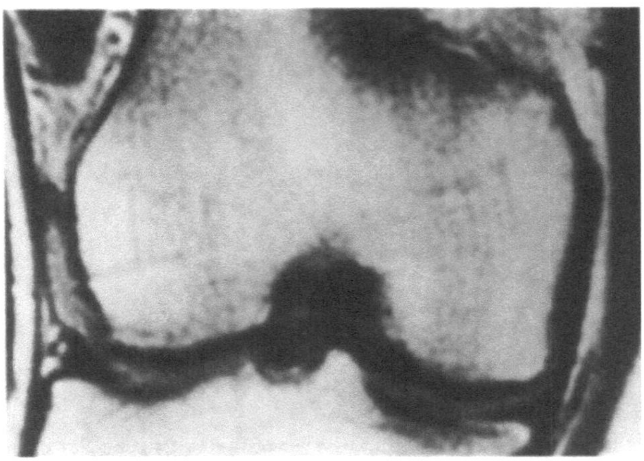

Fig. 7. Bucket-handle medial meniscus tear combined with a chronic anterior cruciate lesion

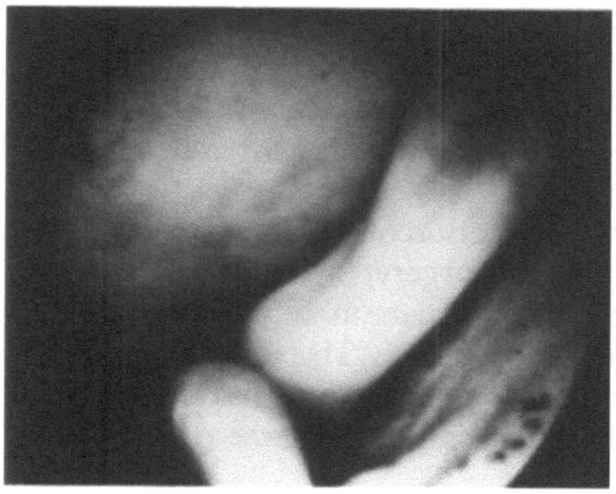

Fig. 8. Arthroscopy of the bucket-handle medial meniscus tear (Fig. 7)

Total and partial, intra- and extrasynovial injuries can be evaluated (Figs. 5, 6). The greatest difficulty in diagnosis and interpretation is undoubtedly associated with menisci, but there will certainly be advances in this field in the future.
We have studied all the types of lesions which occur as spurious signals from the meniscus and concluded that they are probably caused by infiltration of synovial fluid or blood (Figs. 7, 8).

Discussion

The usefulness and diagnostic effectiveness of MRI were assessed by comparing results with clinical diagnoses and with those established surgically or arthroscopically.
At the moment the comparative value of MRI versus clinical diagnosis cannot be expressed. In cases where an intraoperative or arthroscopic diagnosis was available, however, a much more precise comparison was possible. MRI was very reliable in the diagnosis of ligament pathologies, particularly cruciate ones: correct diagnoses were obtained in all the acute or chronic cases.
This group was composed of patients with an indication for surgery and others in whom the clinical picture provided valid support of the MRI findings.
The MRI diagnosis of collateral ligament injury was confirmed arthroscopically (associated with meniscal lesion) or intra-operatively (combined with cruciate lesion).
All the major difficulties were encountered with the diagnosis of meniscal damage. As reported earlier with regard to the normal knee (Denti et al. 1986), clear complete visualisation of the menisci poses a distinct technical problem. This is partly attributable to the size of the menisci themselves, and partly to the objective limits imposed on obtaining very thin layers (less than 4 mm); this difficulty is particularly evident on transverse or axial planes. Flap lesions and small longitudinal lesions prove the most difficult to distinguish.

Conclusions

Since we were faced with the investigation of a totally new set of problems, we found that the first meniscal lesions evaluated by MRI were not always correctly

diagnosed. Further evaluation, however, based on growing experience, especially after arthroscopy, has since enabled us to be more accurate.
MRI proves to have enormous potential for the study of acute injuries of the knee. For example, in situations in which a cruciate lesion may often be appreciated meniscal pathology is not so easily diagnosed.
MRI can thus provide an additional aid in the selection of patients for conservative management or arthroscopic or conventional surgery.
It is of undeniable validity in the case of patients who are athletes for whom the most accurate diagnosis possible must be provided quickly. It is to be emphasised that arthroscopy remains a technique of enormous value due to the possibility of diagnosing the total internal damage to the knee and, at the same time, of performing surgical procedures.
At present, of course, in comparison with arthroscopy, routine employment of MRI in the diagnosis of knee conditions is prevented by its cost and the paucity of apparatuses in current use.

Acknowledgements. The authors are extremely grateful for the assistance of Prof. Lucarelli, Dr. Raiteri and Dr. Landoni of the Nuclear Magnetic Resonance Research Center, San Pio X, Milan, Italy.

Summary

Magnetic resonance imaging (MRI) has proved a sound means of investigating the anatomy of the knee. On examination of acute and chronic injuries of the knee, MRI provides clear diagnostic resolution of meniscal tears and an excellent diagnostic resolution of ligamentous tears.
At present, routine employment of MRI in the diagnosis of knee conditions is prevented by its cost and the paucity of apparatuses in current use compared with arthroscopic techniques.

References

Denti M, Bonizzoni C, Raiteri R, Landoni L (1986) Magnetic resonance imaging of the normal knee. Italian J Sports Traumatology 1:7
Moon K, Genant H, Helms C, Chafetz N, Crooks L, Kaufman L (1983 a) Musculoskeletal application of nuclear magnetic resonance. Radiology 147:161
Moon K, Genant H, Davis P, Chafetz N, Helms C, Morris J, Rodrigo J, Jergens H, Brasch R, Bovill E (1983 b) Nuclear magnetic resonance imaging in orthopedics: principle and applications. J Orthop Res 1:101
McConkey JP, Li D, Davidson RG, Thompson JP (1985) Magnetic resonance imaging in the diagnosis of acute knee ligament rupture. Communication at the 4th Congress of the International Society of the Knee, Salzburg, 14 May
Reicher M, Basset L, Gold R (1985) High-resolution magnetic resonance imaging of the knee joint: pathologic correlations. AJR 145:903
Turner D, Prodromos C, Petasnick J, Clark J (1985) Acute injury of the ligaments of the knee: magnetic resonance evaluation. Radiology 154:717

Diagnosis of Ligamentous and Meniscal Lesions of the Knee Joint with Magnetic Resonance Imaging (MRI)

M. Rodriguez, G. von Schulthess, and N. Friederich

Department of Orthopaedic Surgery, University of Zurich,
Balgrist, 8008 Zurich, Switzerland

We report on our experiences utilising MRI in the diagnosis of acute and old injuries to the menisci, ligaments and capsular structures of the knee joint.

Material and Methods

We compared five normal knees with 30 knees with clinically suspected pathological findings. The knees were imaged with a Philips Gyroscan 1.5 S operated at 1.5 tesla field strength (TR 1100 ms, TE 30 ms). Images were obtained in coronal and sagittal planes. If necessary axial and oblique planes were added. In a few cases the study was performed after instillation of air or of a contrast agent into the joint. A clinical examination with conventional radiographs provided a preliminary diagnosis prior to MRI. Arthrography, arthroscopy and/or arthrotomy completed the diagnostic workup after MRI.

Results

Menisci and their ligamentous attachments to the capsule are easily evaluated in either coronal or sagittal planes. Lesions of the medial and peripheral third of the menisci showed up particularly well. The continuity of the PCL is depicted on sagittal and axial planes. Oblique planes are necessary to demonstrate the ACL completely. Patella, patellar tendon, Hoffa's fat pad and Baker cysts show up nicely in MRI, as do major osteochondral lesions.

Discussion

MRI is still a relatively new diagnostic tool in orthopaedics. Our limited experience suggests that MRI has a broader clinical value in depicting normal and pathological intra- and extra-articular structures of the human knee joint. At present its major drawbacks are long total examination time and the relatively high costs.

Summary

MRI seems to be an excellent alternative to already established tools in the evaluation of suspected structural abnormality of the knee joint. Anticipated improvements in hardware (surface coils) and software (picture generating time) open almost unrestricted diagnostic possibilities.

Surgery and Arthroscopy of the Knee
2nd Congress of the European Society
Editors: Müller/Hackenbruch

The Treatment of Hemophilic Knee Arthropathy and Inferior Limb Discrepancies by the "Ilizarov" Method

G. Torri, G. Bagnoli, E. Lozej, P. Cerea, G. Mistò, R. Ventura, and V. Pietrogrande

Institute Orthopedic Clinic, University of Milan,
Via Bignami n° 1, 20126 Milan, Italy

The aim of this work is the study of the clinical application in hemophilic patients of the "Ilizarov method" to correct varus and valgus knee deviations and to lengthen an inferior limb.

Materials and Methods

Six hemophilic A patients and two hemophilic B patients, aged from 17 to 40 years, entered the study. A cortical bone osteotomy at the proximal metaphysis of the tibia, combined with fibula osteotomy was performed in all cases. In two cases a lengthening of the tibia was also carried out. In another two cases, in which a severe "cystic" resorption of subchondral bone of the femoral condyles was present, a knee joint traction (arthrodiatasis) using the same external apparatus was attempted for about 30 days. All these surgical procedures utilized an "Ilizarov external apparatus" fixed to three or four couples of crossed Kirschner wires placed above and below the cortical bone osteotomy. The requested bone correction was obtained by operating the threaded rod progressively. Weight bearing and muscle and joint reeducation started on the 2nd day after the operation. The correction of varus or valgus knee deviation started on the 2nd day, lengthening of the tibia from the 4th day, and has continued at a rate of one millimeter per day, for about 40 to 60 days on average, until the requested lengthening has been achieved. The "regenerated" bone tissue could be observed radiologically in all patients after 40 days. The so-called stabilisation after lengthening lasted for about 4–6 months. At the end of treatment, the Kirschner wires were removed without anesthesia. Substitutive therapy with factor VIII or IX during the first 30 days was carried out daily in order to keep a factor level of at least 30%. Later, for another 30 days, substitutive therapy was attempted on alternative days, then stopped altogether.

Results and Discussion

The advantages arising from the use of the "Ilizarov method" in the surgical treatment of knee hemophilic arthropathy can be summarized in three ways. (1) Only a small skin incision is required; the thin Kirschner wires are well tolerated, so this technique appears to be the most bloodless of the surgical approaches in patients with a clotting defect. (2) The varus or valgus knee deviation can be corrected gradually. (3) In our experience using this method leads to a reduction of the amount of clotting factor infused during the postoperative period.

Surgery and Arthroscopy of the Knee
2nd Congress of the European Society
Editors: Müller/Hackenbruch

Antibiotic Absorption under Tourniquet

D. P. Johnson, G. C. Bannister, J. M. Auchincloss, and J. H. Newman

Department of Orthopaedics, Bristol Royal Infirmary, Bristol, UK

Introduction

Recent experience at Winford Orthopaedic Hospital, Bristol, has shown that total knee replacement is more susceptible to infection than hip arthroplasty. Deep infection may originate in the joint, or superficial infection may progress to a deep infection. This was found to be the case for at least 28% of deep infections following knee arthroplasty (Johnson and Bannister 1986). There is evidence to suggest that prophylactic antibiotics can significantly reduce the incidence of deep infection in hip arthroplasty, and it seems reasonable to assume that there may be a similar benefit for knee arthroplasty. For proficient prophylaxis adequate concentrations of antibiotic must be present in the tissues at the time of surgery and in the early postoperative phase (Bowers et al. 1973). The concentrations neccessary are assumed to be the minimum bactericidal concentration of any organisms present. This is usually four times the minimum inhibitory concentration (MIC) (Quintillian and Nightingale 1984). The recent study by Johnson (Johnson and Bannister 1986) showed that the most common organism causing infection following knee arthroplasty was *Staphylococcus aureus* (Table 1). Therefore adequate prophylaxis in knee arthroplasty is presumed to be four times the MIC of the *Staphylococcus*.
In order to obtain the maximum level of antibiotic in the tissue it should be administered intravenously a short time prior to surgery (Schurman et al. 1980). However, in knee arthroplasty the limb is exsanguinated, and a tourniquet is applied. If this is performed too soon after antibiotic administration, it may prevent the antibiotics reaching the tissues of the knee. It may also affect the rate of decay of the antibiotics in the tissues in the limb. It was not previously known how much time must elapse between administration of the antibiotic and the application of the tourniquet. This trial was therefore performed to establish the interval neccessary for tissue uptake to occur and to assess the pharmacodynamics of antibiotics in the limb isolated by the tourniquet.

Method

In all, 37 patients who underwent knee replacement were studied. Each patient was randomised to receive 1 g of cefamandole or 1.5 g of cefuroxime administered intravenously prior to application of the tourniquet or after an interval of 30 s, 2 min or 6 min to allow uptake of the antibiotic by the tissue. The limb was exsanguinated using an Esmarch's bandage, and the operation performed in the

Surgery and Arthroscopy of the Knee
2nd Congress of the European Society
Editors: Müller/Hackenbruch

usual way. Samples of tibial and femoral cancellous bone and subcutaneous fat were taken 30, 60 and 90 min after the application of the tourniquet. The samples were frozen and stored until analysis. Analysis was carried out by first crushing the washed bone then extracting the antibiotic with an equal volume of 0.05 μ phosphate buffer (pH 7.0). The bactericidal action of the centrifuged supernatant was tested against known standards of *Bacillus subtilis* NCTC 10400 or *Escherichia coli* 12665 seeded on agar (Leigh et al. 1982). The results were evaluated by comparing the standard error of the means and by the chi-squared test.

Results

The antibiotics were found to be scarcely detectable when administered after the touniquet had been inflated. There was found to be no significant difference between the antibiotic concentrations in the tibia and the femur. In those patients receiving antibiotics 30 s prior to application of the tourniquet, the concentrations detected were grossly inadequate and significantly lower than for the 2-min group ($P<0.05$). The concentrations were lower in the 2-min group than in the 5-min group ($P<0.05$). After 2 min, cefamandole was found in significantly higher concentrations in the bone than cefuroxime ($P<0.01$). After an interval of 5 min higher concentrations of both cefamandole and cefuroxime were obtained. A similar situation was found to occur in the antibiotic concentrations found in the fat samples.

When an adequate prophylactic antibiotic tissue concentration is presumed to be four times the MIC of *Staphylococcus aureus* then there were adequate concentrations of cefamandole in bone samples of all the cases if an interval of 5 min was allowed before applying the tourniquet, whereas the levels of cefuroxime achieved were inadequate in 17% of the cases after the same interval of time.

The half-life decay in the isolated limb was found to be greatly prolonged in bone ($P<0.05$) for both antibiotics. The half-life of cefamandole in bone was found to be 138 min, while that of cefuroxime was found to be 348 min. In contrast the antibiotic concentration in the subcutaneous fat was found not to decay with tourniquet application.

Discussion

The use of prophylactic antibiotics is widely accepted in hip arthroplasty although there is some debate about which is the best agent and method of administration. The recent study performed in Bristol (Table 1) demonstrated that 77% of the organisms implicated in infection following knee arthroplasty are sensitive to cefamandole and cefuroxime. Cefuroxime and cefamandole are the most commonly used antibiotics for joint arthroplasty prophylaxis and were therefore used in this study. The highest tissue concentrations are obtained by intravenous administration, prior to surgery, and thus this was the mode of administration employed here.

Infection rates in knee arthroplasty have consistently been higher than those in hip arthroplasty, but the cause for this is uncertain. One essential difference be-

Table 1. Organisms isolated from cases of infection following knee arthroplasty (Johnson and Bannister 1986)

Organism	Cases of infection		
	Superficial	Deep	Total
Staphylococcus			
Aureus	7 (29%)	13 (45%)	20 (37%)
Albus	12 (50%)	7 (24%)	19 (36%)
Streptococcus			
β-haemolytic	1 (4%)	1 (3%)	2 (4%)
Faecalis (non β-haemolytic)	2 (8%)	4 (14%)	6 (11%)
Diptheroids	0 (0%)	1 (3%)	1 (2%)
Gram-negative			
Proteus	2 (8%)	2 (7%)	4 (8%)
Pseudomonas	0 (0%)	1 (3%)	1 (2%)
Mixed infections	1	4	5
Total number of organisms isolated	24	29	53
Total number of patients	23	25	48

tween the operative procedures is that a tourniquet is used in knee arthroplasty. This study has demonstrated that if the tourniquet is applied too soon after injection of the antibiotics, it can prevent the antibiotics reaching the tissues around the knee during surgery, resulting in inadequate prophylaxis. Cefamandole was found to penetrate the tissues quicker than cefuroxime. For cefamandole an interval of 2 min between injection and application of the tourniquet resulted in adequate prophylaxis in 98% of cases, whereas an interval of 5 min resulted in adequate prophylaxis in all cases. An interval of 2 min resulted in sufficient concentrations of cefuroxime in 40% of cases while an interval of 5 min was adequate in 83%. The half-life of the antibiotic concentration beneath the tourniquet was prolonged both for cefamandole and cefuroxime.

Conclusion

This study has demonstrated that a period of delay is necessary following injection of the antibiotic prophylaxis in knee arthroplasty prior to application of the tourniquet. Too early application of the tourniquet results in inadequate prophylaxis. An interval of 5 min is necessary to obtain high enough tissue concentrations when 1 g of cefamandole is used, though a longer time is required when cefuroxime (1.5 g) is used. Sufficient concentration in the tissue can be ensured for both antibiotics if the antibiotic is injected prior to the induction of anaesthesia.

References

1. Bowers WH, Wilson FC, Green WB (1973) Antibiotic prophylaxis in experimental bone infections. J Bone Joint Surg [Am] 55:795–807

2. Johnson DP, Bannister GC (1986) The outcome of infected arthroplasty of the knee. J Bone Joint Surg [Br] 68:289–291
3. Leigh DA, Martiner J, Nisket D, Powell HDW, Cjurch JCT, Wise K (1982) Bone concentrations of cefuroxime and cefamandole in the femoral head in 96 patients undergoing total hip replacement surgery. J Antimicrob Chemother 18:303–311
4. Quintillian R, Nightingale C (1984) Principles of antibiotic usage. Clin Orthop 190:31–35
5. Schurman DJ, Hirshman HP, Burton DS (1980) Cephalothin and cefamandole penetration into bone, synovial fluid and wound drainage fluid. J Bone Joint Surg [Am] 62:981–985

Pigmented Villonodular Synovitis of the Knee: Advantages of Arthroscopy

B. Locker [1], J. Beguin [1,2], C. Vielpeau [1], and G. Loyau [1,2]

[1] Department of Orthopedics and Traumatology, University Hospital, Caen, France
[2] Department of Rheumatology, University Hospital, Caen, France

Pigmented villonodular synovitis (PVS) is a rare pathology (12 cases in 4000 arthroscopies), found predominantly in the knee. It presents a certain number of diagnostic and therapeutic problems in which our arthroscopic experience has been very useful.

Cases

We describe 12 cases of PVS. Women are in the large majority (nine women to three men), and the average age is 29 years (17–44 years).
The average time of evolution between the first symptoms and the first consultation was 2.5 years (from 3 months to 7 years). The symptoms leading to the consultation are equally divided between blocking (six times) and effusion (six times). No traumatic antecedent has been noted.
During the *physical examination* we have found:
5 palpable tumors
3 effusions
2 limitations of the extension
1 pain in the lateral compartment
1 normal knee
Only once was the *radiographic exam* abnormal (narrowing of the medial femorotibial joint). No arthrography was made.
The tentative diagnosis made after the clinical and radiological examination was of PVS only five times in 12. The other diagnoses made were: meniscus lesion (twice), loose body (twice), plica mediopatellaris (once), and monoarthritis (once). The *diagnosis* was made or confirmed by means of a percutaneous biopsy in one instance and eleven times by arthroscopy.
The arthroscopy was made with local anesthesia (10 times) using two- or four-point incisions [2]. It has allowed the macroscopic evaluation of the synovial membrane, the search for other lesions, multiple and precise synovial biopsies and, sometimes, the teatment of the lesions.
The *pathologic anatomy* of PVS has verified the presence of the histologic elements – cellular proliferation (hystiocytes, giant cells, etc.), ferric pigments, and important vascularisation. This has allowed us, by comparison with the arthroscopic and/or surgical findings, to classify our 12 cases into the following types:
Seven local tumoral (1–7 cm), all localized in the anterior compartment (four near the anterior horn of the medial meniscus, three in the lateral or medial cul-de-sac).

Surgery and Arthroscopy of the Knee
2nd Congress of the European Society
Editors: Müller/Hackenbruch

One diffuse
Four mixed, localized and diffused. One localized type was very peculiar, its symptomatology being sometimes a blocking of the knee and sometimes the sensation of loose body. During arthroscopy, we found a loose body that was identified as a necrotic VPS, the rest of the synovial membrane being normal.

Treatment

The excision of a localized type has been made by arthroscopy twice. The synovectomy has been surgical nine times. Of these, four times by three skin incisions, the only way to make a complete synovectomy (one medial parapatellar, one posterolateral, and one posteromedial for the condyles). In five patients we used one skin incision. In two cases a double meniscectomy was made.
Once, no treatment was carried out due to the refusal of the patient (at the 1-year follow-up, the clinical and radiological examination was normal).
The *follow-up* for our series has been, on average, 3.2 years (6 months to 9 years). It was 1.7 years for the localized forms and, 5.4 years for the others.

Results

Clinical. Eight patients are totally asymptomatic, including the two patients who had a double meniscectomy (9 years follow-up). Four patients experienced some pain (when they tire or with time).

Radiological. There was one external subluxation of the patella associated with a global narrowing of the femorotibial joint – a patient who had a synoviorthesis with osmic acid before and who showed important lesions during the procedure. There was also, one remodeling of the medial femorotibial joint (9 years after a double meniscectomy).

Control Arthroscopy. As performed on four patients, control arthroscopy has shown a posterolateral localization in a knee where there had been an incomplete synovectomy; in spite of no further treatment, the clinical and radiological examination at 3-year follow-up is normal.

Discussion

General Considerations

PVS is almost always monoarticular, particularly in the knee (80% of the localizations in the big articulations, according to Gaubert [6]) between 20 and 40 years of age. The sex distribution is classically the same (in Gaubert's series, 205 males and 236 females). This is not true for Granowitz and Mankin [7] or for us, but these two series are small.
A traumatic history has been reported by several authors [5, 11] and could support the traumatic theory, according to which the repeated hemarthrosis would be the cause of PVS [13]. Our series does not support the arguments in this sense,

and, moreover, this theory does not seem to be able to explain the localized type.

The tolerance time is in general less than 3 years, but it can also be very long, due to the frequent discrete appearance of functional problems and to diagnostic difficulties [4, 11].

Anatomoclinical Study

Following Jaffé et al. [8], we recognize a diffuse form where there is a villous and/or nodular proliferation that affects the entire synovial membrane, giving a symptomatology dominated by an intermittent intra-articular serohematic effusion that later becomes permanent. Similarly, we recognize a localized type where there is only one nodule, generally pediculated, giving a mechanical symptomatology where the blocking is the characteristic element.

Monghal et al. [11] have introduced certain distinctions and propose a classification based on the macroscopic aspect of the nodules. It seems to be better adapted to the two big groups of symptoms already cited, namely for the sessile villonodular types and the pedicled nodular types.

In our series, four cases presented with both an important tumor and diffuse involvement of the synovial membrane. This has led us to create a third type, the so-called "mixed" form, between the diffuse and the localized types. This type associates one or several localized tumors to a diffuse affection of the synovial membrane that is not always hypertrophic or easy to distinguish macroscopically from a reactional synovitis or an ochre synovitis due to blood resorption. Therefore, the biopsies should be systematic and numerous when a localized type is discovered because a mixed type will need a complete synovectomy associated with the elimination of the localized tumor.

Radiology

Standard radiography can show a thickening of the soft tissues or bones (erosions in the zones of synovial reflection, subchondral cysts in the intercondylar notch or under the tibial spines), but is frequently normal as in our series. Bone damage is for Granowitz and Mankin [7] a cause of poor prognosis. We have not had any experience with arthrography in other series that were difficult to interpret.

Arthroscopy

For us arthroscopy is of utmost importance. Its importance with regard to PVS has already been mentioned in the literature [1, 3, 9, 11, 12]. We use local anesthesia, frequently by the superior-lateral path that best enables us to see the localized types which are in the anterior compartment.

Our interest in this technique is threefold. On the diagnostic level, arthroscopy has several advantages. It enables us, first, to perform a better and less traumatic exploration than with traditional arthrotomy, allowing the discovery of small localized types; second, to determine the site, the size, and the aspect of a localized tumor and the morphology of the synovial membrane in different parts of the knee; and third, to take biopsies under direct visual control, allowing the classification of the PVS into one of the types already seen, so that a therapeutic decision can be made.

On the therapeutic level, arthroscopy permits the elimination of localized types of moderate or even great size with the appropriate instruments ("shaver"), allowing a complete synovectomy in the diffuse or mixed types.

Finally, arthroscopy is an excellent method for following these knees in trying to detect an insufficiency requiring surgery. It is a procedure easy to carry out, being not too aggressive and done under local anesthesia [14].

Treatment

PVS requires the complete elimination of the pathologic tissue. The risk of a recidivation is greater in the diffuse type [6, 11] and in the cases with bone involvement. The synovectomy can be arthroscopic, but classically it is surgical. It is performed by three skin incisions in the diffuse or mixed types and by one skin incision in the localized type. It can be completed in the diffuse or mixed types with an isotopic synoviorthesis or with osmic acid, which, for some authors [11], is the first treatment. We have never had to do it after a synovectomy.

Conclusion

Arthroscopy has a fundamental spot in our treatment of PVS. It is the masterpiece of the presurgical evaluation, a guide for the choice of skin incision, eventually a therapeutic tool and, finally, an important follow-up element. Control arthroscopy should be carried out when in doubt.

References

1. Beguin J, Locker B, Souquieres G (1984) Tumeurs bénignes localisées de la synoviale du genou et arthroscopie: à propos de 6 cas. J Méd Lyon 1386:25–28
2. Beguin J, Locker B, Sabatier J-P, Souquieres G, Rochet D, Carmes C (1981) Arthroscopie du genou sous anesthésie locale. J Méd Lyon 1362:7–9
3. Dorfmann H, Debievre J (1976) Arthroscopie et dérangement interne du genou. In: de Seze et al. (eds) L'actualité rhumatologique 1975. Expansion Scientifique, Paris, pp 239–249
4. Dryll A, Lansaman J, Bardin T, Ryckewaert A (1983) Synovite villonodulaire hémopigmentée: étude ultrastructurale et comparaison avec la synovite hémophilique. Rev Rhum Mal Ostéo-articulaires 50(2):87–93
5. Gaubert J (1972) La synovite villonodulaire du genou chez l'enfant. Ann Chir 26:1011–1018
6. Gaubert J, Mazabraud A, Verdie J-C, Cheneau J (1974) Les synovites villonodulaires hémopigmentées des grosses articulations. Rev Chir Orthop 60:265–298
7. Granowitz SP, Mankin HJ (1967) Localised pigmented villonodular synovitis of the knee. J Bone Joint Surg 49-A(1):122–128
8. Jaffe HL, Lichtenstein L, Sutro CJ (1941) Pigmented villonodular synovitis, bursitis and tenosynovitis: discussion of synovial and bursal equivalents of tenosynovial lesions commonly denoted as xanthoma, xanthogranuloma, giant cell tumors of myeloplaxa of tendon sheath, with some consideration of this tendon sheath lesion itself. Arch Pathol 31:731–765
9. Lefranc AM (1983) Arthroscopie et pathologie synoviale rare. Thesis, Caen

10. Locker B, Beguin J, Besnard M, Thomassin C, Agullo M (1986) Techniques anesthésiques en arthroscopie du genou. In: l'Arthroscopie du Genou. Course at C.H.U. Caen, 18–19 April 1986
11. Monghal J-P, Dorfmann H, Caroit M, Ryckewaert A (1981) Contribution à l'étude de la synovite villonodulaire du genou. Revue Rhum 48:397–402
12. Vignon E (1978) L'arthroscopie en rhumatologie. A propos de 102 examens. Lyon 239:119–124
13. Young J-M, Hudacek A-G (1954) Experimental production of pigmented villonodular synovitis in dogs. Am J Path 30:799–811

Villous Lipomatous Proliferation of the Synovia (Lipoma Arborescens)

T. Hallel and S. Lew

Sapir Medical Centre, Kfar Saba, affiliated to Sackler Medical School, Tel-Aviv University, Israel

Introduction

Lipoma arborescens is a rare condition which is usually only found affecting the knee joint. It was first described by Stout in 1953; then, in 1965, Weitzman published a case of lipoma arborescens of the knee, defining its clinical and pathological features, in the *Journal of Bone and Joint Surgery*. The clinical picture is one of slow, progressive development over years, presenting painless synovial swelling and effusion. The macroscopic appearance of the synovia is one of dense, yellowish-white villous proliferation located mainly in the suprapatellar pouch. Microscopy reveals extensive subsynovial hyperplasia of mature fat cells. Since Weitzman's article, only two cases have been subsequently published in detail in the orthopaedic literature [1, 2]. This study examined five cases, seen over a period of 12 years, at three different hospitals. One representative case is presented.

Case Report. In 1972 a 39-year-old male was referred to hospital because of painless swelling of both knees. The symptoms had been present in the right knee for 10 years and in the left, for 5 years. He presented exacerbations of tense effusion, with normal body temperature and normal laboratory tests. Double contrast arthrography of the right knee indicated a huge suprapatellar pouch with villous proliferation. A right knee synovectomy was performed in 1973 followed a year later by a synovectomy of the left knee. On arthrotomy dense, yellowish-white,

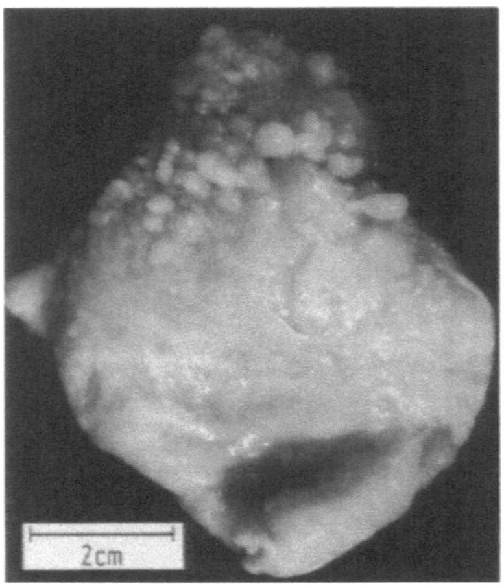

Fig. 1. Dense villous proliferation from the suprapatellar pouch: yellowish-white in colour and elastic in consistency

Surgery and Arthroscopy of the Knee
2nd Congress of the European Society
Editors: Müller/Hackenbruch

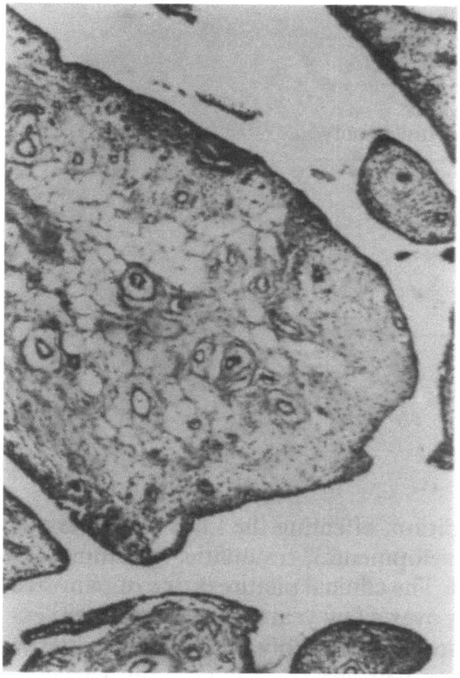

Fig. 2. Villous proliferation with subsynovial fat infiltration and foci of lymphocytes: mild hyperplasia of the synovial lining cells (hematoxylin eosin stain; ×40)

villous proliferation of the synovia was noted (Fig. 1). Microscopic examination of the synovial villi showed extensive mature subsynovial fat cell proliferation with some lymphocytic infiltration (Fig. 2). The preoperative radiological examination showed mild osteoarthritic changes in the right knee and very mild ones in the left. On re-examination in 1984, 10 years after the left knee synovectomy and 11 years after the right, the patient did not present any knee swelling or effusion but had mild knee pain and some crepitations under the patella. Some progression of the radiological osteoarthritic changes as compared with the preoperative radiographes 11 years before was noted.

Results

The five cases treated (Table 1) were all males, aged 39, 66, 56, 56 and 63 years. The duration of the disease before diagnosis was 10 years in the right and 5 years in the left knee in the first case and 30, 6, 5 and 25 years in the other four cases respectively. Synovectomy was performed in all cases. The postoperative follow-up was 11 and 10 years in the first case and 10, 8, 5 and 2 in the other cases, respectively. In cases 2 and 5, in which the condition was present 30 and 25 years before diagnosis, the patients suffered from severe and moderate knee osteoarthritis, respectively. At follow-up (2–11 years, mean 7 years, post-synovectomy) none of the six knees indicated recurrency of effusion or synovial swelling. However, some progression of radiological osteoarthritic changes was noted in the cases with the longer follow-up.

Table 1. Cases of lipoma arborescens examined

Case number	Age (years)	Number of years	
		Before diagnosis	After synovectomy
1	39	10 (R)	11
		5 (L)	10
2	66	30	10
3	56	6	8
4	56	5	5
5	63	25	2

R, right knee; L, left knee.

Discussion

Lipoma arborescens is a rare benign condition, affecting the knee almost exclusively. The aetiology is unknown but developmental, traumatic, inflammatory and neoplastic causes have been suggested. The clinical picture is one of pain-free synovial swelling and effusion, developing over a few years, with intermittent exacerbations of effusion. The exacerbations are probably caused when hypertrophic villi are trapped, from time to time, between the moving joint surfaces. The macroscopic appearance is one of extensive, dense villous proliferation of the synovium, whitish-yellow in colour and elastic in consistency, which is chiefly prominent in the suprapatellar pouch. Microscopy shows marked subsynovial infiltration of mature fat cells. As the name lipoma arborescens implies a tumorous aetiology, it is suggested that the name villous lipomatous proliferation of the synovium is more appropriate for this condition.
The suggested treatment for this condition is synovectomy, however no postoperative follow-up studies have been reported. Synovectomy was performed on six knees in five patients resulting in a full cure with no recurrency of the disease at all over a follow-up period of between 2 and 11 years. The differential diagnosis should include all conditions which cause painless knee effusion and synovial thickening without systemic involvement such as synovial chondromatosis, pigmented villonodular synovitis or synovial haemangiomatosis. The macroscopic and microscopic picture, however, is typical and diagnostic.

Summary

Five cases of villous lipomatous proliferation of the knee synovium were seen over a period of 12 years. The clinical picture is one of painless, long-standing effusion and synovial thickening; however, the precise aetiology is not known. Macroscopically, the synovia show dense whitish-yellow villous proliferation; microscopically, extensive mature subsynovial fat-cell proliferation is seen. Synovectomy was performed in six knees resulting in full recovery at follow-up, 2 and 11 years after treatment.

References

1. David W Burgan (1971) Lipoma arborescens of the knee. Another cause of filling defects on a knee arthrogram. Radiology 101:583–584
2. Hallel (Huller) T, Czernobilsky B, Malkin C (1974) Harefuah 87(6):260–262
3. Stout AP (1953) Atlas of tumor pathology. Section II, Fascicle 5. Tumors of soft tissue. American Registry of Pathology, Armed Forces Institute of Pathology, Washington
4. Weitzman (1965) Lipoma arborescens of the knee. Report of a case. J Bone Joint Surg [Am] 47:1030–1033

Rheumatoid Arthritis: A Result of Different Pathomechanisms

H. G. Fassbender

Zentrum für Rheuma-Pathologie, Kleine Windmuehlenstrasse 2a,
6500 Mainz, FRG

Introduction

Rheumatoid arthritis (RA) is considered to be an inflammatory disease and has been known as such for years (Garrod 1859). The question arises as to whether this definition is still valid in 1987. A precise answer to this question is in fact highly important, because it is necessary to design a therapeutic concept which is flexible enough to adjust to individual pathomechanisms.

Definition of Inflammation. One must not forget that "inflammation" is primarily an empirical, clinical term, which was defined by Celsus in the 1st century A.D. following the known characteristics: calor, rubor, tumor, dolor. The individual symptoms have morphologically been clarified by the proof of tissue changes. The phenomenological definition of inflammation was scientifically delineated and the thereby detected functional and morphological changes of the normal tissue behaviour were canonised as exudation, proliferation and infiltration. These individual processes, which were used to describe inflammation, could authentically be reproduced in animal experiments and became the basis of pharmacological experiments. In this way an obligatory code developed which led to a common agreement on the classification of pathological processes. Naturally pharmacology has been orientated towards this code with the development of antiphlogistic agents.

The fact that rheumatic diseases, above all RA, were seen in terms of a clinical-phenomenological approach to involve pain, swelling and occasional reddening has led to the equating of rheumatic with inflammatory. Consequently, the therapeutic concept in general is an antiphlogistic one. This empirical concept persisted for a long time for the following reasons. First, it was possible to reduce the pain with defined antiphlogistic substances, thus, adhering to the described inflammatory character. Second, the inflammatory nature of rheumatic diseases seemed to be so far beyond question, that test series and experiments were chosen for the analysis of the pathomechanisms, which again produced clearly defined inflammations. In other words, the rheumatic disease in man (RA) was not investigated at all but different variants of the inflammatory process, its immunological and biomechanical mechanisms and its structural changes.

However, disbelief in this global inflammatory conception was then aroused by the recognition that, in spite of a great number of potent antiphlogistic substances, RA was no more influenced by them than by controlling the pain and stiffness caused by swelling. Meanwhile, it was evident that the actual dramatic processes such as joint destruction and vascular necroses continued without being

Surgery and Arthroscopy of the Knee
2nd Congress of the European Society
Editors: Müller/Hackenbruch

inhibited. The limited knowledge of the nosological character of RA other than that of a pure inflammatory concept developed due to the fact that until 1985 it was impossible to reproduce changes of joints and tissue in animal experiments which could be regarded as analogous to the structural changes occurring in RA in man. The only success of countless test series was that inflammatory processes could be produced, which however did not correspond at all to the processes in human RA. Substantial progress is to be expected from the MRL/1-mouse, which shows a genetically conditioned, spontaneous, non-inflammatory joint destruction.

Thus, the study of RA has been carried out under the premise that it is based on inflammatory mechanisms. For a long time this prejudice has stood in the way of an unbiased analysis of the human RA process and thus of new therapeutic concepts. Hitherto existing studies in animal experiments and in vitro models have been performed under these aspects. The extensive immunological and biochemical research has also been carried out considering the concept of inflammation. The activity of the humoral and cellular inflammatory components falls at the end of a long chain of mediators to be examined.

Research to elucidate the nature of RA must, however, examine the results of intensive analysis of RA without any preconceptions.

Materials and Methods

Between 1970 and 1986 more than 24000 tissue specimens with corresponding clinical documentation of patients suffering from different rheumatic diseases from 208 German and Swiss orthopaedic and surgical clinics have been examined at the Zentrum für Rheuma-Pathologie. The tissues were bioptically and surgically obtained. Among them there are approximately 12900 specimens of joint capsule tissue and part of joints from patients with RA aged between 1 and 76 years old. The tissues were predominantly from the interphalangeal, toe and knee joints. In addition, 411 RA necroses from different localisations, e.g. skin, tendon, joint capsule, pleura, myocardium, epicardium, wall of blood vessels, lungs

Table 1. Standardised morphological analysis of synovial tissue

Morphological categories (qualitative)	Possibilities of gradual interpretation (quantitative)
Fibrin	7
Lining cells and stromal cells	5
Giant cells	4
Lymphocytes and plasma cells	9
Polymorphonuclear leucocytes	4
Siderophages	7
Blood vessels	5
Villi	9
Fibrous tissue	3
Cartilage and/or bone remnants	5
RA necroses	4

and sclera of the eye of patients with seropositive rheumatoid arthritis, have been studied.

All tissue specimens were fixed in formalin (5%), routinely processed for paraffin embedding and sectioned. The 5-µm sections were stained with hematoxylin-eosin and van Gieson's stains and examined by light microscopy. The joint capsule specimens were evaluated using a standardised morphological method (Table 1).

In five cases of RA necroses sections were analysed by means of fluorescence microscopy.

Direct Immunofluorescence Method. Sections (5 µm) were taken from paraffin-embedded specimens containing RA necroses. The sections were incubated overnight at 60° C. Paraffin was removed by immersion in xylol (two times for 10 min); the sections were then rehydrated in a descending alcohol series (each concentration for 10 min), rinsed in 10 m*M* phospate buffered salt (PBS) solution for 10 min, and finally with distilled water. Sections were incubated for 15 min at 37° C with pronase. Enzymatic activity was then stopped by incubation in cold 10 m*M* Tris buffer for 30 min followed by 10 m*M* PBS solution for 15 min. Fluorescence staining was achieved by incubation with fluorescein isothiocyanate (FITC)-labelled anti-rabbit and anti-human immunoglobulins (anti-IgG, anti-IgM and anti-fibrinogen) 1:50 in 10 m*M* PBS solution and anti-C3 and anti-C10 1:20 in 10 m*M* PBS solution (by Behring-Werke) for 30 min at room temperature. Stained sections were rinsed for 15 min in 10 m*M* PBS solution and coated with cover slips. Control sections were treated in the same way but without addition of anti-sera to the incubation.

The joint specimens (articular cartilage and juxta-articular bone) were fixed in Bouin's solution [15 parts saturated picrine acid solution, 5 parts formalin (35%), 1 part concentrated acetic acid]. After decalcification the specimens were routinely processed for paraffin embedding and 5-µm sections were taken. Sections were stained with hematoxylin-alcian blue, and Giemsa and van Gieson's stains, and examined by light microscopy.

Analysis of the material revealed the following information on pathomechanisms.

Results

The Inflammatory Process

The most frequent finding in the synovial tissue samples is deposition of fibrin of different ages at the synovial surface (Fig. 1). This also corresponds to electron microscopical evidence of capillary changes in the synovial tissue possibly caused by immunological activity. Initially fibrin deposition leads to destruction of the local lining cell layer followed by a proliferation of stromal cells and hyperplastic regeneration of the lining cells. The exuded fibrin is not laid down in the normal organised way by granulation tissue. The granulation tissue usually consists of a mixture of fibroblasts, macrophages, angioblasts and polymorphonuclear leucocytes (PMN) and repesents the preliminary form of scar tissue. In contrast to this

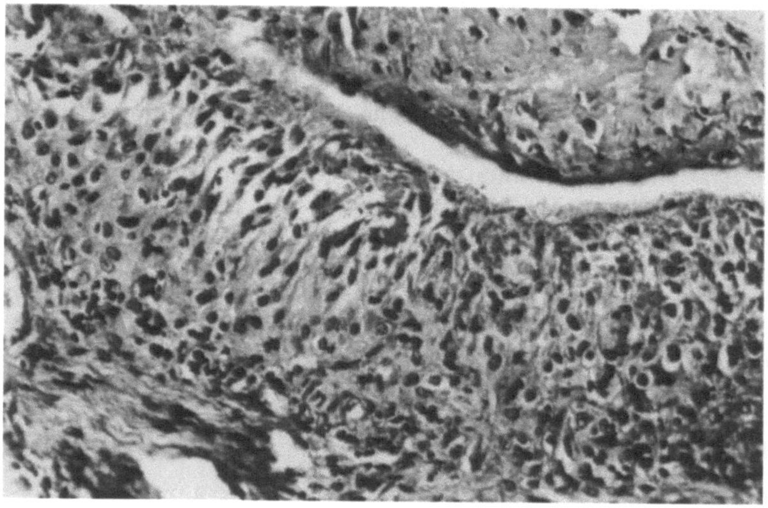

Fig. 1. Fresh fibrinous deposits with reactive proliferation of the synovial cells

process, exclusively in RA, synovial cells transmigrate the fibrin and form a new lining cell layer at the surface. In this way the villous hyperplasia of the synovial tissue (Fassbender 1981) develops subsequent to exudative phases.

During the exudative phase not only plasma extravasates, with subsequent fibrin polymerisation, but also PMN and macrophages emerge from the blood capillaries. While the PMN quickly migrate across the tissue, the macrophages remain visible in the synovial tissue. Lymphocytic infiltration, sometimes with formation of follicles, and plasma cell settlement are also seen to occur in the inflammatory

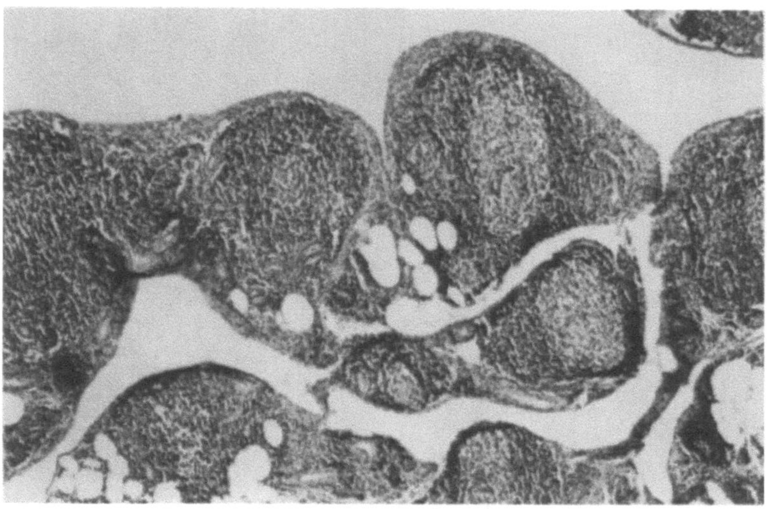

Fig. 2 Newly developed synovial villi with lymphocytic follicles

process. However, these features vanish subsequent to plasma exudation with fibrin precipitation and stromal cell proliferation but can persist in the villi (Botzenhardt 1975) (Fig. 2). After the exudative-productive phase has subsided, lympho-plasmacytic infiltration increases again, signalling the localised inactivity of the process.

Exudation, proliferation and infiltration are the three major components of the synovial inflammatory process. They are qualitatively uncharacteristic and only suggest the presence of RA depending on the degree of their intensity. These components are the cause of pain and swelling. The repeated fibrin exudations lead to an increasing formation of new villi, which may be the major reason for early stiffness (Fassbender 1981). Pain, swelling and stiffness can, therefore, undoubtedly appear to be the cause of inflammation.

The Destructive Process

It was a reasonable assumption that joint destruction was also a result of the inflammatory process. Weissman and co-workers (1971) implicated the enzymes of the PMN, especially elastase. However, an analysis of joints of patients with RA revealed that the articular cartilage in contact with the synovial fluid remains in-

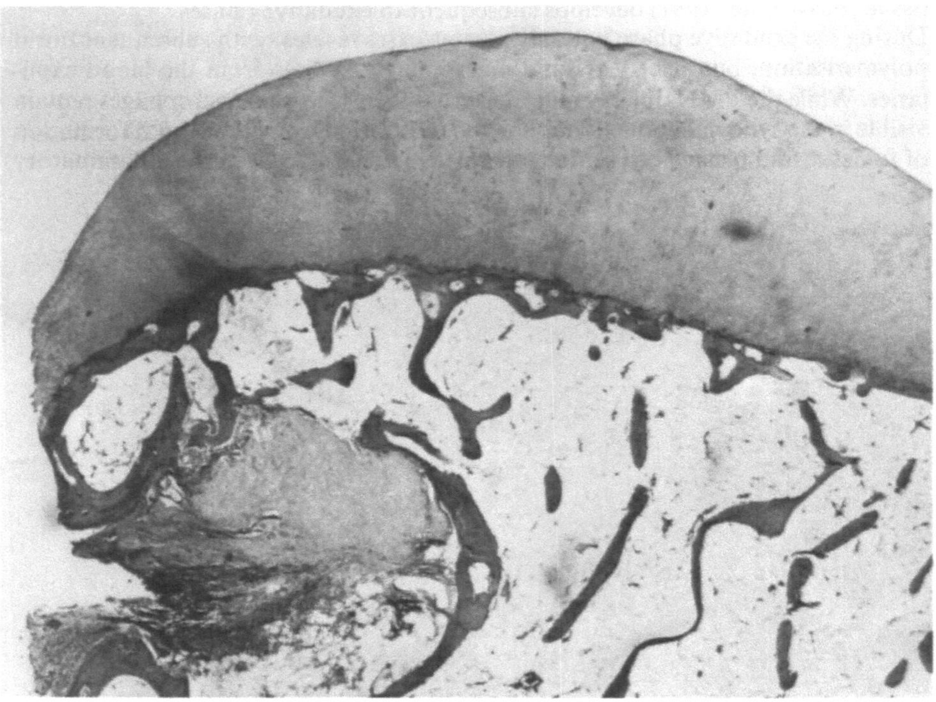

Fig. 3. Completely preserved articular cartilage after long-standing synovitis: old scar pannus indicating termination of subchondral aggression by cells of synovial origin

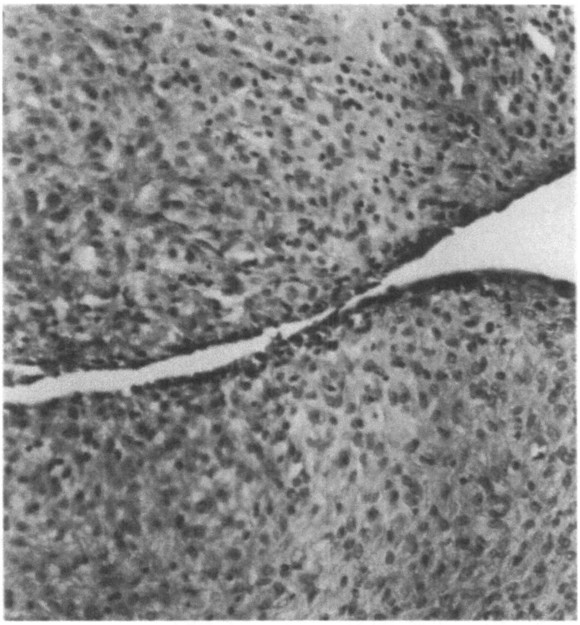

Fig. 4. Highest grade of proliferation of the local synovial cells with numerous mitoses (tumor-like proliferation)

tact and that the destruction develops from both sides, i.e. from the adjacent synovial tissue (Harris 1979; Fassbender 1973) (Fig. 3). The discrepancy with in vitro data can be explained by the fact that the concentration of the PMN enzymes in RA is too low. The number of PMN in the synovial fluid in bacterial purulent arthritis, for example, amounts up to more than 300 000/ml. This results in a rapid disintegration of the hyaline cartilage. However, in RA the number of PMN in the synovial fluid is lower by a factor of ten. With this relatively small number of PMN, the enzymes released are blocked by protease inhibitors and therefore do not degrade the articular cartilage (Kleesiek et al. 1982, 1983). Destruction of the cartilaginous and bony articular parts is rather the work of encroaching synovial cell masses similar to those cell proliferations observed in the synovial tissue following fibrin exudation. These are densely packed, immature, synovial cell elements without a vascular stroma (Fig. 4). These cells are polyhedral or round in shape with round, vesicular nuclei with one or two nucleoli. Scattered mitoses are recognisable within the cells. The cell mass is absolutely homogeneous and contains no inflammatory cell arsenal (Fassbender et al. 1980; Fassbender 1983). This feature represents the invasive phase. It is impossible to mistake these synovial cells for macrophages as they are not disordered cell accumulations but closed cell masses with mitoses. These cell masses have been named tumour-like cell proliferation (TLP; Cancer Research Institute of Switzerland at the Zurich University). TLP degrades the cartilage enzymatically and penetrates the articular surface in a destructive manner. The disintegration of the matrix can

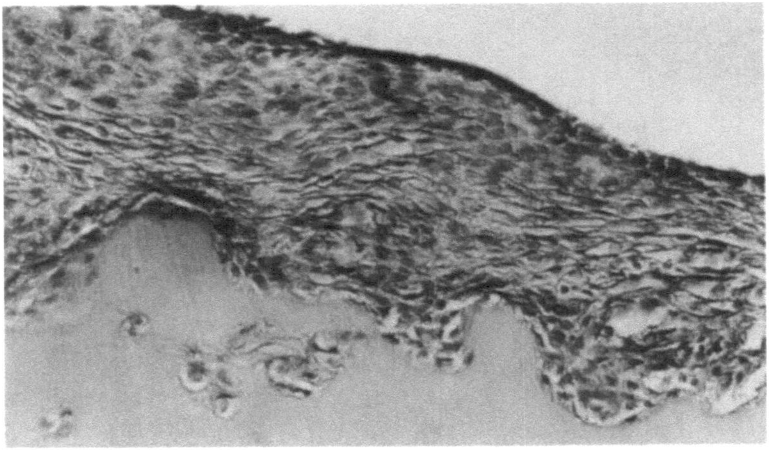

Fig. 5. Encroachment of the aggressive cell mass from the synovial tissue halfway onto the cartilaginous surface in the interphalangeal joint

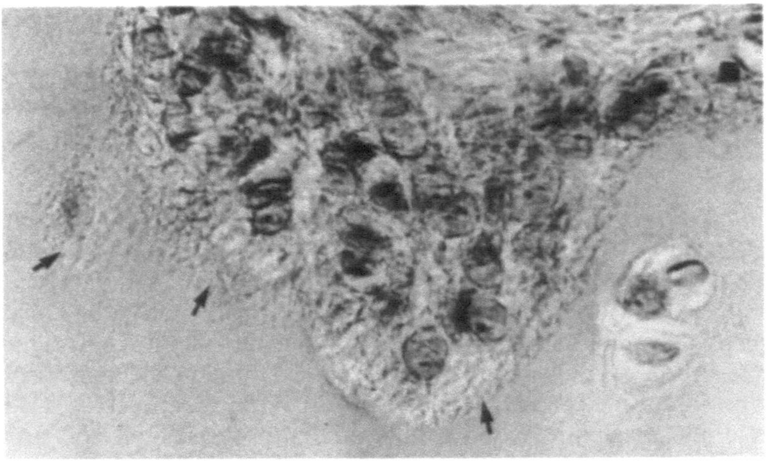

Fig. 6. Invasion of the destructive, compact, avascular, homogeneous cell mass into the articular cartilage: unmasking of collagenous fibres by degradation of proteoglycans is recognisable at the front of the aggression (*arrows*)

be followed step by step: initially the proteoglycans are degraded, then unmasked collagenous fibres are subsequently destroyed (Figs. 5, 6).
An analogous process can be observed at the finger joints in the area of the "bare zone" between the cartilage and the synovial insertion. TLP directly penetrates the medullary space, destroying the bone (Fig. 7). Based on our observations, it appears that the aggressive synovial cell masses destroy the "bare zone" and make an attack on the medullary space rather than directly on the cartilage.
As these aggressive cell masses are primarily avascular, they can only be nourished by diffusion of nutrients through the cells and are thus extremely short-

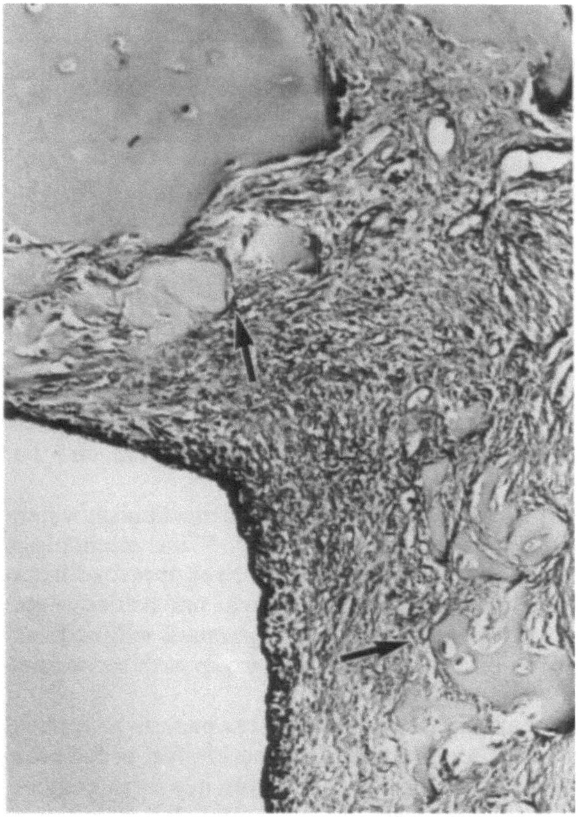

Fig. 7. Invasion of aggressive, tumour-like, proliferative synovial cell mass directly into the subchondral bone in the interphalangeal joint (*arrows* indicate cartilage remnants)

lived. The majority of the cells probably die after a few days. Simultaneous with the breakdown of the cell masses the tissue becomes vascularised. Aggressive synovial cell masses have never been observed after vascularisation. The surviving cell elements "modulate" to fibroblasts, produce collagenous fibres and thus lead to the formation of the pannus.

In general this process repeats after each exudative phase. The development of the persisting for life demonstrable pannus is characterised by vascularisation, migrating lymphocytes, plasma cells and macrophages. PMN can sporadically appear. In contrast to the compact appearance of the aggressive cell mass, a loose vascular cell mixture with some collagenous fibres develops. Thus, the morphological picture corresponds to that of granulation tissue. In this intermediate phase only a sparse number of surviving elements of TLP in contact with the cartilage face and in-between some individual macrophages and multinuclear cells of the osteoclasts type with a degradative function can be found (Bromley and Woolley 1984). During the intermediate phase the destruction of the articular cartilage is minimal. This short "cleaning-up" phase is followed by maturation of the

Table 2. The three phases of joint destruction: structural components

	I Invasive phase (tumour-like)	II Intermediate phase (mixed cells)	III Scar phase (pannus)
Tumor-like cell proliferation	~100%	~20%	0
Fibrocytes	0	~60%	~90%
Macrophages	0	~10%	0
Lymphocytes/plasma cells	0	~10%	~10%
Collagen fibres	0	+	+++
Blood vessels	0	+	+++

synovial tissue into a definite collagenous scar pannus, in which aggression has stopped totally (Table 2). The destruction of cartilage and bone therefore comes to an end; as soon as the compact cell mass degenerates, vessels appear, and the development of the pannus begins.

Thus, analysis shows that joint destruction in RA is based on a mechanism which can no longer be interpreted as being merely inflammatory. If the mechanism should at all be incorporated into known processes, it must be approached from the oncological point of view. There are certain morphological similarities to the giant cell tumours of tendon sheaths and the so-called pigmented villonodular synovitis, which is also a benign tumour of the giant cell group with secondary inflammation.

The frequently seen intermediate phase and the persisting scar pannus have been interpreted to represent the remains of terminated inflammation for, in fact, the actual TLP phase is considered to be extremely short and even in a large number of specimens is not found frequently. However, this misinterpretation has prevented identification of the actual joint-destroying mechanism and has steered the therapeutical effort in the wrong direction.

The occurrence of TLP within RA tissues gains an interesting angle from the investigations of Gran et al. (1984), who found histocompatibility antigen DR 4 (HLA DR 4) in 64.6% of patients presenting a mainly destructive course but only in 34.1% with a mainly inflammatory course.

Primarily Necrotising Processes

The subcutaneous "rheumatoid nodule" (Fig. 8) in mechanically exposed areas of the skin is generally known. Morphologically similar processes also occur in the tendon, joint capsule, pleura, myocardium (Fig. 9) and epicardium, walls of blood vessels, lungs (Kaplan syndrome) and in the sclera of the eye (scleromalacia necroticans). The tissue morphology is characterised by a necrotic centre surrounded by a dense palisade of radially directed fibroblasts and histiocytes. This picture, which can generally only be mistaken for granuloma annulare of the skin, is slightly modified depending on localisation and age of the process.

Here also the question arises, above all from a therapeutical point of view, whether or not this is a primarily inflammatory process. An inflammatory genesis could involve two mechanisms: primary tissue destruction by PMN enzymes

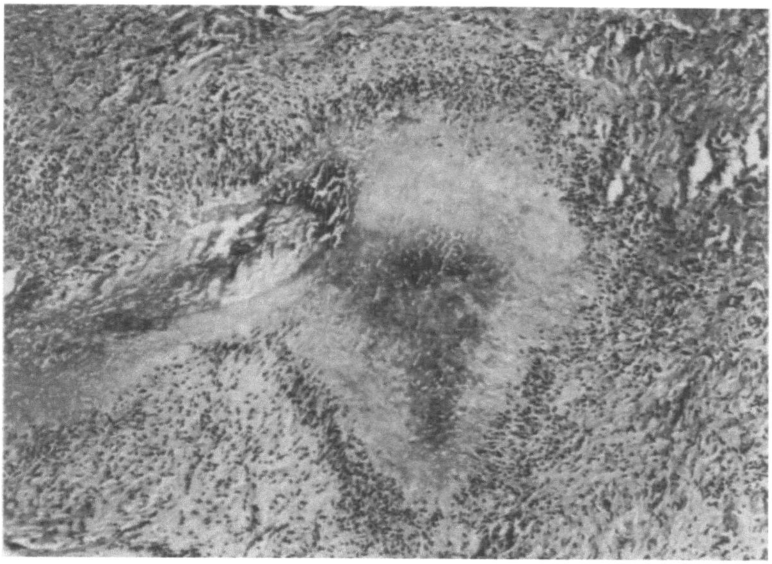

Fig. 8. Subcutaneous rheumatoid nodule

(Mohr and Menninger 1979) or secondary necroses as a consequence of a primary arteritis (Sokoloff 1953).

Analysis of the 411 instances of RA necroses encompassing different age groups and localisations shows that no PMN, or other inflammatory cells, are present in the early stages. In old, disintegrated necroses one can occasionally find sporadic PMN. This is surprising because a large number of PMN would first of all be necessary for such extensive tissue destruction, and it would be expected that an abscess-forming liquefaction would develop, to which RA necrosis bears no analogy.

In the region of RA necrosis together with the remaining collagenous connective tissue the integral blood vessels also degenerate. However, if the necrosis itself were the consequence of a vascular occlusion, a vessel of corresponding calibre would be exterior to the necrosis. Moreover, the necrosis would have the form of an infarct. A necrosis which has developed in this way would also be treated as an infarct by the organism, i.e. a marginal border of inflammatory cells would develop, and the dead area would be taken care of by new capillaries and macrophages.

None of these symptoms applies to regions of RA necrosis. The regions are sealed off, like a sequester, by a dense palisade of fibroblasts and histiocytes and remain unorganised for months. Thus, there is a considerable similarity to a tuberculoma. Neither a direct inflammatory development by PMN enzymes nor an indirect infarction by inflammatory vascular occlusions apply to RA necrosis.

We assume that the deposition of immune complexes and complements leads to primary devitalisation of the collagenous tissue structures and thus is the responsible pathomechanism. The investigations of Nowoslawski and Brsyzosko (1967),

Fig. 9. Old rheumatoid necroses in the myocardium: the necroses are sealed off like a sequestrum and surrounded by a palisade of connective tissue cells and a marginal scar

who detected IgG-IgM immune complexes in the necrotic regions, which bind guinea-pig complement, support this view. IgG and C 3 have already been detected in very small, fresh necroses without vascular involvement. Thus, we have no doubt that RA necroses are caused by a primarily immunologically based destruction of tissue, without any direct or indirect interference of an inflammatory mechanism. This also has been proved by the observation that the rheumatoid factors are detectable in more than 95% of patients in whom RA necroses have been found.

Based on the finding that RA necroses are mainly located in places with an increased mechanical strain, it is believed that IgG-IgM immune complexes and complements penetrate small tissue lesions and destroy the tissue within the localised area.

Conclusions

Even though RA is considered to be an inflammatory disease, its complex picture owes its development to three completely different pathomechanisms namely:

1. Exudative-proliferative processes in the synovial tissue of joints and tendon sheaths and at serous membranes
2. Destruction of articular cartilage and juxta-articular bone by the aggression of non-inflammatory, tumour-like cell elements of synovial origin
3. Primary, non-inflammatory tissue necroses due to deposition of immune complexes and complements

It is therefore understandable that all therapeutic approaches which are of an antiphlogistic character can only influence pain, swelling and matutinal stiffness but not the destruction of the joint and the primarily necrotising processes, which can also destroy vital structures of the heart and vessels.

References

Botzenhardt C (1975) Fibroblastäre Proliferation und lymphozytäre Infiltration des Stratum synoviale bei chronischer Polyarthritis in Beziehung zur klinischen Aktivität der Erkrankung. Dissertation, University of Mainz

Bromley M, Woolley DE (1984) Histopathology of the rheumatoid lesion – identification of cell types at sites of cartilage erosion. Arthritis Rheum 27:417–423

Fassbender HG (1973) Morphologisches Substrat und Pathogenese der rheumatischen Erkrankungen. Therapiewoche 23(8):611–614

Fassbender HG, Simmling-Annefeld M, Stofft E (1980) Transformation der Synovialzellen bei rheumatoider Arthritis. Verh Dtsch Ges Pathol 64:293–312

Fassbender HG (1981) Entstehungsmechanismen synovialer Zottenvegetation. Verh Dtsch Ges Rheumatol 7:484

Fassbender HG (1983) Structural basis of articular cartilage destruction in rheumatoid arthritis. Coll Relat Res 3:141–155

Garrod AB (1859) Nature and treatment of gout and rheumatic gout. Walton and Maberly, London

Gran JT, Husby G, Thorsby E (1984) HLA antigens in palindromic rheumatism, nonerosive rheumatoid arthritis and classical rheumatoid arthritis. J Rheumatol 1147(2):136–140

Harris ED Jr (1979) Recent insights into the pathogenesis of the proliferative lesions in rheumatoid arthritis. Arthritis Rheum 19(1):68–72

Kleesiek H, Neumann S, Greiling H (1982) Determination of the elastase-A 1-proteinase inhibitor-complex. Elastase activity and proteinase-inhibitor in the synovial fluid. Fresenius Z Ann Chem 311:434–435

Kleesiek K, Neumann S, Greiling H (1983) Elastase aus Granulozyten in der Synovialflüssigkeit: Konzentration und Aktivität von Elastase-Inhibitor-Komplexen bei chronischen Gelenkerkrankungen. In: Dettmer N et al. (Hrsg) Theoretische und klinische Befunde der Knorpelforschung, 3rd Symposium für Bindegewebsforschung. Eular, Basel, S. 134–141

Mohr W, Menninger H (1979) Neutrophile Granulozyten bei der entzündlich-rheumatischen Knorpeldestruktion. Bull Schweiz Akad Med Wiss 35:443

Nowoslawski A, Bryzosko WJ (1967) Immunopathology of rheumatoid arthritis. II. The rheumatoid nodule. Pathol Eur 2:203

Sokoloff L (1953) Heart in rheumatoid arthritis. Am Heart J 45:635

Weissmann G, Zurier RB, Spieler PJ, Goldstein JM (1971) Mechanisms of lysosomal release from leucocytes exposed to immune complexes and other particles. J Exp Med 134:149

A Scientific Basis for Surgery in Rheumatoid Arthritis *

T. D. V. Cooke

Surgery Arthritis Laboratory and Clinical Mechanics Group,
Department of Surgery, Division of Orthopaedics, Queen's University,
Apps Medical Research Centre, Kingston General Hospital, Kingston, Ontario,
Canada

Introduction

Understanding a disease provides the scientific basis for specific treatment. In rheumatoid arthritis (RA), the causes of which are unknown, the lack of knowledge about the intrinsic involvement of the articular collagenous tissues (ACT) as a site for the deposition and persistence of immune complexes may have led to poor long-term results of surgery, especially synovectomy. The knowledge that immune complexes in ACT produce inflammatory, immune foci for chronic synovitis responses has led to the introduction of ablative surgery of the damaged cartilaginous tissues as a specific remedy. Such knowledge may be used judiciously with techniques of fusion, resection and arthroplasty to improve and maintain the long-term function of the knee for patients with the crippling seropositive form of the disease.

The optimal management of RA disease(s) of unknown cause requires a knowledge of their epidemiology and the mechanisms of chronic synovitis and cartilage destruction which lead to instability and deformity. Knowledge of pathogenesis provides the scientific basis for both medical and surgical care. Historically, surgery of RA joints has often failed due to recurrence of inflammation after synovectomy; stiffness of adjacent joints follows fusion; progressive deformity develops after resection arthroplasty, and implant loosening attends arthroplasty.

In classical RA [26] severe, unremitting, seropositive, symmetrically distributed disease occurs, resulting in erosive destruction of the joints and leading to progressive loss of function and disability [8]. The requirements for surgery are the highest in this group of arthritics. In classical RA, the chronicity of the disease process and recurrent inflammation after synovectomy are best explained by the persistence of an inflammatory immune stimulus within the joint [29].

Pathogenesis of Chronic Synovitis and Cartilage Loss

Much of the evidence gathered suggests that such a stimulus for chronicity and recurrence of arthritis may relate to the deposition and persistence of antigenic material in the form of immune complexes in the joint ACT [7, 23]. Immune complexes have been identified as aggregates of the immunoglobulins IgG, IgA, IgM, along with complement components (C1q and C3) [7]. The persistence of such

* The research work reviewed in this communication has been supported by grants #6-232-(82) and #6-232-(81) from the Canadian Arthritis Society.

Surgery and Arthroscopy of the Knee
2nd Congress of the European Society
Editors: Müller/Hackenbruch

materials within the joint cavity may be in a large part responsible for the chemotaxis of acute inflammatory cells into the joint fluid, as well as a major reason for the massive local synthesis of antibody and rheumatoid factors (RF) [29].
Although autoimmune reactions [against-self phenomena, i.e. RF, antinuclear antibodies (ANA) and anticollagen antibodies] characterise RA, it is quite likely that the immune complexes deposited in the joint cartilaginous structures are the primary inducers of inflammatory immune reactions. Those local autoantibodies defined as RF, due to their capacity to interact with host IgG, may be directed at immune aggregates in the cartilage [10, 23]. In this light, many of these autoimmune reactions may represent epiphenomena.

Mechanism of Cartilage Destruction

Recent data suggest that deleterious interactions between immune complexes and collagenous matrix components bring about cartilage degradation [9, 10]. These interactions may induce activation of collagenases and/or neutral proteases that reside in a latent form in the cartilaginous matrix [18]. Another important destructive mechanism is the triggering of "frustrated" phagocytosis amongst polymorphonuclear neutrophils (PMN) that collect at the cartilage surfaces, attempting to endocytose the complexes [13, 32]. This PMN activity releases numerous enzymes, including the highly potent elastases. Toxic changes of chondrocytes may also be related to local immune complex activity. The evidence for this comes from observations of necrotic chondrocytes at sites adjacent to immune aggregates. More recently, data indicate an endocytosis capacity of chondrocytes for these immune deposits [9]. Although synovial pannus is responsible for marginal erosion of the joint cartilage [14] current evidence suggests that the mechanisms described above, including immune complex-mediated cartilage destruction, are responsible for the loss of cartilage bulk at the joint margins [10]. Cartilage degradation results in the loss of joint space and of precise centering with resultant instability. Loss of bearing surfaces is probably the most important single factor initiating deformity [8].
The failure of synovectomy to control arthritis consistently in joints [24] indicates that classical RA may be related to the persistence of antigenic immune-complexed material in ACT [6, 7]. The lack of synovitis following arthroplasty procedures, in which ACT-containing antigens are removed or replaced, is most likely because such surgery has removed the local inflammatory immune focus [1, 5, 15].

Surgical Considerations

Bearing in mind these concepts of pathogenesis, it is worth reconsidering the optimal choices of medical and surgical therapy for rheumatoid arthritic joints. In surgery the judicious use of ablative techniques that remove, replace or modify local ACT whilst preserving function will eliminate chronic inflammation and consistently eleviate pain [5].
The mechanical basis for deformity in RA is laxity and loss of centering mechanisms resulting from the destruction of bearing surfaces. When loss of ACT oc-

curs in multiple joints crossed by spanning tendons, Landsmeer's rules provide for complex compensatory zigzag patterns of collapse [16].

Fusion. Although fusion is seldom a first operative choice for the treatment of rheumatoid joints, it may be of great benefit in reconstruction of the hand with zigzag deformities of the thumb, metacarpophalangeal and proximal interphalangeal joints, as well as to stabilise the grossly deformed wrist [2, 12, 27, 29]. In the fore and hindfoot, deformities of the first metatarsophalanged midtarsal and subtalar joints are also effectively treated by fusion [3, 17, 33] as is the unstable cervical spine [20]. After fusion synovitis will not recur and stability with correction of deformity and improved function are predictable long-term outcomes. Fusion is also of great value in the salvage of failed arthroplasties, especially of the knee.

Resection Arthroplasty. Probably the most useful and widely available technique to reduce local recurrence of inflammation in rheumatoid joints is resection arthroplasty. Here removal of parts of the joint articulating system, especially those rich in fibrocartilage tissue, will lessen or eliminate synovial inflammation recurrence and at the same time result in loss of pain and improved function [5]. The major factors to be considered are the correction of deformity by balanced reconstruction of soft tissues so as to balance mechanically the joint complex that has now surgically lost bearing surfaces. Resection arthroplasty is especially useful at the wrist, where it may include decompression of the extensor tendons, excision of the distal ulna, the triangular ligament and a wrist debridement [30]. This resection is best combined with restoration of the correct alignment of the carpus (to eliminate the zigzag pattern) and reconstruction of the distal radioulnar joint with secure suturing of the ulnar collateral structures to the distal radius [12, 27]. Where severe motor imbalance exists in radial deviation, a transfer of a radial wrist extensor to the ulnar aspect of the joint may be added. These measures correct the deformity and maintain the correction. Similarly, at the elbow, resection arthroplasty involving excision of the radial head with its annular fibrocartilaginous ligaments, with joint debridement and synovectomy through a lateral approach, is an excellent early technique to moderate elbow disease [21, 31]. Reconstruction of the lateral collateral ligament is important to prevent subluxation. The acromioclavicular joint is readily dealt with by marginal resection, including the meniscus. Resection arthroplasty at the metacarpal and metatarsophalangeal joints is one of the main techniques for correction of hand and forefoot deformities [12, 31]. At the hand, such procedures are best combined with silastic implants, which provide and maintain alignment [31]. Soft tissue corrections of muscle imbalance are again of mandatory importance, but synovitis will not recur. The most important surgical precepts here are to know that functional improvement will not always accompany improved cosmesis. This means that these techniques for the hand need much improvement [19]. The interphalangeal joint deformities of the rheumatoid hand may also be managed by articular resection and silastic implantation [31]. The key indication for surgical intervention in the hand is the restoration of function [21]. In the foot, resection arthroplasties at the second to fifth metatarsophalangeal joints are combined with fusion of the great toe metatarsophalangeal joint [8]. The role for great toe fusion in maintaining length and functional forefoot balance has been also exemplified by Mann and Oates [17].

Replacement Arthroplasty. Combining arthroplasty with joint resurfacing surgery is a rapidly developing field. These procedures are of proven value at the knee and hip. The arthroplasties will not be associated with recurrent synovitis, and they eliminate pain and provide excellent motion and stability [4, 28]. Interestingly, resurfacings of the knee at the stage when the joint space is lost but the collateral ligaments are still intact restores alignment, joint space and tension to the joint [4]. Arthroplastic techniques have been developed for the ankle, elbow, metacarpal and shoulder, but only those employing implants without major constraints and with resurfacing properties are likely to be useful in the long-term [21]. Prostheses that are mechanically linked together induce major constraints that provide repeatedly high stress on the implant-bone fixation interface. These will invariably cause loosening of the implant with time. Furthermore, the rheumatoid patient's bone stock is always porotic, the result of local disease but which is frequently aggravated by management that includes the use of steroids [25]. Consideration of any form of artificial implant must be carefully balanced with the factors of age, functional requirement, stresses, bone stock, salvageability, etc. Of special note in considering surgery of the RA shoulder is the early destructive involvement of the rotator cuff. Replacement surgery undertaken before there is major degradation of the cuff provides an excellent long-term outcome. Lack of surgical attention to cuff repair and/or integrity will lead to less than optimal function [22]. A long-term bracing and rehabilitation program in the postoperative management may yield better results when the cuff is reconstructed.

Special Problems

Special techniques are required for implant fixation in osteopenic bone. This is most often a requirement at the hip and knee. In the hip, the use of bone grafts, mesh and retention rings are used to deal with acetabular protrusion [11]; the knee requires coverage of the tibia to the cortical margins of the plateau, special cementing techniques by injection and/or stemmed prostheses [8, 28].

Synovectomy

The last point, but not least, is to reconsider the use of synovectomy. Synovectomy is of great benefit in decompression of rheumatoid disease involving flexor and extensor tendons and in the release of tendon entrapments [8]. It may also provide excellent disease control of joint synovitis where the patient has a more benign disease (seronegative with a remissive pattern). Synovectomy is especially useful in the management of rheumatoid joints complicated by sepsis [8].

Conclusion

Ablative surgery of cartilaginous tissues which harbor immune complexes offers the only currently available local cure for classical RA. This treatment is optimally used accompanied by an understanding of the epidemiology of the disease. Each procedure must be integrated with good medical management, selecting the right time, at the optimal location and order, always aiming for the restoration of the patient to functional independence [8, 21].

Acknowledgement. The author acknowledges the excellent secretarial support of Mrs P. Brennan in the preparation of this manuscript.

References

1. Boerbooms AM, van den Broek WJM, van Rens TJG, van de Putte LBA (1982) ^{99M}Tc-Pertechnetate uptake after total knee replacement in rheumatoid arthritis. Acta Orthop Scand 53:125
2. Carroll RE, Dick HM (1971) Arthrodesis of the wrist for rheumatoid arthritis. J Bone Joint Surg [Am] 53:1365
3. Clayton ML (1971) Surgery of the lower extremity in rheumatoid arthritis. J Bone Joint Surg [Am] 53:1465
4. Cloutier JM (1983) Results of total knee arthroplasty with a nonconstrained prosthesis. J Bone Joint Surg [Am] 65(7):906
5. Cooke TDV (1980) How does the mechanism of cartilage destruction influence surgical management of the rheumatoid joint? J Rheumatol 7:119
6. Cooke TDV (1981) The interactions and local disease manifestations of immune complexes in articular collagenous tissues. In: Holborrow EJ, Maroudas A (eds) Studies of joint disease, vol 1. Pitman, London, p 158
7. Cooke TDV, Hurd ER, Jasin HE, Bienenstock J, Ziff M (1975) Identification of immunoglobulins and complement in rheumatoid articular collagenous tissues. Arthritis Rheum 18:541
8. Cooke TDV, Scudamore RA (1984) Rheumatoid arthritis and allied conditions. In: Cruess RL, Rennie W (eds) Adult orthopaedic surgery, vol 1. Churchill, New York, p 271
9. Cooke TDV, Sumi M, Elliott S, Maeda M (1983) Immune complex mediated destruction of cartilage in antigen-induced arthritis of rabbits. J Rheumatol 10(11):103
10. Cooke TDV, Sumi M, Maeda M Deleterious interactions of immune complexes in cartilage of experimental immune arthritis. I. The erosion of pannus-free hyaline cartilage. Clin Orthop Rel Res (in press)
11. Hastings D, Parker SM (1975) Protrusion acetabuli in rheumatoid arthritis. Clin Orthop Rel Res 108:76
12. Hastings WD, Welsh RP (1979) Surgical reconstruction of the rheumatoid hand. Medisport and Orthopaedic Management Corporation, Toronto
13. Henson PM (1971) The immunologic release of constituents from neutrophil leucocytes. I. The role of antibody and complement on non-phagocytosable surfaces or phagocytosable particles. J Immunol 107:1535
14. Ishikawa H, Ohno O, Hirohata K (1984) An electron microscopic study of the synovial-bone junction in rheumatoid arthritis. Rheumatol Int 4(1):1
15. Lance EM (1982) The effect of joint replacement on rheumatoid synovitis and its significance. Arthritis Rheum 25 (Suppl 4):41–59
16. Landsmeer JMF (1980) Studies in the anatomy of articulation 1. The equilibrium of the "intercolated bone". Acta Morphol Neerl Scand 3:3–4, 288
17. Mann RA, Oates JC (1980) Arthrodesis of the first metatarsophalangeal joint. Foot Ankle 1(3):159–166
18. Martel-Pelletier J, Pelletier JP, Cloutier JM, Howell DS, Ghandur-Mnaymneh L, Woessner JF Jr (1984) Neutral proteases capable of proteoglycan digesting activity in osteoarthritic and normal human articular cartilage. Arthritis Rheum 27(3):305
19. McGraw RW, MacBaine KP (1976) Functional results of arthroplasty in the rheumatoid hand. J Bone Joint Surg [Br] 58:139
20. McGraw RW, Rusch RM (1973) Atlanto-axial arthrodesis. J Bone Joint Surg [Br] 55:482

21. Mowat AG (1978) Surgical management of rheumatoid arthritis. In: Mowat AG (ed) Clinics in rheumatic diseases, vol 4. Saunders, London, p 247
22. Neer CS II (1981) Reconstructive surgery and rehabilitation of the shoulder. In: Kelley WN, Harris ED, Ruddy S, Sledge CB (eds) Textbook of rheumatology. Saunders, Philadelphia, p 1944
23. Ohno O, Cooke TD (1978) Electron microscopic morphology of immunoglobulin aggregates and their interactions in rheumatoid articular collagenous tissues. Arthritis Rheum 21:516
24. Patzakis MJ, Mills DM, Bartholomew BA, Clayton ML, Smyth CJ (1973) A visual histological and enzymatic study of regenerating rheumatoid synovium in the synovectomized knee. J Bone Joint Surg [Am] 55:287
25. Rasker JJ, Cosh JA (1978) Radiological study of cervical spine and hand in patients with rheumatoid arthritis of 15 years' duration: an assessment of the effects of corticosteroid treatment. Ann Rheum Dis 37:529
26. Ropes MW, Bennett GA, Cobb S, Jacox R, Messar RA (1959) 1958 revision of diagnostic criteria for rheumatoid arthritis. Arthritis Rheum 2:16
27. Shapiro JS (1968) The etiology of ulnar drift: a new factor. J Bone Joint Surg [Am] 50:634
28. Sledge CB (1981) Introduction to the surgical management of arthritis. In: Kelley WN, Harris ED, Ruddy S, Sledge CB (eds) Textbook of rheumatology, vol 2. Saunders, Philadelphia, p 1867
29. Smiley JD, Sachs C, Ziff M (1968) In vitro synthesis of immunoglobulin by rheumatoid synovial membrane. J Clin Invest 47:624
30. Straub LR, Ranawat CS (1969) The wrist in rheumatoid arthritis. Surgical treatment and results. J Bone Joint Surg [Am] 51:1
31. Swanson AR (1973) Flexible inplant rejection arthroplasty in the hand and extremities. Mosby, St. Louis
32. Ugai K, Ishikawa H, Hirohata K, Shirane H (1983) Interaction of polymorphonuclear leukocytes with immune complexes trapped in rheumatoid articular cartilage. Arthritis Rheum 26(12):1434
33. Vianio K (1956) The rheumatoid foot. A clinical study with pathological and roentgenological comments. Ann Chir Gynaecol 45 [Suppl 1]

Arthroscopic Knee Synovectomy

E. S. Bittar

Department of Orthopedics, University of Florida,
Gainesville, Florida, USA

Open knee synovectomy requires lengthy hospitalization and staged anterior and posterior arthrotomies with significant postoperative morbidity. This paper describes the indications, techniques, and results of arthroscopic synovectomy of the knee.

Materials and Methods

For the past 2 years the Department of Orthopedics at the University of Florida has performed 28 knee synovectomies on 20 patients who underwent arthroscopic subtotal synovectomy for diagnoses that include rheumatoid arthritis, hemophilia, and posttraumatic chronic synovitis. Usually six arthroscopic portals were utilized to perform a subtotal arthroscopic synovectomy using a powered suction shaving device. The patients were hospitalized for 3 to 5 days postoperatively and after 1 or 2 days of immobilization with suction drainage of their knees, the patients were mobilized with a continuous passive motion machine during the remainder of their hospitalization. The patients were not discharged until at least 90° of pain-free range of motion had been obtained, and they were fully ambulatory with the aid of external support.

Results

Of 26 synovectomies 22 were graded as excellent or good and 4 as fair. None required either repeat synovectomy or postoperative manipulation in order to improve their motion, or to provide a range of motion free of pain. No infections have resulted. By the 3rd postoperative week 75% of the patients were ambulatory without the need of external supports unless they had been required pre-operatively for unrelated disabilities.

Discussion

This review indicates that subtotal arthroscopic knee synovectomy is a valuable alternative to an open procedure. A 1–2-day period of postoperative immobilization followed by continuous passive motion for a period of 2–3 days provides a functional range of motion with only minimal discomfort. Patients with rheumatoid arthritis and florid synovitis beyond medical management benefit most from this procedure when it is performed early in the course of the disease, prior to destruction of interarticular structures.

Surgery and Arthroscopy of the Knee
2nd Congress of the European Society
Editors: Müller/Hackenbruch

Weight-bearing plain X-rays are valuable in predicting postoperative results. The arthroscopic procedure is technically demanding but permits a more complete synovectomy to be performed, including submeniscal and posterior compartment areas not easily accessible even with wide arthrotomies.

Summary

Arthroscopic knee synovectomy presents an alternative method of surgical treatment for patients who are beyond medical management of synovitis. The procedure is technically demanding but has yielded excellent or good results in over 80% of patients studied over a 2-year period at the University of Florida Department of Orthopedics. The procedure provides an alternative to open synovectomy with low morbidity, excellent postoperative motion, and modest postoperative pain. The technique is most effective in rheumatoid arthritis when performed early in the course of the disease, prior to destructive changes of interarticular structures.

Synovectomy Applied to Inflammatory Rheumatic Diseases of the Knee under Arthroscopic Control

M. Moraldo, R. Schleberger, and A. Hedtmann

Orthopaedic Department, University of Bochum, St. Josef-Hospital, 4630 Bochum, FRG

Synovectomy in connection with the therapy of inflammatory rheumatic diseases arises from the relative indications of operation (Vainio, Geschwend). It would not be possible without the regeneration faculty of the synovial membrane. Regeneration takes 6 weeks, and for so long is the mutual effect of synovia and articular cartilage disturbed. The regenerated substance must provide sufficient functional compensation. Examinations justify synovectomy.
It is agreed that an operation within the first 6 months after involvement of the joint can be called "early" synovectomy.
What is not so well-known is the morphologically defined early synovectomy (Hijmans et al. 1969), according to which only the stratum synoviale may be afflicted. Nowadays, the early operation is generally recommended, though in connection wit the juvenile affliction one rather tends to exercise restraint. Especially in this context there will be specific secondary alterations if the knee joint is afflicted (Schleberger and Raunio 1981).
Together with the invention of suitable instruments in connection with the development of arthroscopically controlled operation, the arthroscopical synovectomy seems promising (Fig. 1). We have employed it since January 1985 and have compared the patients concerned with a similar group of those who had undergone an "open synovectomy" in the years 1983–1984. The common criterion was the existence of a mono- or oligoarthritis in all cases.

Method

The operation is conducted under complete or local anaesthesia and with tourniquet. While in open operations the medial and lateral approaches were chosen, in operations under arthroscopic control the anterolateral standard approach is

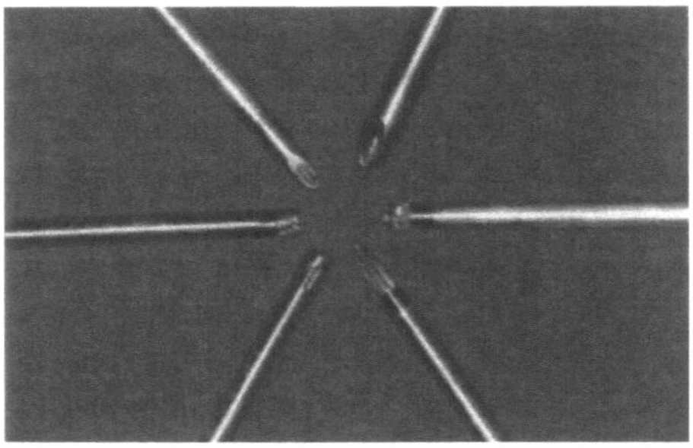

Fig. 1

Surgery and Arthroscopy of the Knee
2nd Congress of the European Society
Editors: Müller/Hackenbruch

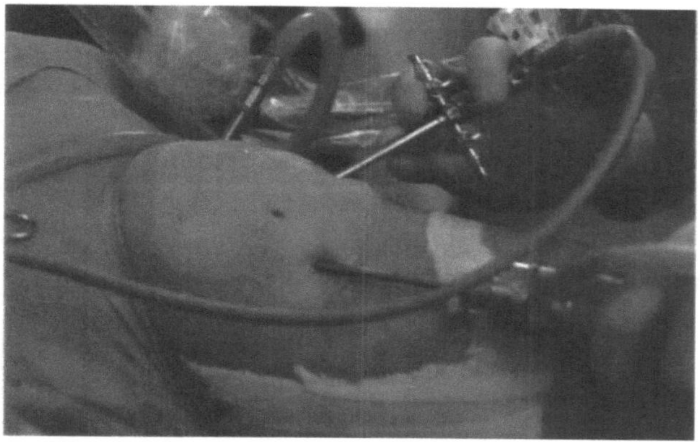

Fig. 2

chosen for the scope, and the medial approach is at first made for the probe hook, later serving for the scope and the shaver (Fig. 2). In- and outflow may work via scope; as a special inflow, one of the suprapatellar portals may be made immediately, the second is needed later (Fig. 3). Permanent in- and outflow demands 10 liters of liquid at least. According to our experience these accesses allow a more complete synovectomy than is possible in connection with an open procedure.

In the arthroscopy group 20 knees of 20 patients were synovectomised; the average age was 32.6 years. There was a group of five "late synovectomies". The male sex and oligoarticular patterns of affliction prevailed. The affliction of the knee joint had lasted from 4 months to 12 years in the arthroscopy group and 6 months to 4 years in the open operation group (Table 1).

In every case at least three ARA (American Rheumatism Association) criteria were made, one of which was always typical histology. In connection with this criterion the arthroscopy group showed 12 times a highly florid or distinct rheumatoid arthritis (RA), and 4 times heavily recidivous forms. The group treated by open operation included 12 RA and 7 recidivous forms. Seronegative forms and HLA B27-associated arthritides prevailed.

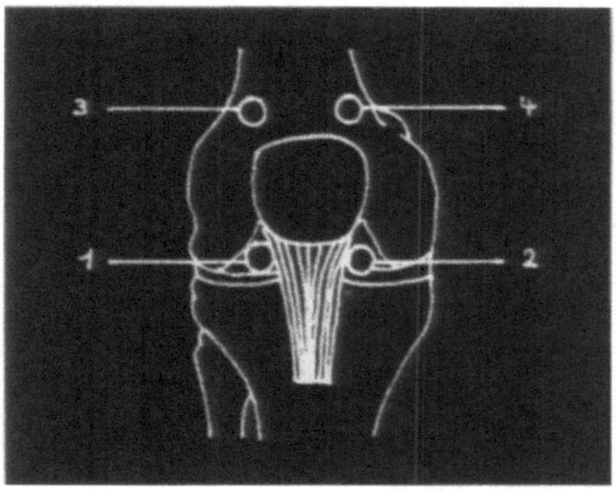

Fig. 3

Table 1. Characteristics of patients in study

	Arthoscopically treated group	Open surgery group
n	16	15
Number of knee joints	20	19
Age (years)	32.6	33.4
Postoperative rehabilitation	14–20 months	26–43 months
Period of affliction before treatment	4 months–12 years	6 months–4 years
Seronegative	8	3
Seropositive	4	3
HLA-B27	4	9
Florid (histology)	15	12
Recidivous (histology)	5	7

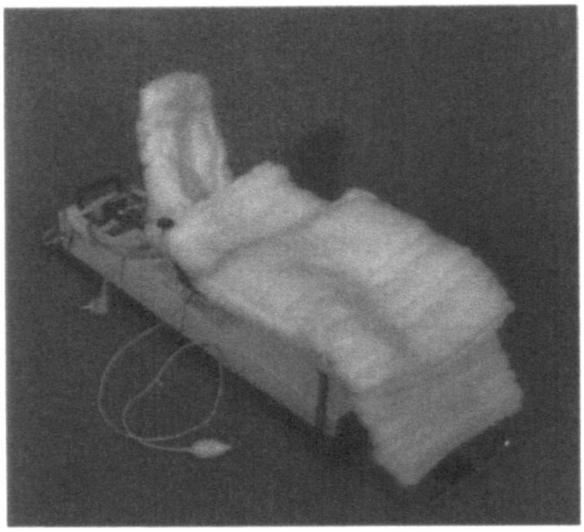

Fig. 4

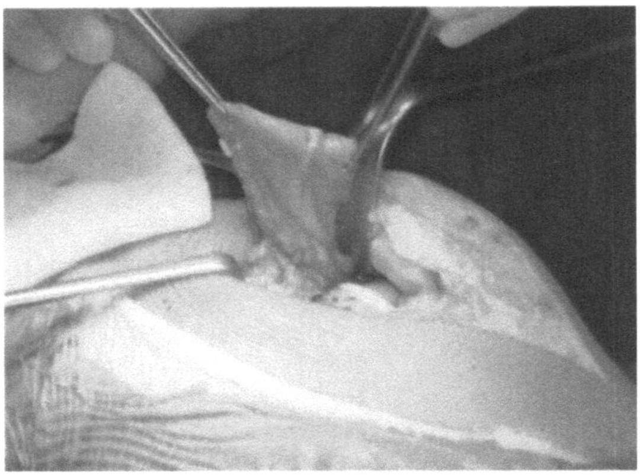

Fig. 5

Postoperative Care and Rehabilitation

The synovectomised patients could bear weight with crutches in 6 weeks. The first critical phase after the operation, when the passive motion machine was applied to the knee (Fig. 4), was decisively different in both groups. This could be seen from the consumption of analgesics: the openly synovectomised groups received morphium and derivates or analogous drugs in all cases, but the arthroscopic group, only in 50% of the cases and just on the day of operation. After that this group could partly do without any analgesics. Within the open operation group prolonged medication frequently proved necessary (Figs. 5–7). All patients got up on the first day after the operation in connection with the general prophylaxis against thrombosis. Physiotherapy for the arthroscopy group usually started with the training of the quadriceps muscles; this was parallelled on the side of the open operation group by efforts of the physiotherapist supporting the bending of the knee (Figs. 8, 9). In the arthroscopic group the passive motion machine was no longer required after 5 days at the latest (Table 2). The time of postoperative control was extremely different in both groups. Up to now no recidivous cases have been found in the arthroscopy group. There was only a slight painless effusion

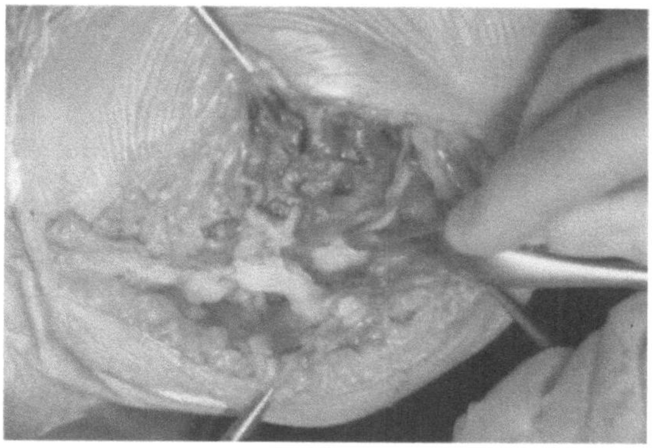

Fig. 6

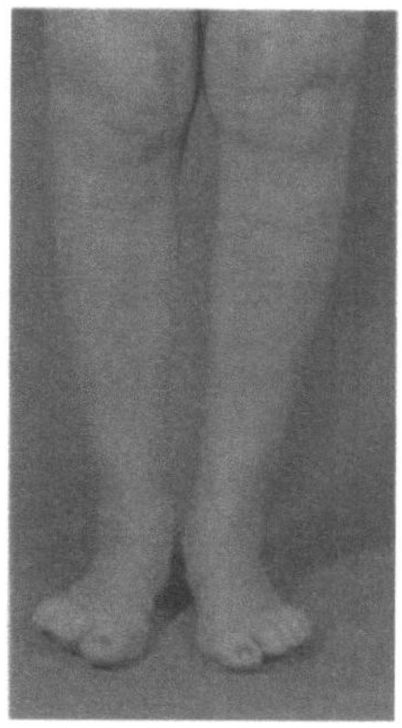

Fig. 7

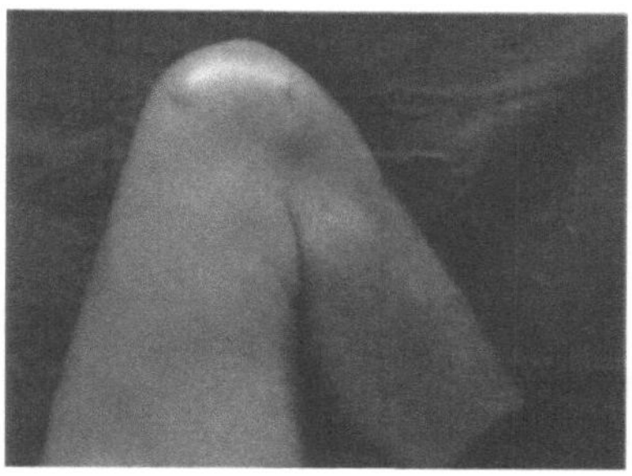

Fig. 8

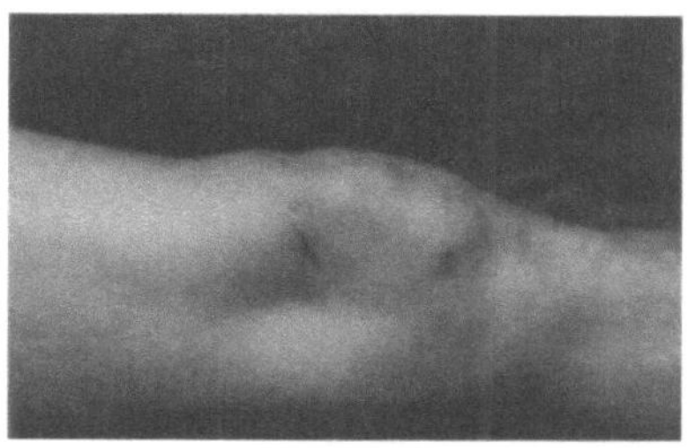

Fig. 9

Table 2. Details of postoperative care

	Arthroscopically treated group	Open surgery group
Stay in hospital (days)	10.6	16.1
Morphium or analogous drugs	8	19
Analgesics	12	–
Flexion over 120°	20	–
Flexion under 90°	0	4

in three cases; in the open operation group we observed one recidivous and one re-recidivous case of which both were histologically highly florid. Pains caused by strain depend on the initial diagnosis; temporary problems with chondromalacia patellae sometimes occur.

Comparing the methods, arthroscopically controlled synovectomy is more effective concerning an anterior access in arthrotomy and the postoperative care is by far less problematical for the patients; an average duration of stay of 7–10 days becomes normal. Further ambulant rehabilitation is also less complicated although a reduced compliance of the patients required distinct instructions concerning the protection of the cartilage for 6 weeks. This method advances the orthopaedically desirable, so-called early synovectomy.

References

Hijmans W, Paul WD (1969) Early synovectomy in rheumatoid arthritis. Excerpta Medica, Amsterdam

Schleberger R et al. (1981) Conditions of unilateral leg lengthening in JRA. Rev Rhum Juin 1981:701

Radical Versus Partial Synovectomy: Pathophysiological Considerations

K. Tillmann and U. Dümmer

Rheumaklinik, Bad Bramstedt, FRG

The question of whether the results of synovectomy depend on the technique of the operation or not is important for the surgeon as well as for the patient. There are some clinical hints and also pathophysiological findings, especially regarding metabolic alterations inside the joint, which seem to show that the results are partly related to the completeness of the operation.

We compared the outcome of "partial" synovectomies by the ventral only approach (MORI) with "subtotal" operations in which we used an additional dorsal approach. Our first short-term follow-up demonstrated a statistically significant superiority of the subtotal technique.

Later on we examined two groups of patients who underwent these two different operations (52 patients had a "partial", 64 a "subtotal"). Like the first, it was a retrospective not a randomized, study. The time of follow-up was 6.6 ± 1.8 years and 5.8 ± 1.2 years respectively. The "subtotal" group had a higher systemical activity and a later local stage of the disease compared with the "partial" group.

The results of the subtotal operation proved to be superior to the partial one, but the difference did not reach statistical significance. The later the stage of the local disease at the time of operation, the smaller the difference between the two techniques. The recurrence rate after the "partial" synovectomy was more than twice as high as the "subtotal" operation.

Surgery and Arthroscopy of the Knee
2nd Congress of the European Society
Editors: Müller/Hackenbruch

Rehabilitation after Knee Injuries – Importance of Pain Relief

E. Eriksson and I. Arvidsson

Trauma Section, Department of Surgery, Karolinska Hospital
and Department of Physical Therapy Education, Karolinska Institute,
Stockholm, Sweden

The rapid development of quadriceps atrophy observed following knee surgery is well-known to physicians and physiotherapists. One explanation for this could be that postoperative pain makes it impossible to maintain any training.

Ten patients, operated for reconstruction of the anterior cruciate ligament were studied the day after surgery. Registration of integrated EMG during maximal voluntary contraction of the quadriceps muscle was done, before and 5, 12 and 20–25 min after epidural injection of 20 ml of 0.25% lidocaine. As the pain gradually subsided, IEMG increased and reached a mean almost 30 times the initial value. This indicates that pain relief plays a significant role in the ability to activate normally the quadriceps muscle after open knee surgery.

In order to compare the pain-relieving effect of placebo transcutaneous nerve stimulation (TNS), high frequency TNS, and epidural analgesia with dilute local anesthetics, a similar technique was used. The patients' ability to contract their quadriceps muscle was measured with IEMG before and after the different treatments and also compared with their subjective assessment of pain. The results showed that placebo-TNS had no significant effect on either pain perception or on IEMG. High frequency TNS given for 20 min decreased pain perception and increased muscle contraction ability, but it was not as effective as epidural analgesia. Since, however, TNS is easy to administer and lacks side effects, it undoubtedly has a place in postoperative pain treatment.

Nonsteroidal anti-inflammatory drugs (NSAIDs) have been claimed to reduce postoperative pain and exudate and might thus offer the possibility of shortening rehabilitation time. Forty patients participated in a double-blind study of piroxicam 20 mg daily or placebo, given for 10 days following open meniscectomy. The patients were studied for 3 weeks postoperatively using these parameters:

- quadriceps torque measured isokinetically
- patient's assessment of pain
- active and passive range of motion
- degree of swelling and irritation of the knee
- efficacy of the medication

The mean values for all parameters, except for pain assessment, were in favour of the piroxicam group, which indicates that the use of an NSAID does improve early postoperative rehabilitation after open meniscectomy.

It is our belief that alleviation of pain and reduction of the inflammatory process postoperatively could give patients the possibility to start training earlier, thus avoiding the usual significant muscle atrophy.

Surgery and Arthroscopy of the Knee
2nd Congress of the European Society
Editors: Müller/Hackenbruch

Recent Results on Post-traumatic Osteoarthrosis after Internal Fixation of Fractures of the Tibial Plateau

H.-M. Märklin and U. Holz

Department of Traumatology and Reconstructive Surgery,
Katharinenhospital, Stuttgart, FRG

The aim of our clinical study has been to analyse factors important in the development of post-traumatic gonarthrosis. These include the type of fracture, post-traumatic radiological findings, the function of the knee and the complaints of the patients themselves.

Material and Methods

Between 1974 and 1980 we conducted a postoperative follow-up study on 48 of 146 patients with fractures of the tibial head at an average time of 5.5 years after operation of these patients: 64.6% were male and 35.4% female. Average age was 44.2 years. Of the 146 fractures 21.2% were open and 78.8% were closed. To provide a conceptual basis for the follow-up we classified the fractures of the tibial plateau in anatomical terms [4]:

1. Monocondylar fractures
2. Bicondylar fractures
3. Split fractures
4. Depression fractures
5. Impression fractures
6. Combined fractures
7. Epiphyseal fractures (Aitken)

Of all fractures 54.1% were monocondylar and 45.9% bicondylar. Fractures of the lateral condyle occurred in 78.5% of cases. Lesions of the knee joint presented commonly; 21.2% showed lesions of the lateral meniscus and 19.2% lesions of the intercondylar eminetia with lesions of the cruciate ligaments. Additionally, 25.5% had fractures of the head of the fibula. Lesions of the collateral ligaments were present in 6.2% of cases. In 39.7% of patients, immediate surgical treatment was performed, while 60.2% were treated within hours or days, depending on their general condition.

Internal fixation according to AO principles was used to treat 139 patients. Primary amputation was necessary in 6 cases due to additional serious vessels or nerve lesions or to multiple comminuted fractures of the leg. In one case a unicondylar knee prosthesis was implanted. Transplantations of autogenous cancellous bone grafts were used in 85 patients [11].

The follow-up involved, first, the analysis of X-ray exposures in terms of incongruence, axial deviations and signs of post-traumatic arthrosis. Cases of incongruence were classified as wave-like, trough-shaped and step-like deformities

Surgery and Arthroscopy of the Knee
2nd Congress of the European Society
Editors: Müller/Hackenbruch

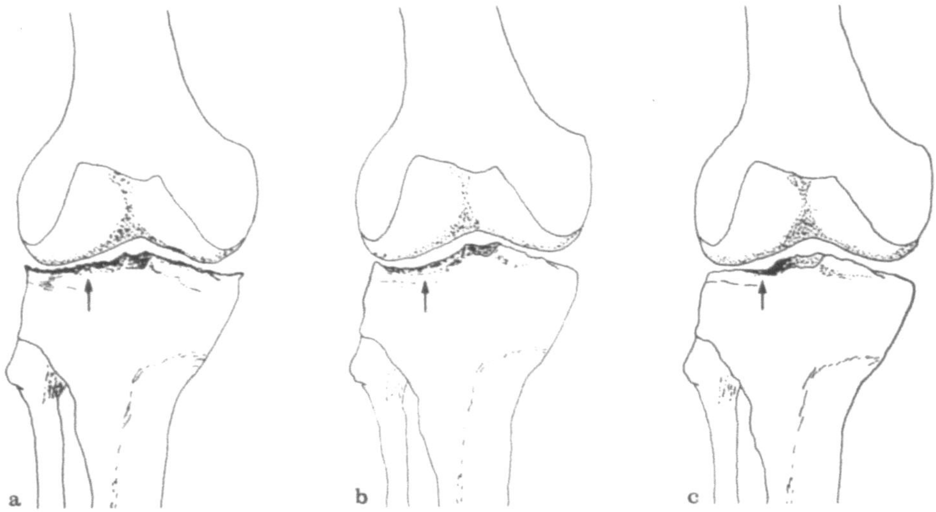

Fig. 1 a–c. Types of incongruence in the tibial plateau. **a** Wave-like, **b** trough-shaped, **c** step-like

(Fig. 1). Wave-like deformities are degenerative in nature. Trough-shaped incongruences result from insufficient transplantation of cancellous bone and/or from unsatisfactory internal fixation of fragments and the consecutive lowering of parts of the tibial plateau. Step-like deformities are caused by insufficient primary reduction.

Second, radiological signs of arthrosis were graded according to the 4-grade scale of Jonasch and Mohing [6, 7]:

Grade I : Slight marginal formation of osteophytes
Slight unrounding of the ambon
Fine extension of intercondylar eminentia

Grade II : Marginal bulging (< 5 mm)
Unrounding of articular surface
Lanciform extension of intercondylar eminentia

Grade 1III : Marginal bulging (> 5 mm)
Joint cavity not reduced

Grade IV : Reduction of joint cavity, irrespective of extent of marginal bulging
Enormous console formation
Subchondral sclerosis

Clinical examination furthermore involved manner of walking, function of the knee joint and ligaments stability. The patients were interviewed about their complaints, their evaluation of the therapeutic procedure and their ability in walking.

Results

Among all patients 66% were satisfied with the results of treatment; half had few or no complaints. Patients with combined fractures complained more frequently.

Also, 46.8% were able to walk more than 5 km without discomfort. Concerning manner of walking, patients with unicondylar depression fractures indicated the most discomfort. Upon examination we found an arthrotic friction murmur in 50%; in 41.7% of knees thickening of the capsule could be observed.

Normal functioning of the knee joint was found in 34 patients (30%). There was no significant difference in knee function between those with combined fractures and those with monocondylar fractures. Ligament instability was found in 39.6%, loosening of lateral ligaments in 25% and loosening of medial ligaments in 20.8%. In 91.7% of patients signs of post-traumatic arthrosis were present; 16.7% had already shown signs of gonarthrosis before the trauma (Table 1). Step-like deformities of up to 3 mm were evidenced in 14.6%. There were no step-like deformities of more than 3 mm. Trough-shaped articular deformities were observed in 25% and wave-like ones in 43%. Step-like and trough-shaped deformities were encountered frequently accompanying impression fractures and combined fractures. Wave-like incongruence was predominant in arthrosis of grade IV. Axial deviations of more than 5° could be found in 27% of cases, valgus deviation being twice as frequent as varus deviation.

Axial deviations occurred most frequently in combined fractures where they accelerate the development of gonarthrosis. Valgus deformation tends to take a more severe course than does varus deformation. Impression fractures and combined fractures are followed by a more severe grade of arthrosis than are depression fractures.

Incongruence of the tibial plateau caused arthrotic deformities in every case. Wave-like incongruence was always accompanied by severe arthrosis. Trough-shaped deformities always presented arthrotic deformities but in a lower grade as

Table 1. Post-traumatic arthrosis in fractures of the tibial head ($n=48$)

	Number of cases	Percent of cases
Post-traumatic arthrosis	44	91.7
Arthrosis grade I–II	28	58.3
Arthrosis grade III–IV	16	33.3
Preoperative arthrosis	8	16.7

Table 2. Arthrosis and tibial plateau incongruence

Arthrosis	Articular surface		Type of incongruence		
	Congruent ($n=16$)	Incongruent ($n=32$)	Step-like ($n=7$)	Trough-shaped ($n=12$)	Wave-like ($n=13$)
Grade 0	4	0	0	0	0
Grade I	11	5	2	2	1
Grade II	1	11	2	2	7
Grade III	0	5	1	4	0
Grade IV	0	11	2	4	5

compared with the wave-like form (Table 2). Step-like differences in the tibial plateau of less than 3 mm did not lead to severe arthrosis.
There is no significant correlation between the degree of arthrosis and the time to onset of operative treatment after trauma. The patients with the most severe complaints were those with a combination of post-traumatic incongruence, axial deviations and ligament instabilities.

Discussion

Aside from the alteration of the joint by bone destruction, the destruction of cartilage is the most important feature in the manifestation and progress of post-traumatic gonarthrosis [1, 3, 5]. Accompanying ligament injuries could not be defined as a significant factor in this series, although these seem to favour the development of arthrosis [8, 10].
Development of arthrosis can take place even after ideal reconstruction of the bone [11]. Primary bone destruction and operative intervention into the biological integrity of the joint stimulate a degenerative process in the cartilage [7]. Of decisive importance for the size and degree of cartilagenous defect is the destruction of the subchondral capillary network [3, 7]. Impression fractures and combined fractures with trough-shaped incongruence lead to more severe arthrotic deformities than do other types of fractures. Autogenous cancellous bone grafts in combination with an anatomical reduction and a rigid stabilisation are of great importance in the operative treatment of these fractures.
Incongruence of less than 3 mm does not significantly influence the development of a post-traumatic arthrosis. A rigid osteosynthesis favouring early functional treatment avoids further degeneration of cartilage [5, 9, 11]. The aim of operative treatment is the restoration of the biochemical balance of the knee joint. Muscular rehabilitation through early functional exercise is almost as important as bone and ligament reconstruction.

References

1. Cotta H, Puhl W (1976) Pathophysiologie des Knorpelschadens. Unfallheilkunde 127:1–32
2. Fryjordet A Jr (1967) Operative treatment of tibial condyle fractures. Acta Chir Scand 133:17–25
3. Hoaglund FT (1967) Experimental hemarthrosis. J Bone Joint Surg 49:285–293
4. Holz U (1975) Form und Einteilung der Tibiakopffrakturen. Chirurg 46:341–344
5. Holz U, Welte G, Märklin HM, Weller S (1985) Ergebnisse nach operativer Versorgung von Tibiakopffrakturen. Unfallchirurg 88:519–527
6. Jonasch E (1964) Das Kniegelenk. De Gruyter, Berlin
7. Mohing W (1966) Die Arthrosis deformans des Kniegelenks. Springer, Berlin Heidelberg New York
8. Moore TM (1981) Fracture dislocation of the knee. Clin Orthop 156:128–132
9. Muggler E, Huber D, Burri C (1975) Ergebnisse nach operativer Versorgung von 225 Tibiakopffrakturen. Chirurg 46:348–352
10. Wilpulla E, Bakalim G (1972) Ligamentous tear concomitant with tibial condylar fracture. Acta Orthop Scand 43:292–300
11. Schweiberer L, Eitel F (1977) Bone transplantation in animals and in man. In: Masshoff JW (ed) Transplantation. Springer, Berlin Heidelberg New York (Handbuch der allgemeinen Pathologie, vol 6/8)

Traumatic Loss of a Condyle of the Femur: Replacement by an Ipsilateral Patellar Graft

V. Bühren, H. Seiler, and H. Niemeyer

Chirurgische Universitätsklinik – Abteilung Unfallchirurgie,
6650 Homburg/Saar, FRG

Due to the shape and size of the patella, it can be used as a replacement for a destroyed tibial plateau or a missing condyle of the femur. Autologous grafts of this type have been described by Wilson and Jacobs (1952) for recent fractures of the head of the tibia and by Karpinski and Botting (1983) for old ones. Merle d'Aubigne reported on two war injuries, during 1945 and 1946, each involving a missing medial condyle of the femur which was replaced by using a petiolated patellar graft.

A larger number of cases can be found for replacement after a tumour resection. Campanacci et al. (1985) described 19 cases chiefly involving osteoclastomata in which a tibial plateau (12 cases) and a condyle of the femur (7 cases) were replaced by a freed ipsilateral patellar graft. The patella graft was either obtained by way of a sharp, radical ectomy or by a sawed cut, retaining the thin anterior osseous lamella. The advantage of an autologous transplantation in contrast to a homologous one which fits better, for example, the condyle of the femur taken from a bone bank, is perceived to lie chiefly in the decreased risk of infection and the more rapid revascularisation, thus resulting in reliable incorporation.

In the first of two cases involving an open knee injury, fragments of a patella, which was also traumatised, were used as a graft. In the second case the uninjured ipsilateral patella was used as a replacement for the lost condyle of the femur.

A 22-year-old patient was involved in an automobile accident while driving and sustained an open luxation of the right knee along with a fracture of the condyle and patella. The patient presented with multiple injuries including fracture of the right and left humerus shafts, gross posterior knee instability, and bilateral fracture of the calcaneum. The right knee was treated, and the wound was thoroughly debrided with removal of embedded glass and paint splinters. The posterior cruciate ligament, which had been torn out of the tibia, and the anterior cruciate ligament (ACL), which was ruptured at the femur, were reinserted. The patella was freed using a sharp cut, and the larger fragments were sculptured in such a manner so they could be used to replace the missing lateral condyle of the femur. Fixation was done using an external tightener bridging the joint.

Rehabilitation and knee mobilisation could not be initiated until 10 weeks after the accident due to the extensive injuries sustained; an extension deficit of 20° still remained. Thus a deflective, supracondylar osteotomy was performed 10 months after the accident. The range of mobility of the knee 2 years after the accident amounts to 0/0/100° (full extension with flexion to 100°). The leg can be subjected to full loading but with an accompanying predominantly anterior instability, and the patient has been able to go back to his job in the post office.

Surgery and Arthroscopy of the Knee
2nd Congress of the European Society
Editors: Müller/Hackenbruch

In the second case a 16-year-old moped rider was involved in an accident. The injuries sustained were a catenoid injury of the right leg with a closed fracture of the femoral shaft, an open second-degree fracture of the lower leg, and an injury to the knee joint which was ripped wide open, accompanied by loss of a large portion of the lateral femoral condyle. The uninjured patella was freed using a sawed cut to retain the anterior osseous lamella. The accompanying fractures were treated by way of osteosynthesis at the same time.
During postoperative treatment, muscle flap and split-skin grafts were used to compensate for extensive soft tissue defects. Radiological checkups 14 months after the accident revealed that the patella had become incorporated and the fractures mended. Arthroscopic examination indicated good cartilaginous grafting of the transplant as well as substitutive formation of fibrocartilage in the patellar bearing. The leg could accommodate maximum loading with an accompanying acceptable range of motion. The patient has now returned to work.
As a rule in a load-bearing joint arthrodesis or an alloplastic replacement has to be performed for cases in which a large section covered by cartilage is lost. The use of autologous osseous grafting in reconstruction of such injuries is recommended to reduce the high risk of infection.
The possibility of using the ipsilateral patella should always be considered when treating such knee traumas and presents the almost ideal mode of treatment in cases for which patellectomy is indicated. However, the sacrifice of the uninjured patella can only be justified when a part of a joint, which is essential for stability and function, can be replaced. The known consequences of a patellectomy can be safely accepted in view of satisfactory global functions of the knee.
Operative strategy and technical details of surgical treatment have to be planned individually for each case, due to the rare occurrence of such cases and the possible wide range of findings. Petiolated removal of the patella does not appear to be necessary. Partial open resection also results in a reliable incorporation, and the extent of surgical intervention involved is smaller. In addition, complete patella removal leaves the extension apparatus of the already seriously damaged knee relatively intact, particularly if the patella is freed without using a sharp cut, retaining the anterior osseous lamella.

References

Campannaci M, Cervellati C, Donati U (1985) Autogenous patella as replacement for a resected femoral or tibial condyle. J Bone Joint Surg [Br] 67:557
Karpinski MRK, Botting TDJ (1983) Patellar graft for late disability following tibial plateau fractures. Injury 15:197
Merle d'Aubigné R, Alexandere G (1966) Reconstruction condylienne fémorale par greffe rotulienne pédiculée. Rev Chir Orthop 52:611
Wilson WJ, Jacobs JE (1952) Patellar grafts for severely depressed comminuted fractures of lateral tibial condyle. J Bone Joint Surg [Am] 34:436

Electron Microscopical Studies on Human Meniscal Tissue Preserved for Transplantation

A. Schmid, F. Schmid, and T. Tiling

Department of Surgery, University Clinic, Robert-Koch-Strasse 40,
3400 Göttingen, FRG

Introduction

Since arthrosis of the knee joint develops at an accelerated rate after meniscectomy (Burr and Radin 1982; Espley and Wangh 1981; Hansen 1983), attempts have been made to preserve human meniscus tissue for homologous transplantation to delay this process. In this study we investigated the ultrastructure of preserved human menisci, to evaluate their suitability as homologous transplants.

Material and Methods

The ultrastructure of ten solvent-dried human menisci of Pfrimmer-Viggo (Fig. 1) and ten lyophilised human menisci of B. Braun, Melsungen (Fig. 2) was examined by light microscopy and transmission electron microscopy (TEM). The ultrastructural results were compared with the morphology of native meniscal tissue removed from organ donors, aged between 18 and 35 years, who had died from an accident and who revealed no previous knee damage.
The meniscal tissue was imbedded in Epon. Thin sections for light microscopical examination were stained with toluidine blue. Ultrathin sections were contrasted with uranyl acetate and lead citrate and viewed by a Zeiss EM 9 electron microscope.

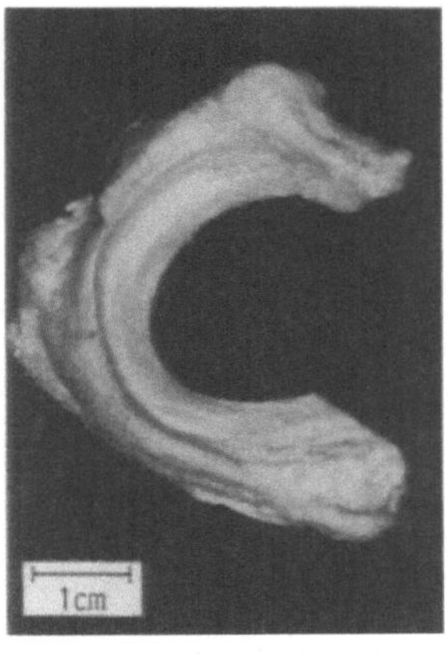

Fig. 1. A solvent-dried human meniscus

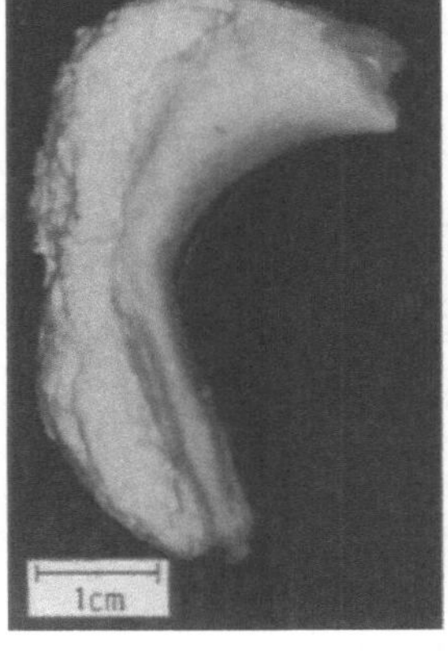

Fig. 2. A lyophilised human meniscus

Surgery and Arthroscopy of the Knee
2nd Congress of the European Society
Editors: Müller/Hackenbruch

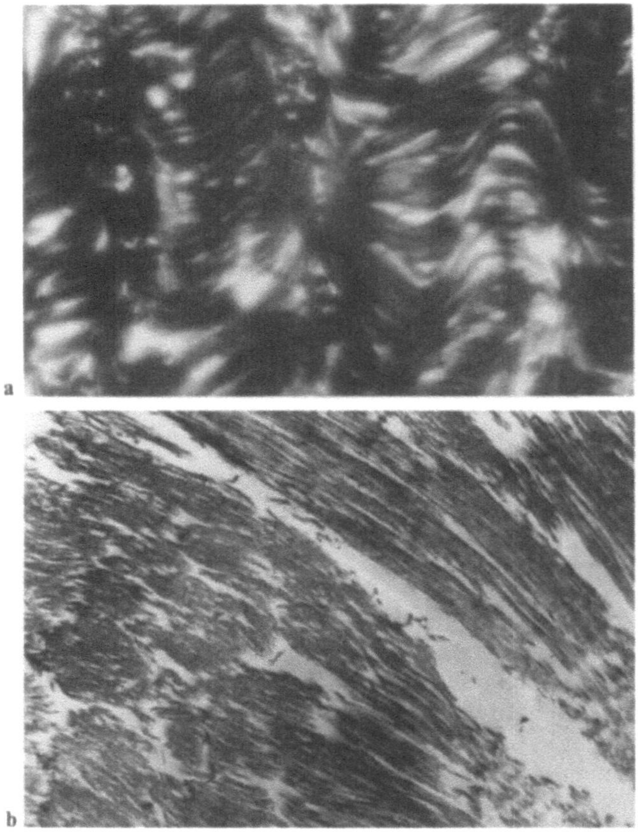

Fig. 3. **a** Light micrograph (× 16) and **b** electron micrograph (× 4500) showing a regular pattern of collagen fibres in solvent-dried menisci

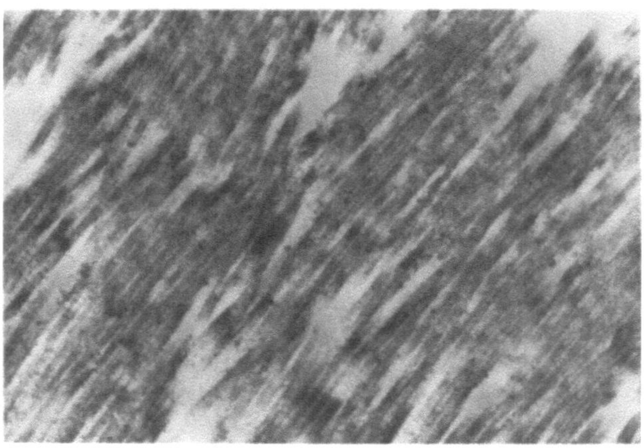

Fig. 4. Uniformly calibrated collagen fibres in solvent-dried menisci, × 19 500

Results

In the solvent-dried menisci the pattern of the collagenous fibres is regular, forming large parallel bundles typical of meniscal tissue (Fig. 3). Bundles are less densely arranged than in native meniscal tissue, but the structure of the individual fibres does not show any features deviating from the native tissue. The fibres are uniformly calibrated and smoothly delineated (Fig. 4). Electron micrographs indicate the presence of cavities with electron dense material at the edge of the bundles (Fig. 5). Since no intact cells are seen in any of the preparations, the cavities were interpreted as representing the site of former cells that had not been preserved during fixation.

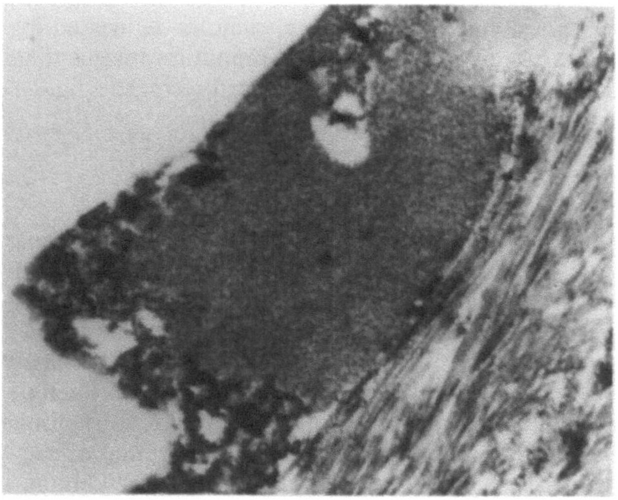

Fig. 5. Electron dense material found in solvent-dried menisci which is thought to be destroyed cells, ×19500

a 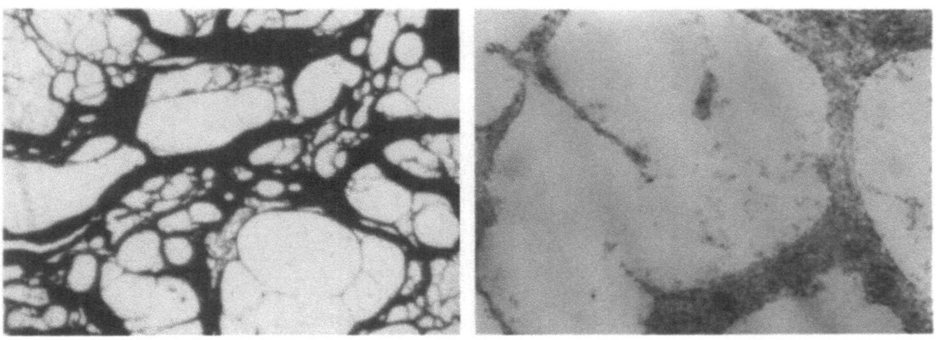 b

Fig. 6. a Light micrograph (×16) and **b** electron micrograph (×4500) showing that the meniscal tissue has been broken up by lyophilisation

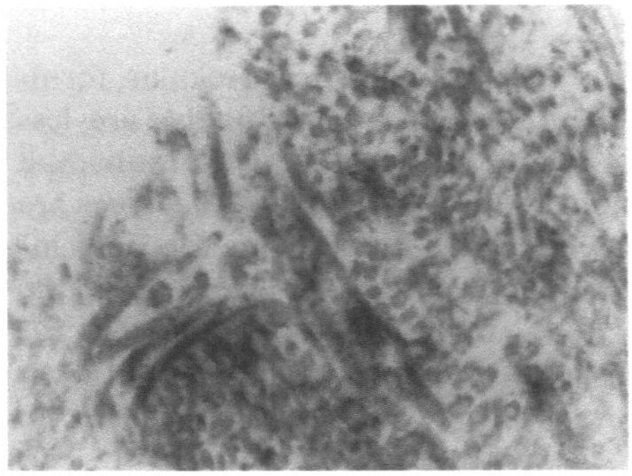

Fig. 7. Tissue structure in lyophilised menisci consisting of irregular calibrated fibres, ×30000

Light micrographs of the lyophilised menisci show large vacuoles delineated by narrow regions of tissue (Fig. 6). Electron microscopal examination reveals that these tissue paths consist of fibres varying in thickness and partly fringed (Fig. 7). No intact cells are seen in any of the preparations.

Discussion

Both preservation methods destroy the cellular components of the meniscal tissue so that a functional incorporation of the transplants by means of an allovital transplant cannot be expected. From follow-up arthroscopic and ultrastructural investigations of recovered tissue samples, it is found that after total meniscectomy cell-rich connective tissue grows in from the capsule and forms a small meniscus wedge (Fig. 8). Since this wedge does not develop to reach the size of a normal meniscus, it cannot replace the biological and biomechanical functions of the meniscus. With a meniscus transplant implanted after meniscectomy, the connective tissue growing in from the capsule could be protected against early biomechanical stress until a wedge of a sufficient size has been formed.

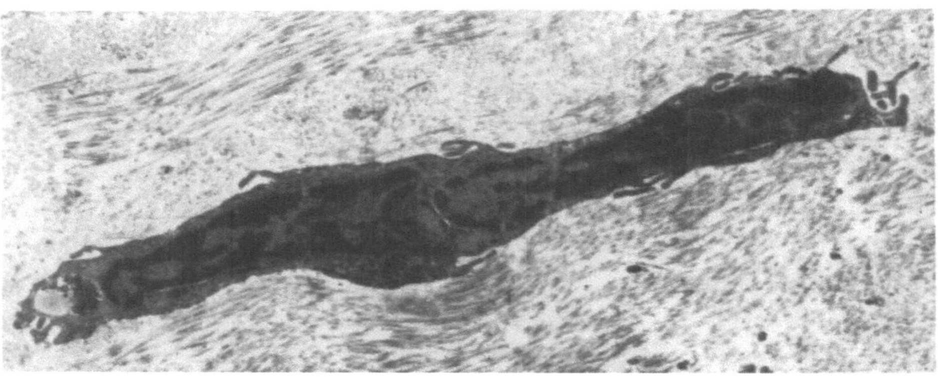

Fig. 8. Cell-rich connective tissue grows in from the capsule after total meniscectomy, ×7500

It is known that alloplastic material, especially collagen, favours the growth of granulation tissue and finally becomes replaced by the ingrowing regenerative tissue (Bedacht 1969; Cherapil and Krajicek 1967; Cutright et al. 1973; Katthagen 1986). With reference to meniscus transplantation, this means that the graft would not only serve as a guide line and substitute until the ingrowing tissue is able to stand biomechanical stress itself, but it would also stimulate the ingrowth of the connective tissue.

Our ultrastructural studies have shown that after solvent drying the spatial and fine structure of the meniscus fibres are almost preserved, whereas after lyophilisation the spatial and fine architecture of the meniscus are mainly destroyed. Although both types of transplants could be used for growth stimulation we consider the solvent-dried menisci to be more capable of functioning as a temporary substitute and protector for the ingrowing tissue.

References

Bedacht R (1969) Tierexperimentelle und klinische Untersuchungen über die Anwendung von heterologem Kollagen als Implantat in der Knochenhöhle von Röhrenknochen. Dissertation, University of Munich

Burr DB, Radin EL (1982) Meniscal function and the importance of meniscal regeneration in preventing late medial compartment osteoarthrosis. Clin Orthop 171:121

Cherapil M, Krajicek M (1967) Die Analyse einiger theoretischer Probleme bei der Verwendung von Erzeugnissen aus Kollagen in der Medizin. Langenbecks Arch Klin Chir 318:80

Cutright DE, Posey WR, Bhaskar SN, Larson WJ (1973) Collagen sponge subcutaneous implantation. Oral Surg 35:144

Espley AJ, Wangh W (1981) Regeneration of menisci after total knee replacement. J Bone Joint Surg [Br] 63:387

Hausen B (1983) Der Einfluß der Meniskusoperation auf das Kniegelenk. Eine Studie der Spätergebnisse. Beitr Orthop Traumatol 30:81

Katthagen BD (1986) Knochenregeneration mit Knochenersatzmaterialien. Hefte Unfallheilkd 178:30

Ultrastructural Analysis of the Cruciate Ligaments and of Tissue Used for their Plastic Repair

H. Arnold-Schmiebusch [1], W. Hackenbruch [2], and J. Staubesand [1]

[1] Department of Anatomy I, 7800 Freiburg, FRG
[2] Department of Orthopaedic Surgery, Regionalspital, 4900 Langenthal, Switzerland

Introduction

Rupture of the anterior cruciate ligament is more and more frequently encountered, probably due both to better diagnosis and increasingly demanding sports. Recovery without surgical intervention is, on purely anatomical grounds, impossible.

In spite of the improved operative techniques that have become available during the past decade, the results of surgical treatment are still not always satisfactory. This is partly accounted for by the precarious blood supply and partly by the biomechanical complexity of the ligament (Hackenbruch and Henche 1981; Müller 1982; Schmiebusch and Hackenbruch 1984; Hackenbruch et al. 1985).

Material and Methods

In a few cases for which a second operation was necessary – or during arthroscopy – we removed very small pieces of tissue at different times after the operation in

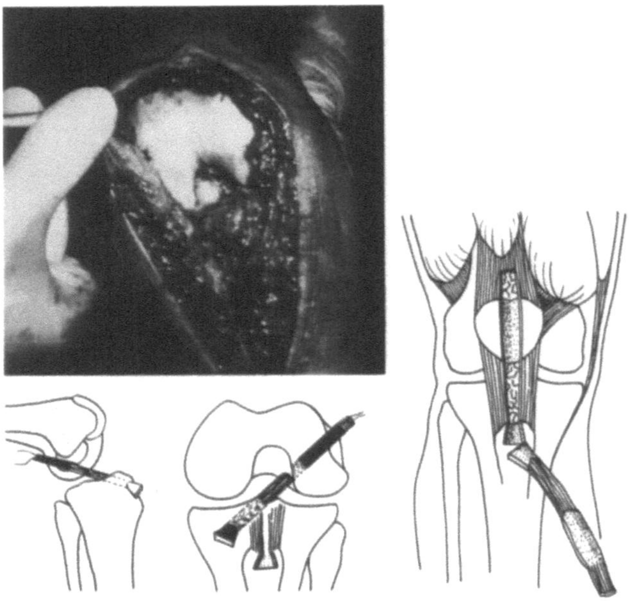

Fig. 1. Operative procedure (Brückner and Jones)

Surgery and Arthroscopy of the Knee
2nd Congress of the European Society
Editors: Müller/Hackenbruch

the hope that ultrastructural investigation would provide us with objective criteria of operative success.

The operation consisted of primary suture or replacement of the anterior cruciate ligament with a strip from the middle third of the ligamentum patellae (Jones 1963; Brückner 1972). This strip was removed, together with a piece of bone from the patella and tibial tuberosity, and, as a free transplant, pulled distally through a bore-hole into the normal anatomical position of the anterior cruciate ligament (Fig. 1 b, c).

Results

Either because of a slight remaining discomfort or because of the patient's urgent desire to take part in some sport, arthroscopy was undertaken in about 30 cases. In nearly every case a vital transplant that could not be distinguished from a normal cruciate ligament was found (Fig. 1 a). Of some 100 such operations carried out over 4 years, 3 required a second operation to enable a satisfactorily functioning cruciate transplant to be put into position.

Control tissue is taken from a macroscopically healthy cruciate ligament. The collagen fibrils are uniformly distributed in transverse section and are not all of the same diameter (Fig. 2).

One year after plastic repair of the cruciate ligament with part of the *ligamentum patellae,* the collagen fibrils display a regular, orderly arrangement with little variation in the individual diameters (Fig. 3).

The arrangement of the collagen fibrils into whorls after successful *primary suture* of the anterior cruciate ligament suggests that the normal texture of this tissue has been destroyed (Fig. 4).

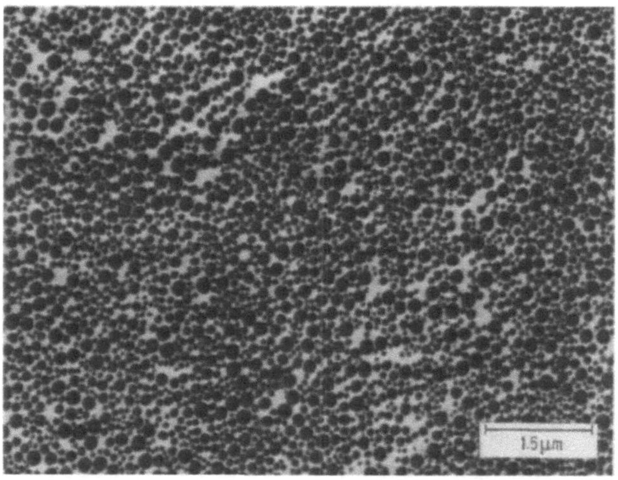

Fig. 2. Control tissue taken from a macroscopically healthy cruciate ligament (biopsy from a 27-year-old male)

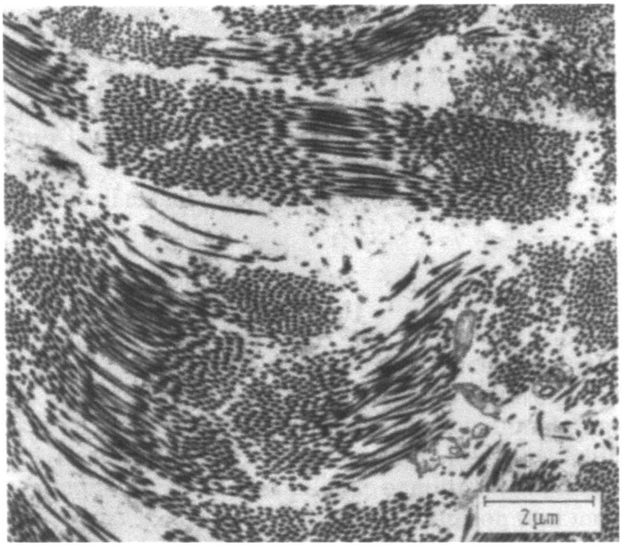

Fig. 3. One year after plastic repair. The clinical result in the case of this 35-year-old male was very good

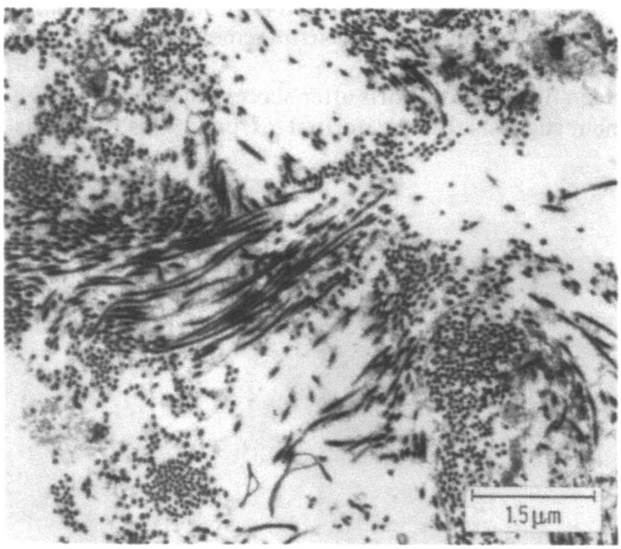

Fig. 4. Successful primary suture of the anterior cruciate ligament in a 62-year-old male

A strip of *fascia lata* used as a transplant shows an essentially irregular orientation of the collagen fibrils (Fig. 5).

The biopsy from the remains of a cruciate ligament, taken from a patient for whom *no surgical repair* had been attempted (and whose medial meniscus had also been torn), shows a complete disorder and irregularity in the shape and structure of the fibrils, which is characteristic (Fig. 6).

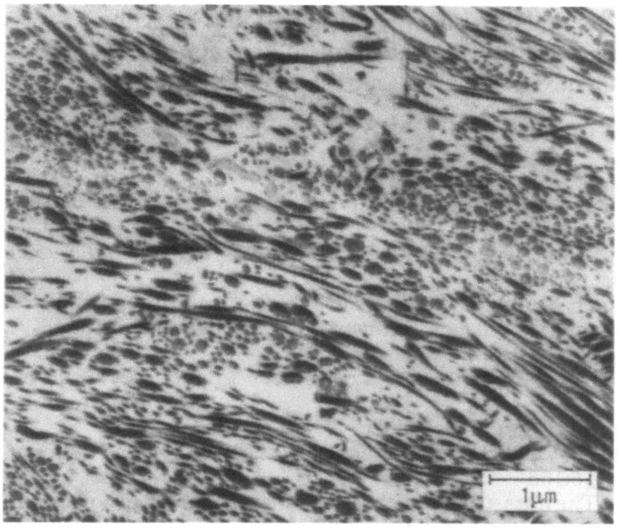

Fig. 5. A strip of fascia lata was used as a transplant in this 20-year-old male

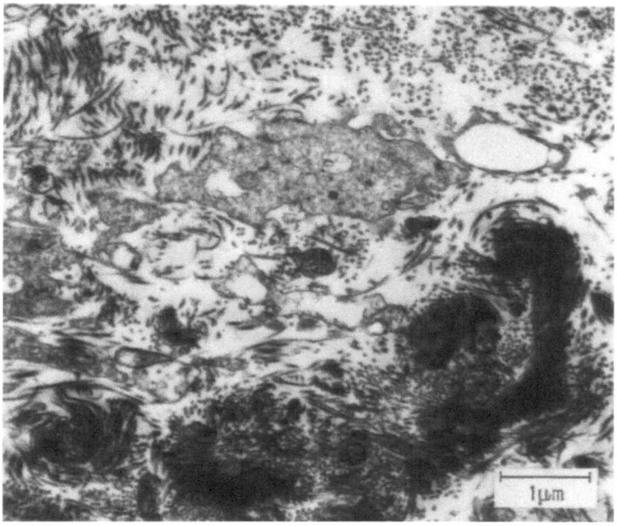

Fig. 6. Biopsy from the remains of a cruciate ligament taken from a patient for whom no surgical repair had been attempted (and whose medial meniscus had also been torn)

Summary

Arthroscopic follow-up of patients who have undergone primary suture of an anterior cruciate ligament confirms that they do less well in terms of ultrastructural change than when part of the patellar ligament is used as a transplant. This, of course, confirms what is already known about the prognosis for joint stability.

Following primary suture the collagen shows both intra- and interfibrillar disorder (Staubesand and Fischer 1980), whereas after transplantation the orderly arrangement of fibers and the texture of the tissue remain fairly normal.

References

Brückner H (1972) Bandplastiken im Kniebereich nach dem Baukastenprinzip. Zentralbl Chir 3:65

Hackenbruch W, Henche HR (1981) Diagnostik und Therapie von Kapselbandläsionen am Kniegelenk. Eular, Basel

Hackenbruch W, Schmiebusch H, Müller W, Staubesand J (1985) Arthroscopy and electron microscopy following anterior cruciate reconstruction. Kongreßband 1, ESKA. Springer, Berlin Heidelberg New York

Jones KG (1963) Reconstruction of the anterior cruciate ligament. J Bone Joint Surg [Am] 45:925

Müller W (1982) Das Knie. Springer, Berlin Heidelberg New York

Schmiebusch H, Hackenbruch W (1984) Operative Rekonstruktion des Ligamentum cruciatum anterius. Anat Anz 156:166

Staubesand J, Fischer N (1980) The ultrastructural characteristics of abnormal collagen fibrils in various organs. Connect Tissue Res 7:213–217

What is a Meniscal Insufficiency?

M. F. Fischmeister

Unfallkrankenhaus Linz der Allgemeinen Unfallversicherungsanstalt,
Blumauerplatz 1, 4020 Linz, Austria

Introduction

Load bearing is thought to be the most important function of the menisci. Axial loading by the femur creates a circumferential tensile stress that accompanies the axial compressive stress in the meniscal tissue (Figs. 1, 2). An analysis of the microstructure of human knee menisci by light and scanning electron microscopy techniques has shown two structurally different regions: a mesial part containing collagen bundles in a radial pattern and a peripheral part consisting of larger collagen bundles in a circumferential pattern. There seems to be a correlation between microstructure and biomechanics [1]. The function of the menisci depends not only on the preserved anatomic shape but also on the integrity of the peripheral fibres. If under load circumferential stress is not built up, e.g. in injuries of the meniscal horns, subluxation of the affected menisci occurs (Fig. 3). This can be described as a functional insufficiency.

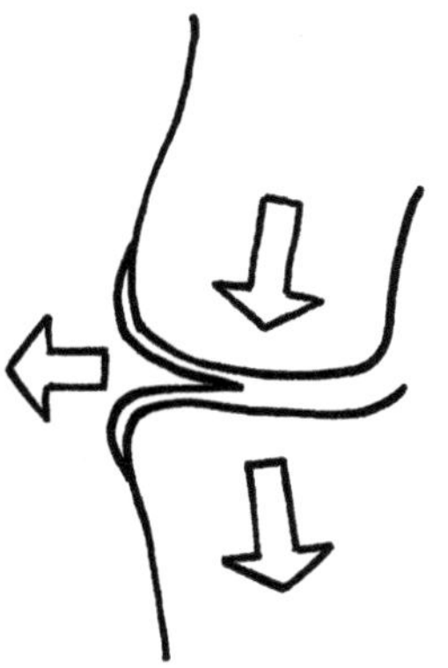

Fig. 1. Flow of forces in the knee joint under conditions of load bearing

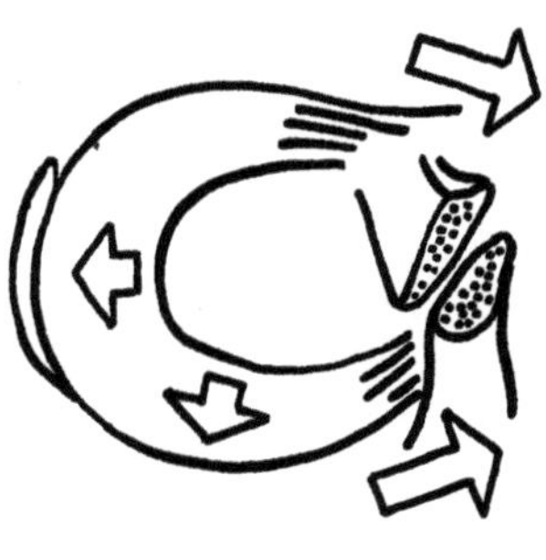

Fig. 2. Circumferential tensile stress in the meniscal tissue is neutralised at the meniscal horns

Surgery and Arthroscopy of the Knee
2nd Congress of the European Society
Editors: Müller/Hackenbruch

Fig. 3. Posterior and medial subluxation of the medial meniscus in avulsion injury of the posterior horn

Methods

To test this hypothesis the posterior horns of otherwise intact menisci were examined during arthroscopy. Arthroscopies were carried out through an anterolateral approach with the use of a legholder and done under general anaesthesia after a clinical examination. Routine radiography consisted of anteroposterior and lateral views.

Results

Two types of injuries were found: (1) transverse rupture of the posterior meniscal horn at the point of insertion and (2) bony avulsion of the posterior meniscal horn. From 1.1.1985 to 31.12.1985 five patients with type 1 injuries and three patients with type 2 injuries were seen. Ligamentous instability was observed in four of them. The other four patients had anterior knee pain and swelling but no typical meniscal signs. Type 2 injuries showed a small round shadow posteroparamedian to the intercondylar eminence on anteroposterior and lateral radiographs (Fig. 4). Diagnosis was made by arthroscopy (Figs. 5, 6). Subluxation of the meniscus posteriorly and medially has also been seen. The treatment of these mensical tears was conservative.

Discussion

This concept of a functional insufficiency of the mensici could have an impact on therapeutic goals of treatment. In type 1 and 2 injuries of the meniscal horns, for

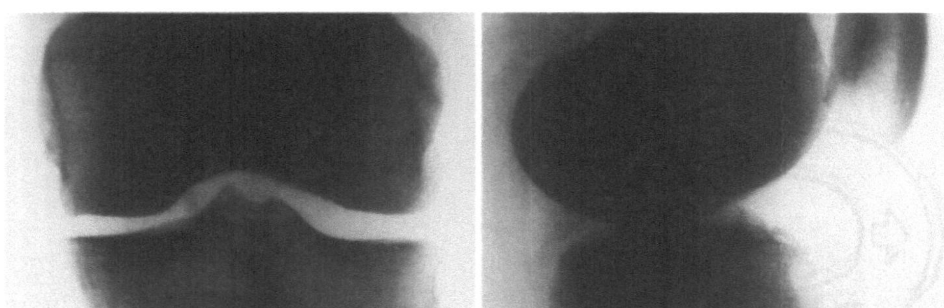

Fig. 4. Type 2 lesion of a medial meniscus, anteroposterior and lateral view of a left knee

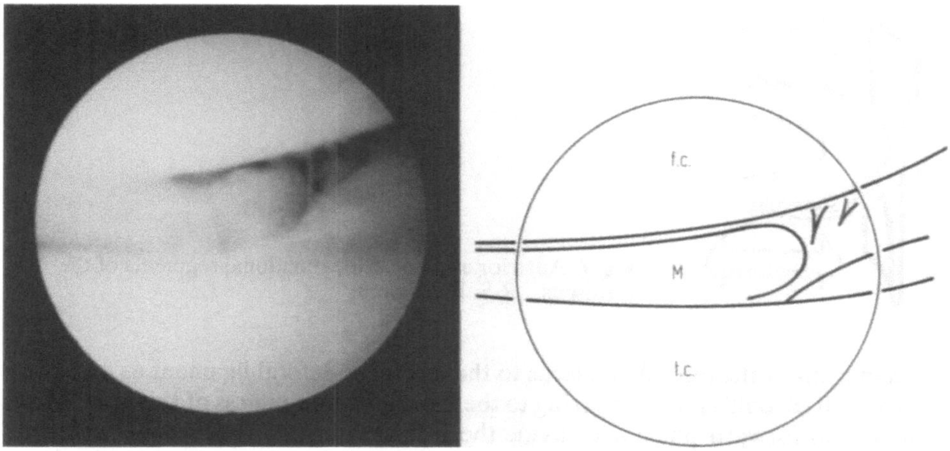

Fig. 5. Type 1 injury of a medial meniscus, left knee, anterolateral approach (4 mm, 70 optic)

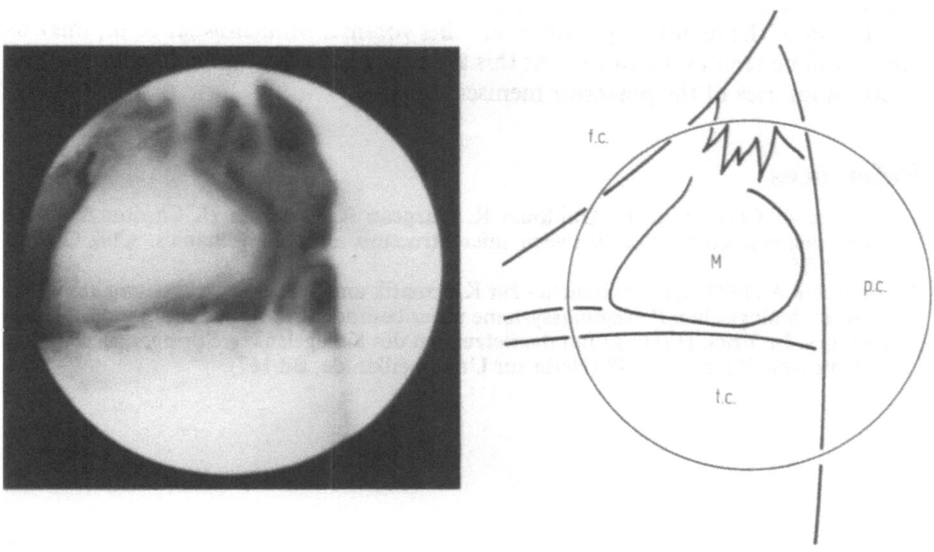

Fig. 6. Old type 2 injury of a medial meniscus, left knee, anterolateral approach, modified Gillquist technique (4 mm, 30 optic)

example, one should try to anchor these horns firmly at their point of insertion. How to do this technically has not yet been solved. It is also of importance to preserve the outer fibrous rim of the meniscus in other operations, e.g. open repair of meniscal tears or open reduction and internal fixation of tibial plateau fractures.

This concept of a functional insufficiency of the menisci is compatible with Menschik's analysis of the kinematics of the knee joint [2]. He describes the area of

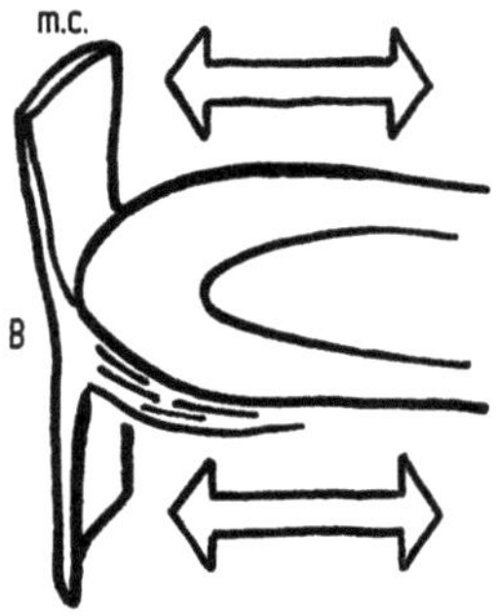

Fig. 7. Anterior and posterior functional segments of the meniscus, "R.S. Ball's point"

attachment of the medial meniscus to the medial collateral ligament as an equivalent of R. S. Ball's point, referring to the kinematic uniqueness of this area. Moving a step forward one could divide the meniscus in two functionally independent parts: a posterior segment from the point of insertion of the posterior horn to Ball's area, and an anterior segment from Ball's area to the attachment of the anterior horn (Fig. 7).

Whether these concepts are of further clinical relevance, for example, in transplantation of the meniscus [Wulff et al., this volume; Milachowski et al., this volume], will be seen in the future. At this time they have led to the detection of the avulsion injuries of the posterior meniscal horns.

References

1. Beaupre A, Choukroun R, Guidouin R, Garneau R, Gerardin H, Cardou A (1986) Knee menisci, correlation between microstructure and biomechanics. Clin Orthop 208:72–75
2. Menschik A (1984) Grundsätzliches zur Kinematik und Selbstverwirklichung der unbekannten biologischen Bewegungssysteme unter besonderer Berücksichtigung des Kniegelenkes. In: Frick H (Hrsg) Bandverletzungen des Kniegelenkes. Springer, Berlin Heidelberg New York, S. 23–48 (Hefte zur Unfallheilkunde, Bd 167)

Reconstruction of the Lateral Femoral Condyle with Homologous Bone Graft after a Third-Degree Compound Comminuted Fracture in Polytrauma after Motorcycle Accident

P. Lauber, A. Kentsch, and W. Müller

Department of Orthopedic Surgery, Kantonsspital Bruderholz, 4104 Bruderholz, Switzerland

Anamnesis. In July 1982 a motorcycle accident (collision with a car) resulted in a massive contusion of the whole left leg.

Diagnosis

Third-degree, compound, comminuted fracture of the lateral femoral condyle with excessive contamination and soft tissue lesions. Nondislocated, lateral, tibial plateau fracture with cartilage damage, sagittal splitting fracture of the patella, rupture of the ligamentum patellae and the quadriceps tendon (Figs. 1, 2). Third-degree, compound luxatio pedis subtalo in equinus position of the foot and completely denuded talus, malleolar fracture type A (Figs. 3, 4). First-degree compound fractures with slight dislocation of the metatarsal bones II–V, fractures of the proximal phalangeal bones II–V. Sensibility, motion, and microcirculation of the toes were intact. The pulses of tibialis anterior and posterior could not be palpated.

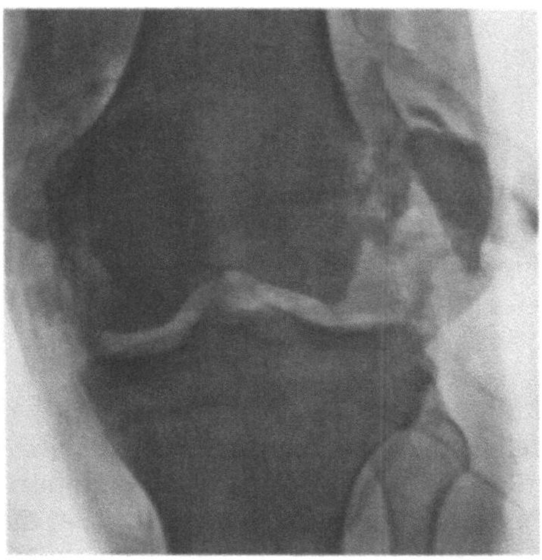

Fig. 1

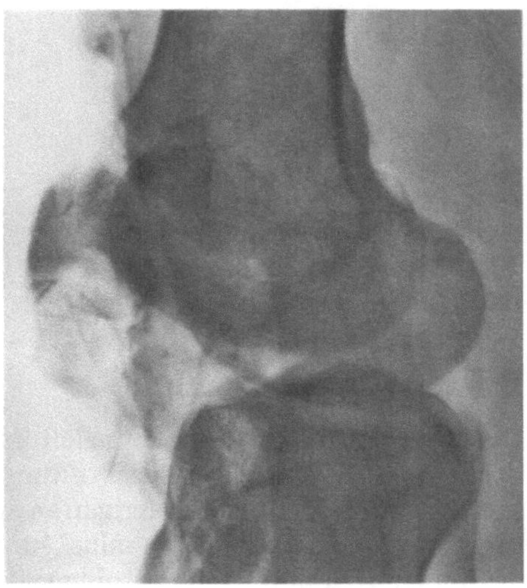

Fig. 2

Surgery and Arthroscopy of the Knee
2nd Congress of the European Society
Editors: Müller/Hackenbruch

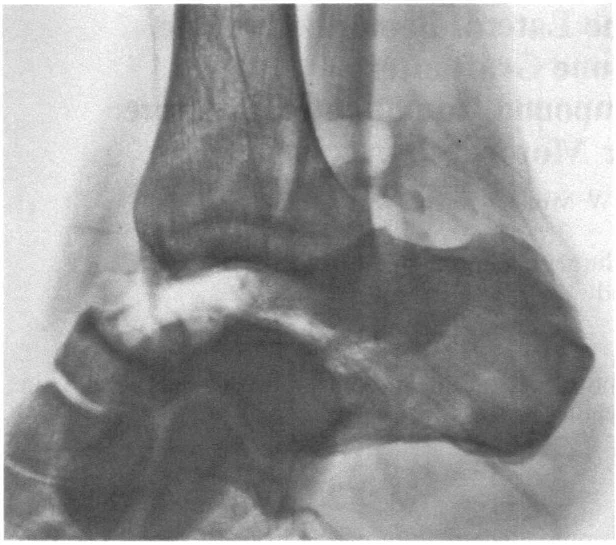

Fig. 3

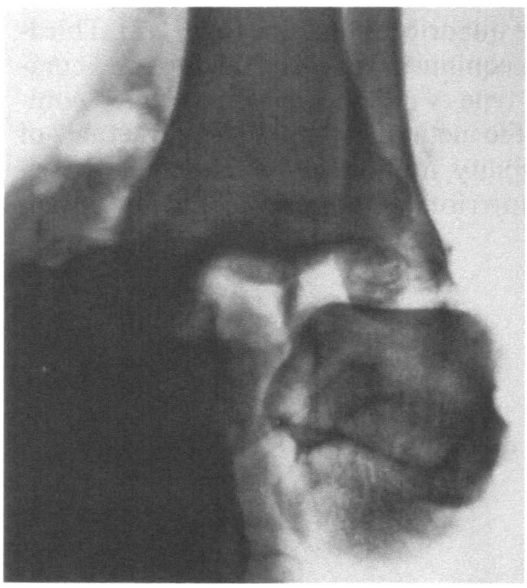

Fig. 4

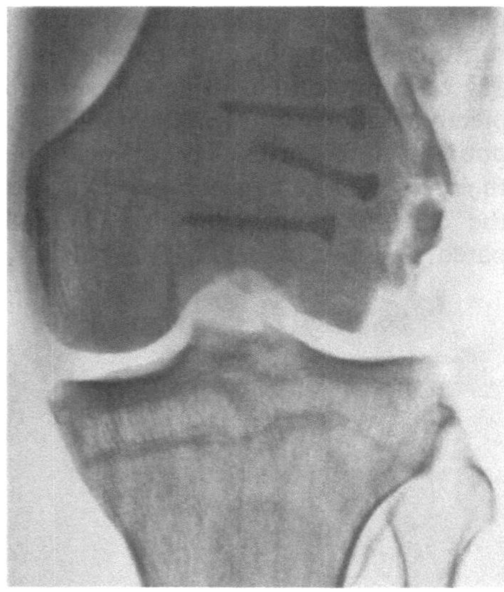

Fig. 5

Primary Reconstruction

Debridement and cleaning of the left knee joint. Minimal internal fixation of the lateral condylus. Osteosynthesis of the patella, suture of the ligamentum patellae and the quadriceps tendon. Irrigation-suction drainage for 48 h, skin adaptation (Figs. 5, 6). Reduction of the talus, sutures of the ligaments, internal fixation of the fibula, irrigation-suction drainage for 48 h (Figs. 7, 8). Open wound treatment of the forefoot.

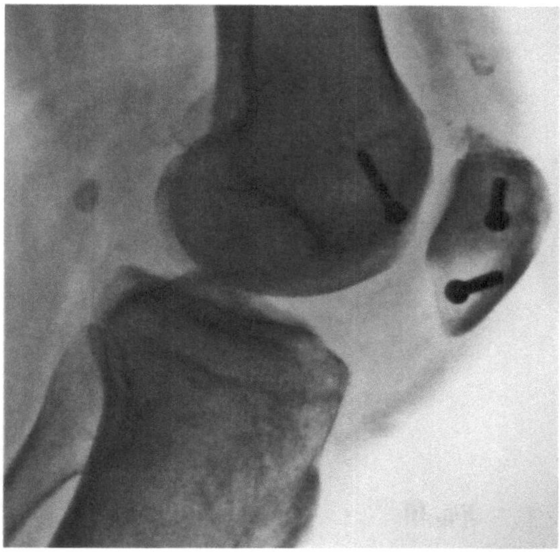

Fig. 6

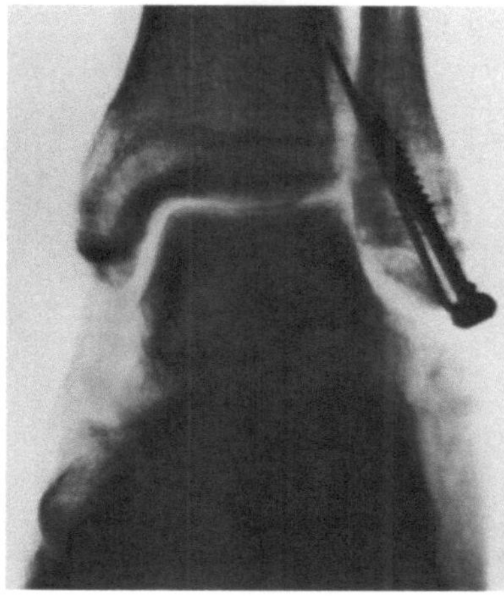

Fig. 7

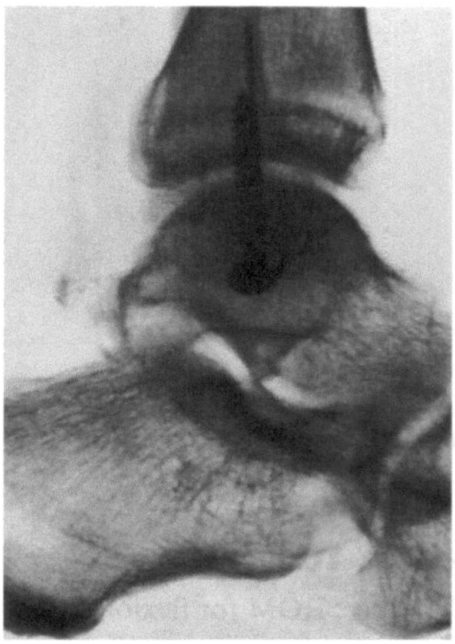

Fig. 8

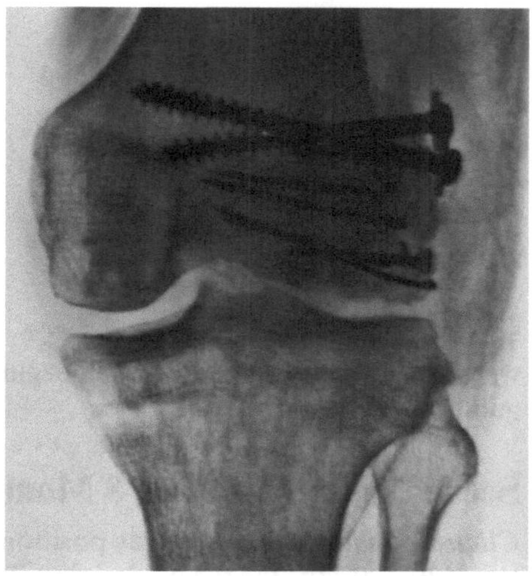

Fig. 9

Continuing Treatment. Antibiotics for 3 weeks, continuous passive motion of the knee and ankle joints on a kinetic machine, wound care. After the wound has healed protect it by a plaster splint.

Postoperative Course. The patient was discharged 9 weeks after the accident. Primary wound healing. Range of motion (ROM) 85°–20°–0° of the knee and 25°–5°–0° of the ankle joint for flexion/extension. Exercise daily with therapist. No

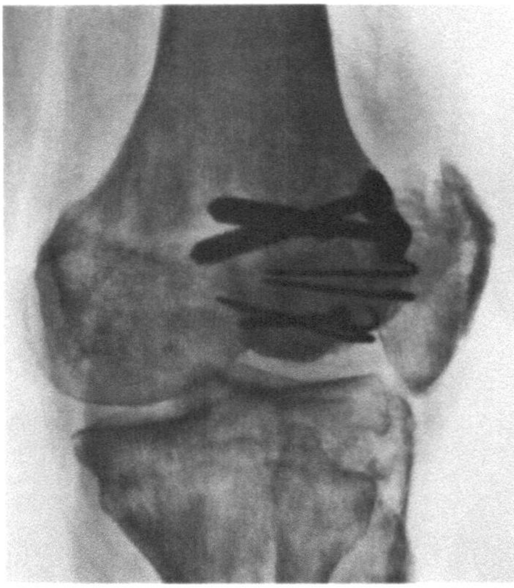

Fig. 10

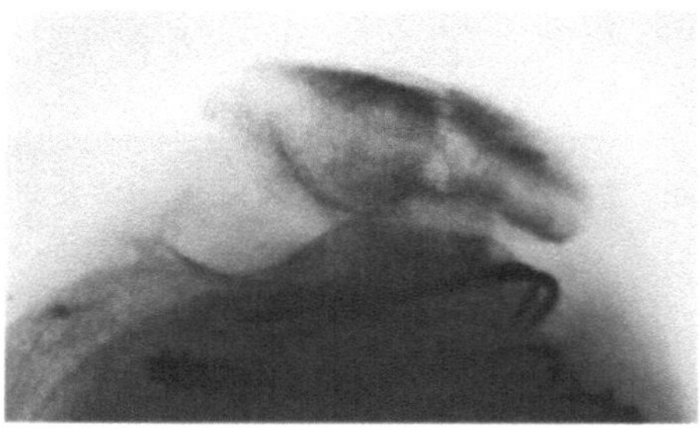

Fig. 11

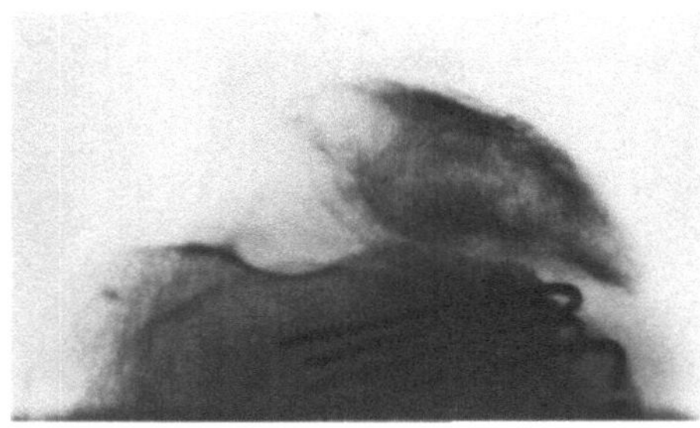

Fig. 12

weight bearing for 5 months, then beginning with slight ground contact. Anticoagulation medication given.

Second Hospitalisation, 8 Months After the Accident

Clinical Findings. Slight valgus position of the left knee; ROM for flexion/extension 130°–20°–0°, for the ankle joint, 25°–10°–0°.

Second Operation. Straight anterolateral incision, debridement of subcutaneous asphalt pieces. The lateral meniscus was intact. The tibial cartilage was partially damaged by fraction of the lateral edge of the lateral femoral condyle. The scarred origins of the lateral collateral ligament system were osteotomized. Sandwich plasty with three 1.5-cm wide bone blocks from a homologous femoral head. Refixation of the ligaments (Figs. 9, 10).

Postoperative Therapy. Adhesion tape and longitudinal traction between 3 and 5 kg in order to extend the retracted posterior capsule, continuous passive motion

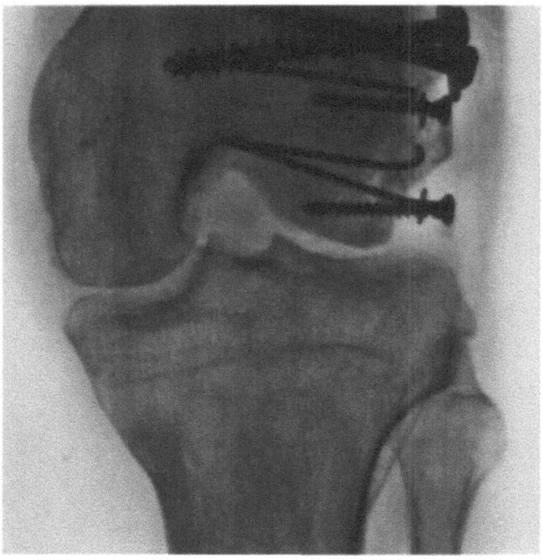

Fig. 13

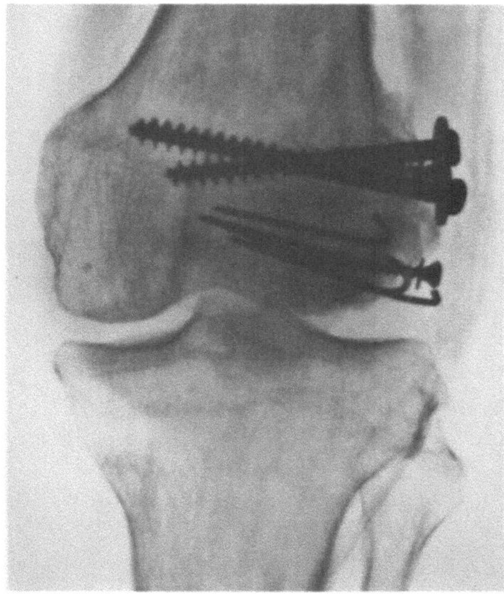

Fig. 14

between maximum extension possible and 90° of flexion. Exercise daily. Anticoagulation medication given. Knee brace in slight varus, when the incision was healed. No weight bearing permitted.

Postoperative Course. Ten weeks later the patient had reached a ROM of 135°–10°–0° for flexion/extension. No refusion, pain free movement. Chronic subluxation of the patella. Still no incorporation of the homologous bone block.

Third Hospitalisation, 1 Year After Trauma

Clinical Findings. Lateral subluxation of the patella (possibly due to adhesions and insufficiency of the quadriceps muscle; Fig. 11).

Third Operation. Again anterolateral straight incision and debridement of asphalt pieces. Resection of adhesions (especially on the medial side). Transosseous fixation of the medial vastus at the medial edge of the patella. Lateral release. No arthrotomy.

Postoperative Course. Primary wound healing. Nonweight bearing with slight ground contact and knee brace. Persisting lateral subluxation of the patella (Fig. 12). Beginning incorporation of the bone block but also partial resorption in the posterolateral compartment (Figs. 13–15). Remarkable crepitation and cartilage loss of the patella. ROM 135°–20°–0° for flexion/extension.

Further Therapy. Hometrainer, increasing weight bearing up to 20 kg. Instep-raisers with medial support.

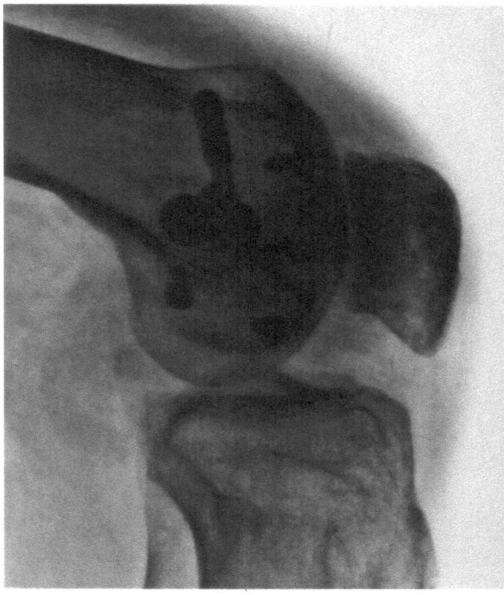

Fig. 15

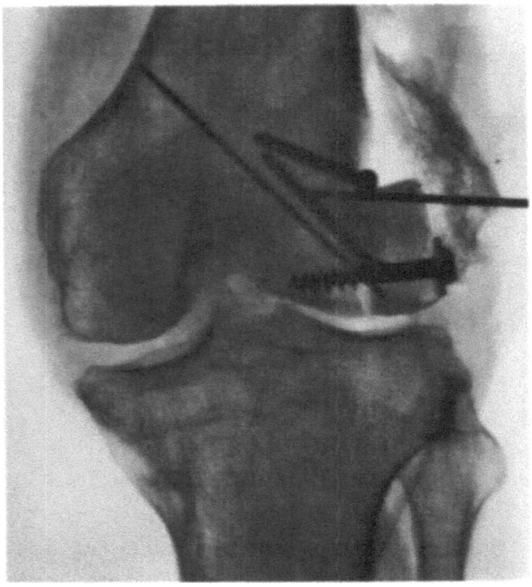

Fig. 16

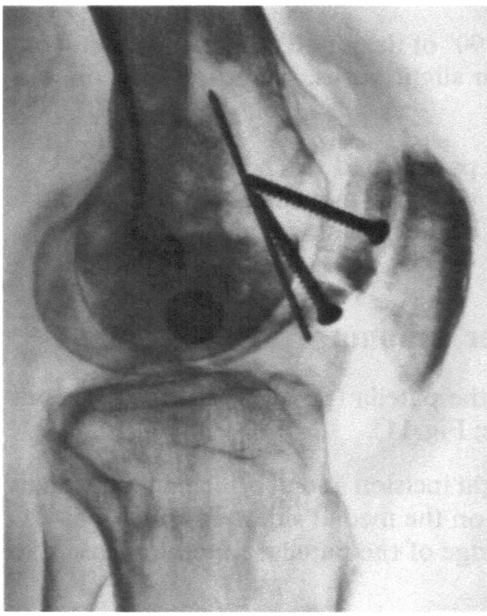

Fig. 17

Fourth Hospitalisation, 22 Months After Trauma

Fourth Operation. Sagittal osteotomy of the lateral femoral condyle and again sandwich apoplasty with a segment of a homologous femoral head. The trochlea femoris was also osteotomized and lifted for 1.5 cm in order to create a new outer bridge edge (Figs. 16, 17).

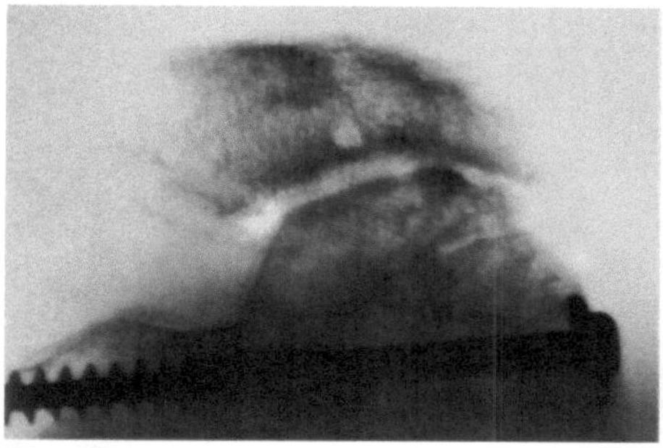

Fig. 18

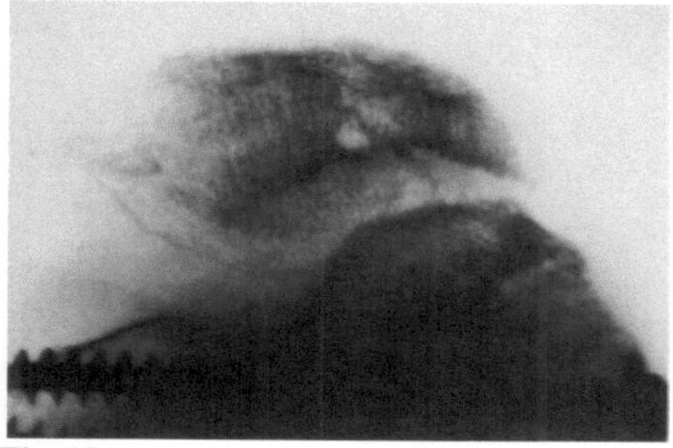

Fig. 19

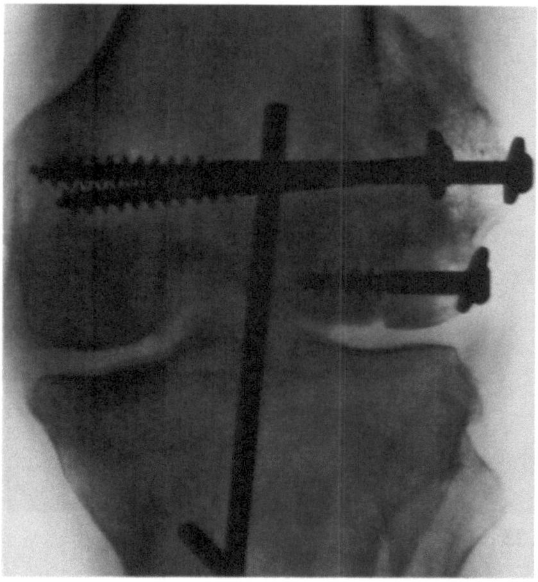

Fig. 20

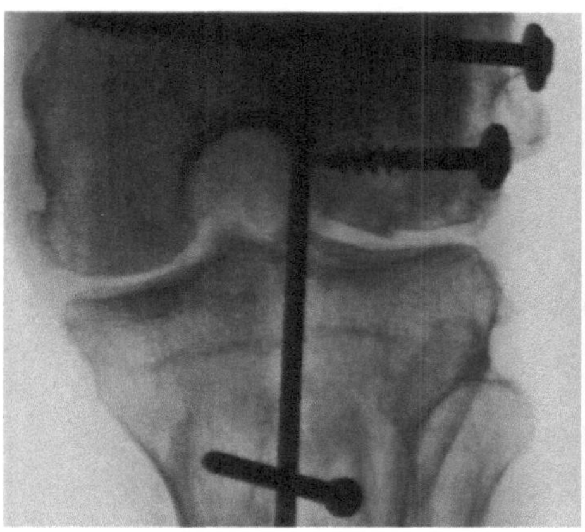

Fig. 21

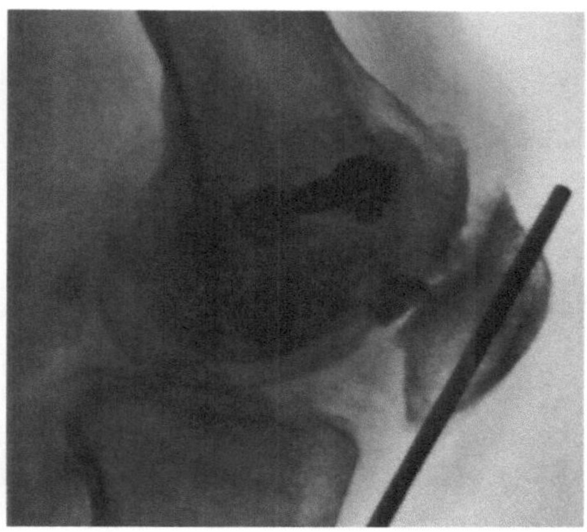

Fig. 22

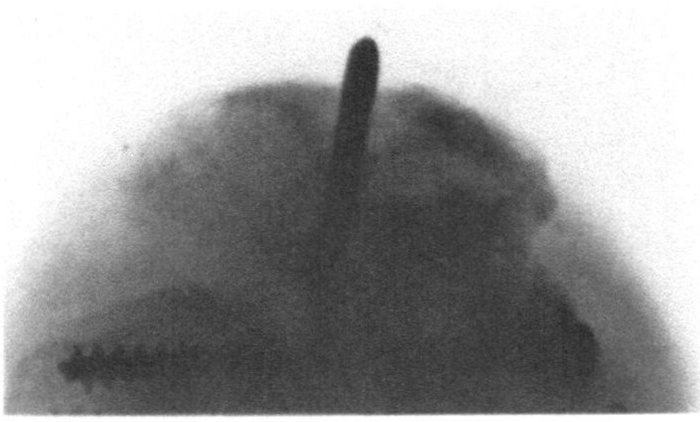

Fig. 23

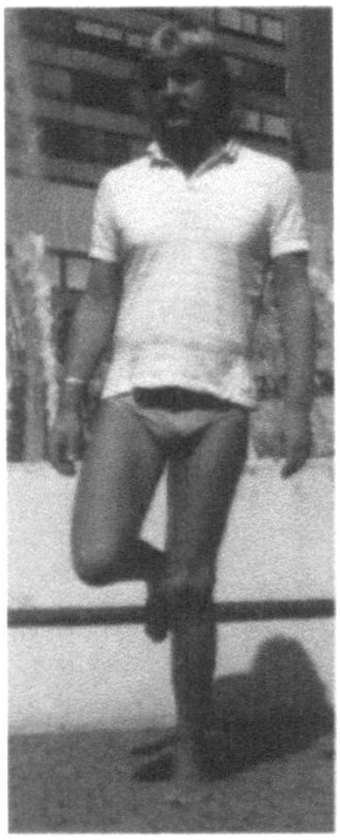

Fig. 24

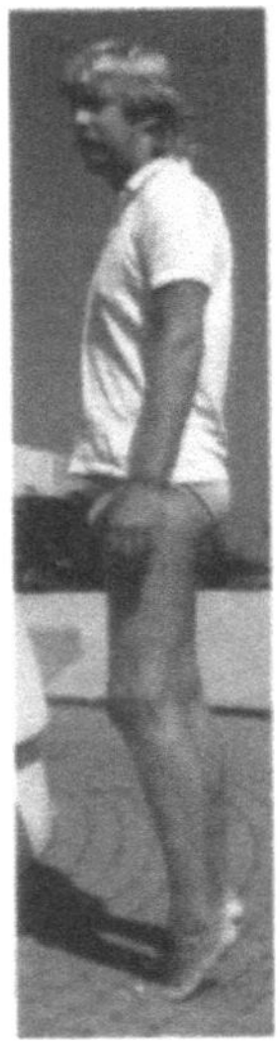

Fig. 25

Fig. 26

Fig. 27

Fig. 28

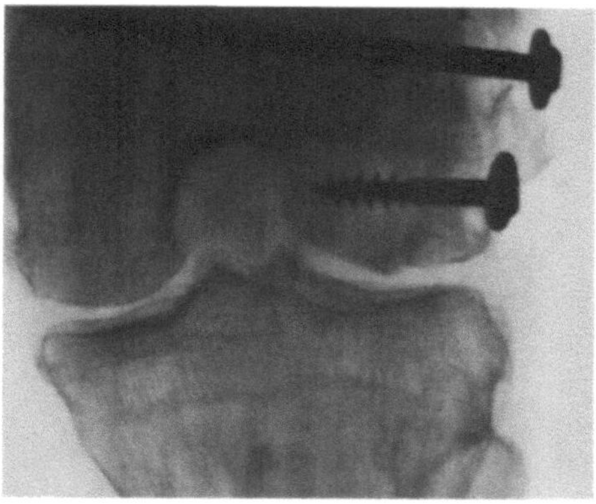

Fig. 29

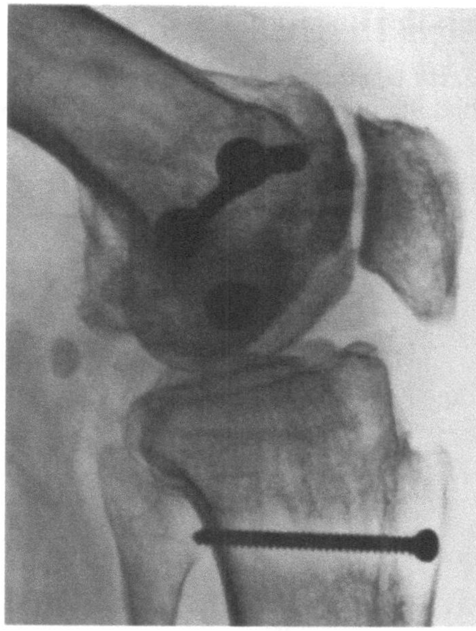

Fig. 30

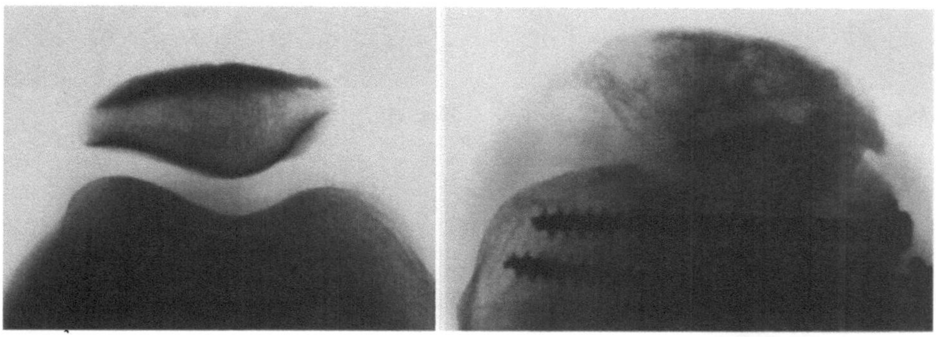

Fig. 31

Postoperative Course. Dorsal protection splint (plaster) in 20° of flexion. Ground contact 7 weeks after the intervention. Still lateral position of the patella, but this position could be corrected by a lateral support (Figs. 18, 19).

Fifth Hospitalisation, 23 Months After Trauma

Correction of the lateral femoral condyle plasty: lifting of the outer bridge edge and curtailment of the lateral condyle. Excessive medial transposition of the tibial tuberosity.
There was improvement during the postoperative course but still insufficient correction of the position of the patella.

Sixth Hospitalisation, 26 Months After Trauma

Sixth Operation. Dissection of lateral scarred tissue and insertion of a patella-tibial Steinmann pin in the correct position of the patella (Figs. 20–23).

Postoperative Course. The patient felt better, and there was pain-free joint play. Partial weight bearing, hometrainer, and intensive exercises. ROM 130°–30°–0°,

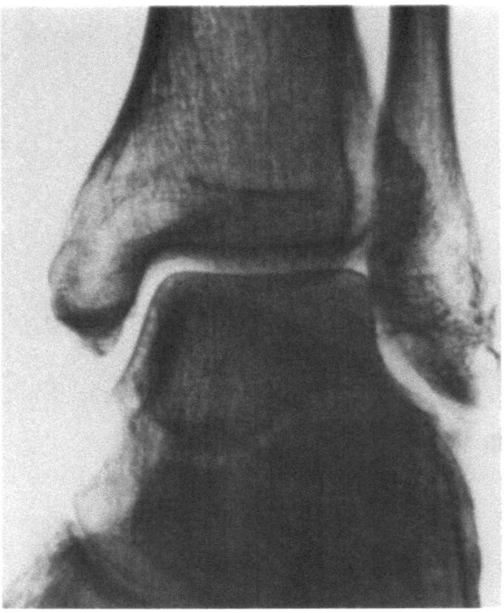

Fig. 32

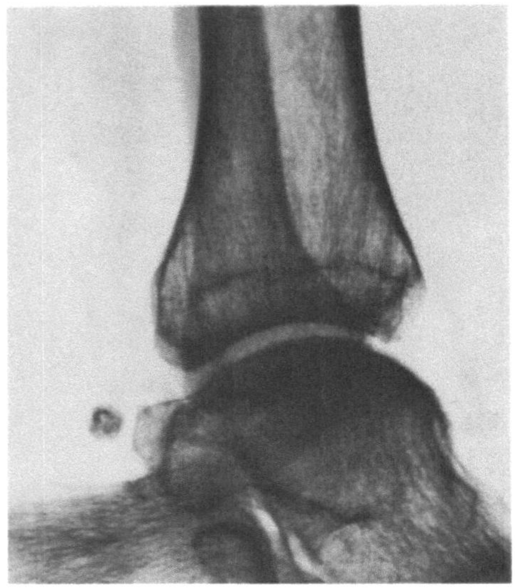

Fig. 33

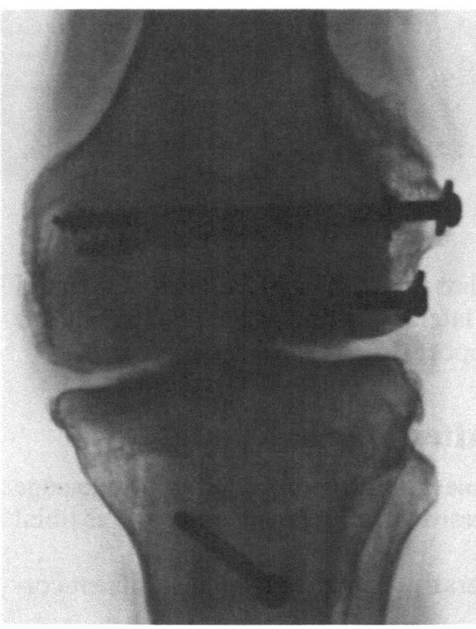

Fig. 34

stretching. To correct the lack of extension, a Campbell cast was applied twice for 4 weeks.
Removal of the Steinmann pin 8 weeks later.

Situation in August 1985, 3 Years After Trauma (Figs. 24–28)

Slight weather and effort-dependent lateral knee pain. Some trouble going upstairs owing to reduced function of the knee (ROM 130°–20°–0°). Dorsal splint with passive stretching during the night. No refusion. Still dystrophy of the thigh, but good action of the quadriceps muscle. No signs of infection and good trophic situation of the knee. No further lateral subluxation of the patella. Incorporated bone blocks (Figs. 29–31).
Ankle joint pain free, moderate reduction in the superior and inferior ankle joint play: flexion/extension 30°–0°–5°, eversion/inversion 10°–0°–25°. Good trophic situation of the left foot with intact sensibility, motor function and pulses, no swelling (Figs. 32, 33). Moderate reduction in the superior and inferior ankle joint play.
The monopedal (Fig. 34) showed a valgus alignment of 12°.
That winter, the patient started with alpine and cross country skiing.

Further Treatment. Removal of the screws and possibly a lateral supracondylar varus and extension osteotomy in order to correct the valgus alignment and the persisting knee flexion.

Subject Index

Abrasion 342
Antibiotics 573, 657
Antiphlogistics 389
Aplasia 116
Arthrodesis 595
Arthrography 313
Arthroplasty
  cementless 640, 646
  complication 607, 617, 645ff.
  fractures 629
  infection 607, 630
  loosening 609, 623, 645
  results 609, 634, 657, 684
Arthrodesis 367, 415, 476, 487

Bone
  graft 619, 629, 643
  reconstruction 585
  resection 544
  stock 527, 612
Braces 499

Cartilage lesion 397
  shaving 427
Center of rotation 8
Centrode, tibial 14
Chondromalacia 399, 443–490
Chondropathy 57, 61
Collagen fibres 334
Contact forces 18
Cruciate ligament 5, 13, 38
  anterior 20, 75
  posterior 19

Disability, score 124, 174, 507–517, 530, 538

Effusion 390, 482, 612
Electromyography 55
Electron-microscopy 150, 329, 380, 408, 415, 424, 427, 705, 713
European Multicenter Study 519

Fibres, neural 18
Fracture 398, 697, 717
Gait movement 31
Geometry of the joint 12
  four-bar linkage 12, 71
Gliding-rolling mechanics 72
Guide system 201, 519

Haemarthrosis 172, 185
Hamstrings 20, 57
Histology 311, 353

Iliotibial tract 45
Incongruence 4
Infection 517
Innervation 53
Instant centre 14
Isometric lines 258
  point 252

Joint, geometry 12

Kinetic centre 99, 554

Lachman test 47, 168, 176, 185, 232
Laser CO_2 349
Lever arm
  anterior cruciate 696
  posterior cruciate 650
Ligament
  augmentation device 156, 166, 237
  collateral 650
Local anaesthesia 351

Magnetic resonance imaging (MRI) 209, 649
Malformation 503
Mechanoreceptors 50
Medial condyle, rolling and gliding 95

Meniscus 38, 655, 703, 716
repair 306, 328
tear 653
Microangiography 381
Microscopic analysis 407
Microtrauma 422
Movement
gliding 23
rolling 3, 579
sliding 579
Muscle, antagonistic 28

Osteoarthritis 57, 102, 176, 185, 246, 305, 390, 471, 557
Osteochondral fracture 398
Osteophyte 372, 507
Osteotomy 41, 182, 294, 463, 604, 656, 722

Patellar dysplasia 458
movement 437
tendinitis 493, 498
Patellectomy 476, 486
Pivot shift 69, 185, 232, 358
Popliteal muscle 38, 58
Popliteus bypass 256, 338
Prosthesis
cemented 613
gliding 591
infection 592
loosening 535ff., 570, 591ff.
uncemented 83, 613

Q-angle 444
Quadriceps tendon 20

Radiographic measurement 82
Rating system 523
Rehabilitation 43, 211, 218, 229, 260, 287, 328, 696
Revascularisation 225
Revision arthroplasty 595
endoprosthesis 517
Roentgen stereophotogrammetric examination 104
Rolling during flexion 40
Rotation, automatic 37

Steinmann pin, transpatellar fixation 260, 271, 275, 727
Stereophotogrammetric analysis 434
Stress radiography 251, 599
Subluxation 255, 360, 476, 567, 589
Synovectomy 664ff., 683ff.
Synovitis 248

Technetium 99m bone scan 93
Tenodesis 79
Tension force 26
Three-dimensional space 6
Tibia, rolling backwards 15
Torsion 64
Tourniquet 659
Two-bundle plasty 37

Ultrasonography 495

Vascularisation 449

MIX
Papier aus verantwortungsvollen Quellen
Paper from responsible sources
FSC® C105338

If you have any concerns about our products,
you can contact us on
ProductSafety@springernature.com

In case Publisher is established outside the EU,
the EU authorized representative is:
Springer Nature Customer Service Center GmbH
Europaplatz 3, 69115 Heidelberg, Germany

Printed by Libri Plureos GmbH
in Hamburg, Germany